Brief Contents

Contents

Preface

Textbook authors are motivated by the feeling that they can write a book that not only serves their own purposes better than others already on the market, but that should also serve other similar teachers and their students. We believe we have achieved this goal for our own classes; wider use depends on the judgment of others.

While the choice of topics included in this book is not markedly different from others, we have chosen to emphasize molecular orbitals and symmetry in nearly every aspect of bonding and reactivity. An early chapter, Chapter 4, is devoted to a discussion of molecular symmetry and introductory group theory, with applications to molecular vibrations and chirality. In later chapters, we have used group theory in a variety of other applications: obtaining molecular orbitals in Chapter 5, predicting the *d* orbital splitting in transition metal complexes in Chapter 8, and interpreting infrared spectra of organometallic compounds in Chapter 12. Additional applications of group theory are included in problems at the end of these and other chapters. At the same time that we are convinced that an introduction to group theory is important in an upper level course in inorganic chemistry, we recognize that not all who teach such a course will agree with us. An alternate, more pictorial, approach to molecular orbitals is also provided in Chapter 5, and we have written the other chapters in such a way that they can be used by those who do not share our enthusiasm for group theory.

After the introduction in Chapter 1, Chapters 2 and 3 provide a review of atomic theory and simple concepts of chemical bonding. We have included a description of the nucleogenesis of the elements in Chapter 1 and have incorporated solid state chemistry into several chapters where appropriate. Following the introduction to group theory, this theory is applied to the construction of molecular orbitals in Chapter 5. Chapter 6 provides a detailed discussion of the various acid-base concepts, with numerous examples emphasizing applications of molecular orbitals to acid-base chemistry. Important additional properties of the main group elements are then described in Chapter 7. This chapter summarizes some of the best known and most important properties of these elements and their compounds; references are cited for those wishing additional information on the main group elements.

In Chapters 8 through 13, attention is turned to the chemistry of the transition elements. The first four of these chapters deal, respectively, with bonding, electronic spectra, structures, and reactions of classical transition metal complexes. A separate chapter on electronic spectra provides a particularly straightforward avenue of presentation. Chapters 12 and 13 provide an introduction to organometallic compounds, their spectra, and reactions. Special attention has been given to catalytic cycles and their application to problems of chemical and industrial significance.

Seldom is sufficient attention paid to parallels between main group and organometallic compounds. We have therefore discussed some of these important parallels in Chapter 14 and have placed particular emphasis on the isolobal analogy developed by Roald Hoffmann. We believe that seeking similarities in the chemistry of different types of compounds can be an extremely valuable exercise. This chapter provides a useful key to these similarities, while pointing out some of the differences, and will encourage readers to investigate additional parallels. No text would be complete without a discussion of the role of inorganic compounds in biological processes. In Chapter 15, bioinorganic chemistry has been combined with the related, and increasingly important, topic of environmental inorganic chemistry.

In addition to selecting the most appropriate topics for inclusion in such a text, we wanted to make our text as accessible to students as possible. We have therefore included examples and exercises within the chapters, with answers to the examples in the chapters, and answers to the exercises in Appendix A. To encourage use of the literature in inorganic chemistry, we have included some end-of-chapter questions from the literature, and have cited many useful references in the text. In addition, there are suggestions for more general supplemental reading at the end of each chapter. We have also included more introductory material than is common in inorganic texts for those students wishing a quick summary of previously covered material.

We want to acknowledge our debt to all previous textbook authors; no one can write a text without being influenced by the books used as texts or references in the past. More specifically, we offer our thanks to those who reviewed this book in preparation and offered many helpful suggestions; first among these is our colleague from across the river, Jim Finholt. We also appreciate the constructive reviews of various chapters by Beth Abdella, Robert Doedens, Al Finholt, Russell Grimes, Mitsuru Kubota, Gary Spessard, and Amy Shachter. We also acknowledge the comments of Kevin Roesselet, who tested a draft version of portions of the manuscript in his classes at Bates College; Kate Doan, who checked the galley proofs; and Dominick Casadonte, who checked the page proofs. We are also grateful for the ability and patience of our editors, Dan Joraanstad and Debra Wechsler, in shepherding us through the writing and production process.

To our students who suffered through the inconveniences and uncertainties of a book in process and offered many helpful suggestions, our thanks for their good humor and continual support. And to Becky and Marge, who learned more about the process than any non-author needs to know, and put up with our physical and mental absences, our thanks and love.

Gary L. Miessler
Donald A. Tarr

1

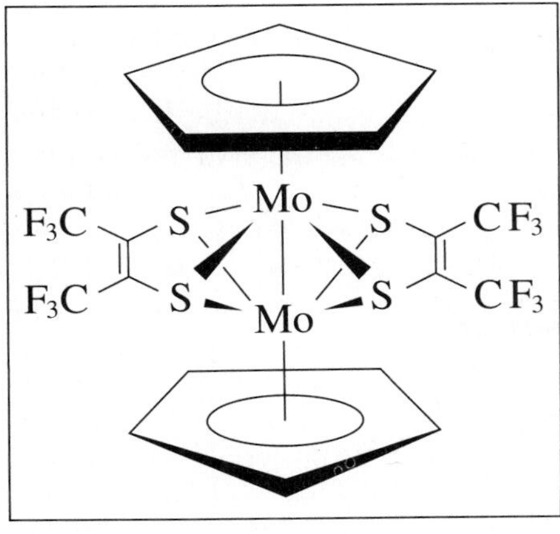

Introduction to Inorganic Chemistry

1-1 WHAT IS INORGANIC CHEMISTRY?

If organic chemistry is defined as the chemistry of compounds of carbon, primarily those containing hydrogen or halogens plus other elements, inorganic chemistry can be described broadly as the chemistry of "everything else." This includes all the remaining elements in the periodic table, as well as carbon, which plays a major role in many inorganic compounds. Organometallic chemistry, a very large and rapidly growing field, bridges both areas by considering compounds containing direct metal–carbon bonds. As can be imagined, the inorganic realm is extremely broad, providing essentially limitless areas for investigation.

1-2 CONTRASTS WITH ORGANIC CHEMISTRY

Some comparisons between organic and inorganic compounds are in order. In both areas, single, double, and triple covalent bonds are found, as shown in Figure 1-1; for inorganic compounds, these include direct metal–metal bonds as well as metal–carbon bonds. However, while the maximum number of bonds between carbon atoms is three, there are many compounds containing quadruple bonds between metal atoms. In addition to the sigma and pi bonds common in organic chemistry, quadruply bonded metal atoms contain a delta (δ) bond (Figure 1-2); a combination of one sigma bond, two pi bonds, and one delta bond makes up the quadruple bond. The delta bond is possible in these cases because metal atoms have *d* orbitals to use in bonding, in contrast to carbon, which has only *s* and *p* orbitals available.

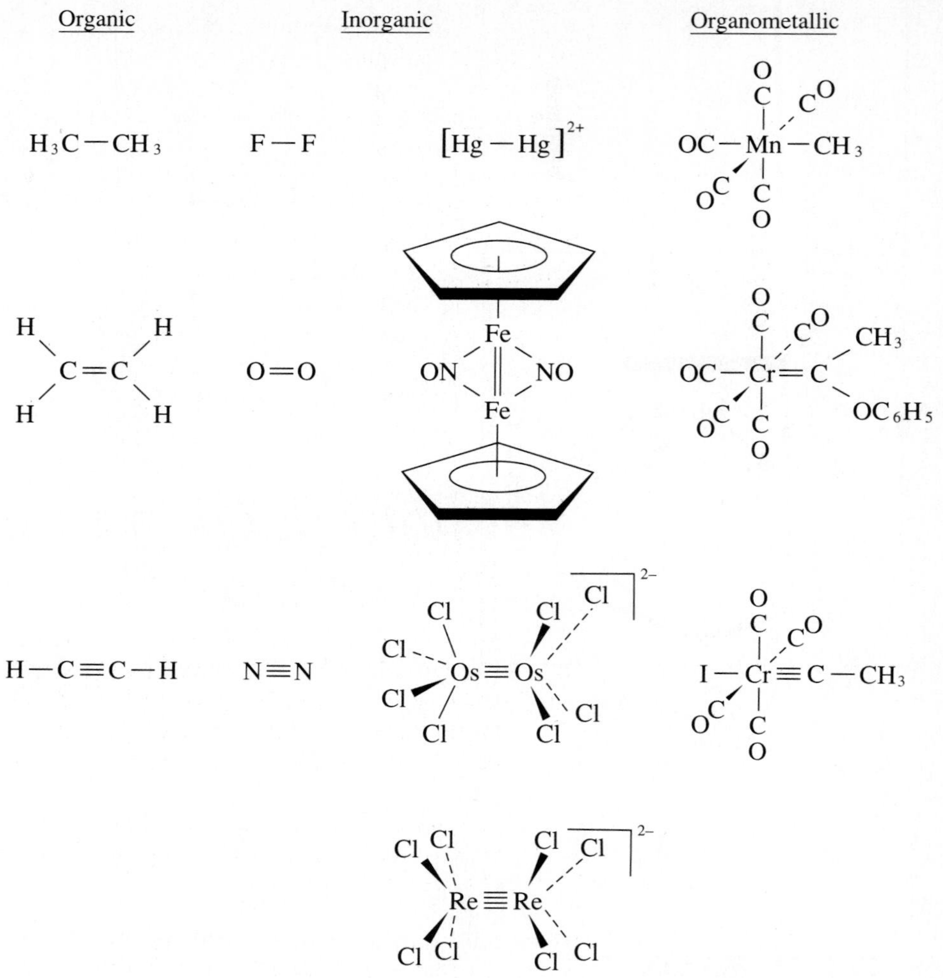

Organic	Inorganic		Organometallic
$H_3C{-}CH_3$	$F{-}F$	$[Hg{-}Hg]^{2+}$	$(OC)_5Mn{-}CH_3$
$H_2C{=}CH_2$	$O{=}O$	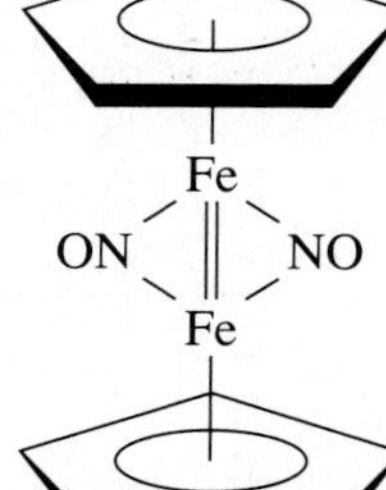	$(OC)_5Cr{=}C(CH_3)(OC_6H_5)$
$H{-}C{\equiv}C{-}H$	$N{\equiv}N$	$[Cl_4Os{\equiv}OsCl_4]^{2-}$	$I{-}Cr(CO)_4{\equiv}C{-}CH_3$
		$[Cl_4Re{\equiv}ReCl_4]^{2-}$	

FIGURE 1-1 Single and Multiple Bonds in Organic and Inorganic Molecules.

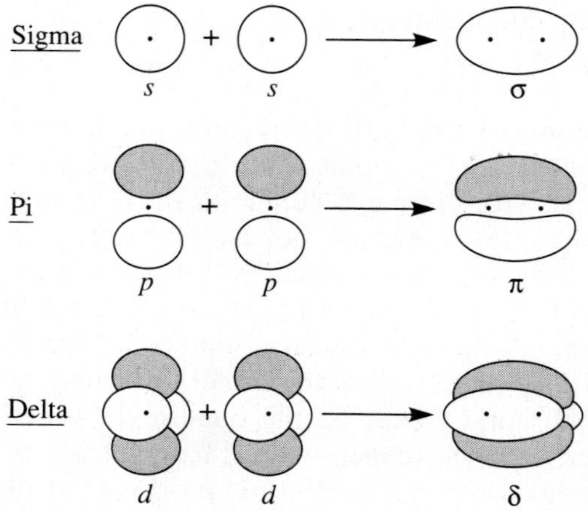

FIGURE 1-2 Examples of Bonding Interactions.

In organic compounds, hydrogen is nearly always bonded to a single carbon. In inorganic compounds, especially of the Group 3 (IIIA) elements, hydrogen is frequently encountered as a bridging atom between two or more other atoms. Bridging hydrogens can also occur in metal cluster compounds; in these clusters hydrogen forms bridges across edges or faces of complex polyhedra of metal atoms. Alkyl groups may also act as bridges in inorganic compounds, a function rarely encountered in organic chemistry (except in reaction intermediates). Examples of terminal and bridging hydrogens and alkyl groups in inorganic compounds are shown in Figure 1-3.

Some of the most striking differences between the chemistry of carbon and that of many other elements are in coordination number and geometry. While carbon is usually limited to a maximum coordination number of four (a maximum of four atoms bonded to carbon, as in CH_4), inorganic compounds having coordination numbers of five, six, seven, and more are very common; the most common coordination geometry is an octahedral arrangement around a central atom, as shown for TiF_6^{3-} in Figure 1-4. Furthermore, inorganic compounds present coordination geometries different from those found for carbon. For example, while 4-coordinate carbon is nearly always tetrahedral, both tetrahedral and square planar shapes occur for 4-coordinate compounds of both metals and nonmetals.

The tetrahedral geometry usually found in 4-coordinate compounds of carbon also occurs in a different form in some inorganic molecules. Methane contains four hydrogens in a regular tetrahedron around carbon. Elemental phosphorus is tetratomic (P_4) and also is tetrahedral, but with no central atom. Examples of some of the geometries found for inorganic compounds are shown in Figure 1-4.

Aromatic rings are very common in organic chemistry. Aryl groups can also form sigma bonds to metals, as in phenyllithium, C_6H_5Li. However,

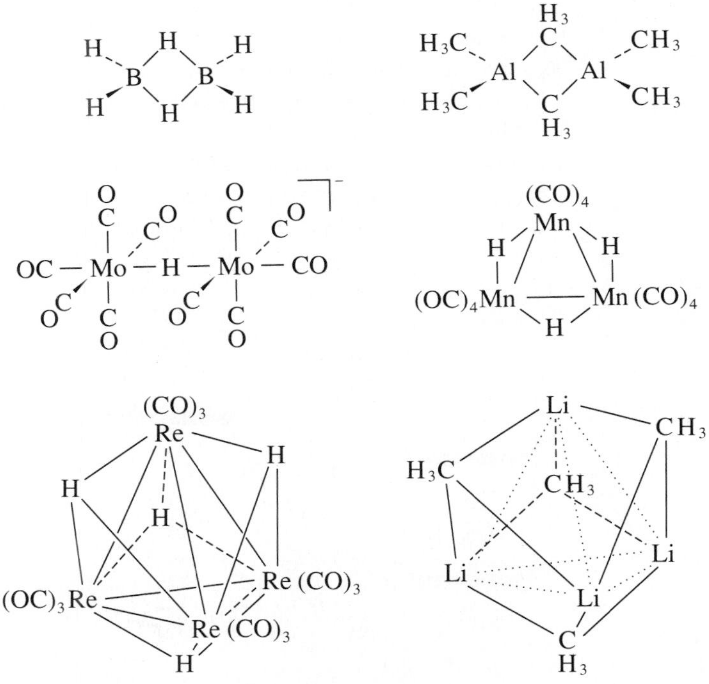

FIGURE 1-3 Examples of Inorganic Compounds Containing Terminal and Bridging Hydrogens and Alkyl Groups.

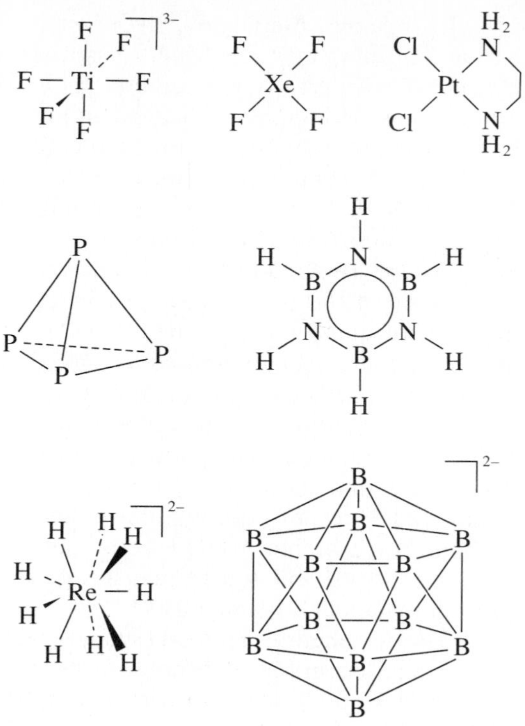

FIGURE 1-4 Examples of Geometries of Inorganic Compounds.

$B_{12}H_{12}^{2-}$ (not shown: one hydrogen on each boron)

aromatic rings can also bond to metals in a dramatically different fashion using their pi orbitals, as shown in Figure 1-5. The result is a metal atom bonded above the center of the ring, almost as if suspended in space. In many cases, metal atoms are sandwiched between two aromatic rings; multiple-decker sandwiches of metals and aromatic rings are also known.

Carbon plays a very unusual role in a number of metal cluster compounds, in which a carbon atom is at the center of a polyhedron of metal atoms. Examples of carbon at the center of clusters of five, six, or more metals are known; several of these are shown in Figure 1-6. The contrast of the role that carbon plays in these clusters to its usual role in organic compounds is striking;

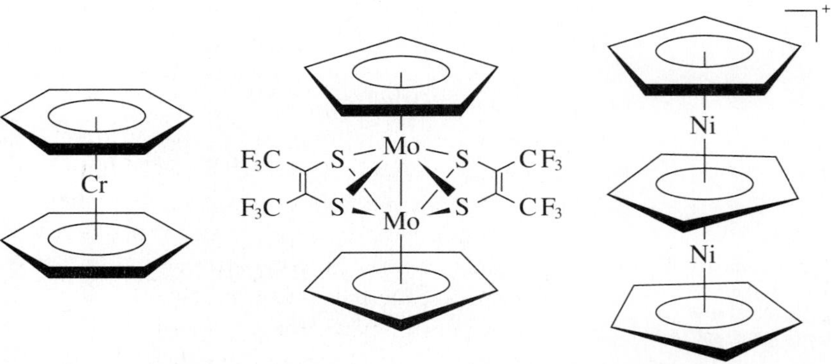

FIGURE 1-5 Inorganic Compounds Containing Pi-bonded Aromatic Rings.

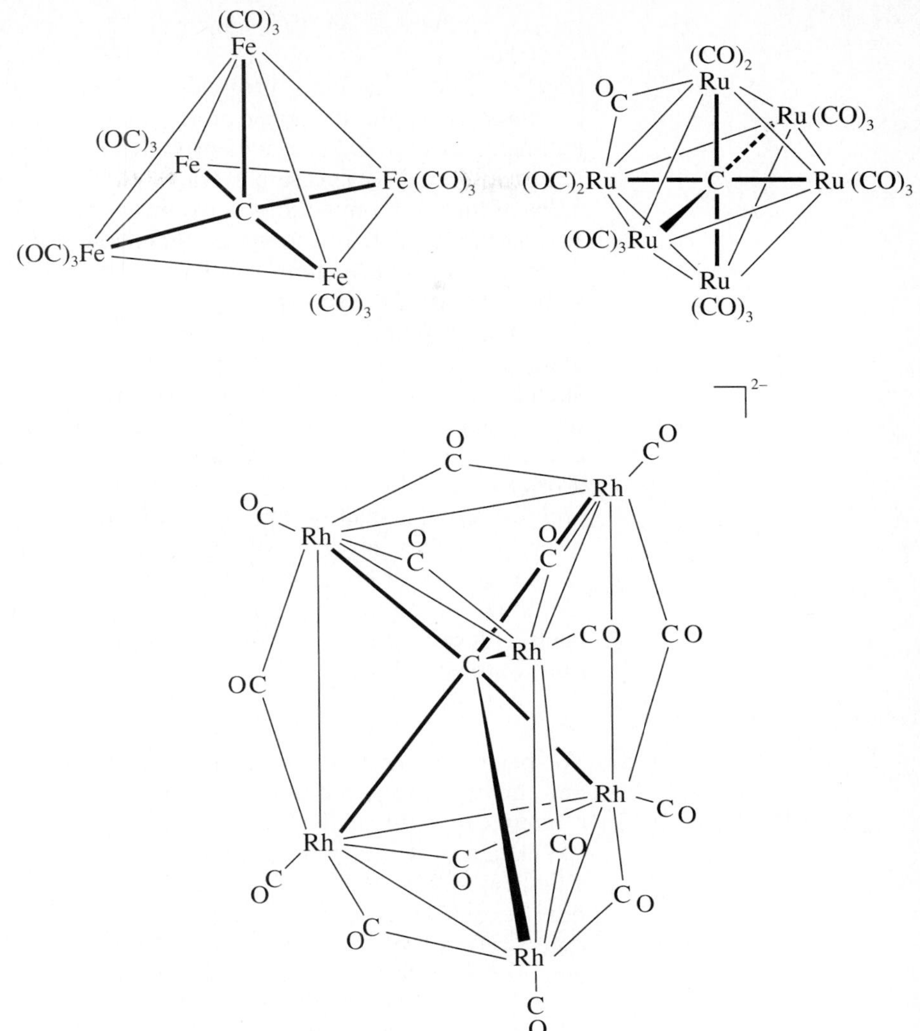

FIGURE 1-6 Carbon-centered Metal Clusters.

attempting to explain how carbon can form bonds to the surrounding metal atoms in clusters has provided an interesting challenge to theoretical inorganic chemists.

As is true of all fields of study, there are no sharp dividing lines between the specialty subfields in chemistry. This is particularly true of inorganic chemistry, which is expanding rapidly into new areas of research. In spite of these changes, there is some general agreement on the topics usually included in the study of inorganic chemistry. This text will include most of them, with more thorough treatment of those that seem to the authors particularly significant as background to further study.

Physical chemistry has been described as "all the parts of chemistry that are interesting." Naturally, all chemists would claim the same thing for their specialty, and that feeling leads to widely different groupings of topics. Many of the subjects included in this book, such as acid–base chemistry and organometallic reactions, are of vital interest to many organic chemists. Others, such as oxidation–reduction reactions, spectra, and solubility relations, also interest analytical chemists; subjects related to structure determination, spectra, and theories of bonding appeal to physical chemists. Finally, the use of organometallic catalysts provides a connection to petroleum and polymer chemistry, and the presence of coordination compounds such as hemoglobin and metal-containing enzymes provides a similar tie to biochemistry. This list is not intended to describe a fragmented field of study, but rather to show some of the interconnections between the different specialty fields of chemistry. Just as material from previous chemistry courses is used in this book, the information provided here can be used to illuminate topics studied earlier or to be studied in the future.

The remainder of this chapter is devoted to the history of inorganic chemistry, from the creation of the elements to the present. It is a quick, very sketchy history intended only to provide the reader with a sense of connection to the past and with a means of putting some of the topics of inorganic chemistry into the context of larger historical events. In many later chapters, a brief history of each topic is given, with the same intention. Although time and space do not allow for much attention to history, we want to avoid the impression that any part of chemistry has sprung full-blown from any one person's work or has appeared suddenly. Although certain events can later be identified as marking a dramatic change of direction in inorganic chemistry, such as a new theory or a new kind of compound or reaction, they were all built on past achievements.

1-3 GENESIS OF THE ELEMENTS (THE BIG BANG)

We begin our study of inorganic chemistry with the genesis of the elements and the creation of the universe. Among the difficult tasks facing anyone who attempts to explain the origin of the universe are the inevitable questions: "What about the time just before the creation? Where did the starting material, whether energy or matter, come from?" The idea of an origin at a specific time requires that such questions be set aside, at least for the time being. No theory attempting to explain the origin of the universe can be expected to extend infinitely far back in time; there is controversy enough in attempting to describe the time after the origin.

Current opinion favors the big bang theory[1] over other creation theories, although many controversial points are yet to be explained. Other theories, such as the steady-state or oscillating theories, have their advocates, and the creation of the universe is certain to remain a source of controversy and study for many years.

According to the big bang theory, the universe began 1.8×10^{10} years ago with an extreme concentration of energy in a very small space (in fact, zero volume, which in turn requires infinite temperature), which then exploded, beginning the process of the creation of atoms. Neutrons were formed initially and decayed quickly (half-life = 11.3 min) into protons, electrons, and antineutrinos:

$$n \longrightarrow p + e^- + \overline{\nu}_e$$

In this and subsequent equations,

p = proton of charge +1 and mass 1.007 atomic mass unit (amu)
γ = gamma ray (high-energy photon) with zero mass

[1] J. Selbin, *J. Chem. Educ.*, **1973,** *50,* 306, 380; A. A. Penzias, *Science,* **1979,** *205,* 549.

e^- = electron of charge -1 and mass $\frac{1}{1823}$ amu (also known as a β particle)

e^+ = positron with charge $+1$ and mass $\frac{1}{1823}$ amu

ν_e = neutrino with no charge and a very small mass

$\overline{\nu}_e$ = antineutrino with no charge and a very small mass

n = neutron with no charge and a mass of 1.009 amu

Nuclei are described by the convention

$$^{\text{nuclear mass}}_{\text{nuclear charge}}\text{symbol}$$

Within one half-life of the neutron, half the matter of the universe was protons and the temperature was near 500×10^6 K. The nuclei formed in the first 30 to 60 min were those of deuterium (^{2}H), ^{3}He, ^{4}He, and ^{5}He. (Helium 5 has a very short half-life [2×10^{-21} s] and decays back to helium 4, effectively limiting the mass of the nuclei formed by these reactions to 4 amu.) The following reactions show how these nuclei can be formed in a process called *hydrogen burning:*

$$^1_1\text{H} + {}^1_1\text{H} \longrightarrow {}^2_1\text{H} + e^+ + \nu_e$$

$$^2_1\text{H} + {}^1_1\text{H} \longrightarrow {}^3_2\text{He} + \gamma$$

$$^3_2\text{He} + {}^3_2\text{He} \longrightarrow {}^4_2\text{He} + 2\,{}^1_1\text{H}$$

The expanding material from these first reactions began to gather together into galactic clusters and then into more dense stars, where the pressure of gravity kept the temperature high and promoted further reactions. The combination of hydrogen and helium with many protons and neutrons led rapidly to the formation of heavier elements. In stars with internal temperatures at 10^7 to 10^8 K, the reactions forming ^{2}H, ^{3}He, and ^{4}He continued, along with reactions that produced heavier nuclei. The following *helium burning* reactions are among those known to take place under these conditions:

$$2\,{}^4_2\text{He} \longrightarrow {}^8_4\text{Be} + \gamma$$

$$^4_2\text{He} + {}^8_4\text{Be} \longrightarrow {}^{12}_6\text{C} + \gamma$$

$$^{12}_6\text{C} + {}^1_1\text{H} \longrightarrow {}^{13}_7\text{N} \longrightarrow {}^{13}_6\text{C} + e^+ + \nu_e$$

In heavier stars (temperatures of 6×10^8 K or higher), the carbon–nitrogen cycle is possible:

$$^{12}_6\text{C} + {}^1_1\text{H} \longrightarrow {}^{13}_7\text{N} + \gamma$$

$$^{13}_7\text{N} \longrightarrow {}^{13}_6\text{C} + e^+ + \nu_e$$

$$^{13}_6\text{C} + {}^1_1\text{H} \longrightarrow {}^{14}_7\text{N} + \gamma$$

$$^{14}_7\text{N} + {}^1_1\text{H} \longrightarrow {}^{15}_8\text{O} + \gamma$$

$$^{15}_8\text{O} \longrightarrow {}^{15}_7\text{N} + e^+ + \nu_e$$

$$^{15}_7\text{N} + {}^1_1\text{H} \longrightarrow {}^4_2\text{He} + {}^{12}_6\text{C}$$

The net result of this cycle is the formation of helium from hydrogen, with gamma rays, positrons, and neutrinos as byproducts. In addition, even heavier elements are formed:

$$^{12}_{6}C + ^{12}_{6}C \longrightarrow ^{20}_{10}Ne + ^{4}_{2}He$$

$$2\ ^{16}_{8}O \longrightarrow ^{28}_{14}Si + ^{4}_{2}He$$

$$2\ ^{16}_{8}O \longrightarrow ^{31}_{16}S + ^{1}_{0}n$$

At still higher temperatures, further reactions take place:

$$\gamma + ^{28}_{14}Si \longrightarrow ^{24}_{12}Mg + ^{4}_{2}He$$

$$^{28}_{14}Si + ^{4}_{2}He \longrightarrow ^{32}_{16}S + \gamma$$

$$^{32}_{16}S + ^{4}_{2}He \longrightarrow ^{36}_{18}Ar + \gamma$$

Even heavier elements can be formed, with the actual amounts depending on a complex relationship between their inherent stability, the temperature of the star, and the lifetime of the star. The curve of inherent stability of nuclei has a maximum at $^{56}_{26}Fe$, accounting for the high relative abundance of iron in the universe. However, the lighter elements are also abundant, leading to the conclusion that the process is continuing, rather than having reached a stable conclusion. Formation of elements of higher atomic number takes place by addition of neutrons to a nucleus, followed by electron emission decay. In environments of low neutron density, this addition of neutrons is relatively slow, one neutron at a time; in high neutron density environments, 10 to 15 neutrons may be added in a very short time, and the resulting nucleus is then neutron rich:

$$^{56}_{26}Fe + 13\ ^{1}_{0}n \longrightarrow ^{69}_{26}Fe \longrightarrow ^{69}_{27}Co + e^-$$

The very heavy elements are also formed by reactions such as this. After addition of the neutrons, β decay (loss of electrons from the nucleus) leads to nuclei with larger atomic numbers.

1-4 FORMATION OF THE EARTH

Gravitational attraction combined with rotation gradually formed the expanding cloud of material into relatively flat spiral galaxies containing thousands of stars each. Complex interactions within the stars led to black holes and other kinds of stars, some of which exploded as supernovas and scattered their material widely. Further gradual accretion of some of this material into planets followed. At the lower temperatures found in planets, the buildup of heavy elements stopped, and decay of the unstable radioactive isotopes of the elements became the predominant reactions.

1-5 NUCLEAR REACTIONS AND RADIOACTIVITY

In addition to the overall curve of nuclear stability, which has its most stable region near atomic number $Z = 26$, combinations of protons and neutrons at each atomic number exhibit different stabilities. In some elements, like fluorine, there is only one stable **isotope** (a specific combination of protons and neutrons), ^{19}F. In others, such as chlorine, there are two or more isotopes. ^{35}Cl has a natural abundance of 75.77%, and ^{37}Cl has a natural abundance of 24.23%. Both are stable, as are all the natural isotopes of the lighter elements (^{3}H, ^{14}C, and a few others are radioactive nuclei that are continually being formed by cosmic rays and have a low, constant concentration). The radioactive isotopes of these elements have short half-lives and have had more than enough time to decay to more stable elements.

Heavier elements ($Z = 40$ or more) may also have radioactive isotopes with longer half-lives. As a result, some of these radioactive isotopes have not had time to decay completely, and the natural substances are radioactive.

Further discussion of isotopic abundances and radioactivity is available in other larger or more specialized sources.[2] We will conclude our discussion with the most widely used and hotly debated reactions, those for **fission** of uranium or plutonium and for **fusion** of hydrogen. ^{235}U naturally splits into two not quite equal parts when it absorbs a neutron. Representative reactions are

$$^{235}_{92}U + ^{1}_{0}n \longrightarrow ^{139}_{56}Ba + ^{94}_{36}Kr + 3\,^{1}_{0}n$$

$$^{235}_{92}U + ^{1}_{0}n \longrightarrow ^{133}_{51}Sb + ^{99}_{41}Nb + 4\,^{1}_{0}n$$

Because each reaction releases more neutrons than it consumes, the reactions can proceed and grow when there is a critical mass of uranium present, large enough to trap the neutrons before they can escape. Under controlled circumstances, the neutrons are slowed by collisions with water molecules or carbon in the form of graphite, and these slower "thermal" neutrons are more readily absorbed to continue the reaction. The heat resulting from the reactions is used to boil water into high-pressure steam to be used for the generation of electricity. In other conditions, when a critical mass is formed suddenly by forcing smaller pieces of uranium metal together with carefully engineered explosives, the reactions are so fast that a nuclear explosion occurs. In either the fast or slow reaction, a small fraction of the mass of the atoms is converted directly into energy according to Einstein's famous equation:

$$E = mc^2,$$

where E = energy released
m = mass being converted
c = velocity of light

Conversion of even a small amount of mass results in an enormous release of energy because the velocity of light is so large.

Fusion reactions are similar to those for the first reactions of the big bang:

$$^{2}_{1}H + ^{2}_{1}H \longrightarrow ^{3}_{2}He + ^{1}_{0}n$$

$$^{2}_{1}H + ^{2}_{1}H \longrightarrow ^{3}_{1}H + ^{1}_{1}H$$

$$^{2}_{1}H + ^{3}_{1}H \longrightarrow ^{4}_{2}He + ^{1}_{0}n$$

[2] N. N. Greenwood and A. Earnshaw, *Chemistry of the Elements,* Pergamon, Elmsford, N.Y., 1984; J. Silk, *The Big Bang, The Creation and Evolution of the Universe,* W. H. Freeman, San Francisco, 1980.

These reactions require temperatures of 10^6 to 10^7 K to continue (and even higher temperatures for ignition). Such temperatures are achieved on the earth only in nuclear explosions, which are used as triggers for fusion bombs (commonly called hydrogen bombs), and in laboratories where small amounts of material can be heated very quickly by lasers and the resulting highly ionized plasma contained by magnetic forces. Efforts to generate and maintain fusion reactions for the generation of power have so far been unsuccessful, but the idea is still being studied intensively.

Theories that attempt to explain the formation of the specific structures of the earth are at least as numerous as those for the formation of the universe. Although the details of these theories differ, there is general agreement that the earth was much hotter during its early life, and the materials fractionated into the gaseous, liquid, and solid states at that time. As the surface of the earth cooled, the lighter materials in the crust solidified and still float on a molten inner layer, according to the plate tectonics explanation of geology. There is also general agreement that the earth has a core of iron and nickel, solid at the center and liquid above that. The outer half of the earth's radius is composed of silicate minerals in the mantle; silicate, oxide, and sulfide minerals in the crust; and a wide variety of materials at the surface, including much water and the gases of the atmosphere.

The different kinds of forces apparent in the early planet earth are seen indirectly in the distribution of minerals and elements. In locations where liquid magma broke through the crust, compounds that are readily soluble in such molten rock were carried along and deposited as ores. Fractionation of the minerals then depended on their melting points and solubilities in the magma. In other locations, water was the source of the formation of ore bodies. At these sites, water leached minerals from the surrounding area and later evaporated, leaving the minerals behind. The solubilities of the minerals in either magma or water depend on the elements, their oxidation states, and the other elements they are combined with. A rough division of the elements can be made according to their ease of reduction to the element and their combination with oxygen and sulfur. *Siderophiles* are reduced to the element in molten iron, *lithophiles* combine primarily with oxygen, and *chalcophiles* combine more readily with sulfur. These divisions are shown in the periodic table of Figure 1-7.

As an example of the action of water, we can explain the formation of bauxite (Al_2O_3) deposits by the leaching away of more soluble salts from aluminosilicate deposits. The silicate portion is enough more soluble in water that it can be leached away, leaving a high concentration of the aluminum ore. This mechanism provides at least a partial explanation for the presence of bauxite deposits in tropical areas or areas that were tropical in the past, with large amounts of rainfall.

Further explanations of these geological processes must be left to more specialized sources.[3] Such explanations are based on many concepts treated later in this book. For example, the ideas of hard and soft acids and bases (HSAB) help explain the different solubilities of minerals in water or molten rock and their resulting deposit in specific locations. The divisions given in Figure 1-7 can also be partly explained by HSAB theory, which is discussed in Chapter 6 and used in later chapters.

[3] J. E. Fergusson, *Inorganic Chemistry and the Earth,* Pergamon, Elmsford, N.Y., 1982.

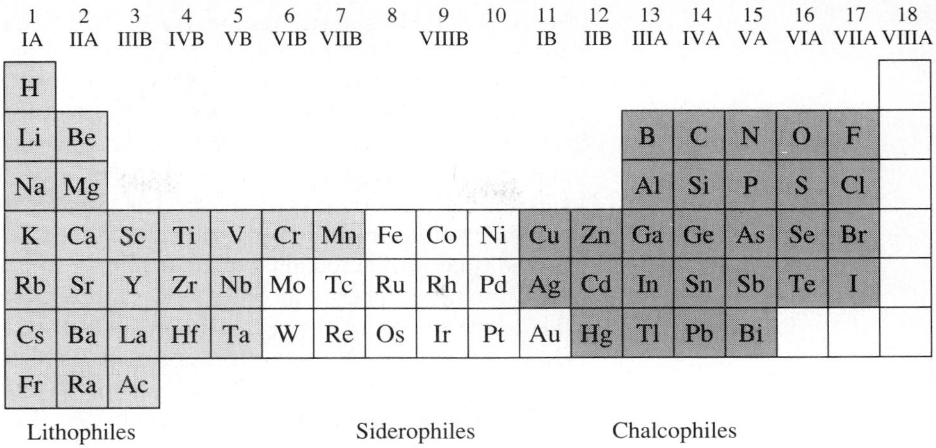

1 IA	2 IIA	3 IIIB	4 IVB	5 VB	6 VIB	7 VIIB	8	9 VIIIB	10	11 IB	12 IIB	13 IIIA	14 IVA	15 VA	16 VIA	17 VIIA	18 VIIIA
H																	
Li	Be											B	C	N	O	F	
Na	Mg											Al	Si	P	S	Cl	
K	Ca	Sc	Ti	V	Cr	Mn	Fe	Co	Ni	Cu	Zn	Ga	Ge	As	Se	Br	
Rb	Sr	Y	Zr	Nb	Mo	Tc	Ru	Rh	Pd	Ag	Cd	In	Sn	Sb	Te	I	
Cs	Ba	La	Hf	Ta	W	Re	Os	Ir	Pt	Au	Hg	Tl	Pb	Bi			
Fr	Ra	Ac															

Lithophiles Siderophiles Chalcophiles

FIGURE 1-7 Geochemical Classification of the Elements.

1-6 THE HISTORY OF INORGANIC CHEMISTRY

We make no attempt here to give a complete history of inorganic chemistry. Rather, we describe some interesting developments as examples of the kind of chemistry being studied at specific times.

Even before alchemy became a subject of study, many chemical reactions were used and the products applied to daily life. For example, the first metals used were probably gold and copper, which can be found in the metallic state. Copper can also be formed readily by the reduction of malachite [basic copper carbonate, $Cu_2(CO_3)(OH)_2$] in charcoal fires. Silver, tin, antimony, and lead were also known as early as 3000 B.C. Iron appeared in classical Greece and in other areas around the Mediterranean Sea by 1500 B.C. At about the same time, colored glasses and ceramic glazes, largely composed of silicon dioxide (SiO_2, the major component of sand) and other metallic oxides that had been melted and allowed to cool to amorphous solids, were introduced.

Alchemists were active in China, Egypt, and other centers of civilization early in the Christian era. Although much effort went into attempts to "transmute" base metals into gold, the treatises of these alchemists also describe many other chemical reactions and operations. Distillation, sublimation, crystallization, and other techniques were developed and used in these studies. Because of the political and social changes of the time, alchemy shifted into the Arab world and later (about A.D. 1000 to 1500) reappeared in Europe. Gunpowder was used in Chinese fireworks as early as 1150, and alchemy was also widespread in China and India at this time. Alchemists appeared in art, literature, and science until at least 1600, by which time chemistry was beginning to take shape as a science. Roger Bacon (1214–1294) is recognized as one of the first great experimental scientists; he also wrote extensively about alchemy.

By the seventeenth century, the common acids (nitric, sulfuric, and hydrochloric) were known, and more systematic descriptions of common salts and their reactions were being accumulated. The combination of acid and base to form salts was appreciated by some chemists. As experimental techniques improved, quantitative study of reactions and the properties of gases became more common, atomic and molecular weights were determined more accurately, and the groundwork was laid for what later became the periodic table. By 1869, the concepts of atoms and molecules were well established, and it was possible for Mendeleev and Meyer to describe different forms of the periodic table.

The chemical industry, which had been in existence since very early times in the form of factories for the purification of salts and the smelting and refining of metals, expanded as methods for the preparation of relatively pure materials became more common. In 1896, Becquerel discovered radioactivity, and another area of study was opened. Studies of subatomic particles, spectra, and electricity finally led to the atomic theory of Bohr in 1913, soon modified by the quantum mechanics of Schrödinger and Heisenberg.

Inorganic chemistry as a field of study was extremely important during the early years of the exploration and development of mineral resources. Qualitative analysis methods were developed to help in identifying minerals, and combined with quantitative methods to assess their purity and value. As the industrial revolution progressed, the chemical industry progressed in parallel. By the early twentieth century, plants for the production of ammonia, nitric acid, sulfuric acid, sodium hydroxide, and many other large-scale inorganic chemicals were common.

In spite of the work of Werner and Jørgensen on coordination chemistry near the turn of the century and the discovery of a number of organometallic compounds, the popularity of inorganic chemistry as a field of study gradually declined through most of the first half of the twentieth century. The need for inorganic chemists to work on military projects during World War II began a rejuvenation of the field. As more work was done on many varied projects (not least, the Manhattan Project, which developed the fission and fusion bombs), new areas of research appeared, old areas were found to have missing information, and new theories were proposed, which in turn led to further experimental work. The enthusiasm and ideas generated during this time resulted in a great expansion of inorganic chemistry, beginning in the 1950s.

A great contribution to inorganic chemistry at this time was the extension[4] of the same mathematical method used to explain the spectra of metal ions in ionic crystals[5] to the explanation of the spectra of coordination compounds, in which metal ions are surrounded by ions or molecules that donate electron pairs. This crystal field theory was modified into ligand field theory, which gave an even more complete picture of the bonding in these compounds through molecular orbitals. With this theoretical framework, the new instruments developed at about this same time, and the generally reawakened interest in inorganic chemistry, the field developed rapidly.

In 1955, Ziegler[6] and Natta[7] discovered organometallic compounds that could catalyze the polymerization of ethylene at lower temperatures and pressures than the common industrial method used up to that time. In addition, the polyethylene formed was more likely to be made up of linear, rather than branched, molecules and, as a consequence, had more durability and strength. Other catalysts were soon developed, and their study contributed to the rapid expansion of organometallic chemistry, now one of the fastest growing areas of chemistry.

In a similar way, but with less obvious milestones, the study of biological materials containing metal atoms has progressed rapidly in the last 30 years.

[4] J. S. Griffith and L. E. Orgel, *Quart. Rev.*, **1957,** *11,* 381.

[5] H. Bethe, *Ann. Physik,* **1929,** *3,* 133.

[6] K. Ziegler, E. Holzkamp, H. Breil, and H. Martin, *Angew. Chem.*, **1955,** *67,* 541.

[7] G. Natta, *J. Polymer Sci.,* **1955,** *16,* 143.

Again, the development of new experimental methods allowed more thorough study of these compounds, and the related theoretical work provided connections to other areas of study. Attempts to make *model* compounds that have chemical and biological activity similar to natural compounds have also led to many new synthetic techniques. Two of the many biological molecules containing metals are shown in Figure 1-8. Although these molecules have very different roles, they share similar ring systems.

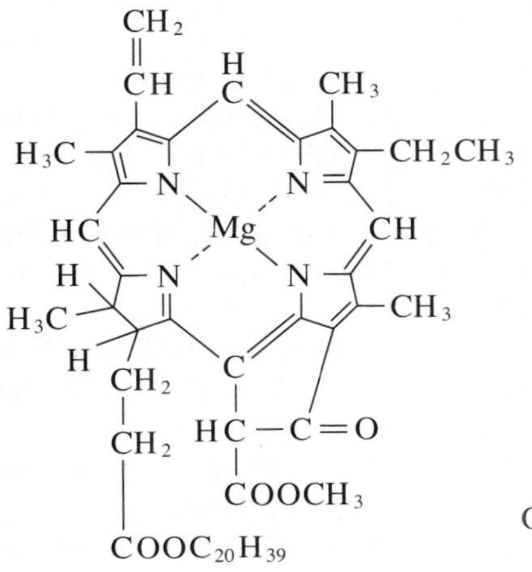

(a)

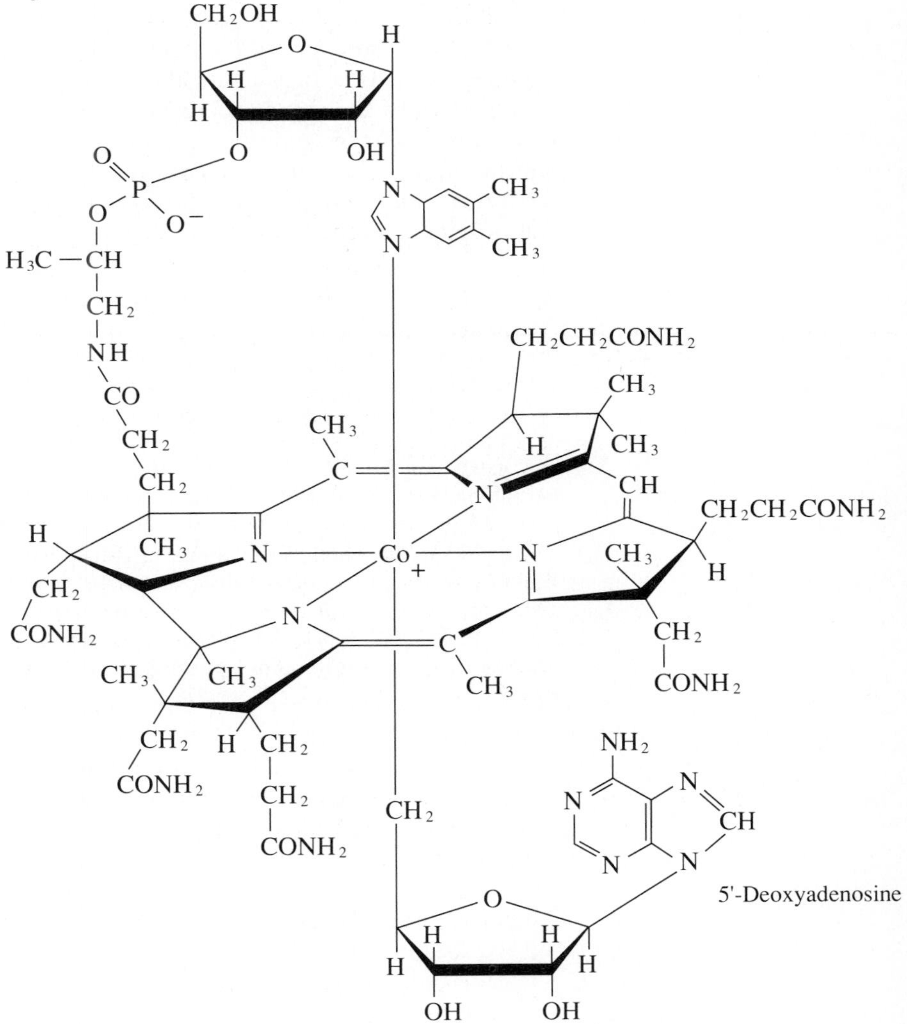

(b)

FIGURE 1-8 Biological Molecules Containing Metal Ions. (a) Chlorophyll *a*, the active agent in photosynthesis. Other chlorophylls are also found with different side chains. A similar ring structure, with Fe replacing Mg, and with different side chains, is called heme, the active site of oxygen-carrying hemoglobin and myoglobin. Similar structures are present in the cytochromes, which are oxidation-reduction catalysts. (b) Vitamin B_{12} coenzyme, a naturally occurring organometallic compound. During isolation from liver, CN^- replaces the 5′-deoxyadenosyl group attached to the cobalt at the bottom of the figure. The coenzyme prevents pernicious anemia and catalyzes many reactions.

A combination of organometallic and bioinorganic chemistry is required for research on a current topic of considerable interest, the conversion of nitrogen to ammonia:

$$N_2 + 3\ H_2 \longrightarrow 2\ NH_3$$

This reaction is one of the most important industrial processes, with about 100 million tons of ammonia now produced each year worldwide. However, in spite of metal oxide catalysts introduced in the Haber–Bosch process in 1913 and improved since then, it is also a reaction that requires temperatures near 400°C and 200 atmospheres pressure and still results in a yield of only 15% ammonia. Bacteria, however, manage to fix nitrogen (convert it to ammonia and then to nitrite and nitrate) at 0.2 atm at room temperature in nodules on the roots of legumes. The nitrogenase enzyme that catalyzes this reaction is a complex iron–molybdenum–sulfur protein that has so far resisted complete characterization. Efforts to synthesize a similar subunit have been moderately successful, but progress has been frustratingly slow.

With this brief survey of the marvelously complex field of inorganic chemistry, we now turn to the details in the remainder of this book. The topics included provide a broad introduction to the field. However, even cursory examination of a chemical library or one of the many inorganic journals will show some of the important aspects of inorganic chemistry that must be omitted in a short textbook. The references cited in the text suggest sources for further study, including historical sources, texts, and reference works that can provide useful additional material.

GENERAL REFERENCES

For those interested in further discussion of the physics of the big bang and related cosmology, a nonmathematical treatment is in S. W. Hawking, *A Brief History of Time,* Bantam, New York, 1988. The inorganic chemistry of minerals, their extraction, and environmental impact at a level understandable to anyone with some understanding of chemistry can be found in J. E. Fergusson, *Inorganic Chemistry and the Earth,* Pergamon, Elmsford, N.Y., 1982. Among the many general reference works available, three of the most useful and complete are N. N. Greenwood and A. Earnshaw, *Chemistry of the Elements,* Pergamon, Elmsford, N.Y., 1984; F. A. Cotton and G. Wilkinson, *Advanced Inorganic Chemistry,* 5th ed., Wiley-Interscience, New York, 1988; and A. F. Wells, *Structural Inorganic Chemistry,* 5th ed., Oxford University Press, New York, 1984. An interesting study of inorganic reactions with a different organization from these is G. Wulfsberg, *Principles of Descriptive Inorganic Chemistry,* Brooks/Cole, Belmont, Calif., 1987.

2

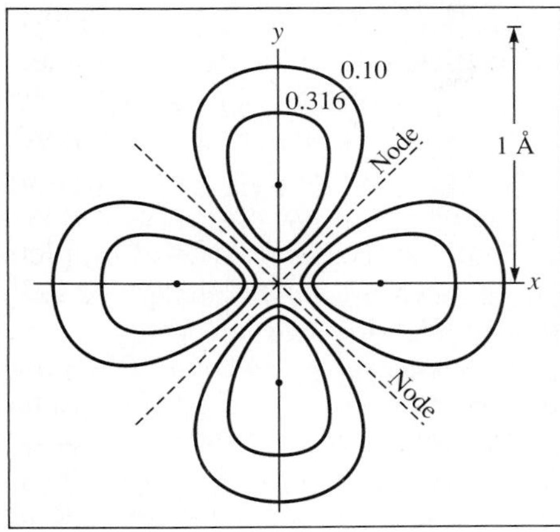

Atomic Structure

The theme of this book is the use of the theories of molecular structure to explain why certain inorganic compounds form and how they react. These theories depend on quantum mechanics to describe atoms and molecules in mathematical terms. While the details of quantum mechanics require considerable mathematical sophistication, it is possible to understand the principles involved with only a moderate amount of mathematics. This chapter presents the fundamentals needed to explain atomic and molecular structure in qualitative or semiquantitative terms. Much of the material may be review, but it is essential background for the remainder of the book.

2-1 HISTORICAL DEVELOPMENT OF ATOMIC THEORY

Although the Greek philosophers Democritus (460–370 B.C.) and Epicurus (341–270 B.C.) presented views of nature that included atoms, many hundreds of years passed before experimental studies could establish the quantitative relationships needed for a coherent atomic theory. In 1808, John Dalton published *A New System of Chemical Philosophy,*[1] in which he proposed that

> the ultimate particles of all homogeneous bodies are perfectly alike in weight, figure, etc. In other words, every particle of water is like every other particle of water, every particle of hydrogen is like every other particle of hydrogen, etc.[2]

[1] John Dalton, *A New System of Chemical Philosophy,* 1808, reprinted with an introduction by Alexander Joseph by Peter Owen Limited, London, 1965.

[2] Ibid., p. 113.

and that atoms combine in simple numerical ratios to form compounds. The terminology he used has since been modified, but he presented clearly the ideas of atoms and molecules, many observations about heat (or caloric, as it was called), and quantitative observations of the masses and volumes of substances combining to form new compounds. Because of confusion about elemental molecules such as H_2 and O_2 (assumed to be monatomic H and O), he did not find the correct formula for water. Dalton said,

> When two measures of hydrogen and one of oxygen gas are mixed, and fired by the electric spark, the whole is converted into steam, and if the pressure be great, this steam becomes water. It is most probable then that there is the same number of particles in two measures of hydrogen as in one of oxygen.[3]

In fact, he then changed his mind about the number of molecules in equal volumes of different gases:

> At the time I formed the theory of mixed gases, I had a confused idea, as many have, I suppose, at this time, that the particles of elastic fluids are all of the same size; that a given volume of oxygenous gas contains just as many particles as the same volume of hydrogenous; or if not, that we had no data from which the question could be solved. . . . I [later] became convinced . . . That every species of pure elastic fluid has its particles globular and all of a size; but that no two species agree in the size of their particles, the pressure and temperature being the same.[4]

Only a few years later, Avogadro used data from Gay-Lussac to argue that equal volumes of gas at equal temperatures and pressures contain the same number of molecules, but uncertainties about the nature of sulfur, phosphorus, arsenic, and mercury vapors delayed acceptance of this idea. Widespread confusion about atomic weights and molecular formulas contributed to the delay; in 1861, Kekulé gave 19 different possible formulas for acetic acid![5] In the 1850s, Cannizzaro revived the argument of Avogadro and argued that everyone should use the same set of atomic weights, rather than the many different sets then being used. At a meeting in Karlsruhe in 1860, he distributed a pamphlet describing his views.[6] His proposal was eventually accepted, and a consistent set of atomic weights and formulas gradually evolved. In 1869, Mendeleev[7] and Meyer[8] independently proposed periodic tables nearly like those used today, and from that time the development of atomic theory progressed rapidly.

2-1-1 THE PERIODIC TABLE

The idea of a periodic table had been considered by many chemists, but either the data to support the idea were insufficient or the classification schemes were incomplete. Mendeleev and Meyer organized the elements in order of atomic weight and then identified families of elements with similar properties.

[3] Ibid., p. 133.

[4] Ibid., pp. 144–45.

[5] J. R. Partington, *A Short History of Chemistry,* 3rd ed., Macmillan, London, 1957, reprinted, 1960, Harper & Row, New York, p. 255.

[6] Ibid., pp. 256–58.

[7] D. I. Mendeleev, *J. Russ. Phys. Chem. Soc.,* **1869,** *i,* 60.

[8] L. Meyer, *Ann.,* **1870,** *Suppl. vii,* 354.

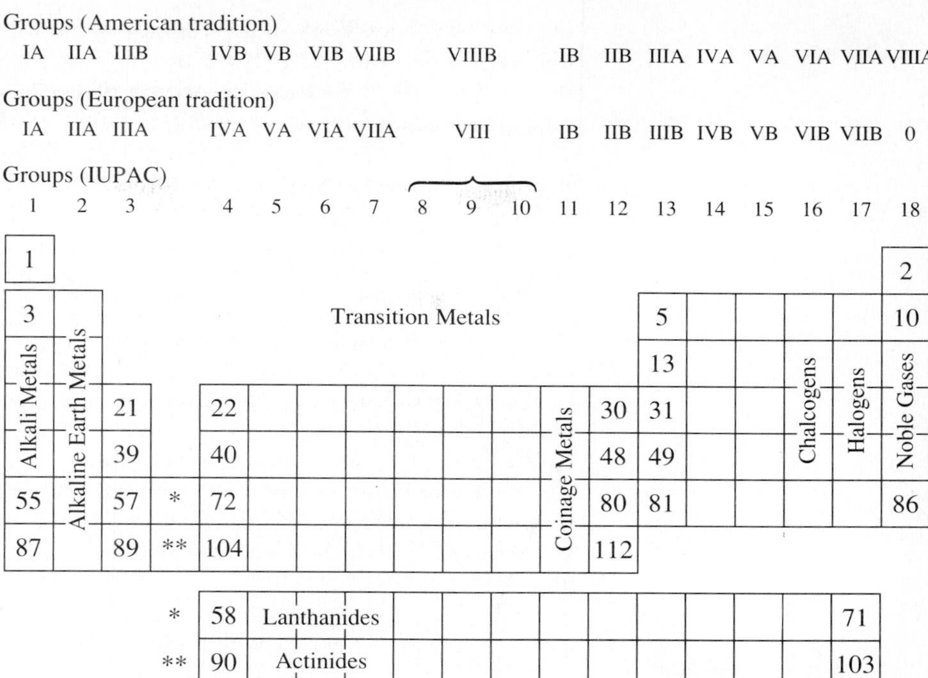

FIGURE 2-1 Names for Parts of the Periodic Table.

By arranging these families in rows or columns, and by considering similarities in chemical behavior as well as atomic weight, Mendeleev found vacancies in the table and was able to predict the properties of several elements (gallium, scandium, germanium, polonium) that had not yet been discovered. When his predictions proved accurate, the concept of a periodic table was quickly established. The discovery of additional elements not known in Mendeleev's time and the synthesis of heavy elements by physical means have led to the more complete periodic table in its modern form, as shown inside the front cover of this text.

Before proceeding with the development of atomic theory, we will define some terms for different parts of the periodic table that will be needed later for the discussion of periodicity and trends in physical and chemical properties. Some of these terms are identified in Figure 2-1. A horizontal row of elements is called a **period,** and a vertical column is a **group** or **family.** Note that the traditional designations of groups in the United States differ from those used in Europe. Recently, the International Union of Pure and Applied Chemistry (IUPAC) has recommended that the groups be numbered 1 through 18, a recommendation that has generated considerable controversy. Rather than argue the point, we will use the IUPAC group numbers, with the traditional American numbers in parentheses. Larger sections of the table are frequently called **blocks;** the names used for some of these are given in Figure 2-1.

2-1-2 DISCOVERY OF SUBATOMIC PARTICLES AND THE BOHR ATOM

Electrons were discovered and studied during the time when the periodic table was being formulated, and in 1897 J. J. Thomson showed they were negatively charged particles very much lighter than the hydrogen atom.[9] Becquerel discovered the radioactivity of uranium in 1896, which in the next 20 years led to the idea of isotopes as atoms with the same chemical properties but different nuclear masses. The combination of these ideas showed a complex structure for the atom, but did not describe it adequately. In 1911, students of Rutherford discovered that a small fraction of alpha particles (helium nuclei) was deflected at large angles on passing through gold foil, while most of them passed directly through. They quickly realized that this required much empty space in the atom and a heavy, but tiny, nucleus carrying a positive charge.

Soon after, Moseley[10] completed this part of the picture by measuring the wavelengths of X-rays emitted when the elements were made the target of a high voltage electron beam. He found that the square root of the frequency of the emitted X-rays varied almost linearly with atomic number Z, approximately half the atomic mass. This experiment confirmed the conclusions of Rutherford, that the atomic number had a more fundamental importance than the mass and that the charge on the nucleus matched the atomic number. This result also explained some discrepancies in the periodic table, where the atomic weight order is not the same as the atomic number order of the elements (Co and Ni, Te and I).

Parallel discoveries in atomic spectra showed that each element emits light of specific energies when excited by an electric discharge or heat. In 1885, Balmer showed that the energies of visible light emitted by the hydrogen atom are given by the equation

$$E = R_H\left(\frac{1}{2^2} - \frac{1}{n_h^2}\right),$$

where

$$\begin{aligned} n_h &= \text{integer, with } n_h > 2 \\ R_H &= \text{Rydberg constant for hydrogen} \\ &= 1.09678 \times 10^7\ \text{m}^{-1} \\ &= 1312.0\ \text{kJ mol}^{-1} \end{aligned}$$

The equation was later made more general, as spectral lines in the ultraviolet and infrared regions of the spectrum were discovered, by replacing 2^2 by n_l^2, with the condition that $n_l < n_h$. The origin of this light energy was unknown until Niels Bohr's quantum theory of the atom,[11] first published in 1913 and refined over the following ten years. This theory assumed that negative electrons in atoms move in stable circular orbits around the positive nucleus with no absorption or emission of energy. However, electrons may absorb light of certain specific energies and be excited to orbits of higher energy; they may also emit light of specific energies and fall to orbits of lower energy. The energy of the light emitted or absorbed can be found, according to the Bohr model of the hydrogen atom, from the equation

$$E = R\left(\frac{1}{n_l^2} - \frac{1}{n_h^2}\right)$$

[9] Partington, op. cit., p. 357.

[10] H. G. J. Moseley, *Phil. Mag.*, **1913,** *xxvi,* 102; **1914,** *xxvii,* 703.

[11] N. Bohr, *Phil. Mag.*, **1913,** *26,* 1.

where $$R = \frac{2\pi^2 \mu Z^2 e^4}{(4\pi\epsilon_0)^2 h^2}$$

μ = reduced mass of the electron–nucleus combination ($1/\mu = 1/m_e + 1/m_{nucleus}$)

m_e = mass of the electron

$m_{nucleus}$ = mass of the nucleus

Z = charge of the nucleus

e = electronic charge

h = Planck's constant

n_h = quantum number describing the higher-energy state

n_l = quantum number describing the lower-energy state

ϵ_0 = permittivity of a vacuum, $4\pi\epsilon_0 = 1.11265 \times 10^{-10}\ C^2\ N^{-1}\ m^{-2}$

This equation shows that the Rydberg constant depends on the mass of the nucleus as well as the fundamental constants.

Examples of the emission spectrum observed for the hydrogen atom and the energy levels responsible are shown in Figure 2-2. As the electrons drop from level n_h to n_l, energy is released in the form of light. Conversely, if light of the correct energy is absorbed by an atom, electrons are raised from level n_l to level n_h (l for lower level, h for higher). The inverse-square dependence of energy on n_i results in energy levels that are far apart in energy at small n_i and become much closer in energy at larger n_i. In the upper limit, as n_i approaches infinity, the energy approaches a limit of zero. Individual electrons can have more energy, but above this point they are no longer part of the atom; an infinite quantum number means that the nucleus and the electron are separate entities.

When applied to hydrogen, Bohr's theory worked well; when atoms with more electrons were considered, the theory failed. Complications such as elliptical rather than circular orbits were introduced in an attempt to fit the data to Bohr's theory. The developing experimental science of atomic spectroscopy provided extensive data for testing of the Bohr theory and its modifications and forced the theorists to work hard to explain the spectroscopists' observations. In spite of their efforts, the Bohr theory eventually proved unsatisfactory. An important characteristic of the electron, its wave nature, still needed to be considered.

All moving particles have wave properties according to the de Broglie equation[12]

$$\lambda = \frac{h}{mv}$$

where

λ = wavelength of the particle

h = Planck's constant

m = mass of the particle

v = velocity of the particle

[12] L. de Broglie, *Phil. Mag.*, **1924**, *47*, 446; *Ann. Phys.*, **1925**, *3*, 22.

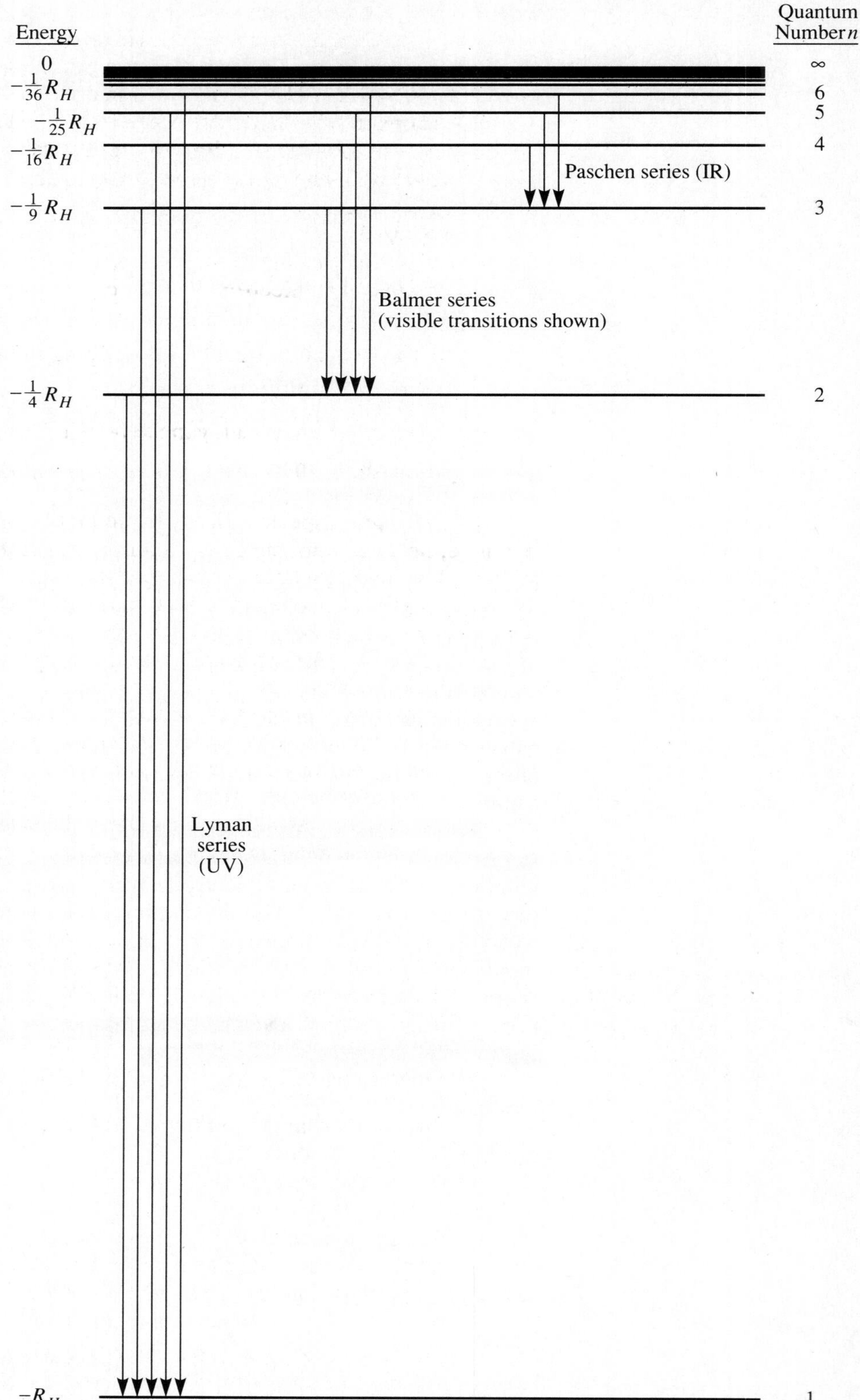

FIGURE 2-2 The Hydrogen Atom Spectrum and Energy Levels.

Particles large enough to be visible (and even those much smaller) have very short wavelengths because of their large masses; the wavelengths are too small to be measured. Electrons, on the other hand, have very distinct wave properties because of their very small mass.

Electrons moving in circles around the nucleus, as in Bohr's theory, can be thought of as forming standing waves according to the de Broglie equation. However, we no longer believe that it is possible to describe the motion of an electron in an atom so precisely. This is a consequence of another fundamental principle of modern physics, **Heisenberg's uncertainty principle,**[13] which states that there is a relationship between the inherent uncertainties in the location and momentum of an electron described by the equation:

$$\Delta x \, \Delta p_x \geq \frac{h}{4\pi}$$

where Δx = uncertainty in the position of the electron

Δp_x = uncertainty in the momentum of the electron

The energy of spectral lines can be measured with great precision (as an example, note the number of significant figures in the Rydberg constant), which in turn allows precise determination of the energy of electrons in atoms. This precision in energy also implies precision in momentum (Δp_x is small); therefore, according to Heisenberg, there is a large uncertainty in the location of the electron (Δx is large). These concepts mean that we cannot treat electrons as simple particles with their motion described precisely, but we must instead consider the wave properties of electrons, characterized by a degree of uncertainty in their location. In other words, instead of being able to describe precise **orbits** of electrons, as in the Bohr theory, we can only describe **orbitals,** regions that describe the probable location of electrons. The **probability** of finding the electron at a particular point in space (also called the **electron density**) can be calculated, at least in principle.

2-2 THE SCHRÖDINGER EQUATION

In 1926 and 1927, Heisenberg[13] and Schrödinger[14] published papers on wave mechanics (descriptions of the wave properties of electrons in atoms) that used very different mathematical techniques. In spite of the different approaches, it was soon shown that their theories were equivalent. Schrödinger's differential equations are more commonly used to introduce the theory, and we will follow that practice. Heisenberg's matrix methods are very useful in many quantum mechanics problems, but are less pictorial and are therefore not used in this introduction.

The Schrödinger equation is based on the usual concepts of potential and kinetic energy, but with major changes to account for the wave properties of the electron. It describes energy levels and other properties of electrons in atoms and molecules, and, when all the interactions between particles can be adequately included, gives results that match experimental observations. The

[13] W. Heisenberg, *Z. Phys.*, **1927,** *43,* 172.

[14] E. Schrödinger, *Ann. Phys.* (Leipzig), **1926,** *79,* 361, 489, 734; **1926,** *80,* 437; **1926,** *81,* 109; *Naturwissenschaften,* **1926,** *14,* 664; *Phys. Rev.*, **1926,** *28,* 1049.

equation is based on the **wave function Ψ,** which describes the behavior of the electron. In its simplest notation, the equation is

$$H\Psi = E\Psi$$

where
H = Hamiltonian operator
E = energy of the electron
Ψ = wave function

The **Hamiltonian operator** (frequently just called the Hamiltonian) includes derivatives that **operate** on the wave function (an operator changes the function it operates on in some way; it can be as simple as multiplication by a constant or much more complex than the Hamiltonian). In the forms used for calculating energy levels, the equation is

$$\left[\frac{-h^2}{8\pi^2 m}\left(\frac{\partial^2}{\partial x^2} + \frac{\partial^2}{\partial y^2} + \frac{\partial^2}{\partial z^2}\right) + V(x, y, z)\right]\Psi(x, y, z) = E\Psi(x, y, z)$$

where

$$H = \frac{-h^2}{8\pi^2 m}\left(\frac{\partial^2}{\partial x^2} + \frac{\partial^2}{\partial y^2} + \frac{\partial^2}{\partial z^2}\right) + V(x, y, z)$$

h = Planck's constant
m = mass of the particle
E = total energy of the system
$\Psi(x, y, z)$ = wave function
$V(x, y, z)$ = potential energy (from electrostatic interactions)

The wave function Ψ is the solution of the equation, which describes the wave properties of the electron and contains all the information about the system that can be obtained by quantum mechanics. Each solution of the Schrödinger equation for an atom describes the behavior of an electron in a particular orbital. The product of Ψ and its complex conjugate,[15] Ψ^*, at any point is the probability of finding the particle at that point, and Ψ can also be used to find other properties of the system. A number of conditions are required for a physically realistic solution for Ψ in atoms:

1. The wave function Ψ must be single valued. (There can only be one value for the electron density at a specific point in space.)
2. The wave function Ψ and its first derivatives must be continuous. (The probability cannot change abruptly from one point to the next.)
3. The wave function Ψ must be quadratically integrable. This means that $\int \Psi\Psi^* \, d\tau$ (where $\int d\tau$ means integrating all dimensions over all space) is finite. In fact, the magnitude of Ψ is adjusted so that $\int \Psi\Psi^* \, d\tau = 1$. This is called **normalizing** the wave function, so the probability of finding the particle summed over all space is 100%. In other words, the total probability of finding the electron somewhere in space is equal to unity; it must be somewhere.

[15] Ψ may be a complex function, containing i ($= \sqrt{-1}$). The complex conjugate, Ψ^*, is the same function with $-i$ replacing i. The product, $\Psi\Psi^*$, is real.

4. Any two solutions to the wave function must be **orthogonal,** which means $\int \Psi_A \Psi_B \, d\tau = 0$. This condition is sometimes ignored in approximate calculations, but it is a condition for a totally accurate wave function.

2-2-1 THE PARTICLE IN A BOX

The usual first example of the wave equation, the one-dimensional particle in a box, shows how these conditions are used. We will give an outline of the method; the details are available elsewhere.[16] The "box" is shown in Figure 2-3. The potential energy $V(x)$ inside the box, between $x = 0$ and $x = a$, is zero. Outside the box, the potential energy is infinite. This means that the particle (which would therefore need to acquire an infinite amount of energy to leave the box) is completely trapped in the box, but has no forces acting on it within the box. The wave equation for locations within the box is

$$\frac{-h^2}{8\pi^2 m}\left(\frac{\partial^2 \Psi(x)}{\partial x^2}\right) = E\Psi(x), \qquad \text{because } V(x) = 0$$

Sine and cosine functions have the properties that we associate with waves—a well-defined wavelength and amplitude—and we may therefore propose that the wave characteristics of our particle may be described by a combination of sine and cosine functions. A general solution to describe the possible waves in the box would then be

$$\Psi = A \sin rx + B \cos sx$$

where A, B, r, and s are constants. Substitution into the wave equation allows solution for r and s (see problem 4 at the end of the chapter):

$$r = s = \sqrt{2mE}\,\frac{2\pi}{h}$$

Since Ψ must be continuous and must equal zero at $x < 0$ and $x > a$ (because the particle is confined to the box), Ψ must go to zero at $x = 0$ and $x = a$. Since $\cos sx = 1$ for $x = 0$, Ψ can equal zero in the general solution above only if $B = 0$. This reduces the expression for Ψ to

$$\Psi = A \sin rx$$

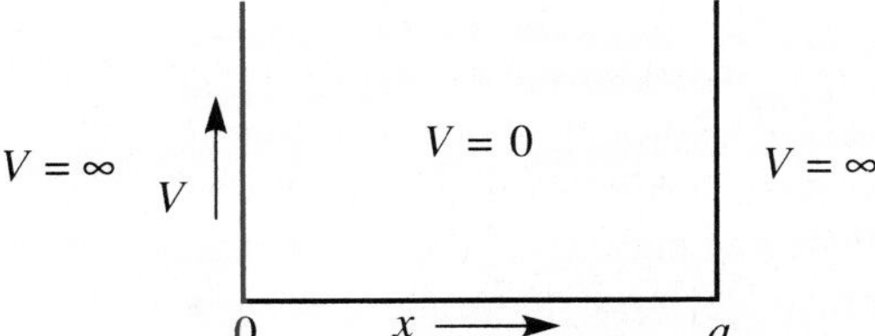

FIGURE 2-3 Potential Energy Well for the Particle in a Box.

[16] G. M. Barrow, *Physical Chemistry,* 5th ed., McGraw-Hill, New York, 1988, pp. 64–70, calls this the "particle on a line" problem. Many other physical chemistry texts also include solutions.

At $x = a$, Ψ must also equal zero; therefore, $\sin ra = 0$, which is possible only if ra is an integral multiple of π:

$$ra = \pm n\pi \quad \text{or} \quad r = \frac{\pm n\pi}{a}$$

where n = any integer $\neq 0$.[17] Substituting the positive value (since both positive and negative values give the same results) for r into the solution for r on the preceding page gives

$$r = \frac{n\pi}{a} = \sqrt{2mE}\,\frac{2\pi}{h}$$

This expression may be solved for E:

$$E = \frac{n^2h^2}{8ma^2}$$

These are the energy levels predicted by the particle in a box model for any particle in a one-dimensional box of length a. The energy levels are *quantized* according to **quantum numbers** $n = 1, 2, 3, \ldots$.

Substituting $r = n\pi/a$ into the wave function gives

$$\Psi = A \sin \frac{n\pi x}{a}$$

and applying the normalizing requirement $\int \Psi\Psi^*\, d\tau = 1$ gives

$$A = \sqrt{\frac{2}{a}}$$

The total solution is then

$$\Psi = \sqrt{\frac{2}{a}} \sin \frac{n\pi x}{a}$$

The resulting wave functions and their squares for the first three states are plotted in Figure 2-4.

The squared wave functions are the probability densities and show the difference between classical and quantum mechanical behavior. Classical mechanics predicts that the electron has equal probability of being at any point in the box. The wave nature of the electron gives it the extremes of high and low probability at different locations in the box. Larger particles (atoms or molecules in a box of visible dimensions) have very short wavelengths and closely spaced energy levels; when they have any significant amount of energy, they have very large quantum numbers (magnitudes in the thousands or higher) and essentially classical probabilities.

[17] If $n = 0$, then $r = 0$ and $\Psi = 0$ at all points. The probability of finding the electron is $\int \Psi\Psi^*\, dx = 0$, and there is no electron at all.

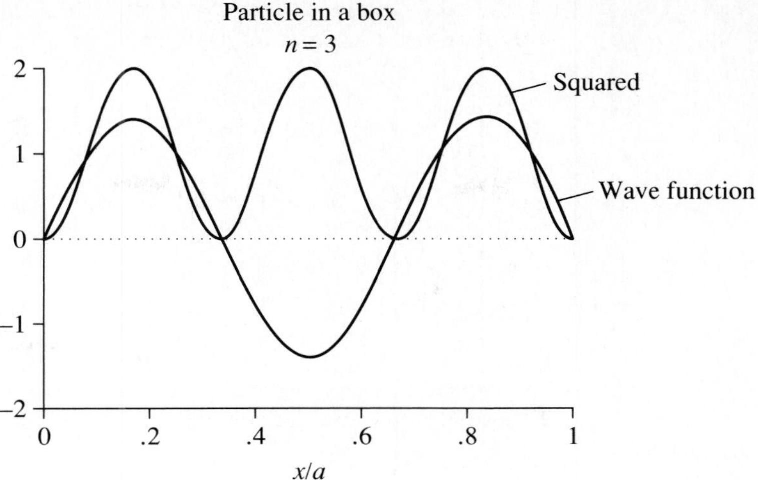

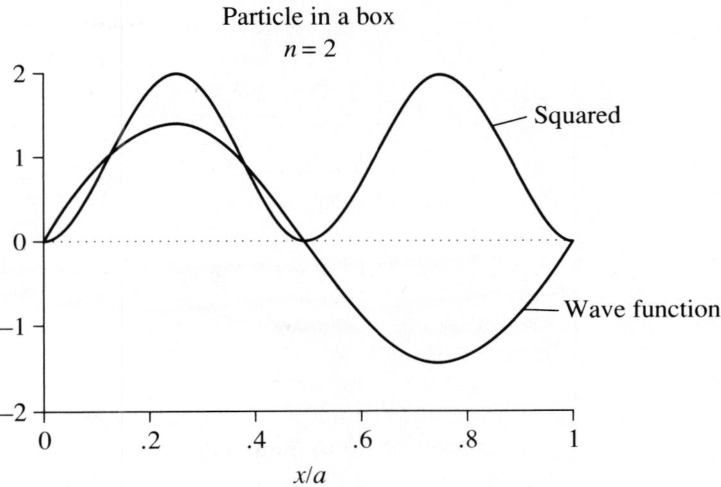

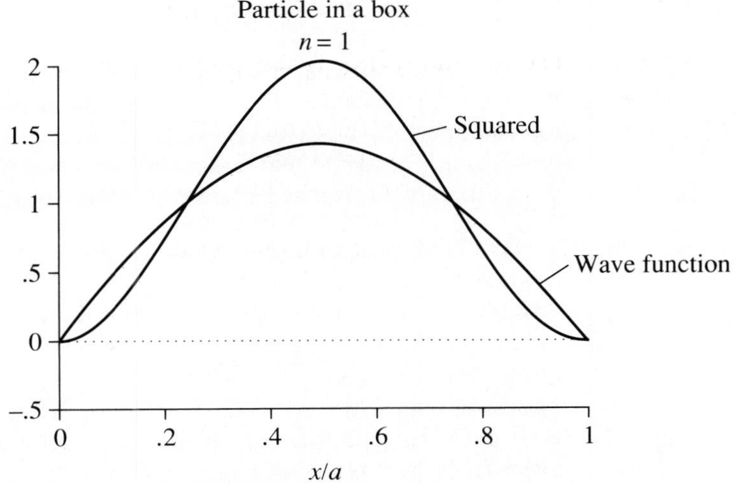

FIGURE 2-4 Wave Functions and Their Squares for the Particle in a Box with $n = 1, 2,$ and 3.

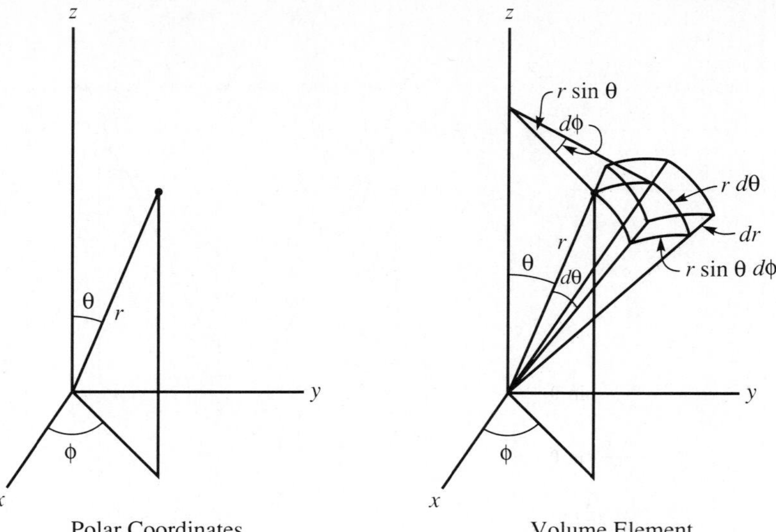

FIGURE 2-5 Polar Coordinates and Volume Element for a Spherical Shell in Polar Coordinates.

2-2-2 ATOMIC WAVE FUNCTIONS

The same methods, with considerably more complexity, can be expanded to three dimensions for atoms. Polar coordinates, as shown in Figure 2-5, are used because of the spherical nature of the potential energy in atoms. The x, y, and z coordinates are changed to r, the distance from the nucleus; θ, the angle from the z axis, varying from 0 to π; and ϕ, the angle from the x axis, varying from 0 to 2π. In polar coordinates, the volume element $dx\ dy\ dz$ becomes $r^2 \sin\theta\ d\theta\ d\phi\ dr$, and the volume of the thin shell between r and $r + dr$ becomes $4\pi r^2\ dr$.

The wave functions can be factored readily into a **radial factor *R*** which is a function of r only, and **angular factors Θ** and **Φ** which are functions of the angles θ and ϕ only. The two angular factors are sometimes combined into one factor, called Y:

$$\Psi(r, \theta, \phi) = R(r)\Theta(\theta)\Phi(\phi) = R(r)Y(\theta, \phi)$$

The equations can then be solved separately for the individual factors, which are shown in Tables 2-1 and 2-2. The Bohr radius, a_0, which appears in the radial part of these functions, is a common unit in quantum mechanics. It is the value of r at the maximum of $\Psi\Psi^*$ for a hydrogen 1s orbital and is also the radius of a 1s orbital according to the Bohr model.

The hydrogen atom has a potential energy (in joules/atom) of

$$V = \frac{-Ze^2}{4\pi\epsilon_0 r}$$

representing the electrostatic attraction between the nucleus and the electron. The negative potential energy is a characteristic of each electronic energy level of the hydrogen atom. An electron near the nucleus (small r) has a large

TABLE 2-1
Hydrogen atom wave functions: Angular factors

		Angular factors			*Real wave functions*		
		Related to angular momentum		*Functions of θ*	*In Polar coordinates*	*In Cartesian coordinates*	*Shapes*
l	m_l	Φ	Θ		$\Theta\Phi(\theta, \phi)$	$\Theta\Phi(x, y, z)$	
0(*s*)	0	$\frac{1}{\sqrt{2\pi}}$	$\frac{1}{\sqrt{2}}$		$\frac{1}{2\sqrt{\pi}}$	$\frac{1}{2\sqrt{\pi}}$	
1(*p*)	0	$\frac{1}{\sqrt{2\pi}}$	$\frac{\sqrt{6}}{2}\cos\theta$		$\frac{1}{2}\sqrt{\frac{3}{\pi}}\cos\theta$	$\frac{1}{2}\sqrt{\frac{3}{\pi}}\frac{z}{r}$	
	+1	$\frac{1}{\sqrt{2\pi}}e^{i\phi}$	$\frac{\sqrt{6}}{2}\sin\theta$		$\frac{1}{2}\sqrt{\frac{3}{\pi}}\sin\theta\cos\phi$	$\frac{1}{2}\sqrt{\frac{3}{\pi}}\frac{x}{r}$	
	−1	$\frac{1}{\sqrt{2\pi}}e^{-i\phi}$	$\frac{\sqrt{6}}{2}\sin\theta$		$\frac{1}{2}\sqrt{\frac{3}{\pi}}\sin\theta\sin\phi$	$\frac{1}{2}\sqrt{\frac{3}{\pi}}\frac{y}{r}$	
2(*d*)	0	$\frac{1}{\sqrt{2\pi}}$	$\frac{1}{2}\sqrt{\frac{5}{2}}(3\cos^2\theta - 1)$		$\frac{1}{4}\sqrt{\frac{5}{\pi}}(3\cos^2\theta - 1)$	$\frac{1}{4}\sqrt{\frac{5}{\pi}}\frac{(2z^2 - x^2 - y^2)}{r^2}$	
	+1	$\frac{1}{\sqrt{2\pi}}e^{i\phi}$	$\frac{\sqrt{15}}{2}\cos\theta\sin\theta$		$\frac{1}{2}\sqrt{\frac{15}{\pi}}\cos\theta\sin\theta\cos\phi$	$\frac{1}{2}\sqrt{\frac{15}{\pi}}\frac{xz}{r^2}$	
	−1	$\frac{1}{\sqrt{2\pi}}e^{-i\phi}$	$\frac{\sqrt{15}}{2}\cos\theta\sin\theta$		$\frac{1}{2}\sqrt{\frac{15}{\pi}}\cos\theta\sin\theta\sin\phi$	$\frac{1}{2}\sqrt{\frac{15}{\pi}}\frac{yz}{r^2}$	
	+2	$\frac{1}{\sqrt{2\pi}}e^{2i\phi}$	$\frac{\sqrt{15}}{4}\sin^2\theta$		$\frac{1}{4}\sqrt{\frac{15}{\pi}}\sin^2\theta\cos 2\phi$	$\frac{1}{4}\sqrt{\frac{15}{\pi}}\frac{x^2 - y^2}{r^2}$	
	−2	$\frac{1}{\sqrt{2\pi}}e^{-2i\phi}$	$\frac{\sqrt{15}}{4}\sin^2\theta$		$\frac{1}{4}\sqrt{\frac{15}{\pi}}\sin^2\theta\sin 2\phi$	$\frac{1}{4}\sqrt{\frac{15}{\pi}}\frac{xy}{r^2}$	

SOURCE: Adapted from G. M. Barrow, *Physical Chemistry*, 5th ed., McGraw-Hill, New York, 1988, with permission.

NOTE: The relations $(e^{i\phi} - e^{-i\phi})/(2i) = \sin\phi$ and $(e^{i\phi} + e^{-i\phi})/2 = \cos\phi$ can be used to convert the exponential imaginary functions to real trigonometric functions, combining the two orbitals with $m_l = \pm 1$ to give two orbitals with $\sin\phi$ and $\cos\phi$. In a similar fashion, the orbitals with $m_l = \pm 2$ result in real functions with $\cos^2\phi$ and $\sin^2\phi$. These functions have then been converted to Cartesian form by using the functions $x = r\sin\theta\cos\phi$, $y = r\sin\theta\sin\phi$, and $z = r\cos\theta$.

TABLE 2-2
Hydrogen atom wave functions: Radial factors

		Radial factors ($\sigma = Zr/a_0$)
n	l	$R(r)$
1	0	$R_{1s} = 2\left[\frac{Z}{a_0}\right]^{3/2} e^{-\sigma}$
2	0	$R_{2s} = \left[\frac{Z}{2a_0}\right]^{3/2} (2 - \sigma)e^{-\sigma/2}$
	1	$R_{2p} = \frac{1}{\sqrt{3}}\left[\frac{Z}{2a_0}\right]^{3/2} \sigma e^{-\sigma/2}$
3	0	$R_{3s} = \frac{2}{27}\left[\frac{Z}{3a_0}\right]^{3/2} (27 - 18\sigma + 2\sigma^2)e^{-\sigma/3}$
	1	$R_{3p} = \frac{1}{81\sqrt{3}}\left[\frac{2Z}{a_0}\right]^{3/2} (6 - \sigma)\sigma e^{-\sigma/3}$
	2	$R_{3d} = \frac{1}{81\sqrt{15}}\left[\frac{2Z}{a_0}\right]^{3/2} \sigma^2 e^{-\sigma/3}$

negative potential energy; electrons farther from the nucleus have potential energies that are less negative. For an electron at infinite distance from the nucleus ($r = \infty$), the attraction between the nucleus and the electron is zero, and the potential energy is zero.

2-2-3 QUANTUM NUMBERS AND ORBITALS

Mathematically, atomic orbitals are discrete solutions of the three-dimensional Schrödinger equation. These orbital equations include three quantum numbers, n, l, and m_l. A fourth quantum number, m_s, completes the description by accounting for the magnetic moment of the electron. This is usually described as the spin of the electron, because a spinning electrically charged particle would have a magnetic moment. The resulting quantum numbers are summarized in Table 2-3.

The quantum number n is primarily responsible for determining the overall energy of an atomic orbital; the other quantum numbers have smaller effects on the energy. The quantum number l determines the angular momentum of the electron and the shape of the orbital, and has a smaller effect on the energy. The quantum number m_l determines the orientation of the angular momentum vector in a magnetic field, as in Figure 2-6. The quantum number m_s determines the orientation of the electron magnetic moment in a magnetic field, either in the direction of the field ($+1/2$) or opposed to it ($-1/2$). When no field is present, all m_l values (all three p orbitals or all five d orbitals) have the same energy, and both m_s values have the same energy. Together the quantum numbers n, l, and m_l define an atomic orbital; the quantum number m_s describes the electron spin within the orbital.

The energy predicted by both the Bohr and Schrödinger equations for an electron with principal quantum number n is

$$E = \frac{-2\pi^2\mu e^4 Z^2}{(4\pi\epsilon_0)^2 h^2 n^2}$$

The radial wave functions [plotted as $a_0^{3/2}R(r)$], and the radial distribution functions [plotted as $a_0 r^2 R(r)^2$] for the $n = 1$, 2, and 3 orbitals are plotted as functions of the radius (in units of r/a_0) in Figure 2-7. The volume element for

TABLE 2-3
Quantum numbers and their properties

Symbol	*Name*	*Values*	*Role*
n	Principal	$1, 2, 3, \ldots$	Determines the major part of the energy
l	Angular momentum	$0, 1, 2, \ldots, n-1$	Describes angular dependence and contributes to the energy
m_l	Magnetic	$0, \pm 1, \pm 2, \ldots, \pm l$	Describes orientation in space
m_s	Spin	$\pm\frac{1}{2}$	Describes orientation of the electron spin in space

Orbitals with different l values are usually known by the following labels, which are derived from early terms for different families of spectroscopic lines:

$l =$	0	1	2	3	4	5, . . .
Label	s	p	d	f	g	continuing alphabetically

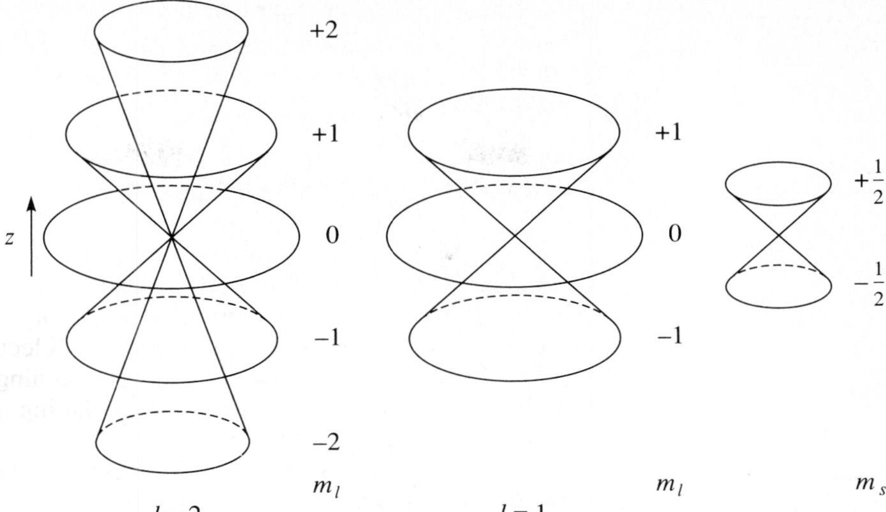

FIGURE 2-6 The Angular Momentum Vector in a Magnetic Field. The cones represent the uncertainty in the direction of the angular momentum vectors. Only the resultant in the field direction can be known.

spherical coordinates is $\sin\theta\, d\theta\, d\phi\, r^2 dr$; for the radial function, the appropriate function is then r^2R^2dr. The angular factors for all the wave functions, when integrated over all angles, have values of 1. The radial distribution function then gives the probability of finding an electron at a given distance from the nucleus, integrated over all angles.

The electron density, or probability of finding the electron, falls off rapidly as the distance from the nucleus increases. The 2*s* orbital also has a **nodal surface,** a surface with zero electron density, in this case a sphere with $r = 2a_0$, where the probability is zero. Nodes appear naturally as a result of the wave nature of the electron and the functions that result from solving the wave equation from Ψ. Although relativistic arguments can be used to explain why the probability at a nodal surface is not precisely zero,[18] the conventional approach is to use these nodes as part of the description of the electronic structures of atoms and molecules. A node is the surface where the wave function is zero as it changes sign (as at $r = 2a_0$ in the 2*s* orbital); this requires that $\Psi = 0$, and the probability of finding the electron on that surface is also zero.

Nodes in orbitals are frequently compared to nodes in vibrating strings. Figure 2-4 shows nodes at $x/a = 0.5$ for $n = 2$ and at $x/a = 0.33$ and 0.67 for $n = 3$ for the particle in a box. The same diagrams could represent the amplitudes of the motion of vibrating strings at the fundamental frequency ($n = 1$) and multiples of 2 and 3. At the nodes, the amplitude of the motion is zero.

As n increases, the number of nodes also increases. Nodes may be spherical in shape as in 2*s*, planar as in 2*p*, conical as in the $3d_{z^2}$, or a mix including several shapes as in 3*p*. The spherical nodes (sometimes called radial nodes) give the atom a layered appearance as n increases. For the wave

[18] A. Szabo, *J. Chem. Educ.*, **1969,** *46*, 678, uses relativistic arguments to explain that the electron probability at a nodal surface has a very small, but finite, value.

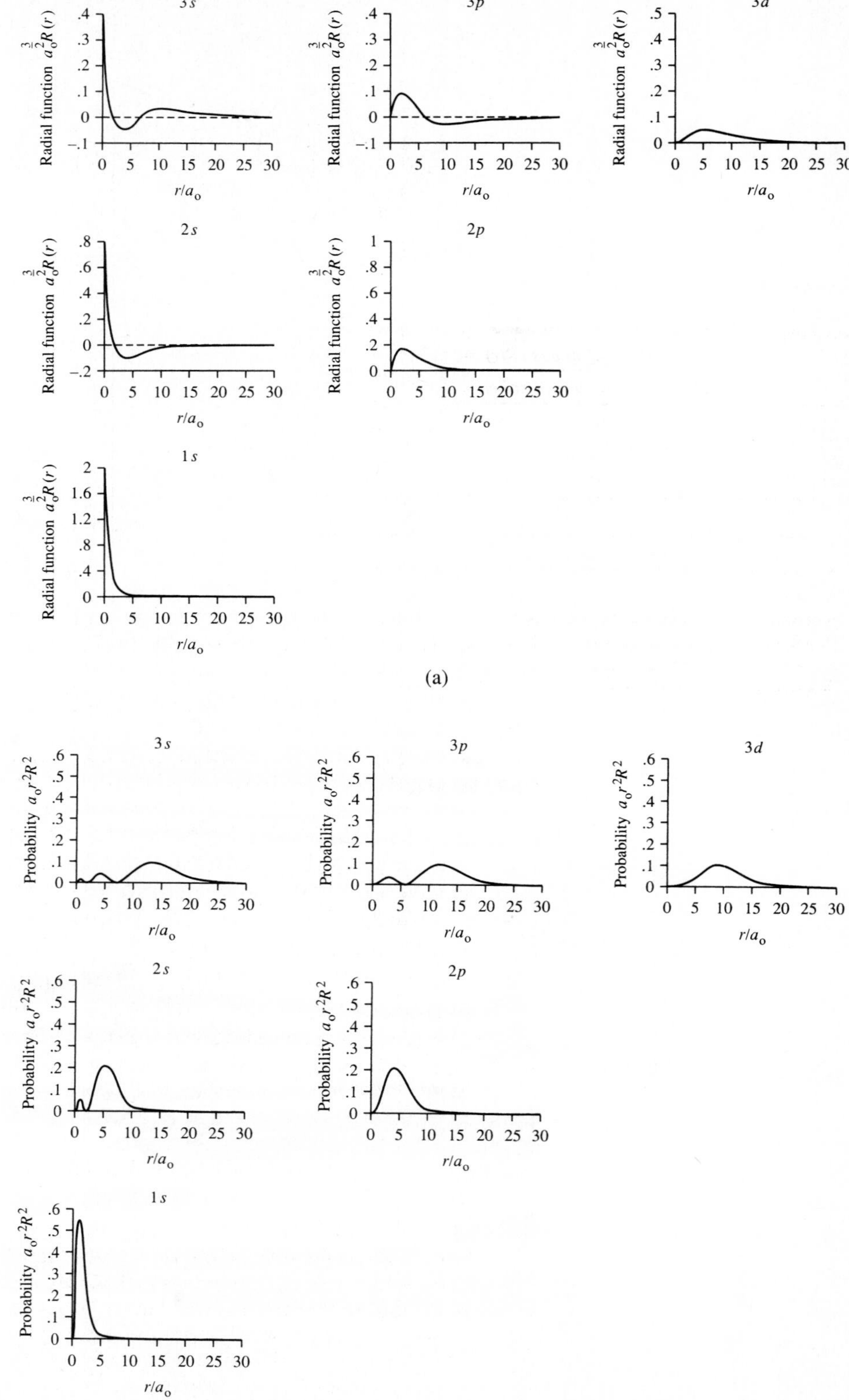

FIGURE 2-7 Radial Wave Functions and Radial Probability Functions. (a) Radial wave functions for hydrogenlike orbitals, $a_0^{3/2}R(r)$. Note the different vertical scales of the graphs. (b) Radial probability functions, $a_0 r^2 R^2(r)$. In these graphs, all the vertical scales are the same, for better comparison of the orbitals.

function $\Psi(r, \theta, \phi) = R(r)Y(\theta, \phi)$, nodes appear when either $R(r) = 0$ or $Y(\theta, \phi) = 0$. Table 2-4 gives the spherical and angular nodes for several orbitals. When the angular part of the wave function is expressed in Cartesian (x, y, z) form (see Table 2-1), the nodal surfaces are easily determined. In addition, the regions where the wave function is positive and where it is negative can be found. This information will be useful in working with molecular orbitals later in Chapter 5. Two examples will show how to determine the nodes.

TABLE 2-4
Nodal surfaces

Spherical nodes [$R(r) = 0$]					
Examples (number of spherical nodes)					
1*s*	0	2*p*	0	3*d*	0
2*s*	1	3*p*	1	4*d*	1
3*s*	2	4*p*	2	5*d*	2

Angular nodes [$Y(\theta, \phi) = 0$]	
Examples (number of angular nodes)	
s orbitals	0
p orbitals	1 plane for each orbital
d orbitals	2 planes for each orbital except d_{z^2} 1 conical surface for d_{z^2}

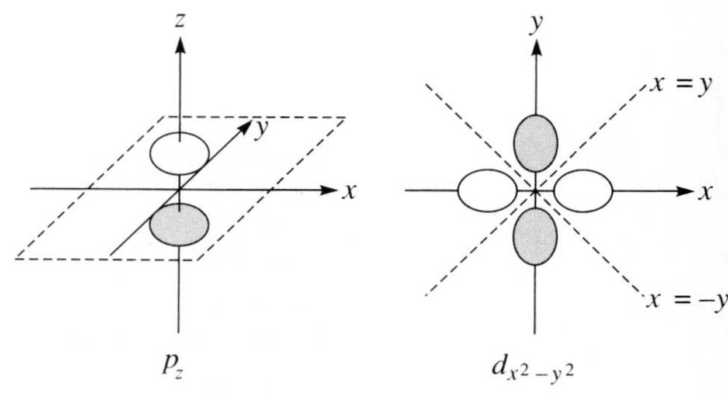

EXAMPLES

p_z The angular factor Y is given in Table 2-1 in terms of Cartesian coordinates:

$$Y = \frac{1}{2}\sqrt{\frac{3}{\pi}}\frac{z}{r}$$

This orbital is designated p_z because z appears in the Y expression. For an angular node, Y must equal zero, which is true only if $z = 0$. Therefore, $z = 0$ (the xy plane) is an angular nodal surface for the p_z orbital. The wave function is positive where $z > 0$ and negative where $z < 0$. A $2p_z$ orbital has no spherical nodes, a $3p_z$ orbital has one spherical node, and so on.

$d_{x^2-y^2}$

$$Y = \frac{1}{4}\sqrt{\frac{15}{\pi}}\frac{x^2 - y^2}{r^2}$$

Here, the expression $x^2 - y^2$ appears in the equation, so the designation is $d_{x^2-y^2}$. Since there are two solutions to the equation $Y = 0$ (or $x^2 - y^2 = 0$), $x = y$ and $x = -y$, the planes defined by these equations are the angular nodal surfaces. They are planes containing the z axis and making 45° angles with the x and y axes. The function is positive where $x > y$ and negative where $x < y$. Table 2-4 shows the nodal surfaces and the positive and negative lobes of these orbitals. A $3d_{x^2-y^2}$ orbital has no spherical nodes, a $4d_{x^2-y^2}$ orbital has one spherical node, and so on.

EXERCISE 2-1

Describe the angular nodal surfaces for a d_{z^2} orbital, whose wave function is

$$Y = \frac{1}{4}\sqrt{\frac{5}{\pi}}\,\frac{2z^2 - x^2 - y^2}{r^2}$$

(Solutions to the exercises are in Appendix A.)

EXERCISE 2-2

Describe the angular nodal surfaces for a d_{xz} orbital, whose angular wave function is

$$Y = \frac{1}{2}\sqrt{\frac{15}{\pi}}\,\frac{xz}{r^2}$$

Students are sometimes puzzled by explanations of nodal surfaces. For example, a p orbital has a nodal plane through the nucleus. How can an electron be on both sides of the node at once without being at the node (at which the probability is zero)? One explanation is the relativistic one cited previously: the probability does not go quite to zero. Another explanation is that such a question really has no meaning for an electron thought of as a wave. A plucked violin string vibrates at a specific frequency, and nodes at which the amplitude of vibration is zero are a natural result. Zero amplitude does not mean that the string does not exist at these points, simply that the magnitude of the vibration is zero. An electron wave exists at the node as well as on both sides of a nodal surface, just as a violin string exists at the nodes and on both sides of points having zero amplitude.

Still another explanation, in a lighter vein, was suggested by R. M. Fuoss to one of the authors in a class on bonding. In a rough quotation from Aquinas,

> Angels are not material beings. Therefore, they can be first in one place and later in another, without ever having been in between.

If the word "electrons" replaces the word "angels," a semitheological interpretation of nodes follows.

The result of the calculations is the set of atomic orbitals familiar to all chemists. Small sketches of the shapes of s, p, and d orbitals are in Table 2-1, with more details in Figure 2-8; similar information on f orbitals can be found in other sources.[19] In the center of Table 2-1 are the shapes for the Θ portion; when they are rotated through $\phi = 0$ to 2π, the three-dimensional shapes in the right column are formed. In the diagrams of orbitals in Table 2-1, the orbital lobes where the wave function is negative are shaded. The probabilities are the same for locations with positive and negative signs for Ψ, but it is useful to distinguish them for bonding purposes, as will be seen in Chapter 5.

Figure 2-8 shows cross-sectional contour diagrams of some selected orbitals. The lines are of constant electron density, $\Psi\Psi^*$, and dotted lines show the nodes. The orbitals we use are the common ones used by chemists; others that are also solutions of the Schrödinger equation can be chosen for special purposes.[20]

[19] H. G. Friedman, Jr., G. R. Choppin, and D. G. Feuerbacher, *J. Chem. Educ.*, **1964,** *41,* 354; C. Becker, *J. Chem. Educ.*, **1964,** *41,* 358.

[20] R. E. Powell, *J. Chem. Educ.*, **1968,** *45,* 45.

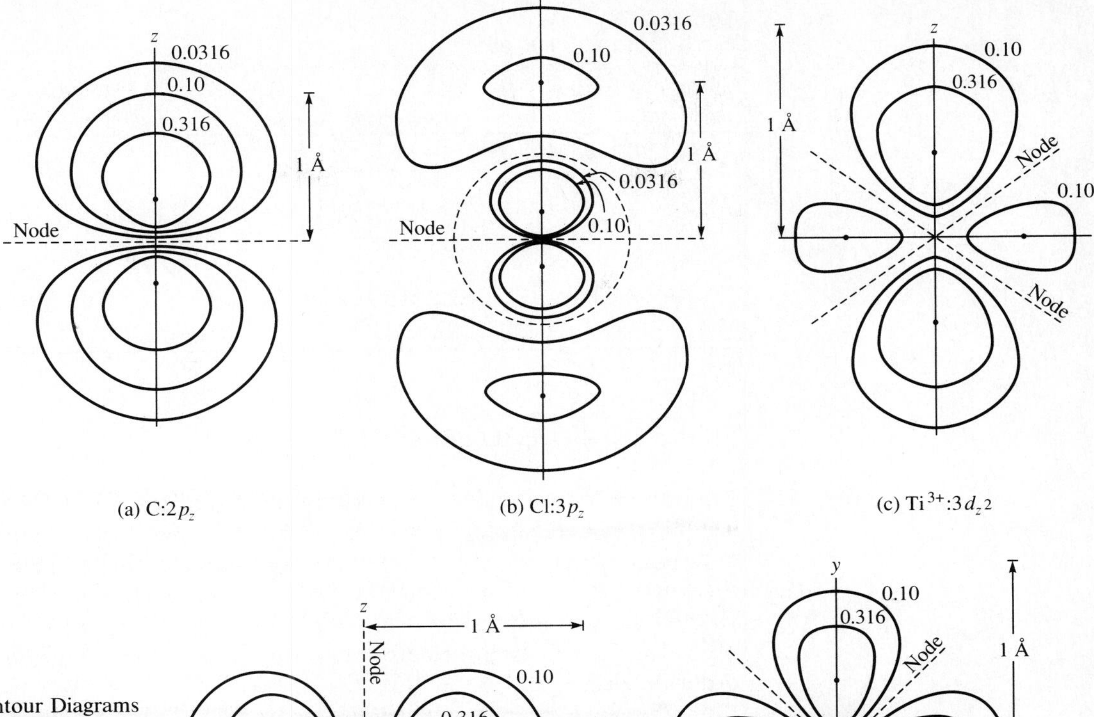

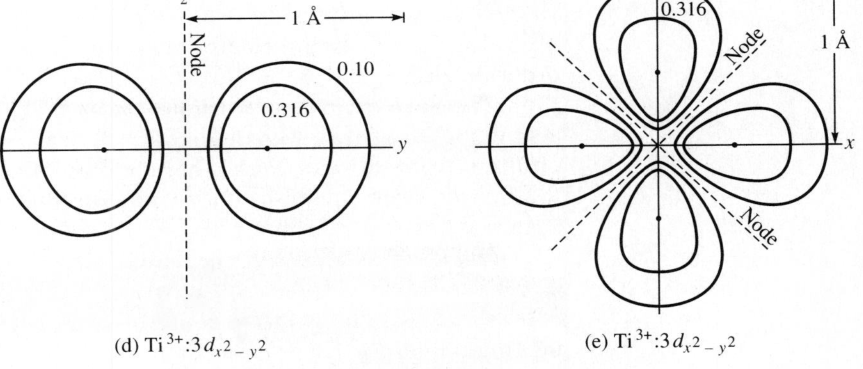

FIGURE 2-8 Contour Diagrams for Selected Atomic Orbitals. For (a) through (c), the cross-sectional plane is any plane containing the z axis. For (d), the cross section is taken through the xz or yz plane, and for (e), the cross section is taken through the xy plane. (From E. A. Orgyzlo and G. B. Porter, *J. Chem. Educ.*, **1963,** *40,* 258. Reproduced with permission.)

One feature that should be mentioned is the appearance of i $(= \sqrt{-1})$ in the p and d orbital wave equations in Table 2-1. Since it is much more convenient to work with real functions than complex functions, we usually take advantage of another property of the wave equation. For equations of this kind, any linear combination of solutions to the equation is also a solution to the equation. The combinations usually chosen for the p orbitals are the sum and difference of the $m_l = +1$ and -1 p orbitals, normalized by multiplying by the constants $i/\sqrt{2}$ and $1/\sqrt{2}$, respectively:

$$\Psi_{2p_x} = -\frac{i}{\sqrt{2}}(\Psi_{+1} - \Psi_{-1}) = \frac{1}{2}\sqrt{\frac{3}{\pi}}[R(r)] \sin\theta \cos\phi$$

$$\Psi_{2p_y} = \frac{1}{\sqrt{2}}(\Psi_{+1} + \Psi_{-1}) = \frac{1}{2}\sqrt{\frac{3}{\pi}}[R(r)] \sin\theta \sin\phi$$

The same procedure used on the d orbital functions for $m_l = \pm 1$ and ± 2 gives the functions in the column headed $\Theta\,\Phi(\theta, \phi)$ in Table 2-1, which are the familiar d orbitals. The d_{z^2} orbital actually uses the function $2z^2 - x^2 - y^2$, which we shorten to z^2 for convenience.

TABLE 2-5
Hund's rule and multiplicity

Number of electrons	*Arrangement*	*Unpaired e^-*	*Multiplicity*
1	↑ ___ ___	1	2
2	↑ ↑ ___	2	3
3	↑ ↑ ↑	3	4
4	⇅ ↑ ↑	2	3
5	⇅ ⇅ ↑	1	2
6	⇅ ⇅ ⇅	0	1

2-2-4 THE AUFBAU PRINCIPLE

Limitations on the values of the quantum numbers lead to the familiar aufbau (building up) principle, where the buildup of electrons in atoms results from continually increasing the quantum numbers. In this process, we start with the lowest n, l, and m_l values (1, 0, and 0, respectively) and either of the m_s values (we will arbitrarily use $-1/2$ first). Two rules then give us the proper order for the remaining electrons as we increase the quantum numbers in the order m_l, m_s, l, and n.

The **Pauli exclusion principle**[21] requires that each electron in an atom have a unique set of quantum numbers. At least one quantum number must be different from those of each other electron. This principle does not come from the Schrödinger equation, but from experimental determination of electronic structures.

Hund's rule of maximum multiplicity[22] requires that electrons be placed in orbitals so as to give the maximum total spin possible (or the maximum number of parallel spins), consistent with the additional requirement for minimum energy. Therefore, when there are 1 to 6 electrons in p orbitals, the required arrangements are those given in Table 2-5. The **multiplicity** is $n + 1$, where n is the number of unpaired electrons; it is explained further in Section 2-3. Any other arrangement of electrons results in fewer unpaired electrons. This is only one of Hund's rules; others are described later in this chapter.

This rule is a consequence of the energy required for pairing electrons in the same orbital. When two electrons occupy the same part of the space around an atom, they repel each other because of their mutual negative charges, with a **Coulombic energy of repulsion, Π_c,** per pair of electrons. As a result, this repulsive force favors electrons in different orbitals (different regions of space) over electrons in the same orbitals.

In addition, there is an **exchange energy, Π_e,** which arises from purely quantum mechanical considerations. This energy depends on the number of electrons at the same energy that could be exchanged while retaining the same overall spins.

For example, the electron configuration of a carbon atom is $1s^2\ 2s^2\ 2p^2$. Three arrangements of the $2p$ electrons can be considered:

(1) ⇅ ___ ___ (2) ↑ ↓ ___ (3) ↑ ↑ ___

[21] W. Pauli, *Z. Physik,* **1925,** *31,* 765.

[22] F. Hund, *Z. Physik,* **1925,** *33,* 345.

The first arrangement involves Coulombic energy, Π_c, since it is the only one that pairs electrons in the same orbital. The energy of this arrangement is higher than that of the other two by Π_c as a result of electron–electron repulsion.

In the first two cases, there is only one possible way to arrange the electrons to give the same diagram, since in each there is only a single electron having + or − spin. However, in the third case there are two possible ways in which the electrons can be arranged:

$\underline{\uparrow_1}\ \underline{\uparrow_2}\ \underline{\quad}$ $\qquad$ $\underline{\uparrow_2}\ \underline{\uparrow_1}\ \underline{\quad}$ $\qquad$ (one exchange of electrons)

The exchange energy is Π_e per possible exchange of parallel electrons and is negative. The more the possible exchanges, the lower the energy. Consequently, the third configuration is lower in energy than the second by Π_e.

The results may be summarized in an energy diagram:

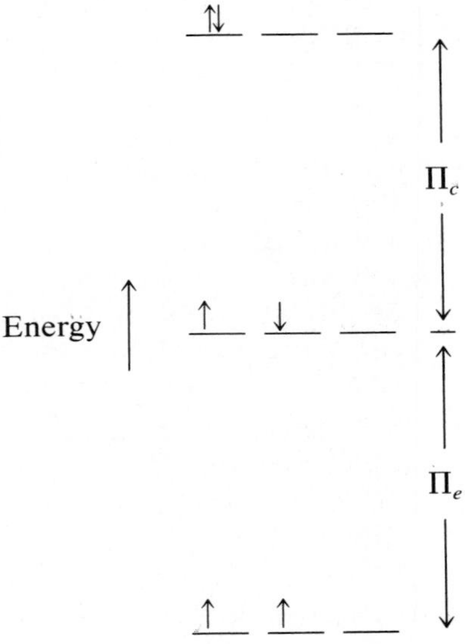

These two pairing terms add to produce the total pairing energy, Π:

$$\Pi = \Pi_c + \Pi_e$$

The Coulombic energy, Π_c, is positive and nearly constant for each pair of electrons. The exchange energy, Π_e, is negative and also nearly constant for each possible exchange of electrons with the same spin. When the orbitals are **degenerate** (have the same energy), both Coulombic and pairing energies favor the unpaired configuration over the paired configuration. If there is a difference in energy between the levels involved, this difference, in combination with the total pairing energy, determines the final configuration. For atoms, this usually means that one set of orbitals is filled before another has any electrons. This breaks down in some of the transition elements, because the 4*s* and 3*d* (or the higher corresponding levels) are so close in energy that the pairing energy is nearly the same as the difference between levels. Section 2-2-5 explains what happens in these cases.

EXAMPLE

Oxygen An oxygen atom with four p electrons could have 2 unpaired electrons (↑↓ ↑ ↑), or it could have no unpaired electrons (↑↓ ↑↓ ___). Find the number of electrons that could be exchanged in each case and the Coulombic and exchange energies for the atom.

↑↓ ↑ ↑ has one electron with ↓ spin, and no possibility of exchange, three electrons with ↑ spin, with 3 exchange possibilities, and one pair. Overall, $3\ \Pi_e + \Pi_c$.

↑↓ ↑↓ ___ has one exchange possibility for each spin pair and two pairs. Overall, $2\ \Pi_e + 2\ \Pi_c$.

EXERCISE 2-3

A nitrogen atom with three p electrons could have 3 unpaired electrons (↑ ↑ ↑), or it could have one unpaired electron (↑↓ ↑ ___). Find the number of electrons that could be exchanged in each case and the Coulombic and exchange energies for the atom.

Many schemes have been used to predict the order of filling of atomic orbitals. One of the simplest, which fits most of the atoms, uses the periodic table blocked out as in Figure 2-9. The electron configurations of hydrogen and helium are clearly $1s^1$ and $1s^2$. After that, the elements in the first two columns on the left (groups 1 and 2 or IA and IIA) are filling s orbitals, with $l = 0$; those in the six columns on the right (groups 13 to 18 or IIIA to VIIIA) are filling p orbitals, with $l = 1$; and the ten in the middle (the transition elements, groups 3 to 12 or IIIB to IIB) are filling d orbitals, with $l = 2$. The lanthanide and actinide series (numbers 58 to 71 and 90 to 103) are filling f

Groups (IUPAC) / (US traditional)

1	2	3		4	5	6	7	8	9	10	11	12	13	14	15	16	17	18
IA	IIA	IIIB		IVB	VB	VIB	VIIB		VIIIB		IB	IIB	IIIA	IVA	VA	VIA	VIIA	VIIIA
1																		2
3	4												5	6	7	8	9	10
11	12												13	14	15	16	17	18
19	20	21		22	23	24	25	26	27	28	29	30	31	32	33	34	35	36
37	38	39		40	41	42	43	44	45	46	47	48	49	50	51	52	53	54
55	56	57	*	72	73	74	75	76	77	78	79	80	81	82	83	84	85	86
87	88	89	**	104	105	106	107	108	109									
			*	58	59	60	61	62	63	64	65	66	67	68	69	70	71	
			**	90	91	92	93	94	95	96	97	98	99	100	101	102	103	

s block   *p* block   *d* block   *f* block

FIGURE 2-9 Atomic Orbital Filling in the Periodic Table.

orbitals, with $l = 3$. This classification is too simple, as shown in the following paragraphs, but does provide a starting point for the electronic configurations.

As a result of shielding (described in the following section) and other more subtle interactions between the electrons, the simple order of orbitals (in order of energy increasing with increasing n) holds only at very low Z and for the innermost electrons of any atom. For the outer orbitals, the increasing split caused by different l values forces the overlap of energy levels with $n = 3$ and $n = 4$, and $4s$ fills before $3d$. In a similar fashion, $5s$ fills before $4d$, $6s$ before $5d$, $4f$ before $5d$, and $5f$ before $6d$. The sudden drop in energy of the d orbitals as they begin filling is at least partly a consequence of the poor shielding of d electrons by other d electrons.

The resulting order of orbital filling for the electrons is shown in Table 2-6. Although the quantum number n is most important in determining the energy, in atoms with more than one electron, l must be included in the calculation of the energy. As the atomic number increases, the electrons are drawn toward the nucleus and the orbital energies become more negative. The pattern that emerges from these factors is shown in Figure 2-10. Although the energies decrease with increasing Z, the changes are irregular due to shielding of outer electrons by inner electrons, as described in the next section.

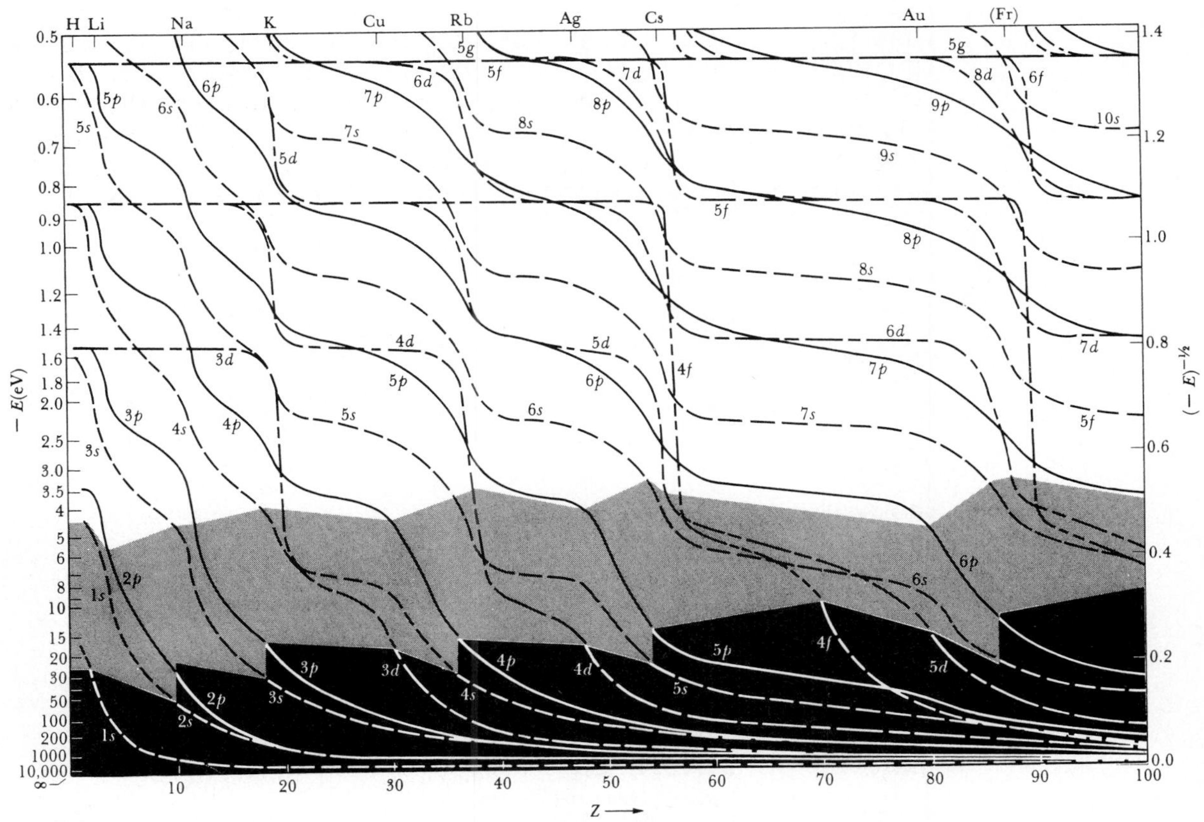

FIGURE 2-10 Energy Levels as a Function of Atomic Number. White background, normally empty subshells; gray background, partly filled subshells; black background, normally full subshells. This diagram is only conceptual and should not be used for detailed electronic configurations. More details are in the text. (From R. L. Rich, *Periodic Correlations*, W. A. Benjamin, Menlo Park, Calif., 1965, p. 6. Reprinted with permission.)

TABLE 2-6
Electron configurations of the elements

Element	*Z*	*Configuration*
H	1	$1s^1$
He	2	$1s^2$
Li	3	[He] $2s^1$
Be	4	[He] $2s^2$
B	5	[He] $2s^2\ 2p^1$
C	6	[He] $2s^2\ 2p^2$
N	7	[He] $2s^2\ 2p^3$
O	8	[He] $2s^2\ 2p^4$
F	9	[He] $2s^2\ 2p^5$
Ne	10	[He] $2s^2\ 2p^6$
Na	11	[Ne] $3s^1$
Mg	12	[Ne] $3s^2$
Al	13	[Ne] $3s^2\ 3p^1$
Si	14	[Ne] $3s^2\ 3p^2$
P	15	[Ne] $3s^2\ 3p^3$
S	16	[Ne] $3s^2\ 3p^4$
Cl	17	[Ne] $3s^2\ 3p^5$
Ar	18	[Ne] $3s^2\ 3p^6$
K	19	[Ar] $4s^1$
Ca	20	[Ar] $4s^2$
Sc	21	[Ar] $4s^2\ 3d^1$
Ti	22	[Ar] $4s^2\ 3d^2$
V	23	[Ar] $4s^2\ 3d^3$
Cr	24	*[Ar] $4s^1\ 3d^5$
Mn	25	[Ar] $4s^2\ 3d^5$
Fe	26	[Ar] $4s^2\ 3d^6$
Co	27	[Ar] $4s^2\ 3d^7$
Ni	28	[Ar] $4s^2\ 3d^8$
Cu	29	*[Ar] $4s^1\ 3d^{10}$
Zn	30	[Ar] $4s^2\ 3d^{10}$
Ga	31	[Ar] $4s^2\ 3d^{10}\ 4p^1$
Ge	32	[Ar] $4s^2\ 3d^{10}\ 4p^2$
As	33	[Ar] $4s^2\ 3d^{10}\ 4p^3$
Se	34	[Ar] $4s^2\ 3d^{10}\ 4p^4$
Br	35	[Ar] $4s^2\ 3d^{10}\ 4p^5$
Kr	36	[Ar] $4s^2\ 3d^{10}\ 4p^6$
Rb	37	[Kr] $5s^1$
Sr	38	[Kr] $5s^2$
Y	39	[Kr] $5s^2\ 4d^1$
Zr	40	[Kr] $5s^2\ 4d^2$
Nb	41	*[Kr] $5s^1\ 4d^4$
Mo	42	*[Kr] $5s^1\ 4d^5$
Tc	43	[Kr] $5s^2\ 4d^5$
Ru	44	*[Kr] $5s^1\ 4d^7$
Rh	45	*[Kr] $5s^1\ 4d^8$
Pd	46	*[Kr] $4d^{10}$
Ag	47	*[Kr] $5s^1\ 4d^{10}$
Cd	48	[Kr] $5s^2\ 4d^{10}$
In	49	[Kr] $5s^2\ 4d^{10}\ 5p^1$
Sn	50	[Kr] $5s^2\ 4d^{10}\ 5p^2$
Sb	51	[Kr] $5s^2\ 4d^{10}\ 5p^3$
Te	52	[Kr] $5s^2\ 4d^{10}\ 5p^4$
I	53	[Kr] $5s^2\ 4d^{10}\ 5p^5$
Xe	54	[Kr] $5s^2\ 4d^{10}\ 5p^6$
Cs	55	[Xe] $6s^1$
Ba	56	[Xe] $6s^2$
La	57	*[Xe] $6s^2\ \ \ 5d^1$
Ce	58	*[Xe] $6s^2\ 4f^1\ 5d^1$
Pr	59	[Xe] $6s^2\ 4f^3$
Nd	60	[Xe] $6s^2\ 4f^4$
Pm	61	[Xe] $6s^2\ 4f^5$
Sm	62	[Xe] $6s^2\ 4f^6$
Eu	63	[Xe] $6s^2\ 4f^7$
Gd	64	*[Xe] $6s^2\ 4f^7\ 5d^1$
Tb	65	[Xe] $6s^2\ 4f^9$
Dy	66	[Xe] $6s^2\ 4f^{10}$
Ho	67	[Xe] $6s^2\ 4f^{11}$
Er	68	[Xe] $6s^2\ 4f^{12}$
Tm	69	[Xe] $6s^2\ 4f^{13}$
Yb	70	[Xe] $6s^2\ 4f^{14}$
Lu	71	[Xe] $6s^2\ 4f^{14}\ 5d^1$
Hf	72	[Xe] $6s^2\ 4f^{14}\ 5d^2$
Ta	73	[Xe] $6s^2\ 4f^{14}\ 5d^3$
W	74	[Xe] $6s^2\ 4f^{14}\ 5d^4$
Re	75	[Xe] $6s^2\ 4f^{14}\ 5d^5$
Os	76	[Xe] $6s^2\ 4f^{14}\ 5d^6$
Ir	77	[Xe] $6s^2\ 4f^{14}\ 5d^7$
Pt	78	*[Xe] $6s^1\ 4f^{14}\ 5d^9$
Au	79	*[Xe] $6s^1\ 4f^{14}\ 5d^{10}$
Hg	80	[Xe] $6s^2\ 4f^{14}\ 5d^{10}$
Tl	81	[Xe] $6s^2\ 4f^{14}\ 5d^{10}\ 6p^1$
Pb	82	[Xe] $6s^2\ 4f^{14}\ 5d^{10}\ 6p^2$
Bi	83	[Xe] $6s^2\ 4f^{14}\ 5d^{10}\ 6p^3$
Po	84	[Xe] $6s^2\ 4f^{14}\ 5d^{10}\ 6p^4$
At	85	[Xe] $6s^2\ 4f^{14}\ 5d^{10}\ 6p^5$
Rn	86	[Xe] $6s^2\ 4f^{14}\ 5d^{10}\ 6p^6$
Fr	87	[Rn] $7s^1$
Ra	88	[Rn] $7s^2$
Ac	89	*[Rn] $7s^2\ \ \ 6d^1$
Th	90	*[Rn] $7s^2\ \ \ 6d^2$
Pa	91	*[Rn] $7s^2\ 5f^2\ 6d^1$
U	92	*[Rn] $7s^2\ 5f^3\ 6d^1$
Np	93	*[Rn] $7s^2\ 5f^4\ 6d^1$
Pu	94	[Rn] $7s^2\ 5f^6$
Am	95	[Rn] $7s^2\ 5f^7$
Cm	96	*[Rn] $7s^2\ 4f^7\ 6d^1$
Bk	97	[Rn] $7s^2\ 5f^9$
Cf	98	*[Rn] $7s^2\ 4f^9\ 6d^1$
Es	99	[Rn] $7s^2\ 5f^{11}$
Fm	100	[Rn] $7s^2\ 5f^{12}$
Md	101	[Rn] $7s^2\ 5f^{13}$
No	102	[Rn] $7s^2\ 5f^{14}$
Lr	103	[Rn] $7s^2\ 5f^{14}\ 6d^1$
	or	[Rn] $7s^2\ 5f^{14}\ 7p^1$
Unq	104	[Rn] $7s^2\ 5f^{14}\ 6d^2$
Unp	105	[Rn] $7s^2\ 5f^{14}\ 6d^3$
Unh	106	[Rn] $7s^2\ 5f^{14}\ 6d^4$
Uns	107	[Rn] $7s^2\ 5f^{14}\ 6d^5$
Uno	108	[Rn] $7s^2\ 5f^{14}\ 6d^6$
Une	109	[Rn] $7s^2\ 5f^{14}\ 6d^7$

SOURCE: Configurations for elements 100 to 109 are predicted, not experimental. Actinide configurations are from J. J. Katz, G. T. Seaborg, and L. R. Morss, *The Chemistry of the Actinide Elements*, 2nd ed., Chapman and Hall, New York and London, 1986.

NOTE: *Indicates configurations that do not follow the simple order of orbital filling.

2-2-5 SHIELDING

Energies of specific levels are difficult to predict quantitatively, but one of the more common approaches is to use the idea of shielding. Each electron acts as a shield for electrons farther out from the nucleus, reducing the attraction between the nucleus and the distant electrons. Slater[23] formulated a set of simple rules that serve as a rough guide to this effect. He defined the effective nuclear charge Z^* as a measure of the nuclear attraction for an electron. Z^* can be calculated from $Z^* = Z - S$, where Z is the nuclear charge and S is the shielding constant. The rules for determining S for a specific electron are as follows:

1. The electronic structure of the atom is written in groupings as follows: $(1s)\ (2s, 2p)\ (3s, 3p)\ (3d)\ (4s, 4p)\ (4d)\ (4f)\ (5s, 5p)$, and so on.
2. Electrons in higher orbitals (to the right in the list above) do not shield those in lower orbitals.
3. For ns or np valence electrons:
   - **a.** Electrons in the same ns, np level contribute 0.35, except the $1s$, where 0.30 works better.
   - **b.** Electrons in the $n - 1$ level contribute 0.85.
   - **c.** Electrons in the $n - 2$ or lower levels contribute 1.00.
4. For nd and nf valence electrons:
   - **a.** Electrons in the same nd or nf level contribute 0.35.
   - **b.** Electrons in groupings to the left contribute 1.00.

The shielding constant S obtained from the sum of the contributions above is subtracted from the nuclear charge Z to obtain the effective nuclear charge Z^* affecting the selected electron.

Some examples follow:

EXAMPLES

Oxygen The electron configuration is $(1s^2)\ (2s^2\ 2p^4)$.

For the outermost electron,

$$Z^* = Z - S$$
$$= 8 - \underbrace{2 \times (0.85)}_{1s} - \underbrace{5 \times (0.35)}_{2s,\, 2p} = 4.55$$

The two $1s$ electrons each contribute 0.85, and the five $2s$ and $2p$ electrons (the last electron is not counted, as we are finding Z^* for it) each contribute 0.35 for a total shielding constant $S = 3.45$. The net effective nuclear charge is then $Z^* = 4.55$. Therefore, the last electron is held with about 57% of the force expected for a +8 nucleus and a −1 electron.

Nickel The electron configuration is $(1s^2)\ (2s^2\ 2p^6)\ (3s^2\ 3p^6)\ (3d^8)\ (4s^2)$.

For a $3d$ electron,

$$Z^* = Z - S$$
$$= 28 - \underbrace{18 \times (1.00)}_{1s,\, 2s,\, 2p,\, 3s,\, 3p} - \underbrace{7 \times (0.35)}_{3d} = 7.55$$

77 [23] J. C. Slater, *Phys. Rev.*, **1930**, *36*, 57.

The 18 electrons in the $1s$, $2s$, $2p$, $3s$, and $3p$ levels contribute 1.00 each, the other 7 in $3d$ contribute 0.35, and the $4s$ contribute nothing. The total shielding is $S = 20.45$ and $Z^* = 7.55$ for the last $3d$ electron.

For the $4s$ electron,

$$Z^* = Z - S$$

$$= 28 - \underset{1s,\,2s,\,2p}{10 \times (1.00)} - \underset{3s,\,3p,\,3d}{16 \times (0.85)} - \underset{4s}{1 \times (0.35)} = 4.05$$

The ten $1s$, $2s$, and $2p$ electrons each contribute 1.00, the sixteen $3s$, $3p$, and $3d$ electrons each contribute 0.85, and the other $4s$ electron contributes 0.35, for a total $S = 23.95$ and $Z^* = 4.05$, considerably smaller than the value for the $3d$ electron above. The $4s$ electron is held less tightly than the $3d$ and should therefore be the first removed in ionization. This is consistent with experimental observations on nickel compounds. Ni^{2+}, the most common oxidation state of nickel, has an electron configuration of $[Ar]3d^8$ (rather than $[Ar]3d^6 4s^2$), corresponding to loss of the $4s$ electrons from nickel atoms. All the transition metals follow this same pattern of losing ns electrons more readily than $(n - 1)d$ electrons.

EXERCISE 2-4

Calculate the effective nuclear charge on a $5s$, a $5p$, and a $4d$ electron in a tin atom.

Justification for Slater's rules (aside from the fact that they work) comes from the electron probability curves for the orbitals. The s and p orbitals have higher probabilities near the nucleus than do d orbitals of the same n. Therefore, the shielding of $3d$ electrons by ($3s$, $3p$) electrons is calculated as 100% effective (a contribution of 1.00). At the same time, shielding of $3s$ or $3p$ electrons by ($2s$, $2p$) electrons is only 85% effective (a contribution of 0.85), because the $3s$ and $3p$ orbitals have regions of significant probability close to the nucleus. Therefore, electrons in these orbitals are not completely shielded by ($2s$, $2p$) electrons.

A complication arises at Cr(24) and Cu(29) in the first transition series and in an increasing number of atoms under them in the second and third transition series. This effect places an extra electron in the $3d$ level and removes one from the $4s$. Cr, for example, has the configuration $[Ar]4s^1 3d^5$ (rather than $[Ar]4s^2 3d^4$). Traditionally, this phenomenon has been explained as a consequence of the special stability of half-filled subshells. However, Rich[24] explains this by specifically considering the difference in energy between an orbital containing one electron and an orbital containing two electrons. Although the orbital itself is usually described as having only one energy, the electrostatic repulsion of the two electrons in one orbital adds the electron pairing energy described as part of Hund's rules. We can visualize two parallel energy levels, each with electrons of only one spin, separated by the electron pairing energy, as in Figure 2-11. As the nuclear charge increases, all the energy levels move down in energy, becoming more stable, with the d orbitals changing more rapidly than the s orbitals. Electrons fill the lowest available energy levels in order up to their capacity, with the results shown in the diagram and in Table 2-6 of electronic structures.

[24] Ronald L. Rich, *Periodic Correlations*, Benjamin, Menlo Park, Calif., 1965, pp. 9–11.

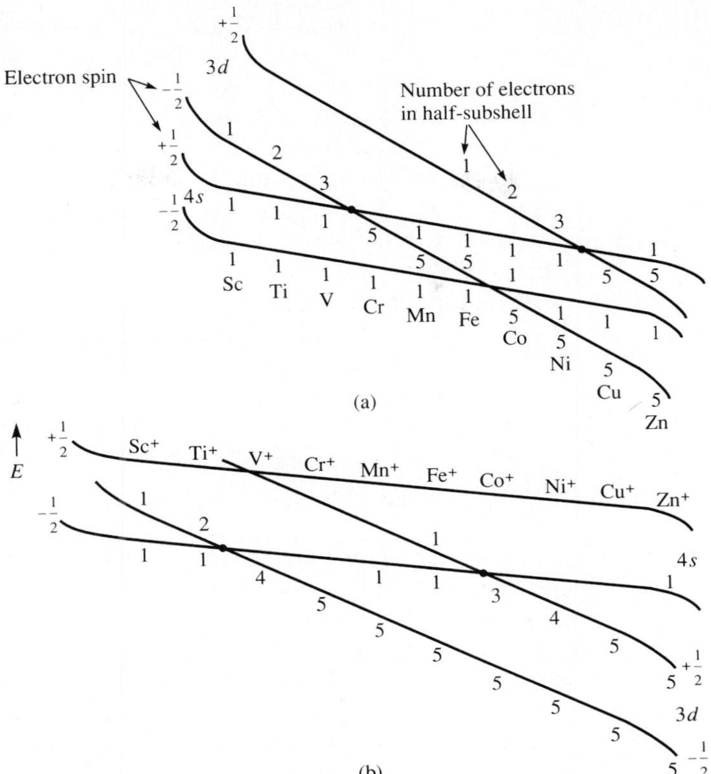

FIGURE 2-11 Schematic Energy Levels for Transition Elements. (a) Schematic interpretation of electron configurations for transition elements in terms of intraorbital repulsion and trends in subshell energies. (b) Similar diagrams for ions, showing the shift in the crossover points on removal of an electron. The diagram shows that *s* electrons are removed before *d* electrons. The shift is even more pronounced for metal ions having 2+ or greater charges. As a consequence, transition metal ions with 2^+ or greater charges have no *s* electrons, only *d* electrons in their outer levels. Similar diagrams, although more complex, can be drawn for the heavier transition elements and the lanthanides. (From R. L. Rich, *Periodic Correlations,* W. A. Benjamin, Menlo Park, Calif., 1965, pp. 9–10. Reprinted with permission.)

The first two electrons enter the $4s$, $-1/2$ and $4s$, $+1/2$ levels; the next three are all in the $3d$, $-1/2$ level, and vanadium has the configuration $4s^2 3d^3$. The $3d$, $-1/2$ level crosses the $4s$, $+1/2$ level between V and Cr. When the 6 electrons of chromium are filled in from the lowest level, chromium has the configuration $4s^1 3d^5$. A similar crossing gives copper its $4s^1 3d^{10}$ configuration. This explanation does not depend on the stability of half-filled shells or other extra factors; those explanations break down for the cases of zirconium ($5s^2 4d^2$) and niobium ($5s^1 4d^4$) and others in the lower periods.

Formation of a positive ion by removal of an electron reduces the overall electron repulsion and lowers the energy of the *d* orbitals more than that of the *s* orbitals. As a result, the remaining electrons occupy the *d* orbitals and we can use the shorthand notion that the electrons with highest *n* (in this case, those in the *s* orbitals) are always removed first in the formation of ions from the transition elements.

A similar, but more complex, crossing of levels appears in the lanthanide and actinide series. The simple picture would have them start filling *f* orbitals at lanthanum (57) and actinium (89), but these atoms have one *d* electron instead. Other elements in these series also show deviations from the "normal" sequence. Rich has shown how these may also be explained by similar diagrams; reference should be made to his book for the details.

2-3 QUANTUM NUMBERS OF MULTIELECTRON ATOMS

Although the quantum numbers and energies of individual electrons can be described in fairly simple terms, interactions between electrons cause this picture to become more complicated. Examples of some interactions have already been described in this chapter: as a result of electron–electron repulsions, electrons tend to occupy separate orbitals; as a result of exchange energy, electrons in separate orbitals tend to have parallel spins. In addition, after the quantum numbers of the individual electrons have been found, there are additional interactions that must be considered. More specifically, the orbital angular momenta combine and the electron spins combine to give new quantum numbers L, S, and J. These quantum numbers, designated by **term symbols,** collectively describe the energy and symmetry of an atom or ion and determine the possible transitions between states of different energies. Examples of these transitions appear in the spectra of coordination compounds, discussed in Chapter 9.

The atomic quantum numbers, describing states of multielectron atoms, are defined as follows:

L = total orbital angular momentum quantum number
S = total spin quantum number
J = total angular momentum quantum number

Consider again the energy levels of a carbon atom. Carbon has the electron configuration $1s^2\ 2s^2\ 2p^2$. We might expect the p electrons to have a single energy, but there are three major energy levels, and, in addition, the lowest energy configuration is also split into three slightly different energies, for a total of five different energy levels. All these can be described in a systematic way by combining the m_l and m_s values of the $2p$ electrons in this configuration. Independently, each of the $2p$ electrons could have

$$n = 2, \qquad l = 1$$

$$m_l = +1, 0, \text{ or } -1 \qquad \text{(three possible values)}$$

$$m_s = +1/2 \text{ or } -1/2 \qquad \text{(two possible values)}$$

or a total of six possible m_l, m_s combinations. The $2p$ electrons do *not* act independently, however; the orbital angular momenta (characterized by m_l values) of the $2p$ electrons interact (couple), and the spin angular momenta (characterized by m_s) also interact in a manner called **Russell–Saunders coupling** or ***LS* coupling.**[25] To describe the types of these interactions, we need to define new quantum numbers describing the atomic states (called **microstates**) that result:

$$M_L = \sum m_l, \qquad \text{total orbital angular momentum}$$

$$M_S = \sum m_s, \qquad \text{total spin angular momentum}$$

We now need to determine how many possible combinations of m_l and m_s values there are for a p^2 configuration.[26] Once these combinations are known, it will be possible to determine the corresponding values of M_L and M_S. For shorthand, we will designate the m_s value of each electron by a superscript +, representing $m_s = +1/2$, or −, representing $m_s = -1/2$. For example, an electron having $m_l = +1$ and $m_s = +1/2$ will be written 1^+.

[25] For a slightly more advanced discussion of coupling and its underlying theory, see M. Gerloch, *Orbitals, Terms, and States,* Wiley, New York, 1986.

[26] Electrons in filled orbitals can be ignored, because their net spin and angular momenta are both zero.

One possible set of values for the two electrons in the p^2 configuration would be

$$\left.\begin{array}{ll}\text{First electron:} & m_l = +1 \text{ and } m_s = +\frac{1}{2} \\ \text{Second electron:} & m_l = 0 \text{ and } m_s = -\frac{1}{2}\end{array}\right\} \text{Notation: } 1^+0^-$$

Each set of possible quantum numbers (such as 1^+0^-) is called a microstate.

The next step is to tabulate the possible microstates. In so doing, we need to take two precautions: (1) to be sure that no two electrons in the same microstate have identical quantum numbers (the Pauli exclusion principle applies), and (2) to avoid duplications. In this connection, for example, the microstates 1^+0^- and 1^-0^+ are different and must both be listed; the microstates 1^+0^- and 0^-1^+ are duplicates and only one will be listed; we must count only the *unique* microstates.

If we determine all possible microstates and tabulate them according to their M_L and M_S values, we obtain a total of 15 microstates.[27] These microstates can be arranged according to their M_L and M_S values and listed conveniently in a microstate table, as shown in Table 2-7.

TABLE 2-7
Microstate table for p^2

		M_S		
		-1	0	$+1$
	$+2$		$1^+\ 1^-$	
	$+1$	$1^-\ 0^-$	$1^+\ 0^-$ $1^-\ 0^+$	$1^+\ 0^+$
M_L	0	$-1^-\ 1^-$	$-1^+\ 1^-$ $0^+\ 0^-$ $-1^-\ 1^+$	$-1^+\ 1^+$
	-1	$-1^-\ 0^-$	$-1^+\ 0^-$ $-1^-\ 0^+$	$-1^+\ 0^+$
	-2		$-1^+\ -1^-$	

EXAMPLE

Determine the possible microstates for an s^1p^1 configuration, and use them to prepare a microstate table.

The s electron can have $m_l = 0$ and $m_s = \pm\frac{1}{2}$.

The p electron can have $m_l = 1, 0,$ or -1 and $m_s = \pm\frac{1}{2}$.

[27] The number of microstates $= i!/[j!(i-j)!]$, where i = number of m_l, m_s combinations (six here) and j = number of electrons.

The resulting microstate table is then:

		M_S		
		-1	0	$+1$
M_L	$+1$	$0^- 1^-$	$0^- 1^+$ $0^+ 1^-$	$0^+ 1^+$
	0	$0^- 0^-$	$0^+ 0^-$ $0^- 0^+$	$0^+ 0^+$
	-1	$0^- -1^-$	$0^- -1^+$ $0^+ -1^-$	$0^+ -1^+$

In this case 0^+0^- and 0^-0^+ are different microstates, since the first electron is an s and the second electron is a p; both must be counted.

EXERCISE 2-5

Determine the possible microstates for a d^2 configuration, and use them to prepare a microstate table. (Your table should contain 45 microstates.)

These microstates can be classified into atomic states having different energies. As mentioned above, atomic states may be described by quantum numbers in a manner similar to describing individual electrons by their quantum numbers; the quantum numbers L and S represent the total orbital and spin angular momenta for these atomic states. The parallels between atomic and electronic quantum numbers should be noted. Just as the quantum number m_l describes the component of the quantum number l of an electron in the direction of a magnetic field, as in Figure 2-6, the quantum number M_L describes the component of L in the direction of a magnetic field for an atomic state. Similarly, m_s describes the component of an electron's spin in a reference direction, while M_S describes the component of S in a reference direction for an atomic state. The possible values of these quantum numbers can also be compared:

Atomic states	*Electrons*
$M_L = 0, \pm 1, \pm 2, \ldots, \pm L$	$m_l = 0, \pm 1, \pm 2, \ldots, \pm l$
$M_S = S, S-1, S-2, \ldots, -S$	$m_s = +1/2, -1/2$

Note that the quantum numbers L and S[28] describe *collections of microstates,* as defined above, M_L and M_S describe microstates, and l, m_l, and m_s describe individual electrons.

TABLE 2-8
Examples of atomic states (free-ion terms) and quantum numbers

Term	*L*	*S*
^1S	0	0
^2S	0	1/2
^3P	1	1
^4D	2	3/2
^5F	3	2

Atomic states are described as S, P, D, F, and higher states in a manner similar to the designation of atomic orbitals as s, p, d, f. The atomic states are further described by their **spin multiplicity,** defined as $2S + 1$; the spin multiplicity is designated as a left superscript. States having spin multiplicities of 1, 2, 3, 4, etc. are described as singlet, doublet, triplet, quartet, etc. states. Examples of atomic states are given in Table 2-8 and in the examples that follow.

Atomic states characterized by S and L quantum numbers are often called **free-ion terms** (sometimes Russell–Saunders terms). Their labels (such

[28] M_S can have values ranging from $+S$ to $-S$; therefore, the largest possible value of M_S is S. Unfortunately, S is used in two ways: to designate the atomic spin quantum number and to designate a state having $L = 0$ (see Table 2-8). Chemists are not always wise in choosing their symbols!

as ^3D, ^2S, and ^5F) are often referred to as **term symbols.**[29] The Russell–Saunders terms are designated as free-ion terms because they describe individual atoms or ions, free of ligands. As will be seen in Chapter 9, the free-ion terms are very important in the interpretation of the spectra of coordination compounds. The following examples show how to determine the values of L, M_L, S, and M_S for a given term and how to prepare microstate tables from them.

EXAMPLES

^1S (singlet S) An S term has $L = 0$ and must therefore have $M_L = 0$. The spin multiplicity (the superscript) is $2S + 1$. Since $2S + 1 = 1$, S must equal 0 (and $M_S = 0$). There can be only one microstate, having $M_L = 0$ and $M_S = 0$, for a ^1S term:

		M_S
		0
M_L	0	0^+0^-

or

		M_S
		0
M_L	0	x

Each microstate is designated by x in the second form of the table.

^2P (doublet P) A P term has $L = 1$; therefore, M_L can have three values: $+1$, 0, and -1. The spin multiplicity is $2 = 2S + 1$. Therefore, $S = 1/2$, and M_S can have two values: $+1/2$ and $-1/2$. There are six microstates in a ^2P term (3 rows $\times$ 2 columns):

		M_S	
		$-\frac{1}{2}$	$+\frac{1}{2}$
	1	1^-	1^+
M_L	0	0^-	0^+
	-1	-1^-	-1^+

or

		M_S	
		$-\frac{1}{2}$	$+\frac{1}{2}$
	1	x	x
M_L	0	x	x
	-1	x	x

The spin multiplicity is equal to the number of possible values of M_S; therefore, the spin multiplicity is simply the number of columns in the microstate table.

EXERCISE 2-6

For each of the following free-ion terms, determine the values of L, M_L, S, and M_S, and diagram the microstate table as in the two examples above: ^2D, ^1P, and ^2S.

At last we are in a position to return to the p^2 microstate table and reduce it to its constituent atomic states (terms). To do this, in general it will be sufficient to designate each microstate simply by x; it will be important to tabulate the number of microstates, but it will not be necessary to write out each microstate in full.

To reduce this microstate table into its component free-ion terms, note that each of the terms described in the examples and Exercise 2-6 consists of

[29] Although "term" and "state" are often used interchangeably, "term" is suggested as the preferred label for the results of Russell–Saunders coupling just described, and "state" for the results of spin–orbit coupling described later, with the quantum number J added. In most cases, there is no problem in using either; the meaning is clear from the context. B. N. Figgis, "Ligand Field Theory," in *Comprehensive Coordination Chemistry,* G. Wilkinson, R. D. Gillard, and J. A. McCleverty, eds., Pergamon, Elmsford, N.Y., 1987, Vol. 1, p. 231.

a rectangular array of microstates. To reduce the p^2 microstate table into its terms, all that is necessary is to find the rectangular arrays. This process is illustrated in Table 2-9.

Therefore, the p^2 electron configuration gives rise to three free-ion terms, designated ^3P, ^1D, and ^1S. In general, each of these terms will have different energy; they represent three states having different degrees of electron–electron repulsions. For our example of a p^2 configuration originating from the electron configuration of a carbon atom, the ^3P, ^1D, and ^1S terms have three distinct energies, the three major energy levels observed experimentally.

The final step in this procedure is to determine which term is of lowest energy. This can be done by using two of **Hund's rules:**

1. The *ground term* (term of lowest energy) is the term of maximum spin multiplicity. In our example of p^2, therefore, the ground term is the ^3P. This is the same rule used earlier in determining electron configurations. This term can be identified as having the following configuration:

$2p$ ↑ ↑ __

$2s$ ⇅

$1s$ ⇅

2. If two or more terms share the maximum spin multiplicity, the ground term will be the one having the highest value of L. For example, if ^4P and ^4F terms are both found for an electron configuration, the ^4F will be of lower energy (^4F has $L = 3$; ^4P has $L = 1$).

The ^1S and ^1D terms have higher energy than the ^3P, but cannot be identified with a single electron configuration. The relative energies of higher-energy terms like these also cannot be determined by simple rules.

EXAMPLE

Reduce the microstate table for the s^1p^1 configuration to its component free-ion terms, and identify the ground-state term.

The microstate table (prepared in the example preceding Exercise 2-5) is the sum of the microstate tables for the ^3P and ^1P terms:

^3P:

M_L \ M_S	−1	0	+1
+1	x	x	x
0	x	x	x
−1	x	x	x

^1P:

M_L \ M_S	−1	0	+1
+1		x	
0		x	
−1		x	

Hund's rule of maximum multiplicity requires ^3P as the ground state.

EXERCISE 2-7

In Exercise 2-5, you obtained a microstate table for the d^2 configuration. Reduce this to its component free-ion terms, and identify the ground-state term.

TABLE 2-9
The microstate table for p^2 and reduction to free-ion terms

M_L \ M_S	−1	0	+1
+2		x	
+1	x	x x	x
0	x	x x (x)	x
−1	x	x x	x
−2		x	

^1D

M_L \ M_S	−1	0	+1
+2		x	
+1		x	
0		x	
−1		x	
−2		x	

^3P

M_L \ M_S	−1	0	+1
+2			
+1	x	x	x
0	x	x	x
−1	x	x	x
−2			

^1S

M_L \ M_S	−1	0	+1
+2			
+1			
0		x	
−1			
−2			

NOTE: The sum of the ^1D, ^3P, and ^1S microstates equals the total p^2 microstates. Therefore, the p^2 electron configuration includes the ^1D, ^3P, and ^1S terms.

2-3-1 SPIN–ORBIT COUPLING

To this point in the discussion of multielectron atoms, the spin and orbital angular momenta have been treated separately. In addition, the spin and orbital angular momenta couple with each other, a phenomenon known as spin–orbit coupling. In multielectron atoms, the S and L quantum numbers combine into the total angular momentum quantum number J. The quantum number J may have the following values:

$$J = L + S, \quad L + S - 1, \quad L + S - 2, \ldots, \quad |L - S|$$

The value of J is given as a subscript.

EXAMPLE

Determine the possible values of J for the carbon terms.

For the term symbols just described for carbon, the ^1D and ^1S terms each have only one J value, while the ^3P term has three slightly different energies, each described by a different J. From the equation above, J can have only the value 0 for the ^1S term $(0 + 0)$ and only the value 2 for the ^1D term $(2 + 0)$. For the ^3P term, J can have the three values 2, 1, and 0 $(1 + 1, 1 + 1 - 1, 1 + 1 - 2)$.

EXERCISE 2-8
Determine the possible values of J for the terms obtained from a d^2 configuration in Exercise 2-7.

Spin–orbit coupling acts to split free-ion terms into states of different energies. The ^3P term therefore splits into states of three different energies, and the total energy level diagram for the carbon atom is

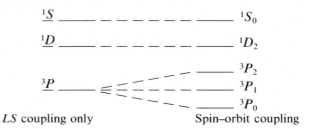

LS coupling only	Spin–orbit coupling
^1S	$^1S_0$
^1D	$^1D_2$
	$^3P_2$
^3P	$^3P_1$
	$^3P_0$

These are the five energy states for the carbon atom referred to at the beginning of this section. The state of lowest energy (spin–orbit coupling included) can be predicted from **Hund's third rule:**

> 3. For subshells (such as p^2) that are less than half-filled, the state having the lowest J value is of lowest energy ($^3P_0$ above); for subshells that are more than half-filled, the state having the highest J value is of lowest energy. Half-filled subshells have only one possible J value.

Spin–orbit coupling can have significant effects on the electronic spectra of coordination compounds, especially those involving fairly heavy metals (atomic number > 40).

2-4 PERIODIC PROPERTIES OF ATOMS

2-4-1 IONIZATION ENERGY

The ionization energy, also known as the ionization potential, is the energy required to remove an electron from a gaseous atom or ion:

$$A^{n+}(g) \longrightarrow A^{(n+1)+}(g) + e^-, \qquad \text{ionization energy} = \Delta U$$

where $n = 0$ (first ionization energy), 1, 2, . . . (second, third, . . .).

As would be expected from the calculation of shielding, the ionization energy varies with different nuclei and different numbers of electrons. Trends for the first ionization energies of the early elements in the periodic table are shown in Figure 2-12. The general trend across a period is an increase in ionization energy as the nuclear charge increases, but a break in the trend appears at the first p electron, at boron. Since the new electron in B is in a new orbital that has most of its electron density farther away from the nucleus, its ionization energy is smaller than that of the $2s^2$ electrons of Be. At the fourth p electron, at oxygen, a similar drop in ionization energy occurs. Here the new electron shares an orbital with one of the previous $2p$ electrons, and the fourth p electron has a higher energy than the trend would indicate because it must be paired with another in the same p orbital. The pairing energy, or repulsion between two electrons in the same region of space, reduces the ionization energy. Similar patterns appear in lower periods, with only small changes in ionization energy for the transition elements, and a generally lower value for heavier atoms in the same family because of increased shielding by inner electrons and increased distance between the nucleus and the outer electrons.

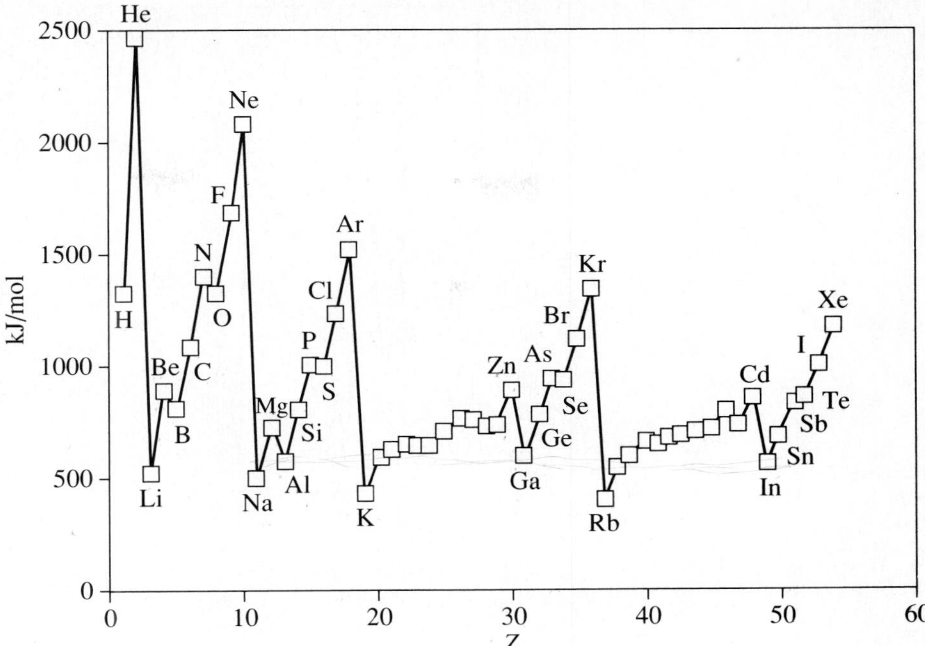

FIGURE 2-12 Ionization Energies; ionization energy = ΔU for $M(g) \longrightarrow M^+(g) + e^-$. [Data from C. E. Moore, "Ionization Potentials and Ionization Limits," *Natl. Stand. Ref. Data Ser. (U.S. Natl. Bur. Stand.)*, **1970**, NSRDS–NBS 34.]

Much larger decreases in ionization energy occur at the start of each new period, because the change to the next major quantum number requires a much higher energy for the new (s) level. The maxima at the noble gases decrease with increasing Z because the outer electrons are farther from the nucleus in the heavier elements.

2-4-2 ELECTRON AFFINITY

Electron affinity is defined as the energy change on adding an electron to an atom, with the sign changed:

$$A(g) + e^- \longrightarrow A^-(g), \qquad \text{electron affinity} = -\Delta U$$

This reaction is exothermic except for the noble gases and the alkaline earth elements, so the electron affinity is usually a positive value. Alternatively, it may be described as the energy change (ΔU) on removal of an electron from a negative ion, $A^-(g) \longrightarrow A(g) + e^-$, usually an endothermic reaction. Because of the similarity of this reaction to the ionization for an atom, electron affinity is sometimes described as the zeroth ionization energy. The pattern of electron affinities with changing Z shown in Figure 2-13 is similar to that of the ionization energies, but for one lower Z value (because of the negative charge of the anions) and with much smaller absolute numbers. In either case, removal of the first electron past a noble gas configuration is easy, so each new period starts with low values. The electron affinities are all much smaller than the corresponding ionization energies because electron removal from a negative ion is easier than from a neutral atom (or addition of an electron to a neutral atom releases less energy than addition of an electron to a positive ion).

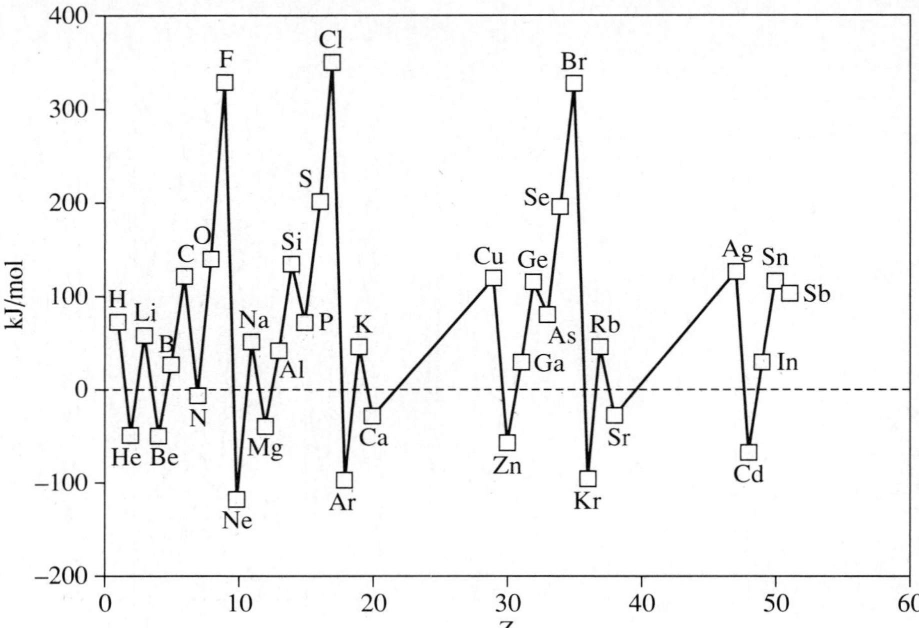

FIGURE 2-13 Electron Affinities: electron affinity = $-\Delta U$ for $M(g) + e^- \longrightarrow M^-(g)$ [or ΔU for $M^-(g) \longrightarrow M(g) + e^-$]. (Data from H. Hotop and W. C. Lineberger, *J. Phys. Chem. Ref. Data*, **1985,** *14,* 731.)

2-4-3 COVALENT AND IONIC RADII

The sizes of atoms and ions are also related to the ionization energies and electron affinities. As the nuclear charge increases, the electrons are pulled in toward the center of the atom, and the size of any particular orbital decreases. On the other hand, as the nuclear charge increases, more electrons are added to the atom and their mutual repulsion keeps the outer orbitals large. The interaction of these two effects (increasing nuclear charge and increasing number of electrons) results in a gradual decrease in atomic size across each period. The numbers used in Table 2-10 are *nonpolar covalent radii* calculated for ideal molecules with no polarity. There are other measures of atomic size, such as van der Waals radii, where collisions with other atoms are used to define the size. It is difficult to obtain consistent data for any such measure, because the polarity, chemical structure, and physical state of molecules change drastically from one compound to another. The numbers shown here are sufficient for a general comparison of one element with another.

There are similar problems in determining the size of ions. Since the stable ions of the different elements have different charges and different numbers of electrons, as well as different crystal structures for their compounds, it is difficult to find a suitable set of numbers for comparison. Earlier data were based on Pauling's approach, in which the ratio of the radii of isoelectronic ions was assumed equal to the ratio of their effective nuclear charges. More recent calculations are based on a number of considerations, including electron density maps from X-ray data, which show larger cations and smaller anions than those previously found. Those in Table 2-11 and Appendix B are called "crystal radii" by Shannon[30] and are generally different from the older values of "ionic radii" by +14 pm for cations and −14 pm for anions, as well as being revised because of more recent measurements.

[30] R. D. Shannon, *Acta Cryst.*, **1976,** *A32,* 751.

TABLE 2-10
Nonpolar covalent radii

Element	*Z*	*Radius (pm)*	*Element*	*Z*	*Radius (pm)*	*Element*	*Z*	*Radius (pm)*	*Element*	*Z*	*Radius (pm)*	*Element*	*Z*	*Radius (pm)*
H	1	32	Li	3	123	Na	11	154	K	19	203	Rb	37	216
He	2	31	Be	4	89	Mg	12	136	Ca	20	174	Sr	38	191
			B	5	82	Al	13	118	Sc	21	144	Y	39	162
			C	6	77	Si	14	111	Ti	22	132	Zr	40	145
			N	7	75	P	15	106	V	23	122	Nb	41	134
			O	8	73	S	16	102	Cr	24	118	Mo	42	130
			F	9	72	Cl	17	99	Mn	25	117	Tc	43	127
			Ne	10	71	Ar	18	98	Fe	26	117	Ru	44	125
									Co	27	116	Rh	45	125
									Ni	28	115	Pd	46	128
									Cu	29	117	Ag	47	134
									Zn	30	125	Cd	48	148
									Ga	31	126	In	49	144
									Ge	32	122	Sn	50	140
									As	33	120	Sb	51	140
									Se	34	117	Te	52	136
									Br	35	114	I	53	133
									Kr	36	112	Xe	54	131

SOURCE: Data from R. T. Sanderson, *Inorganic Chemistry,* Reinhold, New York, 1967, p. 74.

TABLE 2-11
Ionic radii for selected ions

	Z	*Element*	*Radius (pm)*
Alkali metal ions	3	Li^+	90
	11	Na^+	116
	19	K^+	152
	37	Rb^+	166
	55	Cs^+	181
Alkaline earth ions	4	Be^{2+}	59
	12	Mg^{2+}	86
	20	Ca^{2+}	114
	38	Sr^{2+}	132
	56	Ba^{2+}	149
Other cations	13	Al^{3+}	68
	30	Zn^{2+}	88
Halide ions	9	F^-	119
	17	Cl^-	167
	35	Br^-	182
	53	I^-	206
Other anions	8	O^{2-}	126
	16	S^{2-}	170

SOURCE: Data from R. D. Shannon, *Acta Cryst.*, **1976**, *A32*, 751–767. A longer list is available in Appendix B. All the values are for 6-coordinate ions.

The ionic radii in Table 2-11 can be used for rough estimation of the packing of ions in crystals and other calculations, as long as the "fuzzy" nature of atoms and ions is kept in mind. Factors that influence ionic size include the coordination number of the ion, the covalent character of the bonding, distortions of regular crystal geometries, and delocalization of electrons (metallic or semiconducting character, described in Chapter 5). The radius of the anion is also influenced by the size and charge of the cation (the anion exerts a smaller influence on the radius of the cation).[31] The table in Appendix B shows the effect of coordination number.

The values in Table 2-11 show that anions are generally larger than cations with similar numbers of electrons (F^- and Na^+ differ only in nuclear charge, but the radius of fluoride is 37% larger). The radius decreases as nuclear charge increases for ions with the same electronic structure, such as O^{2-}, F^-, Na^+, and Mg^{2+}, with a much larger change with nuclear charge for the cations. Within a family, the ionic radius increases as Z increases because of the larger number of electrons in the ions.

2-5 IONIC BONDING AND CRYSTAL STRUCTURES

It is possible to arrange spheres of the same size in a plane so that each has six nearest neighbors. If the same packing is extended to layers above and below, the predicted coordination number is 12 (3 in a triangle above, 3 in a triangle below, and 6 in a hexagon in the original plane). This geometry, shown in Figure 2-14, exists in many metals, where all the atoms are the same size. It is called **hexagonal close packing** (hcp) when alternate layers are identical (ABAB . . . pattern). When the top layer is rotated 60°, the result is called **cubic close packing** (ccp), with the pattern ABCABC . . . . Both hcp and ccp have the same overall density of packing and the same coordination number; subtle differences in electronic structure determine which metal crystallizes in which form.

In many ionic crystals, the anions are in one of these close-packed geometries, with the cations fitting into the interstices or "holes" between the anions. When the cation fits into a position above the junction of three anions and a fourth anion is directly above it, the coordination number (CN) is 4 and the cation is said to be in a tetrahedral hole. If the next layer of anions is shifted so that the junction of another three anions is directly above the cation, CN = 6 and the cation is in an octahedral hole. In favorable cases, the radius ratio, defined as cation radius divided by anion radius r_+/r_-, can be used to predict which structure is possible when the ions are treated as hard spheres.

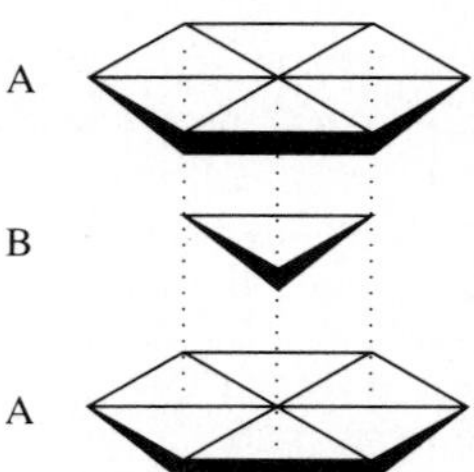

Hexagonal close packed (hcp): Hexagonal close-packed structures have tetrahedral "holes" that run through the entire structure, as shown below.

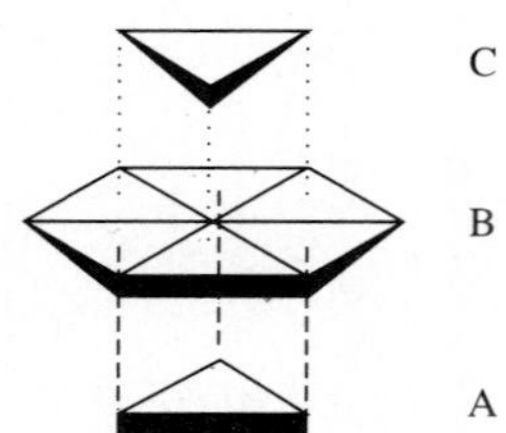

Cubic close packed (ccp): Cubic close-packed structures have all the tetrahedral holes blocked by the next layer.

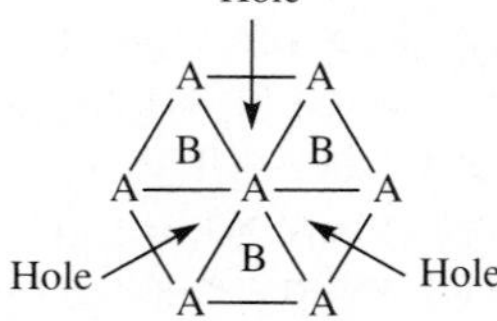

FIGURE 2-14 Close-packed Structures.

[31] O. Johnson, *Inorg. Chem.*, **1973,** *12,* 780.

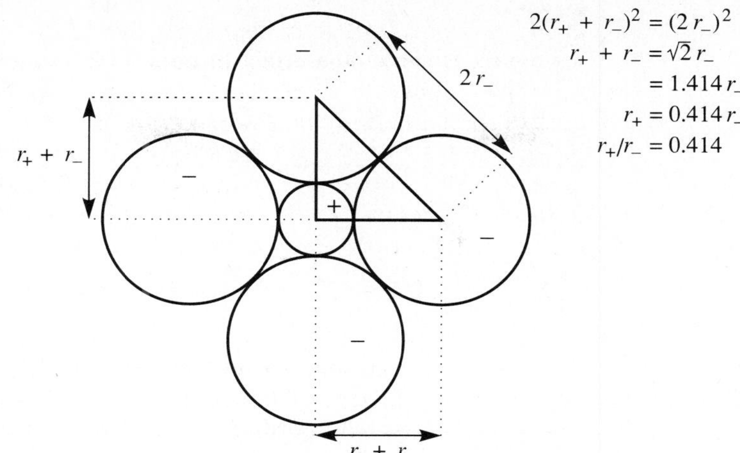

FIGURE 2-15 Calculation of Radius Ratio for Square Planar or Octahedral Geometry.

Figure 2-15 shows the calculation of radius ratio for square planar geometry. Octahedral geometry yields the same numerical value. The coordination number increases as the radius ratio increases, as shown in Table 2-12. Two examples illustrate the use of radius ratios:

EXAMPLES

NaCl Using the radius of the Na^+ cation for either CN = 4 or CN = 6, $r_+/r_- = 113/167 = 0.677$ or $116/167 = 0.695$, both of which predict CN = 6. The Na^+ cation fits easily into the octahedral holes of the Cl^- lattice, which is ccp.

ZnS The zinc ion radius varies more with coordination number. The radius ratios are $r_+/r_- = 74/170 = 0.435$ for the CN = 4 and $r_+/r_- = 88/170 = 0.518$ for the CN = 6 radius. Both predict CN = 6, but the smaller one is close to the tetrahedral limit of 0.414. Experimentally, the Zn^{2+} cation fits in the tetrahedral holes of the S^{2-} lattice, which is either ccp (zinc blende) or hcp (wurtzite).

The predictions match reasonably well with the facts for these two compounds as long as the 4-coordinate radius is used, even though ZnS is largely covalent rather than ionic. However, all radius ratio predictions should be used with caution; ions are not hard spheres and there are many cases where the radius ratio predictions are not correct. One study[32] has found that the actual structure matches the predicted structure in about two-thirds of the cases, with a higher fraction correct at CN = 8 and a lower fraction correct at CN = 4.

EXERCISE 2-9

Calcium ion in CaF_2 (fluorite) is in cubic sites, with $r_+/r_- = 0.96$. Predict the coordination number of Ca^{2+} in $CaCl_2$ and $CaBr_2$.

There are also compounds in which the cations are larger than the anions; in these cases, the appropriate radius ratio is r_-/r_+, determining the CN of the anions in the holes of a cation lattice. Cesium fluoride is an example, with $r_-/r_+ = 119/181$, which places it in the 6-coordinate range, consistent with the NaCl structure observed for this compound.

[32] L. C. Nathan, *J. Chem. Educ.*, **1985,** *62,* 215.

TABLE 2-12
Radius ratios and coordination numbers

Limiting values dividing ranges r_+/r_-	r_-/r_+	*Coordination number*	*Geometry*	*Ionic compounds*
		4	Tetrahedral	ZnS
0.414	2.42			
		4	Square planar	None
		6	Octahedral	NaCl, TiO_2 (rutile)
0.732	1.37			
		8	Cubic	CsCl, CaF_2 (fluorite)
1.00	1.00			
		12	Cubooctahedron	No ionic examples Many metals are 12-coordinate

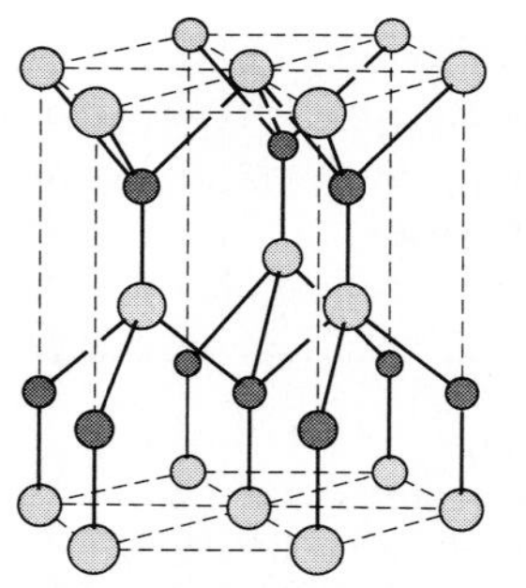
Wurzite (hexagonal ZnS)

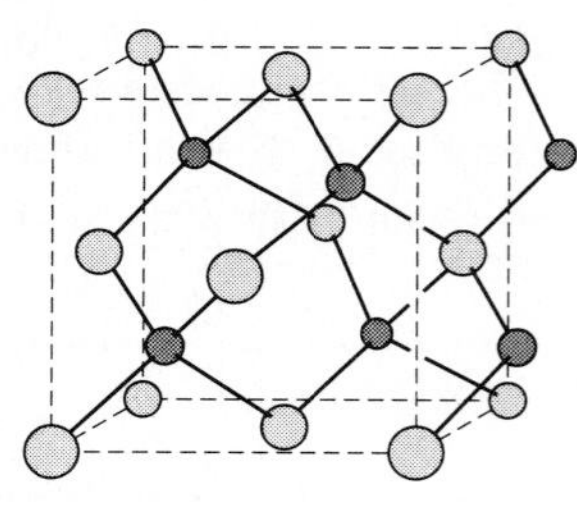
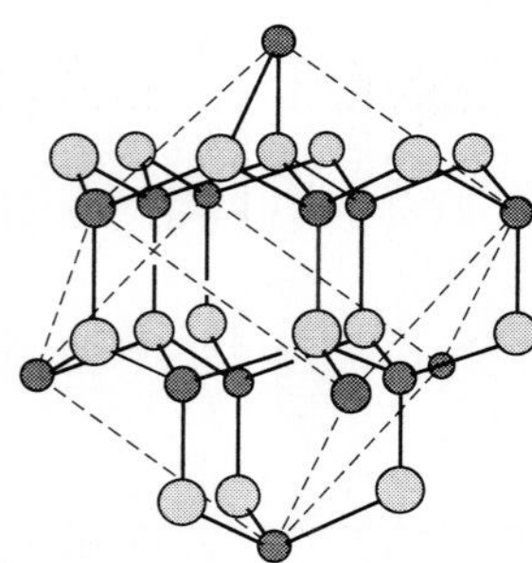
Zinc Blende (cubic ZnS)
Two orientations

Coordination number = 4, tetrahedral (predicted for $r_+/r_- < 0.414$)

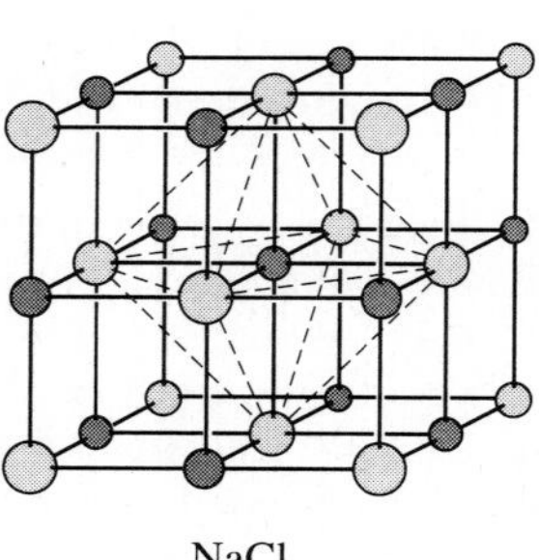
NaCl

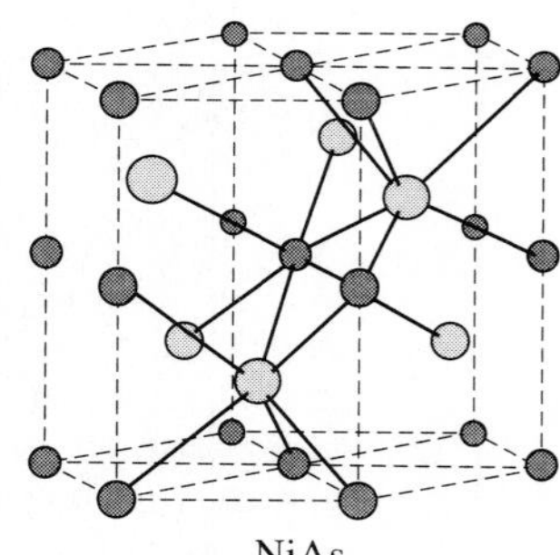
NiAs

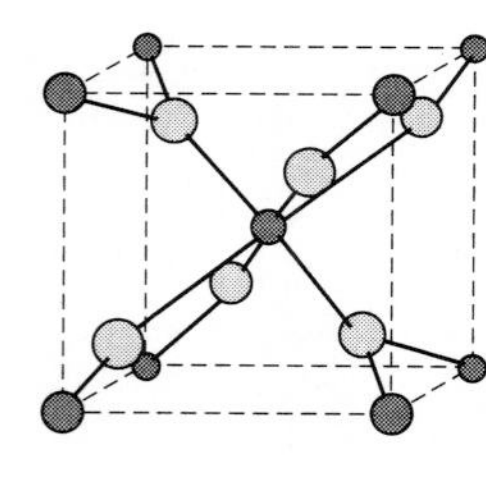
Rutile (TiO_2)

Coordination number = 6, octahedral (predicted for $0.414 < r_+/r_- < 0.732$)

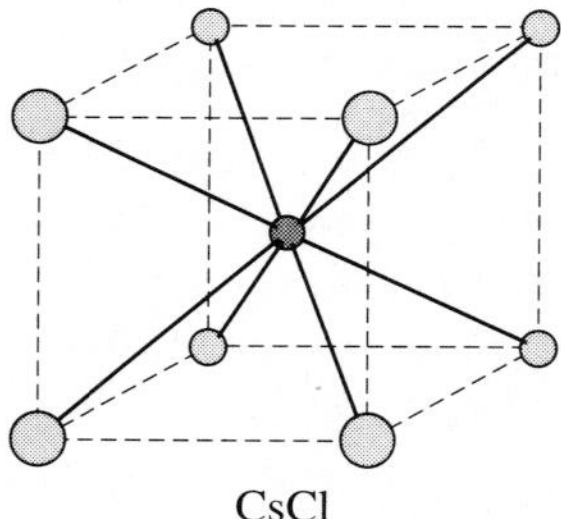
CsCl

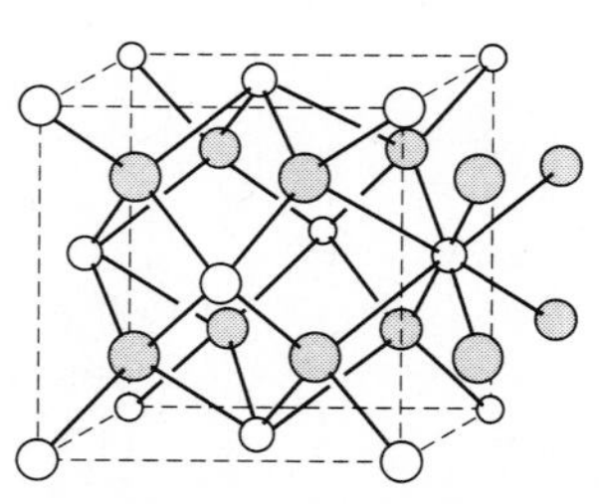
Fluorite (CaF_2)

Coordination number = 8 (predicted for $0.732 < r_+/r_- < 1.00$)

FIGURE 2-16 Crystal Structures and Radius Ratios.

When the ions are nearly equal in size, it seems that the overall structure should be a close-packed structure (hcp or ccp, ignoring the difference between cations and anions). Experimentally, however, equal-sized ions lead to a cubic arrangement with CN = 8, as in cesium chloride. This is a consequence of the charges on the ions and the resulting repulsive forces between ions of the same charge. Figure 2-16 shows some common crystal structures for coordination numbers 4, 6, and 8.

Compounds whose stoichiometry is not 1:1 (such as CaF_2 and Na_2S) may either have different coordination numbers for the cations and anions or structures in which only a fraction of the possible sites are occupied. Details of such structures are available in Wells[33] and other large references.

The electronic structures of metallic and semiconductor solids will be considered in Chapter 5 as special cases of the electronic structures of molecules.

GENERAL REFERENCES

Additional information on the history of inorganic chemistry can be found in *A Short History of Chemistry,* by J. R. Partington, 3rd ed., Macmillan, London, 1957, reprinted by Harper & Row, New York, 1960, and in the *Journal of Chemical Education.* A more thorough treatment of the electronic structure of atoms is in *Orbitals, Terms, and States,* by M. Gerloch, Wiley, New York, 1986.

PROBLEMS

2-1 Determine the de Broglie wavelength of:

a. An electron moving at one-tenth the speed of light.

b. A 400-g Frisbee moving at 10 km/h.

2-2 Using the equation $E = R_H\left(\frac{1}{2^2} - \frac{1}{n_h^2}\right)$, determine the energies and wavelengths of the four visible emission bands in the atomic spectrum of hydrogen (n_h = 3, 4, 5, and 6).

2-3 The transition from the $n = 7$ to the $n = 2$ level of the hydrogen atom is accompanied by the emission of light slightly beyond the range of human perception, in the ultraviolet region of the spectrum. Determine the energy and wavelength of this light.

2-4 The details of several steps in the particle in a box model in this chapter have been omitted. Work out the details of the following steps:

a. Show that if $\Psi = A \sin rx + B \cos sx$ (A, B, r, s = constants) is a solution to the wave equation for the one-dimensional box then $r = s = \sqrt{2mE}\left(\frac{2\pi}{h}\right)$.

b. Show that if $\Psi = A \sin rx$ the boundary conditions ($\Psi = 0$ when $x = 0$ and $x = a$) require that $r = \pm\frac{n\pi}{a}$ where n = any integer other than zero.

c. Show that if $r = \pm\frac{n\pi}{a}$ the energy levels of the particle are given by $E = \frac{n^2h^2}{8ma^2}$.

[33] A. F. Wells, *Structural Inorganic Chemistry,* 5th ed., Oxford University Press, New York, 1984.

d. Show that substitution of the above value of r into $\Psi = A \sin rx$ and applying the normalizing requirement gives $A = \sqrt{\frac{2}{a}}$.

2-5 For the $3s$ and $4d_{x^2-y^2}$ hydrogenlike atomic orbitals, sketch:
a. The radial function R.
b. The radial probability function $a_0 r^2 R^2$.
c. Contour maps of electron density.

2-6 Repeat the exercise in problem 5 for the $4p_z$ and $5d_{xz}$ orbitals.

2-7 Repeat the exercise in problem 5 for the $5s$ and $4d_{z^2}$ orbitals.

2-8 The $4f_{z(x^2-y^2)}$ orbital has the angular function

$$Y = (\text{constant})(x^2 - y^2)z.$$

a. How many spherical nodes does this orbital have?
b. How many angular nodes?
c. Describe the angular nodal surfaces.
d. Sketch the shape of the orbital.

2-9 Repeat the exercise in problem 8 for the $5f_{xyz}$ orbital, which has

$$Y = (\text{constant})xyz$$

2-10 **a.** Find the possible values for the l and m_l quantum numbers for a $5d$ electron, a $4f$ electron, and a $7g$ electron.
b. Find the possible values for all four quantum numbers for a $3d$ electron.

2-11 Give explanations of the following phenomena:
a. The electron configuration of Cr is $[Ar]4s^1 3d^5$ rather than $[Ar]4s^2 3d^4$.
b. The electron configuration of Ti is $[Ar]4s^2 3d^2$, but that of Cr^{2+} is $[Ar]3d^4$.

2-12 Give electron configurations for the following:
a. Sc; **b.** I; **c.** Fe^{3+}; **d.** Bi; **e.** Pt^{2+}

2-13 Which 2^+ ion has six $3d$ electrons? Which has three $3d$ electrons?

2-14 Determine the Coulombic and exchange energies for the following configurations and which configuration is favored (of lower energy):

a. ↑ ↑ and ↑↓ ___

b. ↑ ↑ ↑ and ↑↓ ↑ ___

2-15 Using Slater's rules, determine Z^* for:
a. A $2p$ electron in N, O, F, and Ne. Is the calculated value of Z^* consistent with the relative sizes of these atoms?
b. A $4s$ and a $3d$ electron of Cu. Which type of electron is more likely to be lost when copper forms a positive ion?
c. A $4f$ electron in Ce, Pr, and Nd. There is a decrease in size, commonly known as the *lanthanide contraction* with increasing atomic number in the lanthanides. Are your values of Z^* consistent with this trend?

2-16 For each of the following configurations, construct a microstate table and reduce the table to its constituent free-ion terms. Identify the lowest energy term for each.
a. p^3
b. p^1d^1 (as in a $4p^1 3d^1$ configuration)

2-17 For each of the lowest energy (ground state) terms in problem 16, determine the possible values of J. Which J value describes the state with the lowest energy?

2-18 For each of the following free-ion terms, determine the values of L, M_L, S, and M_S:
a. ^2D; **b.** ^3G; **c.** ^4F

2-19 For each of the free-ion terms in problem 18, choose a possible electron configuration, determine the possible values of J, and decide which is the lowest in energy.

2-20 Select the better choice in each of the following, and explain your selection briefly.
a. Higher ionization energy: Ca or Sr
b. Higher ionization energy: Mg or Al
c. Higher electron affinity: C or N
d. More likely configuration for Mn^{2+}: $[Ar]4s^23d^3$ or $[Ar]3d^5$

2-21 Ionization energies should depend on the effective nuclear charge that holds the electrons in the atom. Calculate Z^* (Slater's rules) for O, S, and Se. Do their ionization energies seem to match these effective nuclear charges? If not, what other factors influence the ionization energies?

2-22 The ionization energies for Cl^-, Cl, and Cl^+ are 343, 1250, and 2300 kJ/mol, respectively. Explain these differences.

2-23 Why are the ionization energies of the alkali metals in the order Li > Na > K > Rb?

2-24 The second ionization of carbon ($C^+ \rightarrow C^{2+} + e^-$) and the first ionization of boron ($B \rightarrow B^+ + e^-$) both fit the reaction $1s^22s^22p^1 \rightarrow 1s^22s^2 + e^-$. Compare the two ionization energies (24.383 and 8.298 eV, respectively) and the effective nuclear charges, Z^*. Is this an adequate explanation of the difference? If not, suggest other factors.

2-25 In each of the pairs below, pick the element with the higher ionization energy and explain your choice.
a. Rb, I; **b.** Cr, Mo; **c.** N, O; **d.** Si, P; **e.** Zn, Ga

2-26 On the basis of electron configurations, explain why:
a. Sulfur has a lower electron affinity than chlorine.
b. Iodine has a lower electron affinity than bromine.
c. Boron has a lower ionization energy than beryllium.
d. Sulfur has a lower ionization energy than phosphorus.
e. Chlorine has a lower ionization energy than fluorine.

2-27 The second ionization energy of He is almost exactly four times the ionization energy of H, and the third ionization energy of Li is almost exactly nine times the ionization energy of H:

	I.E.(MJ mol⁻¹)
$H(g) \longrightarrow H^+(g) + e^-$	1.3120
$He^+(g) \longrightarrow He^{2+}(g) + e^-$	5.2504
$Li^{2+}(g) \longrightarrow Li^{3+}(g) + e^-$	11.8149

Explain this trend on the basis of the Bohr equation for energy levels of single-electron systems.

2-28 The size of the transition metal atoms decreases slightly from left to right. What factors must be considered in explaining this decrease? In particular, why does the size decrease at all, and why is the decrease so gradual?

2-29 All the alkali halides have the NaCl (CN = 6) structure except CsCl, CsBr, and CsI, which have the CsCl structure (CN = 8). Calculate the radius ratios for LiI, KCl, and CsBr. Discuss the agreement (or lack of agreement) between the radius ratio and actual structure for each compound.

3

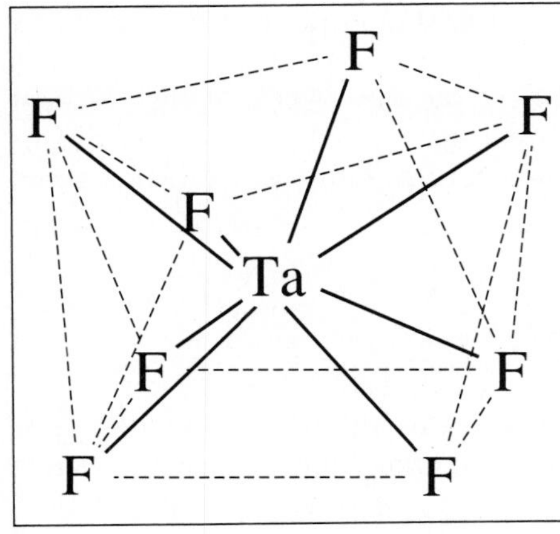

Simple Bonding Theory

We now turn from the use of quantum mechanics and its description of the atom to an elementary description of molecules. Although most of the discussion of bonding in this book uses the molecular orbital approach, simpler methods that provide approximate pictures of the overall shapes and polarities of molecules are also very useful. This chapter provides an overview of Lewis dot structures, valence shell electron-pair repulsion (VSEPR), and related topics. The molecular orbital descriptions of some of the same molecules are presented in Chapter 5 and later chapters, but the ideas of this chapter provide a starting point for that more modern treatment. As in Chapter 2, much of this material may be review, but it is no less important because of that fact. General chemistry texts include discussions of most of these topics; this chapter provides a review for those who have not used them recently.

3-1 LEWIS ELECTRON-DOT DIAGRAMS

Lewis electron-dot diagrams, although very much oversimplified, provide a good starting point for picturing the bonding in molecules. Credit for their initial use goes to G. N. Lewis,[1] an American chemist who contributed much to thermodynamics and chemical bonding in the early years of the twentieth century. In these diagrams a bond between two atoms exists when they share one or more pairs of electrons. In addition, some molecules require nonbonding

[1] G. N. Lewis, *J. Am. Chem. Soc.*, **1916**, *38*, 762; *Valence and the Structure of Atoms and Molecules*, Chemical Catalogue Co., New York, 1923.

pairs (also called lone pairs) of electrons on the atoms. These electrons contribute to the shape and reactivity of the molecule, but do not directly bond the atoms together. Overall, Lewis structures are based on the concept of eight **valence electrons** (those outside the noble gas core) forming a particularly stable arrangement, as in the noble gases with s^2p^6 configurations. Hydrogen, which is stable with two valence electrons, is an exception. Also, some molecules require more than eight electrons around a given atom.

A more detailed approach to electron-dot diagrams is presented in Appendix D. Those who want more background should refer to this appendix or to a general chemistry text.

Simple molecules such as water follow the **octet rule,** in which eight electrons surround the oxygen atom. The hydrogen atoms share two electrons each with the oxygen, forming the familiar picture with two bonds and two lone pairs:

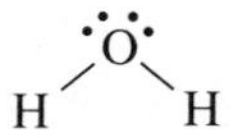

$$\mathrm{H{-}\ddot{\underset{\cdot\cdot}{O}}{-}H}$$

When there are not enough electrons to draw the molecule with single bonds, some bonds must be double bonds, containing four electrons, or triple bonds, containing six electrons:

$$:\ddot{O}{=}C{=}\ddot{O}: \qquad H{-}C{\equiv}C{-}H$$

3-1-1 EXPANDED SHELLS

When it is impossible to draw a structure consistent with the octet rule, it is necessary to increase the number of electrons around the central atom. This is done by expanding the octet on the central atom and using the *d* orbitals that are just above the usual *s* and *p* orbitals in energy. This option is usually limited to elements of the third and higher periods, where the energy of the *d* orbitals is close enough to make it energetically feasible. In most cases, two or four added electrons will complete the bonding, but more can be added if necessary. ClF_3 and SF_6 are examples (Figure 3-1). Ten electrons are required around chlorine in ClF_3 and 12 around sulfur in SF_6.

There are examples with even more electrons around the central atom, such as IF_7, TaF_8^{3-} and XeF_8^{2-}, described later in this chapter. No more than 18 electrons (2 for *s*, 6 for *p*, and 10 for *d* orbitals) are found around a single atom in the top half of the periodic table, and crowding of the outer atoms usually keeps the number below this even for the much heavier atoms with *f* orbitals energetically available.

Section 3-1-4 on formal charge describes some molecules in which multiple bonding raises the number of electrons around the central atom. The final

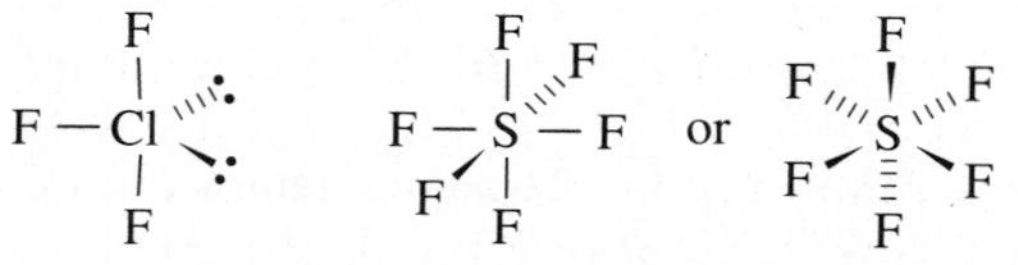

FIGURE 3-1 Structures of ClF_3 and SF_6. The shapes are explained more fully later in this chapter.

description of bonding depends on experimental data on bond lengths and strengths. Bond lengths are most frequently determined by X-ray crystallography, and bond strengths can be measured by determination of vibrational energies using infrared spectroscopy. However, there is sometimes disagreement on the interpretation of the experimental data, particularly when only a few compounds are available for comparison. The beryllium and boron halides are examples discussed in Section 3-1-5.

3-1-2 RESONANCE

In many molecules, the choice of which atoms are connected by the double bond is arbitrary. When several choices exist, all of them should be drawn. For example, three drawings of SO_3 are needed to show the double bond in each of the three possible S—O positions. In fact, experimental evidence shows that all the S—O bonds are identical, with bond lengths between the usual single- and double-bond distances. This phenomenon is called **resonance** to signify that there is more than one possible way in which the valence electrons can be placed in a Lewis structure. (Note that in resonance structures, such as those shown for SO_3 in Figure 3-2, only electrons are being moved; the atomic nuclei remain in fixed positions.) Linus Pauling developed the use of this information to a very high degree in the 1930s. It is more difficult to sketch a molecular orbital picture, but it does not require mentally mixing multiple diagrams. Chapter 5 describes the molecular orbital approach to bonding.

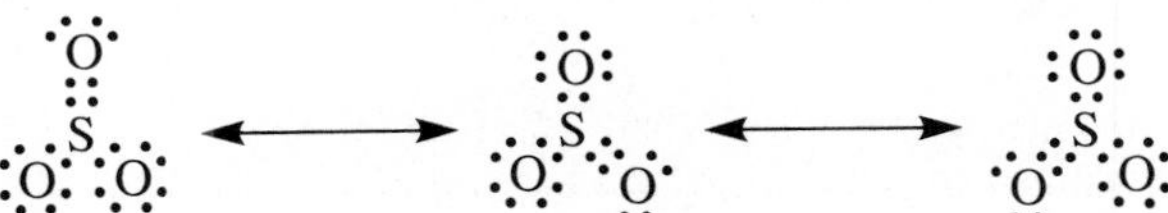

FIGURE 3-2 Lewis Diagrams for SO_3.

The two ions NO_3^- and CO_3^{2-} are **isoelectronic** (have the same electronic structure) with SO_3. Since they are ions, the extra electrons needed for the charge must be added to the electrons from the atoms. The Lewis diagrams are identical to those of SO_3 in Figure 3-2 except for the identity of the central atom.

When a molecule has several resonance structures, its overall electronic energy is lowered, making it more stable. Just as the energy levels of a particle in a box are lowered by making the box larger, the electronic energy levels of the bonding electrons are lowered when the electrons can occupy a larger space. Again, the molecular orbital description of this effect is presented in Chapter 5.

3-1-3 ELECTRONEGATIVITY AND BOND POLARITY

Molecules with identical atoms, such as those in H_2 and N_2, have the same electron density at both ends of the molecule and have **nonpolar bonds.** Molecules with different atoms, such as HCl and CO, have **polar bonds** and slightly higher electron density at one end of the molecule. This phenomenon is a result of **electronegativity,** a measure of the atom's ability to attract electrons

TABLE 3-1
Pauling electronegativities

Element	*Electronegativity*	*Element*	*Electronegativity*
H	2.20	Br	2.96
He		Kr	
Li	0.98	Rb	0.82
Be	1.57	Sr	0.95
B	2.04	Y	1.22
C	2.55	Zr	1.33
N	3.04	Nb	
O	3.44	Mo	2.16
F	3.98	Tc	
Ne		Ru	
Na	0.93	Rh	2.28
Mg	1.31	Pd	2.20
Al	1.61	Ag	1.93
Si	1.90	Cd	1.69
P	2.19	In	1.78
S	2.58	Sn	1.96
Cl	3.16	Sb	2.05
Ar		Te	
K	0.82	I	2.66
Ca	1.00	Xe	
Sc	1.36	Cs	0.79
Ti	1.54	Ba	0.89
V	1.63	La	1.10
Cr	1.66	Hf	
Mn	1.55	Ta	
Fe	1.83	W	2.36
Co	1.88	Ir	2.20
Ni	1.91	Pt	2.28
Cu	1.90	Au	2.54
Zn	1.65	Hg	2.00
Ga	1.81	Tl	2.04
Ge	2.01	Pb	2.33
As	2.18	Bi	2.02
Se	2.55		

SOURCE: Data from A. L. Allred, *J. Inorg. Nucl. Chem.*, **1961,** *17,* 215.

to itself in the molecule. A more complete treatment of this subject is given later in this chapter. For the present, it is sufficient to know that electronegativity is dependent on a complex interplay of electronic energy levels. With the exception of the noble gases, the most electronegative elements (fluorine has the largest value) are in the upper right corner of the periodic table, and the least electronegative (cesium and francium) are in the lower left corner. A table of Pauling electronegativities, recalculated from newer experimental data by Allred, is given in Table 3-1.

3-1-4 FORMAL CHARGE

Although the actual charge on atoms in molecules depends on differences in electronegativity, calculation of the formal charge can help in assigning bonding when there are several possibilities. It can eliminate the least likely forms when we are considering resonance structures and in some cases suggest multiple bonds beyond those required by the octet rule. It is essential, however, to keep in mind that formal charge is only a tool in determining bonding, not a measure of any actual charge on the atoms.

Formal charge is the apparent electronic charge of each atom in a molecule, based on the electron-dot structure. The number of valence electrons available in a free atom of an element minus the total for that atom in the molecule (determined by counting lone pairs as 2 electrons and bonding pairs as one assigned to each atom) is the formal charge on the atom:

$$\text{Formal charge} = \begin{bmatrix}\text{number of valence} \\ \text{electrons in a free} \\ \text{atom of the element}\end{bmatrix} - \begin{bmatrix}\text{number of valence electrons} \\ \text{assigned to the atom in the} \\ \text{molecule or ion}\end{bmatrix}$$

The last term of the equation includes lone electrons, lone pairs, and half the electrons shared with other atoms.

In addition,

$$\text{Charge on the molecule or ion} = \text{sum of all the formal charges}$$

Structures minimizing formal charges and placing negative formal charges on more electronegative elements tend to be favored. Examples of formal charge calculations are given in Appendix D for those who need more review. Three examples, SCN^-, OCN^-, and CNO^-, will illustrate the use of formal charges in describing electronic structures.

EXAMPLES

SCN^- In the thiocyanate ion there are three resonance structures consistent with the electron-dot method, as shown in Figure 3-3. Structure A has only one negative formal charge on the nitrogen atom, the most electronegative atom in the ion, and fits the rules well. Structure B has a single negative charge on the S, which is less electronegative than N. Structure C has charges of 2− on N and 1+ on S, consistent with the relative electronegativities of the atoms but with a larger charge and greater charge separation than the first. Therefore, these structures lead to the prediction that structure A is most likely, structure B is next in importance, and any contribution from C is minor. The bond lengths in Table 3-2 are consistent with this picture, with the S—C bond intermediate in length between a single and double bond and the C—N bond intermediate between a double and triple bond. Structure A, with the negative charge on N, is the predominant form. Protonation occurs at the nitrogen; the acid is HNCS, with N—C 122 pm and C—S 156 pm in length, almost exactly the double bond lengths in both cases, consistent with the Lewis structure H—N=C=S. This also suggests that A is the most important structure of the ion.

TABLE 3-2
Table of S—C and C—N bond lengths

	S—C	*C—N*
SCN^-	165 pm	117 pm
HNCS	156	122
Single bond	181	147
Double bond	155	128 (uncertain)
Triple bond		116

SOURCE: Data from A. F. Wells, *Structural Inorganic Chemistry*, 5th ed., Oxford University Press, New York, 1984, pp. 807, 926, 934–36.

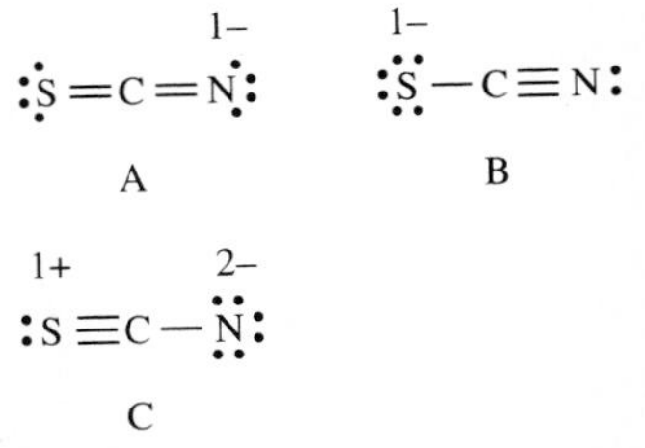

A: $[:\ddot{S}=C=\ddot{N}:]^{1-}$

B: $[:\ddot{S}-C\equiv N:]^{1-}$

C: $:S^{1+}\equiv C-\ddot{N}:^{2-}$

FIGURE 3-3 Resonance Structures of Thiocyanate, SCN^-.

OCN⁻ The analogous cyanate ion (Figure 3-4) has the same possibilities, but the larger electronegativity of O makes structure B more likely. The structure should then be a mix of A and B. The acid, like the sulfur analog, has the proton on the nitrogen, HNCO (about 3% of the HOCN isomer is also present), consistent with protonation of structure A. The bond lengths in OCN^- and HNCO in Table 3-3 are consistent with this picture, but do not agree perfectly.

TABLE 3-3
Table of O—C and C—N bond lengths

	O—C	*C—N*
OCN^-	113 pm	121 pm
HNCO	118	120
Single bond	143	147
Double bond	116	128 (uncertain)
Triple bond	113	116

SOURCE: Data from A. F. Wells, op. cit. pp. 807, 926, 933–34.

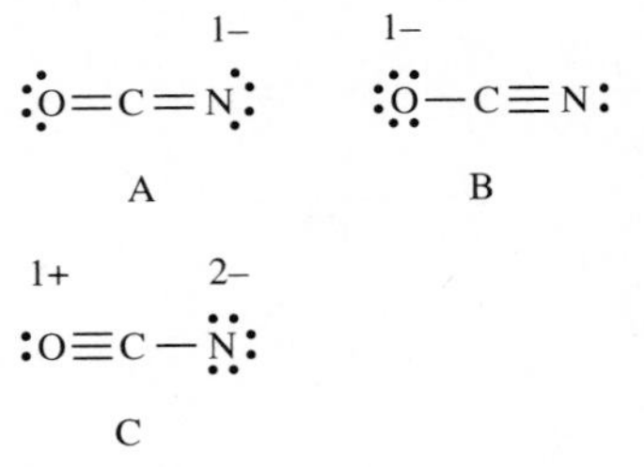

1– :O=C=N: (A) 1– :O—C≡N: (B) 1+ 2– :O≡C—N: (C)

FIGURE 3-4 Resonance Structures of Cyanate, OCN^-.

CNO⁻ The isomeric fulminate ion (Figure 3-5), can be drawn with three similar structures, but the resulting formal charges are unlikely. In A, carbon is 2− and nitrogen 1+; in B, carbon is 3−, and nitrogen and oxygen are each 1+; and in C, carbon and oxygen are each 1− and nitrogen again 1+. Since the order of electronegativities is C < N < O, none of these are plausible structures and the ion is predicted to be unstable. The only common fulminate salts are of mercury and silver; both are explosive. Fulminic acid is linear HCNO in the vapor phase, and coordination complexes of CNO^- with many transition metal ions are known.[2]

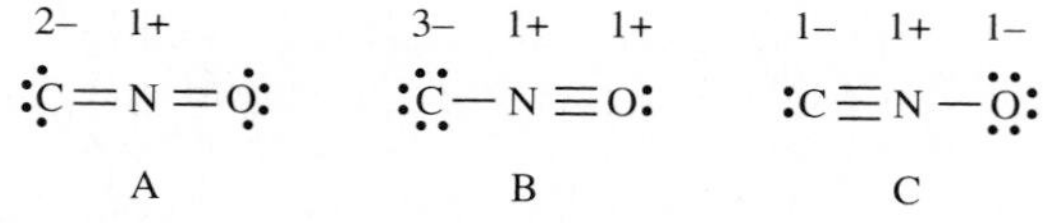

2– 1+ :C=N=O: (A) 3– 1+ 1+ :C—N≡O: (B) 1– 1+ 1– :C≡N—O: (C)

FIGURE 3-5 Resonance Structures of Fulminate, CNO^-.

EXERCISE 3-1

Use electron-dot diagrams and formal charges to find the bond order for each bond in POF_3, SOF_4, and SO_3F^-.

3-1-5 FORMAL CHARGE AND EXPANDED OCTETS

Some molecules seem to have satisfactory electron-dot structures with the usual octets, but have better structures with expanded octets when formal charges are considered. In each of the cases in Figure 3-6, the observed structures are consistent with expanded octets on the central atom and with the resonance structure that uses multiple bonds to minimize formal charges. The multiple bonds may also influence the shapes of the molecules, as described later.

[2] A. G. Sharpe, "Cyanides and Fulminates," in *Comprehensive Coordination Chemistry*, G. Wilkinson, R. D. Gillard, and J. A. McCleverty, eds., Pergamon, Elmsford, N.Y., 1987, Vol. 2, pp. 12–14.

Molecule	Octet: Atom	Octet: Formal charge	Expanded: Atom	Expanded: Formal charge	Expanded to:
SNF_3	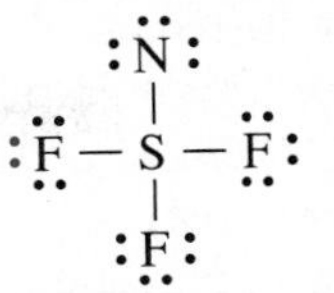 S N	2+ 2−	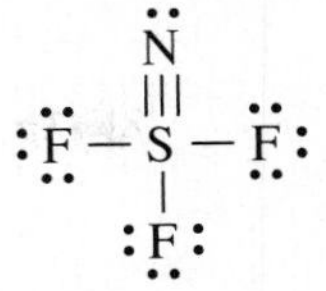 S N	0 0	12
SO_2Cl_2	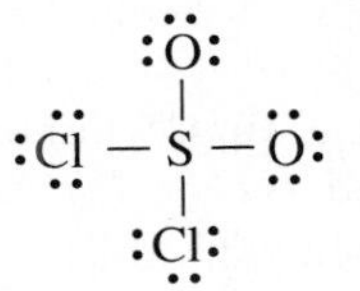 S O	2+ 1−	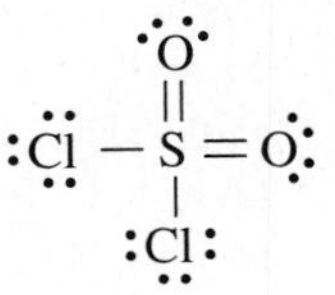 S O	0 0	12
XeO_3	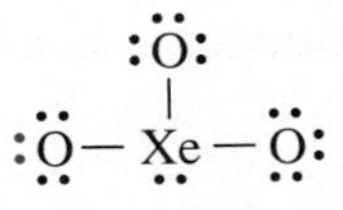 Xe O	3+ 1−	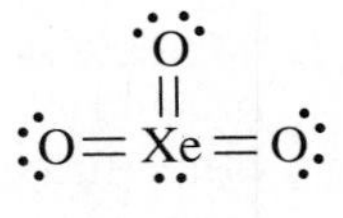 Xe O	0 0	14
SO_4^{2-}	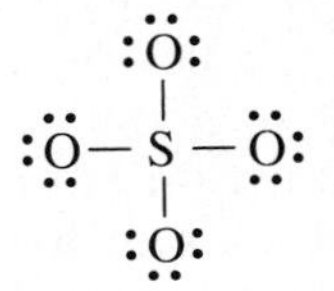 S O	2+ 1−	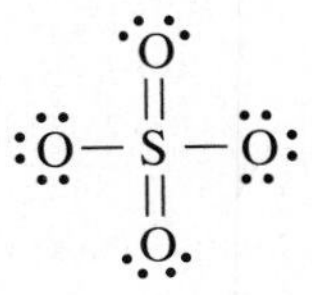 S O	0 0, 1−	12
SO_3^{2-}	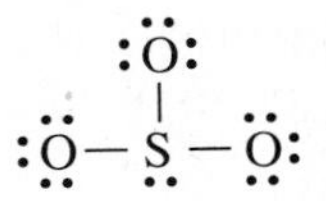 S O	1+ 1−	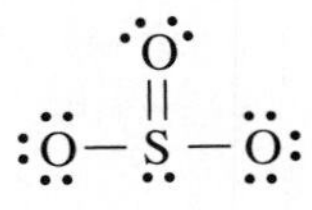 S O	0 0, 1−	10
IOF_5	No octet structure possible	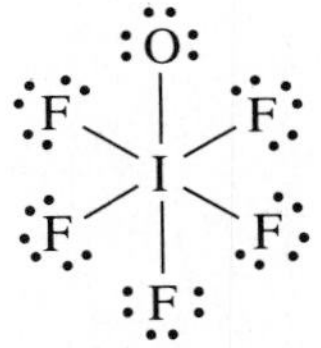	I O	1+ 1−	12
			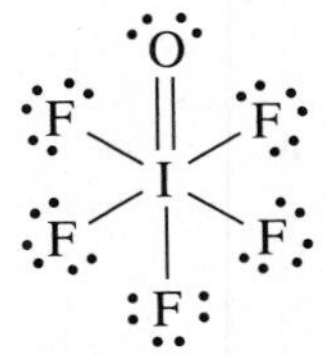 I O	0 0	14

FIGURE 3-6 Formal Charge and Expanded Octets.

3-1-6 MULTIPLE BONDS IN Be AND B COMPOUNDS

A few molecules, such as BeF_2, $BeCl_2$, and BF_3, seem to require multiple bonds, even though we do not usually expect multiple bonds for fluorine and chlorine. Structures minimizing formal charges for these molecules have only four electrons in the valence shell of Be and six electrons in the valence shell of B, in both cases short of the usual octet. The alternative, requiring eight electrons on the central atom, predicts multiple bonds, with BeF_2 analogous to CO_2 and BF_3 analogous to SO_3. These structures, however, result in formal charges (2− on Be and 1+ on F in BeF_2, and 1− on B and 1+ on the double-bonded F in BF_3), which are unlikely by the usual rules.

It has not been experimentally determined whether the bond lengths in BeF_2 and $BeCl_2$ are those of double bonds, because molecules with clear-cut single and double bonds are not available for comparison. In the solid, a complex network is formed with coordination number 4 for the Be atom (see Figure 3-7). $BeCl_2$ tends to dimerize to a 3-coordinate structure in the vapor phase, but the linear monomer is also known at high temperatures. The monomeric structure is unstable; the dimer and polymer share lone pairs from the halogen atoms with the Be atom and bring it closer to the octet structure. The dimer and polymer also have unlikely formal charges. The monomer is still frequently drawn as a singly bonded structure with only four electrons around the beryllium and the ability to add more from lone pairs of other molecules (Lewis acid behavior, discussed in Chapter 6).

Bond lengths in all the boron trihalides are shorter than expected for single bonds, so the partial double-bond character predicted seems reasonable in spite of the formal charges. On the other hand, they combine readily with other molecules that can contribute a lone pair of electrons (Lewis bases), forming a roughly tetrahedral structure with four bonds. Because of this tendency, they are frequently drawn with only six electrons around the boron. Boron trihalides combine with electron donors in the order $BBr_3 > BCl_3 \geq BF_3$. This order is presumed due to π bonding (strongest in BF_3), which is disrupted on formation of the fourth single bond. It is the reverse order that would be expected from electronegativity (F would draw electrons toward itself more effectively, making the boron in BF_3 more positive) or steric crowding (the larger Br atoms would resist compression into a tetrahedral

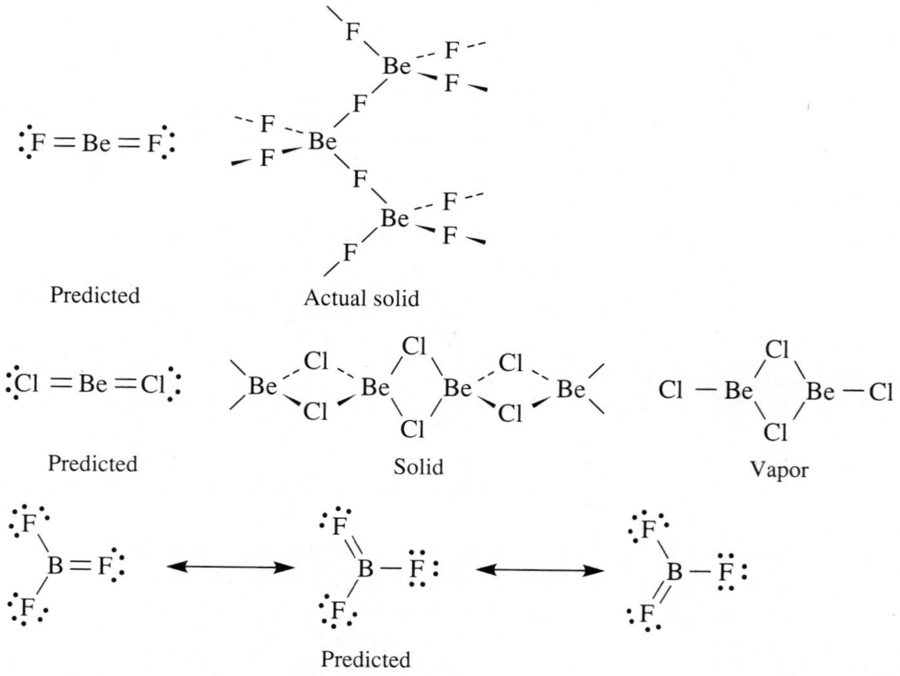

:F̤=Be=F̤:

Predicted

Actual solid

:C̤l=Be=C̤l:

Predicted

Solid

Cl—Be(Cl)₂Be—Cl

Vapor

Predicted

The B—F bond length is 131 pm;
the calculated single-bond length is 152 pm.

FIGURE 3-7 Structures of BeF_2, $BeCl_2$, and BF_3. (From A. F. Wells, *Structural Inorganic Chemistry,* 5th ed., Oxford University Press, New York, 1984, pp. 412, 441, 1047.)

shape most strongly). As can be seen from these examples, although the Lewis electron-dot method often predicts correct structures, a method this simple cannot cover all cases without exceptions.

Other boron compounds that do not fit simple electron-dot structures include the hydrides, such as B_2H_6, and a very large array of more complex molecules. Their structures are discussed in Chapters 7 and 14.

3-2 VALENCE SHELL ELECTRON PAIR REPULSION THEORY

Valence shell electron pair repulsion theory (VSEPR) provides a method for predicting the shapes of molecules, based on the electron-pair diagrams described above and electrostatic repulsion. It was described by Sidgwick and Powell[3] in 1940 and further developed by Gillespie and Nyholm[4] in 1957. The most common method of determining the actual structures is X-ray diffraction, although neutron diffraction and many kinds of spectroscopy are also used.[5]

Since they are negatively charged, electrons repel each other. The quantum mechanical rules force some of them to be fairly close to each other in bonding or lone pairs, but each pair repels all other pairs. According to the VSEPR model, therefore, molecules adopt geometries in which their valence electron pairs position themselves as far from each other as possible.

Carbon dioxide is an example with two bonding positions on the central atom and double bonds in each direction. The electrons in each double bond must be between C and O, and the repulsion between the electrons in the double bonds forces a linear structure on the molecule. Sulfur trioxide has three bonding positions, with partial double-bond character in each. The best position for the oxygens in this molecule is at the corners of an equilateral triangle, with O—S—O bond angles of 120°. The multiple bonding does not affect the geometry because it is shared equally among the three bonds.

The same pattern of finding the Lewis structure and then matching it to a geometry that minimizes the repulsive energy of bonding electrons is followed through four, five, six, seven, and eight directions for bonding, as shown in Figure 3-8.

The structures for two, three, four, and six electron pairs are completely regular, with all bond angles and distances the same. The 5- and 7-coordinate structures cannot have all angles and distances identical, since there are no regular polyhedra with these numbers of vertices. The 5-coordinate molecules have a central triangular plane of three positions plus two other positions above and below the center of the plane. The 7-coordinate molecules have a pentagonal plane of five positions and positions above and below the center of the plane. The regular square antiprism structure (CN = 8) is like a cube with the top and bottom faces twisted 45° into the *antiprism* arrangement, as shown in Figure 3-9. It has three different bond angles for adjacent fluorines: 70.5° for adjacent fluorines in the same square plane, 109.5° for fluorines at opposite corners of the square plane, and 99.6° for one fluorine in the lower plane and one in the upper. TaF_8^{3-} has square antiprism symmetry, but is distorted from this ideal in the solid.[6] (A simple cube has only the 109.5° and

[3] N. V. Sidgwick and H. M. Powell, *Proc. Roy. Soc.*, **1940,** *A176*, 153.

[4] R. J. Gillespie and R. S. Nyholm, *Quart. Revs.*, **1957,** *XI*, 339, a very thorough and clear description of the principles, with many more examples than are included here; R. J. Gillespie, *J. Chem. Educ.*, **1970,** *47*, 18.

[5] G. M. Barrow, *Physical Chemistry*, 5th ed., McGraw-Hill, New York, 1988, pp. 540–680; R. S. Drago, *Physical Methods in Chemistry*, W. B. Saunders, Philadelphia, 1977, pp. 589–625.

[6] J. L. Hoard, W. J. Martin, M. E. Smith, and J. F. Whitney, *J. Am. Chem. Soc.*, **1954,** *76*, 3820.

Bond directions	*Geometry*	*Examples*	*Calculated bond angles*
2	Linear	CO_2	180°
3	Planar triangular (trigonal)	SO_3	120° 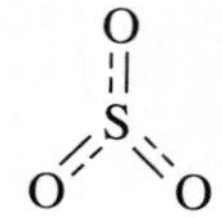
4	Tetrahedral	CH_4	109.5° 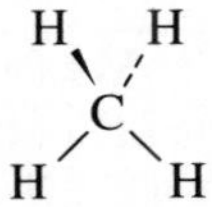
5	Trigonal bipyramidal	PCl_5	120°, 90° 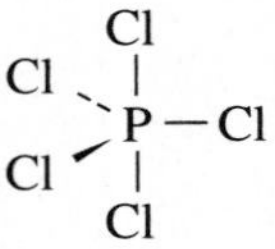
6	Octahedral	SF_6	90° 
7	Pentagonal bipyramidal	IF_7	72°, 90° 
8	Square antiprismatic	TaF_8^{3-}	70.5°, 99.6°, 109.5° 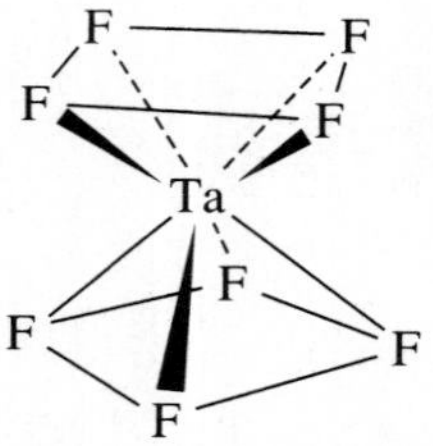

FIGURE 3-8 VSEPR Predictions.

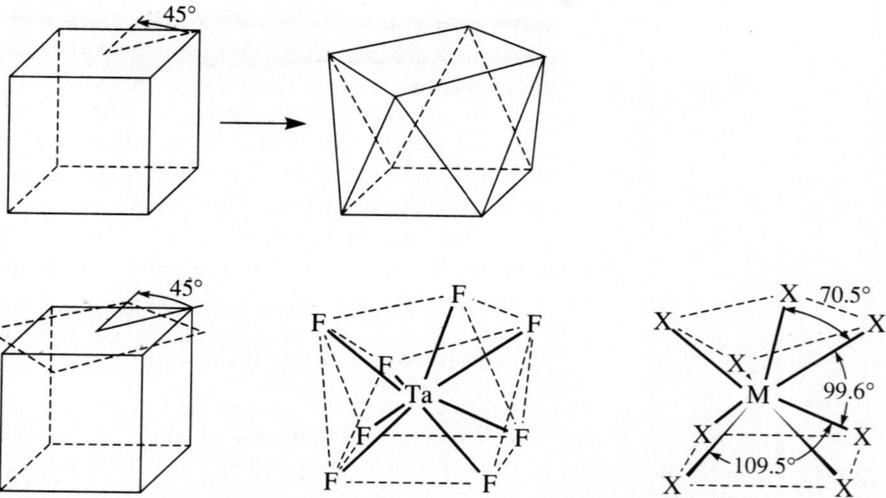

FIGURE 3-9 Conversion of a Cube into a Square Antiprism.

70.5° bond angles measured between two corners and the center of the cube, as all edges are equal and any square face can be taken as the bottom or top.)

3-2-1 LONE PAIR REPULSION

To a first approximation, lone pairs, single bonds, double bonds, and triple bonds can all be treated similarly when predicting molecular shapes. However, better predictions of overall shapes can be made by considering some important differences between lone pairs and bonding pairs.

The molecules CH_4, NH_3, and H_2O (Figure 3-10) illustrate the effect of lone pairs on molecular shape. Methane has four identical bonds between carbon and each of the hydrogens. When the four pairs of electrons are arranged as far from each other as possible, the result is the familiar tetrahedral shape. The tetrahedron, with all H—C—H angles 109.5°, has four identical bonds.

Ammonia also has four pairs of electrons around the central atom, but three are bonding pairs between N and H and the fourth is a lone pair on the nitrogen. The nuclei form a trigonal pyramid; the three bonding pairs and the lone pair make a nearly tetrahedral shape. Since each of the three bonding pairs has a positive nucleus drawing it away from the nitrogen, the electron–electron repulsions near the nitrogen are reduced. The lone pair has a stronger effect on the shape of NH_3 because it is closer to the nitrogen nucleus; this reduces the distance between the lone pair and adjacent bonding pairs and makes the repulsion between them stronger. The balance of these repulsive forces results in larger angles between the lone pair and each of the bonding pairs and smaller angles between the bonding pairs. As a result, the H—N—H angles are 106.6°, nearly 3° smaller than the angles in methane.

The same principles apply to the water molecule, where two lone pairs and two bonding pairs repel each other. Again, the electron pairs have a nearly tetrahedral arrangement, with the atoms arranged in a V-shape. The angle of largest repulsion, between the two lone pairs, is not directly measurable. However, the lone pair–bonding pair (*lp–bp*) repulsion is greater than the bonding pair–bonding pair (*bp–bp*) repulsion, and as a result the H—O—H bond angle is only 104.5°, another 2.1° decrease from the ammonia angles. The net result is that we can predict approximate molecular shapes by assigning more space to lone electron pairs; being attracted to one nucleus rather than two, they are permitted to spread out and thus occupy more space.

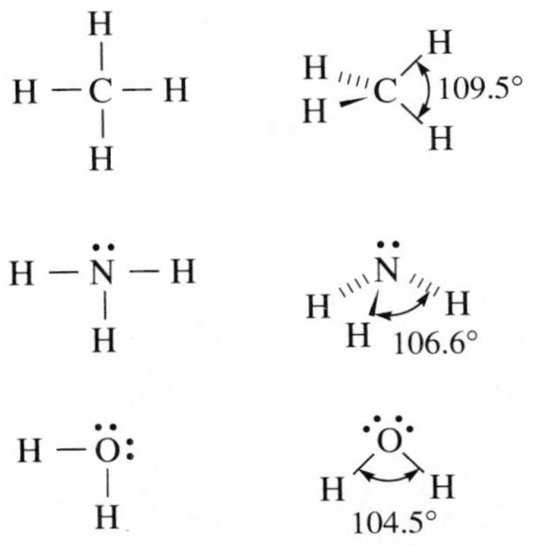

FIGURE 3-10 Shapes of Methane, Ammonia, and Water.

More subtle differences in angle result from differences in atomic size. The series H_2O, H_2S, H_2Se, H_2Te, with the same electron-dot diagram, has angles of 104.5°, 92.1°, 91°, and 90°.[7] The central atoms become larger in this series. This leads to less electron–electron repulsion overall, with a larger effect on the bonding pairs than on the lone pairs. Relatively, the lone pairs remain close to the nucleus, while the bonding pairs are pulled out farther from the larger atoms by the longer bond distance. This reduces the *bp–bp* repulsion, and results in smaller bond angles. Similar effects appear in the NH_3, PH_3, AsH_3, SbH_3 series, with angles of 106.6°, 93.6°, 91.8°, and 91.3°, respectively.[8]

In addition to the effect of size of the central atom, the sizes of atoms around the central atom can also affect bond angles. Some examples are given in Table 3-4. Curiously, the change from hydrogen to fluorine leads to a *decrease* in angle in compounds with nitrogen and oxygen ($NH_3 > NF_3$ and $H_2O > F_2O$). In these compounds, the hydrogen seems to occupy more space near the central atom, perhaps because the small size of the central atom and the low electronegativity of hydrogen leave the bonding pair of electrons close to the nucleus of the central atom. With larger central atoms, these electrons are farther out, smaller angles result, and the bond angles are smallest for hydrogen compounds. For example, the phosphorus series has angles in the order $PH_3 < PF_3 < PCl_3 < PBr_3 < PI_3$. Some argue that fluorine is more electronegative than the other halogens and pulls the electrons away from the central phosphorus atom more effectively, leading to a smaller bond angle. However, the same argument should lead to a smaller angle in PF_3 than in PH_3. Apparently, the size of the halogens has some effect as well.

TABLE 3-4
Bond angles and atomic size

OH_2 104.5°	OF_2 103.3°			
SH_2 92°	SF_2 98°	SCl_2 100°		
NH_3 106.6°	NF_3 102.2°	NCl_3 106.8°		
PH_3 93.8°	PF_3 97.8°	PCl_3 100.3°	PBr_3 101°	PI_3 102°
AsH_3 91.83°	AsF_3 96.2°	$AsCl_3$ 97.7°	$AsBr_3$ 97.7°	AsI_3 99.7°
SbH_3 91.3°	SbF_3 87.3°	$SbCl_3$ 97.2°	$SbBr_3$ 98°	SbI_3 99.1°

An alternative view suggests that the "natural" bond angle is 90°, the angle of the *p* orbitals, and that all the larger angles are a result of repulsion between the outer atoms of the molecule.

We must keep in mind that we are always attempting to match our explanations to experimental data. The explanation that fits the largest amount of data should be the current favorite, but new theories are continually being

[7] N. N. Greenwood and A. Earnshaw, *Chemistry of the Elements,* Pergamon, Elmsford, N.Y., 1984, p. 900.

[8] Ibid., p. 650; L. E. Sutton, *Interatomic Disturbances,* Supplement 1956–1959, Special Publication No. 18, The Chemical Society, London, 1965. The ammonia angle is 106.6° in the absence of vibration, 107.8° including vibration.

suggested and tested. Because we are working with such a wide variety of atoms and molecular structures, it is unlikely that a single, simple approach will work for all of them. While the fundamental ideas of atomic and molecular structure are relatively simple, their application to complex molecules is not.

Another example (ClF_3 again) will illustrate another use of the different repulsions for lone pairs and bonding pairs in predicting shapes. With five electron directions, the lone pairs and bonding pairs form a trigonal bipyramid. Even if all the repulsions were the same, as in PCl_5, the angles cannot all be the same with five directions; three chlorines in PCl_5 are in a triangular plane, with Cl—P—Cl angles of 120°, and the other two are above and below the plane at 90° angles to the plane. There are three possible structures for ClF_3, as shown in Figure 3-11. Examination of the diagrams gives the bond angles in the figure, with lone pairs designated *lp* and bonding pairs *bp*.

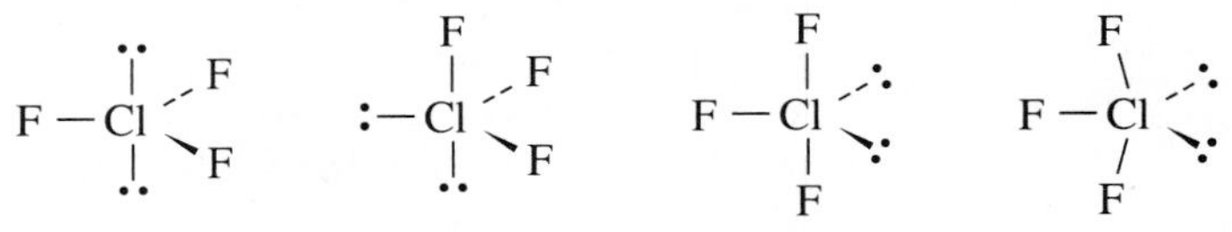

FIGURE 3-11 Possible Structures of ClF_3.

	A	*B*	*C*	*Experimental*
lp–lp	180°	90°	120°	
lp–bp	6 at 90°	3 at 90° 2 at 120°	4 at 90° 2 at 120°	
bp–bp	3 at 120°	2 at 90° 1 at 120°	2 at 90°	2 at 87.5° Axial Cl—F 169.8 pm Equatorial Cl—F 159.8 pm

In determining the structure of molecules, the lone pair–lone pair interactions are most important, with the lone pair–bonding pair interactions next in importance. In addition, interactions at angles of 90° or less are most important; larger angles generally have less influence. Structure B can be eliminated quickly because of the 90° *lp–lp* angle. The *lp–lp* angles are large for A and C, so the choice must come from the *lp–bp* and *bp–bp* angles. Since the *lp–bp* angles are more important, C, which has only four 90° *lp–bp* interactions, is favored over A, which has six such interactions. Experiments confirm that the structure is based on C with slight distortions due to the lone pairs. The lone pair–bonding pair repulsion causes the *lp–bp* angles to be larger than 90° and the *bp–bp* angles less than 90° (actually 87.5°). The Cl—F bond distances show the repulsive effects as well, with the axial fluorines (approximately 90° *lp–bp* angles) at 169.8 pm and the equatorial fluorine (in the plane with two lone pairs) at 159.8 pm.[9]

Additional examples of structures with lone pairs are given in Figure 3-12. Notice that the structures based on a trigonal bipyramidal arrangement of electron pairs around a central atom always place any lone pairs in the equatorial plane, as in SF_4, BrF_3, and XeF_2. These are the shapes that minimize both lone pair–lone pair and lone pair–bonding pair repulsions.

[9] A. F. Wells, *Structural Inorganic Chemistry*, 5th ed., Oxford University Press, New York, 1984, p 390.

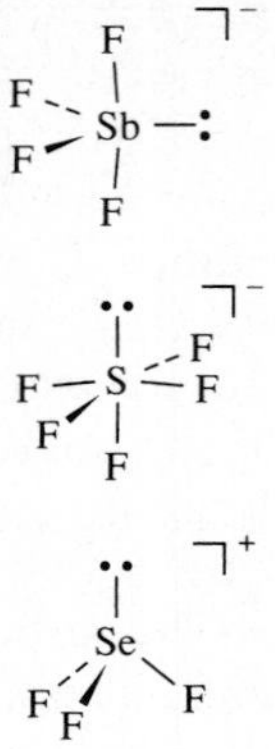

EXAMPLES

SbF_4^- has a single lone pair on Sb. Its structure is therefore similar to SF_4, with a lone pair occupying an equatorial position. This lone pair causes considerable distortion, giving an F—Sb—F (axial positions) angle of 155° and an F—Sb—F (equatorial) angle of 90°.

SF_5^- has a single lone pair. Its structure is based on an octahedron, with the ion distorted away from the lone pair, as in IF_5.

SeF_3^+ has a single lone pair. This lone pair reduces the F—Se—F bond angle significantly, to 94°.

EXERCISE 3-2

Predict the structures of the following ions:

NH_2^- NH_4^+ I_3^- PCl_6^-

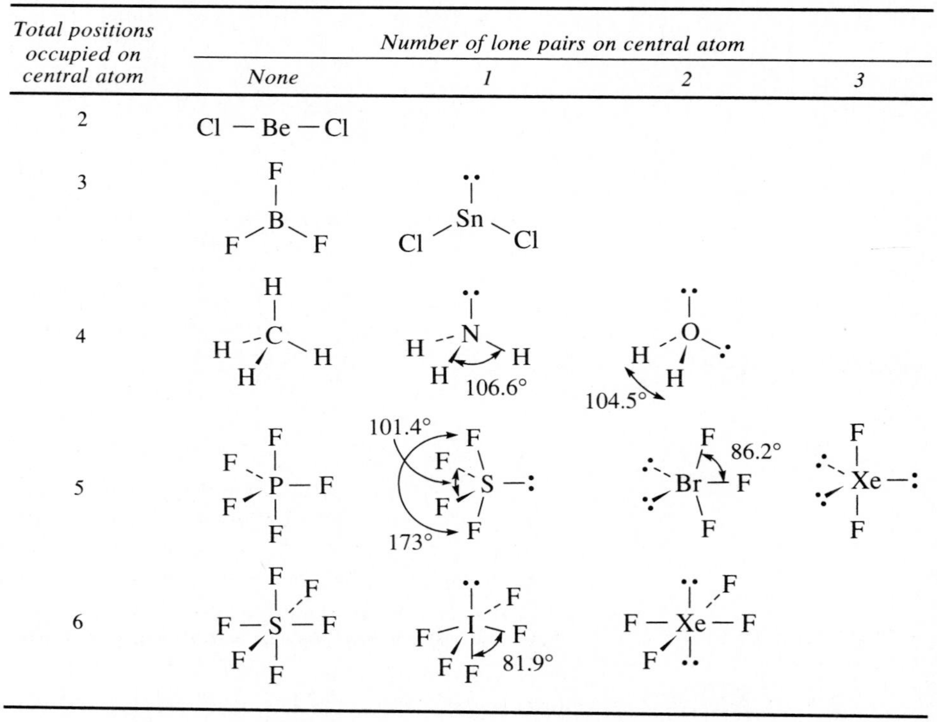

Total positions occupied on central atom	Number of lone pairs on central atom			
	None	1	2	3
2	Cl — Be — Cl			
3	BF_3	$SnCl_2$		
4	CH_4	NH_3 (106.6°)	H_2O (104.5°)	
5	PF_5	SF_4 (101.4°, 173°)	BrF_3 (86.2°)	XeF_2
6	SF_6	IF_5 (81.9°)	XeF_4	

FIGURE 3-12 Structures Containing Lone Pairs.

3-2-2 MULTIPLE BONDS

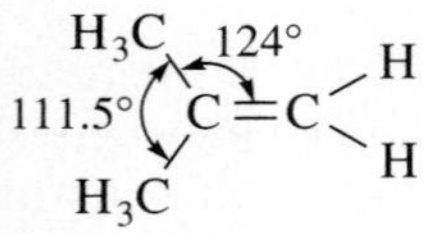

The final refinement of the VSEPR model considers double and triple bonds to have slightly greater repulsive effect than single bonds because of the repulsive effect of pi electrons. For example, the $H_3C—C—CH_3$ angle in $(CH_3)_2C{=}CH_2$ is 111.5° and the C—C=C angle is 124° (both ±2°) (Figure 3-13).[10] In addition, the size of groups attached to a specific atom influences the angles. The arguments used in explaining the chair and boat conformations of cyclohexane in organic chemistry demonstrate that even small differences in repulsions can have a significant effect.

FIGURE 3-13 Bond Angles in $(CH_3)_2C{=}CH_2$.

[10] L. Pauling and L. O. Brockway, *J. Am. Chem. Soc.*, **1937**, *59*, 1223.

In Figure 3-14, we see additional examples of the effect of multiple bonds on molecular geometry. Like lone pairs, the multiple bonds tend to occupy positions that minimize interactions with other electron pairs; for example, both double bonds and lone pairs tend to occupy equatorial positions in molecules with trigonal bipyramidal electronic structures, as shown in Figures 3-12, 3-14, and 3-15. In general, the repulsive effects are in the order *lp* > *bp* (multiple bond) > *bp* (single bond).

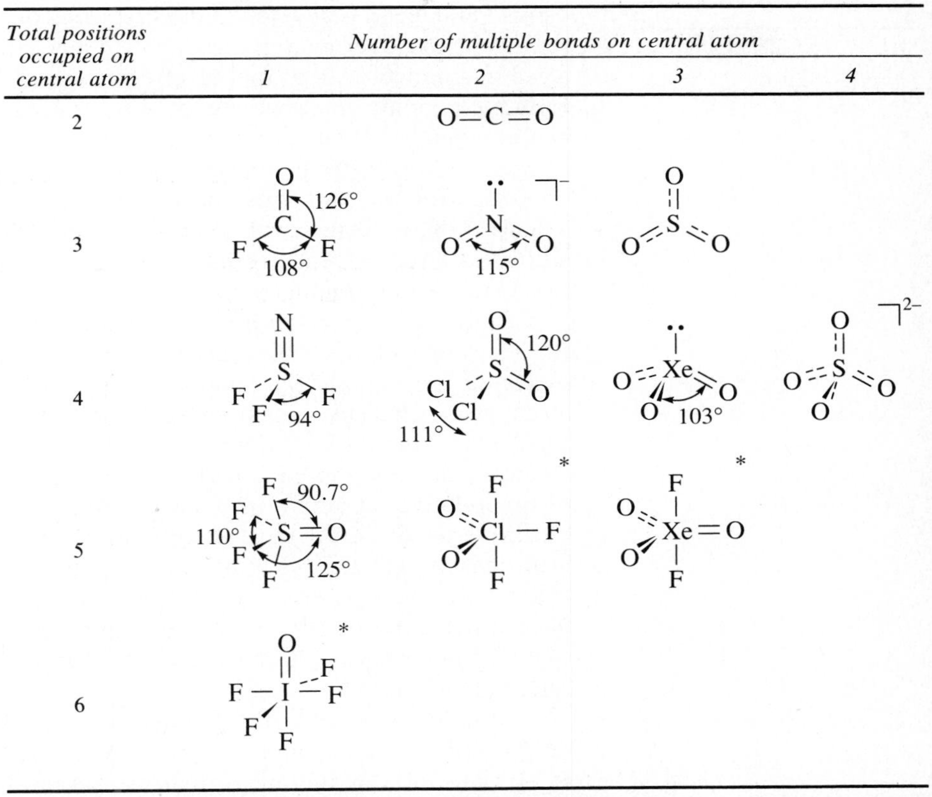

Total positions occupied on central atom	*Number of multiple bonds on central atom*			
	1	*2*	*3*	*4*
2		O=C=O		
3	$F_2C=O$ (126°, 108°)	NO_2^- (115°)	SO_3	
4	NSF_3 (94°)	$SOCl_2$ (120°, 111°)	XeO_3 (103°)	SO_4^{2-}
5	SOF_4 (90.7°, 110°, 125°)	ClO_2F_3*	XeO_3F_2*	
6	IOF_5*			

FIGURE 3-14 Structures Containing Multiple Bonds.

* The bond angles of these molecules have not been determined accurately. However, spectroscopic measurements are consistent with the structures shown.

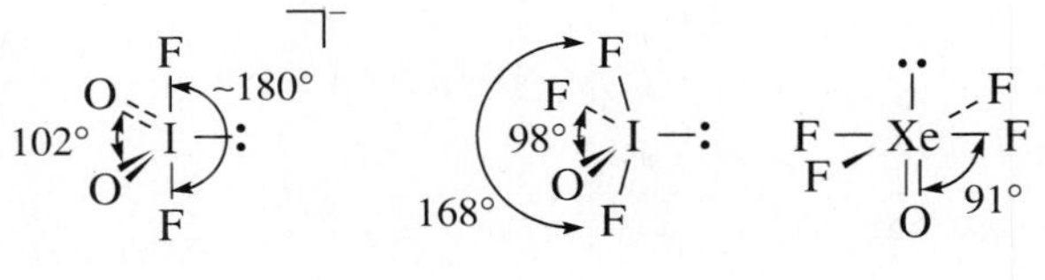

$IO_2F_2^-$ (102°, ~180°) IOF_3 (98°, 168°) $XeOF_4$ (91°)

FIGURE 3-15 Structures Containing Both Lone Pairs and Multiple Bonds.

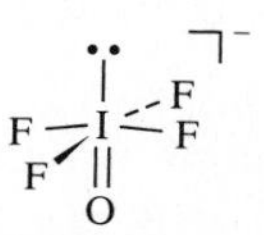

EXAMPLES

HCP like HCN, is linear, with a triple bond: H—C≡P. It is, however, significantly less stable than HCN.

$\mathbf{IOF_4^-}$ has a single lone pair on the side opposite the oxygen. The lone pair has a slightly greater repulsive effect than the double bond to oxygen, as shown by the average O—I—F angle of 89°.

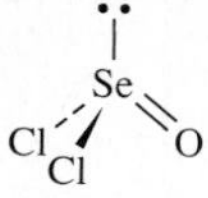

$SeOCl_2$ has both a lone pair and a double bond to the oxygen. The lone pair has a greater effect than the double bond to oxygen; the Cl—Se—Cl angle is reduced to 97° by this effect, and the Cl—Se—O angle is 106°.

EXERCISE 3-3

Predict the structures of the following, including deviations from regular structures:

$XeOF_2$ $ClOF_3$ $SOCl_2$

3-3 ELECTRONEGATIVITY

Linus Pauling, who developed the first electronegativity scale in the 1930s and refined its use over many years, defined electronegativity as "the power of an atom in a molecule to attract electrons to itself."[11] His description involved a form of bonding intermediate between pure covalency, with the electrons shared equally between two atoms, and pure ionic bonding, with the transfer of one or more electrons from one atom to another and the resulting electrostatic bonding between the ions formed. Pauling then used electronegativity to calculate the percent covalent and the percent ionic character for bonds. We will not use the percent covalent and ionic terminology, but do recognize these two extremes of bonding.

In homonuclear diatomic molecules, where the two atoms are identical, the electrons of the chemical bond are shared equally, and the bond and molecule are completely nonpolar. Heteronuclear diatomic molecules have atoms with different nuclear charges and different numbers of electrons. As a result, the atoms have different attractions for the bonding electrons, and the electrons are not shared equally. Electronegativity is a quantitative measure of this relative attraction of bonded electrons to the atoms. Atoms with high electronegativity draw the electrons in closer and thus have a more negative electrical charge. The resulting molecules have centers of positive and negative charge (although generally much less than the charge of an electron) and may have a net molecular polarity. Pauling assigned fluorine an electronegativity of 4, the highest value in his scale. He then calculated other values from this reference value.

Pauling used either the geometric mean (the square root of the product) or the arithmetic mean of the bond dissociation energies (D) of A—A and B—B molecules as the pure covalent energy of the A—B bond. More recent authors have used the arithmetic mean almost exclusively. He called the difference between the experimental A—B bond energy and this calculated covalent bond energy the ionic resonance energy, Δ':

$$\Delta'(\text{A—B}) = D(\text{A—B}) - \frac{D(\text{A—A}) + D(\text{B—B})}{2}$$

He then calculated the difference in electronegativity between A and B:

$$\chi_A - \chi_B = 0.102\sqrt{\Delta'}$$

(The original scale was in electron volts; the factor 0.102 is the constant for energies in kilojoules.)

[11] Linus Pauling, *The Nature of the Chemical Bond,* 3rd ed., Cornell University Press, Ithaca, N.Y., 1960, p. 88.

EXAMPLE

The ionic resonance energy of HCl can be calculated from the electronegativities.

$$\chi_H = 2.20, \quad \chi_{Cl} = 3.16$$

$$\Delta' = \left(\frac{\chi_{Cl} - \chi_H}{0.102}\right)^2 = 88.6\,\frac{\text{kJ}}{\text{mol}}.$$

EXERCISE 3-4

Calculate the bond energy of the H—O bond in water. The H—H bond energy is 436 kJ/mol and the O—O single bond energy is 213 kJ/mol. (Values of the O—O bond energy vary from 142 to 213 kJ, depending on the rest of the molecule. The value for H_2O_2 is 213 kJ/mol.)

The difference in electronegativity between the two atoms in HCl is 0.96, large enough to predict a strongly polar molecule with a negative charge on Cl and positive charge on H. The full table (Table 3-1) of electronegativities requires averaging over a number of compounds to cancel out experimental uncertainties and other minor effects. These average electronegativities are suitable for most uses, but the actual values for atoms in molecules may differ from this average, depending on their electronic environment.

Many others have developed electronegativity scales claimed to be more accurate and useful, usually adjusted to match fairly closely Pauling's values. Mulliken[12] calculated electronegativity from the average of the electron affinity and ionization potential of the atom. Interest in this method has increased recently[13]; further details are in Chapter 6. Allred and Rochow[14] calculated the force of electrostatic attraction, proportional to Z^*/r^2, between electrons and the nucleus at the covalent radius r as the electronegativity. Z^* is the effective nuclear charge calculated using the Slater method described in Chapter 2. Sanderson[15] calculated electronegativities from electron densities of the atoms, making some assumptions about the sizes of the noble gas atoms.

Many of those interested in electronegativity agree that it depends on the structure of the molecule as well as the atom. Jaffé[16] used this idea to develop a theory of the electronegativity of *orbitals* rather than *atoms*. Such theories are useful in detailed calculations of properties that change with subtle changes in structure, but we will not discuss this aspect further.

Since the differences between the different scales are relatively small, we will use Pauling's electronegativity values, as updated by Allred (Table 3-1); remember that these are measures of an atom's ability to attract electrons from a neighboring atom to which it is bonded. Applications of electronegativity are included in the remainder of this chapter and in later chapters.

3-4 POLAR BONDS

Whenever atoms with different electronegativities combine, the resulting molecule has polar bonds, with the electrons of the bond concentrated (perhaps

[12] R. S. Mulliken, *J. Chem. Phys.*, **1934,** *2,* 782.

[13] R. G. Parr, R. A. Donnelly, M. Levy, and W. E. Palke, *J. Chem. Phys.*, **1978,** *68,* 3801; R. G. Pearson, *Inorg. Chem.*, **1988,** *27,* 734; S. G. Bratsch, *J. Chem. Educ.*, **1988,** *65,* 34; 223.

[14] A. L. Allred and E. G. Rochow, *J. Inorg. Nucl. Chem.*, **1958,** *5,* 264.

[15] R. T. Sanderson, *J. Chem. Educ.*, **1952,** *29,* 539; **1954,** *31,* 2, 238; *Inorganic Chemistry*, Van Nostrand-Reinhold, New York, 1967.

[16] J. Hinze and H. H. Jaffé, *J. Am. Chem. Soc.*, **1962,** *84,* 540; *J. Phys. Chem.*, **1963,** *67,* 1501; J. E. Huheey, *Inorganic Chemistry*, 3rd ed., Harper & Row, New York, 1983, pp. 152–56.

very slightly) on the more electronegative atom. As a result, the bonds have permanent dipolar character, with positive and negative ends. This polarity causes specific interactions between molecules, depending on the overall structure of the molecule. Experimentally, the polarity of molecules is measured indirectly by measuring the dielectric constant, which is the ratio of the capacitance of a capacitor filled with the substance to be measured to the capacitance of the same capacitor evacuated. Orientation of the polar molecules in the electric field partially cancels the effect of the field and results in a larger dielectric constant. Measurements at different temperatures allow calculation of the **dipole moment** for the molecule, defined as

$$\mu = Qr$$

where Q is the difference in charge separated by a distance r.[17] In more complex molecules, vector addition of the individual bond dipole moments gives the net molecular dipole moment. It is, however, usually not possible to calculate molecular dipoles directly from bond dipoles. Chloromethane has a dipole moment of 6.24×10^{-30} C m (1.87 D); dichloromethane, a moment of 5.34×10^{-30} C m (1.60 D); and trichloromethane (chloroform), a moment of 3.37×10^{-30} C m (1.01 D).[18] Using the C—H and C—Cl bond dipoles of 1.3 and 4.9×10^{-30} C m, respectively, and approximately tetrahedral bond angles, we can calculate values of 5.90, 6.93, and 6.07×10^{-30} C m (1.77, 2.08, 1.82 D) for the respective molecules. Clearly, some other factors are involved in the experimental dipole moments.

The dipole moments of NH_3 and NF_3 (Figure 3-16) reveal the effect of a lone pair, a much larger effect than those described above. The vector sum of the N—H bond moments is almost as large as the dipole moment of ammonia; the lone-pair polarity reinforces the moment from N—H polarity for a total of 4.9×10^{-30} C m (1.47 D). Similar bond moments with opposite polarity add to give a calculated dipole moment for NF_3 that is much too high. Here the lone-pair moment opposes the N—F moments, resulting in a very small dipole (0.8×10^{-30} C m; 0.23 D). Although the lone-pair electrons are held rather closely by the nuclear charge of the atom, so the distance r is small, there is no compensating positive nucleus to reduce the total negative charge in that part of the molecule, and the product Qr is large. In molecules such as ammonia or water, the bond dipoles add to the lone-pair dipoles to give strongly polar molecules.

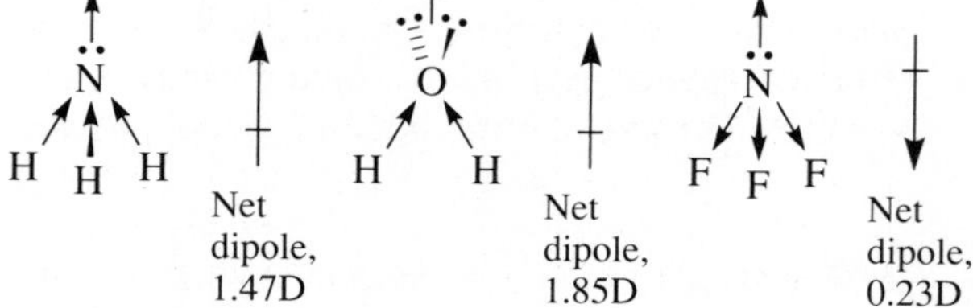

FIGURE 3-16 Bond Dipoles and Molecular Dipoles.

In some cases, therefore, the polarity of the molecule is uncertain because of the competing influences of lone pairs and polar bonds. Another example is SO_2. The dipole of the lone pair on the sulfur opposes the combined dipoles

[17] The units for dipole moments are coulomb meters (C m). Another unit more frequently used in the past was the debye (D) = 3.338×10^{-30} C m.

[18] *Handbook of Chemistry and Physics,* 66th ed., CRC Press, Cleveland, Ohio, 1985–1986, p. E-58 (from NBS table NSRDS-NBS 10).

of the two S—O bonds, but the overall dipole moment (5.5×10^{-30} C m; 1.63 D) is larger than that of ammonia.

Molecules with dipole moments interact electrostatically with each other and with other polar molecules. When the dipoles are large enough, the molecules orient themselves with the positive end of one molecule toward the negative end of another because of these attractive forces, and higher melting and boiling points result. Details of the most dramatic effects are given in the discussion of hydrogen bonding later in this chapter (see also Chapter 6).

On the other hand, if the molecule has a very symmetric structure or if the polarities of different bonds cancel each other, the molecule as a whole may be nonpolar even though the individual bonds are quite polar. Tetrahedral molecules such as CH_4 and CCl_4 and trigonal molecules such as SO_3, NO_3^-, and CO_3^{2-} are all nonpolar. The C—H bond has very little polarity, but the bonds in the other molecules and ions are quite polar. In all these cases, the sum of all the polar bonds is zero because of the symmetry of the molecules, as shown in Figure 3-17.

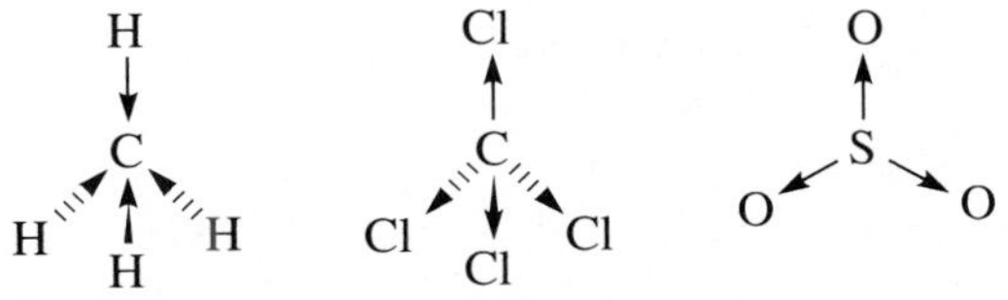

H C H H H Cl C Cl Cl Cl O S O O

Net zero dipole for all three

FIGURE 3-17 Cancellation of Bond Dipoles due to Molecular Symmetry.

In very close proximity to other molecules, there may still be some intermolecular attraction due to the bond dipoles, but at greater distances such interactions are insignificant, since the molecule as a whole is nonpolar. However, attractive forces exist between these molecules, called London or dispersion forces, due to momentary fluctuations in the electron clouds and the resulting polarity. These forces make it possible to liquefy the noble gases and nonpolar molecules. While important in many physical properties, these forces are of minor importance in reactions and we will not consider them further here. Examples of the effects of these forces on the physical properties of molecules are given in Chapter 7.

3-5 HYDROGEN BONDING

Ammonia, water, and hydrogen fluoride all have much higher boiling points than other similar molecules, as shown in Figure 3-18. In water and hydrogen fluoride, these high boiling points are caused by hydrogen bonds, in which hydrogen atoms bonded to O or F also form weaker bonds to a lone pair of electrons on another O or F. Bonds between hydrogen and these strongly electronegative atoms are very polar, with a partial positive charge on the hydrogen. The positive H is strongly attracted to the negative O or F of neighboring molecules. In the past, the attraction between these molecules was considered primarily electrostatic in nature, but an alternative molecular orbital approach, which will be described in Chapters 5 and 6, gives a more complete description of this phenomenon. Regardless of the detailed explanation of the forces involved in hydrogen bonding, the strongly positive H and the strongly negative lone pairs tend to line up and hold the molecules together.

In general, boiling points rise with increasing molecular weight, both because the additional mass requires higher temperature for rapid movement

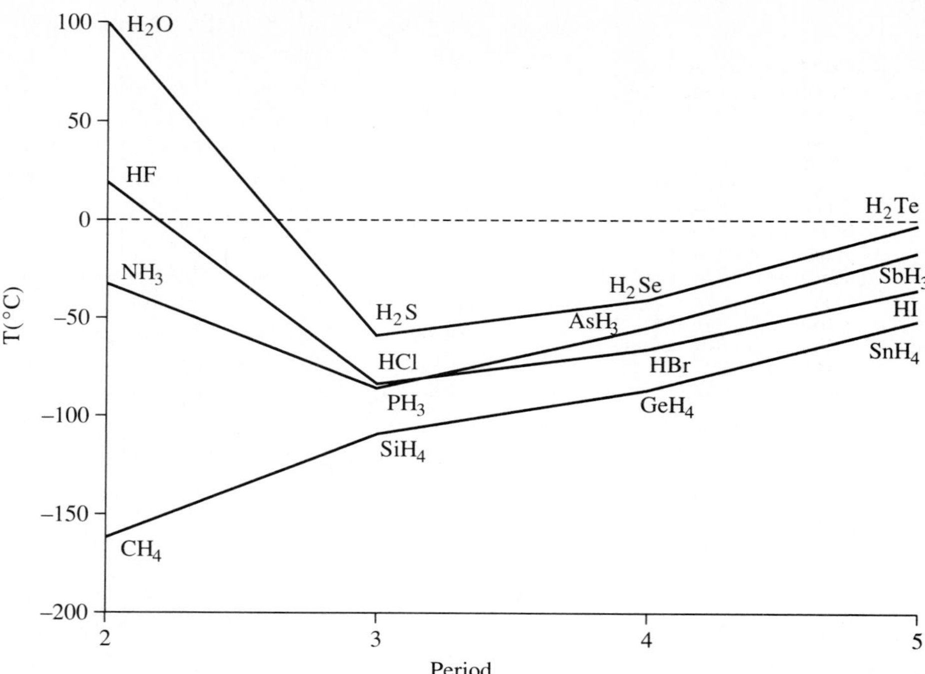

FIGURE 3-18 Boiling Points of Hydrogen Compounds.

of the molecules and because the larger number of electrons in the heavier molecules provide larger London forces. The difference between the actual boiling point of water and the extrapolation of the line connecting the boiling points of the heavier analogous compounds is almost 200°C. Ammonia and hydrogen fluoride have similar, but smaller, differences from the extrapolated values for their families. Water has a much larger effect because each molecule can average four hydrogen bonds (two through the lone pairs and two through the hydrogen atoms), while hydrogen fluoride can only average two (HF has only one H available).

The higher boiling point of ammonia is frequently explained as being due to hydrogen bonding, but the experimental evidence from microwave spectroscopy[19] does not support this reasoning. It does form dimers in the gas phase, and measurement of the rotational motion of these dimers indicates that the molecules are joined at an angle that does not allow a hydrogen to bridge the two nitrogens (shown in Figure 3-19). The N—N distance is also much longer than that of most hydrogen bonds. Ammonia does, however, hydrogen bond to many other compounds that have positive hydrogens, such as HF, HCl, and H_2O. Derivatives such as amines and amides also hydrogen bond through their own hydrogen atoms as well as those of the other component.

Water has other very unusual properties because of hydrogen bonding. The freezing point of water is much higher than that of similar molecules. An even more striking feature is the decrease in density as water freezes. The tetrahedral structure around each oxygen with two regular bonds to hydrogen and two hydrogen bonds to other molecules requires a very open structure with large spaces between ice molecules (Figure 3-20). This makes the solid

[19] D. D. Nelson, Jr., G. T. Fraser, W. Klemperer, *Science*, **1987**, *238*, 1670.

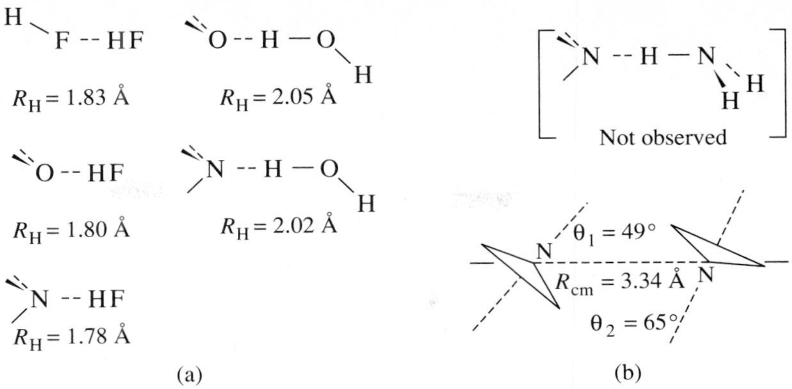

FIGURE 3-19 Hydrogen Bonding and Dimer Structure for Ammonia. (a) Known hydrogen-bonded structures. R_H = hydrogen bond distance. (b) Structure of the NH_3 dimer. The angle between the planes formed by R_{cm} (the line of centers of mass) and the three-fold symmetry axes of the two ammonias is not known. (Redrawn from D. D. Nelson, Jr., G. T. Fraser, W. Klemperer, *Science*, **1987**, *238*, 1670–1674, with permission.)

HF --HF $R_H = 1.83$ Å

O --H—OH $R_H = 2.05$ Å

O --HF $R_H = 1.80$ Å

N --H—OH $R_H = 2.02$ Å

N --HF $R_H = 1.78$ Å

(a)

[N --H—NH₂] Not observed

$\theta_1 = 49°$

$R_{cm} = 3.34$ Å

$\theta_2 = 65°$

(b)

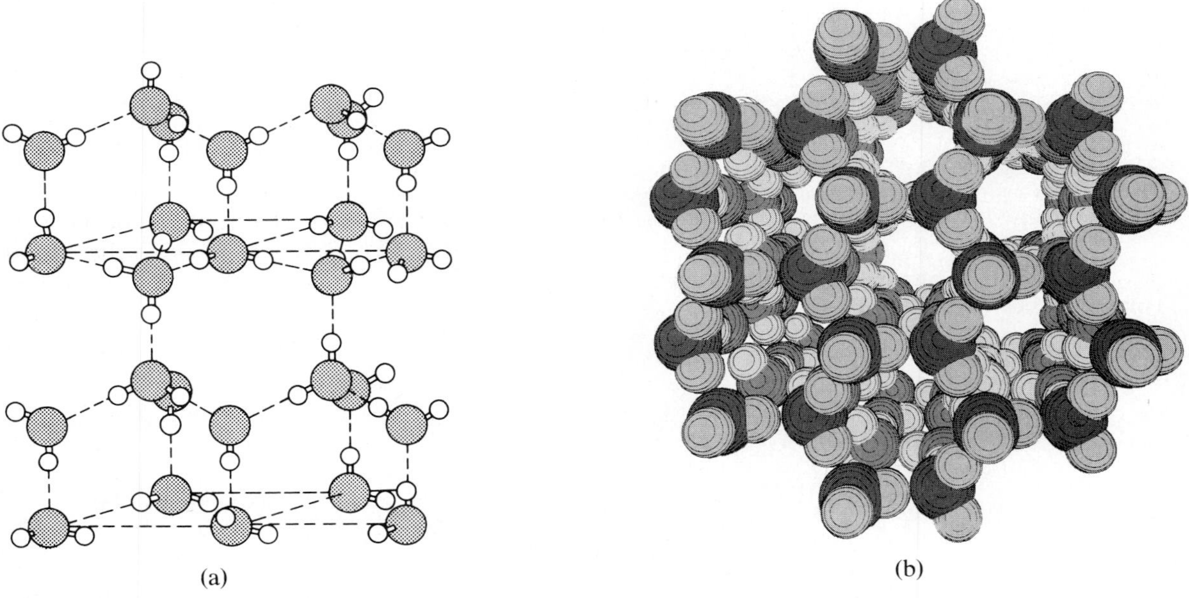

FIGURE 3-20 Two Drawings of Ice. (a) From T. L. Brown and H. E. LeMay, Jr., *Chemistry, the Central Science*, Prentice Hall, Englewood Cliffs, N.J., 1988, p. 628. Reproduced with permission. (b) Copyright © 1976 by W. G. Davies and J. W. Moore, used by permission. All rights reserved.

lighter than the more random liquid water surrounding it, so ice floats. Life on earth would be very different if this were not so. Lakes, rivers, and oceans would freeze from the bottom up, ice cubes would sink, and ice skating would be impossible; the results are difficult to imagine, but would certainly require a much different type of biology and geology. The same forces cause coiling of protein and nucleic acid molecules (Figure 3-21); a combination of hydrogen bonding with other dipolar forces imposes considerable secondary structure on these large molecules.

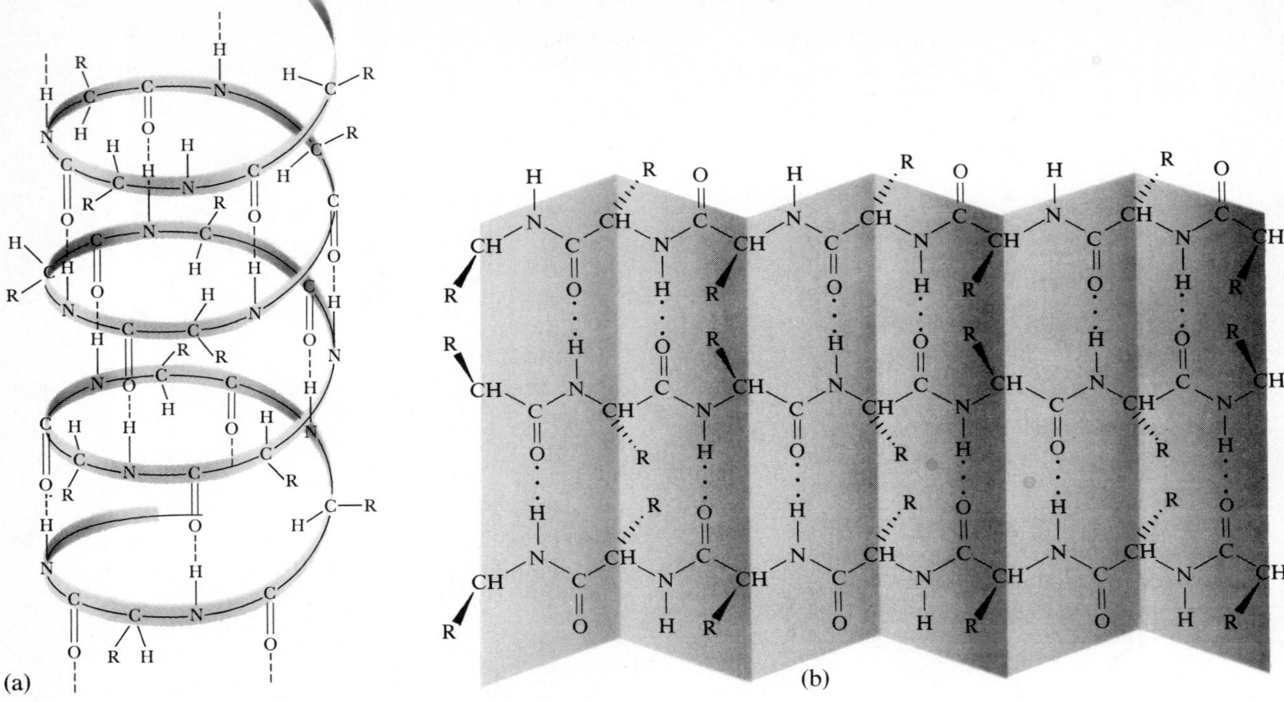

FIGURE 3-21 Hydrogen Bonding in Proteins. (a) A protein α-helix. Peptide carbonyls and N—H hydrogens on adjacent turns of the helix are hydrogen-bonded. (From T. L. Brown and H. E. LeMay, Jr., *Chemistry, the Central Science,* Prentice Hall, Englewood Cliffs, N.J., 1988, p. 946. Reproduced with permission.) (b) The pleated sheet arrangement. Each peptide carbonyl group is hydrogen bonded to a N—H hydrogen on an adjacent peptide chain. (From L. G. Wade, Jr., *Organic Chemistry,* Prentice Hall, Englewood Cliffs, N.J., 1988, pp. 1255–56. Reproduced with permission.)

Another example is a theory of anesthesia by non-hydrogen bonding molecules, such as cyclopropane, chloroform, and nitrous oxide, proposed by Pauling.[20] These molecules are of a size and shape that can fit neatly into a hydrogen-bonded water structure with even larger open spaces than ordinary ice. Such structures, with molecules trapped in a framework of another kind of molecule, are called **clathrates.** Pauling proposed that similar hydrogen-bonded microcrystals form even more readily in nerve tissue because of the presence of other solutes in the tissue. These microcrystals could then interfere with the transmission of nerve impulses.

More specific interactions involving the sharing of electron pairs between molecules can be grouped under acid–base theories. They are discussed in Chapter 6.

GENERAL REFERENCES

Good sources for bond lengths and bond angles are the references by Wells, Greenwood and Earnshaw, and Cotton and Wilkinson cited in Chapter 1. Appendix D provides a review of electron-dot diagrams and formal charges at the level of most general chemistry texts. Alternative approaches to these topics are available in most general

[20] L. Pauling, *Science,* **1961,** *134,* 15.

chemistry texts, as are descriptions of VSEPR theory. One of the best VSEPR references is still the early paper by R. J. Gillespie and R. S. Nyholm, *Quart, Rev.*, **1957**, *XI*, 339.

PROBLEMS

3-1 The dimethyldithiocarbamate ion, $S_2CN(CH_3)_2^-$ has the following skeletal structure:

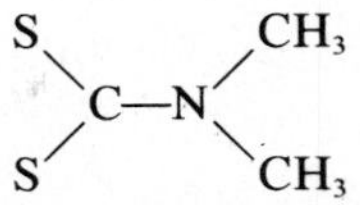

```
S          CH3
  \       /
   C — N
  /       \
S          CH3
```

a. Give the important resonance structures of this ion, including any formal charges where necessary. Select the resonance structure likely to provide the best description of this ion.

b. Repeat for $OSCN(CH_3)_2^-$.

3-2 Several resonance structures are possible for each of the following ions. For each, draw the resonance structures, assign formal charges, and select the resonance structure likely to provide the best description for the ion.

a. Selenocyanate ion, $SeCN^-$

b. Thioformate ion:

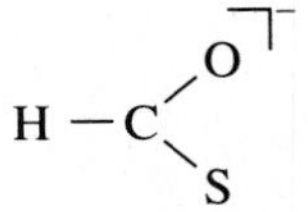

```
         O ⌉⁻
        /
H — C
        \
         S
```

c. Dithiocarbonate, S_2CO^{2-} (C is central)

3-3 Draw the resonance structures for the isoelectronic ions NSO^- and SNO^-, and assign formal charges. Which ion is likely to be more stable?

3-4 Predict the structure of the (as yet) hypothetical ion IF_3^{2-}.

3-5 Select the molecule or ion having the smallest bond angle, and briefly explain your choice:

a. NH_3, PH_3, or AsH_3

b. O_3^+, O_3, or O_3^-

c. Halogen—S—halogen angle:

```
   S            S
 /  \ Cl      /  \ F
O    Cl      O    F
```

d. NO_2^- or O_3

e. ClO_3^- or BrO_3^-

3-6 Predict the most likely structure of PCl_3Br_2 and explain your reasoning.

3-7 Give Lewis dot structures and predict the shapes of the following:

a. $SeCl_4$

b. I_3^-

c. $PSCl_3$ (P is central)

d. IF_4^-

e. PH_2^-

f. TeF_4^{2-}

g. N_3^-

h. $SeOCl_4$ (Se is central)

i. PH_4^+

j. NO^-

3-8 Give Lewis dot structures and predict the shapes of the following:

a. ICl_2^-

b. H_3PO_3 (one H is bonded to P)

c. BH_4^-

d. $POCl_3$

e. IO_4^-

f. $IO(OH)_5$

g. $SOCl_2$

h. $ClOF_4^-$

i. XeO_2F_2

j. $ClOF_2^+$

3-9 Give Lewis dot structures and predict the shapes of the following:

a. SOF_6 (one F is attached to O) **b.** POF_3
c. ClO_2 **d.** NO_2
e. $S_2O_4^{2-}$ (symmetric, with an S—S bond) **f.** N_2H_4 (symmetric, with an N—N bond)

3-10 **a.** Compare the structures of the azide ion, N_3^-, and the ozone molecule, O_3.
b. How would you expect the structure of the ozonide ion, O_3^-, to differ from that of ozone?

3-11 Give Lewis dot structures and predict the shapes for the following:

a. $VOCl_3$ **b.** PCl_3 **c.** SOF_4
d. ClO_2^- **e.** ClO_3^- **f.** P_4O_6

(P_4O_6 is a closed structure with overall tetrahedral shape; an oxygen atom bridges each pair of phosphorus atoms.)

3-12 Consider the series NH_3, $N(CH_3)_3$, $N(SiH_3)_3$, and $N(GeH_3)_3$; these have bond angles at the nitrogen atom of 106.6°, 110.9°, 120°, and 120°, respectively. Explain this trend.

3-13 Explain the bond angles and bond lengths of the following ions:

	X—O (pm)	*O—X—O angle*
ClO_3^-	149	107°
BrO_3^-	165	104°
IO_3^-	181	100°

3-14 Compare the bond orders expected in ClO_3^- and ClO_4^- ions.

3-15 Give Lewis dot structures and predict the shapes for the following:

a. PH_3 **b.** H_2Se **c.** SeF_4
d. PF_5 **e.** ICl_4^- **f.** XeO_3
g. NO_3^- **h.** $SnCl_2$ **i.** PO_4^{3-}
j. SF_6 **k.** IF_5 **l.** ICl_3
m. $S_2O_3^{2-}$ **n.** BF_2Cl

3-16 Which of the molecules or ions in problem 15 are polar?

3-17 Carbon monoxide has a larger bond dissociation energy (1072 kJ mol^{-1}) than molecular nitrogen (945 kJ mol^{-1}). Suggest an explanation.

3-18 **a.** Calculate the expected contribution to the bond energy from the electronegativity differences between oxygen and fluorine, chlorine, bromine, and iodine in the ions OF^-, OCl^-, OBr^-, and OI^-.
b. Using the data in the following table and your answers to part a, discuss the predicted relative stabilities of OF^-, OCl^-, OBr^-, and OI^-.

Bond energies (kJ/mol) for single bonds	
O—O	142
F—F	155
Cl—Cl	240
Br—Br	190
I—I	149

[The O—O bond energy has values from 142 kJ/mol to 213 kJ/mol, depending on the rest of the molecule (J. A. Kerr, *Chem. Rev.*, **1966,** *66,* 465). In this problem, any consistent value can be used; in Exercise 3-4, 213 kJ was more appropriate.]

c. On the basis of their electronic structures, would you expect the charge on each of these ions to contribute to their stability, or would they be more stable as the neutral molecules OF, OCl, OBr, and OI?

3-19 For each of the following bonds, indicate which atom is more negative. Then rank the series in order of polarity.
a. C—N; **b.** N—O; **c.** C—I; **d.** O—Cl; **e.** P—Br; **f.** S—Cl

3-20 Explain the following:
a. PCl_5 is a stable molecule, but NCl_5 is not.
b. SF_4 and SF_6 are known, but OF_4 and OF_6 are not.

3-21 Provide explanations for the following:
a. Methanol, CH_3OH, has a much higher boiling point than methyl mercaptan, CH_3SH.
b. Carbon monoxide has slightly higher melting and boiling points than N_2.
c. The *ortho* isomer of hydroxyaniline [$C_6H_4(NH_2)(OH)$] has a much lower melting point than the *meta* and *para* isomers.
d. The boiling points of the noble gases increase with atomic number.
e. Acetic acid in the gas phase has a significantly lower pressure (approaching a limit of one-half at low temperatures) than predicted by the ideal gas law.
f. Mixtures of acetone and chloroform exhibit significant negative deviations from Raoult's law (which states that the vapor pressure of a volatile liquid is proportional to its mole fraction). For example, an equimolar mixture of acetone and chloroform has a lower vapor pressure than either of the pure liquids.

4

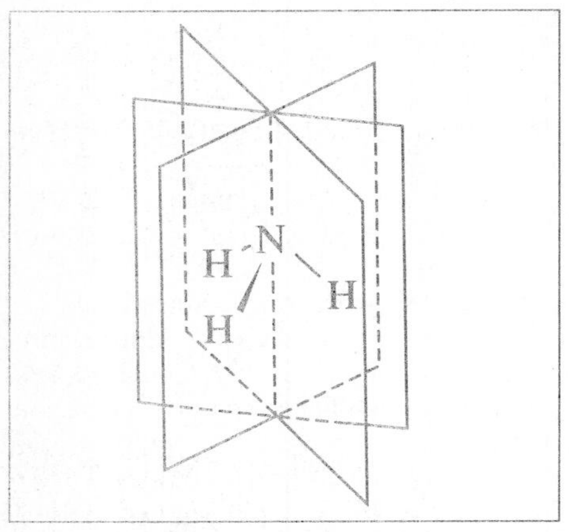

Symmetry and Group Theory

Most readers of this text have at least a qualitative concept of the term symmetry. Symmetry is a phenomenon of the natural world, as well as the world of human invention (Figure 4-1). In nature, many types of flowers and plants, snowflakes, insects, certain fruits and vegetables, and a wide variety of microscopic plants and animals exhibit their characteristic symmetry. Many engineering achievements have as their basis a degree of symmetry, which contributes to their esthetic appeal. Examples include cloverleaf intersections, the pyramids of ancient Egypt, and the Eiffel Tower.

Symmetry concepts can be extremely useful in chemistry. By analysis of the symmetry properties of molecules, we can predict infrared spectra, describe the types of orbitals used in bonding, predict optical activity, interpret electronic spectra, and study a number of additional molecular properties. In this chapter, we first define symmetry very specifically in terms of four fundamental symmetry operations and then describe how molecules can be classified on the basis of the types of symmetry they possess. We conclude this chapter with examples of how symmetry can be used to predict optical activity of molecules and to determine the number and types of infrared-active stretching vibrations.

In later chapters, symmetry will be a valuable tool in the construction of molecular orbitals (Chapters 5 and 8) and in the interpretation of electronic spectra of coordination compounds (Chapter 9) and vibrational spectra of organometallic compounds (Chapter 12).

FIGURE 4-1 Symmetry in Nature, Art, and Architecture. (a) Eiffel Tower from E. B. Feldman, *Varieties of Visual Experience,* Prentice-Hall/Abrams, New York, 1973, p. 398; (b) stalked jellyfish and radiolaria, E. Haeckel, *Art Forms in Nature,* Dover, New York, 1974, pp. 21, 48; (c) Persian medallion, G. Mirow, *A Treasury of Design for Artists and Craftsmen,* Dover, New York, 1969, plate 34; (d) Japanese crest, C. Hornung, *Traditional Japanese Crest Designs,* Dover, New York, 1986, p. 5; (e) Celtic border design, W. and G. Audsley, *Designs and Patterns from Historic Ornament,* Dover, New York, 1968, plate 6. All reproduced with permission.

A molecular model kit is a very useful study aid for this chapter, even for those who can visualize three-dimensional objects easily. We strongly encourage use of such a kit.

4-1 SYMMETRY ELEMENTS AND OPERATIONS

All molecules can be described in terms of their symmetry, even if it is only to say they have none. Molecules or any other objects may contain **symmetry elements** such as mirror planes, axes of rotation, and inversion centers. The actual reflection, rotation, or inversion is called the **symmetry operation.** Most are somewhat familiar, but they must be understood clearly and completely for future use. To contain a given symmetry element, a molecule must have exactly the same appearance after the operation as before. In other words, photographs of the molecule (if such photographs were possible!) taken from the same location before and after the symmetry operation would be indistinguishable. If a symmetry operation yields a molecule that can be distinguished from the original in any way, then that is *not* a symmetry operation of the molecule. The examples in Figures 4-2 through 4-5 illustrate the possible types of molecular symmetry operations and elements.

First is the **identity operation (E),** included for mathematical completeness. This operation causes no change in the molecule. Every molecule has an identity operation, even if it has no symmetry.

Next is the **reflection operation (σ),** where the molecule contains a mirror plane. If details such as hairstyle and location of internal organs are ignored, the human body has a left–right mirror plane. Many molecules have mirror planes, although they may not be immediately obvious. The reflection operation exchanges left and right, as if each point had moved perpendicularly through the plane to a position exactly as far from the plane as when it started. Molecules may have any number of mirror planes. Linear objects such as a round wood pencil or molecules such as acetylene or carbon dioxide have an infinite number, all of which include the center line of the object.

The **rotation operation (C_n)** (also called proper rotation) requires rotation through $360/n$ degrees about a rotation axis. An example of a molecule having a threefold (C_3) axis is $CHCl_3$, with the rotation axis coincident with the C—H bond axis. Two C_3 operations may be performed consecutively to give a new rotation of 240°. The resulting operation is designated C_3^2 and is also a symmetry operation of the molecule. Three successive C_3 operations are the same as the identity operation ($C_3^3 \equiv E$), which is included in all molecules. Many molecules and other objects have multiple rotation axes. Snowflakes are a case in point, with complex shapes that are nearly always hexagonal and nearly planar. The

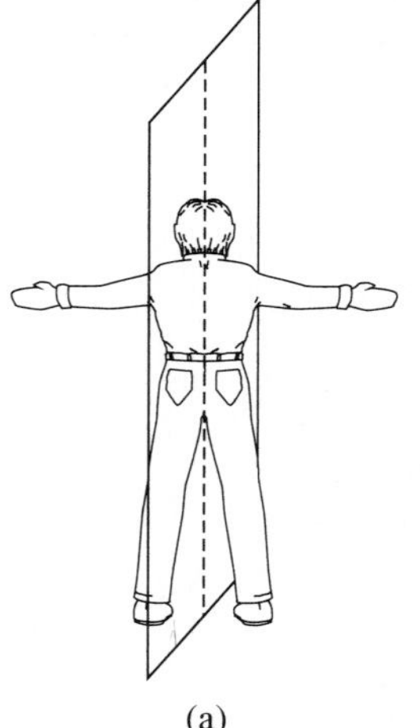
(a)

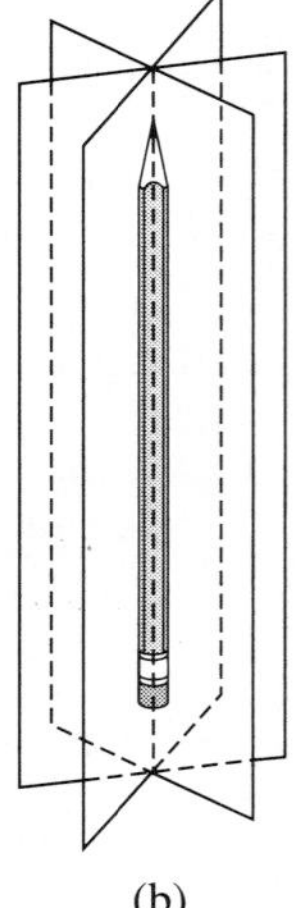
(b)

FIGURE 4-2 Reflections. (a) The human body has a simple reflection through a mirror plane. (b) A round pencil has an infinite number of mirror planes; a hexagonal pencil has six (three through the edges and three through the faces).

Rotation angle	*Symmetry operation*
60°	C_6
120°	C_3 $(= C_6^2)$
180°	C_2 $(= C_6^3)$
240°	C_3^2 $(= C_6^4)$
300°	C_6^5
360°	E $(= C_6^6)$

C_3, H, C, Cl, Cl, Cl

Cl, C_3, C_3^2, H, Cl, Cl

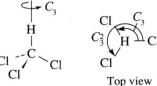

Top view

C_3 rotations of $CHCl_3$

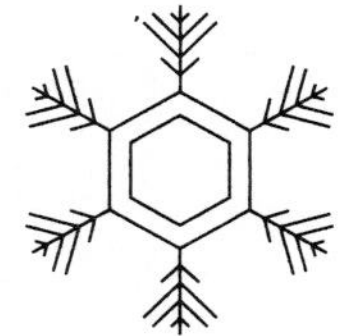

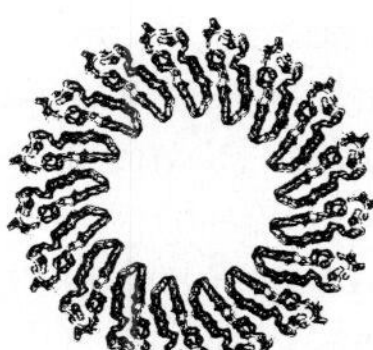

Cross section of protein disk of tobacco mosaic virus

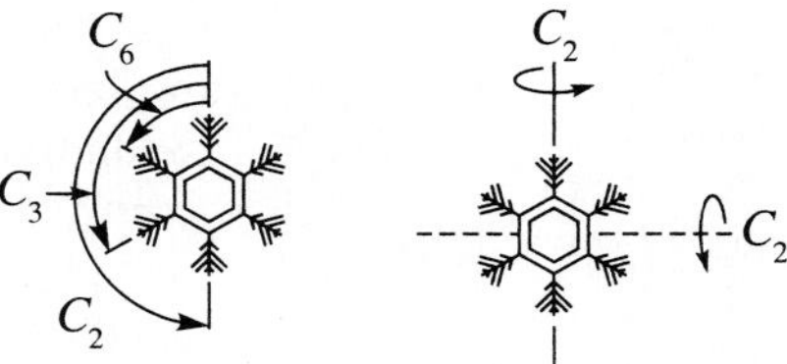

C_2, C_3, and C_6 rotations of a snowflake

FIGURE 4-3 Rotations. The cross section of the tobacco mosaic virus is a cover diagram from *Nature*, **1976**, *259*. Copyright © 1976, Macmillan Journals Ltd. Reproduced with permission of Aaron Klug.

line through the center of the flake perpendicular to the plane of the flake contains a twofold (C_2) axis, a threefold (C_3) axis, and a sixfold (C_6) axis. Note that rotations by 240° (C_3^2) and 300° (C_6^5) are also symmetry operations of the snowflake.

There are also two sets of three C_2 axes in the plane of the snowflake, one set through opposite points and one through the cut-in regions between the points. One of each of these axes is shown in Figure 4-3. In molecules with more than one rotation axis, the C_n axis having the largest value of n may be designated the **highest-order rotation axis** or **principal axis.** The highest-order rotation axis for a snowflake is the C_6 axis. (In assigning Cartesian coordinates, the highest-order C_n axis is usually chosen as the z axis.) When necessary, the C_2 axes perpendicular to the principal axis are designated with primes; a single prime indicates that the axis passes through several atoms of the molecule, while a double prime indicates that it passes between the atoms.

Finding rotation axes for some three-dimensional figures is more difficult, but the same in principle. Remember that nature is not always simple when it comes to symmetry—the protein disk of the tobacco mosaic virus has a 17-fold rotation axis!

Inversion (i) is a somewhat more complex operation. Each point moves through the center of the molecule to a position opposite the original position and as far from the central point as when it started.[1] An example of a molecule having a center of inversion is ethane in the staggered conformation, for which the inversion operation is shown in Figure 4-4.

[1] This operation must be distinguished from the inversion of a tetrahedral carbon in a bimolecular reaction, which is more like that of an umbrella in a high wind.

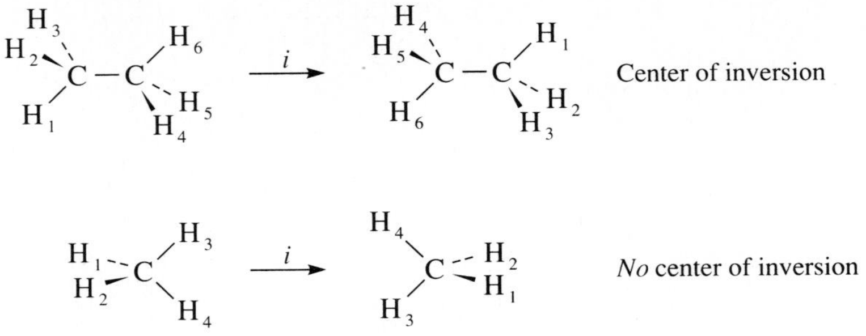

Center of inversion

No center of inversion

FIGURE 4-4 Inversion.

Many molecules that seem at first glance to have an inversion center do not; methane and other tetrahedral molecules are examples. If a methane model is held with two hydrogens in the vertical plane on the right and two hydrogens in the horizontal plane on the left, inversion results in two hydrogens in the horizontal plane on the right and two hydrogens in the vertical plane on the left, as in Figure 4-4. Inversion is therefore *not* a symmetry operation of methane, since the orientation of the molecule following the i operation differs from the original orientation.

Tetrahedra, planar triangles, and pentagons do not have inversion centers; squares, parallelograms, rectangular solids, octahedra, and snowflakes do.

Rotation angle	*Symmetry operation*
90°	S_4
180°	$C_2 \quad (= S_4^2)$
270°	S_4^3
360°	$E \quad (= S_4^4)$

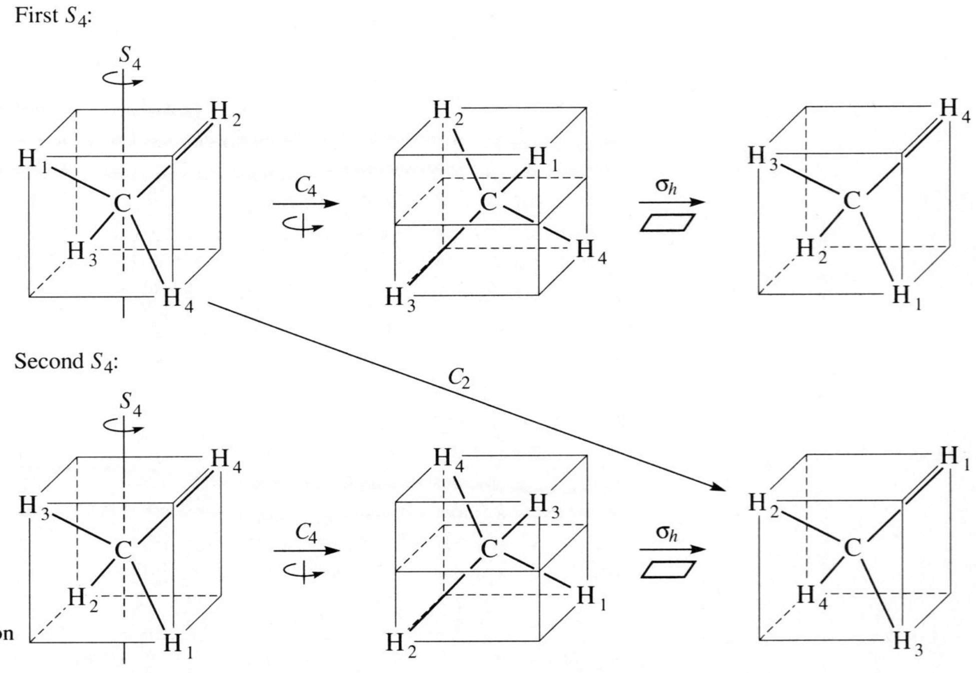

First S_4:

Second S_4:

FIGURE 4-5 Improper Rotation or Rotation–Reflection.

A **rotation–reflection operation** (S_n) (sometimes called **improper rotation**) requires rotation of $360/n$ degrees, followed by reflection through a plane perpendicular to the axis of rotation. In methane, for example, a line through the carbon bisecting the angle between two hydrogens on each side is an S_4 axis. There are three such lines, for a total of three S_4 axes. The operation requires a quarter-turn of the molecule and then reflection through the perpendicular mirror plane. Two S_n operations in succession generate a $C_{n/2}$ axis. In methane, two S_4 operations generate a C_2. These operations are shown in Figure 4-5, along with a table of C and S equivalences for methane.

Sometimes molecules may have an S_n axis that is coincident with a C_n axis. For example, in addition to the rotation axes described above, snowflakes have S_2, S_3, and S_6 axes coincident with the C_6 axis.

Note that an S_2 axis is the same as inversion; an S_1 axis is the same as a mirror plane. The i and σ notation is preferred in each case. Symmetry elements and operations are summarized in Table 4-1.

EXAMPLES

Find all the symmetry elements in the following molecules:

H_2O H_2O has two planes of symmetry, one in the plane of the molecule and one perpendicular to the molecular plane, as shown in Table 4-1. It also has a C_2 axis collinear with the intersection of the mirror planes. H_2O has no inversion center.

O

H H

***p*-Dichlorobenzene** This molecule has three mirror planes: the molecular plane; a plane perpendicular to the molecule, passing through both chlorines; and a plane perpendicular to the first two, bisecting the molecule between the chlorines. It also has three C_2 axes, one perpendicular to the molecular plane (see Table 4-1) and two within the plane: one passing through both chlorines and one perpendicular to the axis passing through the chlorines. Finally, *p*-dichlorobenzene has an inversion center.

Cl Cl

Ethane (staggered conformation) Ethane has three mirror planes, each containing the C-C bond axis and passing through two hydrogens on opposite ends of the molecule. It has a C_3 axis collinear with the carbon-carbon bond and three C_2 axes bisecting the angles between the mirror planes. Ethane also has a center of inversion and an S_6 axis collinear with the C_3 axis (see Table 4-1).

H H H C C H H H

EXERCISE 4-1

Using diagrams as necessary, show that $S_2 = i$ and $S_1 = \sigma$.

EXERCISE 4-2

Find all the symmetry elements in the following molecules:

NH_3 Cyclohexane (boat conformation) Cyclohexane (chair conformation)

TABLE 4-1
Summary table of symmetry elements and operations

Symmetry operation	*Symmetry element*	*Examples*
Identity, E	None	CHFClBr 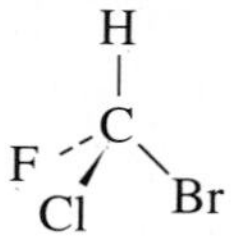
Reflection, σ	Mirror plane	H_2O 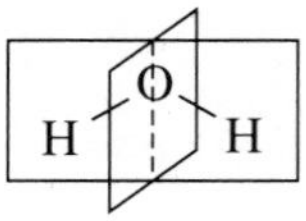
Rotation, C_2	Rotation axis	*p*-dichlorobenzene 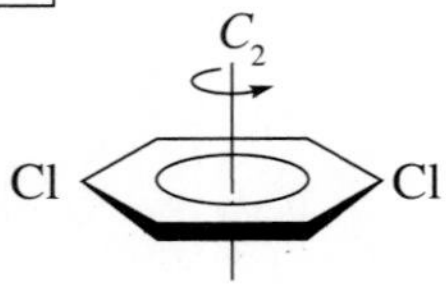
C_3		NH_3 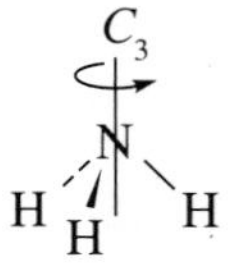
C_4		$[PtCl_4]^{2-}$ 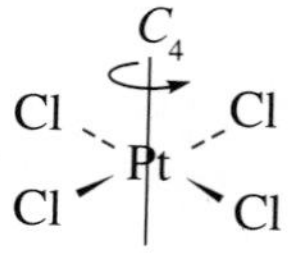
C_5		Cyclopentadienyl group 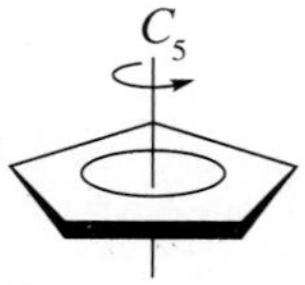
C_6		Benzene 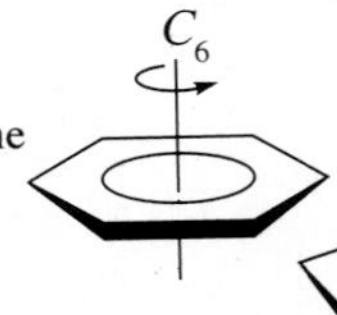
Inversion, i	Inversion center (point)	Ferrocene (staggered) 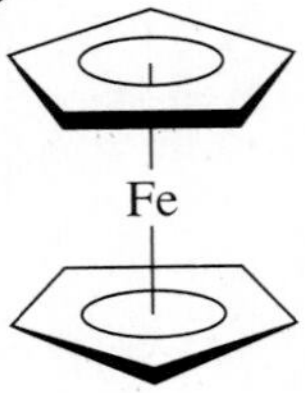
Rotation–reflection, S_4	Rotation–reflection axis (improper axis)	CH_4 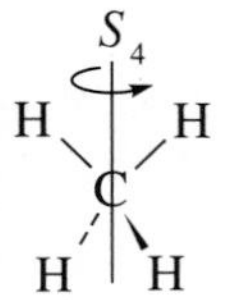
S_6		Ethane (staggered) 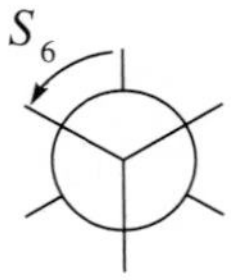
S_{10}		Ferrocene (staggered) 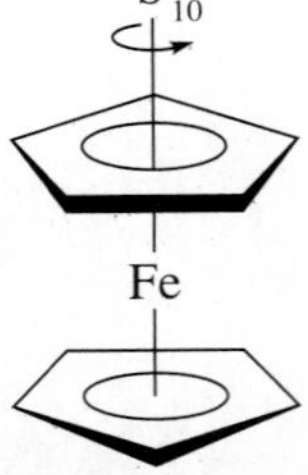

4-2 POINT GROUPS

Each molecule has a set of symmetry operations, which together describes the molecule's overall symmetry. This set of symmetry operations is called the **point group** of the molecule. Several properties of the molecule can then be predicted using **group theory,** which is the mathematical treatment of the properties and behavior of groups. With only a few exceptions, the rules for assigning a molecule to a point group are simple and straightforward. We need only to follow these steps in sequence until a final classification of the molecule is made.

1. Determine whether the molecule belongs to one of the cases of low symmetry (C_1, C_s, C_i) or high symmetry (T_d, O_h, or I_h) described in Tables 4-2 and 4-3.
2. Find the rotation axis with the highest n, the highest-order C_n axis for the molecule.
3. Does the molecule have any C_2 axes perpendicular to the C_n axis? If it does, there will be n of such C_2 axes, and the molecule is in the D set of groups. If not, it is in the C or S set.
4. Does the molecule have a mirror plane (called σ_h, the horizontal plane) perpendicular to the C_n axis? If so, it is classified C_{nh} or D_{nh}. If not, continue with step 5.
5. Does the molecule have any mirror planes that contain the C_n axis? (These are called σ_v, vertical, or σ_d, dihedral.) If so, it is classified C_{nv} or D_{nd}. If not, molecules in the D set are classified D_n; for those in the C or S set, continue with step 6.
6. Is there an S_{2n} axis collinear with the C_n axis? If so, the molecule is classified S_{2n}. If not, it is classified C_n or D_n.

Each step is illustrated below, with the low- and high-symmetry cases treated differently because of their special nature. Molecules that are not in these point groups can be assigned to a point group by questions 2–6. They have at least one C_n axis, but do not belong to one of the high-symmetry

TABLE 4-2
Groups of low symmetry

Group	*Symmetry*	*Examples*
C_1	No symmetry other than the identity operation	CHFClBr
C_s	Only one mirror plane	$H_2C{=}CClBr$
C_i	Only an inversion center; few molecular examples	HClBrC—CHClBr (staggered conformation)

TABLE 4-3
Groups of high symmetry

Group	Description	Examples
T_d	Most (but not all!) molecules in this point group have the familiar tetrahedral geometry. They have eight C_3 axes, three C_2 axes, six S_4 axes, and six σ_d planes. No C_4 axes.	CH_4
O_h	These molecules include those of octahedral structure, although some other geometrical forms, such as the cube, share the same set of symmetry operations. They have four C_3 axes, three C_4 axes, and an inversion center.	SF_6
I_h	Icosahedral structures are best recognized by their six C_5 axes (as well as many other symmetry operations—120 total!).	$B_{12}H_{12}^{2-}$, with a boron at each vertex of an icosahedron

In addition, there are T, T_h, O, and I groups, which are rarely, if ever, seen in nature. For completeness, they are included in the character tables found in Appendix C .

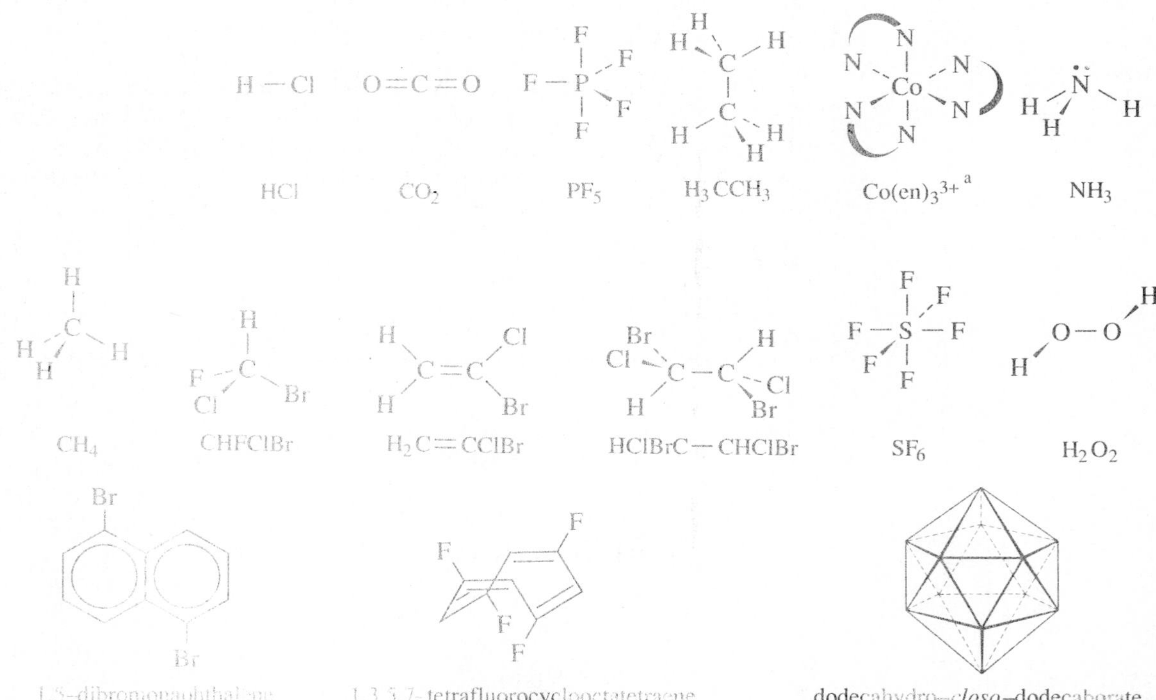

FIGURE 4-6 Molecules to be Assigned to Point Groups.

[a] en = ethylenediamine = $H_2NCH_2CH_2NH_2$, represented by N⌒N.

groups. We will illustrate this procedure by assigning the molecules in Figure 4-6 to their point groups.

Groups of low and high symmetry

> 1. Determine whether the molecule belongs to one of the cases of low or high symmetry.

First, inspection of the molecule will determine if it fits one of the low-symmetry cases. These groups have few or no symmetry operations and are described in Table 4-2.

Molecules with many symmetry operations may fit one of the high-symmetry cases of tetrahedral, octahedral, or icosahedral symmetry with the characteristics described in Table 4-3.

Other groups

> 2. Find the rotation axis with the highest n, the highest-order C_n axis for the molecule.

The rotation axes for the examples are shown in Figure 4-7.

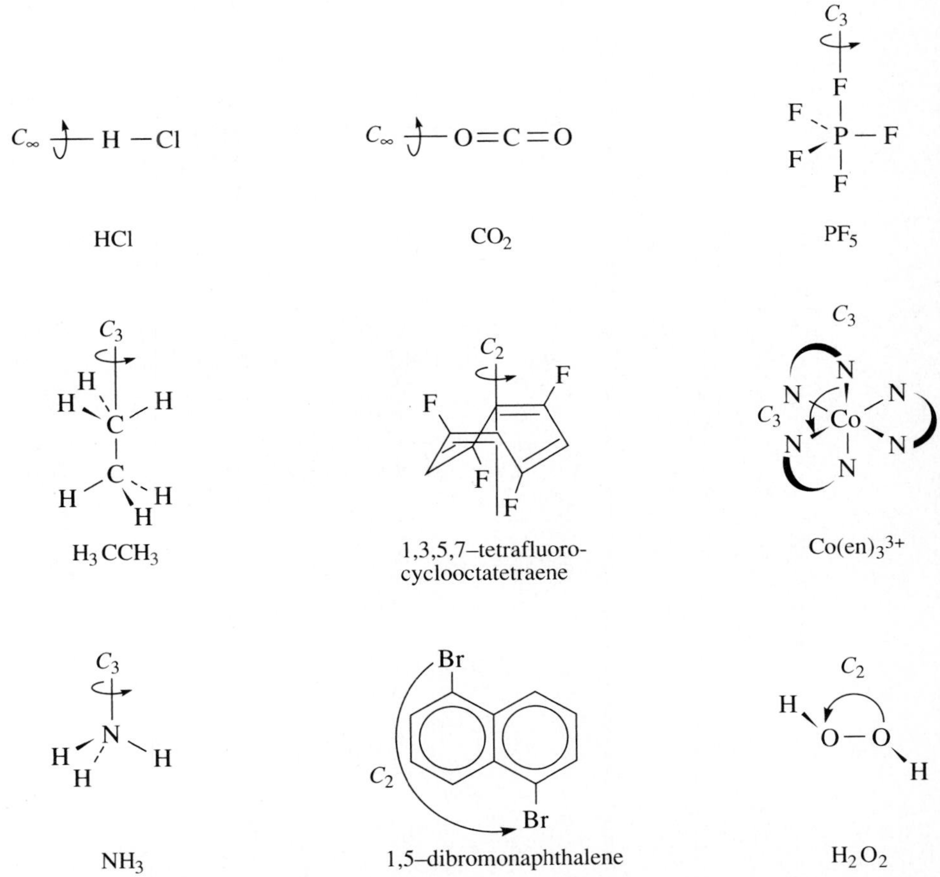

C_∞ H—Cl   HCl

C_∞ O=C=O CO_2

C_3 PF_5

C_3 H_3CCH_3

C_2 1,3,5,7–tetrafluoro-cyclooctatetraene

C_3, C_3 $Co(en)_3^{3+}$

C_3 NH_3

C_2 1,5–dibromonaphthalene

C_2 H_2O_2

FIGURE 4-7 Rotation Axes.

> 3. Does the molecule have any C_2 axes perpendicular to the C_n axis?

The C_2 axes are shown in Figure 4-8.

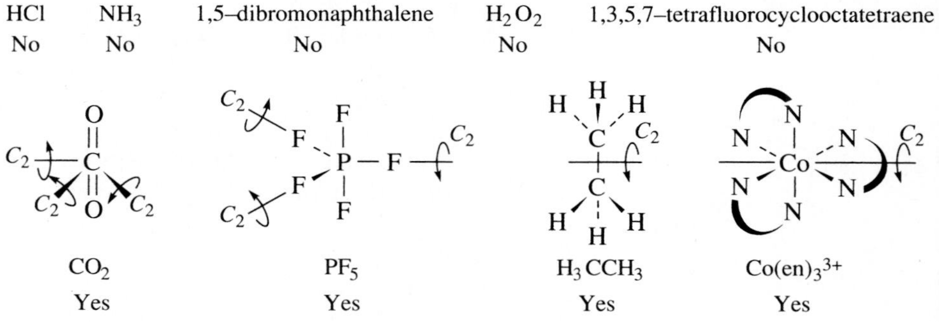

HCl	NH_3	1,5–dibromonaphthalene	H_2O_2	1,3,5,7–tetrafluorocyclooctatetraene
No	No	No	No	No

CO_2	PF_5	H_3CCH_3	$Co(en)_3^{3+}$
Yes	Yes	Yes	Yes

FIGURE 4-8 Perpendicular C_2 Axes.

Yes: D groups

CO_2, PF_5, H_3CCH_3, $Co(en)_3^{3+}$

Molecules with C_2 axes perpendicular to the principal axis are in one of the groups designated by the letter D; there will be n C_2 axes.

No: C or S groups

HCl, NH_3, 1,5-dibromonaphthalene, H_2O_2, 1,3,5,7-tetrafluorocyclooctatetraene

Molecules with no perpendicular C_2 axes are in one of the groups designated by the letters C or S.

No final assignments of point groups have been made, but the molecules have been divided into two major categories, the D set and the C or S set.

> 4. Does the molecule have a mirror plane (σ_h, horizontal plane) perpendicular to the C_n axis?

The horizontal mirror planes are shown in Figure 4-9.

D groups

Yes D_{nh}

CO_2 is $D_{\infty h}$
PF_5 is D_{3h}

C and S groups

Yes C_{nh}

1,5-dibromonaphthalene is C_{2h}

These molecules are now assigned to point groups and need not be considered further. All have horizontal mirror planes.

No: D_n or D_{nd}
H_3CCH_3, $Co(en)_3^{3+}$

No: C_n, C_{nv}, or S_{2n}
HCl, NH_3, H_2O_2, 1, 3, 5, 7-tetrafluorocyclooctatetraene

None of these have horizontal mirror planes; they must be carried further in the process.

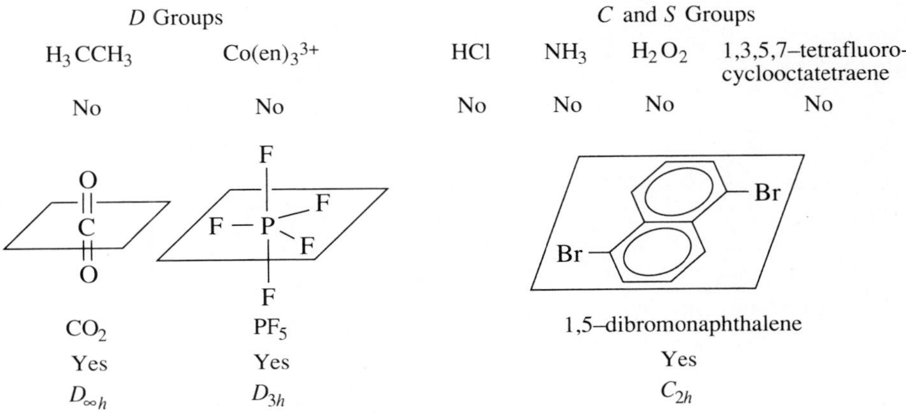

FIGURE 4-9 Horizontal Mirror Planes.

5. Does the molecule have any mirror planes that contain the C_n axis?

These mirror planes are shown in Figure 4-10.

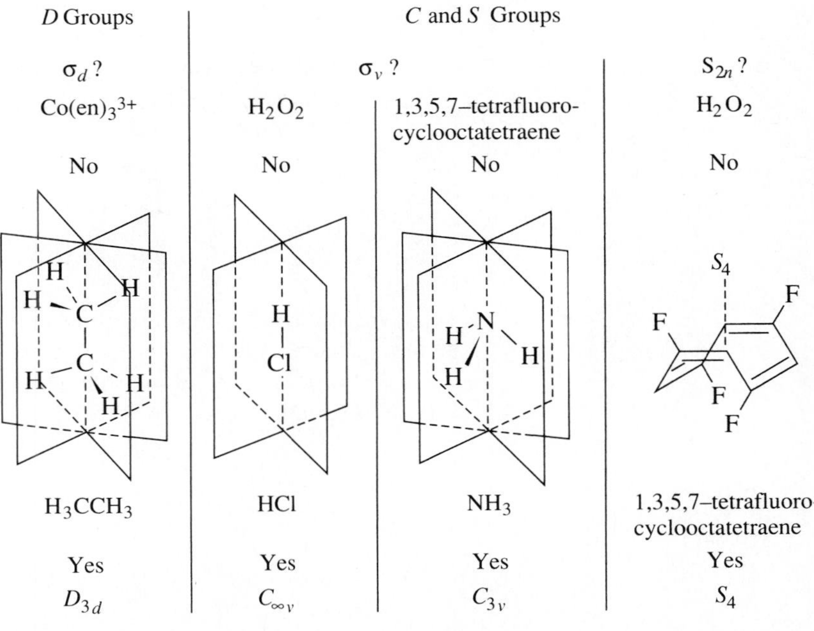

FIGURE 4-10 Vertical or Dihedral Mirror Planes and S_{2n} Axes.

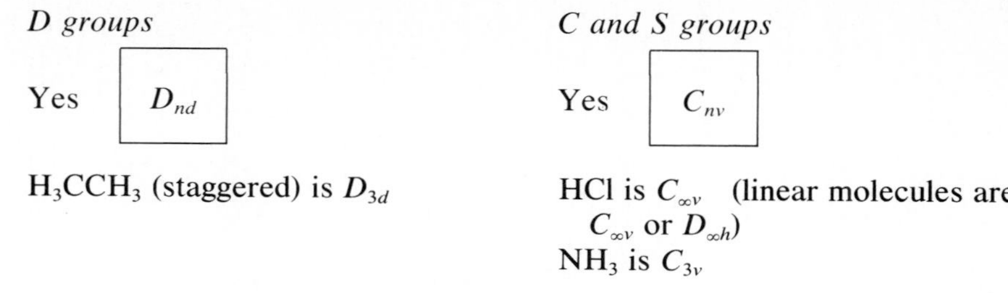

The molecules have mirror planes containing the major C_n axis, but no horizontal mirror planes, and are assigned to the corresponding point groups. There will be n of these planes.

D groups

No: D_n

$Co(en)_3^{3+}$ is D_3

C and S groups

No: C_n or S_{2n}

H_2O_2, 1,3,5,7-tetrafluorocyclo-octatetraene

These molecules are in the simpler rotation groups D_n, C_n, and S_{2n} because they do not have any mirror planes. D_n and C_n point groups have *only* proper rotation axes. S_{2n} point groups have C_n and S_{2n} axes and may have an inversion center. Note that linear molecules have an infinite number of rotation axes and $C_{\infty v}$ or $D_{\infty h}$ symmetry, as shown by HCl and CO_2 above.

6. Is there an S_{2n} axis collinear with the C_n axis?

D groups

While molecules in this category may have S_{2n} axes, they have already been assigned to groups. There are no additional groups to be considered here.

C and S groups

Yes: S_{2n}

1,3,5,7-tetrafluorocyclo-octatetraene is S_4

No: C_n

H_2O_2 is C_2

We have only one example in our list that falls into the S_{2n} groups, as seen in Figure 4-10.

A branching diagram that summarizes this method of assigning point groups is given in Figure 4-11 and more examples are given in Table 4-4.

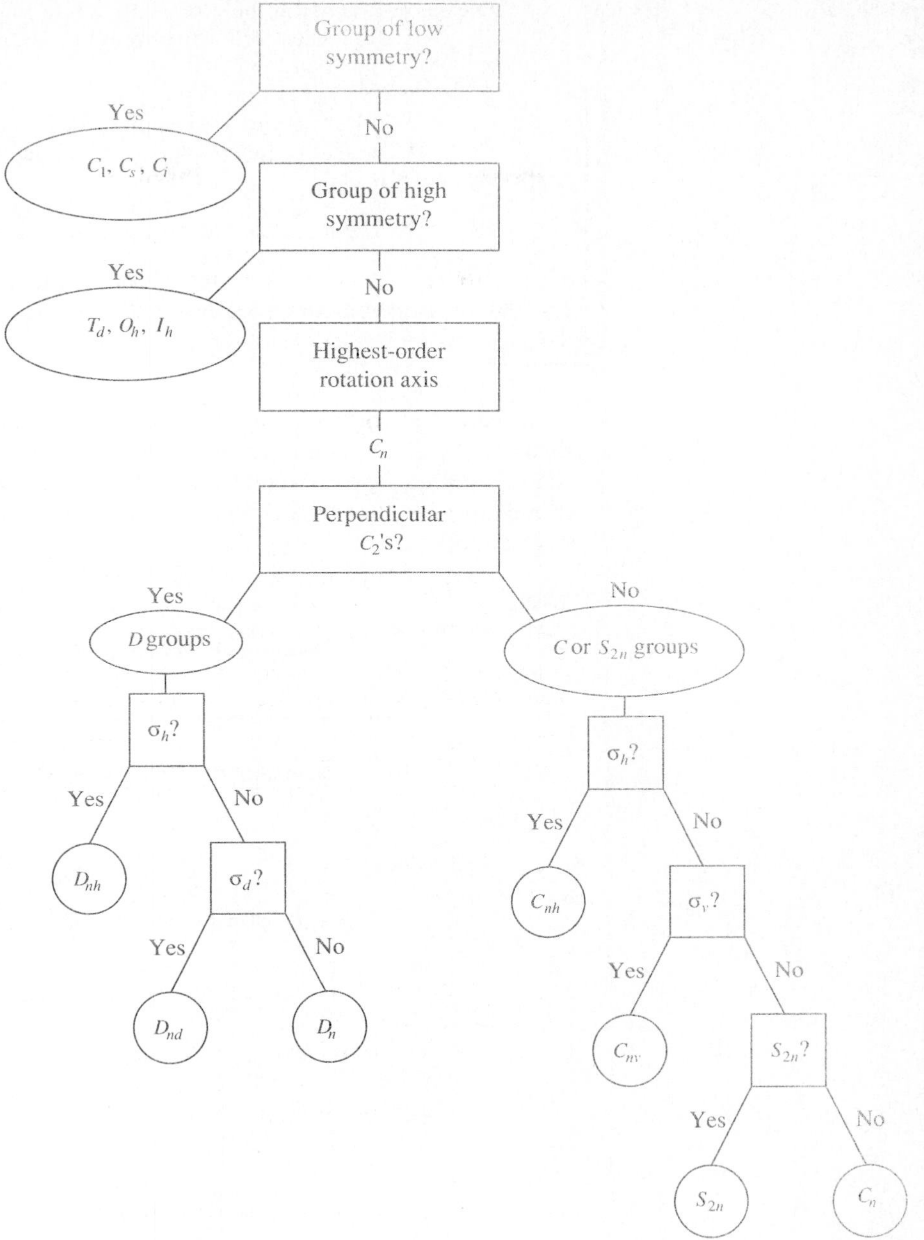

FIGURE 4-11 Diagram of Assignment Method.

EXAMPLES

Determine the point groups of the following molecules and ions from Figures 3-12 and 3-15:

XeF_4

1. XeF_4 is not in the groups of low or high symmetry.
2. Its highest order rotation axis is C_4.
3. It has four C_2 axes perpendicular to the C_4 axis and is therefore in the D set of groups.
4. It has a horizontal plane perpendicular to the C_4 axis. Therefore, its point group is D_{4h}.

$IO_2F_2^-$ 1. It is not in the groups of high or low symmetry.
2. Its highest order (and only) rotation axis is a C_2 passing through the lone pair.
3. The ion has no other C_2 axes and is therefore in the C or S set.
4. It has no mirror plane perpendicular to the C_2.
5. It has two mirror planes containing the C_2. Therefore, the point group is C_{2v}.

IOF_3 1. The molecule has only a mirror plane. Its point group is C_s.

EXERCISE 4-3
Use the procedure described above to verify the point groups of the molecules in Table 4-4.

That's all there is to it! It takes a fair amount of practice, preferably using molecular models, to learn the point groups well; but once you know them, they can be extremely useful. Several practical applications of point groups appear later in this chapter, and additional applications are included in later chapters.

TABLE 4-4
Further examples of C and D point groups

General label	*Group and example*		
C_{nh}	C_{2h}	difluorodiazene	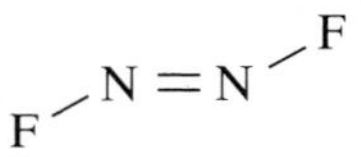 F–N=N–F
	C_{3h}	$B(OH)_3$, planar	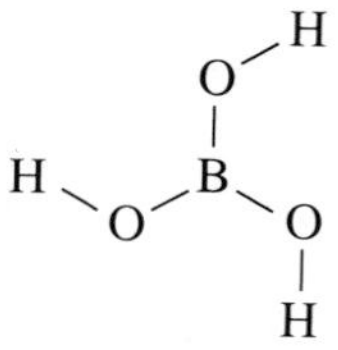
C_{nv}	C_{2v}	H_2O	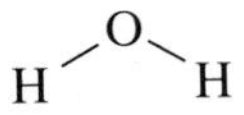
	C_{3v}	PCl_3	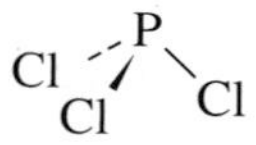
	C_{4v}	BrF_5 (square pyramid)	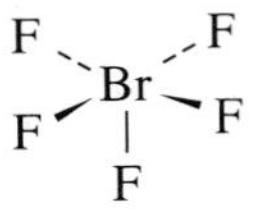
	$C_{\infty v}$	HF, CO, HCN	H–F C≡O H–C≡N
C_n	C_2	N_2H_4, which has a gauche conformation	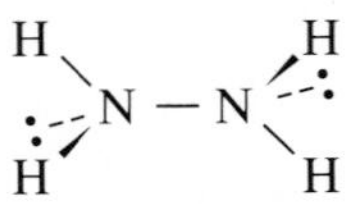
	C_3	$P(C_6H_5)_3$, which is like a three-bladed propeller distorted out of the planar shape by a lone pair on the P	

TABLE 4-4 (Continued)
Further examples of *C* and *D* point groups

General label	Group and example
D_{nh}	D_{3h} BF_3 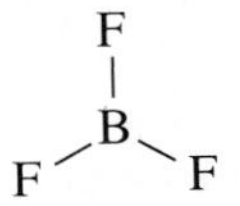
	D_{4h} $PtCl_4^{2-}$ 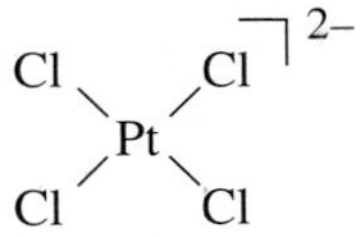
	D_{5h} $Os(C_5H_5)_2$ (eclipsed) 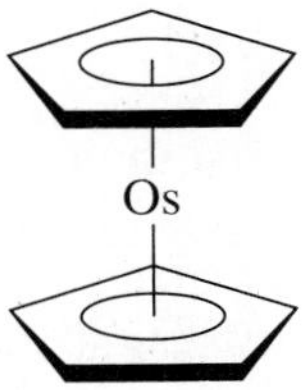
	D_{6h} benzene 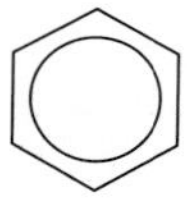
	$D_{\infty h}$ F_2, N_2, F—F N≡N
	acetylene (C_2H_2) H—C≡C—H
D_{nd}	D_{2d} allene ($H_2C{=}C{=}CH_2$) 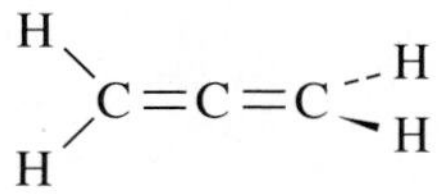
	D_{4d} Ni(cyclobutadiene)$_2$ (staggered) 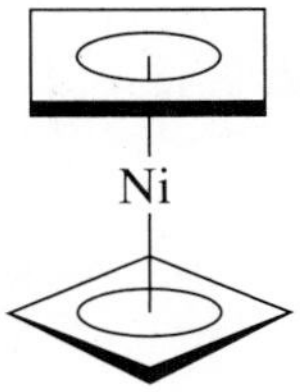
	D_{5d} ferrocene [$Fe(C_5H_5)_2$] (staggered) 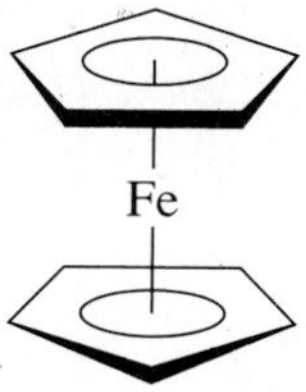
D_n	D_3 $Ru(NH_2CH_2CH_2NH_2)_3^{2+}$ (treating the $NH_2CH_2CH_2NH_2$ group as a planar ring) 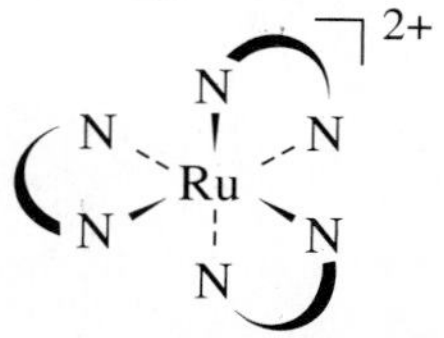

4-3 PROPERTIES AND REPRESENTATIONS OF GROUPS

All mathematical groups (of which point groups are special types) must have certain properties. These properties are listed and illustrated in Table 4-5, using the symmetry operations of NH_3 in Figure 4-12 as an example.

4-3-1 MATRICES

Important information about the symmetry aspects of point groups is summarized in character tables, described later in this chapter. To understand the construction and use of character tables, we first need to consider the properties of matrices, which are the basis for the tables.

By **matrix,** we mean an ordered array of numbers, such as

$$\begin{bmatrix} 3 & 2 \\ 7 & 1 \end{bmatrix} \quad \text{or} \quad [2 \quad 0 \quad 1 \quad 3 \quad 5]$$

To multiply matrices, first, it is required that the number of vertical columns of the first matrix be equal to the number of horizontal rows of the second matrix. To find the product, sum, term by term, the products of each *row* of the first matrix by each *column* of the second (each term in a row must be multiplied by its corresponding term in the appropriate column of the second

TABLE 4-5
Properties of a group

Property of group	*Examples from point group C_{3v}*
1. Each group must contain an **identity** operation that commutes (in other words, $EA = AE$) with all other members of the group and leaves them unchanged ($EA = AE = A$).	C_{3v} molecules (and *all* molecules) contain the identity operation E.
2. Each operation must have an **inverse** that, when combined with the operation, yields the identity operation (sometimes a symmetry operation may be its own inverse). *Note:* By convention, we perform combined symmetry operations *from right to left* as written.	$C_3^2C_3 = E$ 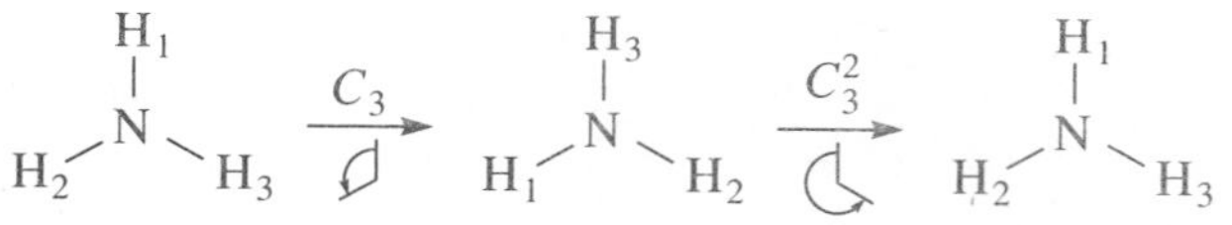 C_3 and C_3^2 are inverses of each other. $\sigma_v\sigma_v = E$ (mirror planes are shown as dashed lines): 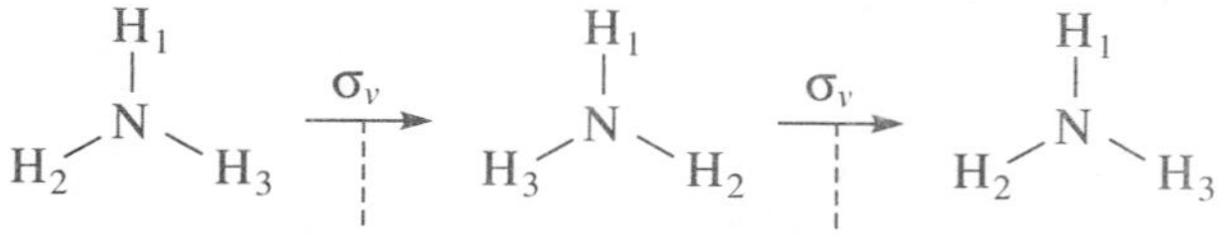
3. The product of any two group operations must also be a member of the group. This includes the product of any operation with itself.	σ_vC_3 has the same overall effect as σ_v''; we therefore write $\sigma_vC_3 = \sigma_v''$: 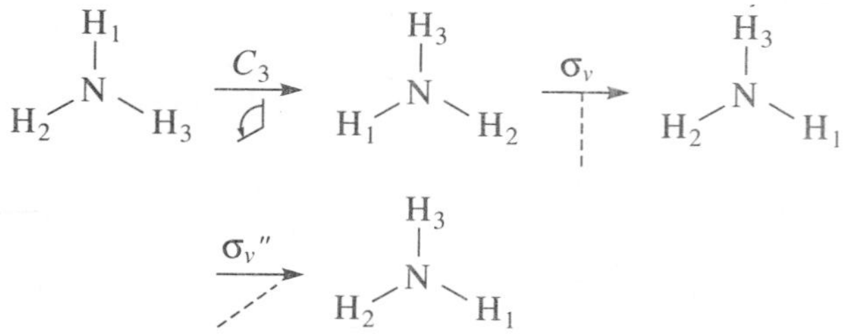 It can be shown that the products of any two operations in C_{3v} are also members of C_{3v}.
4. The associative property of combination must hold. In other words; $A(BC) = (AB)C$.	$C_3(\sigma_v\sigma_v') = (C_3\sigma_v)\sigma_v'$

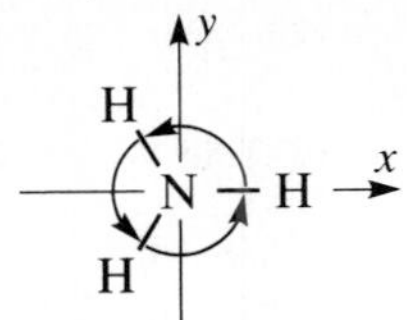

C_3 rotation about the z axis

$\sigma_v\,(xz)$ — One of the mirror planes

H_3–N–H_1, H_2	H_1–N–H_2, H_3	H_2–N–H_1, H_3
NH_3 after E	NH_3 after C_3	NH_3 after $\sigma_v\,(xz)$

FIGURE 4-12 Symmetry Operations for Ammonia. (Top View) NH_3 is of point group C_{3v}, with the symmetry operations E, C_3, C_3^2, σ_v, σ_v', σ_v'', usually written as E, $2C_3$, and $3\sigma_v$ (note that $C_3^3 \equiv E$).

matrix). Place the resulting sum in the product matrix, with the row determined by the row of the first matrix and the column determined by the column of the second matrix:

$$C_{ij} = \Sigma\, A_{ik} \times B_{kj}$$

where
C_{ij} = product matrix, with i rows and j columns
A_{ik} = initial matrix, with i rows and k columns
B_{kj} = initial matrix, with k rows and j columns

EXAMPLES

$$\overset{k}{i\begin{bmatrix} 1 & 5 \\ 2 & 6 \end{bmatrix}} \times \overset{j}{\begin{bmatrix} 2 & 3 \\ 4 & 5 \end{bmatrix}k} = \begin{bmatrix} (1)(2)+(5)(4) & (1)(3)+(5)(5) \\ (2)(2)+(6)(4) & (2)(3)+(6)(5) \end{bmatrix} = \overset{j}{\begin{bmatrix} 22 & 28 \\ 28 & 36 \end{bmatrix}i}$$

$$\overset{k}{i\,[1 \;\; 2 \;\; 3]} \overset{j}{\begin{bmatrix} 1 & 0 & 0 \\ 0 & -1 & 0 \\ 0 & 0 & 1 \end{bmatrix}k} = \overset{j}{[1 \;\; -2 \;\; 3]\,i}$$

$$\overset{k}{i\begin{bmatrix} 1 & 0 & 0 \\ 0 & -1 & 0 \\ 0 & 0 & 1 \end{bmatrix}} \overset{j}{\begin{bmatrix} 1 \\ 2 \\ 3 \end{bmatrix}k} = \overset{j}{\begin{bmatrix} 1 \\ -2 \\ 3 \end{bmatrix}i}$$

EXERCISE 4-4

Do the following multiplications:

a. $\begin{bmatrix} 5 & 1 & 3 \\ 4 & 2 & 2 \\ 1 & 2 & 3 \end{bmatrix} \times \begin{bmatrix} 2 & 1 & 1 \\ 1 & 2 & 3 \\ 5 & 4 & 3 \end{bmatrix}$

b. $\begin{bmatrix} 1 & -1 & -2 \\ 0 & 1 & -1 \\ 1 & 0 & 0 \end{bmatrix} \times \begin{bmatrix} 2 \\ 1 \\ 3 \end{bmatrix}$

c. $[1 \;\; 2 \;\; 3] \times \begin{bmatrix} 1 & -1 & -2 \\ 2 & 1 & -1 \\ 3 & 2 & 1 \end{bmatrix}$

4-3-2 REPRESENTATIONS OF POINT GROUPS

Symmetry operations: Matrix representations

Consider the effects of the symmetry operations of the C_{2v} point group on the set of x, y, and z coordinates. [The set of p orbitals (p_x, p_y, p_z) behaves the same way, so this is a useful exercise.] The water molecule is an example of a molecule having C_{2v} symmetry as shown in Figure 4-13. As mentioned earlier, the z axis is usually chosen as the axis of highest rotational symmetry; for H_2O this is the *only* rotational axis.

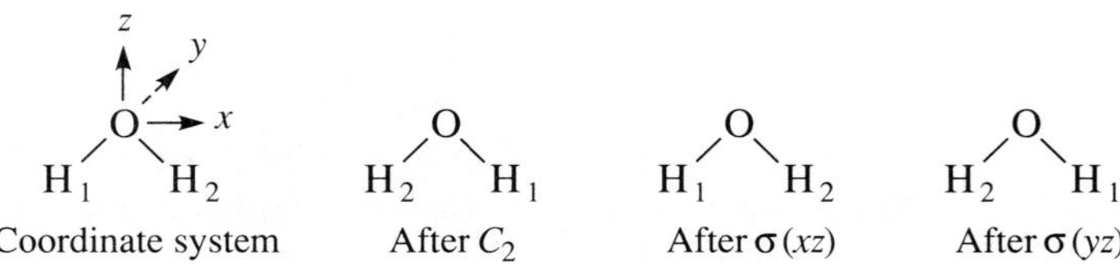

FIGURE 4-13 Symmetry Operations of the Water Molecule.

Each symmetry operation may be expressed as a **transformation matrix** as follows:

[New coordinates] = [transformation matrix][old coordinates]

Examples of symmetry operations of the C_{2v} point group follow.

C_2: Rotate a point having coordinates (x, y, z) about the $C_2(z)$ axis. The new coordinates are given by

$$\begin{aligned} x' &= \text{new } x = -x \\ y' &= \text{new } y = -y \\ z' &= \text{new } z = z \end{aligned} \qquad \begin{bmatrix} -1 & 0 & 0 \\ 0 & -1 & 0 \\ 0 & 0 & 1 \end{bmatrix} \quad \text{transformation matrix for } C_2$$

In matrix notation;

$$\begin{bmatrix} x' \\ y' \\ z' \end{bmatrix} = \begin{bmatrix} -1 & 0 & 0 \\ 0 & -1 & 0 \\ 0 & 0 & 1 \end{bmatrix} \begin{bmatrix} x \\ y \\ z \end{bmatrix} = \begin{bmatrix} -x \\ -y \\ z \end{bmatrix} \quad \text{or} \quad \begin{bmatrix} x' \\ y' \\ z' \end{bmatrix} = \begin{bmatrix} -x \\ -y \\ z \end{bmatrix}$$

$$\begin{bmatrix} \text{New} \\ \text{coordinates} \end{bmatrix} = \begin{bmatrix} \text{transformation} \\ \text{matrix} \end{bmatrix} \begin{bmatrix} \text{old} \\ \text{coordinates} \end{bmatrix} = \begin{bmatrix} \text{new coordinates} \\ \text{in terms of old} \end{bmatrix}$$

$\sigma(xz)$: Reflect a point with coordinates (x, y, z) through the xz plane.

$$\begin{aligned} x' &= \text{new } x = x \\ y' &= \text{new } y' = -y \\ z' &= \text{new } z = z \end{aligned} \qquad \begin{bmatrix} 1 & 0 & 0 \\ 0 & -1 & 0 \\ 0 & 0 & 1 \end{bmatrix} \quad \text{transformation matrix for } \sigma(xz)$$

The matrix equation is

$$\begin{bmatrix} x' \\ y' \\ z' \end{bmatrix} = \begin{bmatrix} 1 & 0 & 0 \\ 0 & -1 & 0 \\ 0 & 0 & 1 \end{bmatrix} \begin{bmatrix} x \\ y \\ z \end{bmatrix} = \begin{bmatrix} x \\ -y \\ z \end{bmatrix} \quad \text{or} \quad \begin{bmatrix} x' \\ y' \\ z' \end{bmatrix} = \begin{bmatrix} x \\ -y \\ z \end{bmatrix}$$

The transformation matrices for the four symmetry operations of the group are.

$$E: \begin{bmatrix} 1 & 0 & 0 \\ 0 & 1 & 0 \\ 0 & 0 & 1 \end{bmatrix} \quad C_2: \begin{bmatrix} -1 & 0 & 0 \\ 0 & -1 & 0 \\ 0 & 0 & 1 \end{bmatrix} \quad \sigma(xz): \begin{bmatrix} 1 & 0 & 0 \\ 0 & -1 & 0 \\ 0 & 0 & 1 \end{bmatrix} \quad \sigma(yz): \begin{bmatrix} -1 & 0 & 0 \\ 0 & 1 & 0 \\ 0 & 0 & 1 \end{bmatrix}$$

This set of matrices satisfies the properties of a mathematical group. We call this a **matrix representation** of the C_{2v} point group. This representation is a set of matrices, each corresponding to an operation in the group, that combines in the same way as the operations. For example, multiplying two of the matrices together parallels carrying out the two corresponding operations and results in a matrix that corresponds to the resulting operation.

These matrices also describe the operations shown in Figure 4-13. The C_2 and $\sigma(yz)$ operations interchange H_1 and H_2, while E and $\sigma(xz)$ leave them unchanged.

Characters

The **character,** defined only for a square matrix, is the sum of the numbers on the diagonal from upper left to lower right. For the C_{2v} point group the following characters are obtained from the preceding matrices:

E	C_2	$\sigma(xz)$	$\sigma(yz)$
3	−1	1	1

These characters can also represent the group, so we can say that they too form a **representation.**

Reducible and irreducible representations

Each transformation matrix in C_{2v} above is "block diagonalized"; it can be broken down into smaller matrices along the diagonal, with all other matrix elements equal to zero:

$$E: \begin{bmatrix} [1] & 0 & 0 \\ 0 & [1] & 0 \\ 0 & 0 & [1] \end{bmatrix} \quad C_2: \begin{bmatrix} [-1] & 0 & 0 \\ 0 & [-1] & 0 \\ 0 & 0 & [1] \end{bmatrix} \quad \sigma(xz): \begin{bmatrix} [1] & 0 & 0 \\ 0 & [-1] & 0 \\ 0 & 0 & [1] \end{bmatrix} \quad \sigma(yz): \begin{bmatrix} [-1] & 0 & 0 \\ 0 & [1] & 0 \\ 0 & 0 & [1] \end{bmatrix}$$

All the nonzero elements become 1×1 matrices along the principal diagonal.

Since each matrix representation for the C_{2v} group can be block diagonalized in this way, it is possible to treat the x, y, and z axes independently. The matrix elements for x form a representation of the group, those for y form a second representation, and those for z form a third representation, all shown in the following table:

		E	C_2	$\sigma(xz)$	$\sigma(yz)$	Coordinate used
Irreducible representations of the C_{2v} point group		1	−1	1	−1	x
		1	−1	−1	1	y
		1	1	1	1	z
	Γ	3	−1	1	1	

These three **irreducible representations** add together to make up the **reducible representation** Γ. The set of 3 × 3 matrices obtained for H_2O is called a reducible representation, because it is the sum of two or more irreducible representations, which cannot be reduced to smaller component parts. In addition, the characters of these matrices form a reducible representation, for the same reason.

4-3-3 CHARACTER TABLES

Three of the representations for C_{2v}, labeled A_1, B_1, and B_2 as shown on the next page, have been determined so far. The fourth, called A_2, can be found by using the properties of a group described in Table 4-6. For the moment, accept it as a necessary addition, which is explained later in this section. A complete set of irreducible representations for a point group is called the **character table** for that group. Each point group has a unique character table; character tables for the common point groups encountered in chemistry are included in Appendix C.

TABLE 4-6
Properties of characters of irreducible representations in point groups

	Example: C_{2v}
1. The total number of symmetry operations in the group is called the **order** (h). To determine the order of a group, simply total the number of symmetry operations listed in the top row of the character table.	Order = 4 [4 symmetry operations: E, C_2, $\sigma(xz)$, and $\sigma(yz)$].
2. Symmetry operations are arranged in **classes.** All operations in a class have identical characters for their transformation matrices and are grouped in the same column in character tables.	Each symmetry operation is in a separate class; therefore, there are 4 columns in the character table.
3. The number of irreducible representations equals the number of classes. This means that character tables have the same number of rows and columns.	Since there are 4 classes, there must also be 4 irreducible representations—and there are.
4. The sum of the squares of the **dimensions** (characters under E) of the irreducible representations equals the order of the group. $h = \sum_i [\chi_i(E)]^2$	$1^2 + 1^2 + 1^2 + 1^2 = 4 = h$, the order of the group.
5. For any irreducible representation, the sum of the squares of the characters equals the order of the group. $h = \sum_R [\chi_i(\mathbf{R})]^2$	For A_2, $1^2 + 1^2 + (-1)^2 + (-1)^2 = 4 = h$.
6. Irreducible representations are **orthogonal** to each other. The sum of the products of the characters for each operation of any pair of irreducible representations is 0. $\sum_R \chi_i(\mathbf{R})\chi_j(\mathbf{R}) = 0$, when $i \neq j$	B_1 and B_2 are orthogonal: $\underset{E}{(1)(1)} + \underset{C_2}{(-1)(-1)} + \underset{\sigma(xz)}{(1)(-1)} + \underset{\sigma(yz)}{(-1)(1)} = 0$
7. A **totally symmetric representation** is included in all groups, with characters of 1 for all operations.	C_{2v} has A_1, which has all characters = 1.

The complete character table for C_{2v} with the irreducible representations in the order commonly used is:

C_{2v}	E	C_2	$\sigma(xz)$	$\sigma(yz)$		
A_1	1	1	1	1	z	x^2, y^2, z^2
A_2	1	1	-1	-1	R_z	xy
B_1	1	-1	1	-1	x, R_y	xz
B_2	1	-1	-1	1	y, R_x	yz

(The labels in the left column used to designate the representations will be described later.)

In the two right columns of the character table are listed several useful functions. In each case, these functions are listed with the irreducible representation of matching symmetry. R_x, R_y, and R_z stand for rotation about the x, y, and z axes, operations useful in working with rotational spectra. Other symbols commonly used in character tables include $\boldsymbol{R}$ for any symmetry operation [such as C_2 or $\sigma(xz)$], χ for the character, i and j for different representations (such as A_1 or A_2), and h for the order of the group (the total number of symmetry operations in the group). Other useful terms are defined in Table 4-6.

The A_2 representation of the C_{2v} group can now be explained. A fourth representation is required for a group of order 4 (property 3 in Table 4-6), and the sum of the products of the characters of any two representations must equal zero (orthogonality, property 6). Therefore, a product of A_1 and the unknown representation must have 1 for two of the characters and -1 for the other two. Since $\chi(E) = 1$ (required by property 4), and no two representations can be the same, A_2 must have $\chi(E) = \chi(C_2) = 1$, and $\chi(\sigma_{xy}) = \chi(\sigma_{yz}) = -1$. This representation is also orthogonal to B_1 and B_2, as required.

Another example: C_{3v} (NH_3)

Full descriptions of the matrices for the operations in this group will not be given, but the characters can be found by using the properties of a group. Consider the C_3 rotation shown in Figure 4-14. Rotation of 120° results in new

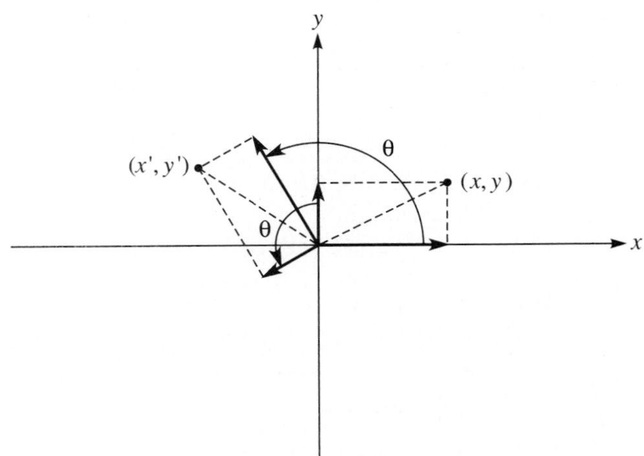

General case: $x' = x \cos\theta - y \sin\theta$
$y' = x \sin\theta + y \cos\theta$

For C_3: $\theta = 2\pi/3 = 120°$

FIGURE 4-14 Effect of Rotation on Coordinates of a Point.

x' and y' as shown, which can be described in terms of the vector sums of x and y by using trigonometric functions:

$$x' = x\cos\frac{2\pi}{3} - y\sin\frac{2\pi}{3} = -\frac{1}{2}x - \frac{\sqrt{3}}{2}y$$

$$y' = x\sin\frac{2\pi}{3} + y\cos\frac{2\pi}{3} = \frac{\sqrt{3}}{2}x - \frac{1}{2}y$$

The transformation matrices for the symmetry operations shown are as follows:

$$E: \begin{bmatrix} 1 & 0 & 0 \\ 0 & 1 & 0 \\ 0 & 0 & 1 \end{bmatrix}, \quad C_3: \begin{bmatrix} \cos\frac{2\pi}{3} & -\sin\frac{2\pi}{3} & 0 \\ \sin\frac{2\pi}{3} & \cos\frac{2\pi}{3} & 0 \\ 0 & 0 & 1 \end{bmatrix} = \begin{bmatrix} -\frac{1}{2} & -\frac{\sqrt{3}}{2} & 0 \\ \frac{\sqrt{3}}{2} & -\frac{1}{2} & 0 \\ 0 & 0 & 1 \end{bmatrix}$$

$$\sigma_{v(xz)}: \begin{bmatrix} 1 & 0 & 0 \\ 0 & -1 & 0 \\ 0 & 0 & 1 \end{bmatrix}$$

Since all characters in a class must be the same, $\chi(C_3^2) = \chi(C_3)$; in addition, the three reflections have identical characters, so it is not necessary to determine the other matrices.

The transformation matrices for C_3 and C_3^2 cannot be block diagonalized into 1×1 matrices because the C_3 matrix has off-diagonal entries; however, the matrices can be block diagonalized into 2×2 and 1×1 matrices as follows:

$$E: \begin{bmatrix} \begin{bmatrix} 1 & 0 \\ 0 & 1 \end{bmatrix} & \begin{matrix} 0 \\ 0 \end{matrix} \\ \begin{matrix} 0 & 0 \end{matrix} & [1] \end{bmatrix} \quad C_3: \begin{bmatrix} \begin{bmatrix} -\frac{1}{2} & -\frac{\sqrt{3}}{2} \\ \frac{\sqrt{3}}{2} & -\frac{1}{2} \end{bmatrix} & \begin{matrix} 0 \\ 0 \end{matrix} \\ \begin{matrix} 0 & 0 \end{matrix} & [1] \end{bmatrix} \quad \sigma_{v(xz)}: \begin{bmatrix} \begin{bmatrix} 1 & 0 \\ 0 & -1 \end{bmatrix} & \begin{matrix} 0 \\ 0 \end{matrix} \\ \begin{matrix} 0 & 0 \end{matrix} & [1] \end{bmatrix}$$

The C_3 matrix must be blocked this way because the (x, y) combination is needed for the new x' and y'; the others must follow the same pattern for consistency across the representation.

The characters of the matrices are the sums of the numbers on the principal diagonal (from upper left to lower right). The set of 2×2 matrices has the characters corresponding to the E representation in the following character table; the set of 1×1 matrices matches the A_1 representation. The third irreducible representation, A_2, can be found by using the defining properties of a mathematical group, as in the C_{2v} example above. Table 4-7 gives the properties of the characters for the C_{3v} point group.

C_{3v}	E	$2C_3$	$3\sigma_v$		
A_1	1	1	1	z	$x^2 + y^2, z^2$
A_2	1	1	-1	R_z	
E	2	-1	0	$(x, y), (R_x, R_y)$	$(x^2 - y^2, xy)(xz, yz)$

TABLE 4-7
Properties of the characters for the C_{3v} point group

Property	C_{3v} *example*
1. Order	6 (6 symmetry operations)
2. Classes	3 classes: E; $2C_3\ (= C_3, C_3^2)$; $3\sigma_v\ (= \sigma_v, \sigma_v', \sigma_v'')$
3. Number of irreducible representations	3 (A_1, A_2, E)
4. Sum of squares of dimensions equals the order of the group	$1^2 + 1^2 + 2^2 = 6$
5. Sum of squares of characters equals the order of the group	$E \quad 2C_3 \quad 3\sigma_v$ A_1: $1^2 + 2(1)^2 + 3(1)^2 = 6$ A_2: $1^2 + 2(1)^2 + 3(-1)^2 = 6$ E: $2^2 + 2(-1)^2 + 3(0)^2 = 6$ (multiply the squares by the number of symmetry operations in the class)
6. Orthogonal representations	The sum of the products of any two representations equals 0. Example of $A_2 \times E$: $(1)(2) + 2(1)(-1) + 3(-1)(0) = 0$
7. Totally symmetric representation	A_1, with all characters = 1

The following notes explain additional features of character tables.

1. When necessary, the C_2 axes perpendicular to the principal axis (in a D group) are designated with primes; a single prime indicates that the axis passes through several atoms of the molecule, while a double prime indicates that it passes between the atoms.
2. When the mirror plane is perpendicular to the principal axis, or horizontal, the reflection is called σ_h. When there is a need to distinguish between the vertical mirror planes, as in the C_{nv} and D_{nh} groups, these planes are labeled σ_v if they pass through outer atoms of the molecule and σ_d if they do not. In the D_{nd} and T_d groups, they are called σ_d.
3. The expressions listed to the right of the characters give the symmetry of mathematical functions of the coordinates x, y, and z and of rotation about the axes (R_x, R_y, R_z). These can be used to find the orbitals that match the representation. For example, x with positive and negative directions matches the p_x orbital with positive and negative lobes, and the product xy with alternating signs on the quadrants matches lobes of the d_{xy} orbital, as in Figure 4-15. In all cases, the totally symmetric s orbital matches the first representation in the group, one of the A set. The rotational functions are used to describe the rotational motions of the molecule.

   In the C_{3v} example above, the x and y coordinates appeared together in the E irreducible representation. The notation for this is to group them as (x, y) in this section of the table. This means that x and y together have the same symmetry properties as the E irreducible representation.
4. Matching the symmetry operations of a molecule with those listed in the top row of the character table will confirm any point group assignment.

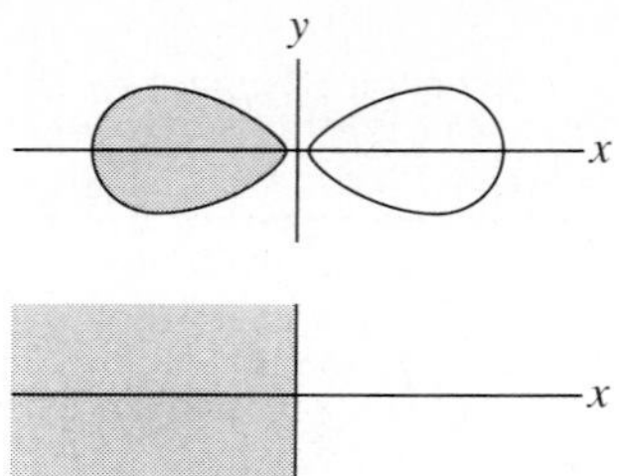

p_x orbitals have the same symmetry as x (positive in half the quadrants, negative in the other half).

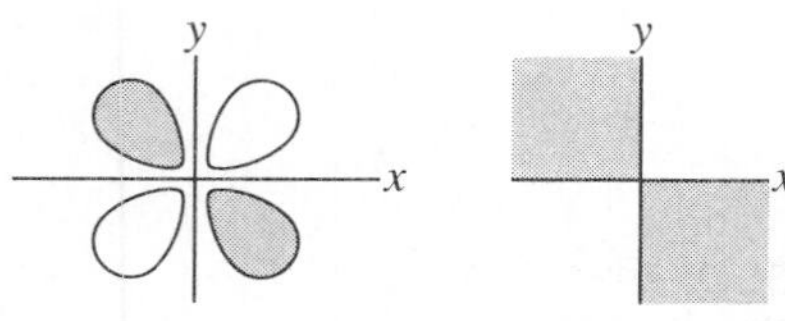

d_{xy} orbitals have the same symmetry as the function xy (sign of the function in the four quadrants).

FIGURE 4-15 Orbitals and Representations.

5. Irreducible representations are assigned labels according to the following rules, where symmetric means a character of 1 and antisymmetric a character of −1 (see the character tables in Appendix C for examples).
   a. Letters are assigned according to the dimension of the irreducible representation (the character for the identity operation).

Dimension	*Symmetry label*	
1	*A*	if the representation is symmetric to the principal rotation operation
	B	if it is antisymmetric
2	*E*	
3	*T*	

   b. Subscript 1 designates a representation symmetric to a C_2 rotation perpendicular to the principal axis; subscript 2, a representation antisymmetric to the C_2. If there are no perpendicular C_2's, 1 designates a representation symmetric to a vertical plane, and 2 a representation antisymmetric to a vertical plane.
   c. Subscript g (*gerade*) designates symmetric to inversion; subscript u (*ungerade*), antisymmetric to inversion.
   d. Single primes are symmetric to σ_h, and double primes are antisymmetric to σ_h, when such distinctions are possible. In other cases, single, double, and triple primes are assigned arbitrarily.

4-4 EXAMPLES AND APPLICATIONS OF SYMMETRY

4-4-1 CHIRALITY

Many molecules are not superimposable on their mirror image. Such molecules, labeled **chiral** or **dissymmetric**, may have important chemical properties as a consequence of this nonsuperimposability. An example of a chiral organic molecule is CBrClFI, and many examples of chiral objects can also be found on the macroscopic scale, as in Figure 4-16.

Chiral objects are termed dissymetric; this terminology does not, however, imply that these objects necessarily have *no* symmetry. For example, the

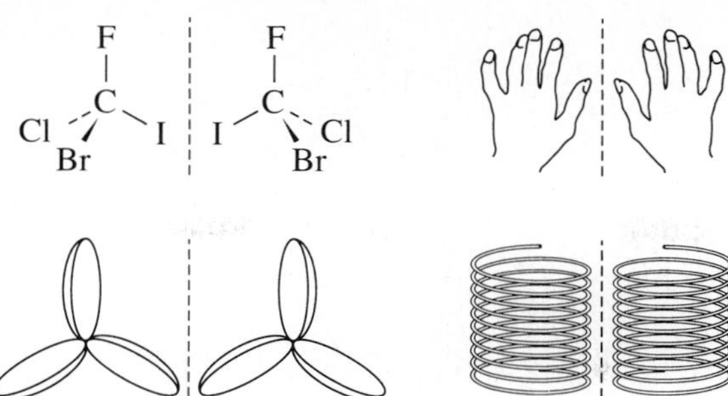

FIGURE 4-16 A Chiral Molecule and Other Chiral Objects.

propellers shown in Figure 4-16 each have a C_3 axis, yet they are nonsuperimposable (if both were spun in a clockwise direction, they would move an airplane in opposite directions!) In general, we can say that a molecule or other object is chiral if it has *no* symmetry operations (other than E) or if it has *only proper rotation axes*.

EXERCISE 4-5
Which point groups are possible for chiral molecules?
[HINT: Refer as necessary to the character tables in Appendix C.]

Air blowing past the stationary propellers in Figure 4-16 will be rotated in either a clockwise or counterclockwise direction. By the same token, plane polarized light will be rotated on passing through chiral molecules (Figure 4-17); clockwise rotation is designated **dextrorotatory**, and counterclockwise rotation **levorotatory**. The ability of chiral molecules to rotate plane polarized light is termed **optical activity** and may be measured experimentally.

Many coordination compounds are chiral and thus exhibit optical activity. One of these is $[Ru(NH_2CH_2CH_2NH_2)_3]^{2+}$, with D_3 symmetry (Figure 4-18). Mirror images of this molecule look much like left- and right-handed three-bladed propellers. Further examples will be discussed in Chapter 8.

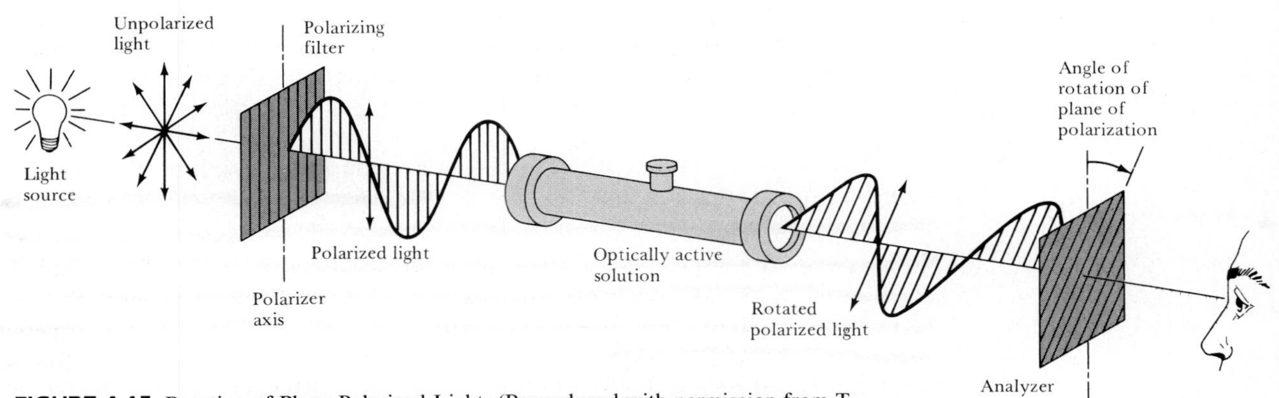

FIGURE 4-17 Rotation of Plane-Polarized Light. (Reproduced with permission from T. L. Brown and E. E. LeMay, Jr., *Chemistry; The Central Science*, Prentice Hall, Englewood Cliffs, N.J., 1988, p. 879.)

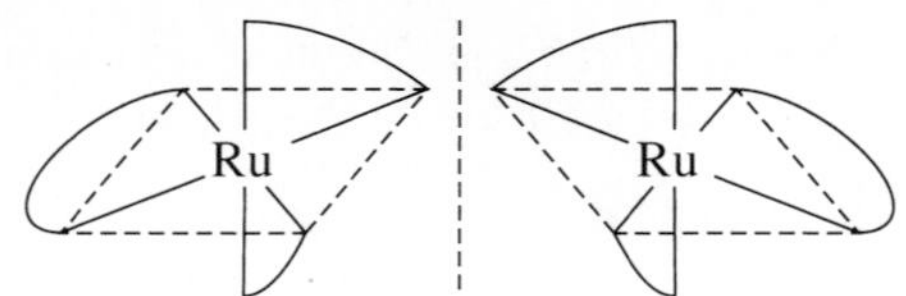

FIGURE 4-18 Chiral Isomers of $[Ru(NH_2CH_2CH_2NH_2)_3]^{2+}$.

4-4-2 MOLECULAR VIBRATIONS

Symmetry can be a great help in determining the modes of vibration of molecules. Vibrational modes of water and the stretching modes of CO in carbonyl complexes are examples that can be treated quite simply, as described in the following pages. Other molecules can be studied using the same methods, but their more complex structure requires more effort.

Water (C_{2v} symmetry)

Since the study of vibrations is the study of motion of the individual atoms in a molecule, we must first attach a set of x, y, and z coordinates to each atom. For convenience, we make the z axes parallel to the C_2 axis of the molecule, the x axes in the plane of the molecule, and the y axes perpendicular to the plane. Each atom can move in all three directions, so a total of nine transformations must be considered. For N atoms in a molecule, there are $3N$ total motions. For water, since there are three atoms, there must be nine different motions (Figure 4-19). Linear molecules have three translational motions, two rotational motions, and $3N - 5$ vibrations, while nonlinear molecules have three translational motions, three rotational motions, and $3N - 6$ vibrations.

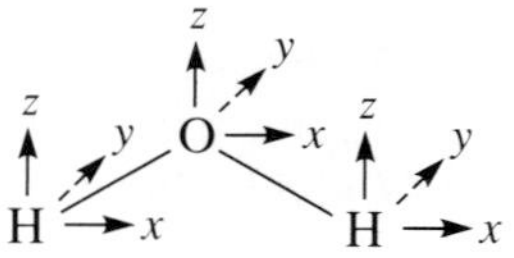

FIGURE 4-19 A Set of Axes for the Water Molecule.

We will determine the symmetry of all nine motions and then assign them to translation, rotation, and vibration. Fortunately, it is only necessary to determine the characters of the transformation matrices, not the individual matrix elements.

In this case, the initial axes make a column matrix with nine elements, and each transformation matrix is 9×9. An entry appears along the diagonal of the matrix only for an atom that does not exchange places with another atom. If the atom changes position during the symmetry operation, a 0 is entered. If the atom remains in its original location and the vector direction is unchanged, a 1 is entered. If the atom remains but the vector direction is reversed, a -1 is entered. (Since all the operations change vector directions by 0° or 180° in the C_{2v} point group, these are the only possibilities.) When all nine vectors are summed, the character of the reducible representation, Γ (gamma), is obtained. (Γ is a common designation for such representations.)

The full 9 × 9 matrix for C_2 is shown as an example; only the diagonal entries are used in finding the character.

C_{2v}	E	C_2	$\sigma_v(xz)$	$\sigma_v'(yz)$		
A_1	1	1	1	1	z	x^2, y^2, z^2
A_2	1	1	−1	−1	R_z	xy
B_1	1	−1	1	−1	x, R_y	xz
B_2	1	−1	−1	1	y, R_x	yz
Γ	9	−1	3	1		

$$
\begin{matrix} \mathrm{O} \\ \\ \\ \mathrm{H}_a \\ \\ \\ \mathrm{H}_b \\ \\ \\ \end{matrix}
\begin{bmatrix} x' \\ y' \\ z' \\ x' \\ y' \\ z' \\ x' \\ y' \\ z' \end{bmatrix}
=
\begin{bmatrix}
-1 & 0 & 0 & 0 & 0 & 0 & 0 & 0 & 0 \\
0 & -1 & 0 & 0 & 0 & 0 & 0 & 0 & 0 \\
0 & 0 & 1 & 0 & 0 & 0 & 0 & 0 & 0 \\
0 & 0 & 0 & 0 & 0 & 0 & -1 & 0 & 0 \\
0 & 0 & 0 & 0 & 0 & 0 & 0 & -1 & 0 \\
0 & 0 & 0 & 0 & 0 & 0 & 0 & 0 & 1 \\
0 & 0 & 0 & -1 & 0 & 0 & 0 & 0 & 0 \\
0 & 0 & 0 & 0 & -1 & 0 & 0 & 0 & 0 \\
0 & 0 & 0 & 0 & 0 & 1 & 0 & 0 & 0
\end{bmatrix}
\begin{bmatrix} x \\ y \\ z \\ x \\ y \\ z \\ x \\ y \\ z \end{bmatrix}
\begin{matrix} \mathrm{O} \\ \\ \\ \mathrm{H}_a \\ \\ \\ \mathrm{H}_b \\ \\ \\ \end{matrix}
$$

E: All nine vectors are unchanged in the identity operation, so the character is 9.

C_2: The hydrogen atoms change position in a C_2 rotation, so all their vectors have zero contribution to the character. The oxygen atom vectors in the x and y directions are reversed, each contributing −1, and in the z direction they remain the same, contributing 1, for a total of −1. The sum of the principal diagonal $= \chi(C_2) = (-1) + (-1) + (1) = -1$.

σ_v: Reflection in the plane of the molecule changes the direction of all the y vectors and leaves the x and z vectors unchanged, for a total of $3 - 3 + 3 = 3$.

σ_v': Finally, reflection perpendicular to the plane of the molecule changes the position of the hydrogens so their contribution is zero; the x vector on the oxygen changes direction, and the y and z vectors are unchanged, for a total of 1.

Since all three directions are included for each of the three atoms, these are the characters of the three translational motions, the three rotational motions, and the three vibrational motions combined.

Reducing representations to irreducible representations

The next step is to separate these representations into their component irreducible representations. This requires another property of groups. The number of times that any irreducible representation appears in a reducible representation is equal to the sum of the products of the characters of the reducible and irreducible representations taken one operation at a time, divided by the order of the group. This may be expressed in equation form with the sum taken over all symmetry operations of the group:[2]

[2] This procedure should yield an integer for the number of irreducible representations of each type; obtaining a fraction in this step indicates a calculation error.

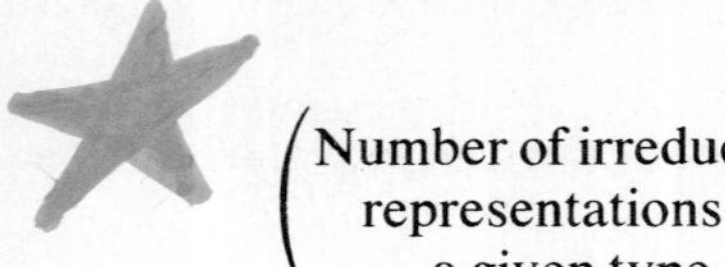

$$\begin{pmatrix}\text{Number of irreducible}\\ \text{representations of}\\ \text{a given type}\end{pmatrix} = \frac{1}{\text{order}} \sum_R \begin{pmatrix}\text{character of}\\ \text{reducible representation}\end{pmatrix} \times \begin{pmatrix}\text{character of}\\ \text{irreducible representations}\end{pmatrix}$$

In the water example,

$$n_{A_1} = \frac{1}{4}[(9)(1) + (-1)(1) + (3)(1) + (1)(1)] = 3$$

$$n_{A_2} = \frac{1}{4}[(9)(1) + (-1)(1) + (3)(-1) + (1)(-1)] = 1$$

$$n_{B_1} = \frac{1}{4}[(9)(1) + (-1)(-1) + (3)(1) + (1)(-1)] = 3$$

$$n_{B_2} = \frac{1}{4}[(9)(1) + (-1)(-1) + (3)(-1) + (1)(1)] = 2$$

The reducible representation for all motions of the water molecule is therefore reduced to $3A_1 + A_2 + 3B_1 + 2B_2$.

Examination of the columns on the far right in the character table shows that translation along the x, y, and z directions is $A_1 + B_1 + B_2$ (translation is motion along the x, y, and z directions, so it transforms in the same way as the three axes) and that rotation in the three directions (R_x, R_y, R_z) is $A_2 + B_1 + B_2$. Subtracting these from the total above leaves $2A_1 + B_1$, the three vibrational modes, as shown in Table 4-8. The number of vibrational modes equals $3N - 6$, as described earlier. Two of the modes are totally symmetric (A_1) and do not change the symmetry of the molecule, while one is asymmetric to C_2 rotation and to reflection perpendicular to the plane of the molecule (B_1). These modes are illustrated as symmetric stretch, symmetric bend, and asymmetric stretch in Table 4-9.

TABLE 4-8
Symmetry of molecular motions of water

All motions	*Translation* (x, y, z)	*Rotation* (R_x, R_y, R_z)	*Vibration (remaining modes)*
$3A_1$	A_1		$2A_1$
A_2		A_2	
$3B_1$	B_1	B_1	B_1
$2B_2$	B_2	B_2	

TABLE 4-9
Vibrational modes of water

A_1	H←O→H (symmetric stretch arrows)	*Symmetric stretch:* Change in dipole moment; more distance between positive H's and negative O *IR active*
B_1	H O H (asymmetric stretch arrows)	*Asymmetric stretch:* Change in dipole moment; change in distances between positive H's and negative O *IR active*
A_1	H O H (bend arrows)	*Symmetric bend:* Change in dipole moment; angles of H–O vectors change *IR active*

A molecular vibration is infrared active (has an infrared absorption) *only if it results in a change in the dipole moment of the molecule.* The three vibrations of water can be analyzed this way to determine their infrared behavior.

Group theory can give us the same information (and can account for the more complicated cases as well; in fact, group theory in principle can account for *all* vibrational modes of a molecule). In group theory terms, a vibrational mode is active in the infrared *if it corresponds to an irreducible representation that has the same symmetry (or transforms) as the Cartesian coordinates x, y, or z.* Otherwise, the vibrational mode is not infrared active.

EXAMPLES

Reduce the following representations to their irreducible representations in the point group indicated: (Refer to character tables in Appendix C.)

C_{2h}	E	C_2	i	σ_h
Γ	4	0	2	2

Solution:

$$n_{A_g} = \frac{1}{4}[(4)(1) + (0)(1) + (2)(1) + (2)(1)] = 2$$

$$n_{B_g} = \frac{1}{4}[(4)(1) + (0)(-1) + (2)(1) + (2)(-1)] = 1$$

$$n_{A_u} = \frac{1}{4}[(4)(1) + (0)(1) + (2)(-1) + (2)(-1)] = 0$$

$$n_{B_u} = \frac{1}{4}[(4)(1) + (0)(-1) + (2)(-1) + (2)(1)] = 1$$

Therefore, $\Gamma = 2A_g + B_g + B_u$

C_{3v}	E	$2C_3$	$3\sigma_v$
Γ	6	3	-2

Solution:

$$n_{A1} = \frac{1}{6}[(6)(1) + (2)(3)(1) + (3)(-2)(1)] = 1$$

$$n_{A2} = \frac{1}{6}[(6)(1) + (2)(3)(1) + (3)(-2)(-1)] = 3$$

$$n_{E} = \frac{1}{6}[(6)(2) + (2)(3)(-1) + (3)(-2)(0)] = 1$$

Therefore, $\Gamma = A_1 + 3A_2 + E$

Be sure to include the number of symmetry operations in a class (column) of the character table. This means that the second term in the C_{3v} calculations must be multiplied by 2 ($2C_3$; there are two operations in this class), and the third term must be multiplied by 3, as shown.

EXERCISE 4-6

Reduce the following representations to their irreducible representations in the point group indicated:

T_d	E	$8C_3$	$3C_2$	$6S_4$	$6\sigma_d$
Γ_1	4	1	0	0	2

D_{2d}	E	$2S_4$	C_2	$2C_2'$	$2\sigma_d$
Γ_2	4	0	0	2	0

C_{4v}	E	$2C_4$	C_2	$2\sigma_v$	$2\sigma_d$
Γ_3	7	−1	−1	−1	−1

EXERCISE 4-7

Vibrational analysis for the NH_3 molecule gives the following representation:

C_{3v}	E	$2C_3$	$3\sigma_v$
Γ	12	0	2

a. Reduce Γ to its irreducible representations.

b. Classify the irreducible representations into translational, rotational, and vibrational modes.

c. Which vibrational modes are infrared active?

Selected vibrational modes

It is often useful to consider a particular type of vibrational mode for a compound. For example, useful information can often be obtained from the C–O stretching bands in infrared spectra of metal complexes containing CO (carbonyl) ligands. The following example of *cis*- and *trans*-dicarbonyl square planar complexes shows the procedure. For these complexes, a simple IR spectrum can distinguish whether a sample is *cis- or trans*-$ML_2(CO)_2$; the number of C–O stretching bands is determined by the geometry of the complex (see Figure 4-20).

***Cis*-$ML_2(CO)_2$, point group C_{2v}.** The principal axis (C_2) is the z axis, with the xz plane the plane of the molecule. Possible C–O stretching motions are shown by arrows in Figure 4-20; either an increase or decrease in the C–O distance is possible. These vectors are used to create the reducible representation

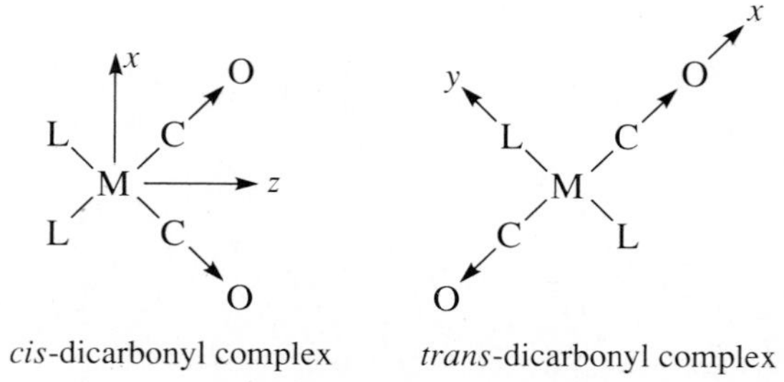

FIGURE 4-20 Carbonyl Stretching Vibrations of *cis*- and *trans*-Dicarbonyl Square Planar Complexes.

shown below using the symmetry operations of the C_{2v} point group. A C–O bond will transform with a character of 1 *if it remains unchanged* by the symmetry operations, and with a character of 0 *if it is changed*. These operations and their characters are shown in Figure 4-21. The reducible representation Γ reduces to $A_1 + B_1$:

C_{2v}	E	C_2	$\sigma_v(xz)$	$\sigma_v'(yz)$		
Γ	2	0	2	0		
A_1	1	1	1	1	z	x^2, y^2, z^2
B_1	1	-1	1	-1	x, R_y	xz

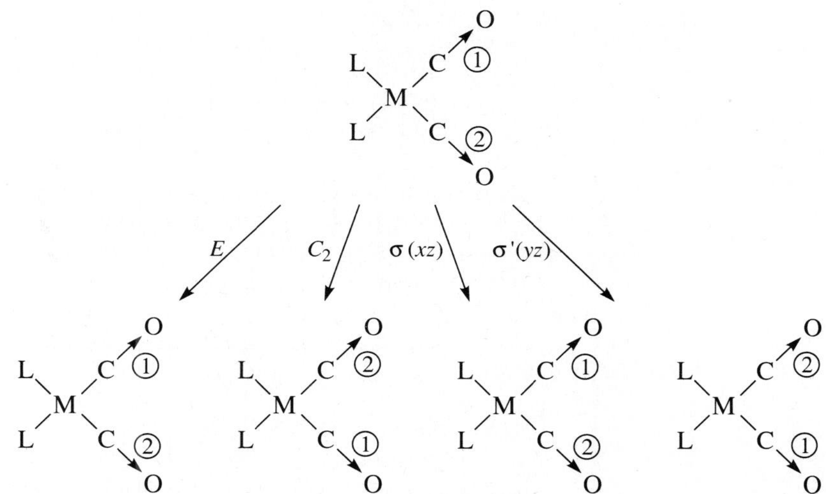

2 vectors unchanged 0 vectors unchanged 2 vectors unchanged 0 vectors unchanged

	E	C_2	$\sigma(xz)$	$\sigma'(yz)$
Γ	2	0	2	0

FIGURE 4-21 Symmetry Operations and Characters for *cis*-$ML_2(CO)_2$

A_1 is an appropriate irreducible representation for an IR active band, since it transforms as (has the symmetry of) the Cartesian coordinate z. Furthermore, the vibrational mode corresponding to B_1 should be IR active, since it transforms as the Cartesian coordinate x.

In summary:
There are two vibrational modes for C–O stretching, one having A_1 symmetry and one B_1 symmetry. Both modes are IR active, and we therefore expect to see two C–O stretches in the IR. This assumes that the C–O stretches are not sufficiently similar in energy to overlap in the infrared spectrum.

***Trans*-$ML_2(CO)_2$, point group D_{2h}.** The principal axis, C_2, is again chosen as the z axis, which this time makes the plane of the molecule the xy plane. Using the symmetry operations of the D_{2h} point group, we obtain a reducible representation for the C–O stretches that reduces to $A_g + B_{3u}$:

D_{2h}	E	$C_2(z)$	$C_2(y)$	$C_2(x)$	i	$\sigma(xy)$	$\sigma(xz)$	$\sigma(yz)$	
Γ	2	0	0	2	0	2	2	0	
A_g	1	1	1	1	1	1	1	1	x^2, y^2, z^2
B_{3u}	1	−1	−1	1	−1	1	1	−1	x

The vibrational mode of A_g symmetry is *not* IR active, since it does not have the same symmetry as a Cartesian coordinate x, y, or z (this is the IR inactive symmetric stretch). The mode of symmetry B_{3u} on the other hand, *is* IR active, since it does have the same symmetry as x.

In summary:
There are two vibrational modes for C–O stretching, one having the same symmetry as A_g and one the same symmetry as B_{3u}. The A_g mode is IR inactive (does not have the symmetry of x, y, or z); the B_{3u} is IR active (has the symmetry of x). We therefore expect to see one C–O stretch in the IR.

It is therefore possible to distinguish *cis*- and *trans*-$ML_2(CO)_2$ by taking an IR spectrum. If one C–O stretching band appears, the molecule is *trans;* if two bands appear, the molecule is *cis*. A significant distinction can be made by a very simple measurement.

EXAMPLE

Determine the number of IR active C–O stretching modes for *fac*-$Mo(CO)_3(NCCH_3)_3$.

This molecule has C_{3v} symmetry. The operations to be considered are E, C_3, and σ_v.

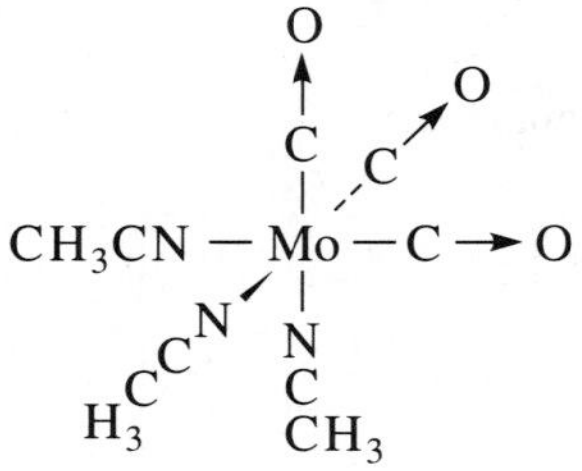

O
C
C O
CH_3CN — Mo — C → O
N N
C C
H_3 CH_3

E leaves the three bond vectors unchanged, giving a character of 3.

C_3 moves all three vectors, giving a character of 0.

Each σ_v passes through one of the CO groups, leaving it unchanged, while interchanging the other two. The resulting character is 1.

The representation to be reduced, therefore, is:

E	$2C_3$	$3\sigma_v$
3	0	1

This reduces to $A_1 + E$. A_1 has the same symmetry as the Cartesian coordinate z and is therefore IR active. E has the same symmetry as the x and y coordinates together and is also IR active. It represents a degenerate pair of vibrations, which appear as one absorption band.

EXERCISE 4-8

Determine the number of IR active C–O stretching modes for $Mn(CO)_5Cl$.

GENERAL REFERENCES

One of the most readable books on this subject is F. A. Cotton, *Chemical Applications of Group Theory,* 3rd ed., Wiley, New York, 1990. Other books are more complete or have different emphases, but this one provides a straightforward approach that moves quickly to the applications most needed by chemists. I. Hargittai and M. Hargittai, *Symmetry through the Eyes of a Chemist,* VCH Publishers, New York, 1986, provides examples of symmetry from nature, art, and other areas in addition to chemistry. It also includes more on symmetry-controlled reactions, the isolobal analogy, and crystalline space groups.

PROBLEMS

4-1 Determine the point groups for:

a. Ethane (staggered conformation)
b. Ethane (eclipsed conformation)
c. Chloroethane (staggered conformation)
d. 1,2-Dichloroethane (staggered conformation)

4-2 Determine the point groups for:

a. Ethylene (H₂C=CH₂ structure shown)

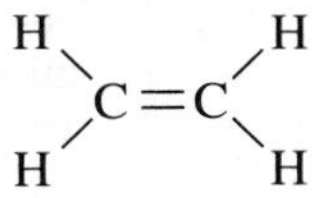

b. Chloroethylene
c. The possible isomers of dichloroethylene

4-3 Determine the point groups for:

a. Acetylene
b. H—C≡C—F
c. H—C≡C—CH_3
d. H—C≡C—CH_2Cl
e. H—C≡C—Ph (Ph = phenyl)

4-4 Determine the point groups for:

a. Naphthalene

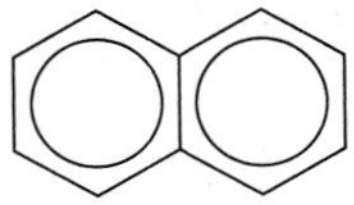

b. 1,8-Dichloronaphthalene

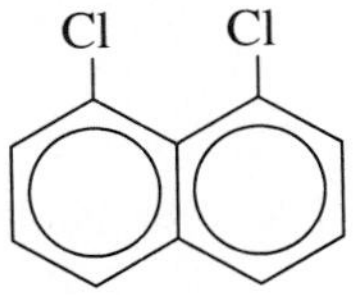

c. 1,5-Dichloronaphthalene

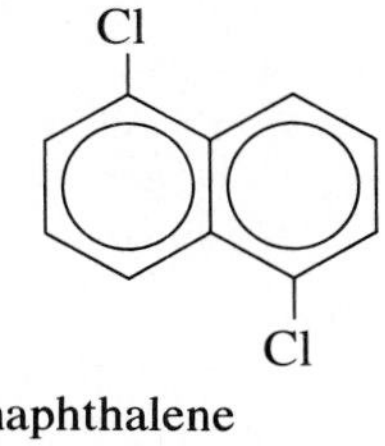

d. 1,2-Dichloronaphthalene

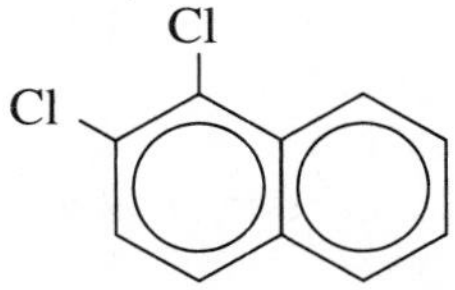

4-5 Determine the point groups for:

a. 1,1′-Dichloroferrocene

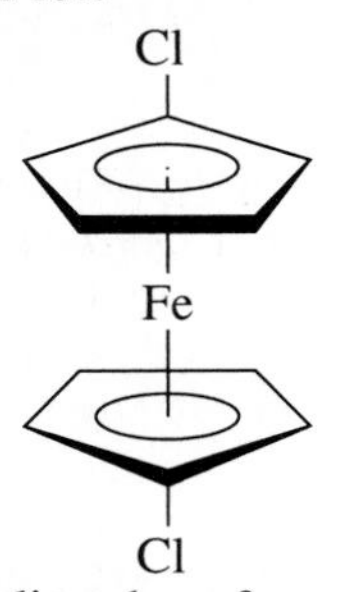

b. Dibenzenechromium (eclipsed conformation)

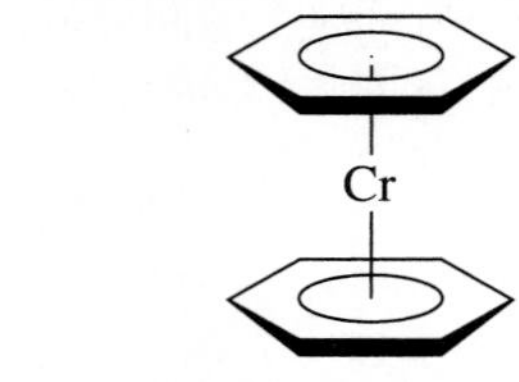

c.

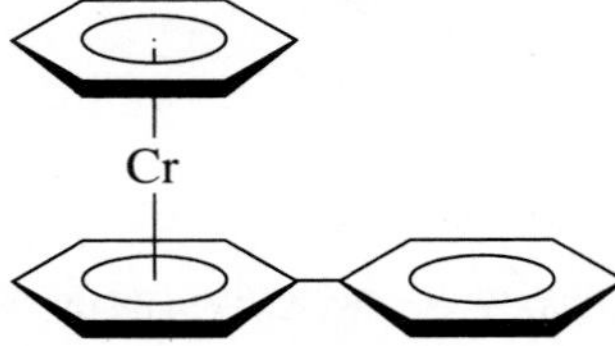

d. H_3O^+

e. O_2F_2 F–O—O–F

f. Formaldehyde

g. S_8 (puckered ring)

h. Borazine (planar)

i. A snowflake

j. A tennis ball (ignoring the label, but including the pattern on the surface)

k. Cyclohexane (chair conformation)

l. Tetrachloroallene $Cl_2C{=}C{=}CCl_2$

m. SO_4^{2-}

n. $Cr(C_2O_4)_3^{3-}$

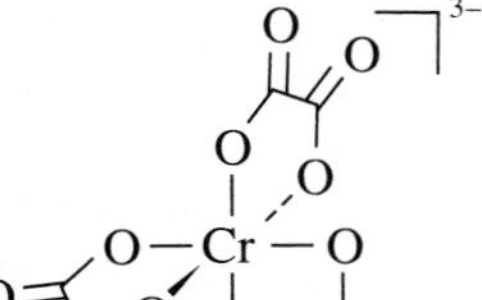

o. Diborane $\begin{matrix} H \\ H \end{matrix}\!\!>\!B\!\!<\!\!\begin{matrix} H \\ H \end{matrix}\!\!>\!B\!\!<\!\begin{matrix} H \\ H \end{matrix}$

p. The possible isomers of tribromobenzene
q. Popeye (remember that he has a perpetual squint in one eye)
r. A 5-pointed star
s. A fork (assuming no decoration)
t. An hour glass
u. A pair of eyeglasses (assuming lenses of equal strength)
v. A sheet of typing paper
w. An Erlenmeyer flask (no label)
x. A screw
y. The number 96
z. Five examples of objects from everyday life; select items from five different point groups.

4-6 Determine the point groups of the molecules in the following problems from Chapter 3:
a. Problem 7
b. Problem 8
c. Problem 9
d. Problem 11
e. Problem 15

4-7 Determine the point groups of the molecules and ions in:
a. Figure 3-8
b. Figure 3-14

4-8 Determine the point groups of the following atomic orbitals:
a. p_x
b. d_{xy}
c. $d_{x^2-y^2}$
d. d_{z^2}

4-9 Show that a cube has the same symmetry elements as an octahedron.

4-10 For *trans*-1,2-dichloroethylene, of C_{2h} symmetry:
a. List all the symmetry operations for this molecule.
b. Obtain a set of transformation matrices to describe the effect of each symmetry operation in the C_{2h} group on a set of coordinates x, y, z for a point. (Your answer should consist of four 3×3 transformation matrices.)
c. Using the terms along the diagonal, obtain as many irreducible representations as possible from the transformation matrices. (You should be able to obtain three irreducible representations in this way, but two will be duplicates.) You may check your results using the C_{2h} character table.
d. Using the C_{2h} character table, verify that the irreducible representations are mutually orthogonal.

4-11 Ethylene is a molecule of D_{2h} symmetry.
a. List all the symmetry operations of ethylene.
b. Obtain a transformation matrix for each symmetry operation to describe the operation of that element on the coordinates of point x, y, z.
c. Using the characters of your transformation matrices, obtain a reducible representation.
d. Using the diagonal elements of your matrices, obtain three of the D_{2h} irreducible representations.
e. Show that your irreducible representations are mutually orthogonal.

4-12 Using the D_{2d} character table:

a. Determine the order of the group.

b. Verify that the E irreducible representation is orthogonal to each of the other irreducible representations.

c. For each of the irreducible representations, verify that the sum of the squares of the characters equals the order of the group.

d. Reduce the following representations to their component irreducible representations:

D_{2d}	E	$2S_4$	C_2	$2C_2'$	$2\sigma_d$
Γ_1	6	0	2	2	2
Γ_2	6	4	6	2	0

4-13 Reduce the following representations to irreducible representations:

C_{3v}	E	$2C_3$	$3\sigma_v$
Γ_1	6	3	2
Γ_2	5	−1	−1

O_h	E	$8C_3$	$6C_2$	$6C_4$	$3C_2$	i	$6S_4$	$8S_6$	$3\sigma h$	$6\sigma d$
Γ_3	6	0	0	2	2	0	0	0	4	2

4-14 For D_{4h} symmetry, show, using sketches, that d_{xy} orbitals have B_{2g} symmetry and that $d_{x^2-y^2}$ orbitals have B_{1g} symmetry. (HINT: You may find it useful to select a molecule that has D_{4h} symmetry as a reference for the operations of the D_{4h} point group.)

4-15 Which items in question 5 are chiral? List three items *not* from this chapter that are chiral.

4-16 For the following molecules, determine the number of IR active C–O stretching vibrations:

a.

b.

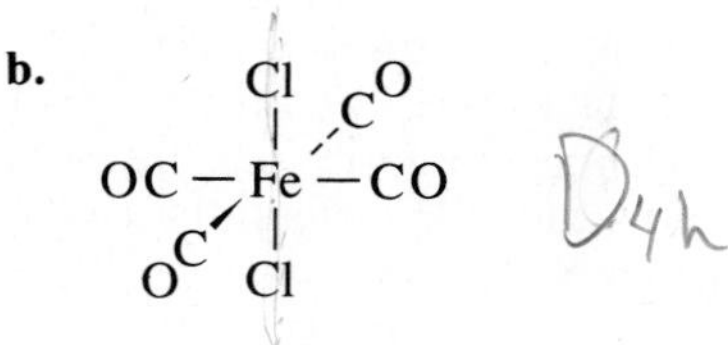

5

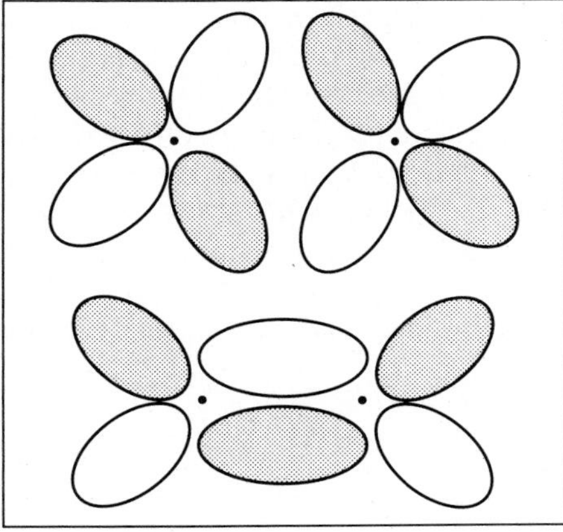

Molecular Orbitals

Molecular orbital theory uses the methods of group theory described in the last chapter to combine atomic orbitals (Chapter 2) into molecular orbitals. This picture of the bonding in molecules complements the simple pictures of bonding introduced in Chapter 3. The symmetry properties of the atomic orbitals and their relative energies determine which combinations of atomic orbitals form useful molecular orbitals. These molecular orbitals are then filled with the available electrons, and the total energy of the electrons in the molecular orbitals is compared with the initial energy of the atoms. If the energy of the electrons in the molecular orbitals is less than that of the atoms, the molecule is stable compared with the atoms; if not, the molecule is unstable and the compound does not form. We will first describe the bonding (or lack of it) of the first ten homonuclear diatomic molecules (H_2 through Ne_2) and then expand the treatment to heteronuclear diatomic molecules and to molecules having more than two atoms.

5-1 FORMATION OF MOLECULAR ORBITALS

As in the case of atomic orbitals, Schrödinger equations can be written for electrons in molecules. Approximate solutions to these molecular Schrödinger equations can be constructed from **linear combinations of the atomic orbitals (LCAO).** These combinations are the sums and differences of the atomic wave functions. For diatomic molecules such as H_2, such wave functions have the form

$$\Psi = c_1\psi_1 + c_2\psi_2$$

where Ψ is the molecular wave function, ψ_1 and ψ_2 are atomic wave functions, and c_1 and c_2 are adjustable coefficients. The coefficients may be equal or unequal, positive or negative, depending on the individual orbitals and their energies. As the distance between two atoms is decreased, their orbitals **overlap,** with significant probability for electrons from both atoms in the overlap region. The result is a **molecular orbital,** the source of bonding between the atoms. Electrons in this molecular orbital occupy the space between the nuclei, and the electrostatic forces between the electrons and the two positive nuclei hold the atoms together. Three conditions are essential for overlap to lead to bonding. First, the symmetry of the orbitals must be such that regions of the same sign overlap. Second, the energies of the atomic orbitals must be similar. When the energies differ by a large amount, the change in energy on formation of the molecular orbitals is small and the net energy change is too small for significant bonding. Third, the overall energy of the electrons in the molecular orbitals must be lower in energy than the overall energy of the electrons in the original atomic orbitals. The methods used in applying molecular orbital theory to these problems and the results obtained are best described by the examples that follow.

5-1-1 BONDING IN H_2

Although it is possible to use the entire set of atomic orbitals in deriving molecular orbitals, whether occupied or not, there is no need to work with any above those usually occupied. For hydrogen, therefore, we need consider only the 1*s* orbitals. For convenience, we label the atoms *a* and *b*, so the atomic orbital wave functions are $\psi(1s_a)$ and $\psi(1s_b)$. We can picture two atoms moving closer to each other until the electron clouds overlap and merge into larger molecular electron clouds. The resulting molecular orbitals are linear combinations of the atomic orbitals, the sum of the two orbitals and the difference between them:

In general terms — *For H_2*

$$\Psi(\sigma) = N[c_a\psi(1s_a) + c_b\psi(1s_b)] = \frac{1}{\sqrt{2}}[\psi(1s_a) + \psi(1s_b)]$$

and

$$\Psi(\sigma^*) = N[c_a\psi(1s_a) - c_b\psi(1s_b)] = \frac{1}{\sqrt{2}}[\psi(1s_a) - \psi(1s_b)]$$

N is the normalizing factor (so $\int \Psi\Psi^*\,d\tau = 1$), and c_a and c_b are adjustable coefficients. In this case, the two atomic orbitals are identical, so $c_a = c_b = 1$ and $N = \frac{1}{\sqrt{2}}$. The results are shown pictorially in Figure 5-1. In this diagram, as in all the orbital diagrams in this book, we indicate the signs of orbital lobes by shading; shaded and unshaded lobes are of opposite sign. (The choice of positive and negative is arbitrary; the important part is how they fit together.) In the contour diagrams on the right in the figure, solid lines and dotted lines show opposite signs of the wave function.

Since the σ molecular orbital is the sum of the two atomic orbitals, $\frac{1}{\sqrt{2}}[\psi(1s_a) + \psi(1s_b)]$, and results in an increased concentration of electrons between the two nuclei where both atomic wave functions contribute, it is a **bonding molecular orbital** and has a lower energy than the starting atomic

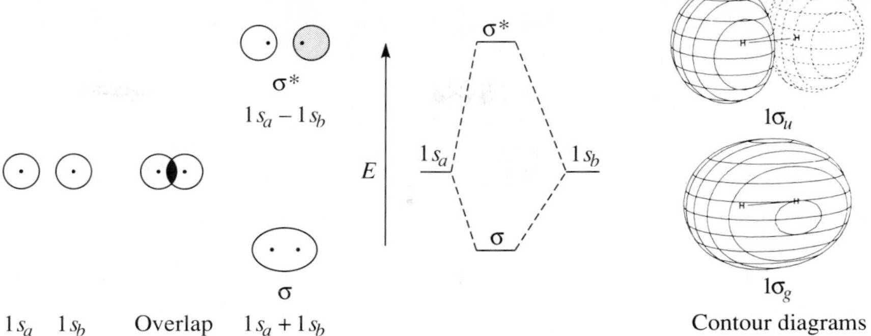

FIGURE 5-1 Molecular Orbitals from Hydrogen 1*s* Orbitals. (The contour diagrams are reproduced with permission from W. L. Jorgensen and L. Salem, *The Organic Chemist's Book of Orbitals,* Academic Press, New York, 1973, p. 61.)

orbitals. The σ^* molecular orbital, which is the difference of the two atomic orbitals, $\frac{1}{\sqrt{2}}[\psi(1s_a) - \psi(1s_b)]$, has a node with zero electron density between the nuclei and a higher energy caused by cancellation of the two wave functions; it is therefore called an **antibonding orbital.** Electrons in bonding orbitals are concentrated between the nuclei and therefore attract the nuclei and hold them together. Electrons in antibonding orbitals avoid the region between the nuclei; the nuclei are thus exposed to each other's positive charge and are mutually repelled.

The σ (sigma) notation indicates orbitals that are symmetric to rotation about the line connecting the nuclei. An asterisk is frequently used to indicate the antibonding orbital, or the orbital of higher energy. Since the bonding, nonbonding, or antibonding nature of a molecular orbital depends on the energies of the atomic orbitals that combine to form it and is sometimes uncertain, this notation will be used only in the simpler cases where the bonding and antibonding character is clear.

This pattern is the usual one for combining two orbitals: two atomic orbitals combine to form two molecular orbitals, one bonding orbital with a lower energy and one antibonding orbital with a higher energy. Regardless of the number of orbitals, the unvarying rule is that the number of resulting molecular orbitals is the same as the number of atomic orbitals initially in the atoms. Another fact to keep in mind is that the increase in energy (destabilization) on formation of antibonding orbitals is greater than the decrease in energy (stabilization) on formation of bonding orbitals. Stable molecules like He_2 are therefore impossible; the total energy of two electrons in the bonding orbital and two electrons in the antibonding orbital is greater than the sum of the energy of two electrons in each of two separate atoms.

5-1-2 MOLECULAR ORBITALS FROM *p* AND *d* ORBITALS

A major factor to be considered for *p* and *d* orbitals is the symmetry of the orbitals, which includes the algebraic sign of the wave functions. When two orbitals overlap and the overlapping regions have the same sign, the sum of the two orbitals has an increased electron probability in the overlap region. When two regions of opposite sign overlap, the combination has a decreased electron probability in the overlap region. Figure 5-1 shows this for the sum and difference of the 1*s* orbitals of H_2; similar effects result from overlapping lobes of *p* and *d* orbitals with their alternating signs. The different possibilities for *p* orbitals and for *s* and *p* orbital combinations are shown in Figure 5-2.

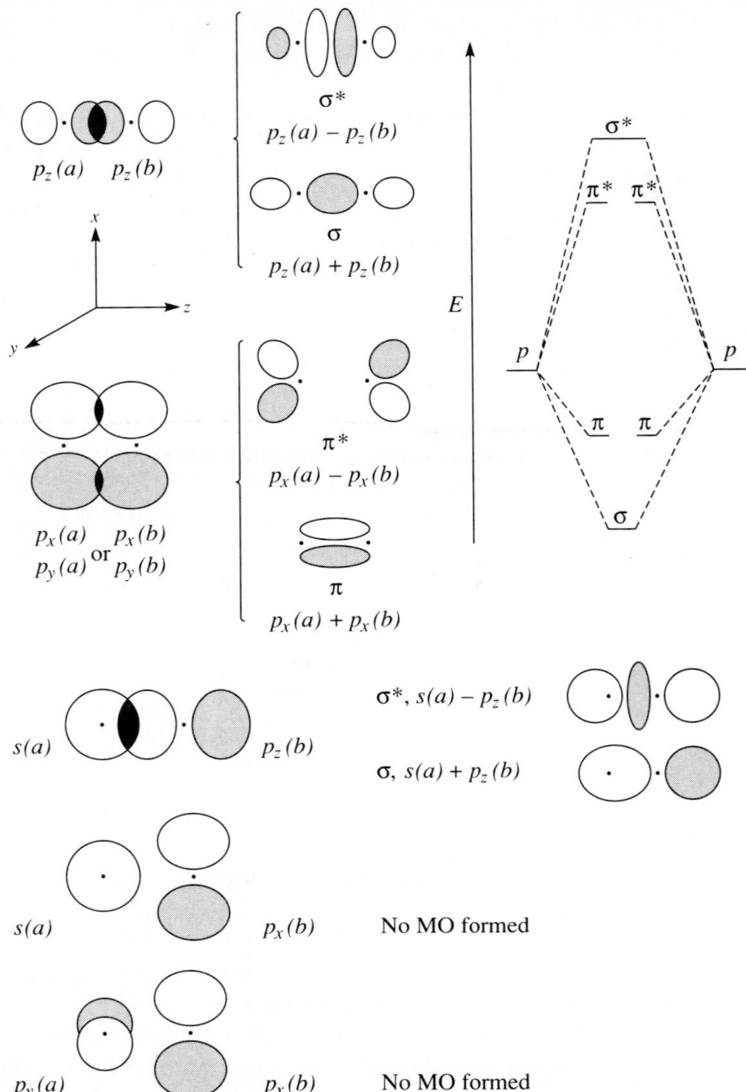

FIGURE 5-2 Some Molecular Orbitals Formed by *p* Orbitals.

When we draw the z axes for the two atoms with positive directions toward each other, the p_z orbitals add to form σ and subtract to form σ^* orbitals, both of which are symmetric to rotation about the z axis, with nodes perpendicular to the line of centers of the nuclei. The same approach applied to the p_x and p_y orbitals lead to π and π^* orbitals as shown. The π (pi) notation indicates a change in sign with C_2 rotation about the bond axis. As in the case of the s orbitals, the overlap of two regions with the same sign leads to an increased concentration of electrons, and the overlap of two regions of opposite sign leads to a node of zero electron density. In this case the nodes of the atomic orbitals lead to similar nodes in the molecular orbitals. In the π^* antibonding case, four lobes result that are similar in appearance to an expanded d orbital.

When orbitals overlap equally with both the same and opposite signs, as in the $s + p_x$ example, the bonding and antibonding effects cancel and no molecular orbital results. Another way to say the same thing is that the symmetry properties of the orbitals do not match and no combination is possible.

In the heavier elements, particularly the transition metals, d orbitals can also be involved in bonding in a similar way. Figure 5-3 shows the possible combinations. Two d_{z^2} orbitals can combine end on for σ bonding, as can $d_{x^2-y^2}$ orbitals. Although the molecular orbitals formed from two $d_{x^2-y^2}$ atomic orbitals have lobes perpendicular to the bond axis, they do not change sign on C_2 rotation about the bond axis; thus the σ notation is retained. The other orbitals, d_{xy}, d_{xz}, and d_{yz}, form either π orbitals or δ (delta) orbitals, depending on their orientation. For example, when the z axes of the two atoms are collinear, the d_{xz} and d_{yz} orbitals form π orbitals. When atomic orbitals meet from two parallel planes and combine side to side, as do the $d_{x^2-y^2}$ and d_{xy} orbitals with collinear z axes, they form δ orbitals. (The δ notation indicates sign changes on C_4 rotation about the bond axis.) Combinations involving

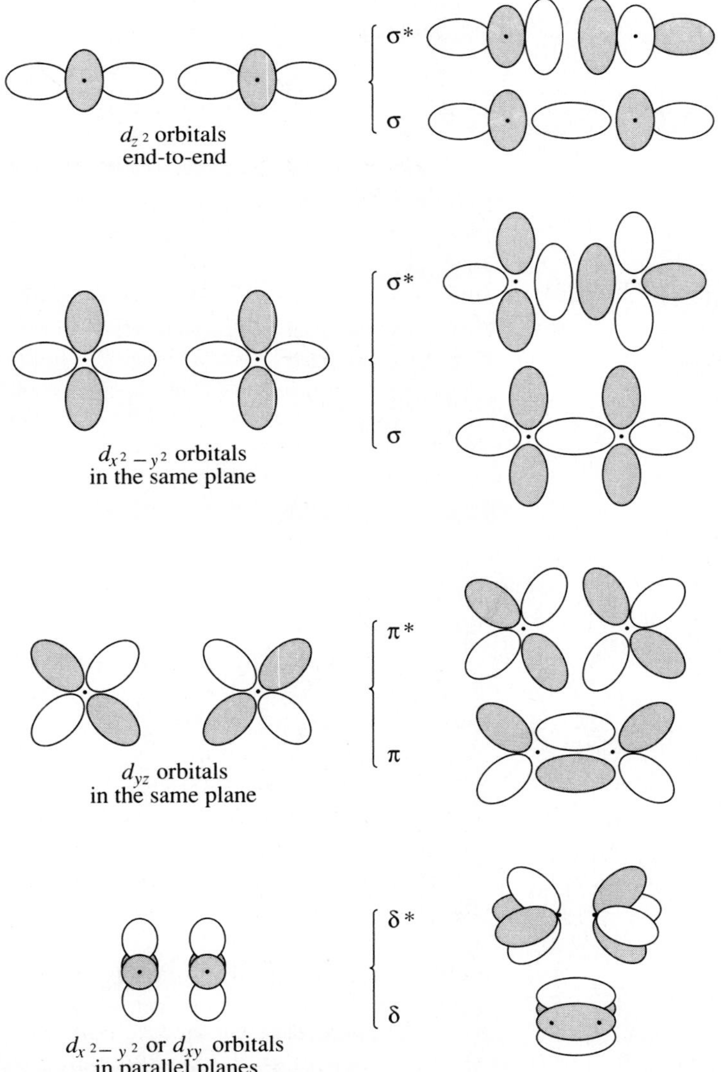

FIGURE 5-3 Some Molecular Orbitals Formed by d Orbitals.

overlapping regions with opposite signs cannot form useful molecular orbitals; for example, p_z and d_{xz} have zero net overlap (one region with overlapping regions of the same sign and another with opposite signs).

EXAMPLE

Sketch the overlap regions of the following combinations of orbitals, all with collinear z axes. Which combinations can form useful molecular orbitals?

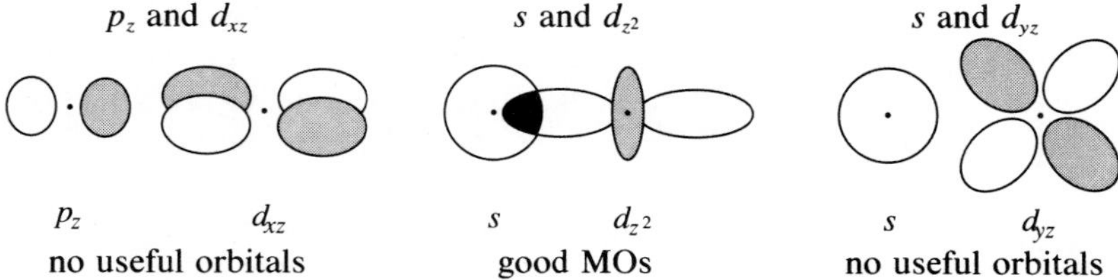

EXERCISE 5-1
Repeat the process for the preceding example, using the following orbital combinations.

p_x and d_{xz} $\quad$ p_z and d_{z^2} $\quad$ s and $d_{x^2-y^2}$

The second major factor that must be considered in forming molecular orbitals is the relative energy of the atomic orbitals. As shown in Figure 5-4, when the two atomic orbitals have the same energy, the resulting interaction is large, and the resulting molecular orbitals have energies well below (bonding) and even farther above (antibonding) that of the original atomic orbitals. When the two atomic orbitals have quite different energies, the interaction is small, and the resulting molecular orbitals have nearly the same energies as the original atomic orbitals. For example, although they have the same symmetry, 1s and 2s orbitals do not combine significantly in diatomic molecules such as N_2 because their energies are too far apart. As we will see, there is some interaction between 2s and 2p, but it is relatively small. The general rule is that the closer the energy match, the stronger the interaction.

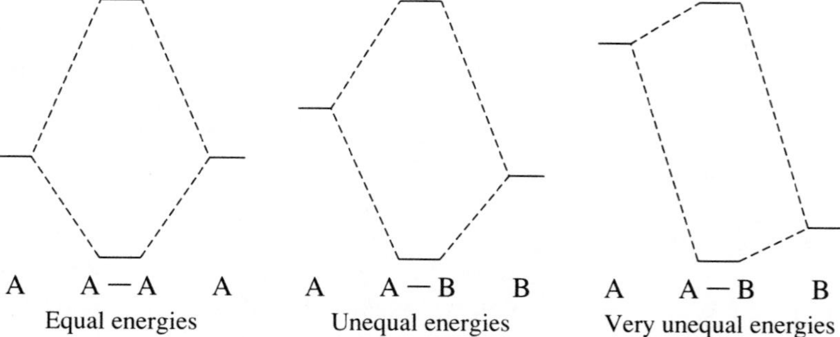

FIGURE 5-4 Energy Match and Molecular Orbital Formation.

5-2 HOMONUCLEAR DIATOMIC MOLECULES

Figure 5-5 shows the full set of molecular orbitals for the homonuclear diatomic molecules of the first ten elements, with N_2 as an example. The diagram shows the order of energy levels for the molecular orbitals, assuming interactions only between atomic orbitals of identical energy. The energies of the molecular

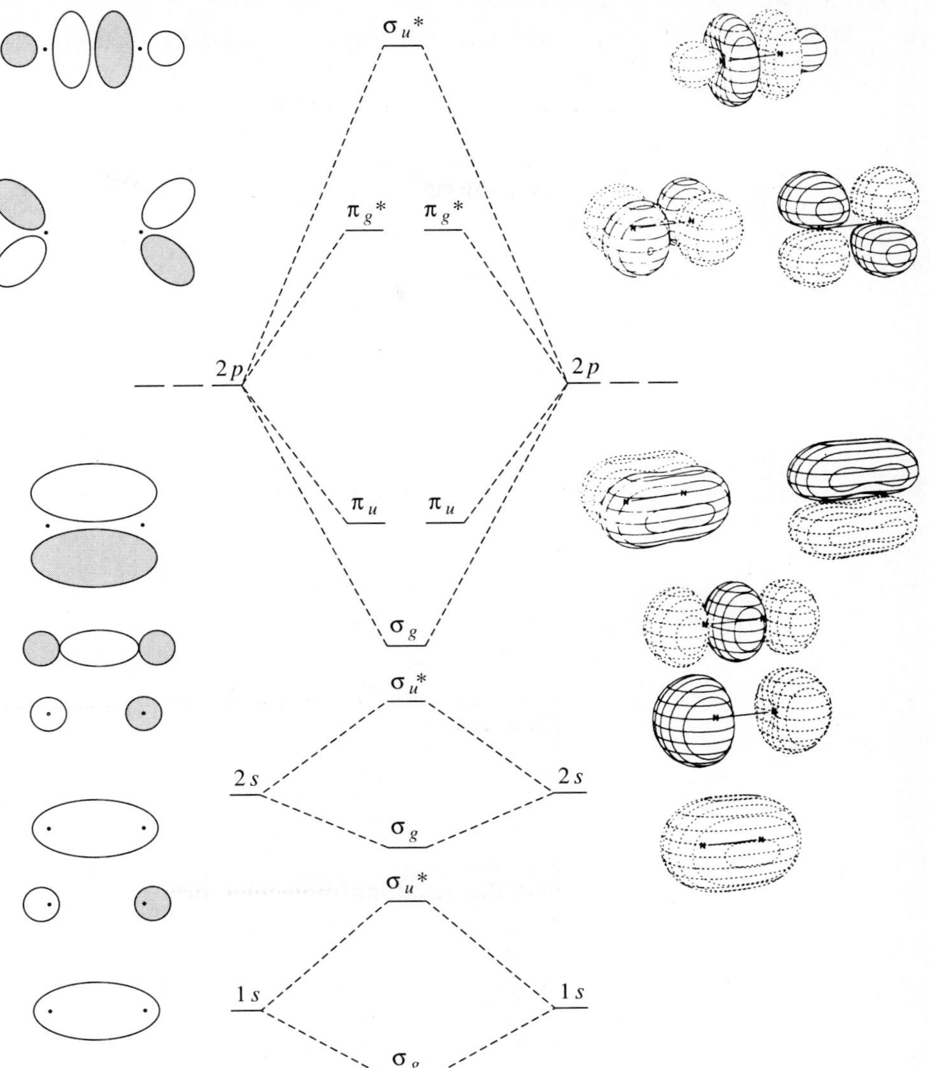

FIGURE 5-5 Molecular Orbitals for the First Ten Elements, with No σ–σ Interaction. (The contour diagrams are for N_2 and are reproduced with permission from W. L. Jorgensen and L. Salem. *The Organic Chemist's Book of Orbitals,* Academic Press, New York, 1973, p. 79.)

orbitals change with increasing atomic number, but the general pattern remains similar (with some subtle changes, as described in several examples below), even for heavier atoms lower in the periodic table. Electrons fill the molecular orbitals according to the same rules governing the filling of atomic orbitals (filling from lowest to highest energy, maximum spin multiplicity consistent with the lowest net energy, and no two electrons with identical quantum numbers).

The overall number of bonding and antibonding electrons determines the number of bonds (bond order):

$$\text{Bond order} = \frac{1}{2}\left[\begin{pmatrix}\text{number of electrons}\\ \text{in bonding orbitals}\end{pmatrix} - \begin{pmatrix}\text{number of electrons}\\ \text{in antibonding orbitals}\end{pmatrix}\right]$$

For example, N_2, with 10 electrons in bonding molecular orbitals and 4 electrons in antibonding orbitals, has a bond order of 3, a triple bond.

Additional symmetry labels have been added to the orbitals in this diagram. Since there are shifts in orbital energies that make the bonding and antibonding labels less useful, we have added g and u subscripts, which are used as described in Chapter 4: g for *gerade,* orbitals symmetric to inversion, and u for *ungerade,* orbitals antisymmetric to inversion (those whose signs change on inversion). The g or u notation describes the symmetry of the orbitals without a judgment on their relative energies. In some later cases, the bonding and antibonding notation will be omitted.

EXAMPLE

Add a g or u label to each of the molecular orbitals in Figure 5-2.

In the upper part of the figure, σ^* and π are antisymmetric to inversion and are labeled u, while σ and π^* are symmetric to inversion, and are labeled g. In the lower part of the figure, the orbitals are inherently unsymmetrical, so no labels are given to them.

EXERCISE 5-2

Add a g or u label to each of the molecular orbitals in Figure 5-3.

When two molecular orbitals of the same symmetry have similar energies, they interact to lower the energy of the lower orbital and raise the energy of the higher. For example, in the homonuclear diatomics, the $\sigma_g(2s)$ and $\sigma_g(2p)$ orbitals both have sigma symmetry, and both are symmetric to inversion; these orbitals interact to lower the energy of the $\sigma_g(2s)$ and to raise the energy of the $\sigma_g(2p)$, as shown in Figure 5-6. Similarly, the $\sigma_u(2s)$ and $\sigma_u(2p)$ orbitals interact to lower the energy of the $\sigma_u(2s)$ and to raise the energy of the $\sigma_u(2p)$. This phenomenon is called **mixing.** Mixing takes into account a factor that has been ignored in Figure 5-5: molecular orbitals with similar energies interact if they have appropriate symmetry. This can also be expressed as the *noncrossing rule,* which states that orbitals of the same symmetry interact so that their energies never cross.[1] When their energies approach each other, they interact to move the higher-energy orbital still higher and the lower-energy orbital lower.

Alternatively, in a more general approach, we can consider the four molecular orbitals (MOs) a result of combining the four atomic orbitals (two $2s$ and two $2p$), because they all have the same symmetry and have similar energies. The method will be demonstrated later in the chapter when the use

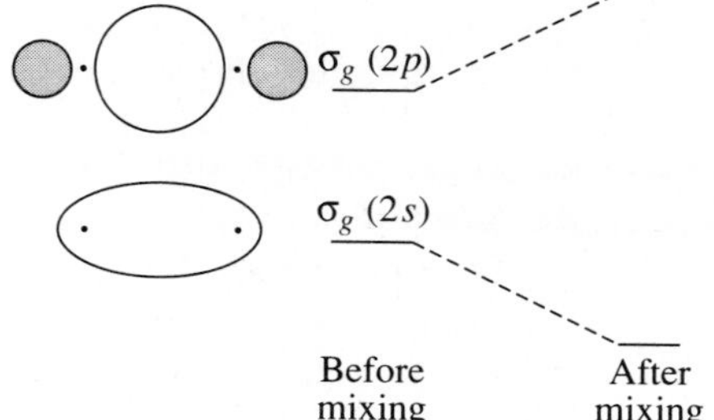

FIGURE 5-6 Interactions between Molecular Orbitals. Mixing molecular orbitals of the same symmetry and the noncrossing rule.

[1] C. J. Ballhausen and H. B. Gray, *Molecular Orbital Theory,* Benjamin, New York, 1965, pp. 36–38.

of group theory to obtain molecular orbitals is explained. The resulting molecular orbitals have the general form

$$\Psi = c_1\psi(2s_a) \pm c_2\psi(2s_b) \pm c_3\psi(2p_a) \pm c_4\psi(2p_b)$$

Since this is a homonuclear molecule, $c_1 = c_2$ and $c_3 = c_4$ in each of the four MOs. The lowest-energy MO has larger values of c_1 and c_2, the highest has larger values of c_3 and c_4, and the two intermediate MOs have intermediate values for all four coefficients.

As we will see, mixing can have an important influence on the energy of molecular orbitals. For example, in the early part of the second period, the σ orbital formed from $2p$ orbitals is higher in energy than the π orbitals formed from the other $2p$ orbitals, an inverted order from that expected without mixing. In some cases, this affects the magnetic properties of molecules. In addition, this mixing changes the bonding–antibonding nature of some of the orbitals. The orbitals with intermediate energies may have either slightly bonding or slightly antibonding character and contribute in minor ways to the bonding, but they will usually be called nonbonding orbitals because of their small contribution and intermediate energy. Each orbital must be considered separately on the basis of the actual energies and electron distributions.

Before proceeding with the examples of homonuclear diatomic molecules shown in Figure 5-7, it is necessary to define two types of magnetic behavior, **paramagnetic** and **diamagnetic.** Paramagnetic compounds are attracted by an external magnetic field, a consequence of one or more unpaired electrons. Diamagnetic compounds, on the other hand, have no unpaired electrons and are repelled slightly by magnetic fields. (An experimental measure of the magnetism of compounds is the **magnetic moment;** this term will be described further in Chapter 8 in the discussion of the magnetic properties of coordination compounds.)

H_2

This is the simplest of the diatomic molecules. As the MO description predicts (Figure 5-1), H_2 has a single bond, a σ bond containing one electron pair. The ionic species H_2^+, having a bond order of one-half, has been detected in low-pressure gas discharge systems. As expected, it is very unstable and has a considerably longer bond distance than H_2.

He_2

The molecular orbital description of He_2 predicts a bond order of zero—in other words, no bond. This is what is found. The noble gas He has no tendency to form diatomic molecules and, with the other noble gases, exists as free atoms.

Li_2

The MO model predicts a single Li—Li bond in Li_2, in agreement with gas phase observations of the molecule.

Be_2

Be_2 would have the same number of antibonding as bonding electrons and consequently a bond order of zero. Hence, like He_2, Be_2 is not a stable chemical species.

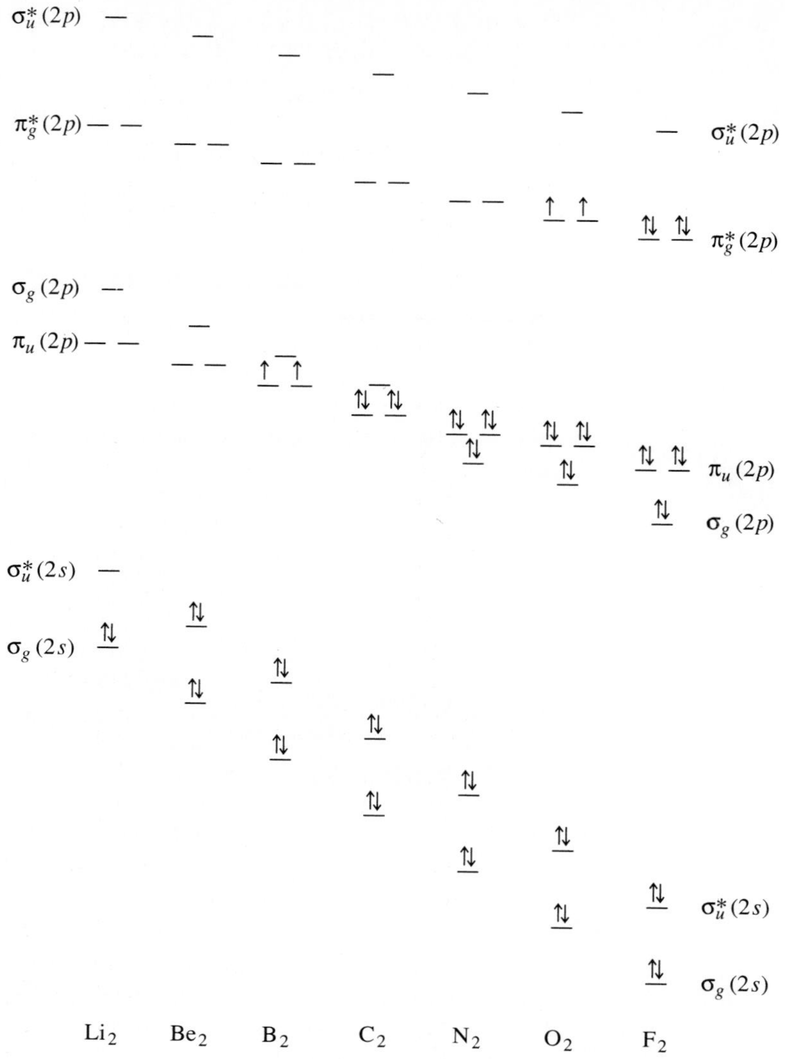

$\sigma_u^*(2p)$

$\pi_g^*(2p)$

$\sigma_g(2p)$

$\pi_u(2p)$

$\sigma_u^*(2s)$

$\sigma_g(2s)$

$\sigma_u^*(2p)$

$\pi_g^*(2p)$

$\pi_u(2p)$

$\sigma_g(2p)$

$\sigma_u^*(2s)$

$\sigma_g(2s)$

Li_2 Be_2 B_2 C_2 N_2 O_2 F_2

FIGURE 5-7 Energy Levels of Homonuclear Diatomic Molecules.

B_2

Here is an example where the MO model has a distinct advantage over the Lewis dot picture. B_2 is known to be paramagnetic. Its magnetic behavior can be explained if its two highest-energy electrons occupy separate π orbitals, as shown. The Lewis dot model cannot account for paramagnetic behavior in this molecule.

B_2 is a good example of the phenomenon of mixing of orbitals. In the absence of mixing, the $\sigma_g(2p)$ orbital is expected to be lower in energy than the $\pi_u(2p)$ orbitals. However, mixing of the $\sigma_g(2s)$ orbital with the $\sigma_g(2p)$ orbital (see Figure 5-6) increases the energy of the $\sigma_g(2p)$ orbital to a higher level than the π orbitals, giving the order of energies shown in Figure 5-7.

C_2

The simple MO picture of C_2 predicts a doubly bonded molecule. The bond dissociation energies of B_2, C_2, and N_2 increase steadily, indicating single, double, and triple bonds with increasing atomic number. While C_2 is not a commonly encountered chemical species (carbon is more stable in two other forms, diamond and graphite, both shown in Chapter 7), the acetylide ion, C_2^{2-}, is well known, particularly in compounds with alkali metals, alkaline earths, and lanthanides. C_2^{2-}, according to the molecular orbital model, should have a bond order of 3. This is supported by a comparison of the C—C distances in acetylene and calcium carbide (acetylide):[2,3]

	C—C distance
H—C≡C—H	120.5 pm
CaC_2	119.1 pm

N_2

N_2 has a triple bond according to the molecular orbital model. This is in agreement with its very short N—N distance (109.8 pm) and extremely high bond dissociation energy (942 kJ/mol). With increasing nuclear charge, all the atomic orbitals move to lower energies, and the separation between the 2*s* and 2*p* levels increases (Section 5-3). Therefore, the $\sigma_g(2s)$ and $\sigma_g(2p)$ levels of N_2 interact (mix) less than the B_2 and C_2 levels, and the σ_g and π_u are very close in energy. The order of energies of these orbitals has been a matter of controversy and will be discussed in more detail in Section 5-2-2 on photoelectron spectroscopy.

O_2

O_2 is paramagnetic. This property, as for B_2, cannot be explained by the traditional Lewis dot structure (:Ö=Ö:), but is evident from the MO picture, which assigns two electrons to the **degenerate** (having the same energy) π_g^* orbitals. The paramagnetism can be demonstrated by pouring liquid O_2 between the poles of a strong magnet; some of the O_2 will be held between the pole faces until it evaporates. Several ionic forms of diatomic oxygen are known, including O_2^+, O_2^-, and O_2^{2-}. The internuclear O—O distance can be conveniently correlated with the bond order predicted by the molecular orbital model, as shown in the following table.

	Bond order	*Internuclear distance (pm)*	*Reference*
O_2^+ (dioxygenyl)	2.5	112.3	4
O_2 (dioxygen)	2.0	121.07	5
O_2^- (superoxide)	1.5	128–132	6
O_2^{2-} (peroxide)	1.0	150–151	6

[2] J. Overend and H. W. Thompson, *Proc. Roy. Soc.*, **1956,** *A234,* 306.

[3] M. Atoji, *J. Chem. Phys.*, **1961,** *35,* 1950.

[4] G. Herzberg, *Molecular Spectra and Molecular Structure I: The Spectra of Diatomic Molecules*, 2nd ed., Van Nostrand Reinhold, New York, 1950, p. 366.

[5] S. L. Miller and C. H. Townes, *Phys. Rev.*, **1953,** *90,* 537.

[6] N.-G. Vannerberg, *Prog. Inorg. Chem.*, **1963,** *4,* 125.

The effect of mixing is not sufficient in O_2 to push the $\sigma_g(2p)$ orbital to higher energy than the π orbitals. The order of molecular orbitals shown is consistent with the photoelectron spectrum discussed below.

F_2

The MO picture of F_2 shows a diamagnetic molecule having a single fluorine–fluorine bond, in agreement with experimental data on this very reactive molecule.

The net bond order in N_2, O_2, and F_2 is the same whether or not mixing is taken into account, but the order of the filled orbitals is different. The switching of the order of the $\sigma_g(2p)$ and $\pi_u(2p)$ orbitals can take place because these orbitals are so close in energy; minor changes in either can switch the order. The energy difference between the $2s$ and $2p$ orbitals of the atoms increases with increasing nuclear charge (see the discussion of valence orbital energies later in this chapter). Because the difference becomes greater, the interaction decreases and the "normal" order of molecular orbitals returns. The higher σ orbital is seen again in CO, described later in this chapter.

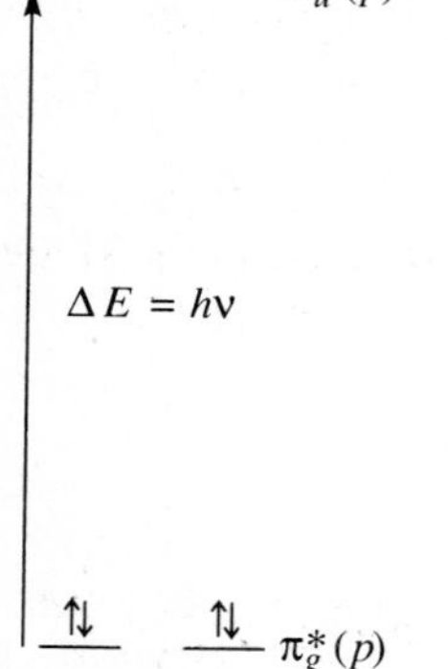

FIGURE 5-8 Halogen Energy Levels and Spectra.

5-2-1 COLORS OF HALOGEN MOLECULES, X_2

An interesting characteristic of the two heavier halogens, Br_2 and I_2, is their vivid colors. These colors arise because the energy difference between the π_g^* and $\sigma_u^*(p)$ orbitals corresponds to the energy of light in the visible spectrum. Thus, absorption of visible light can excite an electron from the π_g^* to the σ_u^* (Figure 5-8) and result in the observed color, deep red for Br_2 and purple for I_2. The energy levels are farther apart in F_2 and Cl_2. The absorption band for F_2 is in the ultraviolet region, resulting in a colorless gas, and Cl_2 absorbs light at the edge of the ultraviolet-visible boundary and, as a result, has a yellow color. Interactions of halogen molecules with donor solvents can have dramatic effects on the color; these will be discussed in Chapter 6.

5-2-2 PHOTOELECTRON SPECTROSCOPY

Photoelectron spectroscopy[7] can be used to help determine the order of molecular orbitals. In this technique, beams of UV light or X-rays dislodge electrons from molecules:

$$O_2 + h\nu \text{ (photons)} \longrightarrow O_2^+ + e^-$$

The kinetic energy of the expelled electrons can be measured; the difference between the energy of the incident photons and this kinetic energy is the ionization energy (binding energy) of the electron:

$$\text{Ionization energy} = h\nu \text{ (photons)} - \text{kinetic energy of expelled electron}$$

Ultraviolet light removes outer electrons, usually from gases; X-rays are more energetic and remove inner electrons as well from any physical state. If

[7] D. N. Hendrickson, in R. S. Drago, *Physical Methods in Chemistry,* Saunders, Philadelphia, 1977, pp. 566–84.

the energy levels of the ionized molecule are assumed to be essentially the same as those of the uncharged molecule, the observed energies can be directly correlated with the molecular orbital energies. This does not hold true in all cases, however. In some molecules, the orbital energies shift when an electron is removed and the order of the orbitals in the positive ion is different from that in the neutral molecule. Since the total energy measurement is from the initial ground state to the final state, any energy of readjustment may appear in the spectrum and must be considered in comparing experimental data with theoretical predictions. In the nitrogen molecule, simple calculations show the calculated energy of the σ_g orbital higher than that of π_u, while more detailed calculations show the π_u higher, with the two orbitals very close in energy.[8] The photoelectron spectrum shows the order $\pi_u(2p)$, $\sigma_g(2p)$. This is explained by the ionization reaction

$$N_2(\sigma_g^2, \pi_u^4) + h\nu \longrightarrow N_2^+(\pi_u^4, \sigma_g^1) + e^-$$

The order of orbitals shown by the photoelectron spectrum is then that of the N_2^+ ion.

The photoelectron spectra of O_2 and of CO (Figure 5-14) show the expected order; apparently the difference in energy between the two molecular orbitals for these molecules is great enough that ionization does not change the order.

Figures 5-9 and 5-10 show photoelectron spectra for N_2 and O_2 and the orbital order of the ions. The lower-energy peaks (at the top in the figures) are for the higher-energy orbitals (less energy required to remove electrons).

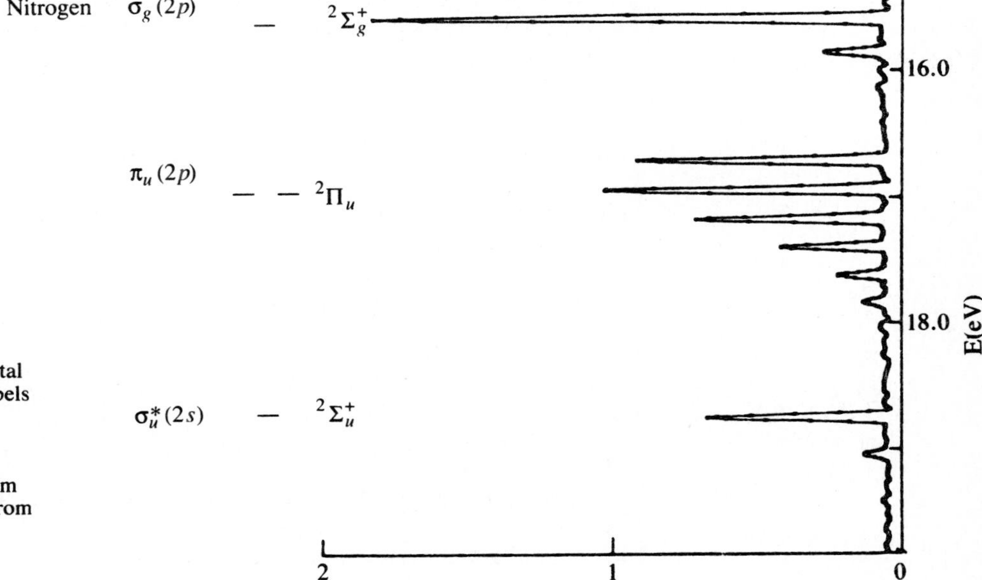

FIGURE 5-9 Photoelectron Spectrum and Molecular Orbital Energy Levels of N_2^+. The labels shown on the spectra are spectroscopic term symbols, similar to those for atomic states. (Photoelectron spectrum reproduced with permission from J. L. Gardner and J. A. R. Samson, *J. Chem. Phys.*, **1975,** *62,* 1447.)

[8] W. C. Emler and A. D. McLean, *J. Chem. Phys.*, **1980,** *73,* 2297.

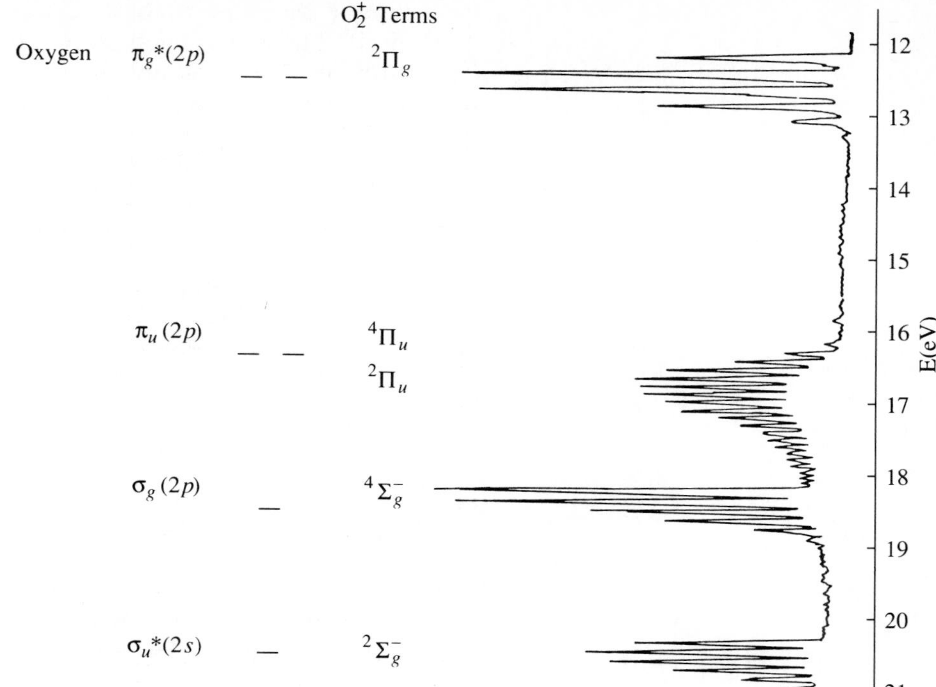

FIGURE 5-10 Photoelectron Spectrum and Molecular Orbital Energy Levels of O_2^+. (Photoelectron spectrum reproduced with permission from J. H. Eland, *Photoelectron Spectroscopy,* Butterworths, London, 1974, p. 11.)

5-2-3 CORRELATION DIAGRAMS

The noncrossing rule and mixing of orbitals of the same symmetry are seen in many other molecules. A more general diagram, called a **correlation diagram,**[9] for this phenomenon is shown in Figure 5-11. This diagram shows the calculated effect of moving two atoms together, from a large interatomic distance on the right, with no interatomic interaction, to zero interatomic distance on the left, where the two nuclei become a single nucleus. The simplest example has two hydrogen atoms on the right and a helium atom on the left. Naturally, such merging of two atoms into one never happens outside the realm of high-energy physics, but it is helpful to consider the orbital changes as if it could.

On the right are the usual atomic orbitals, for atoms separated by a large distance. At smaller distances, they interact to form the molecular orbitals described above, retaining the symmetry required by the atomic orbitals. The s and p orbitals form MOs in order of increasing energy: σ_g and σ_u^* from $1s$, σ_g and σ_u^* from $2s$, and σ_g, π_u, π_g^*, and σ_u^* from $2p$.

On the left are the atomic orbitals for the united atom with twice the nuclear charge. The symmetries of the atomic orbitals of the united atom and of the molecular orbitals are then used to connect the two sides of the diagram. For example, the σ_u^* molecular orbitals have the same symmetry as p_z atomic orbitals (where z is the axis through both nuclei), so they are connected by a line. Similarly, π_u molecular orbitals have the same symmetry as p_x or p_y

[9] R. McWeeny, *Coulson's Valence,* 3rd ed., Oxford University Press, New York, 1979, pp. 97–103.

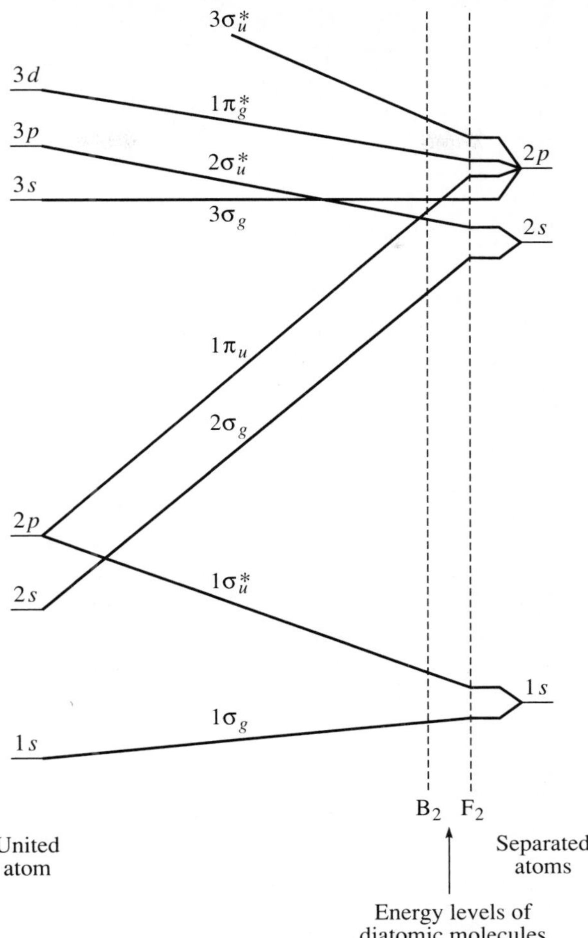

FIGURE 5-11 Correlation Diagram for Homonuclear Diatomic MOs.

atomic orbitals, and π_g^* molecular orbitals have the same symmetry as d_{xz} or d_{yz} atomic orbitals, so they are connected. Figures 5-1 and 5-2 show the shapes of many of these orbitals.

The actual energies of molecular orbitals for diatomic molecules are intermediate between the extremes of this diagram, approximately in the region set off by the vertical lines. Toward the right within this region, closer to the separated atoms, the energy sequence is the "normal" one of N_2 through F_2; farther to the left, the order of molecular orbitals is that of B_2 and C_2, with $\sigma_g(2p)$ above $\pi_u(2p)$.

One triumph of molecular orbital theory was its prediction of two unpaired electrons for O_2. It had long been known that ordinary oxygen is paramagnetic; but the earlier bonding theories required use of a special "three-electron bond"[10] to explain this phenomenon. On the other hand, the order of filling of the molecular orbitals in accordance with Hund's rule results in a paramagnetic molecule with no special treatment needed. In the other cases described

[10] L. Pauling, *The Nature of the Chemical Bond,* 3rd ed. Cornell University Press, Ithaca, N.Y., 1960, pp. 340–54.

above, the experimental facts (paramagnetic B_2, diamagnetic C_2) require a shift of orbital energies, but do not require any different kind of orbitals or bonding. The simplicity of these explanations and the natural way the effects arise lend support to the molecular orbital method.

5-3 HETERONUCLEAR DIATOMIC MOLECULES

5-3-1 POLAR BONDS

Heteronuclear diatomic molecules follow the same general bonding pattern as the homonuclear molecules above, but a greater nuclear charge on one of the atoms lowers its atomic energy levels and shifts the resulting molecular levels. In dealing with heteronuclear molecules, it is necessary to have a way of estimating the energies of the atomic orbitals that may interact. For this purpose, the valence orbital potential energies, given in Table 5-1 and Figure 5-12, are useful. These potential energies are negative, since they represent attraction between valence electrons and atomic nuclei. The values are the average energies for all electrons in the same level (weighted averages of all the terms possible, such as those determined for the p^2 case in Chapter 2). For this reason, the values do not show the variations of the ionization energies, but increase steadily in magnitude from left to right within a period, as the

TABLE 5-1
Valence orbital potential energies

Atomic Number	Element	Orbital potential energy (eV) 1s	2s	2p	3s	3p	4s	4p
1	H	−13.6						
2	He	−24.5						
3	Li		−5.5					
4	Be		−9.3					
5	B		−14.0	−8.3				
6	C		−19.5	−10.7				
7	N		−25.5	−13.1				
8	O		−32.4	−15.9				
9	F		−46.4	−18.7				
10	Ne		−48.5	−21.6				
11	Na				−5.2			
12	Mg				−7.7			
13	Al				−11.3	−6.0		
14	Si				−15.0	−7.8		
15	P				−18.7	−10.0		
16	S				−20.7	−12.0		
17	Cl				−25.3	−13.7		
18	Ar				−29.3	−15.9		
19	K						−4.3	
20	Ca						−6.1	
30	Zn						−9.4	
31	Ga						−12.6	−6.0
32	Ge						−15.6	−7.6
33	As						−17.6	−9.1
34	Se						−20.8	−11.0
35	Br						−24.1	−12.5
36	Kr						−27.5	−14.3

SOURCE: J. G. Verkade, *A Pictorial Approach to Molecular Bonding*, Springer-Verlag, New York, 1988, p. 69.
NOTE: All energies are negative, representing average attractive potentials between the electrons and the nucleus for all terms of the specified orbitals.

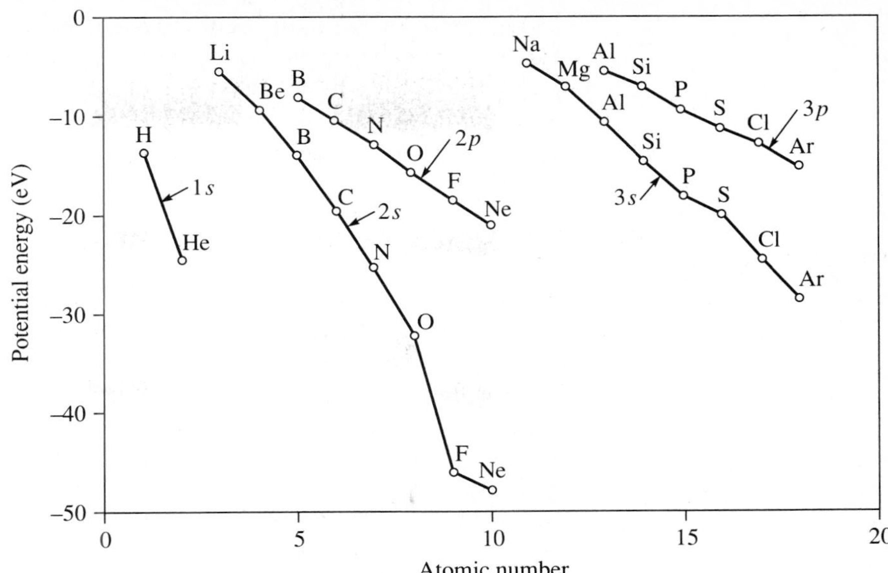

FIGURE 5-12 Valence Orbital Potential Energies.

increasing nuclear charge attracts all the electrons more strongly. The inner orbital values are also those for neutral atoms; the ionization energies are for positive ions and are therefore much larger.

In homonuclear diatomic molecules, the LCAO method results in molecular orbitals that have equal contributions from each of the atoms, so the coefficients for the two atomic orbitals in the molecular orbital are identical. When the atomic orbitals have different energies, the coefficients for orbitals from the two atoms are different. Overall, each atomic orbital contributes the same to the sum of the molecular orbitals formed, but each molecular orbital has a larger contribution from one atomic orbital than the other. As the atomic orbitals on one side of the molecule (atom B in Figure 5-13) drop in energy relative to the orbitals of the other atom, the magnitude of the interaction decreases, the molecular orbitals have coefficients in the LCAO equation that favor the atomic orbital of similar energy, and the resulting molecular orbitals have energies close to those of the contributing atomic levels. Thus, the bonding orbital has more contribution from the lower-energy atomic orbital (atom B), and the antibonding orbital has more contribution from the higher (atom A). In the simplest case, the bonding electrons thus concentrate on the atom with the lower levels, and the antibonding electrons concentrate on the atom with the higher levels, as in Figure 5-13.

An example is carbon monoxide, with $C_{\infty v}$ symmetry (Figure 5-14). Oxygen has lower energy levels than carbon; as a result, the s and p_z orbitals do not match as neatly as in the homonuclear case. These orbitals have the same A_1 symmetry in the $C_{\infty v}$ group (both are symmetric to rotation about the z axis and reflection through any of the infinite number of planes including the z axis) and can be mixed together to form σ orbitals, one bonding and antibonding pair at relatively low energy from the $2s$ orbitals and the other, higher-energy pair from the $2p$. Interaction of the two σ_g levels and the two σ_u levels, like that seen in the homonuclear case, causes a larger split in energy between them, and the 5σ is higher than the π levels. The p_x and p_y orbitals

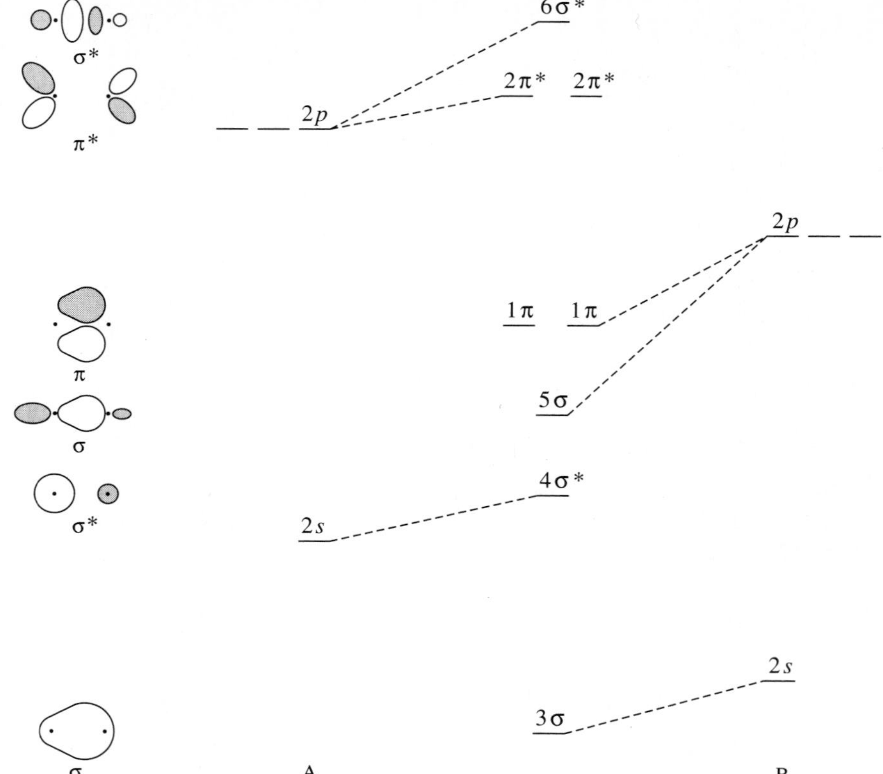

FIGURE 5-13 Heteronuclear Diatomic Molecular Orbitals. Molecular orbitals 1σ and $2\sigma^*$ are from the $1s$ orbitals and are not shown. The g and u labels are not used for the molecular orbitals because the molecules inherently lack a center of inversion, but the same general shapes and signs can be seen.

also form four molecular π orbitals, two bonding and two antibonding. When the electrons are filled in, there are five bonding pairs and two antibonding pairs for a net of three bonds.

EXERCISE 5-3

Use the valence orbital potential energies to explain the bonding in the HF molecule.

The molecular orbitals that will be of greatest interest for reactions between molecules are the **highest occupied molecular orbital (HOMO)** and the **lowest unoccupied molecular orbital (LUMO),** collectively known as **frontier orbitals** (at the occupied–unoccupied frontier). In CO, the HOMO has σ symmetry and is mostly composed of carbon s and p_z orbitals; it is therefore concentrated on the carbon. The LUMO have π symmetry and are mostly composed of the carbon p_x and p_y orbitals. These frontier orbitals are the orbitals that can either contribute electrons (HOMO) or accept electrons (LUMO) in reactions.

5-3-2 IONIC COMPOUNDS AND MOLECULAR ORBITALS

Ionic compounds can be considered the limiting form of polarity in heteronuclear diatomic molecules. As the polarity difference between the two

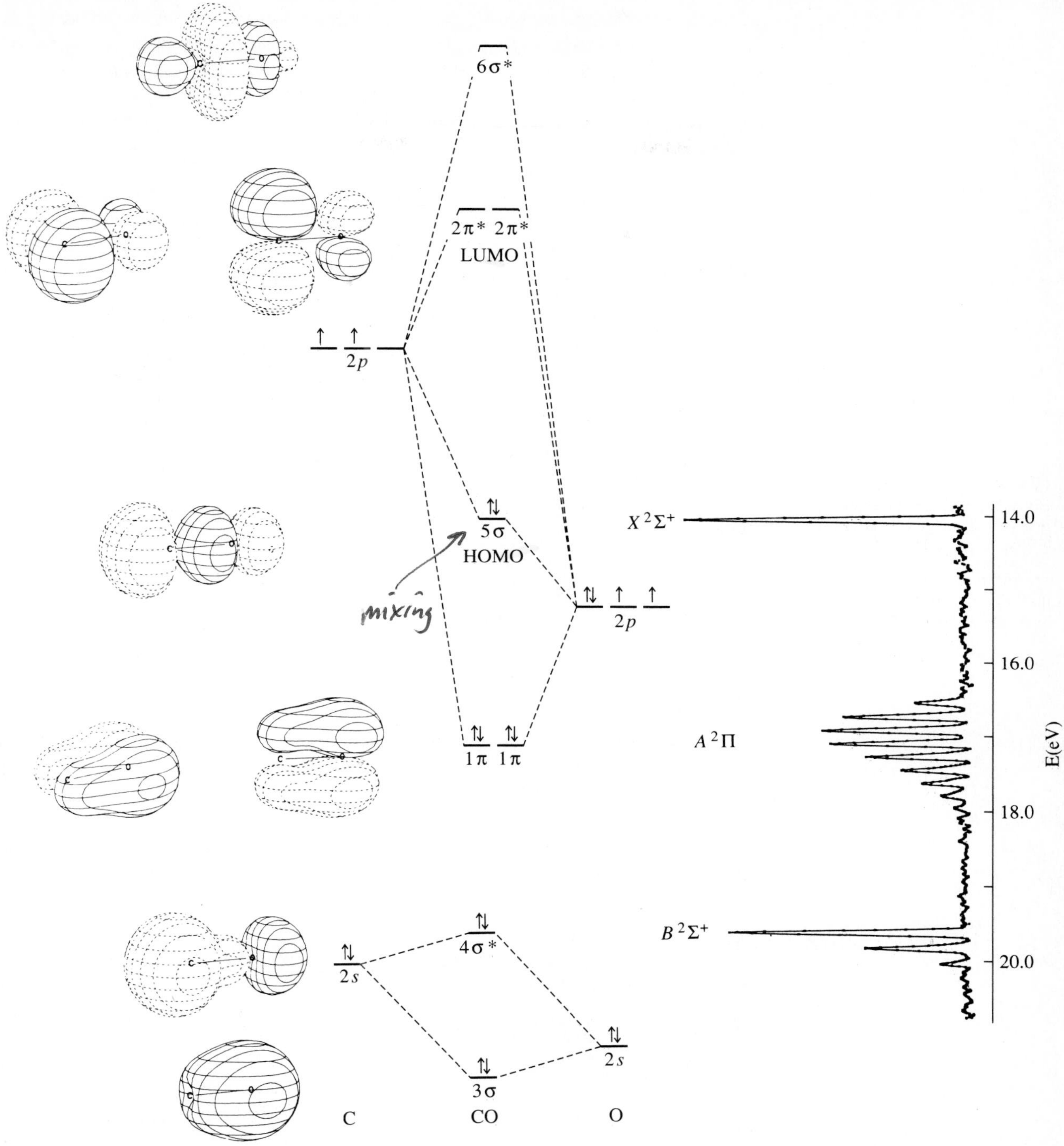

FIGURE 5-14 Molecular Orbitals and Photoelectron Spectrum of CO. The 1σ and 2σ orbitals from the 1s interactions are not shown. (Contour diagrams reproduced with permission from W. L. Jorgensen and L. Salem, *The Organic Chemist's Book of Orbitals,* Academic Press, New York, 1973, p. 78. Photoelectron spectrum reproduced with permission from J. L. Gardner and J. A. R. Samson. *J. Chem. Phys.,* **1975,** *62,* 1447.)

atoms increases, the difference in energy of the orbitals also increases, and the concentration of electrons shifts to favor the more electronegative atom. In the limit, the electron is completely on the negative ion and the positive ion remains with a high-energy vacant orbital. When two elements with a large difference in their electronegativities (such as Li and F) combine, the result is an ionic compound. However, in molecular terms, we can consider an ion pair as if it were a covalent compound. In Figure 5-15, the atomic orbitals and an approximate indication of molecular orbitals for such a diatomic molecule are given. The significant change on formation of the bond from the two isolated atoms is the transfer of an electron from the Li $2s$ orbital to the F $2p$ orbital, and a decrease in the energy of the $2p$ orbital caused by interaction with the Li $2s$ orbital.

In a more accurate picture, the ions are held together in a three dimensional lattice, primarily by electrostatic attraction. Although there is still a small amount of covalent character to the bonding, there are no directional bonds, and each Li^+ ion is surrounded by six F^- ions, each of which in turn is surrounded by six Li^+ ions. The crystal energy cannot be represented easily on any molecular orbital diagram, although it is the major factor in the formation of the compound.

Formation of the ions can be described as a sequence of elementary steps, beginning with solid Li and gaseous F_2:

$$
\begin{array}{rcllr}
Li(s) & \longrightarrow & Li(g) & 161 \text{ kJ/mol} & \text{Vaporization} \\
Li(g) & \longrightarrow & Li^+(g) + e^- & 531 \text{ kJ/mol} & \text{Ionization} \\
\frac{1}{2}F_2(g) & \longrightarrow & F(g) & 79 \text{ kJ/mol} & \text{Dissociation} \\
F(g) + e^- & \longrightarrow & F^-(g) & -328 \text{ kJ/mol} & \text{Ionization} \\
\hline
Li(s) + \frac{1}{2}F_2(g) & \longrightarrow & Li^+(g) + F^-(g) & 443 \text{ kJ/mol} &
\end{array}
$$

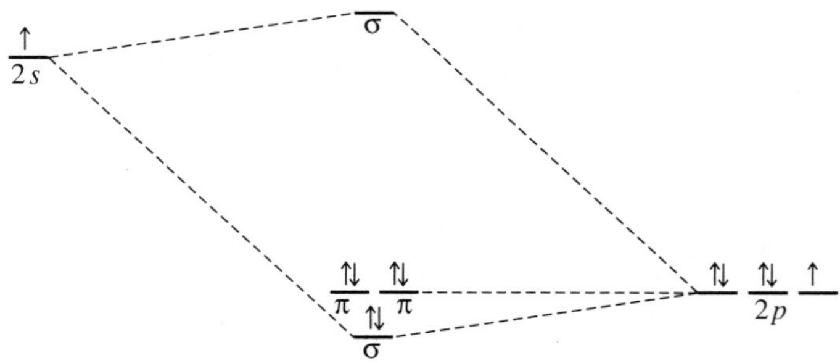

FIGURE 5-15 Approximate LiF Molecular Orbitals.

If this were the final result, Li(s) and F_2 would not react. However, the large attraction between the ions results in release of 709 kJ/mol on formation of a single Li^+F^- ion pair, and 1239 kJ/mol on formation of ordinary crystals:

$$Li^+(g) + F^-(g) \longrightarrow LiF(g) \qquad -709 \text{ kJ/mol}$$

$$Li^+(g) + F^-(g) \longrightarrow LiF(s) \qquad -1239 \text{ kJ/mol}$$

This lattice energy is enough to overcome all the endothermic processes and makes formation of LiF from the elements a very favorable reaction.

5-4 A PICTORIAL APPROACH TO MOLECULAR ORBITALS

The method described above for diatomic molecules can be extended to obtain molecular orbitals for molecules consisting of three or more atoms. We will present two approaches to describing the molecular orbitals of polyatomic species. In the first, we will describe in a pictorial fashion how atomic orbitals on a central atom can interact with orbitals on surrounding atoms; the result will be qualitative, pictorial descriptions of molecular orbitals. In the second approach, we will show how group theory, as described in Chapter 4, can be used to obtain molecular orbitals, in principle, for any molecule.

FHF^-

One way of viewing interactions between atomic orbitals in a polyatomic species is to consider separately the orbitals on a central atom and the orbitals on outer atoms. The orbitals on outer atoms will be labeled **group orbitals,** also sometimes called **symmetry-adapted linear combinations,** or **SALC**s. For example, in the linear FHF^- ion the group orbitals of interest will be the $2s$ and $2p$ orbitals of the fluorine atoms, considered as pairs (in general we will need to consider only the valence atomic orbitals), as in Figure 5-16.

In each case ($2s$, $2p_x$, $2p_y$, and $2p_z$), the atomic orbitals may add and subtract in both a bonding (set 1 in the figure) and antibonding (set 2) sense. For example, the $2s$ orbitals on the fluorine atoms give two group orbitals: ○ · ○ and ○ · ●. The designation "group orbital" does not imply direct bonding between the two fluorine atoms; group orbitals should be viewed merely as collections of similar orbitals. The number of group orbitals is the

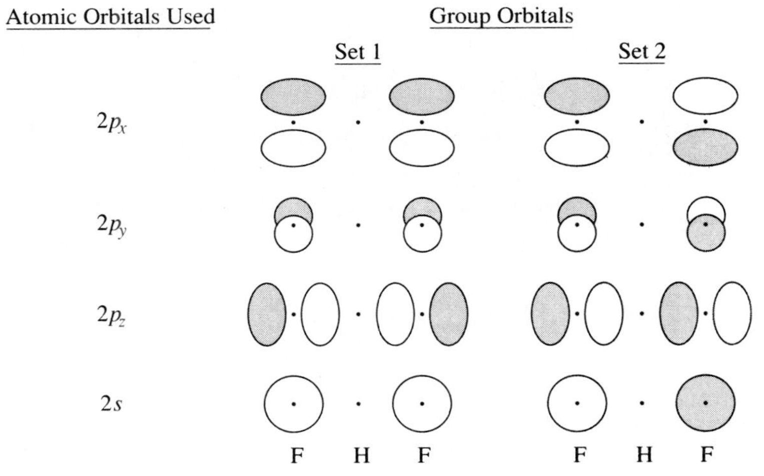

Atomic Orbitals Used | Group Orbitals: Set 1, Set 2

$2p_x$

$2p_y$

$2p_z$

$2s$

F H F   F H F

FIGURE 5-16 Group Orbitals in FHF^-.

same as the number of atomic orbitals combined to form them. Our model will then consider how these group orbitals may interact with atomic orbitals on the central atom, each group orbital being treated in the same manner as an atomic orbital.

In FHF^-, an example of very strong hydrogen bonding,[11] the only orbital available on the central hydrogen atom is the $1s$. Of the eight group orbitals shown, only two are of suitable symmetry for interaction with this $1s$ orbital. These are the group orbitals in set 1 derived from the $2s$ and $2p_z$ orbitals of the fluorines. The possible interactions of these group orbitals with the $1s$ orbital of hydrogen are shown in Figure 5-17.

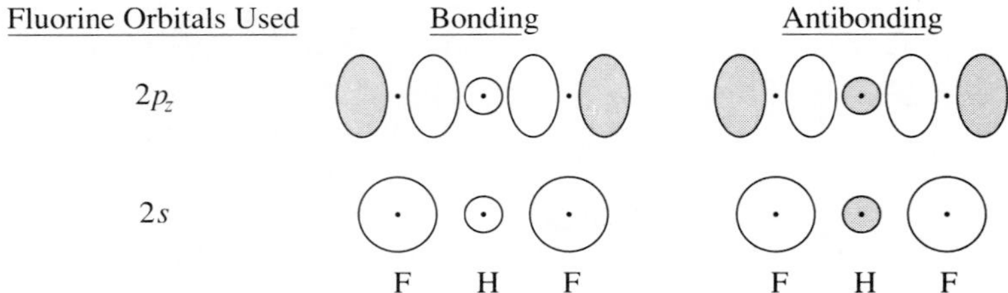

FIGURE 5-17 Interaction of Fluorine Group Orbitals with Hydrogen $1s$ Orbital.

Both sets of interactions are permitted by the symmetry of the orbitals involved. However, the energy match of the $1s$ orbital of hydrogen (orbital potential energy = -13.6 eV) is much better with the $2p_z$ of fluorine (-18.7 eV) than with the $2s$ of fluorine (-46.4 eV). Consequently, the more important interaction of the $1s$ orbital of hydrogen is with the $2p_z$ group orbital of the fluorines.

In sketching the molecular orbital energy diagrams of polyatomic species, we will show the valence orbitals of the central atom on the far left and the group orbitals of the surrounding atoms on the far right. The resulting molecular orbitals will be shown in the middle. For FHF^-, the resulting MO diagram is shown in Figure 5-18.

Five of the six group orbitals derived from the $2p$ orbitals of the fluorines do not interact with the central atom; these orbitals remain nonbonding. The sixth $2p$ group orbital, the $2p_z$ group orbital described above, interacts with the $1s$ orbital of hydrogen to give two molecular orbitals, one bonding and one antibonding. An electron pair occupies the bonding orbital. The group orbitals from the $2s$ orbitals of the fluorines are much lower in energy than the $1s$ orbital of hydrogen; one interacts only slightly with the hydrogen and is essentially nonbonding, and the other is not of appropriate symmetry to interact with the $1s$ of hydrogen at all.

FHF^- may therefore be described in molecular orbital terms as having a 3-center, 2-electron bond; in other words, it has a single molecular orbital containing a pair of bonding electrons delocalized over all three atoms. The lower energy of this orbital illustrates a common phenomenon. Bonding molecular orbitals that include atomic orbitals from three or more atoms have lower energies than those that include atomic orbitals from only two atoms. As a general rule, the more space covered by the orbital, the lower its energy. Conjugated systems (alternating double and single bonds in their electron-dot representation) acquire added stability from this effect.

[11] J. H. Clark, J. Emsley, D. J. Jones, R. E. Overill, *J. Chem. Soc.*, **1981,** 1219.

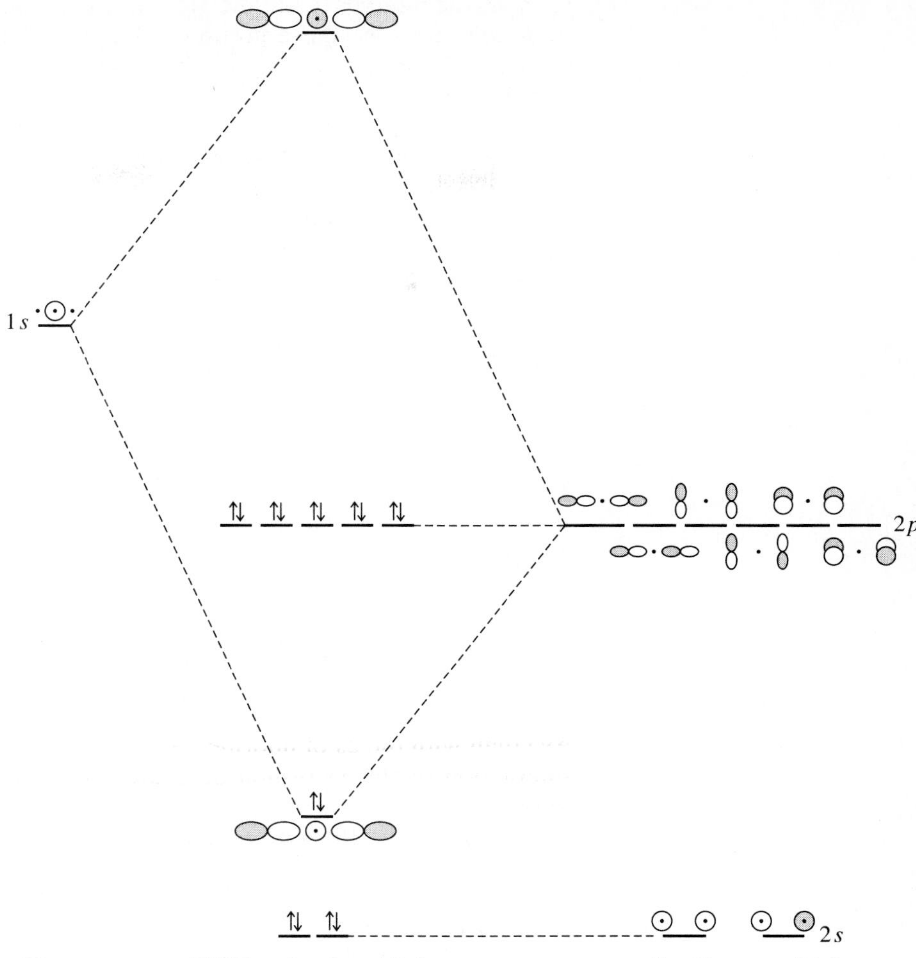

FIGURE 5-18 Molecular Orbital Diagram of FHF^-.

CO_2

Carbon dioxide, another linear molecule, has a more complicated molecular orbital description. In contrast to FHF^-, the central carbon atom in CO_2 now has p orbitals capable of interacting with $2p$ group orbitals on surrounding atoms. To obtain a molecular orbital picture of CO_2, we again begin with the group orbitals. The group orbitals of the oxygen atoms are similar to those for the fluorine atoms shown in Figure 5-16. To determine which atomic orbitals of carbon are of correct symmetry to interact with the group orbitals, it will be useful to consider the group orbitals in turn.

Group orbitals 1 and 2 (the numbers are arbitrary) are formed by adding and subtracting the oxygen $2s$ orbitals. Group orbital 1 is of appropriate symmetry to interact with the $2s$ orbital of carbon, and group orbital 2 is of appropriate symmetry to interact with the $2p_z$ orbital of carbon. Both interactions are very weak, since the $2s$ orbitals of oxygen are much lower in energy than the carbon orbitals; the interaction with the carbon $2s$ is slightly stronger. In the final molecular orbital diagram (Figure 5-19 on page 147), both group orbitals are shown with essentially no interaction with carbon orbitals.

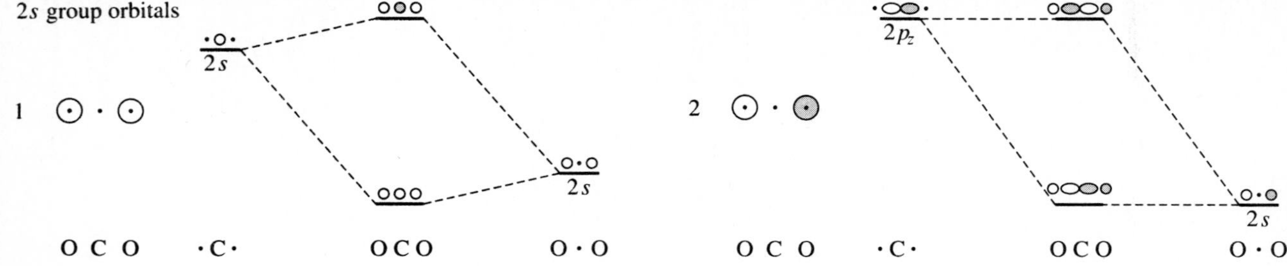

Group orbitals 3 and 4 are formed by adding and subtracting the oxygen $2p_z$ orbitals. Group orbital 3 can interact with the $2s$ of carbon, and group orbital 4 can interact with the carbon $2p_z$.

$2p_z$ group orbitals

3

2s

2pz

O C O   ·C·   OCO   O·O

4

2pz

2pz

O C O   ·C·   OCO   O·O

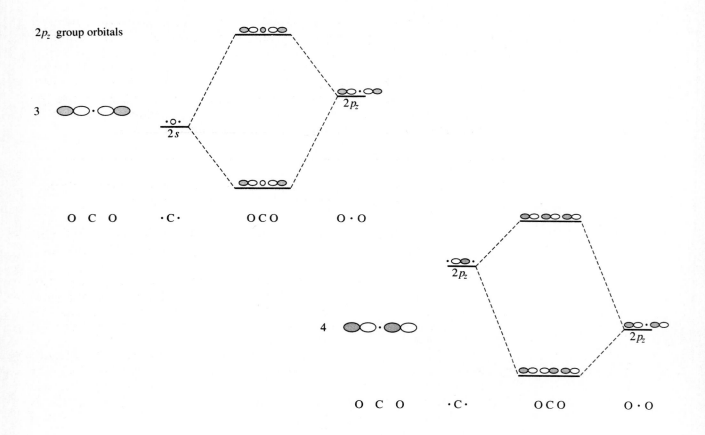

The $2s$ and $2p_z$ orbitals of carbon therefore have two possible sets of group orbitals with which they may interact. In other words, all four interactions described above occur; all four are symmetry allowed. It is then necessary to estimate which interactions can be expected to be the strongest. Useful approximations to the energies of atomic orbitals are the valence orbital potential energies (Table 5-1). For carbon and oxygen the potential energies of the $2s$ and $2p$ orbitals are as follows:

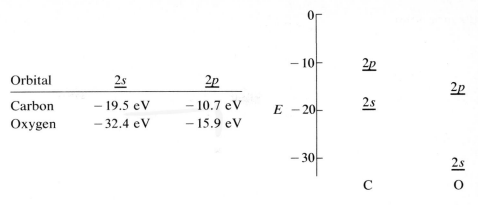

Orbital	$2s$	$2p$
Carbon	−19.5 eV	−10.7 eV
Oxygen	−32.4 eV	−15.9 eV

E 0 −10 −20 −30

C: $2p$, $2s$ O: $2p$, $2s$

Valence orbital potential energies

Interactions are strongest for orbitals having similar energies. For example, consider group orbitals 1 and 3. Both group orbital 1, arising from the $2s$ orbitals of the oxygen, and group orbital 3, arising from the $2p_z$ orbitals, have the proper symmetry to interact with the $2s$ orbital of carbon. However, the energy match between group orbital 3 and the $2s$ orbital of carbon is much better than the energy match between group orbital 1 and the $2s$ of carbon, as can be seen from the diagram above; therefore, the primary interaction is between the $2p_z$ orbitals of oxygen and the $2s$ orbital of carbon.

EXERCISE 5-4

Using valence orbital potential energies, show that group orbital 4 is more likely than group orbital 2 to interact strongly with the $2p_z$ orbital of carbon.

The $2p_x$ orbital of carbon is of appropriate symmetry to interact with group orbital 5. The result is the formation of two π molecular orbitals, one bonding and one antibonding.

However, there is no orbital on carbon of suitable symmetry and energy to interact with group orbital 6, formed by combining $2p_x$ orbitals of oxygen (a d orbital would be required, but they have higher energy); hence group orbital 6 is nonbonding.

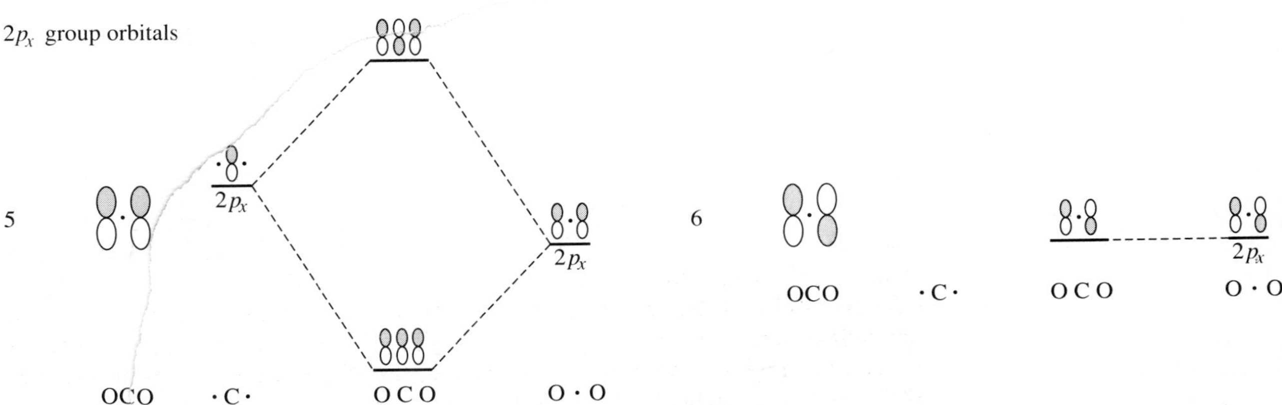

Interactions of the $2p_y$ orbitals are similar to those of the $2p_x$ orbitals. Group orbital 7 interacts with the $2p_y$ orbital of carbon to form π bonding and antibonding orbitals, while group orbital 8 is nonbonding.

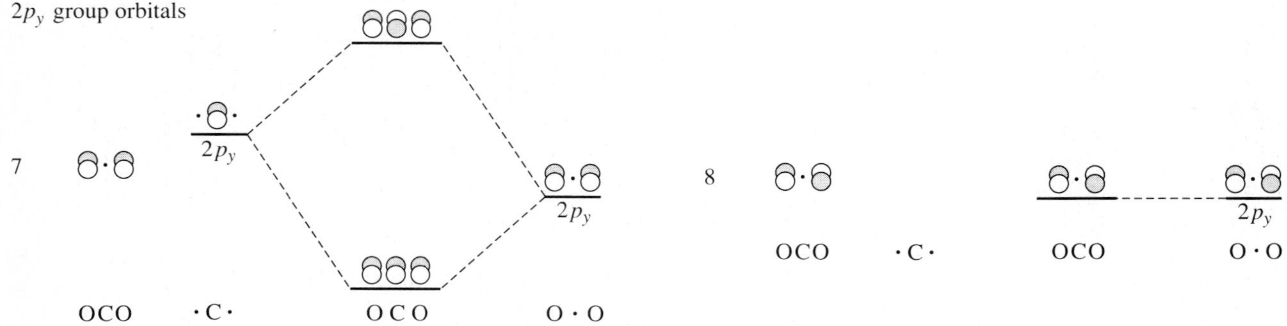

The overall molecular orbital diagram of CO_2, shown in Figure 5-19, has two σ and two π bonding orbitals occupied by pairs of electrons and two σ and two π nonbonding orbitals, each also occupied by pairs of electrons. Thus, half the lone pairs are in σ orbitals and half in π orbitals, and there are four bonds in the molecule, as expected. As in the FHF^- case, all the occupied molecular orbitals are 3-center, 2-electron orbitals; all are more stable (have lower energy) than 2-center orbitals.

The molecular orbital picture of other linear triatomic species, such as N_3^-, CS_2, and OCN^-, can be determined similarly. Likewise, we can describe the molecular orbitals of longer polyatomic species by a similar method. Examples of bonding in linear π systems will be considered in Chapter 12.

EXERCISE 5-5

Prepare a molecular orbital diagram for the azide ion, N_3^-.

H_2O

Molecular orbitals of nonlinear molecules can be determined pictorially by the same procedure. In the case of H_2O, we first sketch the group orbitals, the $1s$ orbitals of the hydrogens, and then match them with appropriate orbitals on the oxygen. The orbital interactions and the resulting molecular orbitals are shown in Figure 5-20 on page 148.

Two orbitals of oxygen, the $2s$ and the $2p_z$, can interact with the first of the group orbitals. The primary interaction is with the $2p_z$ to give the bonding and antibonding molecular orbitals shown (second and fourth highest in energy of the MOs). In addition, the $2s$ orbital of oxygen is stabilized slightly by interaction with the group orbital (forming the lowest-energy molecular orbital in the diagram). Some antibonding character is at the same time imparted to the molecular orbital second highest in energy by the interaction shown in brackets in the diagram. In addition, the second group orbital interacts with the $2p_x$ orbital of oxygen to give a bonding and antibonding pair of orbitals (the second lowest and the highest energy orbitals shown).

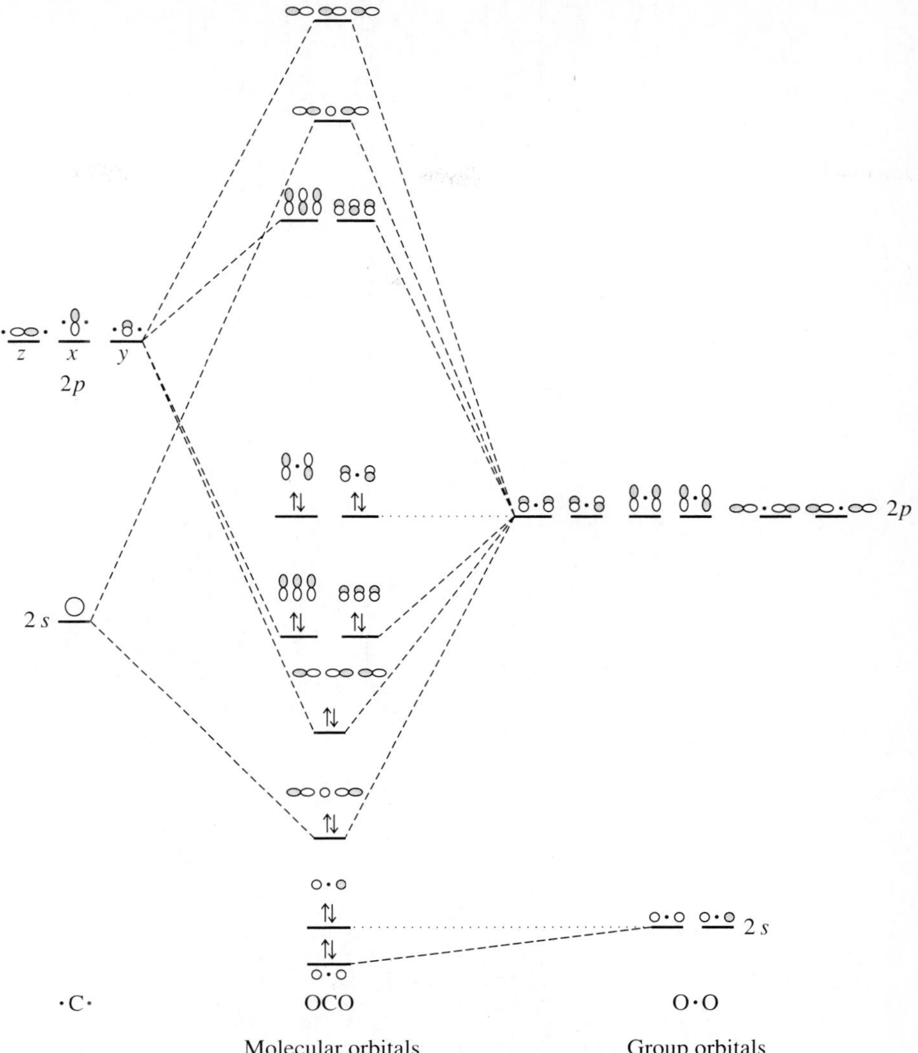

FIGURE 5-19 Molecular Orbital Diagram of CO_2.

The molecular orbital picture is different from the common conception of the water molecule having two equivalent lone electron pairs and two equivalent O—H bonds. In the MO picture, the highest-energy electron pair is truly nonbonding, occupying the $2p_y$ orbital perpendicular to the plane of the molecule. The next two pairs are bonding pairs, resulting from overlap of the $2p_z$ and $2p_x$ orbital with the $1s$ orbitals of the hydrogens. The lowest-energy pair is a lone pair in the essentially unchanged $2s$ orbital of the oxygen. Here, all four occupied molecular orbitals are different.

NH_3

VSEPR arguments describe ammonia as a pyramidal molecule with a single lone electron pair. The symmetry of ammonia is C_{3v}. For the purpose of obtaining a molecular orbital picture of NH_3, it is convenient to view this molecule looking down on the lone pair (down the C_3 axis). Since there are three hydrogen $1s$ orbitals to be considered, there must be three group orbitals.

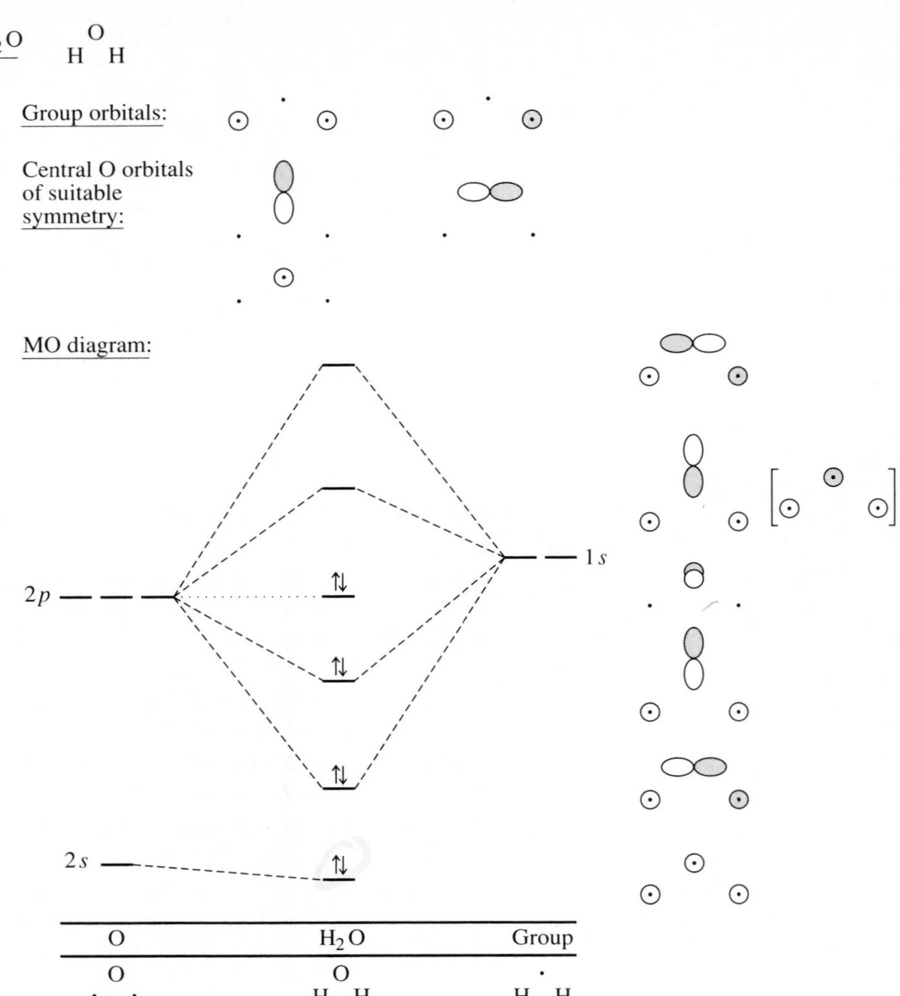

FIGURE 5-20 Molecular Orbitals of H_2O.

To this point it has been a simple matter to obtain a description of the group orbitals. Each polyatomic example considered (FHF^-, CO_2, H_2O) has had two atoms attached to a central atom; the group orbitals could be obtained by matching atomic orbitals on the terminal atoms in both a bonding and antibonding sense. In NH_3 this is no longer possible. One group orbital contains all hydrogen 1*s* orbitals of the same sign. How can the other two group orbitals be determined? (The number of group orbitals must equal the number of atomic orbitals used, in this case three.)

One way of viewing NH_3 is as a three-membered ring of hydrogen atoms, with a nitrogen atom above the ring on the C_3 axis. Group orbitals of regular rings of atoms have well-defined patterns of nodal planes. For example, for three-, four-, and five-membered rings, the group orbitals are shown in Figure 5-21. Several characteristics of these group orbitals should be noted:

1. In each case there is a single group orbital in which all atomic orbitals are of the same sign.
2. The remaining group orbitals have nodal planes cutting through the center of the polygon.

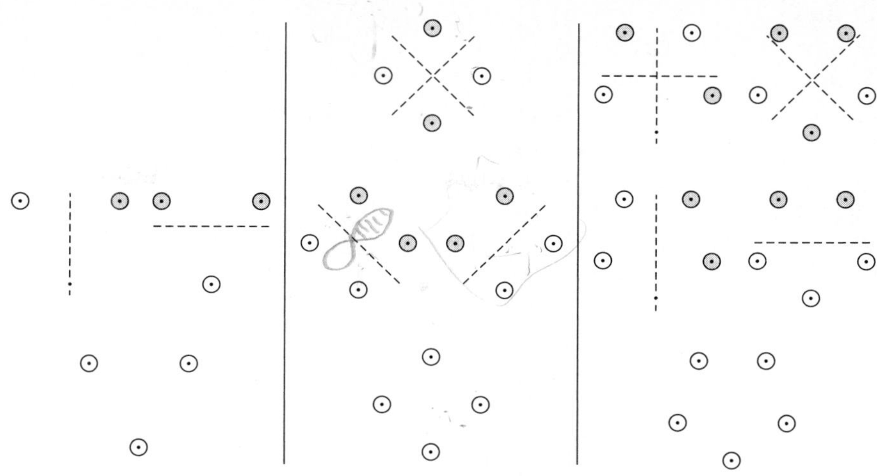

FIGURE 5-21 Group Orbitals for Rings (only s orbitals shown).

3. The nodal planes are oriented symmetrically; the angles between them are all equal. For example, in the five-atom ring the nodal planes in the single-node group orbitals are perpendicular to each other. In each two-node orbital, the nodal planes are perpendicular to each other, and the pair of nodal planes in the first two-node orbital bisects the angles between the nodal planes in the second two-node orbital. In the highest energy orbital for a six-atom ring (Exercise 5-6) there are three nodal planes, at 60° angles to each other.
4. The number of group orbitals equals the number of atomic orbitals used. There are no more than two group orbitals in a set having the same number of nodes. For larger rings than those shown, one group orbital in a set has no nodes, two have a single node, two have two nodes, and so on, until the number of group orbitals equals the number of atomic orbitals used. This method can also be used to determine the molecular orbitals of cyclic organic molecules. Examples of the molecular orbitals in cyclic π systems such as benzene will be discussed in Chapter 12.

EXERCISE 5-6
Sketch the group orbitals of a hexagonal ring of s orbitals.

The group orbitals determined by this method for NH_3 are shown in Figure 5-22. In NH_3 we can begin by first considering the zero-node group orbital. Two atomic orbitals of nitrogen can interact with this group orbital, the $2p_z$ and the $2s$. Each one-node group orbital can interact with one of the remaining $2p$ orbitals of nitrogen, the $2p_x$ or the $2p_y$. The resulting molecular orbital picture is shown in Figure 5-23. (A smaller interaction is shown in brackets.)

The HOMO of NH_3 is slightly bonding; it contains an electron pair in an orbital resulting from interaction of the $2p_z$ orbital of nitrogen with the $1s$ orbitals of the hydrogens (from the zero-node group orbital). This is the lone pair of the electron-dot and VSEPR models. It is also the pair donated by ammonia when it functions as a Lewis base (Lewis acids and bases will be discussed in Chapter 6).

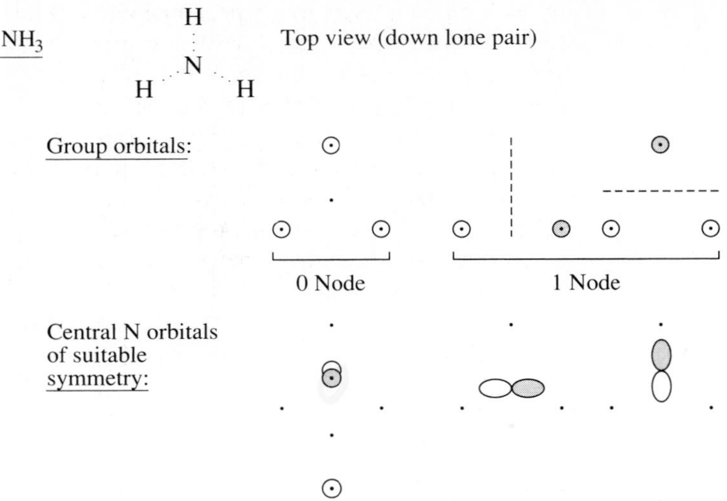

FIGURE 5-22 Group Orbitals of NH_3.

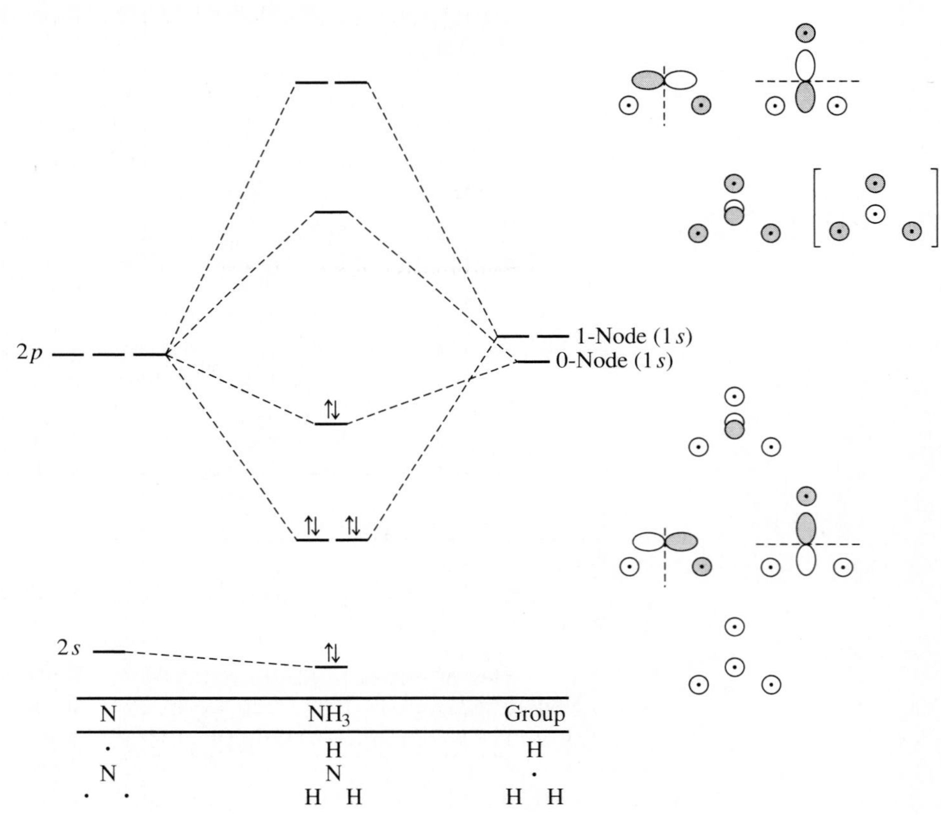

FIGURE 5-23 Molecular Orbitals of NH_3.

BF_3

Boron trifluoride is a classic Lewis acid; an accurate molecular orbital picture of BF_3 should therefore show, among other things, an orbital capable of acting as an electron-pair acceptor.

The procedure for describing molecular orbitals of BF_3 is more complicated than for NH_3, since the fluorines surrounding the central boron have 2*p* as well as $2s$ electrons to be considered. The group orbitals and the central boron atoms that interact with the group orbitals are shown in Figure 5-24; the molecular orbitals are shown in Figure 5-25 (omitting sketches of the five nonbonding $2p$ group orbitals for clarity).

As discussed in Chapter 3, resonance structures may be drawn for BF_3 showing this molecule to have some double-bond character in the B—F bonds. The molecular orbital view of BF_3 shows an electron pair in a bonding π orbital delocalized over all four atoms (this is the orbital slightly below the five nonbonding electron pairs in energy). Overall, BF_3 has three bonding σ orbitals and one slightly bonding π orbital occupied by electron pairs, together with eight nonbonding pairs on the fluorine atoms.

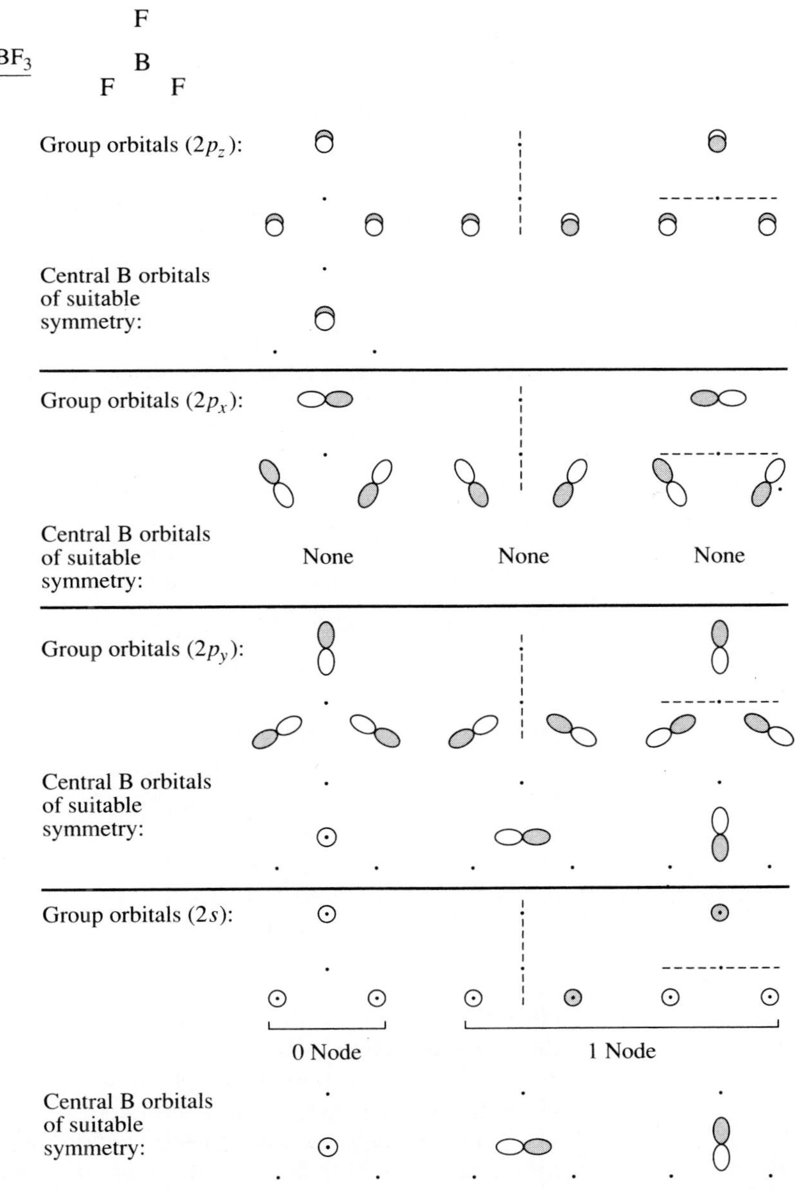

FIGURE 5-24 Group Orbitals for BF_3.

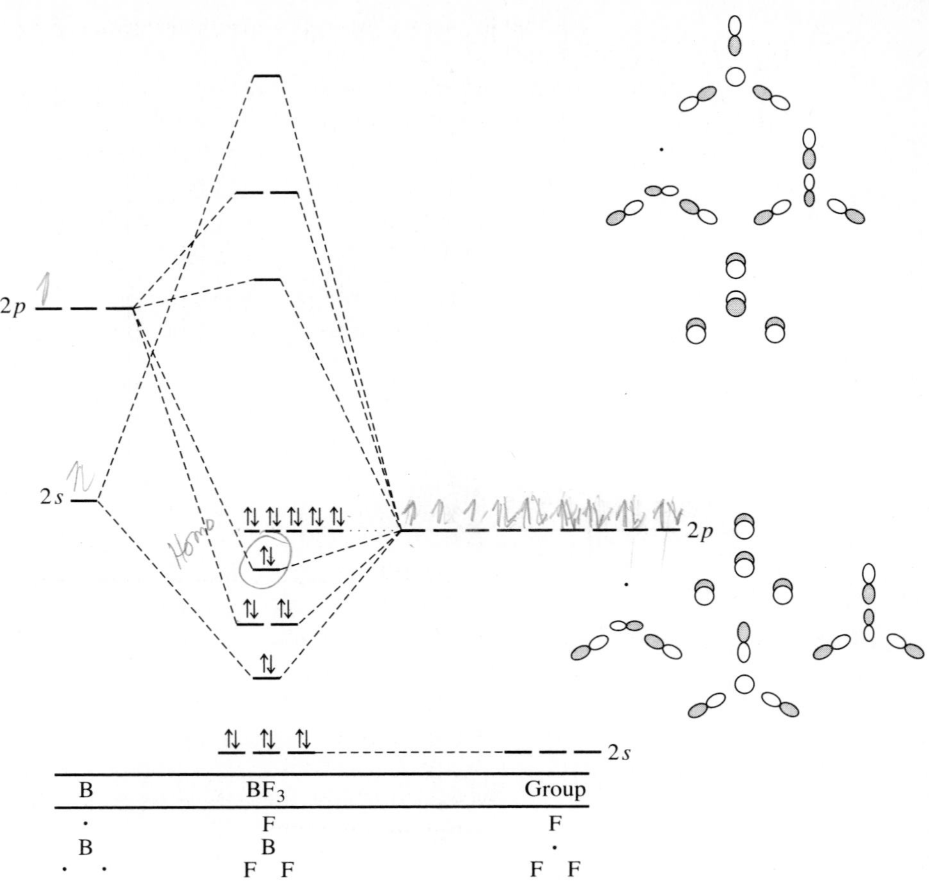

FIGURE 5-25 Molecular Orbitals of BF_3.

The lowest unoccupied molecular orbital (LUMO) of BF_3 is an empty π orbital; it has antibonding interactions between the $2p_z$ orbital on boron and the $2p_z$ orbitals of the surrounding fluorines. This orbital can act as an electron-pair acceptor (as, for example, from the HOMO of NH_3) in Lewis acid–base interactions.

The molecular orbitals of other trigonal planar species, such as SO_3, CO_3^{2-}, and NO_3^-, can be treated by a similar procedure. Group orbitals can also be used to derive molecular orbital descriptions of more complicated molecules. However, while the simple pictorial approach described in these past few pages can lead conveniently to a qualitatively useful description of bonding in simple molecules, the utility of this method decreases with increasing molecular complexity. Furthermore, this method gives only a qualitative, although useful, description of orbital interactions within molecules. More advanced methods, based on computer calculations, are necessary to deal with more complex molecules and to obtain wave equations for the molecular orbitals. These more advanced methods often make use of molecular symmetry and group theory. In the section that follows, we will describe how several of the molecules already described in this chapter can be treated by group theory. The resulting molecular orbital pictures will be similar to those we have already obtained by our more pictorial approach; however, the group theory approach, while somewhat more laborious for simple molecules, provides a foundation for describing molecular orbitals of more complex molecules.

5-5 GROUP THEORY APPROACH TO MOLECULAR ORBITALS

The process in group theoretical terms can be broken into the following steps:

1. Find the point group of the molecule from its shape as described in Chapter 4.
2. Assign *x*, *y*, and *z* coordinates to the atoms, chosen for convenience. Experience is the best guide here. The general rule in all the examples in this book is that the highest-order rotation axis of the molecule is chosen as the *z* axis of the central atom. In nonlinear molecules, the *y* axes of the outer atoms are chosen to point toward the central atom.
3. Find the characters of the representation for the combination of all valence orbitals on the outer atoms. These will later be mixed with the appropriate orbitals of the central atom. As in the case of the vectors described in Chapter 4, any orbital that changes position during a symmetry operation contributes zero to the character of the resulting representation, any orbital that remains in its original position contributes 1, and any orbital that remains in the original position with its signs reversed contributes -1.
4. Reduce each representation from step 3 to its irreducible representations. This is equivalent to finding the group orbitals or the symmetry-adapted linear combinations (SALCs) of the orbitals.
5. Find the atomic orbitals of the central atom with the same irreducible representations as those found in step 4.
6. Combine the atomic orbitals of the central atom and those of the outer atoms with the same symmetry to form molecular orbitals. The number of molecular orbitals formed equals the number of atomic orbitals used from all the atoms.

The process can be carried further to obtain the numerical values of the coefficients of the atomic orbitals used in the molecular orbitals.[12] For the qualitative pictures we will describe, it is sufficient to be able to say that a given orbital is primarily composed of one of the atomic orbitals or that it is composed of roughly equal contributions from each of several atomic orbitals. The coefficients may be small or large, positive or negative, similar or quite different, depending on the characteristics of the orbitals under consideration.

H_2O

1. Water is a simple triatomic bent molecule with a C_2 axis through the oxygen and two mirror planes that intersect in this axis, as shown in Figure 5-26. The point group is therefore C_{2v}.

2. The C_2 axis is chosen as the *z* axis. Since the hydrogen 1*s* orbitals have no directionality, it is not necessary to assign axes to the hydrogens.

3. Since the hydrogen atoms determine the symmetry of the molecule, we will use their orbitals as a starting point. The 1*s* orbitals of the hydrogens, taken as a pair, are tested with the symmetry operations of the C_{2v} group. The matrices showing the effects of the symmetry operations on the hydrogen 1*s* orbitals for the C_{2v} point group are shown in Table 5-2. These matrices yield the representation Γ.

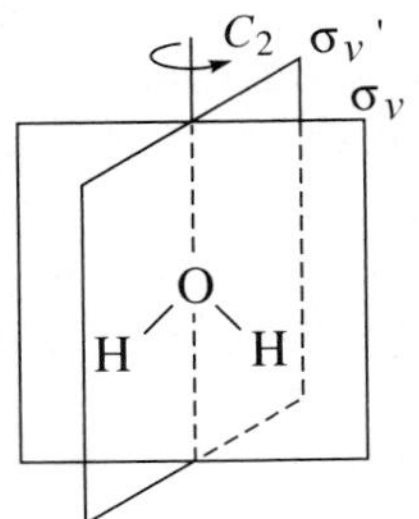

FIGURE 5-26 Symmetry of the Water Molecule.

[12] F. A. Cotton, *Chemical Applications of Group Theory*, 3rd ed., Wiley, New York, 1990, pp. 133–88.

TABLE 5-2
Matrices and representations for C_{2v} symmetry operations for water hydrogens

$$\begin{bmatrix} H'_a \\ H'_b \end{bmatrix} = \begin{bmatrix} 1 & 0 \\ 0 & 1 \end{bmatrix} \begin{bmatrix} H_a \\ H_b \end{bmatrix}, \quad \text{for the identity operation}$$

$$\begin{bmatrix} H'_a \\ H'_b \end{bmatrix} = \begin{bmatrix} 0 & 1 \\ 1 & 0 \end{bmatrix} \begin{bmatrix} H_a \\ H_b \end{bmatrix}, \quad \text{for } C_2 \text{ rotation}$$

$$\begin{bmatrix} H'_a \\ H'_b \end{bmatrix} = \begin{bmatrix} 1 & 0 \\ 0 & 1 \end{bmatrix} \begin{bmatrix} H_a \\ H_b \end{bmatrix}, \quad \text{for } \sigma_v \text{ reflection } (xz \text{ plane})$$

$$\begin{bmatrix} H'_a \\ H'_b \end{bmatrix} = \begin{bmatrix} 0 & 1 \\ 1 & 0 \end{bmatrix} \begin{bmatrix} H_a \\ H_b \end{bmatrix}, \quad \text{for } \sigma'_v \text{ reflection } (yz \text{ plane})$$

The reducible representation is made up of the irreducible representations

C_{2v}	E	C_2	$\sigma_v(xz)$	$\sigma'_v(yz)$	
Γ	2	0	2	0	
A_1	1	1	1	1	z
and					
B_1	1	−1	1	−1	x

C_{2v}	E	C_2	$\sigma_v(xz)$	$\sigma'_v(yz)$		
A_1	1	1	1	1	z	x^2, y^2, z^2
A_2	1	1	−1	−1	R_z	xy
B_1	1	−1	1	−1	x, R_y	xz
B_2	1	−1	−1	1	y, R_x	yz

The characters of the matrices can be obtained without writing out the matrices. The sum of the contributions to the character (1, 0, or −1, as described earlier) for each symmetry operation is the character for that operation, and the complete list for all operations of the group is the reducible representation for the atomic orbitals. The identity operation leaves both hydrogen orbitals unchanged, with a character of 2. Twofold rotation interchanges the orbitals, so each contributes 0 for a total character of 0. Reflection in the plane of the molecule (σ_v) leaves both hydrogens unchanged, for a character of 2; reflection perpendicular to the plane of the molecule (σ'_v) switches the two orbitals, for a character of 0. For simplicity, we will omit the matrix operations in the following examples.

4. The representation Γ can be reduced to the irreducible representations $A_1 + B_2$. Several different approaches can then be used to find the molecular orbitals. The one used here parallels the process used in the more pictorial approach presented earlier. The first step is to combine the two hydrogen 1s orbitals. The sum of the two, $\frac{1}{\sqrt{2}}[\psi(H_a) + \psi(H_b)]$, has symmetry A_1 and the difference, $\frac{1}{\sqrt{2}}[\psi(H_a) - \psi(H_b)]$, has symmetry B_1, as can be seen by examining Figure 5-27. These group orbitals, or symmetry-adapted linear combinations, are each then treated as if they were atomic orbitals. In this case, the atomic orbitals are identical and have equal coefficients, so they contribute equally to the group orbitals. The normalizing factor is $\frac{1}{\sqrt{2}}$. In general, the normalizing factor for a group orbital is

$$N = \frac{1}{\sqrt{\sum c_i^2}}$$

where c_i = coefficients on the atomic orbitals.

Again, each group orbital is treated as a single orbital in combining with the oxygen orbitals.

5. The same kind of analysis can be applied to the oxygen orbitals. This requires only the addition of -1 as a possible character, when a p orbital changes sign. Each orbital can be treated independently.

The s orbital is unchanged by all the operations, so it has A_1 symmetry.
The p_x orbital has the B_1 symmetry of the x axis.
The p_y orbital has the B_2 symmetry of the y axis.
The p_z orbital has the A_1 symmetry of the z axis.

The x, y, and z variables and the more complex functions in the character tables assist in assigning representations to the atomic orbitals.

6. The atomic and group orbitals with the same symmetry are combined into molecular orbitals, as listed in Table 5-3 and shown in Figure 5-28. They are numbered in order of their energy, with 1 the lowest and 6 the highest.

The A_1 group orbital combines with the s and p_z orbitals of the oxygen to form three molecular orbitals, one bonding, one nearly nonbonding (slightly bonding), and one antibonding (three atomic or group orbitals forming three

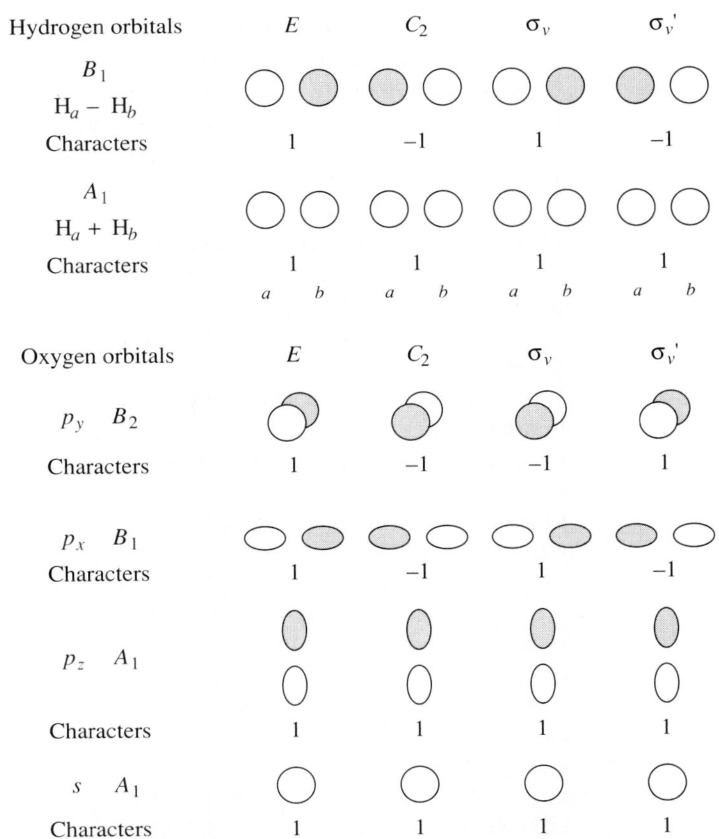

FIGURE 5-27 Symmetry of Atomic and Group Orbitals in the Water Molecule.

TABLE 5-3
Molecular orbitals for water

Symmetry	*Molecular orbitals*		*Oxygen atomic orbitals*		*Group orbitals from hydrogen atoms*	*Description*
B_1	Ψ_6	=	$c_9\psi(p_x)$	+	$c_{10}[\psi(H_a) - \psi(H_b)]$	Antibonding (c_{10} is negative)
A_1	Ψ_5	=	$c_5\psi(s)$	+	$c_6[\psi(H_a) + \psi(H_b)]$	Antibonding (c_6 is negative)
B_2	Ψ_4	=	$\psi(p_y)$			Nonbonding
A_1	Ψ_3	=	$c_3\psi(p_z)$	+	$c_4[\psi(H_a) + \psi(H_b)]$	Nearly nonbonding (slightly bonding; c_4 is very small)
B_1	Ψ_2	=	$c_7\psi(p_x)$	+	$c_8[\psi(H_a) - \psi(H_b)]$	Bonding (c_8 is positive)
A_1	Ψ_1	=	$c_1\psi(s)$	+	$c_2[\psi(H_a) + \psi(H_b)]$	Bonding (c_2 is positive)

molecular orbitals, Ψ_1, Ψ_3, and Ψ_5). The oxygen p_z has only minor contributions from the other orbitals in the weakly bonding Ψ_3 orbital, and the oxygen s and the hydrogen group orbital combine to form the bonding and antibonding Ψ_1 and Ψ_5 orbitals.

The hydrogen B_1 group orbital combines with the oxygen p_x orbital to form two MOs, one bonding and one antibonding (Ψ_2 and Ψ_6). The oxygen p_y (Ψ_4, with B_2 symmetry) does not have the same symmetry as any of the H 1s group orbitals and is a nonbonding orbital. As a result, there are two bonding orbitals, two nonbonding or nearly nonbonding orbitals, and two antibonding orbitals.

When the 8 valence electrons available are added, we have two pairs in bonding orbitals and two pairs in nonbonding orbitals, which can be compared with the two bonds and two lone pairs of the Lewis electron-dot structure. The lone pairs are in molecular orbitals, one b_2 from the p_y of the oxygen, the other a_1 from a combination of s and p_z of the oxygen and the two hydrogen 1s orbitals.

The qualitative methods we have described do not allow us to determine the energies of the orbitals, but we can place them in approximate order from their shapes and the expected overlap. The intermediate energy levels in particular are difficult to place in order. Whether an individual orbital is precisely nonbonding, slightly bonding, or slightly antibonding is likely to make little difference in the overall energy of the molecule. Such intermediate orbitals can be described as essentially nonbonding.

Differences in energy between two clearly bonding orbitals are likely to be more significant in the overall energy of a molecule. The π interactions are generally weaker than σ interactions, so a double bond made up of one σ orbital and one π orbital is not twice as strong as a single bond. In addition, single bonds between the same atoms may have widely different energies. For example, the C—C bond is usually described as having an energy near 345 kJ/mol, a value averaged from a large number of different molecules. These individual values may vary tremendously; numbers as small as 63 and as large

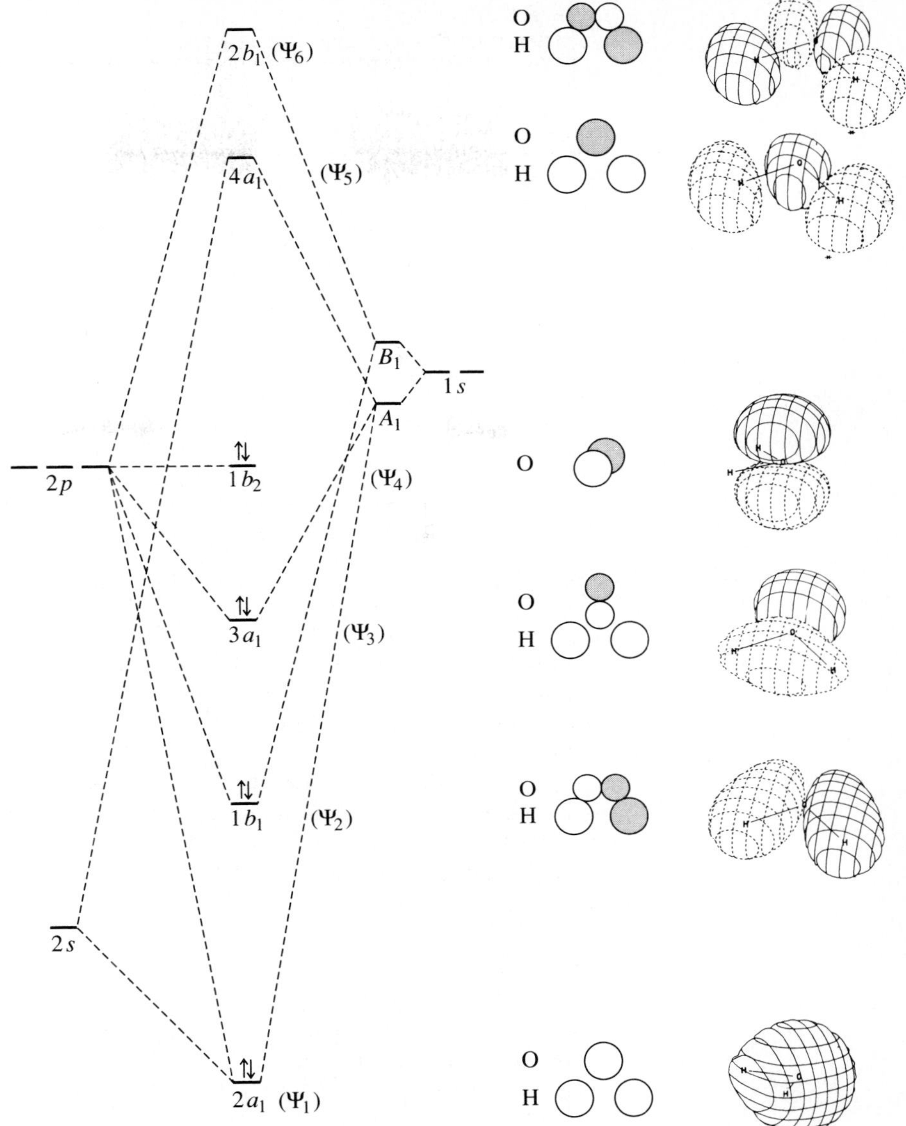

FIGURE 5-28 Water Molecular Orbitals. (Contour diagrams reproduced with permission from W. L. Jorgensen and L. Salem, *The Organic Chemist's Book of Orbitals,* Academic Press, New York, 1973, p. 70.)

as 628 kJ/mol have been reported.[13] The low value is for hexaphenyl ethane, the high for diacetylene (H—C≡C—C≡C—H), extremes in steric crowding and bonding on either side of the C—C bond.

NH_3

Ammonia is a case similar to water. The point group of the molecule is C_{3v}. When the symmetry of the three H atom 1*s* orbitals is considered, we find the reducible representation in Table 5-4. It can be reduced by the methods in Chapter 4 to the A_1 and E irreducible representations, with the orbital combinations in Figure 5-29.

[13] S. W. Benson, *J. Chem. Educ.*, **1965,** *42,* 502.

TABLE 5-4
Representations for atomic orbitals in ammonia

C_{3v} character table

C_{3v}	E	$2C_3$	$3\sigma_v$		
A_1	1	1	1	z	$x^2 + y^2, z^2$
A_2	1	1	-1		
E	2	-1	0	$(x, y)(R_x, R_y)$	$(x^2 - y^2, xy)(xz, yz)$

	C_{3v}	E	$2C_3$	$3\sigma_v$	
The reducible representation	Γ	3	0	1	
is made up of	A_1	1	1	1	z
	and				
	E	2	-1	0	(x, y)

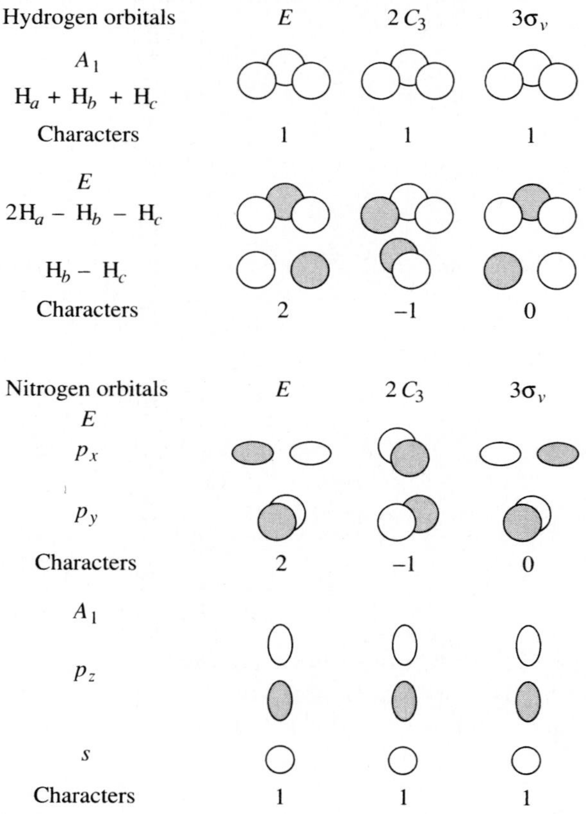

FIGURE 5-29 Symmetry of Atomic Orbitals in the Ammonia Molecule.

The s and p_z orbitals of the nitrogen have A_1 symmetry, and the pair p_x, p_y has E symmetry, exactly the same as the representations of the hydrogen 1s orbitals. Therefore, all orbitals of N are capable of combining with the hydrogen orbitals. As in the water case, we group the orbitals by symmetry and then combine them.

The A_1 symmetry of the sum of the three hydrogen $1s$ orbitals is easily seen, but the two group orbitals of E symmetry require more explanation. (The matrix description of C_3 rotation for the x and y axes in Chapter 4 may also be helpful.) One condition of the equations describing the molecular orbitals is that the square of the coefficients of the atomic orbitals in the LCAOs add up to equal contributions from each atomic orbital. A second condition is that the symmetry of the resulting LCAOs match the symmetry of the representation determined for the orbitals that are being combined (in this case, the E symmetry of the combined x and y axes). Here, this condition requires one node for each of the combined orbitals having E symmetry. With three atomic orbitals, the appropriate combinations are then

$$\frac{1}{\sqrt{6}}[2\psi(H_a) - \psi(H_b) - \psi(H_c)] \quad \text{and} \quad \frac{1}{\sqrt{2}}[\psi(H_b) - \psi(H_c)]$$

The coefficients in these group orbitals result in equal contribution by each atomic orbital when each term is squared (as is done in calculating probabilities) and the terms for each orbital summed.

For H_a, the contribution is $\left(\frac{2}{\sqrt{6}}\right)^2 = \frac{2}{3}$.

For H_b and H_c, the contribution is $\left(\frac{1}{\sqrt{6}}\right)^2 + \left(\frac{1}{\sqrt{2}}\right)^2 = \frac{2}{3}$.

H_a, H_b, and H_c each also have a contribution of $\frac{1}{3}$ in the A_1 group orbital, giving a total contribution of unity by each of the atomic orbitals.

Each group orbital is treated as a single orbital in combining with the nitrogen orbitals, as shown in Figure 5-30. The nitrogen s and p_z orbitals combine with the hydrogen group orbital $\frac{1}{\sqrt{3}}[\psi(H_a) + \psi(H_b) + \psi(H_c)]$ to give three a_1 orbitals, one bonding, one nonbonding, and one antibonding. The nonbonding orbital is almost entirely nitrogen p_z, with the nitrogen s orbital combining effectively with the hydrogen group orbital for the bonding and antibonding orbitals.

The nitrogen p_x and p_y orbitals combine with the group orbitals $\frac{1}{\sqrt{6}}[2\psi(H_a) - \psi(H_b) - \psi(H_c)]$ and $\frac{1}{\sqrt{2}}[\psi(H_b) - \psi(H_c)]$ to form four E orbitals, two bonding and two antibonding (E has a dimension of 2, which requires pairs of orbitals of the same energy).

When we put eight electrons into the lowest energy levels, we get three bonds and one nonbonded lone pair, just as suggested by the electron-dot structure.

CO_2

Carbon dioxide is a linear molecule, with $D_{\infty h}$ symmetry. The oxygens present a more complicated case than the hydrogens in water and ammonia, as we must now consider the p orbitals of these atoms as well as the s orbitals. A useful selection of axes for CO_2 has the z axis for all three atoms running through all three nuclei and the x and y axes parallel to each other within each set, as in Figure 5-31.

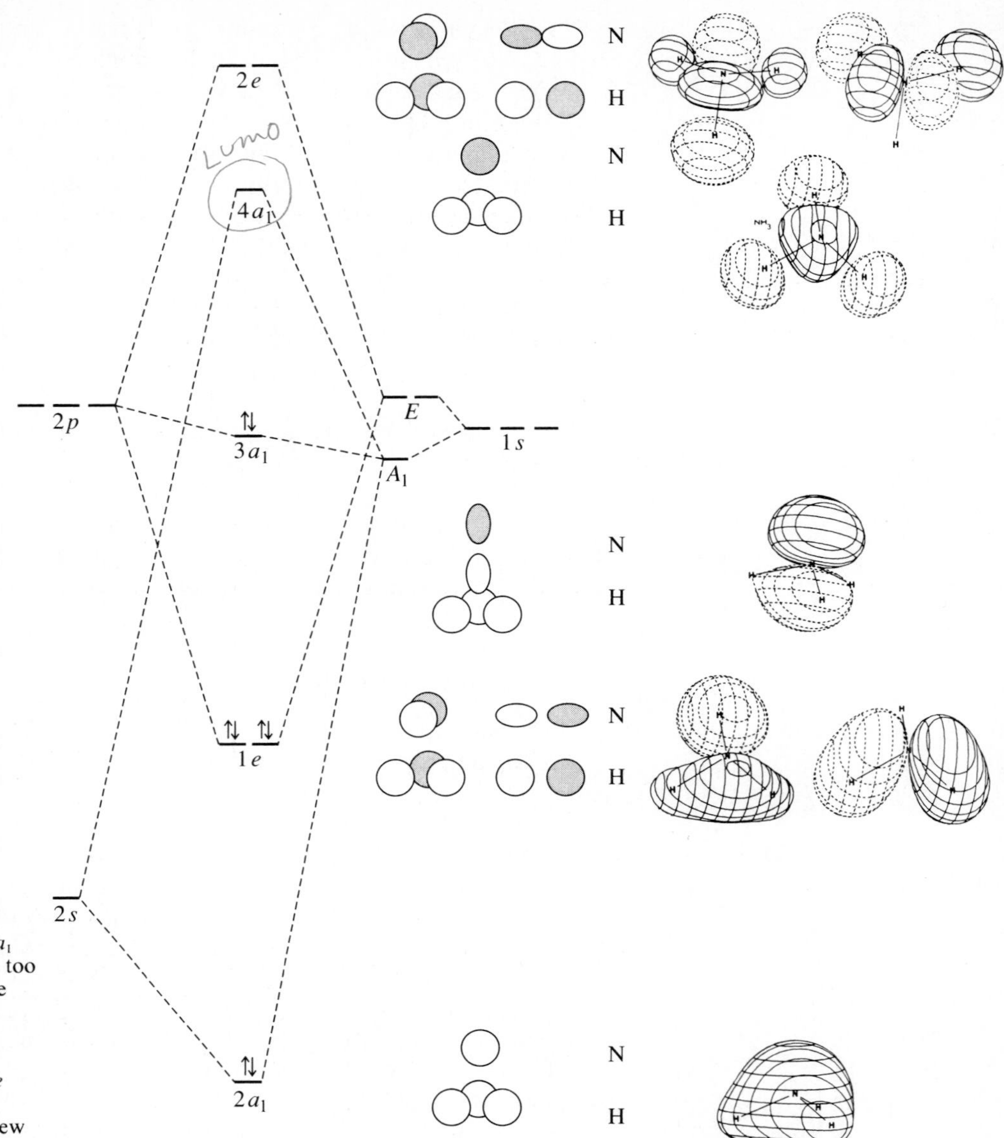

FIGURE 5-30 Molecular Orbitals of Ammonia. The $1a_1$ orbital is the $1s$ orbital of N, too low in energy to mix with the hydrogen orbitals. (Contour diagrams reproduced with permission from W. L. Jorgensen and L. Salem, *The Organic Chemist's Book of Orbitals,* Academic Press, New York, 1973, pp. 130–31.)

Because the molecule has an infinite rotation axis, the angle of rotation is indefinite, and any value will work. To avoid the complications of this indefinite angle, we can choose to work with a simpler group that also fits the symmetry of the atomic orbitals. Since the *p* orbitals have twofold symmetry, we will use a D_{2h} group. The labels for the representations in the remainder of this discussion are for D_{2h}, with those of $D_{\infty h}$ in parentheses.

All nine of the axes could be used together and the resulting representation reduced; but because there are so many orbitals, we can simplify the process by taking each kind of oxygen orbital separately. In effect, this does part of the reduction in advance. The selection is made by examining the individual orbitals in the context of the molecular symmetry. If two orbitals can be

$D_{\infty h}$	E	$2C_\infty^\phi$	...	$\infty\sigma_v$	i	$2S_\infty^\phi$	...	∞C_2		
Σ_g^+	1	1	...	1	1	1	...	1		$x^2 + y^2, z^2$
Σ_g^-	1	1	...	-1	1	1	...	-1	R_z	
Π_g	2	$2\cos\phi$	...	0	2	$-2\cos\phi$	...	0	(R_x, R_y)	(xz, yz)
Δ_g	2	$2\cos 2\phi$	...	0	2	$2\cos 2\phi$	...	0		$(x^2 - y^2, xy)$
...	...	...	...	...	...	...	...	...		
Σ_u^+	1	1	...	1	-1	-1	...	-1	z	
Σ_u^-	1	1	...	-1	-1	-1	...	1		
Π_u	2	$2\cos\phi$	...	0	-2	$2\cos\phi$	...	0	(x, y)	
Δ_u	2	$2\cos 2\phi$	...	0	-2	$-2\cos 2\phi$	...	0		
...	...	...	...	...	...	...	...	...		

D_{2h}	E	$C_2(z)$	$C_2(y)$	$C_2(x)$	i	$\sigma(xy)$	$\sigma(xz)$	$\sigma(yz)$		
A_g	1	1	1	1	1	1	1	1		x^2, y^2, z^2
B_{1g}	1	1	-1	-1	1	1	-1	-1	R_z	xy
B_{2g}	1	-1	1	-1	1	-1	1	-1	R_y	xz
B_{3g}	1	-1	-1	1	1	-1	-1	1	R_x	yz
A_u	1	1	1	1	-1	-1	-1	-1		
B_{1u}	1	1	-1	-1	-1	-1	1	1	z	
B_{2u}	1	-1	1	-1	-1	1	-1	1	y	
B_{3u}	1	-1	-1	1	-1	1	1	-1	x	

FIGURE 5-31 Axes for Bonding and Character Tables for CO_2.

interchanged by a molecular symmetry operation, they must be kept together. If any orbital cannot be changed to another by a molecular symmetry operation, it can be treated separately. In this case, all the orbitals of the oxygen atoms are independent and can be treated separately. Taking each set of oxygen orbitals in turn gives the following representations, all shown pictorially in Figure 5-32:

	D_{2h} representations	*$D_{\infty h}$ representations*
$2p_x$	B_{2g} and B_{3u}	Π_g and Π_u
$2p_y$	B_{3g} and B_{2u}	
$2p_z$	A_g and B_{1u}	Σ_g^+ and Σ_u^+
$2s$	A_g and B_{1u}	Σ_g^+ and Σ_u^+

The representations for the carbon orbitals are

	D_{2h} representations	*$D_{\infty h}$ representations*
$2p_x$	B_{3u}	Π_u
$2p_y$	B_{2u}	
$2p_z$	B_{1u}	Σ_u^+
$2s$	A_g	Σ_g^+

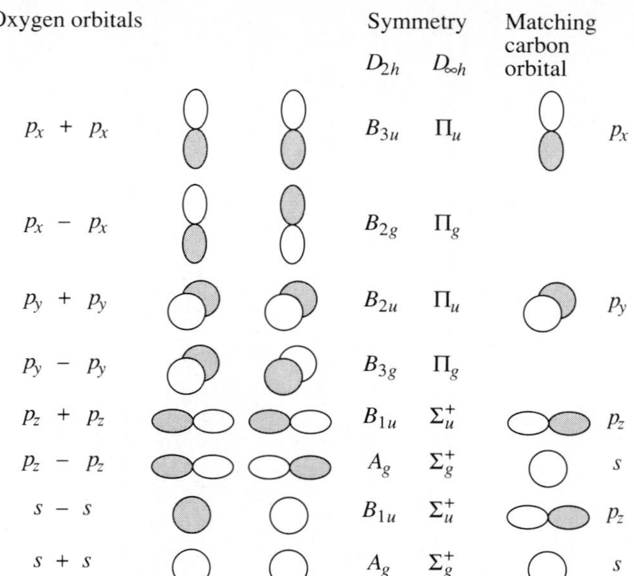

Oxygen orbitals
Symmetry
D_{2h}
$D_{\infty h}$
Matching carbon orbital
$p_x + p_x$
B_{3u}
Π_u
p_x
$p_x - p_x$
B_{2g}
Π_g
$p_y + p_y$
B_{2u}
Π_u
p_y
$p_y - p_y$
B_{3g}
Π_g
$p_z + p_z$
B_{1u}
Σ_u^+
p_z
$p_z - p_z$
A_g
Σ_g^+
s
$s - s$
B_{1u}
Σ_u^+
p_z
$s + s$
A_g
Σ_g^+
s

FIGURE 5-32 Orbital Symmetry in CO_2.

The six s and p_z orbitals combine to form six σ orbitals of A_g and B_{1u} symmetry.

The three $A_g(\Sigma_g^+)$ representations (the sum of the oxygen s orbitals, the carbon s orbital, and the difference of the oxygen p_z orbitals) form three molecular orbitals, one bonding, one essentially nonbonding, and one antibonding ($3\sigma_g$, $4\sigma_g$, and $5\sigma_g$ in the contour diagrams of Figure 5-33). In fact, the "nonbonding" orbital, $4\sigma_g$ in the contour diagrams, has considerable bonding character.

The three $B_{1u}(\Sigma_u^+)$ representations (the difference of the oxygen s orbitals, the sum of the oxygen p_z orbitals and the carbon p_z orbital) also form three σ molecular orbitals, one bonding, one essentially nonbonding, and one antibonding ($2\sigma_u$, $3\sigma_u$, and $4\sigma_u$). As in the $4\sigma_g$, the middle orbital has some bonding character.

The sums of the oxygen p_x orbitals and the carbon p_x form two molecular orbitals of $B_{3u}(\Pi_u)$ symmetry, one bonding and one antibonding, in the xz plane, as do the p_y orbitals, of $B_{2u}(\Pi_u)$ symmetry, in the yz plane. The two bonding orbitals ($1\pi_u$) are degenerate (have the same energy), as are the two antibonding orbitals ($2\pi_u$).

A nonbonding molecular orbital of B_{2g} symmetry uses only the difference of the oxygen p_x orbitals because none of the carbon orbitals has this symmetry, and another of B_{3g} symmetry uses the oxygen p_y orbitals similarly. These orbitals are also degenerate, with π_g notation in the $D_{\infty h}$ point group. We use lowercase labels on the molecular orbitals, with capitals for the atomic orbitals and for representations in general. This is a common, but not universal, practice.

The $D_{\infty h}$ labels can be translated directly to the σ and π orbitals used in the contour diagrams of Figure 5-33.

Adding the 16 valence electrons from the original atoms fills all the bonding orbitals and all the nonbonding orbitals, for a total of four bonds and four lone pairs, just as expected from the Lewis electron-dot picture. As is true in most molecules, the σ molecular orbitals result from the strongest

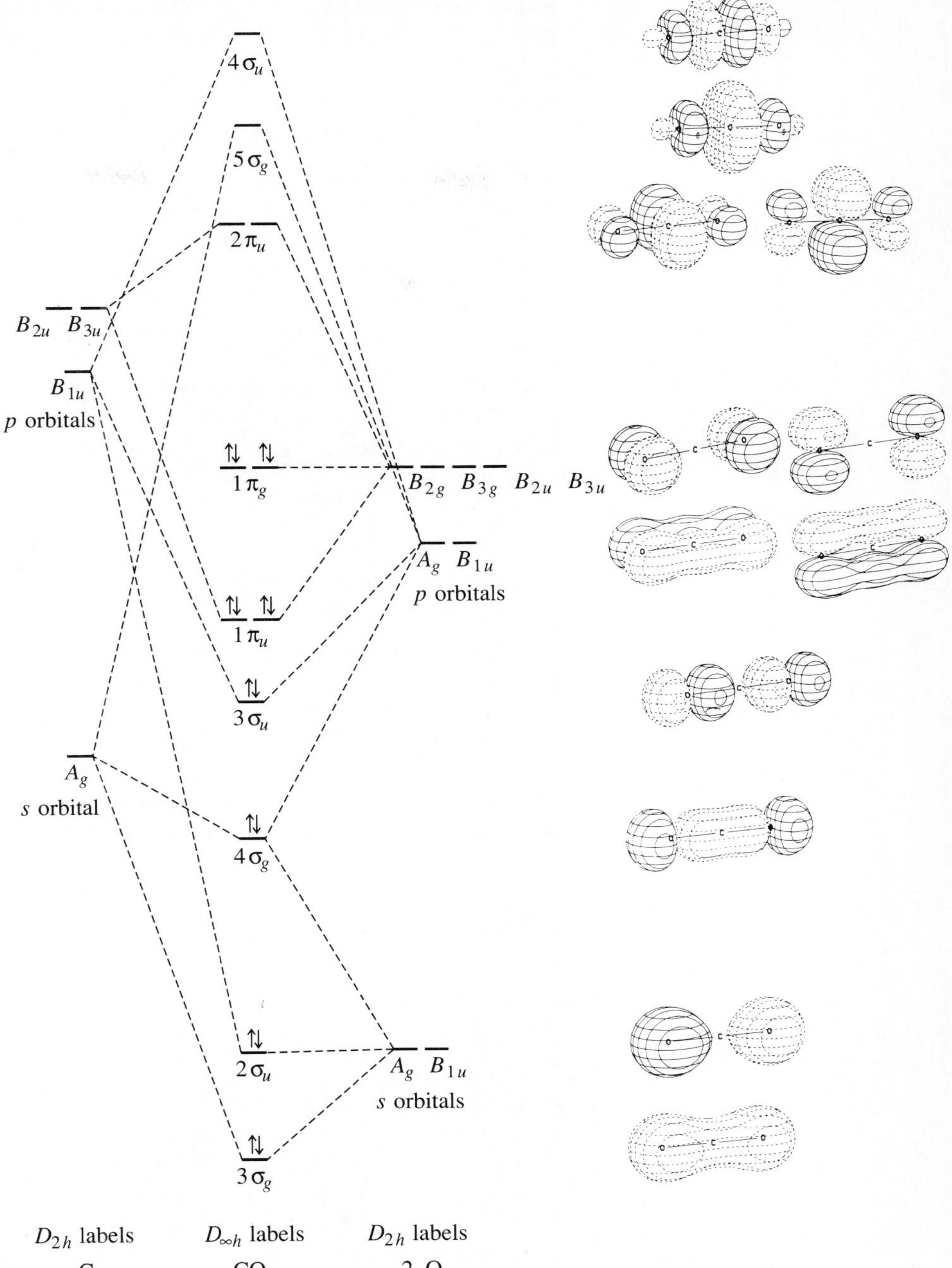

FIGURE 5-33 Molecular Orbitals of Carbon Dioxide. (Contour diagrams reproduced with permission from W. L. Jorgensen and L. Salem, *The Organic Chemist's Book of Orbitals,* Academic Press, New York, 1973, pp. 130–31.)

interactions (largest overlap), so they are both the lowest energy bonding orbitals and the highest energy antibonding orbitals. The contour diagrams show the unequal contributions of different atomic orbitals reflected in the energy levels. Because the oxygen atomic orbital energies are so much below the carbon orbital levels, the lowest b_{1u} level ($2\sigma_u$) is composed mostly of oxygen *s* orbitals, and the corresponding highest b_{1u} level ($4\sigma_u$) has a large contribution from the carbon p_z orbital. Similarly, the carbon 2*s* orbital contributes more to the highest a_g level ($5\sigma_g$) and the oxygen $2s$ orbitals contribute more heavily to the lowest ($3\sigma_g$).

SO_3

Sulfur trioxide is a planar D_{3h} molecule. We choose to place the oxygen axes as shown in Figure 5-34, with y pointed toward the sulfur, x in the plane of the molecule, and z perpendicular to the plane of the molecule. For sulfur, x and y are in the plane of the molecule, and z is perpendicular to the plane. The entire set of orbitals could be taken together in one reducible representation and then broken down, but it is simpler to take smaller subsets. Here, each set (x, y, and z) of the p orbitals is different and cannot be converted to any other by molecular symmetry operations, so each set of three (one from each of the oxygens) can be considered separately.

The resulting representations for the oxygen group orbitals (three orbitals for each type, one from each oxygen atom) are

$2s$	$A_1' + E'$
$2p_x$	$A_2' + E'$
$2p_y$	$A_1' + E'$
$2p_z$	$A_2'' + E''$

The representations for the sulfur orbitals are

$2s$	A_1'
$2p_x$, $2p_y$	E'
$2p_z$	A_2''

The molecular orbitals resulting from combining these orbitals are shown in Figure 5-35. Combining the A_1' orbitals (s and p_y group orbitals of the

D_{3h}	E	$2C_3$	$3C_2$	σ_h	$2S_3$	$3\sigma_v$		
A_1'	1	1	1	1	1	1		$x^2 + y^2$, z^2
A_2'	1	1	−1	1	1	−1	R_z	
E'	2	−1	0	2	−1	0	(x, y)	$(x^2 - y^2, xy)$
A_1''	1	1	1	−1	−1	−1		
A_2''	1	1	−1	−1	−1	1	z	
E''	2	−1	0	−2	1	0	(R_x, R_y)	(xz, yz)

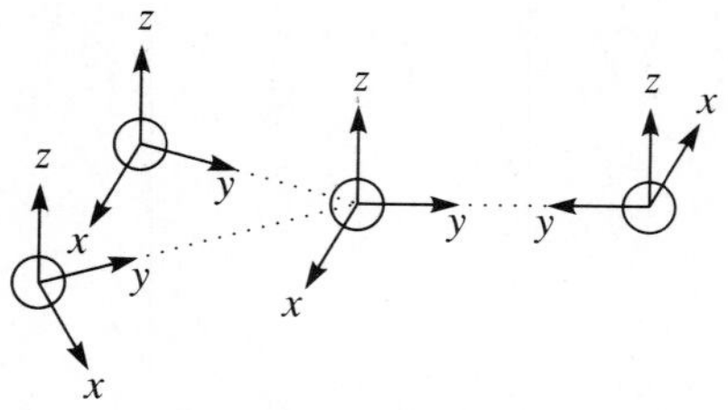

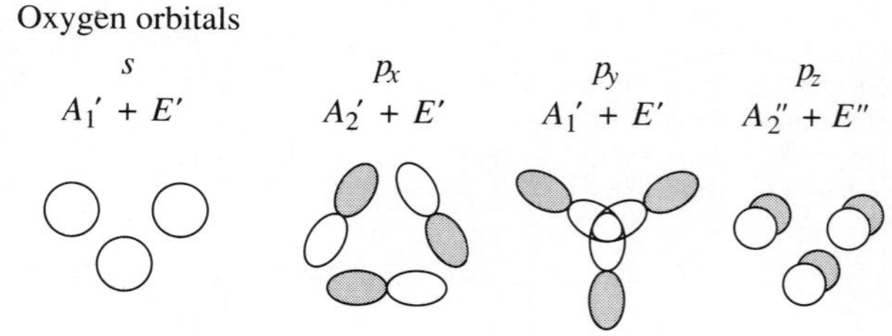

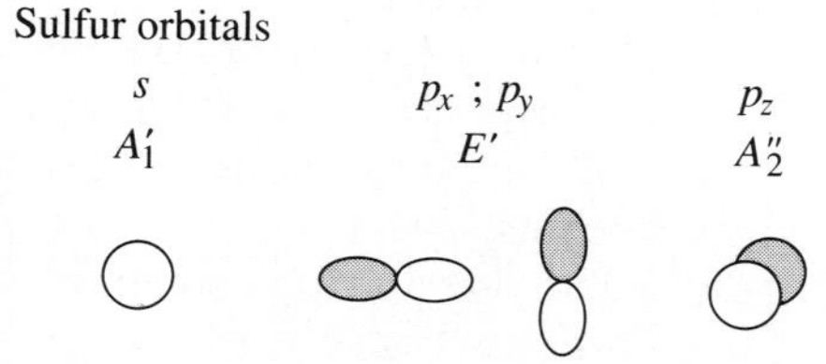

Oxygen orbitals

s	p_x	p_y	p_z
$A_1' + E'$	$A_2' + E'$	$A_1' + E'$	$A_2'' + E''$

Sulfur orbitals

s	p_x ; p_y	p_z
A_1'	E'	A_2''

FIGURE 5-34 Axes, Character Table, and Orbital Symmetry for SO_3. Only the A_1', A_2', and A_2'' combinations are shown for the oxygen orbitals. See Figure 5-36 for mixtures that provide E' and E'' symmetry.

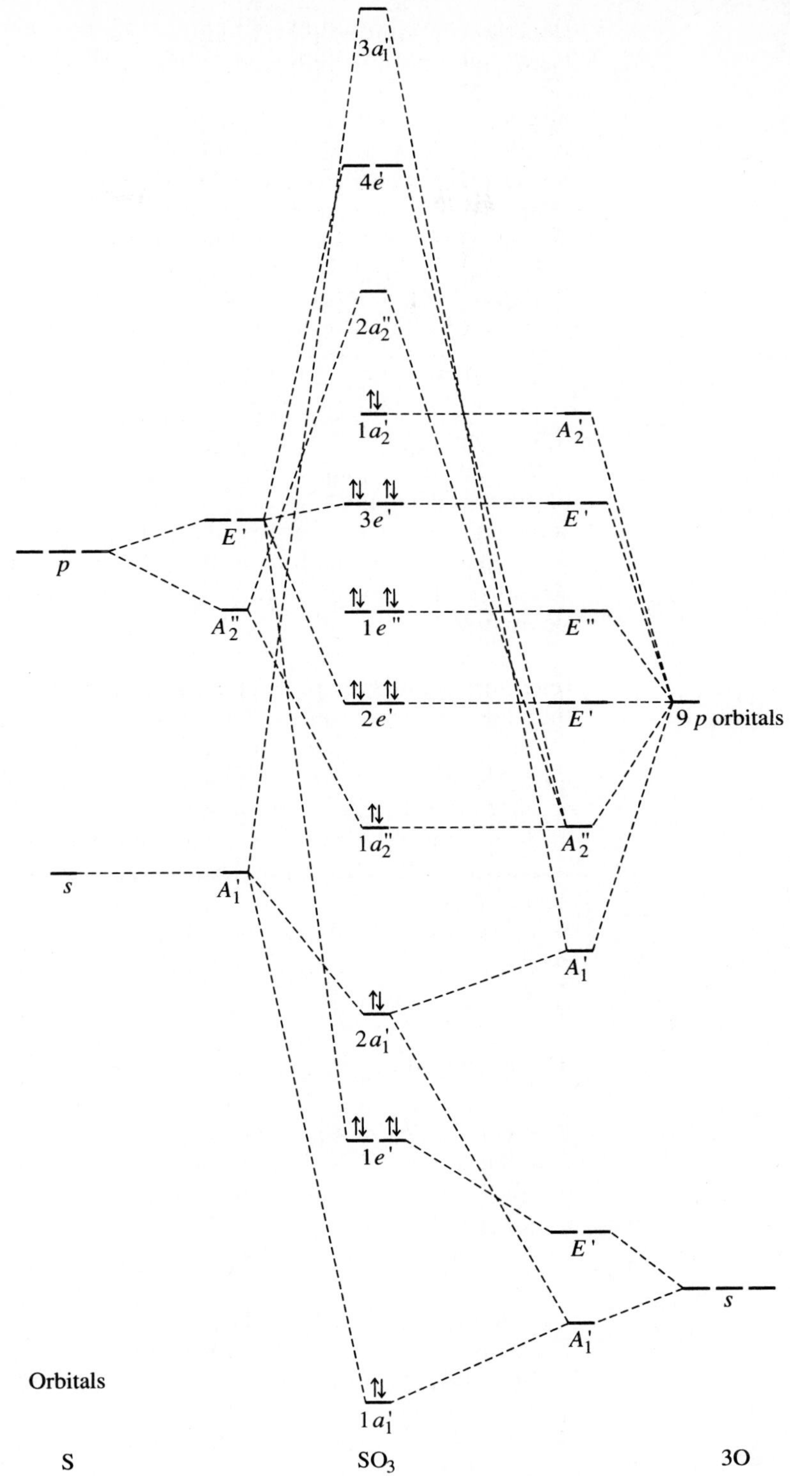

FIGURE 5-35 Energy Levels of Sulfur Trioxide.

oxygens, s of the sulfur), we obtain three molecular orbitals, one bonding, one essentially nonbonding, and one antibonding.

Similarly, the E' orbitals (s, p_y, and p_x group orbitals for the oxygens, p_x and p_y for the sulfur) also combine, with a result of six σ orbitals, two each bonding, nonbonding, and antibonding, and two nonbonding orbitals from the p_x orbitals of the oxygens, which do not overlap well with the sulfur orbitals. (The overall symmetry allows overlap, but the oxygen orbitals do not extend far enough toward the sulfur to allow much overlap.)

The p_z orbitals form one bonding and one antibonding a_2'' and two nonbonding e'' orbitals.

Filling in the 24 valence electrons gives a total of three σ bonds ($1a_1'$ and $1e'$), one π bond ($1a_2''$), and eight lone pairs in the nonbonding orbitals ($2a_1'$, $1a_2'$, $2e'$, $3e'$, and $1e''$), again as expected from the electron-dot model. The difference is in the π (a_2'') orbital, which covers the entire molecule in the molecular orbital model and contributes partial double-bond character to each of the S—O bonds. Each S—O bond therefore has a bond order of $1\frac{1}{3}$, the same as the average from the electron-dot picture. The advantage of the molecular orbital method is that this bond order arises naturally from the overall analysis, with no requirement that three different resonance structures be included.

A more rigorous approach to the bonding mixes all the E' orbitals together, blending s, p_x, and p_y orbitals, and mixes all the A_1' orbitals together, blending s and p_y orbitals. The result of such a calculation leads to the orbitals in Figure 5-36. The atomic orbital combinations shown are those with the largest coefficients; keep in mind that the complete molecular orbitals include smaller contributions from other atomic orbitals of the same symmetry.

EXERCISE 5-7

Work out the molecular orbitals of BF_3 using the group theory approach described above.

5-5-1 HYBRID ORBITALS

It is sometimes more convenient to be able to label the atomic orbitals that combine in molecular orbitals as **hybrid orbitals,** or **hybrids.** In this method, the orbitals of the central atom are combined into hybrids in a process similar to formation of the group orbital combinations used earlier. These hybrid orbitals are then used to form bonds with other atoms whose orbitals overlap properly.

Identification of hybrids is easily done on the basis of molecular symmetry. For example, water can be considered as having nearly tetrahedral symmetry (counting the two lone pairs and the two bonds equally). As a result, all four valence orbitals of oxygen must be used, and the hybrid orbitals are sp^3. The predicted bond angle is then the tetrahedral angle of 109.5°; the experimental value is 104.5°. Repulsion by the lone pairs, as described in the VSEPR section of Chapter 3, is the usual explanation for this smaller angle.

An alternative approach uses the results of the molecular orbital method. The oxygen orbitals used in molecular orbital bonding in water are the $2s$, $2p_x$, and $2p_z$. As a result, the hybrid could be described as sp^2, a mix of one

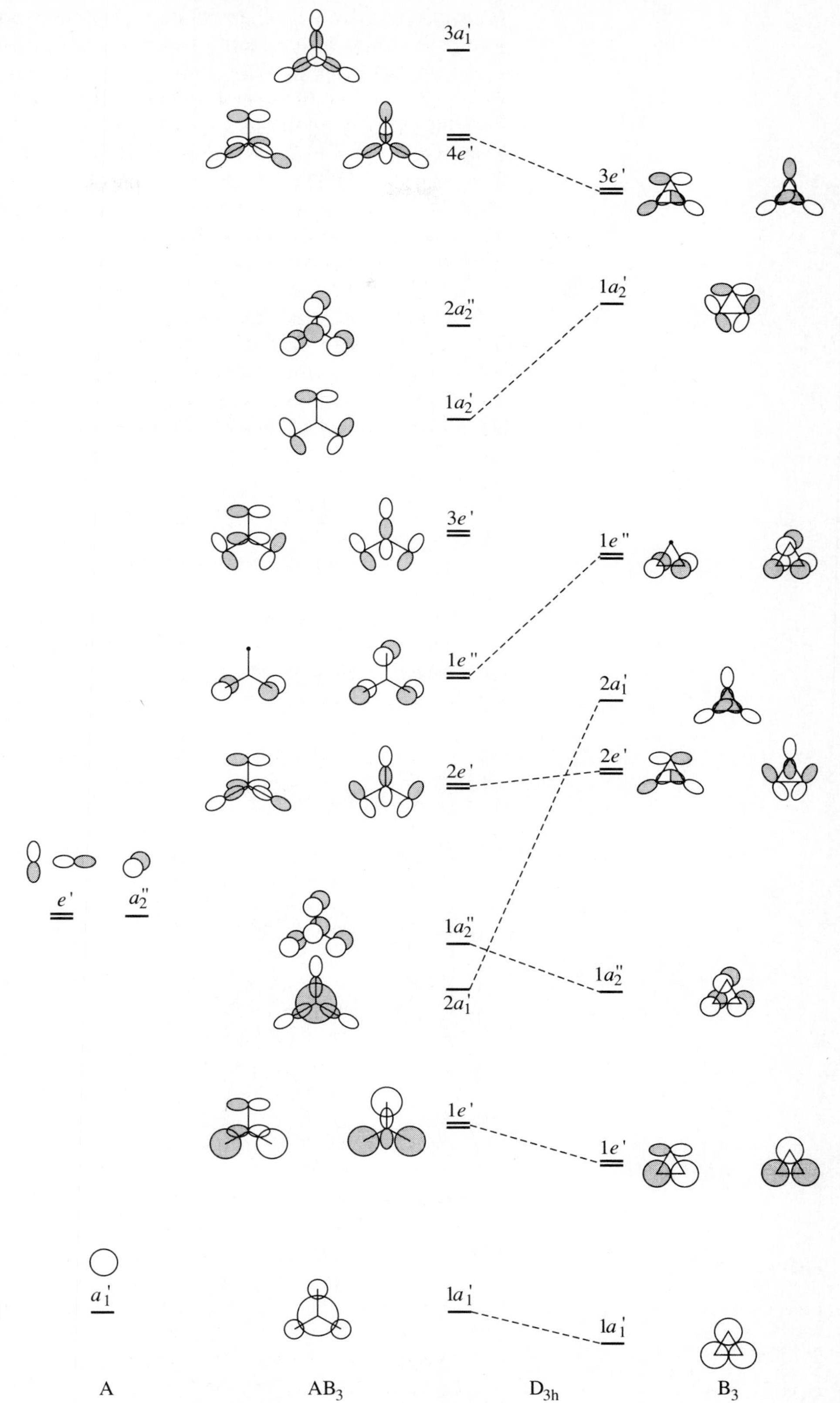

FIGURE 5-36 Molecular Orbitals of Planar AB_3. (Adapted with permission from B. M. Gimarc, *Molecular Structure and Bonding*, Academic Press, New York, 1981, p. 171. The $3a_1'$ orbital was omitted in the original.)

s orbital and two *p* orbitals. Three sp^2 orbitals have trigonal symmetry and a predicted HOH angle of 120°, considerably larger than the experimental value. Repulsion by the lone pairs on the oxygen (one in an sp^2 orbital, one in the remaining p_y orbital) forces the angle to be smaller.

Bonding in NH_3 uses all the nitrogen orbitals, so the hybrids are sp^3, including one *s* orbital and all three *p* orbitals, with overall tetrahedral symmetry. The predicted HNH angle is 109.5°, narrowed to the actual 106.6° by repulsion from the lone pair, also in an sp^3 orbital. Similarly, CO_2 uses *sp* hybrids and SO_3 uses sp^2 hybrids. Only the σ bonding is considered in determining the orbitals used in hybridization, with the *p* orbitals not used in the hybrids available for π interactions. The number of atomic orbitals used in the hybrids is frequently the same as the number of directions counted in the VSEPR method. The MO treatment of water is an exception, as explained above. All these hybrids are summarized in Figure 5-37, along with others using *d* orbitals.

Both the simple descriptive approach and the group theory approach to hybridization are used in the following example.

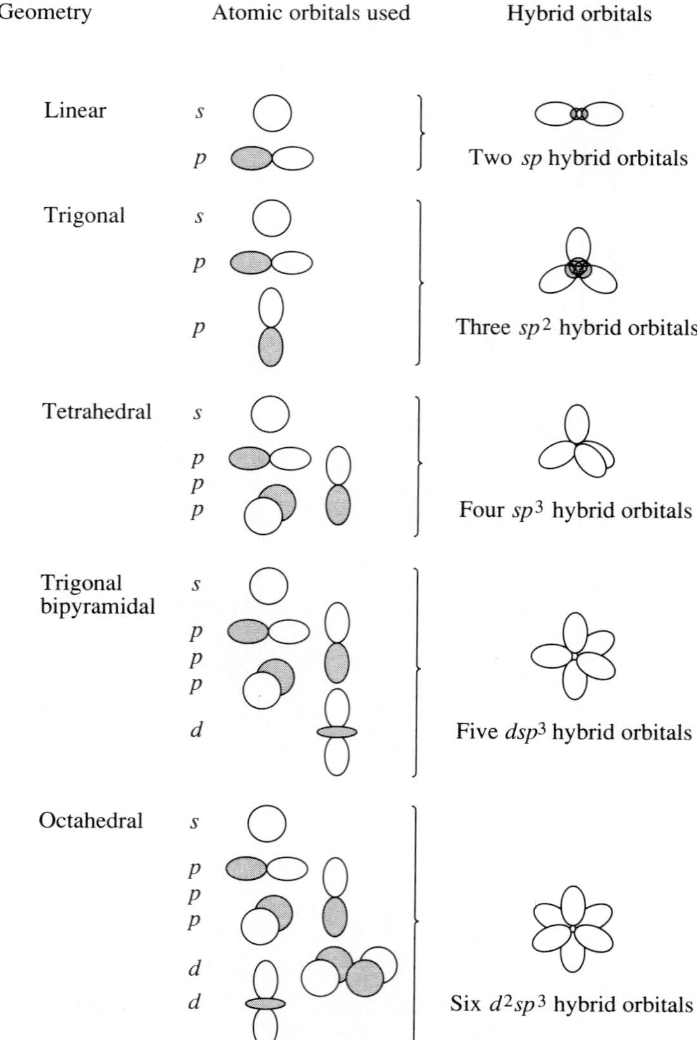

FIGURE 5-37 Hybrid Orbitals. Each single hybrid has the general shape. The figures here show all the resulting hybrids combined, omitting the smaller lobe in the sp^3 and higher orbitals.

EXAMPLE

CH_3^+

VSEPR Shape and Hybrids

We can formally visualize CH_3^+ to result from interactions between a carbon 1+ ion and three hydrogen atoms, and further consider it to include a separate hybridization step, as in Figure 5-38. The $2s$, $2p_x$, and $2p_y$ orbitals of carbon interact to form the three sp^2 hybrids with energy intermediate between the energies of the original atomic orbitals. The $2p_z$ orbital, which is not involved with the hybrids, remains unaffected. The sp^2 hybrids can then interact with the $1s$ orbitals of the hydrogens to form σ and σ^* molecular orbitals, as shown. The net result places three electron pairs in a set of three σ orbitals, the equivalent of three single C—H bonds.

EXERCISE 5-8

Sketch the hybrid orbitals and prepare a molecular orbital diagram for BeH_2.

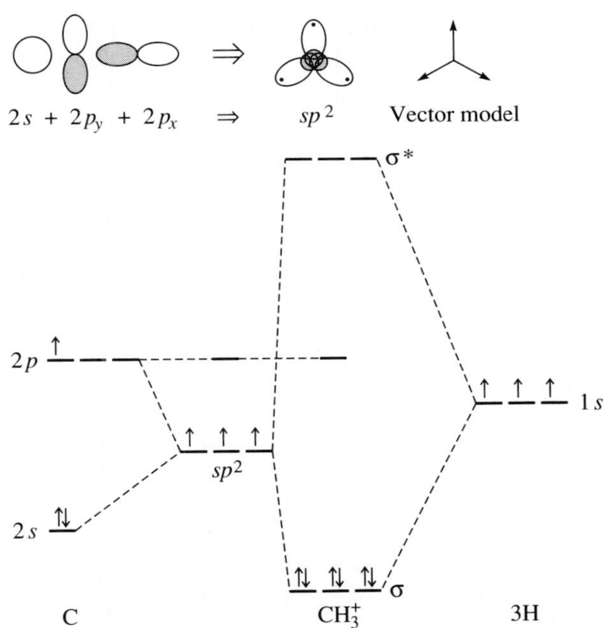

FIGURE 5-38 CH_3^+ Hybrids.

Group theory and hybridization

The group theory methods described earlier can also be used directly to find hybrids used for sigma bonding by the central atom in a molecule. To do this:

1. Determine the shape of the molecule by VSEPR techniques and consider each sigma bond to the central atom and each lone pair on the central atom to be a vector pointing out from the center.
2. Find the reducible representation for the vectors, using the appropriate group and character table, and find the irreducible representations that combine to form the reducible representation.
3. The atomic orbitals that fit the irreducible representations are those used in the hybrid orbital.

Hybrid orbitals are directional. They have a major lobe, plus a smaller lobe of opposite sign, and can be viewed as having the major lobe pointing toward surrounding atoms or lone electron pairs much in the manner of an arrow or vector. Consequently, the symmetry properties of a set of hybrid orbitals are identical to the symmetry properties of a set of vectors with origins at the atomic nucleus of a central atom and pointing toward the surrounding atoms (or toward lone electron pairs). For example, the sp^2 hybrid orbitals of CH_3^+ have the same symmetry as three vectors pointing toward corners of an equilateral triangle (see Figure 5-38). The molecule has D_{3h} symmetry, and these three vectors can be used as the basis of a reducible representation. Using the symmetry operations of D_{3h}, we find the reducible representation $\Gamma = A_1' + E'$:

D_{3h}	E	$2C_3$	$3C_2$	σ_h	$2S_3$	$3\sigma_v$	
Γ	3	0	1	3	0	1	
A_1'	1	1	1	1	1	1	$x^2 + y^2, z^2$
E'	2	−1	0	2	−1	0	$(x, y), (x^2 - y^2, xy)$

This means that the atomic orbitals in the hybrids must have the same symmetry properties as A_1' and E'. More specifically, it means that one orbital must have the same symmetry as A_1' (which is one dimensional) and two orbitals must have the same symmetry, collectively, as E' (which is two dimensional). Looking at the functions listed for each, we see that the s orbital (not listed, but understood to be present for the totally symmetric representation) and the d_{z^2} orbital match the A_1' symmetry. However, the $3d$ orbitals, the lowest possible d orbitals, are too high in energy for bonding in CH_3^+ compared to the $2s$ and $2p$. Therefore, the $2s$ orbital is the contributor with A_1' symmetry.

The functions listed for E' symmetry match the p_x, p_y, or the $d_{x^2-y^2}$, d_{xy} orbitals. The $3d$ are too high in energy for effective bonding, so the $2p_x$ and $2p_y$ orbitals are used by the central atom.

Overall, the orbitals used in the hybridization are the $2s$, $2p_x$, and $2p_y$ orbitals of carbon, comprising the familiar sp^2 hybrids. In other atoms, the d orbitals identified as having the appropriate symmetry might be used in bonding, provided they have energies close to those of the orbitals of the surrounding atoms.

The same procedure can be used for the water and ammonia molecules. Such a procedure results in sp^3 hybridization for water, with two bonds and two lone pairs and sp^3 hybridization for ammonia, with three bonding pairs and the lone pair in hybrid orbitals.

The procedure just described for determining hybrids is very similar to that used in finding the molecular orbitals. Hybridization uses vectors pointing toward the outlying atoms and usually deals only with σ bonding. Once the σ hybrids are known, π bonding is easily added. It is also possible to use hybridization techniques for π bonding, but we will not describe that approach here.[14] Hybridization may be quicker than the molecular orbital approach; the molecular orbital approach uses all the atomic orbitals of the atoms and includes

[14] F. A. Cotton, op. cit., p. 227.

both σ and π bonding directly. Both methods are useful, and the choice of method depends on the particular problem and personal preference.

5-5-2 MORE COMPLEX MOLECULES

The methods described in this chapter also apply to more complex molecules involving d orbitals in bonding. As an example, consider SF_6, strictly octahedral both by VSEPR prediction and experimentally. Although we should consider all the fluorine orbitals for completeness, we will limit this description to a single orbital on each fluorine directed toward the sulfur. The justification for this is the lack of multiple bonds in SF_6. In Chapter 8, we will expand the treatment of octahedral compounds to include all orbitals and allow for π bonding.

The representation of the six fluorine orbitals reduces to $A_{1g} + T_{1u} + E_g$:

O_h	E	$8C_3$	$6C_2$	$6C_4$	$3C_2$	i	$6S_4$	$8S_6$	$3\sigma_h$	$6\sigma_d$	Sulfur orbitals
Γ	6	0	0	2	2	0	0	0	4	2	
A_{1g}	1	1	1	1	1	1	1	1	1	1	s
T_{1u}	3	0	-1	1	-1	-3	-1	0	1	1	(p_x, p_y, p_z)
E_g	2	-1	0	0	2	2	0	-1	2	0	$(d_{x^2-y^2}, d_{z^2})$

On the sulfur, the s orbital, all three p orbitals, and the two d orbitals directed along the axes are used in bonding. The remaining d orbitals, with T_{2g} symmetry, are nonbonding. In hybrid terms, sulfur forms six d^2sp^3 orbitals, all equivalent and pointing along the x, y, and z axes.

EXERCISE 5-9

Find the reducible representation for the fluorine sigma orbitals, reduce it to its irreducible representations, and determine the xenon orbitals used in bonding for XeF_4.

5-6 METALS, SEMICONDUCTORS, AND INSULATORS

In the examples described earlier in this chapter, atomic orbitals combine to form molecular orbitals when their symmetry and energy match. In a homonuclear diatomic molecule, every pair of atomic orbitals forms a pair of molecular orbitals. If the number of atoms is increased, the number of orbitals (both atomic and molecular) grows, but the spread of energy from lowest to highest molecular orbital does not change appreciably, because the interactions between neighboring atoms are still similar. Extending this same description to large aggregates of identical atoms results in a large number of molecular orbitals in the extended crystal, all within a narrow energy span. These energy levels are then known as a **band** of energy levels, essentially continuous rather than sharply defined.

Figure 5-39 shows these bands as a function of interatomic distance in a diamond lattice. At large distances (500 to 600 pm), the $2s$ and $2p$ levels are the usual distinct, narrow energy levels of the individual atoms. At smaller

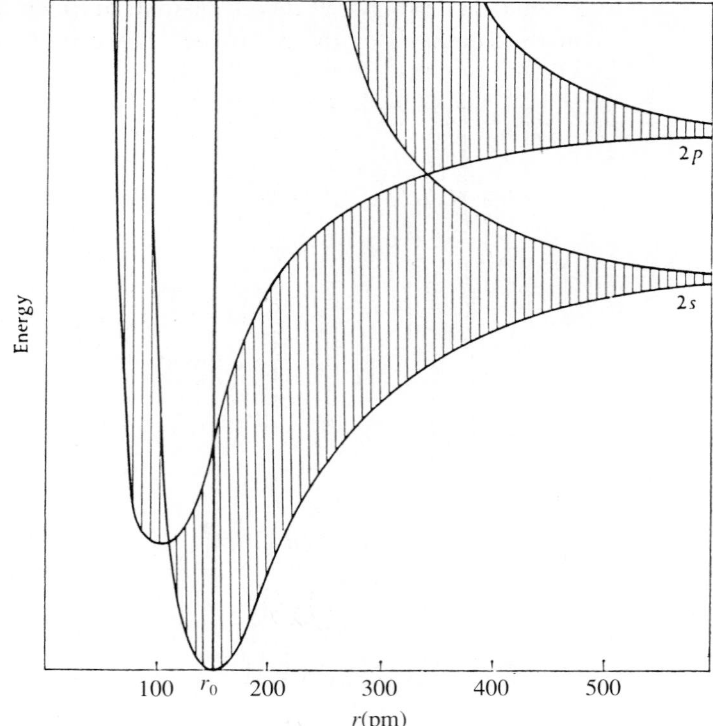

FIGURE 5-39 Energy Bands of Diamond. The energy bands of diamond as a function of internuclear (C—C) distance *r*. All bands are hatched. (Reproduced with permission from C. S. G. Phillips and R. J. P. Williams, *Inorganic Chemistry,* Vol. 1, Oxford University Press, New York, 1965, p. 214.)

distances, the bands broaden until the energy bands originating in *s* and *p* atomic orbitals overlap to form a single continuous set of energy levels at an intermediate distance (330 pm, close enough for strong interactions, but considerably larger than those for usual interatomic distances). At still smaller distances, a gap reappears between the two bands because of mixing interactions similar to those described earlier between molecular orbitals of the same symmetry. At 150 pm, the normal bond distance in diamond, the 2*s* and 2*p* levels of carbon combine in the diamond crystal to form two bands, each with the capacity to hold four electrons per atom, separated by a large energy difference (the upper band is off the scale of Figure 5-39). The four electrons of carbon fill the lower band to capacity, and the large energy gap prevents any easy motion of electrons through the crystal. Therefore, band theory leads to a description of diamond in which four electrons form bonds to surrounding atoms. Although the orbitals belong to the entire crystal in the band theory, the result is still essentially the same as a localized bond picture, with four covalent bonds to each atom. In both cases the electrons are prevented from moving by large energy gaps between levels.

The energies and widths of these bands for other atoms are determined by the symmetry of the orbitals, the symmetry of the crystal, and the interatomic distances in the crystal. When there are large gaps between the energy levels and the lower levels are filled, the electrons cannot move easily within the crystal and it is an **insulator,** which does not conduct electricity.

The alkali metals, such as lithium, have similar band energies, but fewer electrons. As a result, the lower energy band is only partly filled. It requires only a small amount of energy to lift some of these electrons into the empty portion of the band, where they are then free to move through the crystal under the influence of an applied external electric voltage. These elements are **conductors;** this easy conductance of electricity is the major property that defines metals.

Transition metals also conduct electricity readily. Their band structure is more complex, but leads to similar ease of motion by the electrons. The 4*s* levels of the transition metals merge to form a broad band wide enough to extend both above and below the 3*d* band; the 4*s* band also overlaps slightly with the 4*p* band. In Figure 5-40, the energy is plotted horizontally and the number of electrons in the band is shown vertically. The 4*s* and 3*d* bands of transition metal crystals are incompletely filled with electrons, and, as a result, it takes very little energy to move an electron to a higher energy level. A small applied voltage is enough to move such electrons from one end of the crystal to another, and the metal conducts electricity. Even at the end of the transition series, zinc is a conductor because the 4*s* and 4*p* bands overlap enough to allow easy excitation of electrons into empty levels.

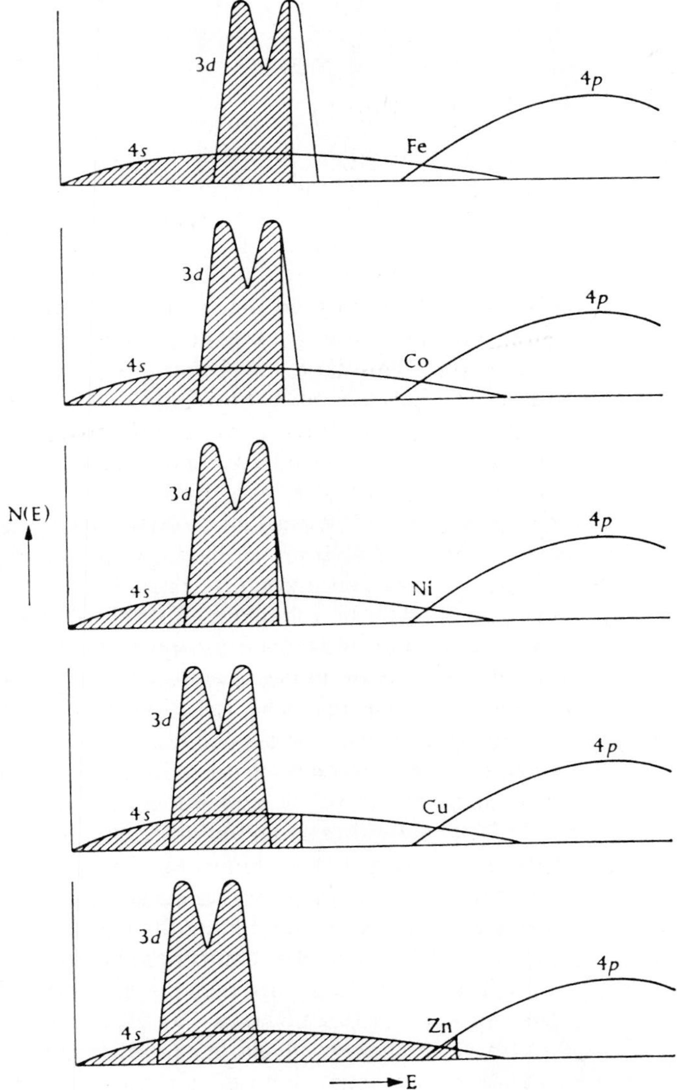

FIGURE 5-40 Energy bands of Transition Metals. Relative positions of 3*d*, 4*s*, and 4*p* bands according to Kiestra (D. Phil. thesis, Groningen, 1956). Occupied levels are hatched. (Reproduced with permission from C. S. G. Phillips and R. J. P. Williams, *Inorganic Chemistry,* Vol. 1, Oxford University Press, New York, 1965, p. 205.)

In fact, the excitation of electrons requires such a small amount of energy that the sharp edge of the top of the occupied portion of a band in a metal is blurred and broadened, even at relatively low temperature. As a result, the energy levels just above and just below the ''normal'' upper limit of filled levels are only partly occupied. Contrast this with the behavior of individual atoms, in which excitation of electrons requires high temperatures (flames or arcs). A related effect appears in the heat capacity. Metals, which have partly filled bands, have an appreciable heat capacity contribution from the excitation of electrons. Insulators, with completely filled bands, require larger amounts of energy to excite electrons to the higher empty levels and have very small electronic contributions to their heat capacities.

Conduction in metals can also be described in an alternative way. In this description, the metal is made up of positive ions in the crystal, all surrounded by a sea of electrons required to balance the overall charge. Since the electrons are relatively free, they can move easily through the crystal and conduct electricity. The temperature dependence of conductance is explained by this model as caused by vibration of the positive ions. As the temperature is raised, the normal vibrations of the metal ions increase, interfering with the easy movement of the electrons. As a result, conductance *decreases* at higher temperatures.

Group 14 (IVA) shows a progression from an insulator (C, in diamond form), through **semiconductors** (Si and Ge), to a conductor (Sn), with conductance increasing by a factor of 100 to 1000 between each of the elements with increasing atomic number. All four have the same crystal structure, and each atom can be thought of as having four tetrahedral bonds to other identical atoms. In diamond, the filled bonding orbitals are well below the empty antibonding orbitals, and the conductance is negligible. In silicon and germanium, the higher atomic number and the addition of *d* orbitals result in closer energies for the orbitals; thermal energy or energy from incident light is enough to raise some electrons to the higher levels and conductance is significant, although still small. In tin, the atomic levels are still closer, and the bands overlap to give a conductance that approaches that of a typical metal (there is also a high-temperature form of tin with even higher conductance).

In contrast to metals, the conductance of **intrinsic semiconductors** such as Si and Ge *increases* with increasing temperature. The conductance is low because a significant amount of energy is needed to excite electrons from the occupied band to the vacant band. As the temperature is raised, more electrons acquire the necessary energy, and the conductance rises. The interference from vibration seen in metal conductance has a much smaller effect; overall, a higher temperature results in a higher conductance for semiconductors.

The properties of semiconductors can be changed by adding a small amount of another element as a **dopant** in the crystal. If arsenic is added to germanium or silicon, it adds an extra electron, and the crystal is then an ***n*-type** (for negative) semiconductor. On the other hand, if boron is added, there is a shortage of electrons, and it is a ***p*-type** (for positive) semiconductor. In either case, the positive or negative charge can move through the crystal and carry current. By combining the two and applying appropriate voltages or currents, it is possible to control the flow of electrons in a circuit and make useful transistors. Recent developments allow the equivalent of thousands of transistors to be formed on a tiny silicon crystal by applying thin layers of different conductors, insulators, and dopants. As a result, single **chips** can contain all the essential circuitry for a computer.

Superconductivity

In contrast to the lower conductivity of semiconductors, there are a number of substances that exhibit **superconductivity.** In these, an electric current experiences no resistance, and a current started in a continuous loop will continue to circulate indefinitely with no loss of energy. One of the most common uses of this phenomenon in chemistry is in the production of high-energy magnetic fields for nuclear magnetic resonance. In the superconductors commonly used, liquid helium is required to attain the temperatures required for the superconducting state (below 20 K). More recently, materials that become superconducting at the temperature of liquid nitrogen (77 K) or higher have been discovered.[15]

Although the theory of superconductivity is still under development, particularly as it relates to high-temperature materials, the general outline is known. Instead of individual electrons moving through a crystal lattice, as in metals, superconductors seem to have pairs of electrons that are able to move with surprising ease through the lattice. These **Cooper pairs** result from vibrational interactions of the positive nuclei with an electron. As these nuclei move toward the electron, they create a local concentration of positive charge that attracts another electron. At the low temperatures required, the ordinary thermal vibrations of the lattice are small enough that they do not interfere with this effect. As the electrons move through the lattice under the influence of an electric field, their mutual interaction keeps them moving together and prevents collisions with stationary atoms from scattering them. Most of the common low-temperature superconductors are niobium-tin or niobium-germanium alloys (mixtures of metals). The high-temperature superconductors are commonly oxides, such as $YBa_2Cu_3O_7$, and are easily prepared by heating the appropriate mixture to 800 or 900°C. The mechanism of conduction in these compounds is not yet clear. For practical use, the problem of forming the brittle oxides into useful shapes is a serious one. The usual methods of drawing wires through a series of smaller and smaller holes are not possible for such brittle materials. Engineering problems such as this are being studied as extensively as the more fundamental questions about the mechanisms of conduction.

GENERAL REFERENCES

There are many books describing bonding and molecular orbitals, with levels ranging from those even more descriptive and qualitative than the treatment in this chapter to those designed for the theoretician interested in the latest methods. A classic that starts at the level of this chapter and includes many more details is R. McWeeny's revision of *Coulson's Valence,* 3rd ed., Oxford University Press, New York, 1979. A different approach, which uses the concept of generator orbitals, is that of J. G. Verkade in *A Pictorial Approach to Molecular Bonding,* Springer-Verlag, New York, 1986. The group theory approach in this chapter is similar to that of F. A. Cotton, *Chemical Applications of Group Theory,* 3rd ed., Wiley, New York, 1990. The contour diagrams in this chapter are from W. L. Jorgensen and L. Salem, *The Organic Chemist's Book of Orbitals,* Academic Press, New York, 1973. It is a useful reference for many other molecules as well.

[15] A. Müller and J. G. Bednorz, *Science,* **1987,** *217,* 1133.

PROBLEMS

5-1 Expand the list of orbitals considered in Figures 5-2 and 5-3 by using all three p orbitals of atom A and all five d orbitals of atom B. Which of these have the necessary match of symmetry for bonding and antibonding orbitals? These combinations are rarely seen in simple molecules, but can be important in transition metal complexes.

5-2 Although KrF^+ and XeF^+ have been studied, $KrBr^+$ has not yet been prepared. For $KrBr^+$:

a. Propose a molecular orbital diagram, showing the interactions of the valence shell s and p orbitals to form molecular orbitals.
b. Toward which atom would the HOMO be polarized? Why?
c. Predict the bond order.
d. Which is more electronegative, Kr or Br? Explain your reasoning.

5-3 **a.** Prepare a molecular orbital energy level diagram for the cyanide ion. Use sketches to show clearly how the atomic orbitals interact to form MOs.
b. What is the bond order, and how many unpaired electrons does cyanide have?
c. Which molecular orbital of CN^- would you predict to interact most strongly with a hydrogen $1s$ orbital to form an H—C bond in the reaction $CN^- + H^+ \longrightarrow HCN$? Explain.

5-4 For the compound XeF_2:

a. Sketch the valence shell group orbitals for the fluorines (with the z axes collinear with the molecular axis).
b. For each of the group orbitals, determine which outermost s, p, and d orbitals of xenon are of suitable symmetry for interaction and bonding.

5-5 **a.** Prepare a molecular orbital energy level diagram for NO, showing clearly how the atomic orbitals interact to form MOs.
b. How does your diagram illustrate the difference in electronegativity between N and O?
c. Predict the bond order and the number of unpaired electrons.
d. NO^+ and NO^- are also known. Compare the bond orders of these ions with the bond order of NO. Which of the three would you predict to have the shortest bond? Why?

5-6 The hypofluorite ion, OF^-, can be observed only with difficulty.

a. Prepare a molecular orbital energy level diagram for this ion.
b. What is the bond order and how many unpaired electrons are in this ion?
c. What is the most likely position for adding H^+ to the OF^- ion? Explain your choice.

5-7 For the ozone molecule, O_3:

a. Prepare a molecular orbital energy level diagram without mixing of the s and p orbitals.
b. Indicate the changes in molecular orbital energies that you would predict on mixing of s and p orbitals.

5-8 The ion H_3^+ has been observed, but its structure has been the subject of some controversy. Prepare molecular orbital energy level diagrams for H_3^+, assuming:

a. A linear structure.
b. A cyclic structure.

5-9 Compare the bonding in O_2^{2-}, O_2^-, and O_2. Include Lewis structures, molecular orbital structures, bond lengths, and bond strengths in your discussion.

5-10 Diborane, B_2H_6, has the structure shown. Using molecular orbitals (and showing appropriate orbitals on B and H from which the MOs are formed), explain how hydrogen can form bridges between two B atoms. (This type of bonding is discussed in Chapter 7.)

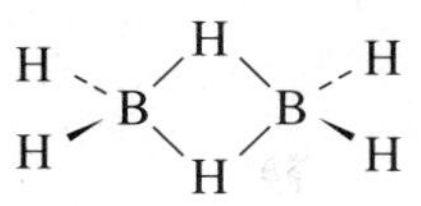

H H H
B B
H H H

5-11 SF_4 has C_{2v} symmetry. Using the group theory method, predict the possible hybridization schemes for the sulfur atom in SF_4.

5-12 Consider a square pyramidal AB_5 molecule. Using the C_{4v} character table, determine the possible hybridization schemes for central atom A. Which of these schemes would you expect to be most likely?

5-13 In coordination chemistry, many square planar species are known (for example, $PtCl_4^{2-}$). For a square planar molecule, use the appropriate character table to determine the types of hybridization possible for a metal surrounded in a square planar fashion by four ligands (consider hybrids used in σ bonding only).

5-14 For the molecule PCl_5:

a. Using the character table for the point group of PCl_5, determine the possible type(s) of hybrid orbitals that can be used by P in forming σ bonds to the five Cl atoms.

b. What type(s) of hybrids can be used in bonding to the axial chlorines? To the equatorial chlorines?

c. Considering your answer to part b, explain the experimental observation that the axial P—Cl bonds (219 pm) are longer than the equatorial bonds (204 pm).

5-15 Determine the types of hybridization possible for atoms in the following environments:

a. Pentagonal bipyramidal

b. Trigonal prismatic

5-16 Use molecular orbital arguments to explain the structures of SCN^-, OCN^-, and CNO^- and compare the results to the electron-dot pictures of Chapter 3.

5-17 Thiocyanate and cyanate ion both bond to H^+ through the nitrogen atoms (HNCS and HNCO), while SCN^- forms bonds with metal ions through either nitrogen or sulfur, depending on the rest of the molecule. What does this suggest about the relative importance of S and N orbitals in the MOs of SCN^-? [HINT: see the discussion of CO bonding.]

5-18 What is the likelihood of cyanide, CN^-, forming bonds to metals through N as well as C, similar to the options of thiocyanate ion?

5-19 The isomeric ions NSO^- (thiazate) and SNO^- (thionitrite) ions have been reported by S. P. So, *Inorg. Chem.*, **1989,** *28,* 2888.

a. On the basis of the resonance structures of these ions, predict which would be more stable.

b. Sketch the approximate shapes of the π and π^* orbitals of these ions.

c. Predict which ion would have the shorter N—S bond and which would have the higher-energy N—S stretching vibration? (Stronger bonds have higher-energy vibrations.)

6

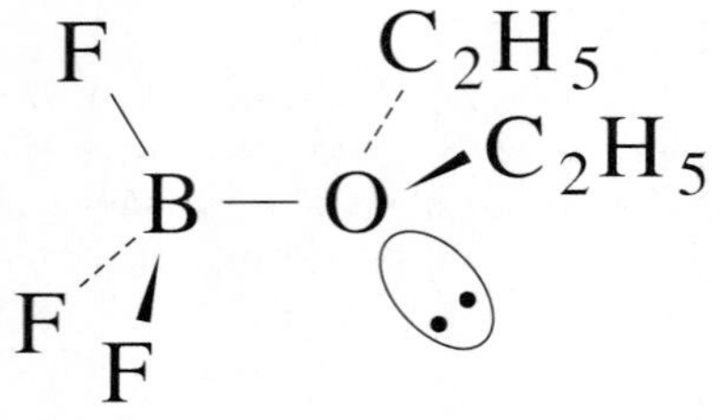

Acid–Base and Donor–Acceptor Chemistry

6-1 ACID–BASE CONCEPTS AS ORGANIZING CONCEPTS

The idea of acids and bases has been a concept of great importance in chemistry since the earliest time, in some cases helping to correlate large amounts of data and in others leading to new predictive ideas. Jensen[1] describes a useful approach in the preface to his book on the Lewis acid–base concept:

> acid–base concepts occupy a somewhat nebulous position in the logical structure of chemistry. They are, strictly speaking, neither facts nor theories and are, therefore, never really "right" or "wrong." Rather they are classificatory definitions or organizational analogies. They are useful or not useful . . . the study of their historical evolution . . . clearly shows that acid–base definitions are always a reflection of the facts and theories current in chemistry at the time of their formulation and that they must, necessarily, evolve and change as the facts and theories themselves evolve and change. . . . the older definitions . . . generally represent the most powerful organizational analogy consistent with the facts and theories extant at the time.

The changing definitions described in this chapter have generally led to a more inclusive and useful approach to acid–base concepts. Most of this chapter is concerned with the Lewis definition, its more recent explanation in terms of molecular orbitals, and its application to inorganic chemistry.

6-2 HISTORY

Practical acid–base chemistry was known in ancient times, developed gradually during the time of the alchemists, and was first satisfactorily explained in

[1] W. B. Jensen, *The Lewis Acid–Base Concepts,* Wiley, New York, 1980, p. vii.

molecular terms after Ostwald and Arrhenius established the existence of ions in aqueous solution in 1880–1890. During the early development of acid–base theory, the experimental observations included the sour taste of acids and the bitter taste of bases, indicator color changes caused by acids and bases, and the reaction of acids with bases to form salts. Partial explanations included the idea that all acids contained oxygen (oxides of nitrogen, phosphorus, sulfur, and the halogens all form acids in water), but by the early nineteenth century many acids that do not contain oxygen were known. By 1838, Liebig defined acids as "compounds containing hydrogen, in which the hydrogen can be replaced by a metal,"[2] a definition that still works well in many instances.

6-2-1 ARRHENIUS CONCEPT

With the discovery of electrolytic dissociation by Ostwald and Arrhenius, a more satisfactory explanation of acid–base reactions became possible. In this concept, **Arrhenius acids form hydrogen ions (or hydronium ions, H_3O^+) in aqueous solution, Arrhenius bases form hydroxide ions in solution,** and the reaction of hydrogen ions and hydroxide ions to form water is the universal aqueous acid–base reaction. The ions accompanying the hydrogen and hydroxide ions form a salt, so the overall Arrhenius acid–base reaction can be written

$$\text{acid} + \text{base} \longrightarrow \text{salt} + \text{water}$$

For example,

$$\text{hydrochloric acid} + \text{sodium hydroxide} \longrightarrow \text{sodium chloride} + \text{water}$$

$$H^+ + Cl^- + Na^+ + OH^- \longrightarrow Na^+ + Cl^- + H_2O$$

This explanation works well in aqueous solution, but is inadequate for nonaqueous solutions and for gas and solid phase reactions where H^+ and OH^- may not exist, and later definitions by Brønsted and Lewis are more appropriate for general use.

6-2-2 BRØNSTED CONCEPT

Brønsted[3] **defined an acid as a species with a tendency to lose a proton and a base as a species with a tendency to add a proton.** These definitions expanded the Arrhenius list of acids and bases to include the gases HCl and NH_3, along with many others. This definition also introduced the concept of **conjugate acids and bases,** differing only in the presence or absence of a proton, and described all reactions as occurring between a stronger acid and base to form a weaker acid and base:

$$\underset{\text{acid 1}}{H_3O^+} + \underset{\text{base 2}}{NO_2^-} \longrightarrow \underset{\text{base 1}}{H_2O} + \underset{\text{acid 2}}{HNO_2}$$

Conjugate acid–base pairs:

Acid	Base
H_3O^+	H_2O
HNO_2	NO_2^-

[2] R. P. Bell, *The Proton in Chemistry,* 2nd ed., Cornell University Press, Ithaca, N.Y., 1973, p. 9.

[3] J. N. Brønsted, *Rec. trav. chim.,* **1923,** *42,* 718.

In water, HCl and NaOH react as the acid H_3O^+ and the base OH^- to form water, which is the conjugate base of H_3O^+ and the conjugate acid of OH^-. Reactions in nonaqueous solvents having ionizable protons parallel those in water. An example of such a solvent is liquid ammonia, where NH_4Cl and $NaNH_2$ react as the acid NH_4^+ and the base NH_2^- to form NH_3, which is both conjugate base and conjugate acid:

$$NH_4^+ + Cl^- + Na^+ + NH_2^- \longrightarrow Na^+ + Cl^- + 2\,NH_3$$

with the net reaction

$$\underset{\text{acid}}{NH_4^+} + \underset{\text{base}}{NH_2^-} \longrightarrow \underset{\substack{\text{conjugate base and}\\ \text{conjugate acid}}}{2\,NH_3}$$

6-2-3 SOLVENT SYSTEM CONCEPT

Aprotic nonaqueous solutions require a similar approach, but with a different definition of acid and base. The solvent system definition applies to any solvent that can dissociate into a cation and an anion (autodissociation), where **the cation resulting from autodissociation of the solvent is the acid and the anion is the base.** The Arrhenius reaction

$$\text{acid} + \text{base} \longrightarrow \text{salt} + \text{water}$$

and the Brønsted acid–base reaction

$$\text{acid 1} + \text{base 2} \longrightarrow \text{base 1} + \text{acid 2}$$

can then become

$$\text{acid} + \text{base} \longrightarrow \text{solvent (both acid and base)}$$

In the solvent BrF_3, for example, the dissociation takes the form

$$2\,BrF_3 \rightleftharpoons BrF_2^+ + BrF_4^-$$

and the acid + base reaction is the reverse:

$$BrF_2^+ + BrF_4^- \rightleftharpoons 2\,BrF_3$$

with BrF_2^+ the acid and BrF_4^- the base. Solutes that increase the concentration of the acid BrF_2^+ are classified as acids, and those that increase the concentration of BrF_4^- are classified as bases. For example, SbF_5 is an acid in BrF_3:

$$SbF_5 + BrF_3 \longrightarrow BrF_2^+ + SbF_6^-$$

Ionic fluorides such as KF are bases in BrF_3:

$$F^- + BrF_3 \longrightarrow BrF_4^-$$

Of course, autoionizing protonic solvents such as H_2O and NH_3 also satisfy the solvent system definition: solutes that increase the concentration of the cation (H_3O^+, NH_4^+) of the solvent are considered acids, and solutes that increase the concentration of the anion (OH^-, NH_2^-) are considered bases.

Table 6-1 gives some of the properties of common solvents.

TABLE 6-1
Properties of solvents

Protic solvents				
Solvent	*Acid cation*	*Base anion*	pK_{ion} *(25°C)*	*Boiling point (°C)*
Ammonia, NH_3	NH_4^+	NH_2^-	27	−33.38
Sulfuric acid, H_2SO_4	$H_3SO_4^+$	HSO_4^-	3.4 (10°)	330
Acetic acid, CH_3COOH	$CH_3COOH_2^+$	CH_3COO^-	14.45	118.2
Hydrogen fluoride, HF	H_2F^+	HF_2^-	~12 (0°)	19.51
Methanol, CH_3OH	$CH_3OH_2^+$	CH_3O^-	18.9	64.7
Water, H_2O	H_3O^+	OH^-	14	100

Aprotic solvents	
Solvent	*Boiling point (°C)*
Dinitrogen tetroxide, N_2O_4	21.15
Sulfur dioxide, SO_2	−10.2
Pyridine, C_5H_5N	115.5
Acetonitrile, CH_3CN	81.6
Diglyme, $CH_3(OCH_2CH_2)_2OCH_3$	162.0
Bromine trifluoride, BrF_3	127.6

SOURCE: Data from W. L. Jolly, *The Synthesis and Characterization of Inorganic Compounds,* Prentice Hall, Englewood Cliffs, N.J., 1970, pp. 99–101. Data for many other solvents are also given by Jolly.

Caution is needed in interpreting acid–base and, indeed, any reaction. For example, $SOCl_2$ and SO_3^{2-} react as acid and base in SO_2 solvent:

$$SOCl_2 + SO_3^{2-} \rightleftharpoons 2\,SO_2 + 2\,Cl^-$$

It was at first believed that $SOCl_2$ dissociated and the resulting SO^{2+} reacted with SO_3^{2-}:

$$SOCl_2 \rightleftharpoons SO^{2+} + 2\,Cl^-$$

$$SO^{2+} + SO_3^{2-} \rightleftharpoons 2\,SO_2$$

However, the reverse reactions should lead to oxygen exchange between SO_2 and $SOCl_2$, but none is observed.[4] The details of the $SOCl_2 + SO_3^{2-}$ reaction are still uncertain.

6-2-4 LEWIS CONCEPT

Lewis[5] defined **a base as an electron-pair donor and an acid as an electron-pair acceptor.** This definition further expands the list to include metal ions and other electron-pair acceptors as acids and provides a handy framework for

[4] W. L. Jolly, *The Synthesis and Characterization of Inorganic Compounds,* Prentice Hall, Englewood Cliffs, N.J., 1970, pp. 108–9; R. E. Johnson, T. H. Norris, and J. L. Huston, *J. Am. Chem. Soc.,* **1951,** *73,* 3052.

[5] G. N. Lewis, *Valence and the Structure of Atoms and Molecules,* Chemical Catalog Co., New York, 1923, pp. 141–42; *J. Franklin Inst.,* **1938,** *226,* 293.

nonaqueous reactions. Most of the acid–base descriptions in this book will use the Lewis definition, which encompasses the Brønsted and solvent system definitions. In addition to all the reactions above, the Lewis definition includes reactions such as

$$Ag^+ + 2\ :NH_3 \longrightarrow H_3N:Ag:NH_3^+$$

with silver ion (or other cation) as an acid and ammonia (or other electron-pair donor) as a base. In reactions such as this, the product is often called an **adduct,** a product of the reaction of a Lewis acid and base to form a new combination. Another example of a Lewis acid–base adduct is a common reagent in synthesis, the boron trifluoride–diethyl ether adduct, $BF_3—O(C_2H_5)_2$. The BF_3 molecule described in Chapters 3 and 5 has a planar triangular structure with some double-bond character in each B—F bond. Since fluorine is the most electronegative element, the boron atom in BF_3 is quite positive. Lone pairs on the oxygen of the diethyl ether are attracted to the boron; the result is that one of the lone pairs bonds to boron, changing the geometry around B from planar to nearly tetrahedral, as shown in Figure 6-1. As a result, BF_3, with a boiling point of $-99.9°C$, and diethyl ether, with a boiling point of 34.5°C, form an adduct with a boiling point of 125° to 126°C (at which temperature it decomposes into its two components).

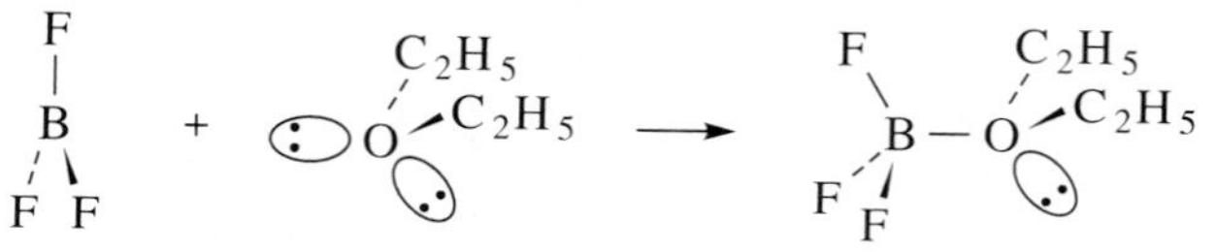

FIGURE 6-1 Boron Trifluoride–Ether Adduct.

Lewis acid–base adducts involving metal ions are **coordination compounds;** their chemistry will be discussed in Chapters 8 to 13. The rest of this chapter will develop the Lewis concept, in which adduct formation is common.

Other acid–base definitions have been proposed. While they are useful in particular types of reactions, none has been widely adopted for general use. The Lux–Flood definition[6] is based on oxide ion, O^{2-}, as the unit transferred between acids (oxide ion acceptors) and bases (oxide ion donors). The Usanovich[7] definition proposes that any reaction leading to a salt (including oxidation–reduction reactions) should be considered an acid–base reaction. This definition could include nearly all reactions and has been criticized for this all-inclusive approach. The Usanovich definition is rarely used today. The electrophile–nucleophile approach of Ingold[8] and Robinson,[9] widely used in organic chemistry, is essentially the Lewis theory with terminology related to reactivity (electrophilic reagents are acids, nucleophilic reagents are bases).

Table 6-2 summarizes these acid–base definitions.

[6] H. Lux, *Z. Electrochem.*, **1939,** *45,* 303; H. Flood and T. Förland, *Acta Chem. Scand.*, **1947,** *1,* 592, 718; W. B. Jensen, *The Lewis Acid–Base Concepts,* Wiley, New York, 1980, pp. 54–55.

[7] M. Usanovich, *Zh. Obshch. Khim.*, **1938,** *9,* 182; H. Gehlen, *Z. Phys. Chem.*, **1954,** *203,* 125; H. L. Finston and Allen C. Rychtman, *A New View of Current Acid–Base Theories,* Wiley, New York, 1982, pp. 140–46.

[8] C. K. Ingold, *J. Chem. Soc.*, **1933,** 1120; *Chem. Rev.*, **1934,** *15,* 225; *Structure and Mechanism in Organic Chemistry,* Cornell University Press, Ithaca, N.Y., 1953, Chap. V; Jensen, op. cit., pp. 58–59.

[9] R. Robinson, *Outline of an Electrochemical (Electronic) Theory of the Course of Organic Reactions,* Institute of Chemistry, London, 1932, pp. 12–15; Jensen, op. cit., pp. 58–59.

TABLE 6-2
Comparison of acid–base definitions

	Definitions		*Examples*	
	Acid	*Base*	*Acid*	*Base*
Lavoisier	Oxide of N, P, S	Reacts with acid	SO_3	NaOH
Liebig	Replaceable H	Reacts with acid	HNO_3	NaOH
Arrhenius	Hydronium ion	Hydroxide ion	H^+	OH^-
Brønsted	Proton donor	Proton acceptor	H_3O^+ H_2O NH_4^+	H_2O OH^- NH_3
Solvent system	Solvent cation	Solvent anion	BrF_2^+	BrF_4^-
Lewis	Electron-pair acceptor	Electron-pair donor	Ag^+	NH_3
Usanovich	Electron acceptor	Electron donor	Cl_2	Na

6-3 ACID AND BASE STRENGTH

6-3-1 MEASUREMENT OF ACID–BASE INTERACTIONS

Interaction between Lewis acids and bases can be measured in many ways in addition to the elevated boiling points of hydrogen-bonded solvents, such as water and hydrogen fluoride (Chapter 3), and the boron trifluoride–ether adduct mentioned in the preceding section. Enthalpies and entropies of acid-base reactions can be measured directly by calorimetric measurements or by temperature dependence of equilibrium constants. The following section gives more details on use of data from these measurements. Similar thermodynamic data can be obtained from gas-phase measurements of the formation of protonated species. Infrared spectra can provide indirect measures of bonding in acid–base adducts by showing changes in bond force constants. For example, free CO has a stretching energy of 2143 cm^{-1}, while CO in $Ni(CO)_4$ has a stretching energy of 2058 cm^{-1}. Nuclear magnetic resonance coupling constants provide a similar indirect measure of changes in bonding on adduct formation, and ultraviolet or visible spectra can show changes in energy levels in the molecules as they combine. Some of these are included in the specific acid–base examples given in the remainder of the chapter.

6-3-2 THERMODYNAMIC MEASUREMENTS

Thermodynamic data can be combined to obtain quantitative data for reactions that are difficult to measure directly. For example, the enthalpy and entropy of ionization of a weak acid HA can be found by measuring (1) the enthalpy of reaction of HA with NaOH, (2) the enthalpy of reaction of a strong acid (such as HCl) with NaOH, and (3) the equilibrium constant for dissociation of the acid (usually determined from the titration curve). Data for some of these functions for acetic acid are given in Table 6-3.

	Reaction	*Enthalpy change*
(1)	$HA + OH^- \longrightarrow A^- + H_2O$	ΔH_1°
(2)	$H_3O^+ + OH^- \longrightarrow 2\,H_2O$	ΔH_2°
(3)	$HA + H_2O \overset{K_a}{\rightleftharpoons} H_3O^+ + A^-$	ΔH_3°

TABLE 6-3
Thermodynamics of acetic acid dissociation

		$\Delta H°$ (kJ mol^{-1})		$\Delta S°$ (J K^{-1} mol^{-1})
$H^+ + OH^- \rightleftharpoons H_2O$		-55.9		80.4
$HOAc + OH^- \rightleftharpoons H_2O + OAc^-$		-56.3		-12.0
		$HOAc \rightleftharpoons H^+ + OAc^-$		
$T(K)$	303	308	313	318	323
$K_a(\times 10^{-5})$	1.750	1.728	1.703	1.670	1.633

NOTE: $\Delta H°$ and $\Delta S°$ for these reactions change rapidly with temperature. Calculations based on these data are valid only over the limited temperature range given above.

From the usual thermodynamic relationships,

(4) $\Delta H_3^\circ = \Delta H_1^\circ - \Delta H_2^\circ$ [since reaction (3) = reaction (1) − reaction (2)]
(5) $\Delta G_3^\circ = -RT \ln K_a = \Delta H_3^\circ - T\Delta S_3^\circ$

Rearranging (5):

(6) $\ln K_a = -\Delta H_3^\circ/RT + \Delta S_3^\circ/R$

Naturally, the final calculation can be more complex than this when HA is already partly dissociated in the first reaction, but the principle remains the same. It is also possible to measure the equilibrium constant at different temperatures and use equation (6) above to calculate $\Delta H°$ and $\Delta S°$. On a plot of $\ln K_a$ versus $1/T$, the slope is $-\Delta H_3^\circ/R$ and the intercept is $\Delta S_3^\circ/R$. This method works as long as $\Delta H°$ and $\Delta S°$ do not change appreciably over the temperature range used. This is sometimes a difficult condition.

EXERCISE 6-1
Use the data in Table 6-3 to calculate the enthalpy and entropy of reaction for dissociation of acetic acid using (a) the combination of equations (4) and (5) above, and (b) the temperature dependence of K_a of equation (6), by graphing $\ln K_a$ versus $1/T$.

Different methods of measuring acid–base strength yield different results, not surprising when the different physical properties being measured are considered. In addition, the different methods are frequently used under different conditions. Different aspects of acid–base strength are explained below, with brief explanations of the experimental methods used included when necessary.

6-3-3 PROTON AFFINITY

One of the purest measures of acid–base strength, but one difficult to relate to solution reactions, is gas-phase proton affinity:[10]

$$B + H^+ \longrightarrow BH^+, \qquad \text{proton affinity} = -\Delta H$$

[10] Finston and Rychtman, op cit., pp. 53–62.

(Like electron affinity, proton affinity is defined as $-\Delta H$ for the combination reaction; it can also be considered as ΔH for the dissociation reaction $BH^+ \longrightarrow B + H^+$.) By convention, proton affinities are tabulated as positive numbers. The larger the value, the weaker the acid BH^+ and the stronger the base B in the gas phase. In favorable cases, mass spectroscopy and ion cyclotron resonance spectroscopy[11] can be used to measure the reaction indirectly by changing the voltage of the ionizing electron beam in mixtures of B and H_2 until BH^+ appears. The enthalpy of formation for BH^+ can then be calculated from the voltage of the electron beam, and combined with enthalpies of formation of B and H^+ to calculate the enthalpy change for the reaction.

In spite of the simple concept, the measured values of proton affinities have uncertainties because the molecules involved frequently are in excited states (with excess energy above their normal ground states), and some species do not yield BH^+ as a fragment. In addition, under more common experimental conditions, the proton affinity is strongly modified by interactions with solvent molecules or other environmental effects. However, gas-phase proton affinities are useful in sorting out the different factors influencing acid–base behavior and their importance. For example, the alkali metal hydroxides that are of equal basicity in aqueous solution have gas-phase basicities in the order $LiOH < NaOH < KOH < CsOH$. This order matches the increase in the electron-releasing ability of the cation in these hydroxides. Proton affinity studies have also shown that pyridine and aniline are stronger bases than ammonia in the gas phase, in contrast to their behavior in aqueous solution.[12] Other comparisons of gas-phase data with solution data allow at least partial separation of the different factors influencing reactions.

6-3-4 ACIDITY AND BASICITY OF BINARY HYDROGEN COMPOUNDS

The binary hydrogen compounds (compounds containing hydrogen and one other element) range from the strong acids HCl, HBr, and HI to the weak base NH_3. Others, such as CH_4, show almost no acid–base properties. Some of these molecules in order of increasing gas-phase acidities from left to right are shown in Figure 6-2.

Two apparently contradictory trends are seen in these data. Within each group (column of the periodic table), acidity increases on going down the series. The strongest acid is the largest, heaviest member, low in the periodic table, containing the nonmetal of lowest electronegativity of the group. However, in any period (row), the strongest acid is the smallest, but heaviest, member, containing the nonmetal of highest electronegativity. Acidity increases with increasing number of electrons in the central atom, either going across the table or down, but the electronegativity effects are opposite for the two directions.

Acidity is greatest with lowest electronegativity in each group, where the comparison is with molecules of the same stoichiometry (for example, acid strength is in the order $H_2Se > H_2S > H_2O$). An explanation of this is that the conjugate bases (SeH^-, SH^-, and OH^-) of the larger molecules have

[11] R. S. Drago, *Physical Methods in Chemistry,* Saunders, Philadelphia, 1977, pp. 552–65.

[12] Finston and Rychtman, op cit., pp. 59–60.

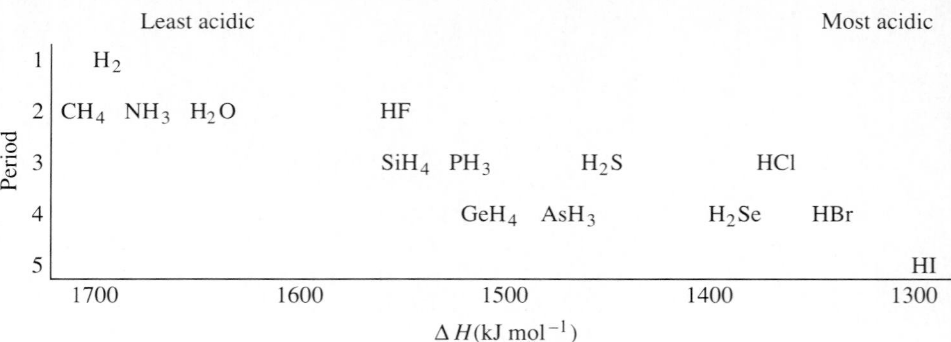

FIGURE 6-2 Acidity of Binary Hydrogen Compounds. Enthalpy of dissociation in kJ/mole for the reaction $AH \longrightarrow A^- + H^+$ (numerically the same as the proton affinity of the anion). (Data from J. E. Bartmess, J. A. Scott, and R. T. McIver, Jr., *J. Am. Chem. Soc.*, **1979,** *101,* 6046; AsH_3 value from J. E. Bartmess and R. T. McIver, Jr. in *Gas-Phase Ion Chemistry,* M. T. Bowers, ed., Academic Press, New York, 1979, p. 87.)

lower charge density and therefore a smaller attraction for protons. As a result, the larger molecules are stronger acids and their conjugate bases are weaker.

On the other hand, within a period, acidity is greatest for the compounds of elements toward the right, with greater electronegativity. The same argument cannot be used, since the more electronegative elements form the stronger acids. Although it may have no fundamental significance, one explanation that assists in remembering the trends divides the 1− charge of each conjugate base evenly between the lone pairs. Thus, NH_2^- has a charge of $-\frac{1}{2}$, OH^- has $-\frac{1}{3}$, and F^- has $-\frac{1}{4}$ on each lone pair. The amide ion, NH_2^-, is the strongest of these three conjugate bases, and NH_3 is therefore the weakest of the acids ($NH_3 < H_2O < HF$).

The same general trends persist when the acidity of these compounds is measured in aqueous solution. The reactions are more complex, forming aquated ions, but the overall effects are similar. The three heaviest hydrohalic acids (HCl, HBr, HI) are equally strong in water, due to the leveling effect of the water. All the other binary hydrogen compounds are weaker acids, with their acid strength decreasing toward the left in the periodic table. Methane and ammonia exhibit no acidic behavior in aqueous solution, nor do silane and phosphine. Details of the leveling effect and other solvent effects are considered in greater detail later in this chapter.

6-3-5 INDUCTIVE EFFECTS

Substitution of electronegative atoms or groups, such as fluorine or chlorine, in place of hydrogen on ammonia or phosphine results in weaker bases. The electronegative atom draws electrons toward itself; as a result, the nitrogen or phosphorus atom has less negative charge and the lone pair is less readily donated to an acid. For example, PF_3 is a much weaker base than PH_3.

A similar effect in the reverse direction results from substitution of alkyl groups for hydrogen. For example, in amines the alkyl groups contribute electrons to the nitrogen, increasing its negative character and making it a stronger base. Additional substitutions increase the effect, with the following resulting order of base strength in the gas phase:

$$NMe_3 > NHMe_2 > NH_2Me > NH_3$$

These **inductive effects** are similar to the effects seen in organic molecules containing electron-contributing or electron-withdrawing groups.

6-3-6 STRENGTH OF OXYACIDS

In the series of oxyacids of chlorine, the acid strength in aqueous solution is in the order

$$HClO_4 > HClO_3 > HClO_2 > HOCl$$

$$\begin{array}{cccc} \text{O} & \text{O} & \text{O} & \\ | & | & | & \\ \text{H—O—Cl—O} & \text{H—O—Cl—O} & \text{H—O—Cl} & \text{H—O—Cl} \\ | & & & \\ \text{O} & & & \end{array}$$

Pauling suggested a rule that predicts the strength of such acids semi-quantitatively, based on n, the number of *nonhydrogenated oxygen atoms* per molecule. This equation describing the acidity at 25°C is $pK_a = 9 - 7n$. Several other equations have been proposed; one that fits some acids better is $pK_a = 8 - 5n$. The pK_a's of the acids above are then

Acid	$HClO_4$ <	$HClO_3$ <	$HClO_2$ <	$HOCl$
n	3	2	1	0
pK_a (calc by $9 - 7n$)	−12	−5	2	9
pK_a (calc by $8 - 5n$)	−7	−2	3	8
pK_a (experimental)	(−10)	−1	2	7.2

where the experimental value of $HClO_4$ is somewhat uncertain. Neither equation is very accurate, but either provides approximate values.

For oxyacids with more than one ionizable proton, the pK_a's change by about 5 units with each successive removal. Thus, phosphoric acid, H_3PO_4, has predicted pK_a's of 2, 7, and 12, using the $9 - 7n$ formula, and 3, 8, and 13, using the $8 - 5n$ formula. Experimentally, the values are 2.15, 7.20, and 12.37, between the two calculated values for each pK. Sulfuric acid, H_2SO_4, has a predicted $pK_a = -5$ and -2 and the hydrogen sulfate ion, HSO_4^-, has a predicted $pK_a = 0$ and 3 by the two formulas. Experimentally, the sulfuric acid pK_a is somewhat uncertain, but below 0, and the hydrogen sulfate $pK_a = 2$.

The molecular explanation for the approximations above is that each nonhydrogenated oxygen draws electrons away from the central atom, increasing its positive charge. This positive charge in turn draws the electrons of the hydrogenated oxygen toward itself. The net result is a weakening of the O—H bond (less electron density in these bonds) and an increase in acid strength, with an increasing number of nonhydrogenated oxygens. An alternative statement of the same argument is that the negative charge of the conjugate base is spread over all the nonprotonated oxygens. The larger the number of highly electronegative atoms (oxygen in this case), the more stable the conjugate base. As a result, there is a smaller attractive force toward the proton, and the anions with more nonhydrogenated oxygens are less basic (the protonated acid is more acidic).

EXERCISE 6-2

a. Calculate approximate pK_a's for H_2SO_3 using both the equations above.
b. H_3PO_3 has one hydrogen bonded directly to the phosphorus. Calculate approximate pK_a's for H_3PO_3 using both the equations above.

6-3-7 ACIDITY OF CATIONS IN AQUEOUS SOLUTION

Many positive ions exhibit acidic behavior in solution. For example, the Fe^{3+} ion in water forms an acidic solution, with yellow or brown iron species formed by reactions such as

$$[Fe(H_2O)_6]^{3+} + H_2O \longrightarrow [Fe(H_2O)_5(OH)]^{2+} + H_3O^+$$

$$[Fe(H_2O)_5(OH)]^{2+} + H_2O \longrightarrow [Fe(H_2O)_4(OH)_2]^{+} + H_3O^+$$

In less acidic (or more basic) solutions, Fe—O—Fe and Fe—O(H)—Fe bonding occurs, and hydrated metal hydroxides precipitate.

Transition metal ions with a charge of 3+ are moderately strong acids; the larger the charge density of the ion and the larger the electronegativity of the metal, the stronger the acid. The alkali metals show no acidity, the alkaline earth metals show it only slightly, 2+ transition metal ions are weakly acidic, 3+ transition metal ions are moderately acidic, and ions with charges of 4+ or higher are strong acids in aqueous solutions.

Solubility of the hydroxide is one measure of cation acidity. Generally, transition metal 3+ ions are acidic enough to precipitate as hydroxides even in the slightly acidic solutions formed when their salts are dissolved in water. The 2+ *d*-block ions and Mg^{2+} precipitate as hydroxides in neutral or slightly basic solutions, and the alkali and remaining alkaline earth ions are so weakly acidic that no pH effects are measured.

At the highly charged extreme, the metal cation is no longer a detectable species. Instead, ions such as permanganate (MnO_4^-) and chromate (CrO_4^{2-}) are formed, with oxidation numbers of 7 and 6 for the metals. Such ions are weak bases, although CrO_4^{2-} does react with acid:

$$CrO_4^{2-} + H^+ \rightleftharpoons HCrO_4^-$$

In more concentrated acid, the dichromate ion is formed by loss of water:

$$2\,HCrO_4^- \rightleftharpoons Cr_2O_7^{2-} + H_2O$$

6-3-8 STERIC EFFECTS

There are also steric effects that influence acid–base behavior. When bulky groups are forced together by adduct formation, their mutual repulsion makes the reaction less favorable. H. C. Brown, who contributed a great deal to these studies, has reviewed the field.[13] Many of the arguments and the data to support them are from his article.

F-strain

Reactions of a series of substituted pyridines with protons show the order of base strengths to be

2,6-dimethylpyridine > 2-methylpyridine > 2-*t*-butylpyridine > pyridine

[13] H. C. Brown, *J. Chem. Soc.*, **1956**, 1248.

which matches the expected order for electron donation (induction) by alkyl groups. Methyl groups tend to push more electron density into the ring, making the nitrogen more negative, a better donor, and therefore a stronger base. Such donation from the 2 and 6 positions is more effective than from the 3 position (and the 4 position is even more effective).

However, reaction with larger acids, such as BF_3 or BMe_3, shows that the methyl groups interfere with the approach of the acid, with the following resulting order of basicity:

pyridine > 2-methylpyridine > 2,6-dimethylpyridine > 2-*t*-butylpyridine

The enthalpy changes on reaction of these four bases with trimethylboron also show greater differences from one base to another than the enthalpy changes for reactions of the same sequence of bases with boron trifluoride or borane. Although BMe_3 is a stronger acid, it shows more steric crowding around the B—N bond. Methylsulfonic acid is the strongest acid shown, with almost no steric effects.

When substituent groups interfere directly with the close approach of the base, as in these examples, the steric effect is called **F-strain** (for front strain). Figure 6-3 illustrates the effect for this series of substituted pyridines.

B-strain

B-strain (for back strain) is less easily established, but shows similar general effects. Gas-phase measurements of proton affinity show the sequence $Me_3N > Me_2NH > MeNH_2 > NH_3$, as predicted on the basis of electron donation (induction) by the methyl groups and resulting increased electron

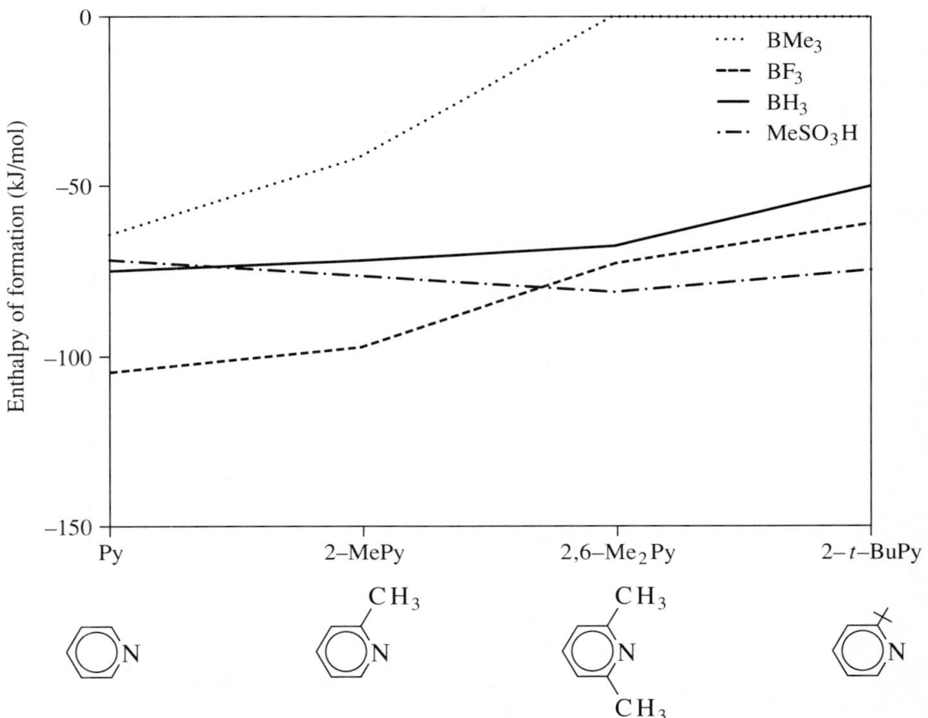

FIGURE 6-3 F-Strain in Substituted Pyridines. (H. C. Brown, *J. Chem. Soc.*, **1956,** 1248.)

density and basicity of the nitrogen.[14] When larger acids are used, the order changes, as shown in Table 6-4. With both BF_3 and BMe_3, Me_3N is a much weaker base, very nearly the same as $MeNH_2$. With the even more bulky acid tri(*t*-butyl)boron, the order is nearly reversed from the proton affinity order, although ammonia is still weaker than methylamine. Brown argues that these effects are from crowding of the methyl groups at the back of the nitrogen as the adduct is formed. Whether this is a complete explanation or whether some F-strain is also present as well may be argued.

TABLE 6-4
Methyl amine reactions

Amine	*ΔH of proton addition (kJ/mol)*	*pK_b (aqueous)*	*ΔH of adduct formation*		
			BF_3 (order)	*BMe_3 (kJ/mol)*	*$B(t\text{-}Bu)_3$ (order)*
NH_3	−846	4.75	4	−57.53	2
CH_3NH_2	−884	3.38	2	−73.81	1
$(CH_3)_2NH$	−912	3.23	1	−80.58	3
$(CH_3)_3N$	−929	4.20	3	−73.72	4
$(C_2H_5)_3N$	−958			−42 (?)	
Quinuclidine	−967			−84	

SOURCES: *Proton addition:* P. Kebarle, *Ann. Rev. Phys. Chem.*, **1977,** *28,* 445. *Aqueous pKs:* N. S. Isaacs, *Physical Organic Chemistry,* Longman/Wiley, New York, 1987, p. 213. *Adduct formation:* H. C. Brown, *J. Chem. Soc.*, **1956,** 1248.

When triethylamine is used as the base, it does not form an adduct with trimethylboron, although the enthalpy change is slightly favorable. Initially, this seems to be another example of B-strain, but examination of molecular models shows that one ethyl group is normally twisted out to the "front" of the molecule, where it interferes with adduct formation. When the alkyl chains are linked into rings, as in quinuclidine 1-azabicyclo[2.2.2]octane, adduct formation is more favorable because the potentially interfering chains are "pinned back" and do not change on adduct formation. The proton affinities of quinuclidine and triethylamine are nearly identical, releasing 967 and 958 kJ/mol. When mixed with trimethylboron, whose methyl groups are large enough to interfere with the ethyl groups of triethylamine, the quinuclidine reaction is twice as favorable as that of triethylamine (−84 versus −42 kJ/mol for adduct formation). Whether one calls the triethylamine effect F- or B-strain is a subtle question, because the interference at the front is indirectly caused by other steric interference at the back between the methyl hydrogens.

Solvation and acid–base strength

A further complication appears in the amine series. In aqueous solution, the methyl-substituted amines have basicities in the order $Me_2NH > MeNH_2 > Me_3N > NH_3$, as given in Table 6-4 (a smaller pK_b indicates a stronger base); ethyl-substituted amines are in the order $Et_2NH > Et_3N = EtNH_2 > NH_3$. In both series, the tri-substituted amines are weaker bases than expected, with

[14] M. S. B. Munson, *J. Am. Chem. Soc.*, **1965,** *87,* 2332; J. I. Brauman and L. K. Blair, *J. Am. Chem. Soc.*, **1968,** *90,* 6561; J. I. Brauman, J. M. Riveros, and L. K. Blair, *J. Am. Chem. Soc.*, **1971,** *93,* 3914.

B-strain given as the reason by Brown.[15] An alternate explanation is that the decreased basicity of the tri-substituted amines must be due to reduced solvation of their protonated cations. Solvation energies for the reaction

$$R_nH_{4-n}N^+ \text{ (g)} + H_2O \longrightarrow R_nH_{4-n}N^+ \text{ (aq)}$$

are in the order $RNH_3^+ > R_2NH_2^+ > R_3NH^+$.[16] Solvation increases on protonation, as water molecules are attracted to the positive charge and form H—O · · · H—N hydrogen bonds. With fewer protons available for such hydrogen bonding, the more highly substituted molecules should be less basic. Competition between the two effects (induction and solvation) gives the scrambled order seen above.

I-strain

The third common steric effect is called **I-strain** (for internal), where the acidic or basic atoms are affected by the size and shape of a ring molecule to which they are attached. Carbonyl groups in small ring compounds are less basic than those in larger rings, an effect that is explained by the smaller bond angles in the ring and the resulting increase in *p* character of the orbitals of the ring and compensating increase in *s* character of the carbonyl carbon. This in turn draws the electrons away from the oxygen, making it less basic. The series in Table 6-5 shows this effect, along with a similar one for nitrogen bases with different ring sizes.

In molecular orbital terms, greater *p*-orbital character in the hybrids of the donor atom leads to increased basicity, due to greater extension of the hybrid orbital from the nucleus and better potential overlap with acceptor orbitals of acids. Comparison of pyridine with piperidine illustrates this effect. In pyridine, the conjugated ring requires an sp^2 hybrid for the lone pair on the nitrogen, while the saturated piperidine ring has sp^3 hybridization for its nitrogen lone pair. Adduct formation with H^+ is slightly more favorable with piperidine ($\Delta H = -964$ versus -941 kJ/mol). The larger *p* contribution to the sp^3 hybrid results in a larger extension of the orbital and thus better overlap with the H^+ orbital. With trimethylboron as acid, the sequence can be extended to smaller rings. Steric strain appears to be least for four-membered rings and greatest for three-membered rings or pyridine, as shown in Table 6-5. Manxine (1-azabicyclo[3.3.3]undecane) has large enough rings that the nitrogen is forced into a nearly planar geometry. Adduct formation, even with a proton, requires a more nearly tetrahedral geometry and is therefore less favorable than for the comparable tri-(*i*-propyl)amine.

6-3-9 NONAQUEOUS SOLVENTS AND ACID–BASE STRENGTH

As mentioned in Section 6-2, different solvents have different acid or base properties. Water is an **amphoteric** solvent, with both acid and base properties, while sulfuric acid is such a strong acid that its basic properties are usually insignificant. The strongest acid possible in any solvent is the cation of that solvent molecule; the strongest base is the anion of the solvent molecule. Any

[15] H. C. Brown, *J. Am. Chem. Soc.*, **1945**, *67*, 374, 378.

[16] J. E. Huheey, *Inorganic Chemistry*, 2nd ed., Harper & Row, New York, 1978, p. 269.

TABLE 6-5
I-strain in acids and bases

Carbonyl base strengths
1-methyl-2-piperidinone (N—CH_3) > 1-methyl-2-pyrrolidinone (N—CH_3) > 1-methyl-2-azetidinone (N—CH_3)

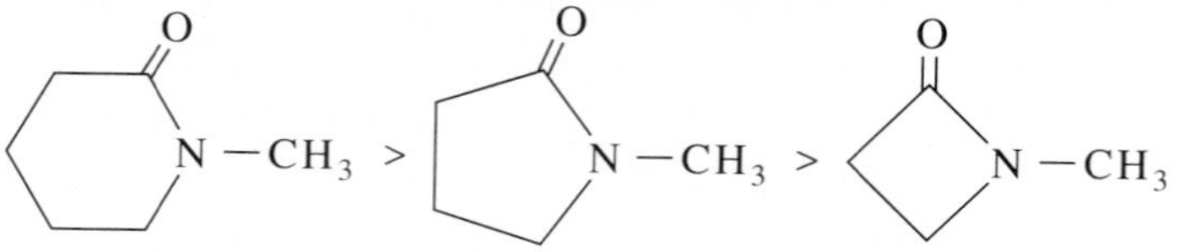

Amine base strengths

	Ring size	*ΔH for adduct formation (kJ/mol)* H^+	$B(CH_3)_3$	
Pyridine	6	−941[b]	−71.1	
$(CH_3)_2NH$	None	−941[b]	−80.6	
Ethyleneimine	3	−921[b]	−73.6	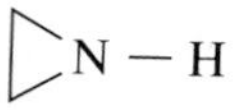 (3-ring)N—H
Trimethyleneimine	4	−952[b]	−94.1	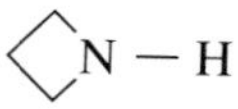 (4-ring)N—H
Pyrrolidine	5	−957[c]	−85.5	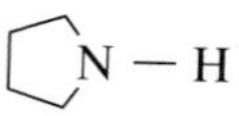 (5-ring)N—H
Piperidine	6	−964[b]	−82.2	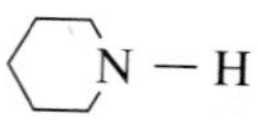 (6-ring)N—H
Manxine	8	−995[a]		 (bicyclic)N
$(n\text{-}Pr)_3N$	None	−968[c]		

SOURCES: Carbonyl base strengths from C. A. L. Figueiras and J. E. Huheey, *J. Org. Chem.*, **1976,** *41,* 49. Trimethylboron data from H. C. Brown, *J. Chem. Soc.*, **1956,** 1248. Proton data from D. H. Aue, H. M. Webb, and M. T. Bowers, *J. Am. Chem. Soc.*, *(a)* **1975,** *97,* 4136; (b) **1975,** *97,* 4137; (c) **1976,** *98,* 311.

stronger acid or base will react to form the weaker solvent acid or base, as described earlier for the Brønsted concept.

Solutes in different solvents also have different properties. Nitric, sulfuric, perchloric, and hydrochloric acids are all equally strong in water, where the strongest acid is H_3O^+. In acetic acid solvent, where the strongest acid is H_2OAc^+, even these strong acids do not dissociate completely, and the acid strengths are in the order $HClO_4 > HCl > H_2SO_4 > HNO_3$. Since acetic acid is less easily protonated than water, the difference in acid strength becomes measurable. On the other hand, all but the weakest bases act as strong bases in acetic acid, forming OAc^- and the protonated base. Conversely, even weak acids appear strong when in basic solvents such as liquid ammonia. The strongest acid in this case is NH_4^+, and even weak acids can protonate ammonia readily. In a basic solvent, all but the weakest acids act as strong acids. The compression of a range of acid or base strengths to the strengths of the solvent acid or base is called the **leveling effect.**

Inert solvents, with neither acidic nor basic properties, allow a wider range of acid–base behavior. For example, hydrocarbon solvents do not limit acid or base strength because they do not form solvent acid or base species. In such solvents, the acid or base strengths of the solutes determine the

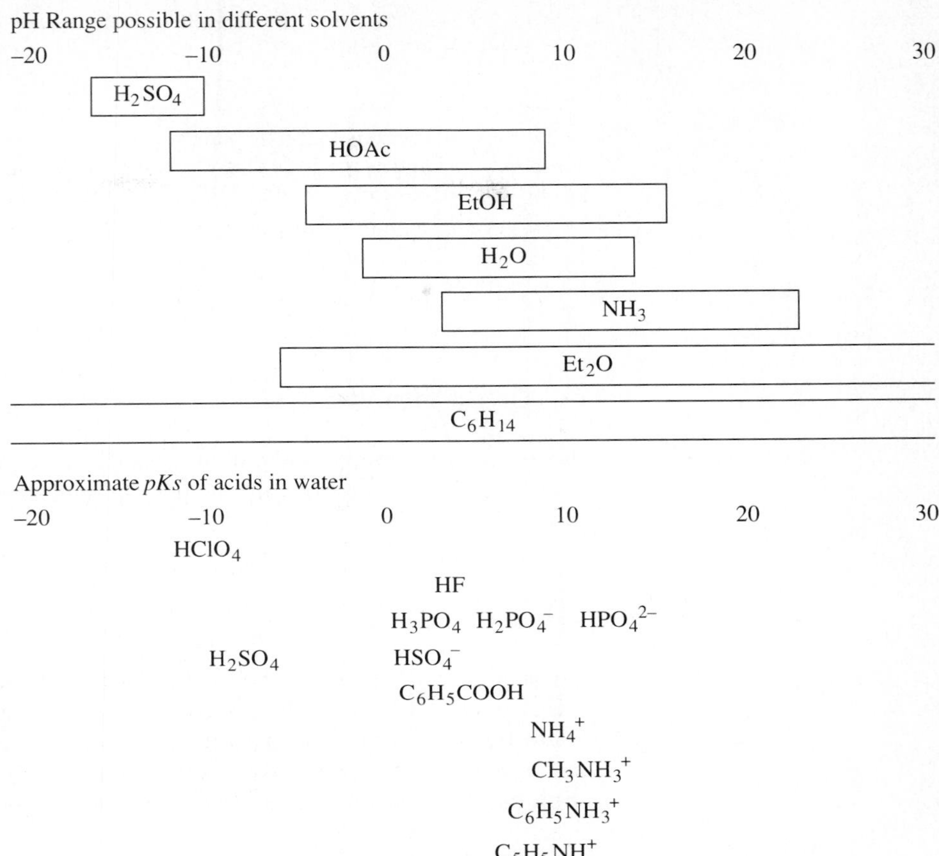

FIGURE 6-4 The Leveling Effect and Solvent Properties. (Adapted from R. P. Bell, *The Proton in Chemistry*. First edition, copyright © 1959 by Cornell University Press. Second edition, copyright © 1973 by R. P. Bell. Used by permission of the Publisher, Cornell University Press.)

reactivity and there is no leveling effect. Balancing the possible acid–base effects of a solvent with requirements for solubility, safety, and availability is one of the challenges for experimental chemists.

Figure 6-4 shows the range of pK values that are measurable in different solvents and the pK's of selected acids in water.

EXAMPLE

What are the reactions that take place and the major species in solution at the beginning, midpoint, and end of the titration of a solution of ammonia in water by hydrochloric acid in water?

Beginning NH_3, and a very small amount of NH_4^+ and OH^- are present. As a weak base, ammonia dissociates very little.

Midpoint The reaction taking place during the titration is $H_3O^+ + NH_3 \longrightarrow NH_4^+ + H_2O$, since HCl is a strong acid and completely dissociated. At the midpoint, equal amounts of NH_3 and NH_4^+ are present, along with about 5×10^{-10} M H_3O^+ and 2×10^{-5} M OH^- (pH = pK_a at the midpoint). Cl^- is the major anion present.

End Point All NH_3 has been converted to NH_4^+, so NH_4^+ and Cl^- are the major species in solution, along with about 2×10^{-6} M H_3O^+ (pH about 5.7).

After the End Point Excess HCl has been added, so the H_3O^+ concentration is now larger, and the pH lower. NH_4^+ and Cl^- are still the major species.

EXERCISE 6-3

What are the reactions that take place and the major species in solution at the beginning, midpoint, and end of the following titrations? Include estimates of the extent of reaction (does the acid dissociate completely, to a large extent, or very little?).

a. Titration of a solution of acetic acid in water by sodium hydroxide in water.
b. Titration of a solution of acetic acid in pyridine by tetramethylammonium hydroxide in pyridine.

6-3-10 SUPERACIDS

Acid solutions more acidic than sulfuric acid are called superacids.[17] The acidity of such solutions is frequently measured by the Hammett acidity function:[18]

$$H_0 = pK_{BH^+} - \log\frac{[BH^+]}{[B]}$$

where B and BH^+ are a nitroaniline indicator and its conjugate acid. On this scale, pure sulfuric acid has an H_0 of -11.9. Fuming sulfuric acid (oleum) is made by dissolving SO_3 in sulfuric acid. This solution contains $H_2S_2O_7$ and higher polysulfuric acids, all of them stronger acids than H_2SO_4. Other superacid solutions and their acidities are given in Table 6-6.

TABLE 6-6
Superacids

Acid		H_0
Sulfuric acid	H_2SO_4	−11.9
Hydrofluoric acid	HF	−11.0
Perchloric acid	$HClO_4$	−13.0
Fluorosulfonic acid	HSO_3F	−15.6
Trifluoromethanesulfonic acid (triflic acid)	HSO_3CF_3	−14.6
Magic Acid*	HSO_3F–SbF_5	−21.0 to −25 (depending on concentration)
Fluoroantimonic acid	HF–SbF_5	−21 to −28 (depending on concentration)

NOTE: * Magic Acid is a registered trademark of Cationics, Inc., Columbia, S.C.

The Lewis superacids formed by the fluorides are a result of transfer of anions to form complex fluoro anions:

$$2\,HF + 2\,SbF_5 \rightleftharpoons H_2F^+ + Sb_2F_{11}^-$$

$$2\,HSO_3F + 2\,SbF_5 \rightleftharpoons H_2SO_3F^+ + Sb_2F_{10}(SO_3F)^-$$

[17] G. Olah, G. K. S. Prakash, J. Sommer, *Science*, **1979,** *206,* 13.
[18] L. P. Hammett and A. J. Deyrup, *J. Am. Chem. Soc.*, **1932,** *54,* 2721.

These acids are very strong Friedel–Crafts catalysts. For this purpose, the term superacid is applied to any acid stronger than $AlCl_3$, the most common Friedel–Crafts catalyst. Other fluorides, such as those of arsenic, tantalum, niobium, and bismuth, also form superacids. Many other compounds exhibit similar behavior; recent additions to the list of superacids include HSO_3F–$Nb(SO_3F)_5$ and HSO_3F–$Ta(SO_3F)_5$, synthesized by oxidation of niobium and tantalum in HSO_3F by $S_2O_6F_2$.[19] Their acidity is explained by reactions similar to those for SbF_5 in fluorosulfonic acid.

6-3-11 DRAGO'S *E*, *C* EQUATION

A quantitative system of acid–base parameters proposed by Drago and Wayland[20] uses the equation

$$-\Delta H = E_A E_B + C_A C_B$$

where ΔH is the enthalpy of the reaction $A + B \longrightarrow AB$ in the gas phase or in an inert solvent, and E and C are parameters calculated from experimental data. Drago has separated the enthalpy into two components, where E is a measure of the capacity for electrostatic interactions and C a measure of the tendency to form covalent bonds. The subscripts refer to values assigned to the acid and base, with I_2 chosen as the reference acid and *N,N*-dimethylacetamide and diethyl sulfide as reference bases. The defined values (in units of kcal/mol) are

	E_A	C_A	E_B	C_B
I_2	1.00	1.00		
N,N-dimethylacetamide			1.32	
Diethyl sulfide				7.40

Values of E_A and C_A for selected acids and E_B and C_B for selected bases are given in Table 6-7. Combining the values of these parameters for acid–base pairs gives the enthalpy of reaction in kcal/mol; multiplying by 4.184 J/cal converts to joules (although we use joules in this book, these numbers were originally derived for calories and we have chosen to leave them unchanged).

Examination of the table shows that most acids have lower C_A values and higher E_A values than I_2. Since I_2 has no permanent dipole, it has little electrostatic attraction for bases and therefore has a low E_A. On the other hand, it has a strong tendency to bond with some other bases, accounted for by a relatively large C_A. Because 1.00 was chosen as the reference value for both parameters for I_2, C_A values are mostly below 1 and E_A values are mostly above 1. For C_B and E_B, this relationship is reversed.

[19] W. V. Cicha and F. Aubke, *J. Am. Chem. Soc.*, **1989,** *111,* 4328.

[20] R. S. Drago and B. B. Wayland, *J. Am. Chem. Soc.*, **1965,** *87,* 3571; R. S. Drago, G. C. Vogel, and T. E. Needham, *J. Am. Chem. Soc.*, **1971,** *93,* 6014; R. S. Drago, *Struct. Bond.*, **1973,** *15,* 73; R. S. Drago, L. B. Parr, and C. S. Chamberlain, *J. Am. Chem. Soc.*, **1977,** *99,* 3203.

TABLE 6-7
C_A, E_A, C_B, and E_B values

Acid	C_A	E_A
Trimethyl boron, $B(CH_3)_3$	1.70	6.14
Boron trifluoride (gas), BF_3	1.62	9.88
Trimethylaluminum, $Al(CH_3)_3$	1.43	16.9
Iodine (standard), I_2	*1.00	*1.00
Trimethylgallium, $Ga(CH_3)_3$	0.881	13.3
Iodine monochloride, ICl	0.830	5.10
Sulfur dioxide, SO_2	0.808	0.920
Phenol, C_6H_5OH	0.442	4.33
tert-Butyl Alcohol, C_4H_9OH	0.300	2.04
Pyrrole, C_4H_4NH	0.295	2.54
Chloroform, $CHCl_3$	0.159	3.02
Base	C_B	E_B
1-Azabicyclo[2.2.2] octane, $HC(C_2H_4)_3N$ (quinuclidine)	13.2	0.704
Trimethylamine, $(CH_3)_3N$	11.54	0.808
Triethylamine, $(C_2H_5)_3N$	11.09	0.991
Dimethylamine, $(CH_3)_2NH$	8.73	1.09
Diethyl sulfide, $(C_2H_5)_2S$	*7.40	0.339
Pyridine, C_5H_5N	6.40	1.17
Methylamine, CH_3NH_2	5.88	1.30
Pyridine-N-oxide, C_5H_5NO	4.52	1.34
Tetrahydrofuran, C_4H_8O	4.27	0.978
7-Oxabicyclo[2.2.1] heptane, $C_6H_{10}O$	3.76	1.08
Ammonia, NH_3	3.46	1.36
Diethyl ether, $(C_2H_5)_2O$	3.25	0.963
Dimethyl sulfoxide, $(CH_3)_2SO$	2.85	1.34
N,N-dimethylacetamide, $(CH_3)_2NCOCH_3$	2.58	*1.32
p-Dioxane, $O(C_2H_4)_2O$	2.38	1.09
Acetone, CH_3COCH_3	2.33	0.987
Acetonitrile, CH_3CN	1.34	0.886
Benzene, C_6H_6	0.681	0.525

SOURCE: Data from R. S. Drago, *J. Chem. Educ.*, **1974,** *51,* 300.
NOTE: * Reference values.

The example of iodine and benzene shows how these tables can be used.

$$\underset{\text{acid}}{I_2} + \underset{\text{base}}{C_6H_6} \longrightarrow I_2 \cdot C_6H_6$$

$$-\Delta H = E_A E_B + C_A C_B \quad \text{or} \quad \Delta H = -(E_A E_B + C_A C_B)$$

$$\Delta H = -(1.00 \times 0.681 + 1.00 \times 0.525) = -1.206 \text{ kcal/mol or } -5.046 \text{ kJ/mol}$$

The experimental value of ΔH is -1.3 kcal/mol, or -5.5 kJ/mol, 10% larger.[21] This is a weak adduct (other bases combining with I_2 have enthalpies as exothermic as -12 kcal/mol, or -50 kJ/mol), and the calculation does not agree with experiment as well as many. Because there can be only one set of numbers for each compound, Drago has developed statistical methods for averaging experimental data from many different combinations. In many cases, the agreement between calculated and experimental enthalpies is within 5%.

One phenomenon not well accounted for by other approaches is seen in Table 6-8.[22] It shows a series of four acids and five bases in which both E and

[21] R. M. Keefer and L. J. Andrews, *J. Am. Chem. Soc.*, **1955,** *77,* 2164.

[22] R. S. Drago, *J. Chem. Educ.*, **1974,** *51,* 300.

TABLE 6-8
Acids and bases with parallel changes in E and C

Acids	C_A	E_A
$CHCl_3$	0.154	3.02
C_6H_5OH	0.442	4.33
m-$CF_3C_6H_4OH$	0.530	4.48
$B(CH_3)_3$	1.70	6.14
Bases	C_B	E_B
C_6H_6	0.681	0.525
CH_3CN	1.34	0.886
$(CH_3)_2CO$	2.33	0.987
$(CH_3)_2SO$	2.85	1.34
NH_3	3.46	1.36

C increase. In most descriptions of bonding, as electrostatic (ionic) bonding increases, covalent bonding decreases, but these data show both increasing at the same time. Drago argues that this means that the E and C approach explains acid–base adduct formation better than an alternative, the HSAB theory described in the next section.

EXAMPLE

Calculate the enthalpy of adduct formation predicted by Drago's E, C equation for the reaction of I_2 with diethyl ether and diethyl sulfide.

	C_A C_B E_A E_B	ΔH (kcal/mol)	Experimental ΔH
Diethyl ether	$-(1.00 \times 3.25 + 1.00 \times 0.963) =$	-4.21	-4.2
Diethyl sulfide	$-(1.00 \times 7.40 + 1.00 \times 0.339) =$	-7.74	-7.8

Agreement is very good, with the product $C_A \times C_B$ by far the dominant factor. The softer sulfur reacts more strongly with the soft I_2.

EXERCISE 6-4

Calculate the enthalpy of adduct formation predicted by Drago's E, C equation for the following combinations:

a. BF_3 reacting with ammonia, methylamine, dimethylamine, and trimethylamine.
b. Pyridine reacting with trimethyl boron, trimethyl aluminum, and trimethyl gallium.
c. Explain the trends in terms of the electrostatic and covalent contributions.

6-3-12 HARD AND SOFT ACIDS AND BASES

The hard and soft acids and bases (HSAB) concept was developed by Ralph Pearson[23] as an explanation of data concerning reactions of metal ions and

[23] R. G. Pearson, *J. Am. Chem. Soc.*, **1963,** *85*, 3533; *Chem. Brit.*, **1967,** *3*, 103; R. G. Pearson, ed., *Hard and Soft Acids and Bases*, Dowden, Hutchinson & Ross, Inc., Stroudsburg, Pa., 1973.

anions; the concept has since been expanded to include many other reactions and has recently been placed on a more mathematical foundation.[24]

For many years, chemists tried to explain experimental observations such as the insolubility of silver halides and other salts that can be used to separate the metal ions into groups for identification in qualitative analysis schemes. Fajans[25] proposed that insolubility of a salt in water was a consequence of the degree of covalent bonding in these compounds, since water dissolves ionic compounds more readily than covalent compounds. He proposed the following correlations:

1. Covalent character increases with increase in size of the anion and decrease in size of the cation.
2. Covalent character increases with increasing charge on either ion.
3. Covalent character is greater for cations with non-noble gas electronic configurations.

For example, $Fe(OH)_3$ is much less soluble than $Fe(OH)_2$ (rule 2), AgS is much less soluble than AgO (rule 1), FeS is much less soluble than $Fe(OH)_2$ (rules 1 and 2), Ag_2S is much less soluble than AgCl (rule 2), and salts of the transition metals in general are less soluble than those of the alkali and alkaline earth metals (rule 3). These rules are helpful in predicting the behavior of specific cation–anion combinations in relation to others, although they are not enough to explain all such reactions. The HSAB concept provides a more general approach that covers some of the exceptions.

Ahrland, Chatt, and Davies[26] classified some of the same phenomena (as well as others) by dividing the metal ions into class (a) ions, including most metals, and class (b) ions, a smaller group including Cu^+, Pd^{2+}, Ag^+, Pt^{2+}, Au^+, Hg_2^{2+}, Hg^{2+}, Tl^+, Tl^{3+}, Pb^{2+}, and heavier transition metal ions. The members of class (b) are located in a small region in the periodic table at the lower right side of the transition metals. The class (b) ions form halides whose solubility is in the order $F > Cl > Br > I$, the reverse of the solubility order of class (a) halides. The class (b) metal ions also have a larger enthalpy of reaction with phosphorus donors than with nitrogen donors, again the reverse of the class (a) metal ion reactions. In the periodic table of Figure 6-5, the elements that are always in class (b) and those that are commonly in class (b) when they have low or zero oxidation states are identified. In addition, the transition metals have class (b) character in compounds where their oxidation state is zero.

Ahrland, Chatt, and Davies explained the class (b) metals as having d electrons available for π bonding (a discussion of metal–ligand bonding is included in Chapter 8). Therefore, high oxidation states of elements to the right of the transition metals have more class (b) character than low oxidation states. For example, thallium(III) and thallium(I) are both class (b) in their reactions with halides, but Tl(III) shows stronger class (b) character because Tl(I) has two $6s$ electrons that screen the $5d$ electrons and keep them from being fully available for π bonding. Elements farther left in the table have

[24] R. G. Parr and R. G. Pearson, *J. Am. Chem. Soc.*, **1983,** *105,* 7512; R. G. Pearson, *J. Am. Chem., Soc.*, **1985,** *107,* 6801; R. G. Pearson, *J. Am. Chem. Soc.*, **1986,** *108,* 6109; R. G. Pearson, *Inorg. Chem.*, **1988,** *27,* 734.

[25] K. Fajans, *Naturwissenschaften*, **1923,** *11,* 165.

[26] S. Ahrland, J. Chatt, and N. R. Davies, *Quart. Rev. Chem. Soc.*, **1958,** *12,* 265.

1	2	3	4	5	6	7	8	9	10	11	12	13	14	15	16	17	18
1																	2
3												B	C				10
11																	18
19		21				Mn	Fe	Co	Ni	Cu							36
37		39			Mo	Tc	Ru	Rh	Pd	Ag	Cd				Te		54
55			72		W	Re	Os	Ir	Pt	Au	Hg	Tl	Pb	Bi	Po		86
87			104														
			57													71	
			89													103	

FIGURE 6-5 Location of Class (b) Metals in the Periodic Table. Those in the outlined region are class (b) acceptors. Others indicated by their symbols are borderline elements, whose behavior depends on their oxidation state and the donor. The remainder (blank) are class (a) acceptors. (Adapted with permission from S. Ahrland, J. Chatt, and N. R. Davies, *Quart. Rev. Chem. Soc.*, **1958,** *12,* 265.)

more class (b) character in low or zero oxidation states, when more *d* electrons are present.

Donor molecules or ions that have the most favorable enthalpies of reaction with class (b) metals are those that are more readily polarizable and have vacant *d* or π^* orbitals available for π bonding.

Pearson has designated the class (a) ions **hard acids** and class (b) ions **soft acids.** Bases are also classified as hard or soft. For example, the halide ions range from F^-, a very hard base, through less hard Cl^- and Br^- to I^-, a soft base. Reactions are more favorable for hard–hard and soft–soft interactions than for a mix of hard and soft in the reactants. Much of the hard–soft distinction depends on polarizability, the degree to which a molecule or ion is easily distorted by interaction with other molecules. Electrons in polarizable molecules can be attracted or repelled by charges on other molecules, forming slightly polar species that can then combine with the other molecule. Hard acids and bases are relatively small, compact, and nonpolarizable, while soft acids and bases are larger and more polarizable (therefore softer). The hard acids are therefore any cations with large positive charge (3+ or larger) or those whose *d* electrons are relatively unavailable for π bonding. Soft acids are those whose *d* electrons or orbitals are readily available for π bonding. In addition, the more massive the atom, the softer it is likely to be, because the large number of inner electrons shield the outer ones and make the atom more polarizable. This description fits the class (b) ions well; they are primarily 1+ and 2+ ions with filled or nearly filled *d* orbitals, and most are in the second and third rows of the transition elements, with 45 or more electrons. Tables 6-9 and 6-10 list bases and acids in terms of their hardness or softness.

The trends in bases are even easier to see, with $F^- > Cl^- > Br^- > I^-$ the hardness order of the halides. Again, more electrons and larger size lead to softer behavior. In another example, S^{2-} is softer than O^{2-} because it has more electrons spread over a slightly larger volume, making S^{2-} more polarizable. Within a group, such comparisons are easy; as the electronic structure and size change, comparisons become more difficult but still possible. Thus, S^{2-} is softer than Cl^-, which has the same electronic structure, because S^{2-} has a smaller nuclear charge and a slightly larger size. As a result, the negative charge is more available for polarization. Numerical values have been

TABLE 6-9
Hard and soft bases

Hard bases	*Borderline bases*	*Soft bases*
		H^-
F^-, (Cl^-)	Br^-	I^-
H_2O, OH^-, O^{2-}		H_2S, HS^-, S^{2-}
ROH, RO^-, R_2O, CH_3COO^-		RSH, RS^-, R_2S
NO_3^-, ClO_4^-	NO_2^-, N_3^-	SCN^-, CN^-, RNC, CO
CO_3^{2-}, SO_4^{2-}, PO_4^{3-}	SO_3^{2-}	$S_2O_3^{2-}$
NH_3, RNH_2, N_2H_4	$C_6H_5NH_2$, C_5H_5N, N_2	R_3P, $(RO)_3P$, R_3As
		C_2H_4, C_6H_6

SOURCE: Adapted from R. G. Pearson, *J. Chem. Educ.*, **1968,** *45,* 581.

TABLE 6-10
Hard and soft acids

Hard acids	*Borderline acids*	*Soft acids*
H^+, Li^+, Na^+, K^+	Fe^{2+}, Co^{2+}, Ni^{2+}, Cu^{2+}, Zn^{2+}	$Co(CN)_5^{3-}$, Pd^{2+}, Pt^{2+}, Pt^{4+}
Be^{2+}, Mg^{2+}, Ca^{2+}, Sr^{2+}		
BF_3, BCl_3, $B(OR)_3$	$B(CH_3)_3$	BH_3, Tl^+, $Tl(CH_3)_3$
Al^{3+}, $Al(CH_3)_3$, $AlCl_3$, AlH_3		
Sc^{3+}, Ga^{3+}, In^{3+}, La^{3+}	GaH_3	$Ga(CH_3)_3$, $GaCl_3$, $GaBr_3$, GaI_3
Cr^{3+}, Mn^{2+}, Fe^{3+}, Co^{3+}	Rh^{3+}, Ir^{3+}, Ru^{3+}, Os^{2+}	Cu^+, Ag^+, Au^+, Cd^{2+}, Hg^+
		Hg^{2+}, CH_3Hg^+
CO_2, RCO^+, CH_3Sn^{3+}, $(CH_3)_2Sn^{2+}$	R_3C^+, $C_6H_5^+$, Sn^{2+}, Pb^{2+}	CH_2, carbenes
N^{3+}, RPO_2^+, $ROPO_2^+$, As^{3+}	NO^+, Sb^{3+}, Bi^{3+}	
SO_3, RSO_2^+, $ROSO_2^+$	SO_2	
Ions with oxidation states of 4 or higher		Br_2, I_2
HX (hydrogen-bonding molecules)		Metals with zero oxidation oxidation state
		π acceptors: trinitrobenzene, choroanil, quinones, tetracyanoethylene, etc.

SOURCE: Adapted from R. G. Pearson, *J. Chem. Educ.*, **1968,** *45,* 581.

assigned to hardness parameters and related data; these are described in Section 6-4. The equations defining the parameters are also described in that section.

More detailed comparisons are possible, but another factor, called the **inherent acid–base strength,** must also be kept in mind in these comparisons. An acid or a base may be either hard or soft and at the same time either strong or weak. The strength of the acid or base may be more important than the hard–soft characteristics; both must be considered at the same time. For example, if two soft bases are in competition for the same acid, the one with more inherent base strength may be favored unless there is considerable difference in softness. Such comparisons require care; seldom is one factor totally responsible for the reaction, and the reaction is nearly always a competition between acid–base pairs. As an example, consider the following reaction. Two hard–soft combinations react to give a hard–hard and a soft–soft combination, although ZnO is composed of the strongest acid (Zn^{2+}) and the strongest base (O^{2-}).

$$\underset{\text{soft–hard}}{ZnO} + \underset{\text{hard–soft}}{2\,LiC_4H_9} \longrightarrow \underset{\text{soft–soft}}{Zn(C_4H_9)_2} + \underset{\text{hard–hard}}{Li_2O}$$

As a general rule, the hard–hard combinations are more favorable energetically than soft–soft combinations. When in doubt, this explanation may be helpful in deciding the determining factor in a reaction.

EXAMPLE

Qualitative Analysis

TABLE 6-11
HSAB and qualitative analysis

	Qualitative analysis separation				
	Group 1	Group 2	Group 3	Group 4	Group 5
HSAB Acid	Soft	Borderline and soft	Borderline	Hard	Hard
Reagent	HCl	H_2S (acidic)	H_2S (basic)	$(NH_4)_2CO_3$	Soluble
Precipitated	AgCl $PbCl_2$ Hg_2Cl_2	HgS CdS CuS SnS_2 As_2S_3 Sb_2S_3 Bi_2S_3	MnS FeS CoS NiS ZnS $Al(OH)_3$ $Cr(OH)_3$	$CaCO_3$ $SrCO_3$ $BaCO_3$	Na^+ K^+ NH_4^+

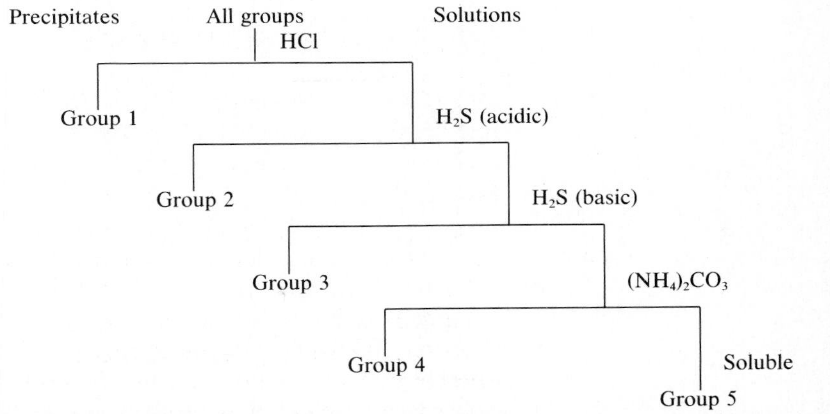

Precipitates — All groups — Solutions

All groups → HCl → Group 1 (precipitate); solution → H_2S (acidic) → Group 2 (precipitate); solution → H_2S (basic) → Group 3 (precipitate); solution → $(NH_4)_2CO_3$ → Group 4 (precipitate); Soluble → Group 5

The traditional qualitative analysis scheme can be used to show how the HSAB theory can be used to correlate solubility behavior; it also can show some of the difficulties with such correlations. In qualitative analysis for metal ions, the cations are successively separated into groups by precipitation for further detailed analysis. The details differ with the specific reagents used, but generally fall into the categories in Table 6-11. In the usual analysis, the reagents are used in the order given from left to right. The cations Ag^+, Pb^{2+}, and Hg_2^{2+} precipitate with chloride even though they are considered soft acids and chloride is a marginally hard base. Apparently, the sizes of the ions permit strong bonding in the crystal lattice in spite of this mismatch, partly because their interaction with water (another hard base) is not strong enough to prevent precipitation. The reaction

$$M^{n+}(H_2O)_m + n\ Cl^-(H_2O)_p \longrightarrow MCl_n \downarrow + (m + p)\ H_2O$$

is favorable (although $PbCl_2$ is appreciably soluble in hot water).

Group 2 is made up of borderline and soft acids that are readily precipitated in acidic H_2S solution, where the S^{2-} concentration is very low because the equilibrium

$$H_2S \rightleftharpoons 2\,H^+ + S^{2-}$$

lies far to the left in acid solution. Group 3 cleans up the remaining transition metals in the list, all of which are borderline acids. In the basic H_2S solution, the equilibrium above lies far to the right, and the high sulfide ion concentration precipitates even these cations. Al^{3+} and Cr^{3+} are hard enough that they prefer OH^- over S^{2-}. Another hard acid could be Fe^{3+}, but it is reduced by S^{2-}. Group 4 is a clear-cut case of hard–hard interactions, and Group 5 cations are larger, with only a single electronic charge, and thus have small electrostatic attractions for anions. For this reason they do not precipitate except with certain highly specific reagents, such as perchlorate, ClO_4^-, for potassium and tetraphenylborate, $B(C_6H_5)_4^-$, or zinc uranyl acetate, $[Zn(UO_2)_3(C_2H_3O_2)_9]^-$, for sodium.

This quick summary of the analysis scheme shows where hard–hard or soft–soft combinations lead to insoluble salts, but also shows that the rules have limitations. Some cations considered hard will precipitate under the same conditions as others that are clearly soft. For this reason, any predictions based on HSAB must be considered tentative, and solvent and other interactions must be considered carefully.

6-3-13 COMPARISON OF HSAB AND *E, C*

Drago's system emphasizes the two factors involved in acid–base strength (electrostatic and covalent) in the two terms of his equation for enthalpy of reaction, while Pearson's puts more obvious emphasis on the "covalent" factor. Pearson[27] has proposed the equation $\log K = S_A S_B + \sigma_A \sigma_B$, with the inherent strength S modified by a softness factor σ. Larger values of strength and softness then lead to larger equilibrium constants or rate constants. Although Pearson attached no numbers to this equation, it does show the need to consider more than just hardness or softness in working with acid–base reactions. However, his more recent development of absolute hardness based on orbital energies (Section 6-4) returns to a single parameter and considers only gas-phase reactions. Both systems (Drago's E and C parameters, Pearson's HSAB) are useful, but neither covers every case, and it is usually necessary to make judgments about reactions for which information is incomplete. With E and C numbers available, quantitative comparisons can be made. When they are not, the HSAB approach can provide a rough guide for predicting reactions. Examination of the tables also shows little overlap of the examples chosen. Neither approach is completely satisfactory; both can be of considerable help in classifying reactions and predicting which reactions will proceed and which will not.

6-3-14 SOLVENT CONTRIBUTION

When reactions in solution are considered, the effects of solvent interactions become very important, sometimes to the extent of controlling the reaction.

[27] R. G. Pearson, *J. Chem. Educ*, **1968,** *45*, 581.

Drago's E and C parameters specifically ignore solvation, in order to separate the acid–base properties of the reactants from solvation effects. Because solvation varies hugely as the solvent is changed, it gives differing values for both enthalpy and entropy changes on reaction. The inevitable changes in solvation during adduct formation are large and frequently difficult to evaluate. For example, reaction of Fe(III) with halide ions in aqueous solution requires not only that the attraction of water molecules to iron and halide change as the reactant charges of 3+ and 1− change to the product charge of 2+, but also that water molecules be removed from each ion to allow them to come into contact. Particularly in the case of Fe(III), the water molecules are strongly attracted and difficult to remove. In comparing this series of halide ion reactions, only the difference in solvation between the halide ion and the product is important, because the other changes are the same for all halides. The difference between these two determines the contribution of solvation to the overall enthalpy and entropy of the reaction. The fluoride ion is the most strongly hydrated of the halides, and FeF^{2+}, as the smallest product ion, should also be the most strongly hydrated. Competition between the two hydrations and the Fe^{3+}–X^- interactions finally balances out to favor formation of FeX^{2+} in the order $F^- > Cl^- > I^- > Br^-$, as in Table 6-12.

TABLE 6-12
FeX^{2+} stability constants

$Fe^{3+} + X^- \rightleftharpoons FeX^{2+}$ $\qquad K_{stab} = \dfrac{[FeX^{2+}]}{[Fe^{3+}][X^-]}$

X^-	*Approximate* K_{stab}
F^-	$10^{5.18}$
Cl^-	$10^{0.64}$
Br^-	$10^{-0.2}$
I^-	$10^{0.36}$

SOURCE: Data from R. M. Smith and A. E. Martell, *Critical Stability Constants*, Vol. 4, Plenum, New York, 1976.

6-4 FRONTIER ORBITALS AND ACID–BASE REACTIONS[28]

A molecular orbital description of acid–base reactions uses **frontier molecular orbitals** (those at the occupied–unoccupied frontier) and can be illustrated by the simple reaction $NH_3 + H^+ \longrightarrow NH_4^+$. In this reaction, the orbital containing the lone pair electrons of the ammonia molecule combines with the empty $1s$ orbital of the hydrogen ion to form bonding and antibonding orbitals. The lone pair in the a_1 orbital of NH_3 is stabilized by this interaction, as shown in Figure 6-6. Ammonia has three bonding pairs of electrons in the e and a_1 orbitals, a lone pair in a nonbonding a_1 orbital, and three empty antibonding e and a_1 orbitals, as described in Chapter 5. The NH_4^+ ion has the same molecular orbital structure as methane, CH_4, with four bonding orbitals (a_1 and t_2) and four antibonding orbitals (also a_1 and t_2). Combining the seven NH_3 orbitals and the one H^+ orbital, with the change in symmetry from C_{3v} to T_d, gives the eight orbitals of the NH_4^+. When the 8 valence electrons are placed in these orbitals, one pair enters the bonding a_1 orbital and three pairs enter bonding t_2 orbitals. The net result is a lowering of energy as the nonbonding a_1 becomes a bonding t_2, making the combined NH_4^+ more stable than the

[28] Jensen, op. cit., Chap. 4, pp. 112–55.

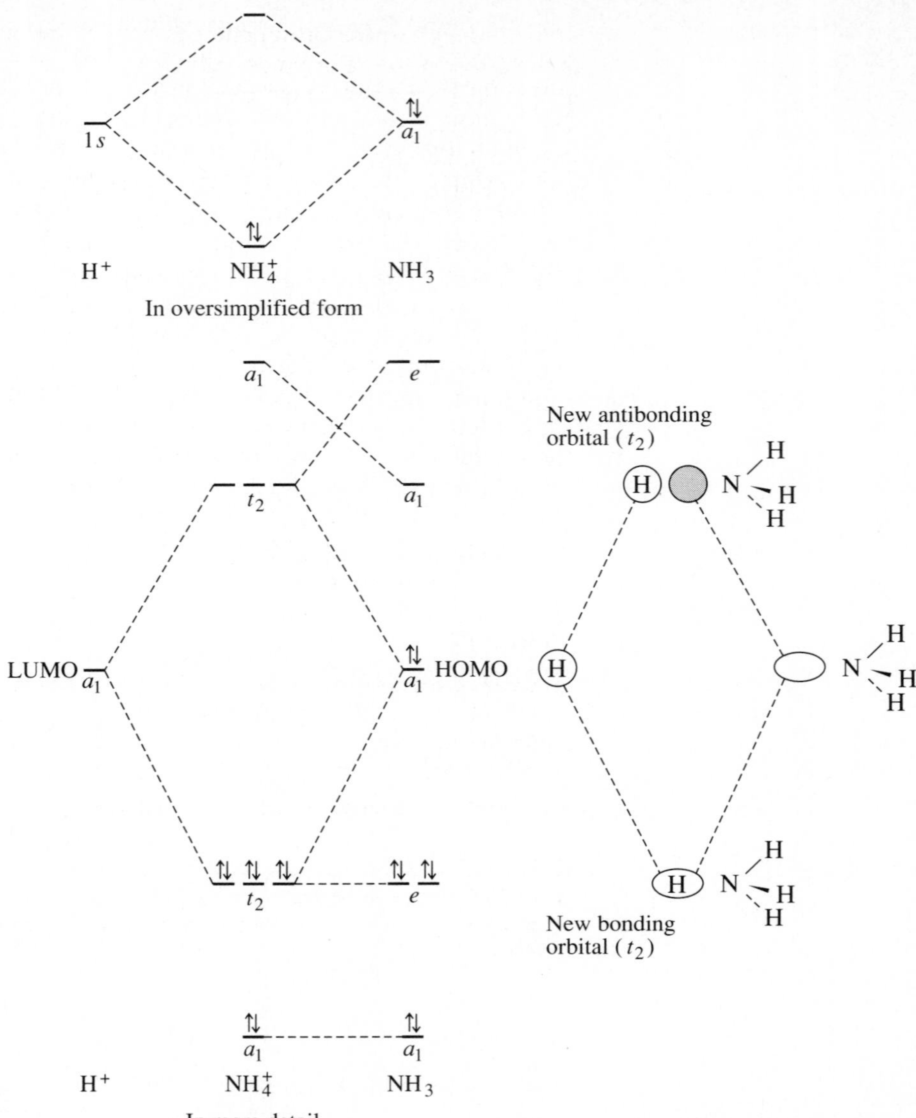

FIGURE 6-6 $NH_3 + H^+ \longrightarrow NH_4^+$ Molecular Energy Levels.

separated $NH_3 + H^+$. This is an example of the combination of the highest occupied molecular orbital (HOMO) of the base NH_3 and the lowest unoccupied molecular orbital (LUMO) of the acid H^+ accompanied by a change in symmetry to make the new t_2 orbitals, one bonding and one antibonding. Energy levels and orbital overlap for this reaction are shown in Figure 6-6.

In most acid–base reactions, **a HOMO–LUMO combination forms new HOMO and LUMO orbitals of the product.** We can see that orbitals whose shapes allow significant overlap and whose energies are similar form useful bonding and antibonding orbitals. On the other hand, if the orbital combinations have no useful overlap, no net bonding is possible (as shown in Chapter 5) and they cannot form acid–base products.[29]

[29] In a few cases the match of symmetry and energy does not involve the HOMO; although rare, this possibility should be kept in mind.

Even when the orbital shapes match, several reactions may be possible, depending on the relative energies. A single species can act as an oxidant, an acid, a base, or a reductant, depending on the other reactant. These possibilities are shown in Figure 6-7. While predictions on the basis of these arguments may be difficult (and frequently impossible) when the orbital energies are not known, they still provide a useful background to these reactions.

Reactant A is taken as a reference; water is a good example. The first combination, A + B, has all the B orbitals at a much higher energy than those of water (Ca, for example; the alkali metals react similarly but have only one electron in their highest *s* orbital). The energies are so different that no adduct can form, but a transfer of electrons can take place from B to A. From simple electron transfer, we might expect formation of H_2O^-, but reduction of water yields hydrogen gas instead. As a result, water is reduced to H_2 and OH^-, while Ca is oxidized to Ca^{2+}:

$$2\ H_2O + Ca \longrightarrow Ca^{2+} + 2\ OH^- + H_2 \qquad \text{(water as oxidant)}$$

If orbitals with matching shapes have similar energies, the resulting bonding orbitals will have lower energy than the reactant HOMOs, and a net decrease in energy (stabilization of electrons in the new HOMOs) results. An adduct is formed, with its stability dependent on the difference between the total energy of the product and the total energy of the reactants. An example with water as acceptor (with lower–energy orbitals) is the reaction with chloride ion (C in Figure 6-7):

$$n\ H_2O + Cl^- \longrightarrow Cl(H_2O)_n^- \qquad \text{(water as acid)}$$

In this reaction, water is the acceptor, and the LUMO used is an antibonding orbital centered primarily on the hydrogen atom (the chloride HOMO is one of its lone pairs from a 3*p* orbital). A reactant with orbitals lower in energy

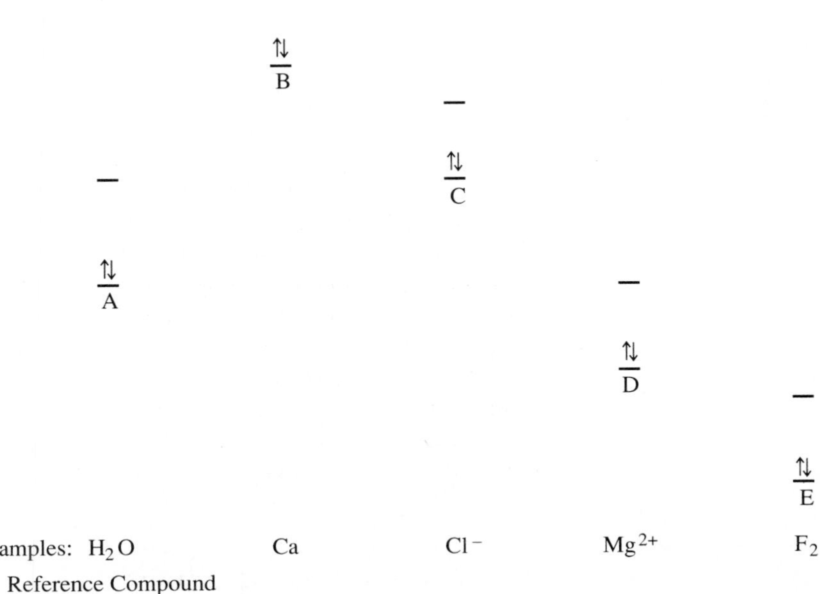

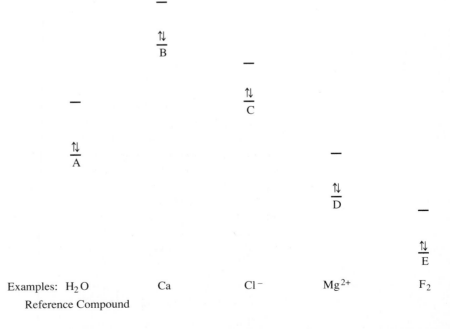

FIGURE 6-7 HOMO–LUMO Interactions. (Adapted from Figure 4-6 of W. B. Jensen. *The Lewis Acid–Base Concepts,* Wiley-Interscience, New York, 1980, p. 140. Copyright © 1980, John Wiley & Sons, Inc. Reprinted by permission of John Wiley & Sons, Inc.)

than those of water (for example, Mg^{2+}, D in Figure 6-7) allows water to act as a donor:

$$6\ H_2O + Mg^{2+} \longrightarrow Mg(H_2O)_6^{2+} \qquad \text{(water as base)}$$

Here, water is the donor, contributing a lone pair primarily from the HOMO, which has a large contribution from the oxygen atom (the magnesium ion LUMO is the vacant 3*s* orbital). The molecular orbital levels that result from reactions with B or C are similar to those in Figures 6-9 and 6-10 (Section 6-4-2) for hydrogen bonding.

Finally, if the reactant has orbitals much lower than the water orbitals (F_2, for example, E in Figure 6-7), water can act as a reductant and transfer electrons to the other reactant. The product is not the simple result of electron transfer (H_2O^+), but the result of breakup of the water molecule to oxygen and hydrogen ions:

$$2\ H_2O + 2\ F_2 \longrightarrow 4\ F^- + 4\ H^+ + O_2 \qquad \text{(water as reductant)}$$

Similar reactions can be described for other species, and the adducts formed in the acid–base reactions can be quite stable or very unstable, depending on the exact relationship between the orbital energies. Naturally, there can be many possible energies between the ones shown here, and the reaction possibilities for each are dependent on other factors as well. Proton transfer is illustrated later as one extreme of hydrogen bonding.

We are now in a position to reformulate the Lewis definition of acids and bases in terms of frontier orbitals: **A base has an electron pair in a HOMO of suitable symmetry to interact with the LUMO of the acid.** The better the energy match between the base's HOMO and the acid's LUMO, the stronger the interaction.

6-4-1 CARBON MONOXIDE

In the case of CO, electronegativity predicts that the oxygen end of the molecule is more negative, and combination of CO with metal atoms might seem likely to result in M—O—C geometry. In fact, nearly all known metal–carbon monoxide compounds (called **carbonyls**) bond through the carbon, with M—C—O geometry.[30] This behavior is supported by formal charge arguments, which place a formal charge of 1− on carbon in CO, and by the molecular orbital pictures. Examination of the CO molecular orbitals from Chapter 5 (duplicated in Figure 6-8) shows that the HOMO is a bonding orbital derived primarily from the 2*s* and 2*p* orbitals of carbon and a 2*p* orbital of oxygen. As the HOMO, this lone-pair orbital is the preferred donor for the Lewis base, and the M—C—O geometry follows naturally from the concentration of the orbital on carbon. Further details of metal carbonyl bonding are given in Chapter 12.

6-4-2 HYDROGEN BONDING

The molecular orbitals for the symmetric FHF^- ion were described in Chapter 5 as combinations of the atomic orbitals. They are also shown in Figure 6-9

[30] Examples of isocarbonyl ligands are also known, when CO forms bridges between two metal atoms. In these compounds, the C is bonded to one metal atom, the O to the other.

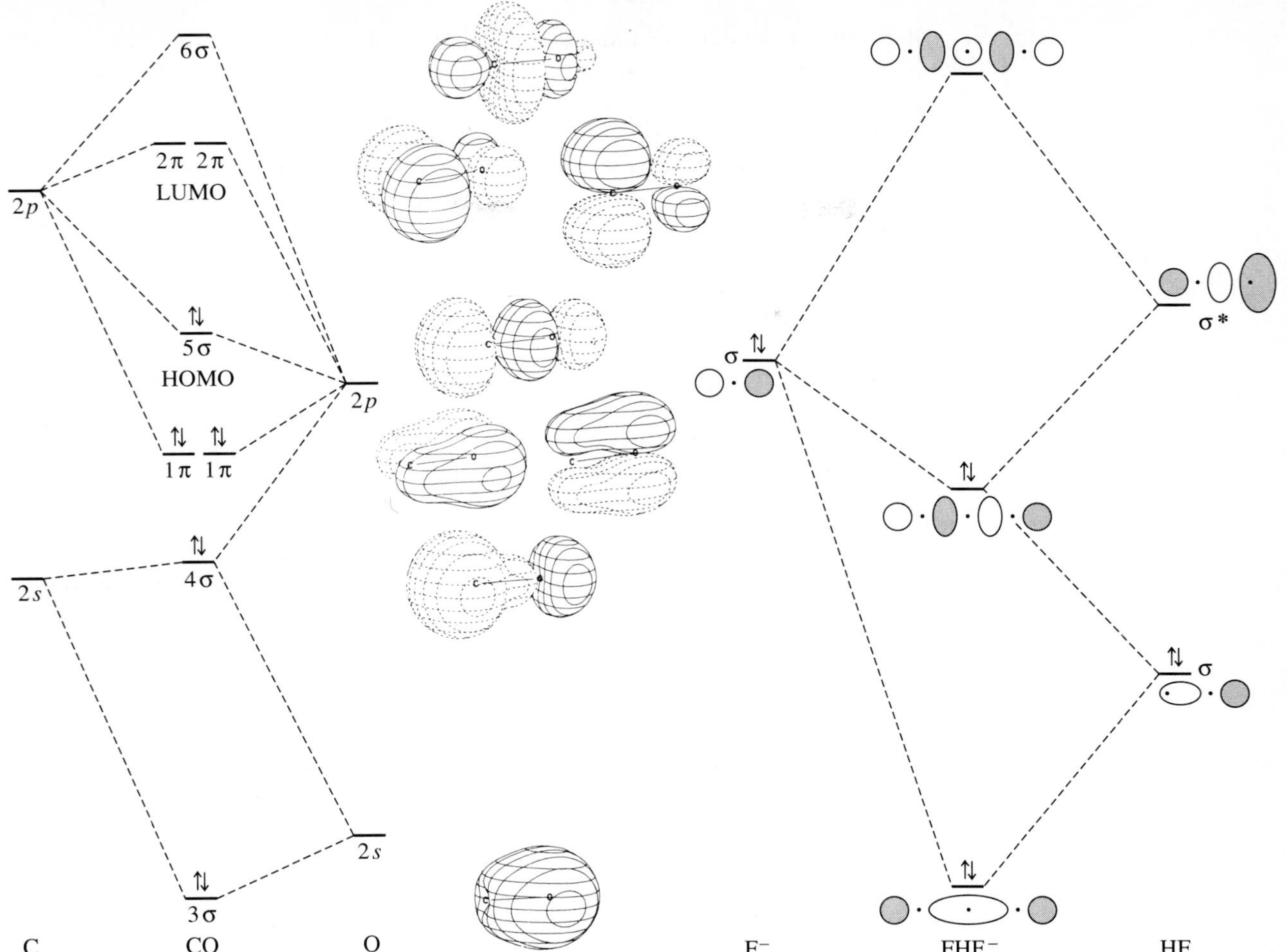

FIGURE 6-8 Molecular Orbitals of CO. (Orbital diagrams reproduced with permission from W. L. Jorgensen and L. Salem, *The Organic Chemist's Book of Orbitals,* Academic Press, New York, 1973, p. 78.)

FIGURE 6-9 Molecular Orbitals for Hydrogen Bonding in FHF^-.

to result from combination of HF with F^-. The shapes of the orbitals are appropriate for bonding, overlap of the F^- orbital with the HF orbitals forming the three product orbitals shown in the middle. The three orbitals that result are all symmetric about the central H nucleus. The lowest orbital is distinctly bonding, with all three component orbitals contributing and no nodes between the atoms. The middle orbital is essentially nonbonding, with nodes through each of the nuclei, and as a result has no contribution from the H orbital. The highest-energy orbital is antibonding, with nodes between each pair of atoms. The symmetry of the molecule dictates the nodal pattern, increasing from two to three to four nodes with increasing energy. In general, orbitals with nodes between adjacent atoms are antibonding; orbitals with nodes through atoms may be either bonding or nonbonding, depending on the orbitals involved. When three atomic orbitals are used (a $2p$ orbital from each F^- and the $1s$ orbital from H^+), the resulting pattern is one low-energy molecular orbital, one high-energy molecular orbital, and one intermediate-energy molecular orbital. The intermediate orbital may be slightly bonding, slightly antibonding, or nonbonding; we describe such orbitals as essentially nonbonding.

For unsymmetrical hydrogen bonding, such as that of B + HA $\rightleftharpoons$ BHA shown in Figure 6-10, the node of the middle orbital would be on one side or the other of the hydrogen nucleus, and the orbital would stress bonding toward one or the other of the donor atoms. At the same time, the lowest orbital would favor bonding in the other direction.

Regardless of the exact energies and location of the nodes, the general pattern is the same. The resulting FHF^- or BHA structure has a total energy lower than the sum of the energies of the reactants. The stability of the hydrogen bond depends strongly on the energies of the contributing orbitals; the total energy of the occupied product orbitals must be lower than that of the occupied MOs of the reactants in order for the combination to be stable. For the general case of B + HA, three possibilities exist, but with a difference from the earlier HOMO–LUMO illustration (Figure 6-7) created by the possibility of proton transfer. These possibilities are illustrated in Figure 6-11. First, for a poor match of energies when the occupied reactant orbitals are lower in total energy than those of the possible hydrogen-bonded product, no new product will be formed; there will be no hydrogen bonding. Second, for a good match of energies, the occupied product orbitals may be lower in

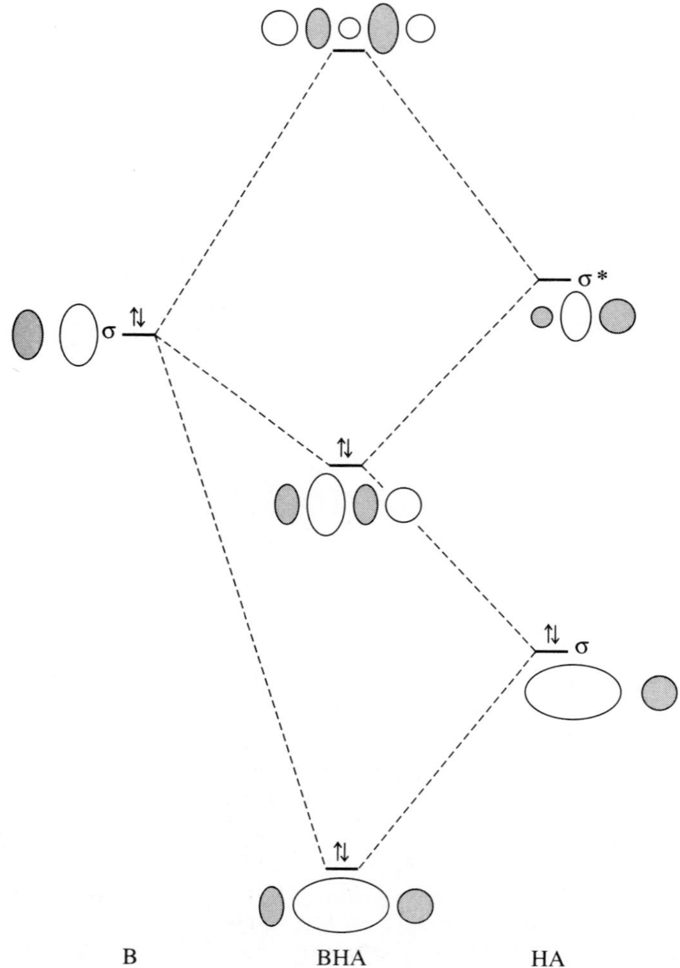

FIGURE 6-10 Molecular Orbitals for Unsymmetrical Hydrogen Bonding.

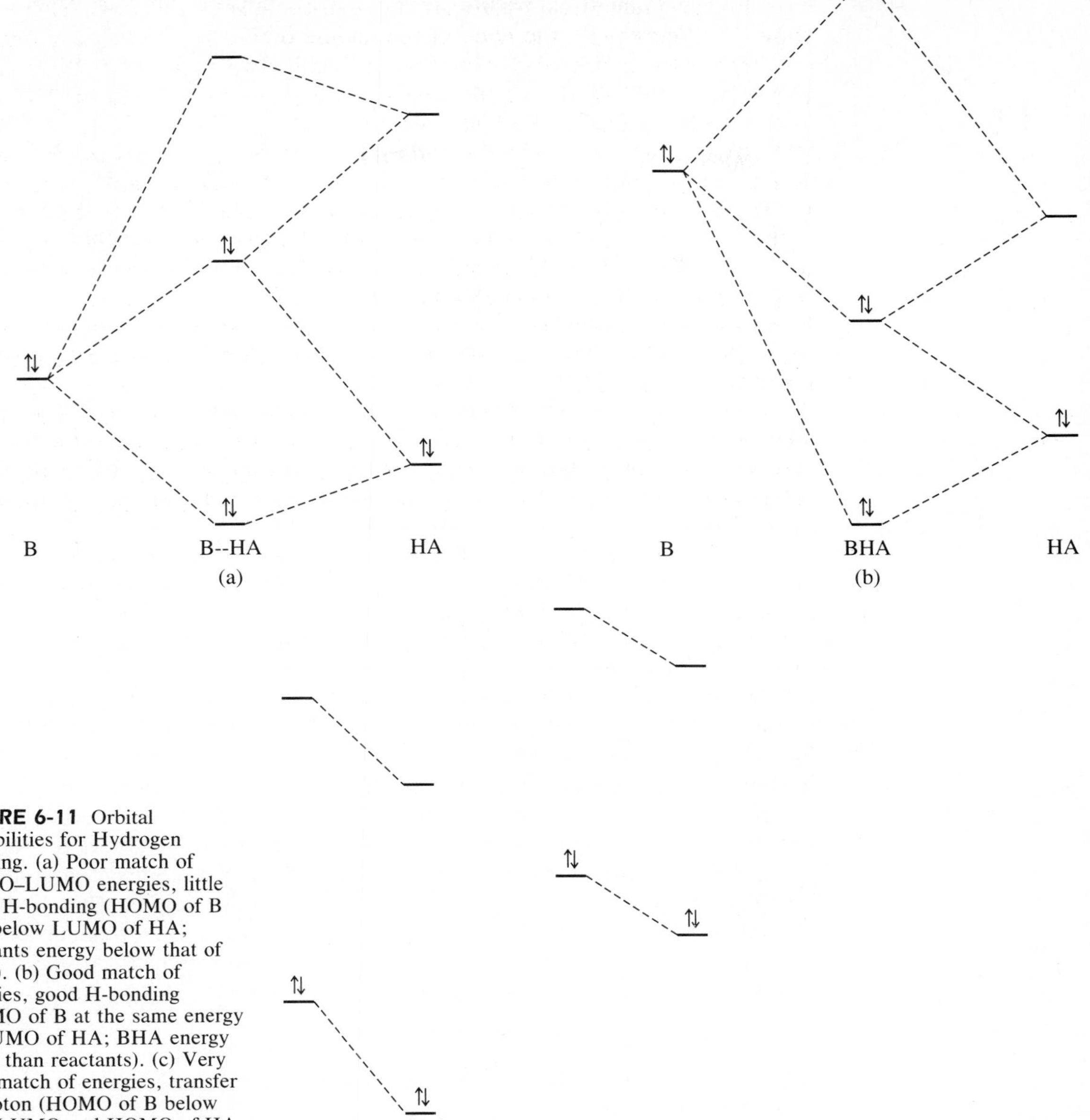

FIGURE 6-11 Orbital Possibilities for Hydrogen Bonding. (a) Poor match of HOMO–LUMO energies, little or no H-bonding (HOMO of B well below LUMO of HA; reactants energy below that of BHA). (b) Good match of energies, good H-bonding (HOMO of B at the same energy as LUMO of HA; BHA energy lower than reactants). (c) Very poor match of energies, transfer of proton (HOMO of B below both LUMO and HOMO of HA; BH + A energy lower than B + HA or BHA).

energy and a hydrogen-bonded product forms. The greater the lowering of energies of these orbitals, the stronger the hydrogen bonding. Finally, for a poor match of energies, occupied orbitals of the species BH + A may be lower than those of B + HA; in this case, complete proton transfer occurs.

In Figure 6-11a, the HOMO of B is well below that of the LUMO of HA. Since the lowest orbital is only slightly lower than the HA orbital, and the middle orbital is higher than the B orbital, little or no reaction occurs. In aqueous solution, interactions between water and molecules with almost no acid–base character like CH_4 fit this group. There is little interaction between the hydrogens of the methane molecule and the lone pairs of surrounding water molecules.

In Figure 6-11b, the LUMO of HA and the HOMO of B have similar energies, both occupied product orbitals are lower than the respective reactant orbitals, and a hydrogen-bonded product forms. The node of the product HOMO is near H, and the hydrogen-bonded product has a B—H bond similar in strength to the H—A bond. If the B HOMO is slightly higher than the HA LUMO, as in the figure, the H—A portion of the hydrogen bond is stronger. If the B HOMO is lower than the HA LUMO, the B—H portion is stronger (the product HOMO consists of more B than A orbital). Weak acids like acetic acid are examples of hydrogen-bonding solutes in water. Acetic acid will hydrogen-bond strongly with water (and to some extent with other acetic acid molecules), with a small amount of proton transfer to water to give hydronium and acetate ions.

In Figure 6-11c, HOMO–LUMO energy match is so poor that no useful adduct orbitals can be formed. The product MOs here are those of A and BH, and the proton is transferred from A to B. Strong acids like HCl will donate their protons completely to water, after which the H_3O^+ formed will hydrogen bond strongly with other water molecules.

In all these diagrams, either HA or BH (or both) may have a positive charge and either A or B (or both) may have a negative charge, depending on the circumstances.

When A is a highly electronegative element such as F, O, or N, the highest occupied orbital of A has lower energy than the hydrogen 1*s* orbital, and the H—A bond is relatively weak, with most of the electron density near A and with H somewhat positively charged. This favors the hydrogen-bonding interaction by lowering the overall energy of the HA bonding orbital and improving overlap with the B orbital. In other words, when the reactant HA has a structure close to H^+——A^-, hydrogen bonding is more likely. This explains the strong hydrogen bonding in cases with hydrogen bridging between F, O, and N atoms in molecules and the much weaker or nonexistent hydrogen bonding between other atoms. Although some have argued that hydrogen bonds are purely electrostatic,[31] others lean more toward a covalent model.[32] The description above can be described as a 3-center 4-electron model[33] that results in a bond angle at the hydrogen within 10° to 15° of a linear 180° angle.

6-4-3 ELECTRONIC SPECTRA (INCLUDING CHARGE TRANSFER)

One reaction that shows the effect of adduct formation dramatically is the reaction of I_2 as an acid with different solvents and ions that act as bases. The changes in spectra and visible color caused by changes in electronic structure (shown in Figures 6-12 and 6-13) are striking. The upper molecular orbitals of I_2 are similar to those of F_2 shown in Figure 5-7, with a net single bond and lone pairs in the $5\pi_g^*$ HOMO orbitals. In the gas phase, I_2 is violet, absorbing light near 500 nm due to promotion of an electron from the $4\pi_g^*$ level to the

[31] L. Pauling, *The Nature of the Chemical Bond,* 3rd ed., Cornell University Press, Ithaca, N.Y., 1960, pp. 449–52.

[32] G. C. Pimentel and A. L. McClellan, *The Hydrogen Bond,* W. H. Freeman Co., San Francisco, Calif., 1959, pp. 236–38.

[33] R. L. DeKock and W. B. Bosma, *J. Chem. Educ.*, **1988,** *65,* 194.

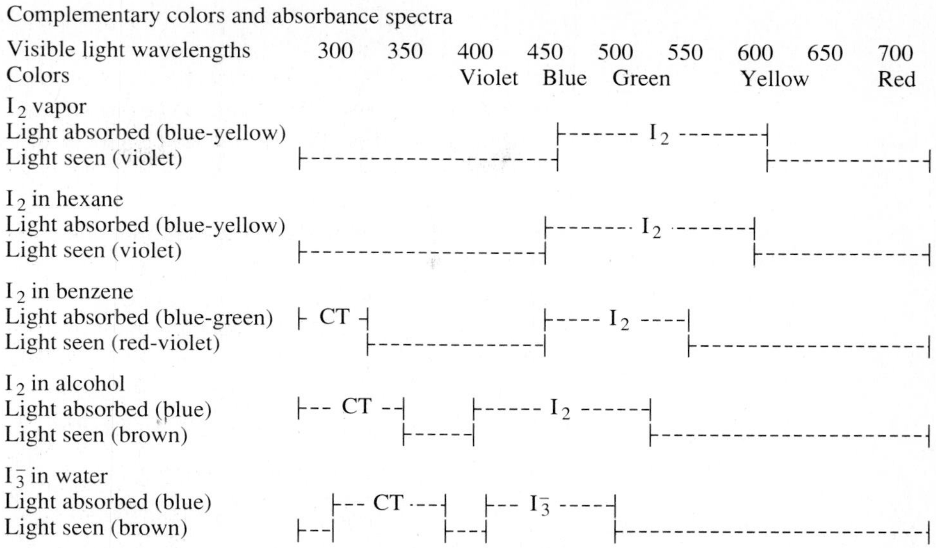

Complementary colors and absorbance spectra

Visible light wavelengths	300	350	400	450	500	550	600	650	700
Colors			Violet	Blue	Green		Yellow		Red

I_2 vapor
Light absorbed (blue-yellow) ├------ I_2 -------┤
Light seen (violet) ├------------------┤ ├-----------┤

I_2 in hexane
Light absorbed (blue-yellow) ├------- I_2 -------┤
Light seen (violet) ├------------------┤ ├-----------┤

I_2 in benzene
Light absorbed (blue-green) ├ CT ┤ ├---- I_2 ----┤
Light seen (red-violet) ├--------------┤ ├------------------┤

I_2 in alcohol
Light absorbed (blue) ├-- CT --┤ ├----- I_2 ------┤
Light seen (brown) ├-----┤ ├---------------------┤

I_3^- in water
Light absorbed (blue) ├-- CT ---┤ ├-- I_3^- ----┤
Light seen (brown) ├--┤ ├--┤ ├------------------------┤

FIGURE 6-12 Spectra of I_2 with Different Bases.

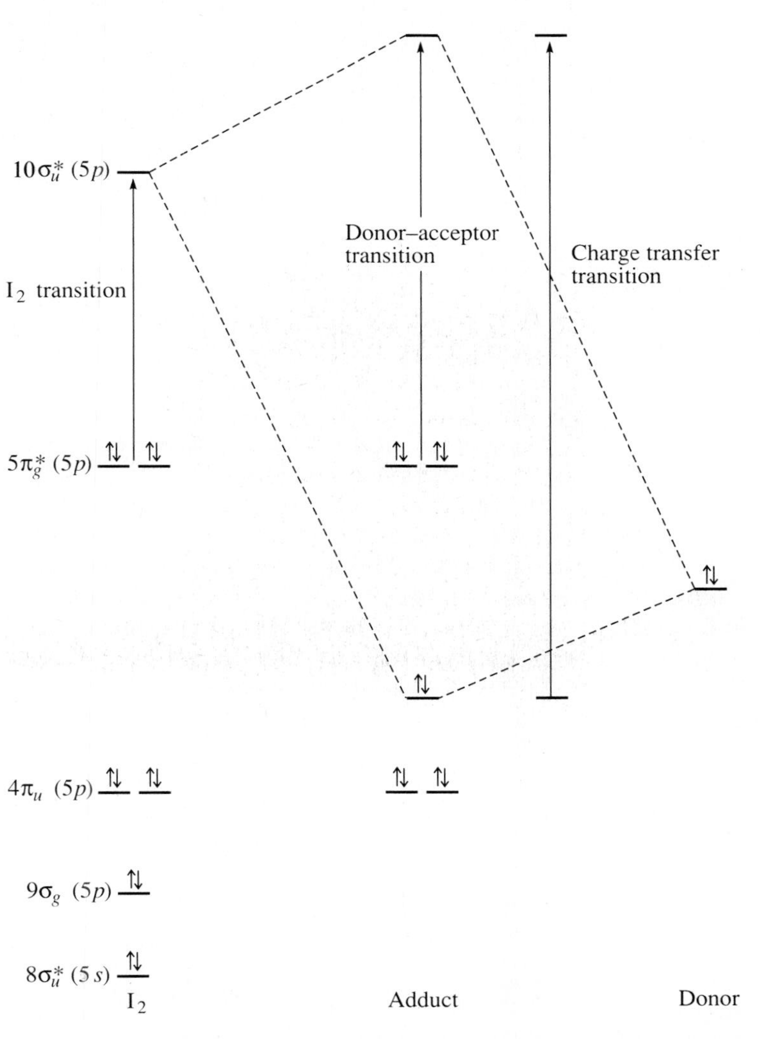

FIGURE 6-13 Electronic Transitions in I_2 Adducts.

$10\sigma_u^*$ level (see Figure 6-13). This absorption removes the middle yellow, green, and blue part of the visible spectrum, leaving red and violet at opposite ends of the spectrum to combine in the violet color that is seen.

In nondonor solvents such as heptane, the iodine color remains essentially the same violet, but in benzene and other pi-electron solvents it becomes more red-violet, and in good donors like ethers, alcohols, and amines the color becomes distinctly brown. The solubility of I_2 also increases with increasing donor character of the solvent. Interaction of the donor orbital of the solvent with the $10\sigma_u^*$ orbital results in a lower occupied bonding orbital and a higher unoccupied antibonding orbital. As a result, the $\pi_g^* \longrightarrow \sigma_u^*$ transition for I_2 + donor (Lewis base) has a higher energy and a maximum absorbance shifted toward the blue. The transmitted color shifts toward brown (combined red, yellow, and green), as more of the yellow and green light passes through. Water is also a donor, but not a very good one; I_2 is only slightly soluble in water, and the solution is yellow-brown. Adding I^- results in formation of I_3^-, which is very soluble in water and brown in color.

In addition to the shifts described above, a new **charge-transfer band** appears at the edge of the ultraviolet (230 to 400 nm). This band is due to the transition $\sigma \longrightarrow \sigma^*$, between the two new orbitals formed by the interaction. Because the σ orbital has a larger proportion of the donor (solvent or I^-) orbital and the σ^* orbital has a larger proportion of the I_2 orbital, the transition transfers an electron from an orbital that is primarily of donor composition to one that is primarily of acceptor composition; hence the name charge transfer for this transition. The transition may be shown schematically as

$$I_2 \cdot \text{donor} \xrightarrow[\text{CT}]{h\nu} [I_2]^- \cdot [\text{donor}]^+$$

As the energies of the donor and acceptor orbitals become more similar and the energy of the donor–acceptor transition ($\pi_g^* \longrightarrow \sigma_u^*$) increases, the energy of the charge-transfer transition decreases. In poor donor–acceptor combinations, the σ orbital is essentially that of the donor, and the σ^* orbital is essentially that of the I_2 acceptor. With better donors, the change in energy from the reactant orbitals to the product orbitals is larger, but the $\sigma \longrightarrow \sigma^*$ energy difference is smaller.

The charge-transfer phenomenon also appears in many other adducts. If the charge-transfer transition actually transfers the electron permanently, the result is an oxidation–reduction reaction; the donor is oxidized and the acceptor is reduced. The sequence of reactions of Fe^{3+} with the ions F^-, Cl^-, Br^-, and I^- [actually $Fe(H_2O)_6^{3+}$ and aquated halide ions forming $Fe(H_2O)_5X^{2+}$, described in Table 6-12 and the associated text] illustrates the whole range of possibilities. FeF^{2+} is a very stable combination ($Fe^{3+} + F^- \rightleftharpoons FeF^{2+}$ has a large equilibrium constant) that is colorless, $FeCl^{2+}$ is somewhat less stable and has a yellow color, $FeBr^{2+}$ is still less stable and more strongly yellow-brown, and FeI^{2+} is unstable and in high concentrations reacts to form Fe^{2+} and molecular iodine as products of a redox reaction. In these examples, the acid is Fe^{3+} and the base is the halide ion.

In the series F^-, Cl^-, Br^-, I^-, the highest occupied (p) orbital of the halide ion increases in energy with the increase in ionic size and the number of electrons. This increase in energy and size is a consequence of more effective screening by the inner electrons (described in Chapter 2) in the heavier elements. All the halide HOMOs are also well above the Fe^{3+} LUMO, with F^- forming the closest energy match. In FeF^{2+}, the interaction is strong enough to result in a large energy difference between the bonding and antibonding orbitals. Therefore, the transition between them requires high energy ultraviolet photons. As the halide orbital energy increases for $FeCl^{2+}$ and $FeBr^{2+}$, the interaction decreases, the difference between bonding and antibonding orbitals decreases, and the absorbance moves into the visible region. Finally, with I^-, the bonding orbital becomes nearly the same as the lowest unoccupied orbital of the Fe^{3+}, and an electron is easily transferred to it, forming Fe^{2+}.

6-4-4 MOLECULAR ORBITALS AND HSAB

It is desirable to be able to relate acidity and basicity to fundamental properties of the system being studied; with sufficient information about the physical properties of potential reactants, we might therefore be able to predict if a reaction might occur. In his more recent papers, Pearson[34] has related hardness and softness to the **absolute electronegativity,** as defined by Mulliken (Chapter 3):

$$\chi = \frac{I + A}{2}, \qquad I = \text{ionization energy (in eV)}, \quad A = \text{electron affinity (in eV)}$$

The **absolute hardness,** η, is defined (see Figure 6-14) as

$$\eta = \frac{I - A}{2}$$

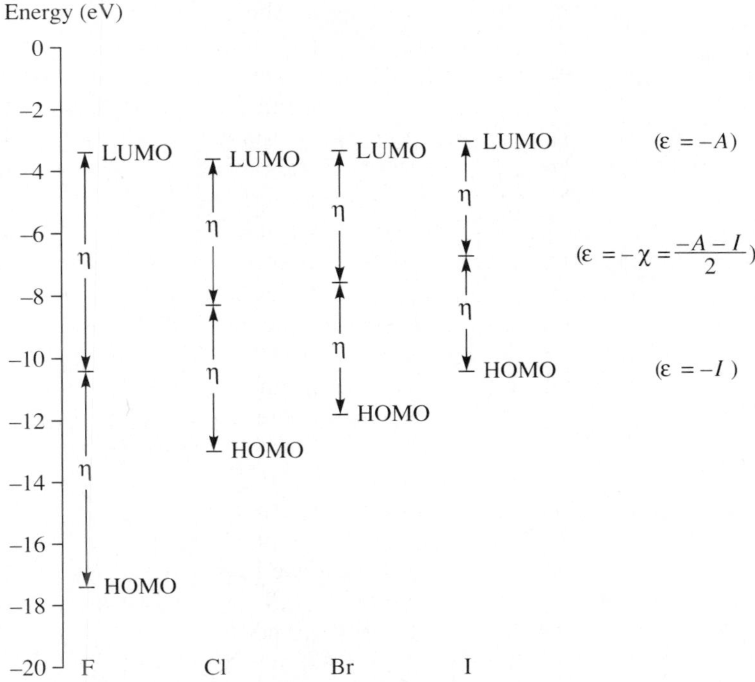

FIGURE 6-14 Energy Levels for Halogens. Relation between absolute electronegativity (χ), absolute hardness (η), and HOMO and LUMO energies for the halogens.

[34] R. G. Pearson, *Inorg. Chem.*, **1988,** *27,* 734.

and the softness, σ, is the inverse of the hardness:

$$\sigma = \frac{1}{\eta}$$

Ionization energy is assumed to measure the energy of the HOMO, and electron affinity is assumed to measure the energy of the LUMO for a given molecule:

$$E_{HOMO} = -I, \qquad E_{LUMO} = -A$$

Tables 6-13 through 6-15 give the ionization energy, electron affinity, electronegativity, and absolute hardness for cations, molecules, atoms, and radicals, all in order of decreasing hardness. Because ionization energies are much larger in magnitude than electron affinities, electronegativity as a function of

TABLE 6-13
Hardness parameters for cations (all in eV)

Ion	*I*	*A*	χ	η
B^{3+}	259.37	37.93	148.65	110.72
Be^{2+}	153.89	18.21	86.05	67.84
Al^{3+}	119.99	28.45	74.22	45.77
Li^{+}	75.64	5.39	40.52	35.12
Mg^{2+}	80.14	15.04	47.59	32.55
Na^{+}	47.29	5.14	26.21	21.08
Ca^{2+}	50.91	11.87	31.39	19.52
Sr^{2+}	43.6	11.03	27.3	16.3
K^{+}	31.63	4.34	17.99	13.64
Fe^{3+}	54.8	30.65	42.73	12.08
Rb^{+}	27.28	4.18	15.77	11.55
Rh^{3+}	53.4	31.1	42.4	11.2
Zn^{2+}	39.72	17.96	28.84	10.88
Cs^{+}	25.1	3.89	14.5	10.6
Cd^{2+}	37.48	16.91	27.20	10.29
Cr^{3+}	49.1	30.96	40.0	9.1
Mn^{2+}	33.67	15.64	24.66	9.02
Mn^{3+}	51.2	33.67	42.4	8.8
Co^{3+}	51.3	33.50	42.4	8.8
V^{3+}	46.71	29.31	38.01	8.70
Ni^{2+}	35.17	18.17	26.67	8.50
Pb^{2+}	31.94	15.03	23.49	8.46
Au^{3+}	54.1	37.4	45.8	8.4
Cu^{2+}	36.83	20.29	28.56	8.27
Co^{2+}	33.50	17.06	25.28	8.22
Pt^{2+}	35.2	19.2	27.2	8.0
Sn^{2+}	30.50	14.63	22.57	7.94
Ir^{3+}	45.3	29.5	37.4	7.9
Hg^{2+}	34.2	18.76	26.5	7.7
V^{2+}	29.31	14.65	21.98	7.33
Fe^{2+}	30.65	16.18	23.42	7.24
Cr^{2+}	30.96	16.50	23.73	7.23
Ag^{+}	21.49	7.58	14.53	6.96
Ti^{2+}	27.49	13.58	20.54	6.96
Pd^{2+}	32.93	19.43	26.18	6.75
Rh^{2+}	31.06	18.08	24.57	6.49
Cu^{+}	20.29	7.73	14.01	6.28
Sc^{2+}	24.76	12.80	18.78	5.98
Ru^{2+}	28.47	16.76	22.62	5.86
Au^{+}	20.5	9.23	14.9	5.6

SOURCE: Data from R. G. Pearson, *Inorg. Chem.*, **1988**, *27*, 734.

TABLE 6-14
Hardness parameters for molecules (all in eV)

Molecule	*I*	*A*	χ	η
BF_3	15.81	−3.5	6.2	9.7
H_2O	12.6	−6.4	3.1	9.5
N_2	15.58	−2.2	6.70	8.9
NH_3	10.7	−5.6	2.6	8.2
CH_3CN	12.2	−2.8	4.7	7.5
C_2H_2	11.4	−2.6	4.4	7.0
PF_3	12.3	−1.0	5.7	6.7
$(CH_3)_3N$	7.8	−4.8	1.5	6.3
C_2H_4	10.5	−1.8	4.4	6.2
PH_3	10.0	−1.9	4.1	6.0
O_2	12.2	.4	6.3	5.9
$(CH_3)_3P$	8.6	−3.1	2.8	5.9
$(CH_3)_3As$	8.7	−2.7	3.0	5.7
SO_2	12.3	1.1	6.7	5.6
SO_3	12.7	1.7	7.2	5.5
C_6H_6	9.3	−1.2	4.1	5.3
C_5H_5N	9.3	−.6	4.4	5.0
Butadiene	9.1	−.6	4.3	4.9
PCl_3	10.2	.8	5.5	4.7
PBr_3	9.9	1.6	5.6	4.2

SOURCE: Data from R. G. Pearson, *Inorg. Chem.*, **1988,** *27,* 734.

TABLE 6-15
Hardness parameters for atoms and radicals (all in eV)

Atom or radical	*I*	*A*	χ	η
F	17.42	3.40	10.41	7.01
H	13.60	.75	7.18	6.43
OH	13.17	1.83	7.50	5.67
NH_2	11.40	.74	6.07	5.33
CN	14.02	3.82	8.92	5.10
CH_3	9.82	.08	4.96	4.87
Cl	13.01	3.62	8.31	4.70
C_2H_5	8.38	−.39	4.00	4.39
Br	11.84	3.36	7.60	4.24
C_6H_5	9.20	1.1	5.2	4.1
NO_2	>10.1	2.30	>6.2	>3.9
I	10.45	3.06	6.76	3.70
SiH_3	8.14	1.41	4.78	3.37
C_6H_5O	8.85	2.35	5.60	3.25
$Mn(CO)_5$	8.44	2.	5.2	3.2
CH_3S	8.06	1.9	5.0	3.1
C_6H_5S	8.63	2.47	5.50	3.08

SOURCE: Data from R. G. Pearson, *Inorg. Chem.*, **1988,** *27,* 734.
NOTE: The hardness values approximate those of the corresponding anions.

atomic number parallels the shape of the ionization energy curve given in Chapter 2. There are no electron affinities for anions, so Pearson uses the values for atoms to calculate electronegativities and hardness, arguing that these functions will be as accurate as can be obtained. Examination of these tables and the hard and soft acids and bases mentioned earlier in this chapter shows discrepancies in the assignments. For example, hard, borderline, and soft metal ions are intermixed in Table 6-13. This shows the difficulty of using only one parameter to describe their behavior.

The halogen molecules offer good examples of the use of these orbital arguments. Fluorine is the most electronegative halogen. It is also the smallest and least polarizable halogen and is therefore the hardest. In orbital terms, the LUMOs of all the halogen molecules are nearly identical, and the HOMOs increase in energy from F_2 to I_2, as in Figure 6-14. The electronegativities decrease in order $F_2 > Cl_2 > Br_2 > I_2$ as the HOMO energies increase; the hardness also decreases in the same order as the difference between the HOMO and LUMO decreases.

A somewhat oversimplified way to look at the hard–soft question considers the hard–hard interactions as simple electrostatic interactions, with the LUMO of the acid far above the HOMO of the base and relatively little change in orbital energies on adduct formation.[35] The corresponding soft–soft interaction then involves HOMO and LUMO energies that are much closer and give a large change in orbital energies on adduct formation. The larger separation between HOMO and LUMO for hard species leads to a smaller change in orbital energies on adduct formation. Diagrams of such interactions are in Figure 6-15. Diagrams like these need to be used with caution, however. The

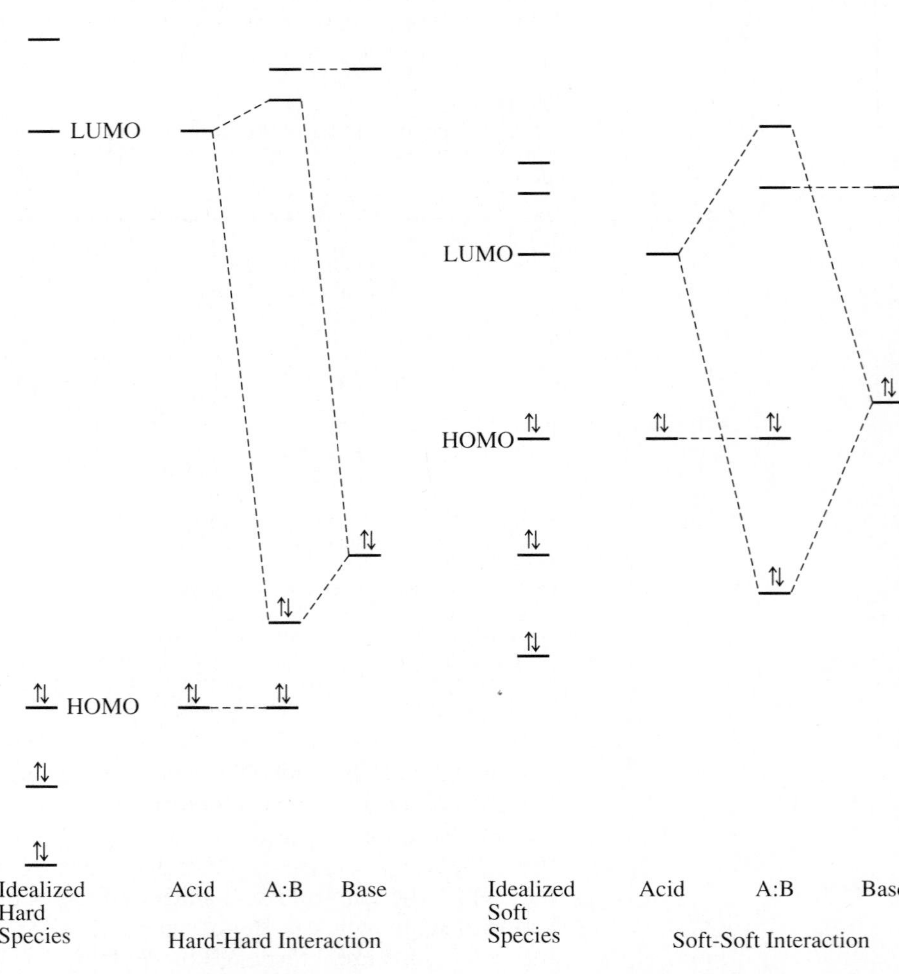

FIGURE 6-15 HOMO–LUMO Diagrams for Hard–Hard and Soft–Soft Interactions. (Adapted from W. B. Jensen, *The Lewis Acid–Base Concepts,* Wiley-Interscience, New York, 1980, pp. 262–63. Copyright © 1980, John Wiley & Sons, Inc. Reprinted by permission of John Wiley & Sons, Inc.)

[35] Jensen, op. cit., pp. 262–65; C. K. Jørgensen, *Struct. Bonding Berlin,* **1966,** *1,* 234.

small drop in energy in the hard–hard case that seems to indicate only small interactions is not necessarily the entire story. The hard–hard interaction depends on a longer-range electrostatic force, and this interaction can be quite strong. Many comparisons of hard–hard and soft–soft interactions indicate that the hard–hard combination is stronger and is the primary driving force for the reaction. The contrast between the hard–hard product and the hard–soft reactants in such cases provides the net energy difference that leads to the products. One should also keep in mind that many reactions to which the HSAB approach is applied involve competition between two different conjugate acid–base pairs; only in a limited number of cases is one interaction large enough to overwhelm the others and determine whether the reaction will proceed or not.

GENERAL REFERENCES

W. B. Jensen, *The Lewis Acid–Base Concept: An Overview,* Wiley, New York, 1980, and H. L. Finston and Allen C. Rychtman, *A New View of Current Acid–Base Theories,* Wiley, New York, 1982, give good overviews of the history of acid–base theories and critical discussions of the different theories. R. G. Pearson's *Hard and Soft Acids and Bases,* Dowden, Hutchinson, & Ross, Stroudsburg, Pa., 1973, is a review by one of the leading exponents of HSAB. For other viewpoints, the references provided in the chapter should be consulted.

PROBLEMS

Additional acid–base problems may be found at the end of Chapter 7.

6-1 List the following acids in order of acid strength in aqueous solution:
a. $HMnO_4$ H_3AsO_4 H_2SO_3 H_2SO_4
b. $HClO$ $HClO_4$ $HClO_2$ $HClO_3$

6-2 List the following acids in order of their strength when reacting with NH_3:

BF_3 $B(CH_3)_3$ $B(C_2H_5)_3$ $B(C_6H_2(CH_3)_3)_3$

[$C_6H_2(CH_3)_3$ is 2,4,6-trimethylphenyl]

6-3 Solvents can change the acid–base behavior of solutes. Compare the acid–base properties of dimethylamine in water, acetic acid, and 2-butanone.

6-4 Choose the stronger acid or base in the following pairs and explain your choice.
a. CH_3NH_2 or NH_3 in reaction with H^+
b. Pyridine or 2-methylpyridine in reaction with trimethylboron
c. Triphenylboron or trimethylboron in reaction with ammonia

6-5 Predict the reactions of the following hydrogen compounds with water and explain your reasoning.
a. CaH_2 **b.** HBr **c.** H_2S **d.** CH_4

6-6 Rationalize the following data in HSAB terms:

Reaction	ΔH
$CH_3CH_3 + H_2O \longrightarrow CH_3OH + CH_4$	12 kcal
$CH_3COCH_3 + H_2O \longrightarrow CH_3COOH + CH_4$	−13 kcal

6-7 Predict the order of solubility in water of each of the following series and explain the factors involved.

a. $MgSO_4$ $CaSO_4$ $SrSO_4$ $BaSO_4$

b. $PbCl_2$ $PbBr_2$ PbI_2 PbS

6-8 Choose and explain:

a. Strongest Brønsted acid: SnH_4 SbH_3 TeH_2

b. Strongest Brønsted base: NH_3 PH_3 SbH_3

c. Strongest base to H^+ (gas phase):
NH_3 CH_3NH_2 $(CH_3)_2NH$ $(CH_3)_3N$

d. Strongest base to BMe_3: pyridine 2-methylpyridine 4-methylpyridine

6-9 B_2O_3 is acidic, Al_2O_3 is amphoteric, and Sc_2O_3 is basic. Why?

6-10 Baking powder is a mixture of aluminum sulfate and sodium hydrogen carbonate, which generates a gas and makes bubbles in biscuit dough. Explain what the reactions are.

6-11 AlF_3 is insoluble in liquid HF, but dissolves if NaF is present. When BF_3 is added to the solution, AlF_3 precipitates. Explain.

6-12 CsI is much less soluble in water than CsF, and LiF is much less soluble than LiI. Why?

6-13 For each of the following reactions, identify the acid and the base. Also indicate which acid–base definition (Lewis, solvent system, Brønsted) applies. In some cases, more than one definition may apply.

a. $BF_3 + 2\,ClF \longrightarrow [Cl_2F]^+ + [BF_4]^-$

b. $AsF_5 + 2\,ClF \longrightarrow [Cl_2F]^+ + [AsF_6]^-$

c. $PCl_5 + ICl \longrightarrow [PCl_4]^+ + [ICl_2]^-$

d. $NOF + ClF_3 \longrightarrow [NO]^+ + [ClF_4]^-$

e. $2\,ClO_3^- + SO_2 \longrightarrow 2\,ClO_2 + SO_4^{2-}$

f. $Pt + XeF_4 \longrightarrow PtF_4 + Xe$

g. $XeO_3 + OH^- \longrightarrow [HXeO_4]^-$

h. $2\,HF + SbF_5 \longrightarrow [H_2F]^+ + [SbF_6]^-$

i. $2\,NOCl + Sn \longrightarrow SnCl_2 + 2\,NO$ (in N_2O_4 solvent)

j. $PtF_5 + ClF_3 \longrightarrow [ClF_2]^+ + [PtF_6]^-$

6-14 The conductivity of BrF_3 is enhanced by adding either AgF or SnF_4. Explain this enhancement, using the appropriate chemical equations.

6-15 The conductivity of ICl is enhanced by adding either NaCl or $AlCl_3$.

a. Suggest an equation to describe the autoionization of ICl.

b. Account for the enhancement of conductivity by the two solutes.

6-16 Titration of NH_4Cl with $SnCl_4$ in ICl requires 2 moles of NH_4Cl for every mole of $SnCl_4$ to reach the end point. Explain, using chemical equations.

6-17 Dissolution of KF in IF_5 increases the conductivity of the latter. Suggest an explanation.

6-18 Anhydrous H_2SO_4 and anhydrous H_3PO_4 both have high electrical conductivities. Explain.

6-19 HF has $H_0 = -11.0$. Addition of 4% SbF_5 lowers H_0 to -21.0. Explain why this should be true and why the resulting solution is so strongly acidic that it can protonate alkenes:

$$(CH_3)_2C{=}CH_2 + H^+ \longrightarrow (CH_3)_3C^+$$

6-20 a. Use Drago's E and C parameters to calculate ΔH for the reactions of pyridine and BF_3 and of pyridine and $B(CH_3)_3$. Compare your results with the experimental values in Table 6-5.

b. Explain the differences found in part a in terms of the structures of BF_3 and $B(CH_3)_3$.
c. Explain the differences in terms of HSAB theory.

6-21 Repeat the calculations of the previous problem with NH_3 as the base, and put the four reactions in order of the magnitudes of their ΔH values.

6-22 Compare the results of problems 20 and 21 with the absolute hardness parameters of Table 6-14 for BF_3, NH_3, and pyridine (C_5H_5N). What value of η would you predict for $B(CH_3)_3$? [Compare NH_3 and $N(CH_3)_3$ as a guide.]

7

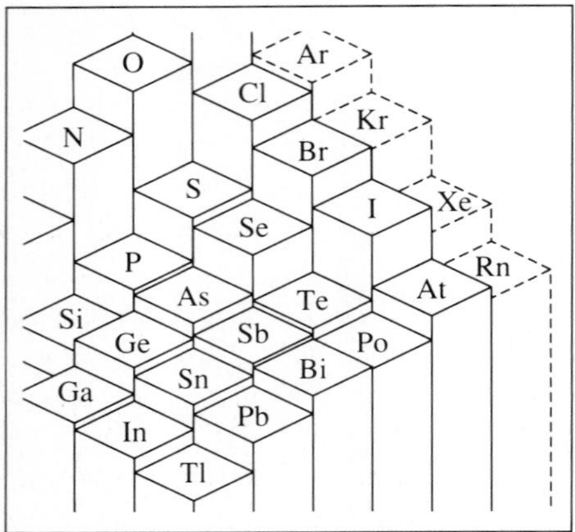

Chemistry of the Main Group Elements

A discussion of main group chemistry provides a useful context in which to introduce a variety of topics not covered previously in this text. These topics are particularly characteristic of main group chemistry but may have application to the chemistry of other elements as well. For example, numerous examples are known in which atoms form bridges between other atoms; main group examples include:

In this chapter we will discuss in some detail one important type of bridge, the hydrogens that form bridges between boron atoms in boranes. A similar approach can be used to describe bridges by halogen atoms, other atoms, and groups such as CO (CO bridges between transition metal atoms will be discussed in Chapter 12).

This chapter also provides examples of how modern chemistry has developed in ways contrary to previously held ideas (including misconceptions persisting in some general chemistry texts). Examples include the synthesis of alkali metal *anions*, compounds in which carbon is bonded to more than four atoms, and the now fairly extensive chemistry of noble gas elements. Much of the information in this chapter is included for the sake of handy reference; for more details, the interested reader should consult the references listed at the end of this chapter. Many examples of main group chemistry have already been discussed in this text, especially in connection with the bonding and structures of main group compounds (Chapters 3 and 5) and acid–base reactions involving these compounds (Chapter 6). Problems at the end of this chapter provide further illustrations of these topics as well as the new topics introduced in this chapter.

Main group chemistry is extremely important. The 20 top industrial chemicals produced in the United States are main group elements or compounds (see Table 7-1), and eight of the top nine may be classified as inorganic; many other compounds of these elements are of great commercial importance.

This chapter presents some of the most significant physical and chemical data on each of the main groups of elements (also known as the representative elements), treating hydrogen first and continuing in sequence through groups 1, 2, and 13 to 18 (groups IA to VIIIA in common American notation).

TABLE 7-1
Top 20 industrial chemicals produced in United States, 1988

Rank	*Chemical*	*Production* ($\times 10^9$ *kg*)
1	H_2SO_4	38.81
2	N_2	23.63
3	$H_2C{=}CH_2$	16.58
4	O_2	16.82
5	NH_3	15.37
6	CaO (lime)	14.67
7	NaOH	10.87
8	H_3PO_4	10.63
9	Cl_2	10.28
10	$H_2C{=}CH{-}CH_3$ (propylene)	9.06
11	Na_2CO_3	8.66
12	HNO_3	7.16
13	H_2NCONH_2 (urea)	7.15
14	NH_4NO_3	6.52
15	$ClH_2C{-}CClH_2$ (ethylene dichloride)	6.19
16	C_6H_6 (benzene)	5.37
17	$C_6H_5C_2H_5$ (ethylbenzene)	4.51
18	p-$C_6H_4(COOH)_2$ (terephthalic acid)	4.35
19	CO_2	4.25
20	$H_2C{=}CHCl$ (vinyl chloride)	4.11

SOURCE: Data from *Chem. Eng. News*, June 19, **1989**, 39.

7-1 GENERAL TRENDS IN MAIN GROUP CHEMISTRY

7-1-1 PHYSICAL PROPERTIES

The main group elements are characterized by the completion of their electron configurations using *s* and *p* electrons; the total number of such electrons in the outermost shell is conveniently given by the traditional American group numbers. These elements cover the full range of classifications, from the most metallic to the most nonmetallic, with elements of intermediate properties, the semimetals (also known as metalloids), in between. On the far left, the alkali metals and alkaline earths exhibit the expected metallic characteristics of luster, high ability to conduct heat and electricity, and malleability. The distinction between metals and nonmetals is best made on their electrical

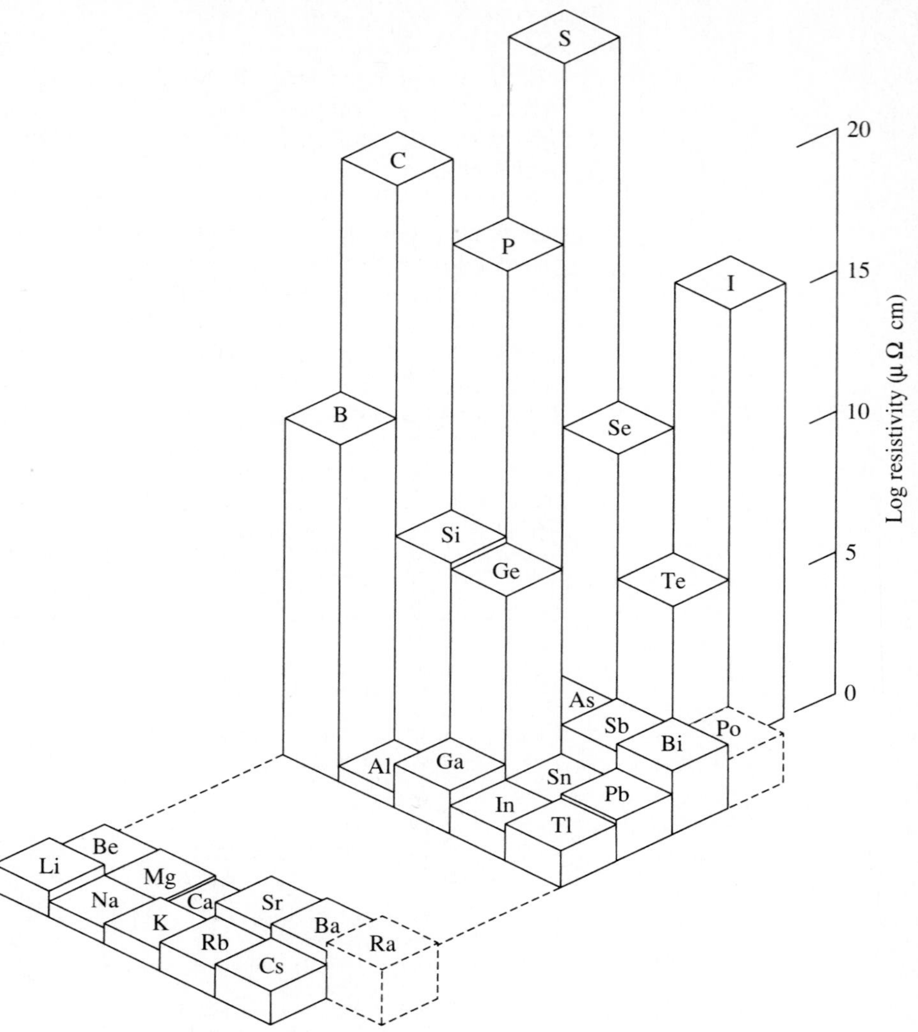

FIGURE 7-1 Electrical Resistivities of the Main Group Elements. Dashed lines indicate estimated values. (From J. Emsley, *The Elements,* Oxford University Press, New York, 1989.)

conductivity. In Figure 7-1 are plotted electrical resistivities (inversely proportional to conductivity) of the solid main group elements. At the far left are the alkali metals, having low resistivities (high abilities to conduct); at the far right are the nonmetals. As discussed in Chapter 5, these dramatic differences in conductivity between metals and nonmetals can be explained on the basis of the bonding and structure of crystals of these elements. In short, metals contain loosely bound valence electrons that are relatively free to move and thereby conduct current; nonmetals contain much more localized lone electron pairs and covalently bonded pairs, which are much less free.

Elements along a rough diagonal from boron to polonium are intermediate in behavior, in some cases having both metallic and nonmetallic allotropes (elemental forms); these elements are designated **metalloids** or **semimetals**. Some, such as silicon and germanium, are capable of having their conductivity finely tuned by the addition of small amounts of impurities and are consequently of enormous importance in the manufacture of semiconductors in the computer industry.

7-1-2 ELECTRONEGATIVITY

Electronegativity, shown in Figure 7-2, also provides a guide to the chemical behavior of the main group elements. The extremely high electronegativity of the nonmetal fluorine is evident, with a steady decline in electronegativity toward the left and the bottom of the periodic table. The semimetals form a diagonal of intermediate electronegativity. Definitions of electronegativity are given in Section 3-3, and tabulated values for the elements are given in Table 3-1 and Appendix B; values for the main group elements are included in tables for each group later in this chapter.

Hydrogen, although usually classified with group 1 (IA), is quite dissimilar from the alkali metals in its electronegativity, as well as in many other properties, both chemical and physical. Hydrogen's chemistry is distinctive from all the groups, so this element will be discussed separately in this chapter.

Electronegativities of the noble gases have not been measured experimentally. However, the noble gases have higher ionization energies than the halogens (one method of calculating electronegativity, proposed by Mulliken, defines the electronegativity as the average of the ionization energy and the electron affinity; see Chapter 3), and calculations have suggested that the electronegativities of the noble gases may match or even exceed those of the

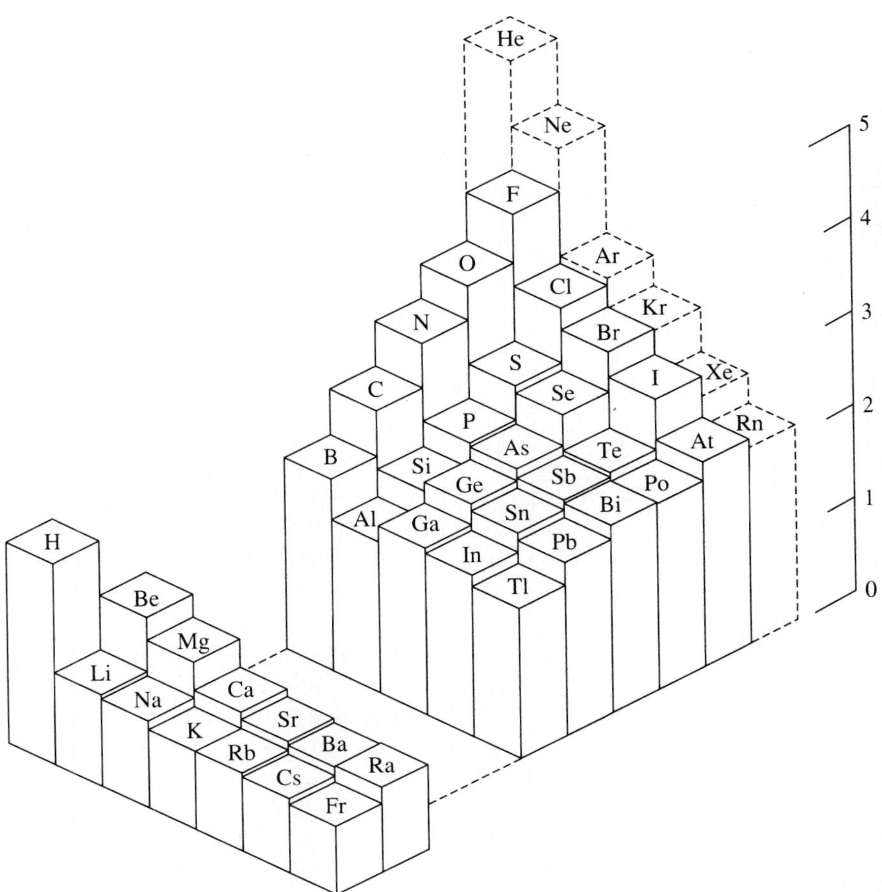

FIGURE 7-2 Pauling Electronegativities of the Main Group Elements. (From A. L. Allred, *J. Inorg. Nucl. Chem.*, **1961,** *17,* 215. Dashed lines indicate estimated values from L. C. Allen and J. E. Huheey, *J. Inorg. Nucl. Chem.*, **1980,** *42,* 1523.)

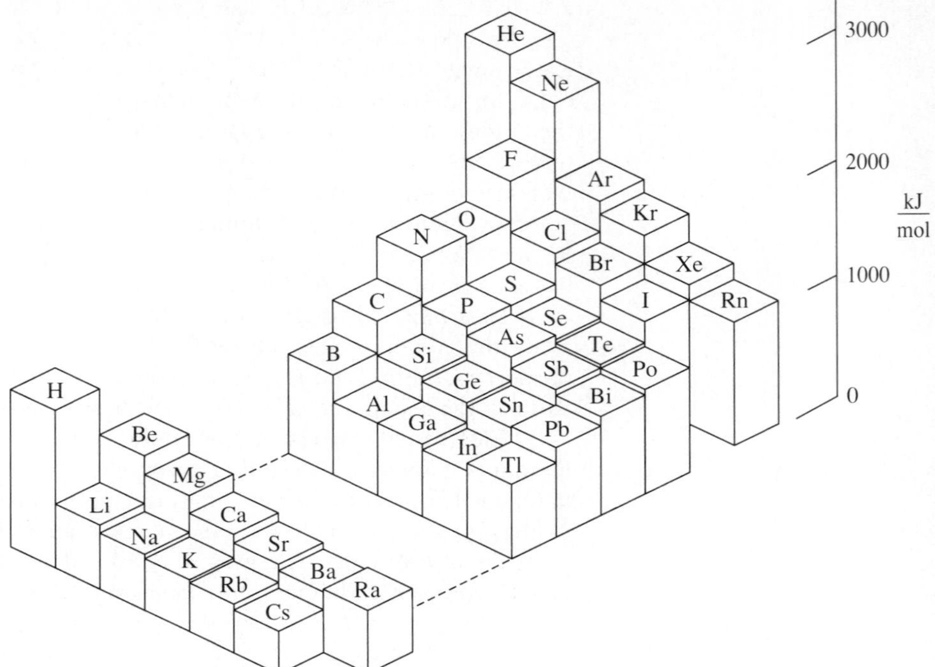

FIGURE 7-3 Ionization Energies of the Main Group Elements. (From C. E. Moore, *Ionization Potentials and Ionization Limits Derived from the Analyses of Optical Spectra*, National Standard Reference Data Series, U.S. National Bureau of Standards (NSRDS–NBS 34), 1970.)

halogens.[1] The noble gas atoms are somewhat smaller than the neighboring halogen atoms (for example, Ne is smaller than F) as a consequence of a greater effective nuclear charge. This charge, which is able to attract noble gas electrons strongly toward the nucleus, is also likely to exert a strong attraction on electrons of neighboring atoms; hence high electronegativities predicted for the noble gases are reasonable. Estimated values of these electronegativities are included in Figure 7-2.

7-1-3 IONIZATION ENERGY

Ionization energies of the main group elements exhibit rather similar trends to electronegativity, as shown in Figure 7-3. There are some subtle differences, however. As discussed in Section 2-4-1, while there is a general increase in ionization energy toward the upper-right corner of the periodic table, several of the group 13 (IIIA) elements have lower ionization energies than the preceding group 12 (IIA) elements, and several group 16 (VIA) elements have lower ionization energies than the preceding group 15 (VA) elements. For example, the ionization energy of boron is lower than that of beryllium, and oxygen is lower than nitrogen (see also Figure 2-12). The high ionization energies of Be and N are associated with their electron configurations, with these atoms having electron subshells that are completely filled ($2s^2$ for Be) or half-filled ($2p^3$ for N). The next atoms (B and O) have an additional electron that is lost with comparative ease. In boron the outermost electron, a $2p$, has significantly higher energy (higher quantum number l) than the filled $1s$ and $2s$ orbitals and is thus more easily lost than a $2s$ electron of Be. In oxygen, the

[1] L. C. Allen and J. E. Huheey, *J. Inorg. Nucl. Chem.*, **1980**, *42*, 1523.

fourth $2p$ electron must pair with another $2p$ electron; occupation of this orbital by two electrons is accompanied by an increase in electron–electron repulsions, which facilitates loss of an electron. Additional examples of this phenomenon can be seen in Figures 7-3 and 2-12 (tabulated values of ionization energies are included in Appendix B).

7-1-4 CHEMICAL PROPERTIES

Efforts to find similarities in the chemistry of the main group elements began well before the formulation of the modern periodic table. The strongest parallels, of course, are within each group: the alkali metals most strongly resemble other alkali metals, halogens resemble other halogens, and so on. In addition, certain similarities have been recognized between some elements along diagonals (upper left to lower right) in the periodic table. One example is that of electronegativities. As can be seen from Figure 7-2, electronegativities along diagonals are quite similar; for example, values along the diagonal from B to Te are in the range 1.9 to 2.2. Other "diagonal" similarities include the unusually low solubilities of LiF and MgF_2 (a consequence of the small sizes of Li^+ and Mg^{2+}, which lead to high lattice energies in these ionic compounds); similarities in solubilities of carbonates and hydroxides of Be and Al; and the formation of complex three-dimensional structures based on SiO_4 and BO_4 tetrahedra. These parallels are interesting but somewhat limited in scope; they can often be explained on the basis of similarities in sizes and electronic structures of the compounds in question.

Another characteristic of main group elements frequently cited is the "first row anomaly" (counting as first row the elements Li through Ne). In many instances, properties of elements in this row are significantly different from properties of other elements in the same group. For example, F_2 has a much lower bond energy than expected by extrapolation of bond energies of Cl_2, Br_2, and I_2; HF is a weak acid in aqueous solution, while HCl, HBr, and HI are all strong; multiple bonds between carbon atoms are much more common than between other elements in group 14 (IVA); and hydrogen bonding is much stronger for compounds of F, O, and N than for compounds of other elements in their groups. There is no single explanation to account for all the differences between elements in this row and other elements. However, in many cases the distinctive chemistry of the first row elements is related to the small atomic size and high electronegativity of these elements. Examples are described in several places in this chapter.

7-2 HYDROGEN

The most appropriate position of hydrogen in the periodic table has been a matter of dispute among chemists. Its electron configuration, $1s^1$, is similar to the valence electron configurations of the alkali metals (ns^1); hence hydrogen is most commonly listed in the periodic table at the top of group 1 (IA). However, it has little chemical similarity to the alkali metals. Hydrogen is also one electron short of a noble gas configuration and could conceivably be classified with the halogens; yet, while hydrogen has some similarities with the halogens, for example in forming a diatomic molecule and an ion of 1− charge, these similarities are limited. A third possibility is to place hydrogen in group 14 (IVA) above carbon: both elements have half-filled valence electron shells, are of similar electronegativity, and usually form covalent rather than ionic bonds. We prefer not to attempt to fit hydrogen into any particular group in the periodic table; it is a unique element in many ways and deserves separate consideration.

Hydrogen is extremely abundant; it is by far the most abundant element in the universe (and on the sun) and, primarily in its compounds, is the third most abundant element in the earth's crust. The element occurs in three isotopes, ordinary hydrogen, ^{1}H; deuterium, ^{2}H or D; and tritium, ^{3}H or T. Both ordinary hydrogen and deuterium have stable nuclei; tritium undergoes beta decay,

$$^{3}_{1}H \longrightarrow {}^{3}_{2}He + e^{-}$$

and has a half-life of 12.35 years. Naturally occurring hydrogen is 99.985% ^{1}H, and essentially all the remainder is ^{2}H; only traces of the radioactive ^{3}H are found on earth. Deuterium compounds are used extensively as solvents for NMR spectroscopy and in kinetic studies on reactions involving bonds to hydrogen (deuterium isotope effects). Tritium is produced in nuclear reactors by bombardment of ^{6}Li nuclei with neutrons:

$$^{6}_{3}Li + {}^{1}_{0}n \longrightarrow {}^{4}_{2}He + {}^{3}_{1}H$$

It is used as a tracer, especially in metallurgy, in studying the movement of ground waters, and to study the *ab*sorption of hydrogen by metals and the *ad*sorption of hydrogen on metal surfaces. Numerous deuterated and tritiated compounds have been synthesized and studied. Some of the important physical properties of the isotopes of hydrogen are listed in Table 7-2.

TABLE 7-2
Properties of hydrogen, deuterium, and tritium

			Properties of molecules, X_2			
Isotope	*Abundance (%)*	*Atomic mass*	*Melting point (K)*	*Boiling point (K)*	*Critical temperature (K)**	*Enthalpy of dissociation (kJ mol^{-1} at 25°C)*
Hydrogen, H	99.985	1.007825	13.957	20.39	33.19	435.88
Deuterium, D	0.015	2.014102	18.73	23.67	38.35	443.35
Tritium, T	$\sim 10^{-16}$	3.016049	20.62	25.04	40.6 (calc)	446.9

SOURCES: Abundance and atomic mass data from *Quantities, Units, and Symbols in Physical Chemistry*, International Union of Pure and Applied Chemistry, Blackwell Scientific Publications, Oxford, England, 1988. Other data from N. N. Greenwood and A. Earnshaw, *Chemistry of the Elements*, Pergamon Press, Elmsford, N.Y., 1984.
NOTE: * The highest temperature at which gas can be condensed to a liquid.

7-2-1 CHEMICAL PROPERTIES

As mentioned above, hydrogen can gain an electron to achieve a noble gas configuration in forming the hydride ion, H^{-}. Many metals, for example the alkali metals and alkaline earths, form hydrides that are essentially ionic and contain discrete H^{-} ions. In many other cases, bonding to hydrogen atoms is essentially covalent, as in compounds with carbon and the other nonmetals. Examples of these compounds will be discussed later in this chapter. Hydride ions may also act as ligands in bonding to metals, with as many as nine hydrogens on a single metal, as in ReH_9^{2-}. Numerous complex hydrides such as BH_4^- and AlH_4^- serve as important reagents in organic and inorganic synthesis. Although such complexes may be described formally as hydrides, their bonding is primarily covalent.

Reference to the hydrogen ion, H^+, is also common. However, in the presence of solvent, the extremely small size of the proton (radius approximately 1.5×10^{-3} pm) requires that it be associated with solvent molecules or other dissolved species. In aqueous solution a more correct description is H_3O^+ (aq), although larger species such as $H_9O_4^+$ are also likely. Another important characteristic of H^+ that is a consequence of its small size is its ability to form hydrogen bonds. Protonic acids and hydrogen bonds have been discussed in Chapter 6.

The ready combustibility of hydrogen, together with the lack of potentially polluting byproducts has led to the suggestion of using hydrogen extensively as a fuel. For example, as a potential fuel for automobiles, H_2 can provide a greater amount of energy per unit mass than gasoline without producing such environmentally damaging byproducts as carbon monoxide, sulfur dioxide, and nitrogen oxides. A challenge for chemists is to develop practical thermal or photochemical processes for generating hydrogen from its most abundant source, water.

Molecular hydrogen is also an important reagent, especially in the industrial hydrogenation of unsaturated organic molecules. Examples of such processes, involving transition metal catalysts, are discussed in Chapter 13.

7-3 GROUP 1 (IA): THE ALKALI METALS

Alkali metal salts, in particular sodium chloride, have been known and used since antiquity. In early times, long before the chemistry of these compounds was understood, they were used in the preservation and flavoring of food and even as a medium of exchange. However, because of the difficulty of reducing the alkali metals, they were not isolated until comparatively recently, well after many other elements. Two of the alkali metals, sodium and potassium, are essential for life; their careful regulation is often important in treating a variety of medical conditions.

7-3-1 THE ELEMENTS

Potassium and sodium were first isolated within a few days of each other in 1807 by Humphry Davy as products of the electrolysis of molten KOH and NaOH. In 1817, J. A. Arfvedson, a young chemist working with J. J. Berzelius, recognized similarities between the solubilities of compounds of lithium and those of sodium and potassium. The following year Davy also became the first to isolate lithium, this time by electrolysis of molten Li_2O. Cesium and rubidium were discovered with the help of the spectroscope in 1860 and 1861, respectively; they were named after the colors of their most prominent emission lines, from the Latin *caesius* for sky blue and *rubidus* for deep red. Francium was not identified until 1939 as a short-lived radioactive isotope from the nuclear decay of actinium.

The alkali metals are silvery, highly reactive solids having low melting points. They are ordinarily stored under nonreactive oil to prevent air oxidation;

TABLE 7-3
Properties of the group 1 (IA) elements: the alkali metals

Element	*Ionization energy (kJ mol⁻¹)*	*Electron affinity (kJ mol⁻¹)*	*Melting point (°C)*	*Boiling point (°C)*	*Electronegativity (Pauling)*	$\mathscr{E}^\circ$ $(M^+ \longrightarrow M)$ (V)[a]
Li	520	60	180.5	1347	0.98	−3.04
Na	496	53	97.8	881	0.93	−2.71
K	419	48	63.2	766	0.82	−2.92
Rb	403	47	39.0	688	0.82	−2.92
Cs	376	46	28.5	705	0.79	−2.92
Fr	400[b,c]	60[b,d]	27[b]		0.7[b]	−2.9[b]

SOURCES: Ionization energies cited in this chapter are from C. E. Moore, *Ionization Potentials and Ionization Limits Derived from the Analyses of Optical Spectra*, National Standard Reference Data Series, U.S. National Bureau of Standards (NSRDS-NBS 34), 1970, unless noted otherwise. Reported values of electron affinities vary considerably. Values listed in this chapter are from H. Hotop and W. C. Lineberger, *J. Phys. Chem. Ref. Data*, **1985**, *14*, 731. Standard electrode potentials listed in this chapter are from *Standard Potentials in Aqueous Solutions*, A. J. Bard, R. Parsons, and J. Jordan, eds., Marcel Dekker (for IUPAC), New York, 1985. Electronegativities cited in this chapter are from A. L. Allred, *J. Inorg. Nucl. Chem.*, **1961**, *17*, 215. Approximate values are from L. Pauling, *The Nature of the Chemical Bond*, 3rd ed., Cornell University Press, Ithaca, N.Y., 1960, p. 93. Other data are from N. N. Greenwood and A. Earnshaw, *Chemistry of the Elements*, Pergamon Press, Elmsford, N.Y., 1984 except where noted.
NOTES: [a] Aqueous solution, 25°C.
[b] Approximate value.
[c] J. Emsley, *The Elements*, Oxford University Press, New York, 1989.
[d] S. G. Bratsch, *J. Chem. Educ.*, **1988**, *65*, 34.

they are soft enough to be easily cut with a knife or spatula. Physical properties of the alkali metals are summarized in Table 7-3.

7-3-2 CHEMICAL PROPERTIES

Elements in group 1 are remarkably similar in their chemical properties, which are governed in large part by the ease with which these metals can lose one electron (the alkali metals have the lowest ionization energies of all the elements) and thereby achieve a noble gas configuration. All elements in this group are highly reactive metals and excellent reducing agents. The metals react vigorously with water to form hydrogen; for example,

$$2\ Na + 2\ H_2O \longrightarrow 2\ NaOH + H_2$$

This reaction is highly exothermic, and the hydrogen formed may ignite, sometimes explosively. Consequently, special precautions must be taken to prevent these metals from coming into contact with water when they are stored.

Alkali metals dissolve in liquid ammonia and other donor solvents, such as aliphatic amines and $P(NMe_2)_3$ (hexamethylphosphoramide), to give blue solutions believed to contain solvated electrons:[2]

$$Na + x\,NH_3 \longrightarrow Na^+ + e(NH_3)_x^-$$

Because of these solvated electrons, dilute solutions of alkali metals in ammonia conduct electricity far better than completely dissociated ionic compounds in aqueous solutions. As the concentration of the alkali metals is increased, the conductivity first declines, and then increases; at sufficiently high concentration

[2] J. L. Dye, "Electrides, Negatively Charged Metal Ions, and Related Phenomena," in *Progress in Inorganic Chemistry*, Vol. 32, S. J. Lippard, ed., Wiley, New York, 1984, p. 327.

the solution acquires a metallic luster and a conductivity comparable to a molten metal. Dilute solutions are paramagnetic, with approximately one unpaired electron per metal atom. At higher concentrations the paramagnetism decreases, possibly as a consequence of formation of solvated electron pairs. One interesting aspect of these solutions is that they are less dense than liquid ammonia itself; the solvated electrons may be viewed as occupying cavities (estimated radius of approximately 300 pm) in the solvent, thus increasing the volume significantly. The blue color, corresponding to a broad absorption band centered near 1500 nm, is attributed to the solvated electron (alkali metal ions are colorless). At higher concentrations, these solutions have a coppery color and contain alkali metal anions, M^- (see discussion below).

Not surprisingly, solutions of alkali metals in liquid ammonia are excellent reducing agents. Examples of reductions that can be effected by these solutions follow:

$$RC{\equiv}CH + e^- \longrightarrow RC{\equiv}C^- + \tfrac{1}{2}H_2$$
$$NH_4^+ + e^- \longrightarrow NH_3 + \tfrac{1}{2}H_2$$
$$S_8 + 2\,e^- \longrightarrow S_8^{2-}$$
$$Fe(CO)_5 + 2\,e^- \longrightarrow Fe(CO)_4^{2-} + CO$$

The solutions of alkali metals are unstable and undergo slow decomposition to form amides:

$$M + NH_3 \longrightarrow MNH_2 + \tfrac{1}{2}H_2$$

Other metals, especially the alkaline earths Ca, Sr, and Ba and the lanthanides Eu and Yb (both of which can form 2+ ions), can also dissolve in liquid ammonia to give the solvated electron; however, the alkali metals undergo this reaction more efficiently and have been used far more extensively for synthetic purposes.

Alkali metal atoms readily lose their outermost (ns^1) electron to form their common ions of 1+ charge. These ions can form complexes with a variety of Lewis bases (ligands, discussed more fully in Chapters 8 and following), although such complexes are generally not as stable as complexes of transition metals. Of particular interest are cyclic Lewis bases that have several donor atoms that can surround, or trap, cations. Examples of such molecules are shown in Figure 7-4. The first of these is one of a large group of cyclic ethers, commonly known as *crown ethers*, which are able to donate electron density to metals through their oxygen atoms. The second, one of a family of cryptands (or cryptates), can be even more effective as a cage with eight donor atoms surrounding a central metal.

As might be expected, the ability of a cryptand to trap an alkali metal cation depends on the sizes of both the cage and the metal ion; the better the

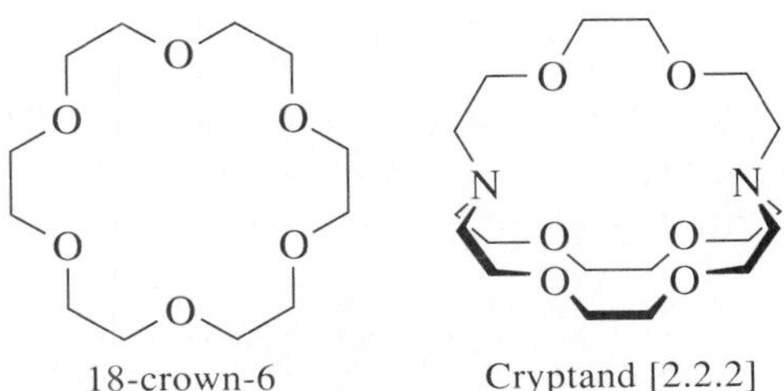

FIGURE 7-4 A Crown Ether and a Cryptand.

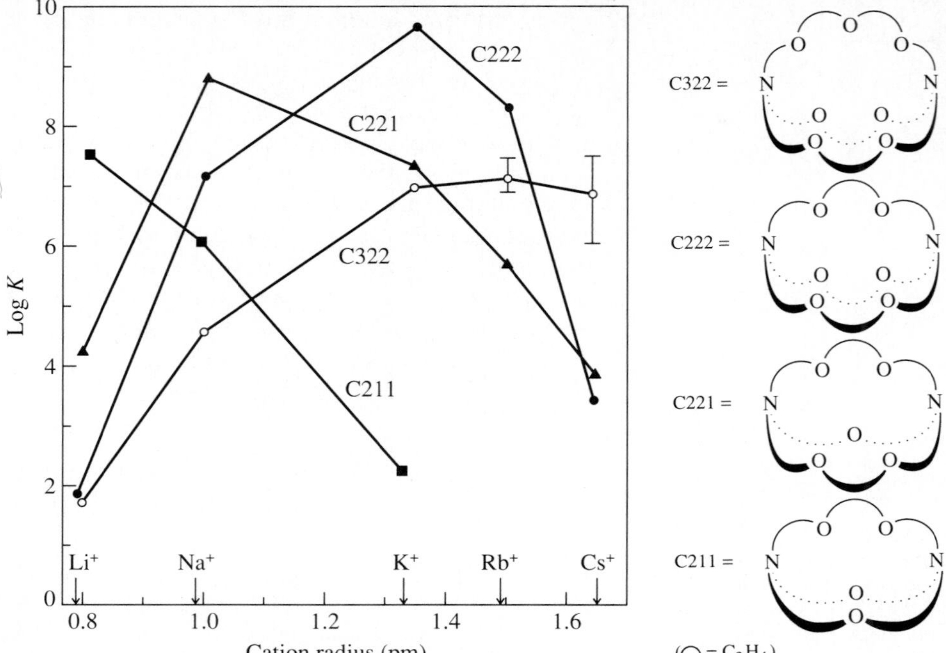

FIGURE 7-5 Formation Constants of Alkali Metal Cryptands. (Reproduced from J. L. Dye, "Electrides, Negatively Charged Metal Ions, and Related Phenomena," in *Progress in Inorganic Chemistry,* Vol. 32, S. J. Lippard, ed., Wiley, New York, 1984, p. 337. Copyright © 1984, John Wiley & Sons, Inc. Reprinted by permission of John Wiley & Sons, Inc.)

match between these sizes, the more effectively the ion can be trapped. This effect is shown graphically for the alkali metal ions in Figure 7-5. The largest of the alkali metal cations, Cs^+, is trapped most effectively by the largest cryptand ([3.2.2]), and the smallest, Li^+, by the smallest cryptand ([2.1.1]).[3] Other correlations can easily be seen in Figure 7-5. Cryptands have also played an important role in the study of a characteristic of the alkali metals that was not recognized until rather recently, the capacity to form negatively charged ions.

Although the alkali metals are known primarily for their formation of unipositive ions, since 1974 many examples of alkali metal anions (alkalides) have been reported. The first of these was the sodide ion, Na^-, formed from the reaction of sodium with the cryptand $N\{(C_2H_4O)_2C_2H_4\}_3N$ in the solvent ethylamine:

$$2\ Na + \underset{\text{cryptand [2.2.2]}}{N\{(C_2H_4O)_2C_2H_4\}_3N} \longrightarrow [Na(N\{(C_2H_4O)_2C_2H_4\}_3N)]^+ + Na^-$$

In this complex the Na^- occupies a site sufficiently remote from the coordinating N and O atoms of the cryptand that it can be viewed as a separate entity; it is formed as the result of what may be viewed as disproportionation of Na into Na^+ (surrounded by the cryptand) plus Na^-. Alkalide ions are also known for the other members of group 1 (IA) and for other metals, especially those for which a 1− charge gives rise to an $s^2\ d^{10}$ electron configuration. As might be expected, alkalide ions are powerful reducing agents; this means that the cryptand or other cyclic group to surround the associated positive ion must be highly resistant to reduction. Even if such groups are carefully chosen, most alkalides are rather unstable and subject to irreversible decomposition.

[3] The numbers indicate the number of oxygen atoms in each bridge between the nitrogens. Thus cryptand [3.2.2] has one bridge with three oxygens and two bridges with two oxygens, as shown in Figure 7-5.

7-4 GROUP 2 (IIA): THE ALKALINE EARTHS

7-4-1 THE ELEMENTS

Compounds of magnesium and calcium have been used since antiquity. For example, the ancient Romans used mortars containing lime (CaO) mixed with sand, and the ancient Egyptians used gypsum ($CaSO_4 \cdot 2\ H_2O$) in the plasters used to decorate their tombs. These two alkaline earths are among the most abundant elements in the earth's crust and occur extensively in a wide variety of minerals. Strontium and barium are less abundant but, like magnesium and calcium, commonly occur as sulfates and carbonates in their mineral deposits. Beryllium is fifth in abundance of the alkaline earths and is obtained primarily from the mineral beryl [$Be_3Al_2(SiO_3)_6$]. Radium is radioactive in all its isotopes (longest lived isotope: ^{226}Ra, half-life = 1600 years); it was first isolated by Pierre and Marie Curie from the uranium ore pitchblende in 1898. Selected physical properties of the alkaline earths are given in Table 7-4.

TABLE 7-4
Properties of the group 2 (IIA) elements: the alkaline earths

Element	*Ionization energy (kJ mol⁻¹)*	*Electron affinity (kJ mol⁻¹)*[b]	*Melting point (°C)*	*Boiling point (°C)*	*Electro-negativity (Pauling)*	$\mathscr{E}°$ $M^{2+} + 2e^- \longrightarrow M$ *(V)*[a]
Be	899	−50	1287	2500[b]	1.57	−1.97
Mg	738	−40	649	1105	1.31	−2.36
Ca	590	−30	839	1494	1.00	−2.84
Sr	549	−30	768	1381	0.95	−2.89
Ba	503	−30	727	1850[b]	0.89	−2.92
Ra	509	−30	700[b]	1700[b]	0.9[b]	−2.92

SOURCES: See Table 7-3.
NOTES: [a] Aqueous solution, 25°C.
[b] Approximate values.

Atoms of the group 2 (IIA) elements are smaller than the neighboring group 1 (IA) elements as a consequence of the greater nuclear charge of the former. The observed result of this decrease in size is that the group 2 elements are more dense and their atoms pack together more tightly than the group 1 elements; the group 2 elements therefore have higher melting and boiling points (see Tables 7-3 and 7-4) and higher enthalpies of fusion and vaporization. Beryllium, the lightest of the alkaline earth metals, is widely used in alloys with copper, nickel, and other metals. When added in small amounts to copper, for example, beryllium increases the strength of the metal dramatically and improves the corrosion resistance, while preserving high conductivity and other desirable properties. Beryllium is also among the most effective neutron reflectors and moderators and is therefore used extensively in nuclear reactors. It is interesting to note that emeralds and aquamarine are two types of beryl, the mineral source of beryllium. (The vivid colors of these stones are due to small amounts of chromium and other impurities.) Magnesium, with its alloys, is used widely as a strong, but very light, construction material; its density is less than one-fourth that of steel. The other alkaline earth metals are used occasionally, but in much smaller amounts, in alloys. Radium has been used in treatment of cancerous tumors, but has largely been superseded by other radioisotopes.

7-4-2 CHEMICAL PROPERTIES

As for the group 1 (IA) elements, the elements in this group, with the exception of beryllium, display a remarkable degree of similarity in their chemical properties, with much of their chemistry governed by their tendency to lose two electrons to achieve a noble gas electron configuration. In general, therefore, elements in this group are good reducing agents. While not as violently reactive toward water as the alkali metals, the alkaline earths react readily with acids to generate hydrogen:

$$Mg + 2H^+ \longrightarrow Mg^{2+} + H_2$$

The reducing ability of these elements increases with atomic number. As a consequence, calcium and the heavier alkaline earths react directly with water in a reaction that can conveniently generate small quantities of hydrogen:

$$Ca + 2\,H_2O \longrightarrow Ca(OH)_2 + H_2$$

Beryllium is distinct from the other alkaline earths in its chemical properties. Because of its small size it participates primarily in covalent rather than ionic bonding; although the ion $Be(H_2O)_4^{2+}$ is known, free Be^{2+} ions are rarely, if ever, encountered. Beryllium and its compounds are extremely toxic, and consequently special precautions are required in their handling. As discussed in Chapter 5, although beryllium halides of formula BeX_2 may be monomeric and linear in the gas phase, in the crystal the molecules polymerize to form halogen-bridged chains, with tetrahedral coordination around beryllium, as shown in Figure 7-6. Beryllium hydride, BeH_2, is also polymeric in the solid, with bridging hydrogens. The 3-center bonding involved in bridging by halogens, hydrogen, and other atoms and groups is also commonly encountered in the chemistry of the group 3 (IIIA) elements and will be discussed more fully with those elements.

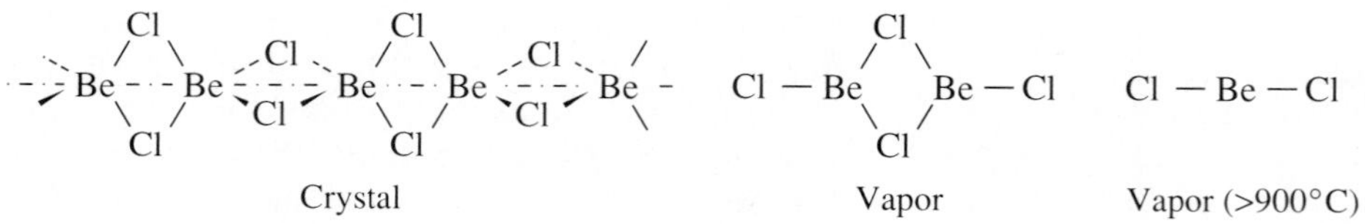

Crystal — Vapor — Vapor (>900°C) Cl—Be—Cl

FIGURE 7-6 Structure of $BeCl_2$.

The most chemically useful magnesium compounds are the Grignard reagents, of general formula RMgX (R = alkyl or aryl). These reagents have been found to be extremely complex in their structure and function. It has been demonstrated that Grignard reagents contain a variety of species in solution, linked by equilibria. The relative positions of these equilibria, and hence the concentrations of the various species, are affected by the nature of the R group and the halogen, the solvent, and the temperature. Equilibria believed to be important in solutions of Grignard reagents are summarized in Figure 7-7.

Grignard reagents are extraordinarily versatile and can be used to synthesize a vast range of organic compounds, including alcohols, aldehydes, ketones, carboxylic acids, esters, thiols, amines, and other classes of com-

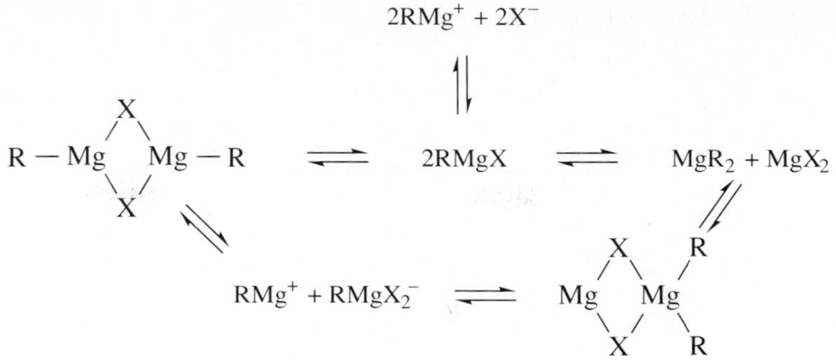

$$2RMg^+ + 2X^-$$

$$\rightleftharpoons$$

$$R-Mg(\mu\text{-}X)_2Mg-R \rightleftharpoons 2RMgX \rightleftharpoons MgR_2 + MgX_2$$

$$R-Mg(\mu\text{-}X)_2Mg-R \rightleftharpoons RMg^+ + RMgX_2^- \rightleftharpoons Mg(\mu\text{-}X)_2MgR_2 \rightleftharpoons MgR_2 + MgX_2$$

FIGURE 7-7 Grignard Reagent Equilibria.

pounds. Details of these syntheses are presented in organic chemistry texts. The development of these reagents since their original discovery by Victor Grignard in 1900 has been reviewed.[4]

Chlorophylls contain magnesium coordinated by chlorin groups. These compounds, essential in photosynthesis, will be discussed in Chapter 15.

Calcium fluoride (fluorite) is the principal source of fluorine; the crystal structure of this compound is the prototype for many structures of compounds of formula MX_2; each calcium is surrounded by a cube of fluorides and each fluoride by a tetrahedron of calciums, as shown in Figure 7-8 (this structure is compared with other structures in Figure 2-16).

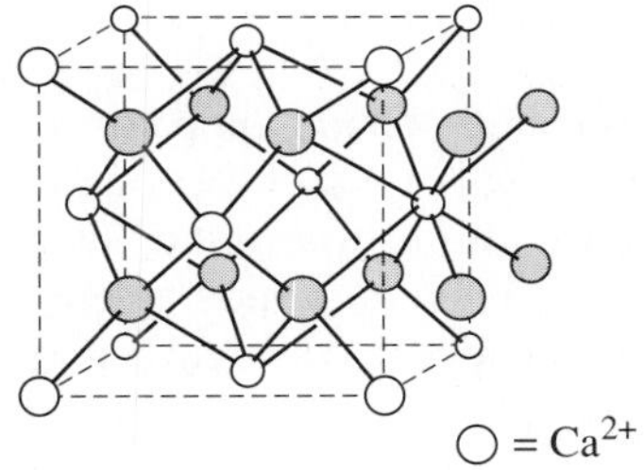

FIGURE 7-8 Unit Cell of Calcium Fluoride (Fluorite).

7-5 GROUP 13 (IIIA)

7-5-1 THE ELEMENTS

Elements in this group include one nonmetal, boron, and four elements that are primarily metallic in their properties. Physical properties of these elements are shown in Table 7-5.

Boron

Boron's chemistry is so different from that of the other elements in this group that it deserves separate discussion. Chemically, boron is a nonmetal; in its tendency to form covalent bonds it shares more similarities with carbon and silicon than with aluminum and the other group 13 elements. Like carbon, boron forms many hydrides; like silicon, it forms oxygen-containing minerals of complex structure (borates). Compounds of boron have been used since ancient times in the preparation of glazes and borosilicate glasses, but the element itself has proved to be extremely difficult to purify. In recent decades, crystallographic studies have begun to reveal a fascinating diversity of allo-

[4] *Bull. Soc. Chim. France*, **1972**, 2127 (several papers).

TABLE 7-5
Properties of the group 13 (IIIA) elements

Element	*Ionization energy (kJ mol⁻¹)*	*Electron affinity (kJ mol⁻¹)*	*Melting point (°C)*	*Boiling point (°C)*	*Electronegativity (Pauling)*
B	801	27	2180	3650[b]	2.04
Al	578	43	660	2467	1.61
Ga	579	30[a]	29.8	2403	1.81
In	558	30[a]	157	2080	1.78
Tl	589	20[a]	304	1457	2.04[b]

SOURCES: See Table 7-3.
NOTES: [a] Approximate value.
[b] Value for Tl (III). The electronegativity of Tl (I) is reported as 1.62.

tropes, with most based on the icosahedral B_{12} structural unit. As described in Chapter 14, the icosahedron is an important structural unit of many boron compounds.

In the boron hydrides, called **boranes**, hydrogen often serves as a bridge between boron atoms, a function rarely performed by hydrogen in carbon chemistry. How is it possible for hydrogen to serve as a bridge? To answer this question, we need to consider the bonding in diborane, B_2H_6:

```
         H
H.     /   \    .H
   `B       B´
H´     \   /    `H
         H
```

The geometry around each boron is approximately tetrahedral, with sp^3 hybrid orbitals. Two hybrid orbitals of each boron are involved in sigma bonds with the terminal hydrogens. The total number of valence electrons in B_2H_6 is 12, with 8 of these involved in sigma bonds to the terminal hydrogens; this leaves only 4 electrons to hold the borons together via the bridging hydrogens.

The possible interactions between the hybrid orbitals and the bridging hydrogens are shown in Figure 7-9. By the approach used in Chapter 5, each set of B—H—B interactions may be viewed as involving the $1s$ orbital on hydrogen with group orbitals arising from the sp^3 hybrids on the boron atoms, as shown. The first group orbital, in which both orbital lobes pointing toward the bridging hydrogen are of the same sign, can interact with the $1s$ orbital of hydrogen in either a bonding or antibonding fashion. The second group orbital is not of appropriate symmetry for interaction with the $1s$ of hydrogen and is therefore nonbonding. The overall result is the formation of three molecular orbitals, one bonding, one nonbonding, and one antibonding, from the two sp^3 hybrids of the borons and the $1s$ orbital of hydrogen. Two electrons (half of the available four) are available to occupy the bonding molecular orbital; stabilization of these electrons by formation of the bond holds these atoms together. An identical set of interactions accounts for the bonding in the other B—H—B bridge. This type of bonding, involving three atoms and two bonding electrons, is described as 3-center, 2-electron bonding.[5]

The bonding in diborane may also be described according to an equivalent group theory approach, as presented in Chapter 5. The point group of B_2H_6 is D_{2h}. Using the set of internal sp^3 hybrid orbitals of each boron to obtain group orbitals, we obtain group orbitals of A_g, B_{3u}, B_{1u}, and B_{2g}, symmetry, as shown

[5] W. N. Lipscomb, *Boron Hydrides*, W. A. Benjamin, New York, 1963.

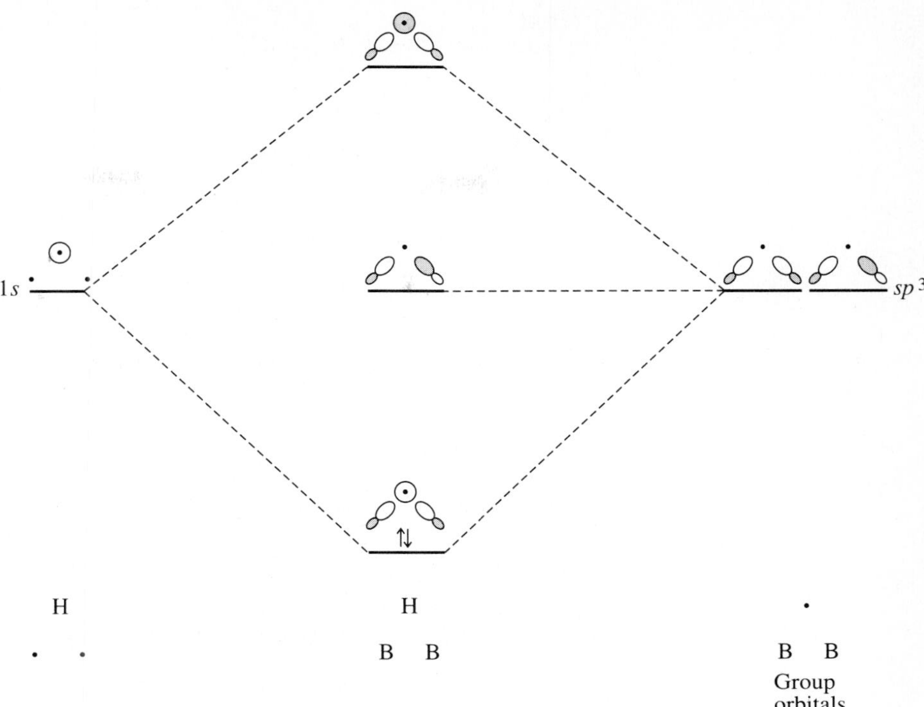

FIGURE 7-9 Bonding in B—H—B Bridge in Diborane.

in Figure 7-10. Similarly, combination of the 1*s* orbitals of the bridging hydrogens gives two possibilities, of A_g and B_{3u} symmetry. Interactions of these group orbitals, as shown in Figure 7-11, produces two bonding orbitals (A_g and B_{3u}), two antibonding orbitals (also A_g and B_{3u}), and two nonbonding orbitals (B_{1u} and B_{2g}). The four electrons occupy the two bonding orbitals, each of which can be described as a 3-center, 2-electron bond.

Similar bridging hydrogen atoms occur in many other boranes, as well as in carboranes, which contain both boron and carbon atoms arranged in clusters. In addition, bridging hydrogens and alkyl groups are frequently encountered in aluminum chemistry and in the chemistry of elements of other groups. A few examples of these compounds are shown in Figure 7-12 on page 238.

The boranes, carboranes, and related compounds are also of interest in the field of cluster chemistry, the chemistry of compounds containing metal–metal bonds. The bonding in these compounds will be discussed and compared with the bonding in transition metal cluster compounds in Chapter 14.

^{10}B has a very high neutron absorption cross section (it is a good absorber of neutrons). This property has recently been developed for use in the treatment of cancerous tumors: boron-containing compounds having a strong preference for attraction to tumor sites can be irradiated with beams of neutrons; the subsequent nuclear decay can kill the cancerous tissue.

7-5-2 OTHER CHEMISTRY OF THE GROUP 13 (IIIA) ELEMENTS

Elements in this group, especially boron and aluminum, form 3-coordinate Lewis acids capable of accepting an electron pair and increasing their coordination number. Some of the most commonly used Lewis acids are the boron

Reducible representation for sp^3 hybrid orbitals involved in bonding with bridging hydrogens:

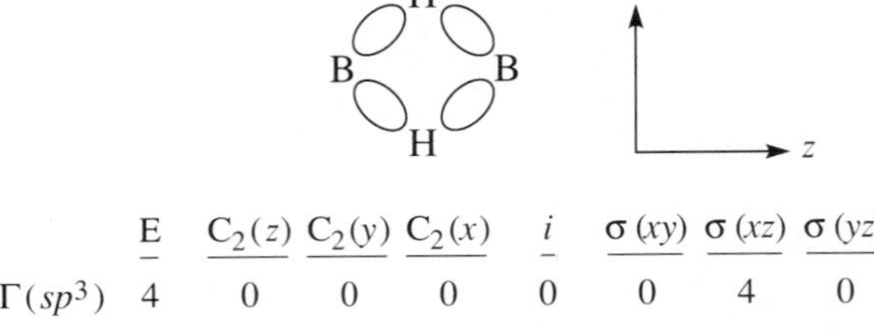

	E	$C_2(z)$	$C_2(y)$	$C_2(x)$	i	$\sigma(xy)$	$\sigma(xz)$	$\sigma(yz)$
$\Gamma(sp^3)$	4	0	0	0	0	0	4	0

(Only orbitals that remain unchanged by symmetry operations contribute to the character)

This representation reduces to $A_g + B_{2g} + B_{1u} + B_{3u}$. These irreducible representations have the following symmetries:

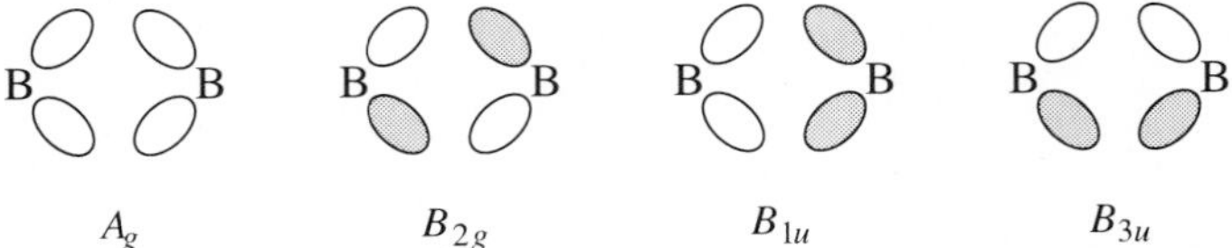

A_g B_{2g} B_{1u} B_{3u}

Reducible representation for $1s$ orbitals of bridging hydrogens:

H

B B

H

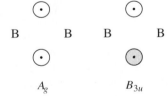

	E	$C_2(z)$	$C_2(y)$	$C_2(x)$	i	$\sigma(xy)$	$\sigma(xz)$	$\sigma(yz)$
$\Gamma(1s)$	2	0	0	2	0	2	2	0

This reduces to $A_g + B_{3u}$, which have the following symmetries:

B B B B

A_g B_{3u}

FIGURE 7-10 Orbital Combinations for Diborane.

trihalides, BX_3. These compounds are monomeric (unlike diborane, B_2H_6, and aluminum halides, Al_2X_6) and, as discussed in Chapter 3, are planar molecules having significant π bonding. In their Lewis acid role, they can accept an electron pair from a halide ion to form tetrahaloborate ions, BX_4^-. The Lewis acid behavior of these compounds has been discussed in Chapter 6.

Aluminum, gallium, indium, and thallium commonly form 3+ ions, with the metallic nature of the elements increasing somewhat going down the group. Thallium also exhibits a 1+ ion, the first case we have encountered of the *inert pair effect*. The inert pair effect is the occurrence of oxidation states of metals 2 less than the group number (for example, Pb, in group 14 (IVA), exhibits a 2+ ion as well as a 4+ ion); the effect has frequently been ascribed to the stability of the electron configuration having all filled subshells (including an outermost s^2 subshell), which results from loss of p electrons from the outermost subshell.

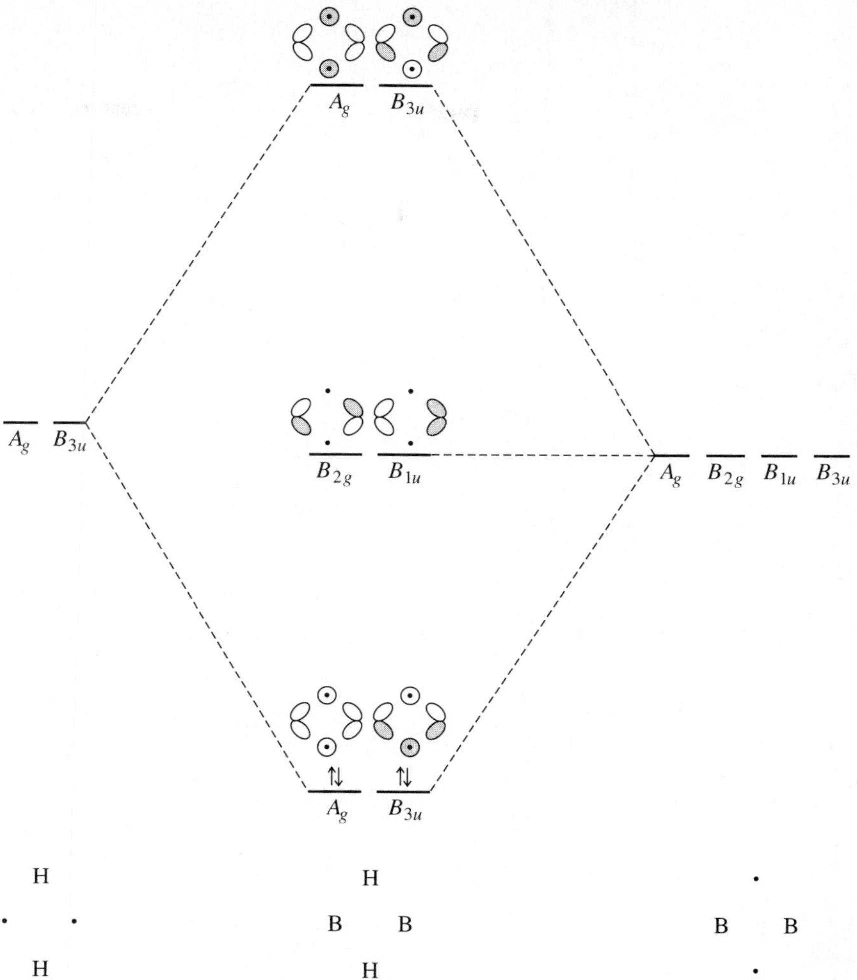

FIGURE 7-11 Orbital Interactions in Diborane.

Parallels between main group and organic chemistry can be extremely interesting. One of the best known of these parallels is between the organic molecule benzene and the isoelectronic borazine (alias "inorganic benzene"), $B_3N_3H_6$. Some of the similarities in physical properties between these two are striking, as shown in Table 7-6. Despite these parallels, the chemistry of these two compounds is quite different. In borazine the difference in electronegativity between boron (2.04) and nitrogen (3.04) adds considerable polarity to the B—N bonds and makes the molecule much more susceptible to attack by nucleophiles (at the more positive boron) and electrophiles (at the more negative nitrogen) than benzene.

Parallels between benzene and isoelectronic inorganic rings remain of interest. Recent examples include reports of boraphosphabenzenes (containing B_3P_3 rings)[6] and $[(CH_3)AlN(2,6\text{-diisopropylphenyl})]_3$ containing an Al_3N_3 ring.[7]

[6] H. V. R. Dias and P. P. Power, *Angew, Chem. Int. Ed. Engl.*, **1987**, *26*, 1270; *J. Am. Chem. Soc.*, **1989**, *111*, 144.

[7] K. M. Waggoner, H. Hope, and P. P. Power, *Angew. Chem. Int. Ed. Engl.*, **1988**, *27*, 1699.

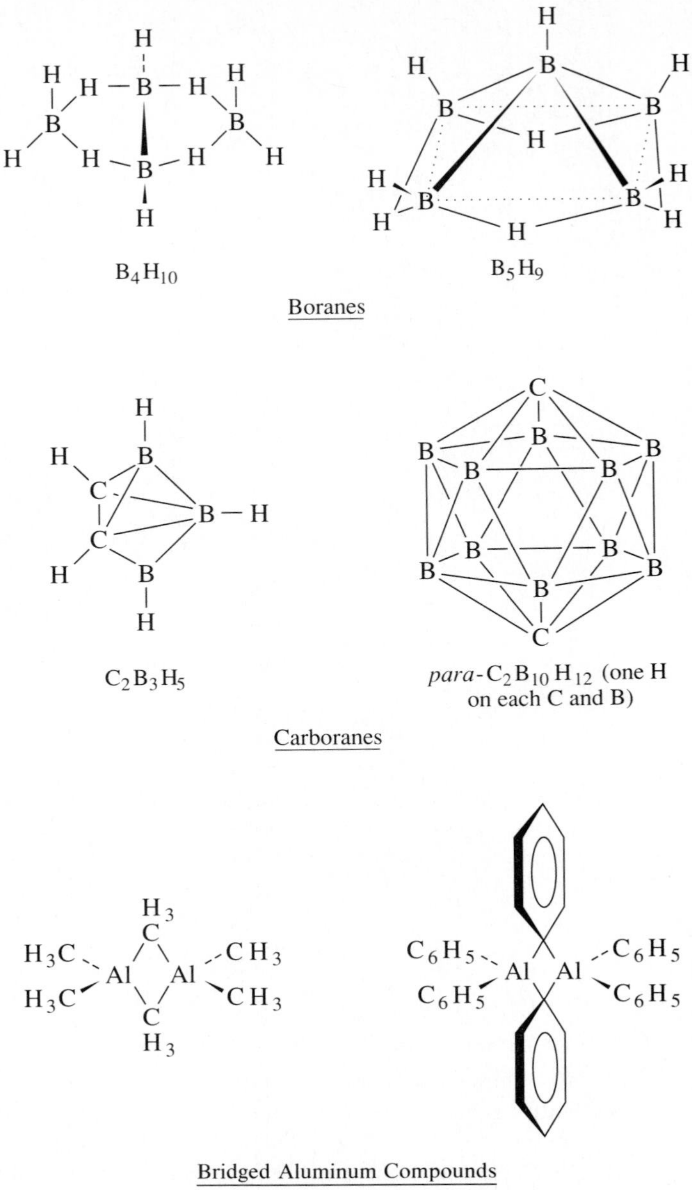

B_4H_{10}

B_5H_9

Boranes

$C_2B_3H_5$

para-$C_2B_{10}H_{12}$ (one H on each C and B)

Carboranes

Bridged Aluminum Compounds

FIGURE 7-12 Boranes, Carboranes, and Bridged Aluminum Compounds.

Another interesting parallel between boron–nitrogen chemistry and carbon chemistry is offered by boron nitride, BN. Like carbon (Sections 7-6-1 and 7-6-2), boron nitride exists in two forms, a diamondlike form and a form similar to graphite. In the diamondlike (cubic) form, each nitrogen is coordinated tetrahedrally by four borons and each boron by four nitrogens; as in diamond, such coordination gives high rigidity to the structure and makes BN comparable to diamond in hardness. In the graphitelike hexagonal form, BN also occurs in extended fused ring systems. However, there is much less delocalization of π electrons in this form and, unlike graphite, hexagonal BN is a poor conductor. As in the case of diamond, the harder, more dense form (cubic) can be formed from the less dense form (hexagonal) under high pressures.

TABLE 7-6
Benzene and borazine

```
    H       H              H       H
     \     /                \     /
      C — C                  B — N
     /     \                /     \
H — C   ○   C — H      H — N   ○   B — H
     \     /                \     /
      C — C                  B — N
     /     \                /     \
    H       H              H       H
```

	Benzene	Borazine
Melting point (°C)	6	−57
Boiling point (°C)	80	55
Density* (g cm^{-3})	0.81	0.81
Surface tension* (Nm^{-1})	0.0310	0.0311
Dipole moment	0	0
Internuclear distance in ring (pm)	142	144
Internuclear distance, bonds to H (pm)	C—H: 108	B—H: 120 N—H: 102

SOURCE: Data from N. N. Greenwood and A. Earnshaw, *Chemistry of the Elements*, Pergamon Press, Elmsford, N.Y., 1984, p. 238.
NOTE: * At melting point.

7-6 GROUP 14 (IVA)

7-6-1 THE ELEMENTS

Elements in this group span the range from a nonmetal, carbon, to the metals tin and lead, with the intervening elements showing semimetallic behavior. Carbon has been known from prehistory as the soot or charcoal resulting from combustion of organic matter. In recorded history, diamonds have been prized as precious gems for thousands of years. Neither form of carbon, however, was recognized as a chemical element until late in the eighteenth century. Tools made of flint (primarily SiO_2) were used throughout the Stone Age. However, free silicon was not isolated until 1823, when J. J. Berzelius obtained it by reducing K_2SiF_6 with potassium. Tin and lead have also been known since ancient times. A major early use of tin was in combination with copper in the alloy bronze; weapons and tools containing bronze date back more than 5000 years. Lead was used by the ancient Egyptians in pottery glazes and by the Romans for plumbing and other purposes. Germanium, on the other hand, was a "missing" element for a number of years. Mendeleev accurately predicted the properties of this then unknown element in 1871 ("eka-silicon"), but it was not discovered until 1886, by C. A. Winkler. Properties of the group 14 (IVA) elements are summarized in Table 7-7.

Although carbon occurs primarily as the isotope ^{12}C (whose atomic mass serves as the basis of the modern system of atomic mass), two other isotopes, ^{13}C and ^{14}C are of importance as well. ^{13}C, which has a natural abundance of 1.10%, has a nuclear spin of $\frac{1}{2}$, in contrast to ^{12}C, which has zero nuclear spin. This means that ^{13}C, even though it comprises only about 1 part in 90 of naturally occurring carbon, can be used as the basis of NMR observations for the characterization of carbon-containing compounds. With the advent of Fourier transform technology, ^{13}C NMR has become a valuable tool in both organic and inorganic chemistry. Uses of ^{13}C NMR in organometallic chemistry are described in Chapter 12.

TABLE 7-7
Properties of the group 14 (IVA) elements

Element	*Ionization energy (kJ mol⁻¹)*	*Electron affinity (kJ mol⁻¹)*	*Melting point (°C)*	*Boiling point (°C)*	*Electronegativity (Pauling)*
C	1086	122	4100	[a]	2.55
Si	786	134	1420	3280[b]	1.90
Ge	762	120	945	2850	2.01
Sn	709	120	232	2623	1.96[c]
Pb	716	35	327	1751	2.33[d]

SOURCES: See Table 7-3.
NOTES: [a] Sublimes.
[b] Approximate value.
[c] Value for Sn (IV). The electronegativity of Sn (II) is reported as 1.80.
[d] Value for Pb (IV). The electronegativity of Pb (II) is reported as 1.87.

^{14}C is formed in the atmosphere from nitrogen by thermal neutrons from the action of cosmic rays:

$$^{14}_{7}N + ^{1}_{0}n \longrightarrow ^{14}_{6}C + ^{1}_{1}H$$

^{14}C is formed by this reaction in comparatively small amounts (approximately $1.2 \times 10^{-10}\%$ of atmospheric carbon); it is incorporated into plant and animal tissues by biological processes. When a plant or animal dies, the process of exchange of its carbon with the environment by respiration and other biological processes ceases, and the ^{14}C in its system is effectively trapped. However, ^{14}C decays by beta emission, with a half-life of 5730 years:

$$^{14}_{6}C \longrightarrow ^{14}_{7}N + e^-$$

Therefore, by measuring the remaining amount of ^{14}C, we can determine to what extent this isotope has decayed and, in turn, the time elapsed since death. Often called simply *radiocarbon dating*, this procedure has been used to estimate the ages of many archeological samples, including Egyptian remains, charcoal from early campfires, and the Shroud of Turin.

Carbon occurs primarily in two allotropes, diamond and graphite. The diamond structure is very rigid, with each atom surrounded tetrahedrally by four other atoms in a structure that has a cubic unit cell; as a result diamond is extremely hard, the hardest of all naturally occurring substances. Graphite, on the other hand, consists of layers of fused six-membered rings of carbon atoms. The carbon atoms in these layers may be viewed as being sp^2 hybridized. The remaining, unhybridized p orbitals are perpendicular to the layers and participate in extensive π bonding, with π electron density delocalized over the layers. Because of the relatively weak interactions between the layers, the layers are free to slip with respect to each other, and π electrons are free to move within each layer, making graphite a good lubricant and electrical conductor. The structures of diamond and graphite are shown in Figure 7-13, and important physical properties are shown in Table 7-8.

At room temperature, graphite is thermodynamically the more stable form. However, the density of diamond (3.514 g cm^{-3}) is much greater than that of graphite (2.266 g cm^{-3}), and graphite can be converted to diamond at very high pressure (high temperature and molten metal catalysts are also

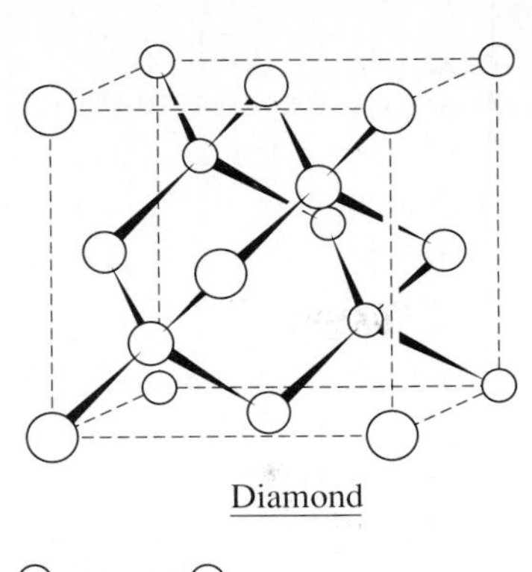

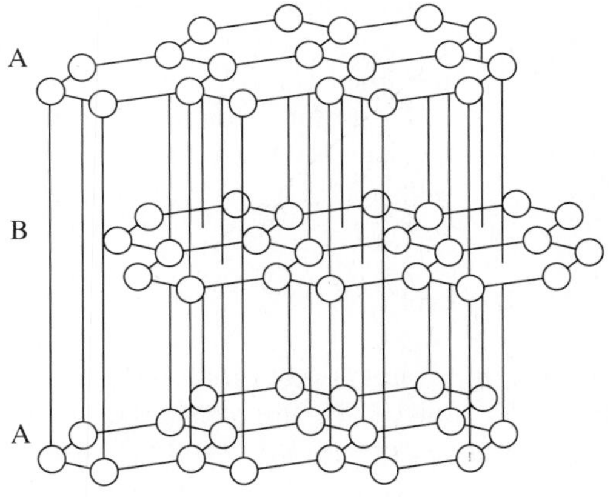

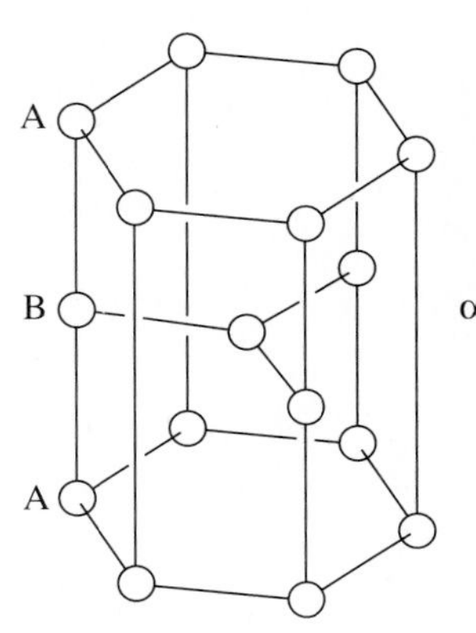

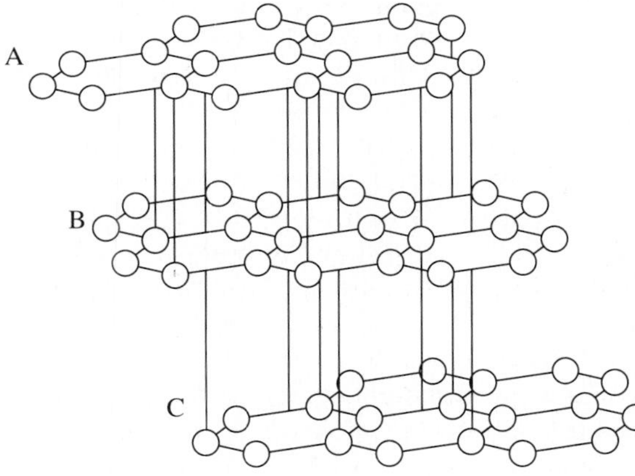

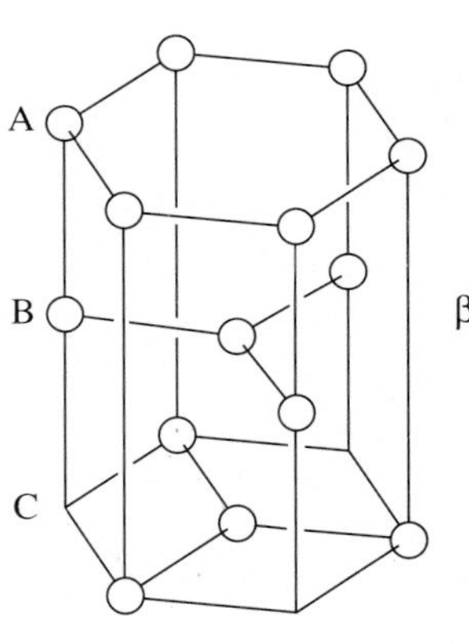

Graphite

FIGURE 7-13 Diamond and Graphite.

TABLE 7-8
Physical properties of diamond and graphite

Property	*Diamond*	*Graphite*
Density (g cm^{-3})	3.513	2.260
Electrical resistivity (μohm cm)	10^{19}	1375
Standard molar entropy (J mol^{-1} K^{-1})	2.377	5.740
C_p at 25°C (J mol^{-1} K^{-1})	6.113	8.527
C—C distance (pm)	154.4	141.5 (within layer) 335.4 (between layers)

Source: J. Emsley, *The Elements*, Oxford University Press, New York, 1989, p. 44.

used to facilitate this conversion). Since the first successful synthesis of diamonds from graphite in the mid 1950s, the manufacture of industrial diamonds has developed rapidly, and nearly half of all industrial diamonds are now produced synthetically. Additional allotropes of carbon, involving long —C≡C—C≡C—C≡C— chains, have been identified in nature; they may also be prepared from graphite at high temperatures and pressures.

Silicon crystallizes in the diamond structure. However, it has somewhat weaker covalent bonds than carbon as a consequence of less efficient orbital overlap. These weaker bonds result in a lower melting point for silicon (1420°C, compared with 4100°C for diamond) and a greater chemical reactivity. Germanium also has the diamond structure and is a semiconductor, with somewhat weaker bonding than silicon and consequently lower melting point and enthalpy of vaporization.

Tin, on the other hand, has two allotropes, a diamond form (α) favored below 13.2°C and a metallic form (β) favored at higher temperatures. Lead is entirely metallic and is among the most dense (and most poisonous) of the metals.

7-6-2 COMPOUNDS

A common misconception is that carbon can, at most, be 4-coordinate. While in the vast majority of compounds, carbon is bonded to four or fewer atoms, many examples are now known in which carbon has coordination numbers of 5, 6, or higher. Five-coordinate carbon is actually rather common, with methyl and other groups frequently forming bridges between two metal atoms, as in $Al_2(CH_3)_6$ (see Figure 7-12). Many organometallic cluster compounds contain carbon atoms surrounded by polyhedra of metal atoms. Such compounds, often designated carbide clusters, are discussed in Chapter 14. Examples of carbon atoms having coordination numbers of 5, 6, 7, and 8 are shown in Figure 7-14 (other examples were shown in Chapter 1).

The two most familiar oxides of carbon, CO and CO_2, are colorless, odorless gases. Carbon monoxide is a rarity of sorts, a stable compound in which carbon formally has only three bonds. It is extremely toxic, forming a bright red complex with the iron in hemoglobin. As described in Chapter 5, the highest occupied molecular orbital of CO is concentrated on carbon; this provides the molecule an opportunity to interact strongly with a variety of metal atoms, which in turn can donate electron density through their *d* orbitals to empty π^* orbitals (LUMOs) on CO. The details of such interactions are described more fully in Chapter 12.

Carbon dioxide is familiar as a major component of the earth's atmosphere (although only fifth in abundance, after nitrogen, oxygen, argon, and water vapor) and as the product of respiration, combustion, and other natural and industrial processes. It was the first gaseous component to be isolated from air, the "fixed air" isolated by Joseph Black in 1752. In recent years, CO_2 has gained international attention because of its role in the greenhouse effect and the potential atmospheric warming and other climatic consequences of an increase in CO_2 abundance. Carbon dioxide, because of its vibrational energy levels, is capable of absorbing a significant amount of thermal energy and acting as a sort of atmospheric blanket. Since the beginning of the industrial revolution, the carbon dioxide concentration in the atmosphere has increased substantially, an increase that will continue indefinitely in the absence of major

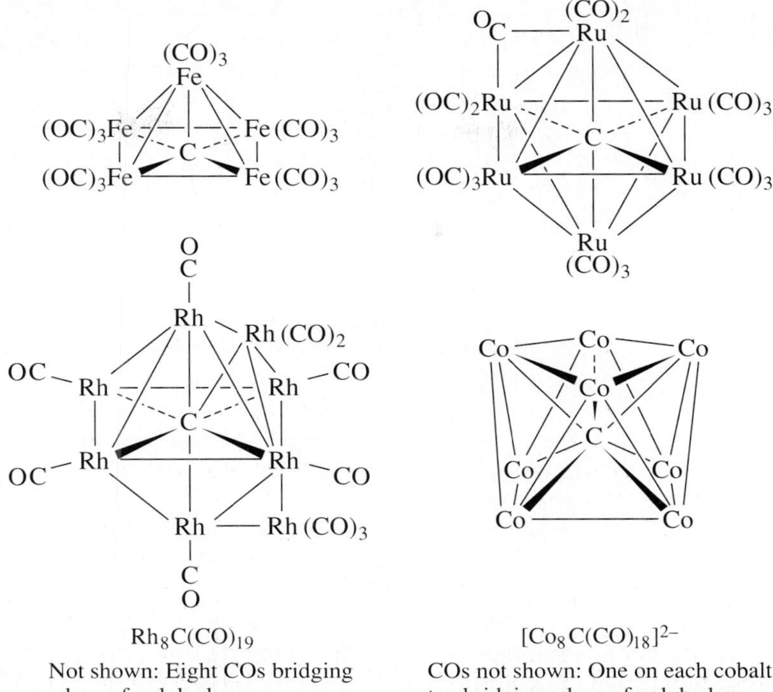

$Rh_8C(CO)_{19}$
Not shown: Eight COs bridging edges of polyhedron

$[Co_8C(CO)_{18}]^{2-}$
COs not shown: One on each cobalt; ten bridging edges of polyhedron

FIGURE 7-14 High Coordination Numbers of Carbon.

policy changes by the industrialized nations. The consequences of a continuing increase in atmospheric CO_2 are extremely difficult to forecast; the dynamics of the atmosphere are extremely complex, and the interplay between atmospheric composition, human activity, the oceans, solar cycles, and other factors is not yet well understood.

Although only two forms of elemental carbon are common, carbon forms several anions, especially in combination with the most electropositive metals. In these compounds, collectively called the **carbides**, there is considerable covalent as well as ionic bonding, with the proportion of each depending on the metal. The best characterized carbide ions are:

Ion	*Common name*	*Systematic name*	*Example*	*Major hydrolysis product*
C^{4-}	Carbide or methanide	Carbide	Al_4C_3	CH_4
C_2^{2-}	Acetylide	Dicarbide (2−)	CaC_2	H—C≡C—H
C_3^{4-}		Tricarbide (4−)	Mg_2C_3*	H_3C—C≡C—H

NOTE: * This is the only known compound containing the C_3^{4-} ion.

These carbides, as indicated, liberate organic molecules on reaction with water; for example,

$$Al_4C_3 + 12\,H_2O \longrightarrow 4\,Al(OH)_3 + 3\,CH_4$$
$$CaC_2 + 2\,H_2O \longrightarrow Ca(OH)_2 + HC{\equiv}CH$$

It may seem surprising that carbon, with its vast range of literally millions of compounds, is not the most abundant element in this group. By far the most abundant of group 14 (IVA) elements on earth is silicon, which comprises 27% of the earth's crust (by mass) and is second in abundance (after oxygen); carbon is only seventeenth in abundance. Silicon, with its semimetallic properties, is of enormous importance in the semiconductor industry, with wide applications in such fields as computers and solar energy collection.

In nature, silicon occurs almost exclusively in combination with oxygen, with many ores containing tetrahedral SiO_4 structural units. Silicon dioxide, SiO_2, occurs in a variety of forms in nature, the most common of which is α-quartz, a major constituent of sandstone and granite. SiO_2 is of major industrial importance as the major component of glass, in finely divided form as a chromatographic support (silica gel) and catalyst substrate, in filtration plants (as diatomaceous earth, the remains of diatoms, tiny unicellular algae), and in numerous other applications.

The SiO_4 structural units occur in nature in silicates, compounds in which these units may be fused together by sharing corners, edges, or faces in an extremely diverse number of ways. Discrete silicate ions, chains, nets, and three-dimensional networks are all known. Selected examples of the first three types of silicate structures are shown in Figure 7-15, along with the structure of SiO_2 (quartz). The interested reader can find extensive discussions of these structures in the chemical literature.[8]

With carbon forming the basis for the colossal number of organic compounds, it is interesting to consider whether silicon or other members of this group can form the foundation for an equally vast array of compounds. Unfortunately, such does not seem the case; the ability to catenate (form bonds with other atoms of the same element) is much less for the other group 14 (IVA) elements than for carbon, and the hydrides of these elements are also much less stable.

Silane, SiH_4, is stable and, like methane, tetrahedral. However, although silanes (of formula Si_nH_{2n+2}) up to eight silicon atoms in length have been synthesized, their stability decreases markedly with chain length; Si_2H_6 (disilane) undergoes only very slow decomposition, but Si_8H_{18} decomposes rapidly. In recent years a few compounds containing Si═Si bonds have been synthesized, but there is no promise of a chemistry of multiply bonded Si species comparable at all in diversity with the chemistry of unsaturated organic compounds. Germanes of formulas GeH_4 to Ge_5H_{12} have been made, as have SnH_4 (stannane), Sn_2H_6, and, possibly, PbH_4 (plumbane), but the chemistry in these cases is even more limited than that of the silanes.

Why are the silanes and other analogous compounds less stable (more reactive) than the corresponding hydrocarbons? First, the Si—Si bond is slightly weaker than the C—C bond (approximate bond energies of 340 and 368 kJ mol^{-1}, respectively), and Si—H bonds are weaker than C—H bonds (393 versus 435 kJ mol^{-1}). Silicon is less electronegative (1.90) than hydrogen (2.20) and is therefore more susceptible to nucleophilic attack, in contrast to carbon, which is *more* electronegative (2.55) than hydrogen. Silicon atoms are also larger and therefore provide greater surface area for attack by nucleophiles; in addition, silicon atoms have low-lying *d* orbitals that can act as acceptors of electron pairs from nucleophiles. Similar arguments can be used to describe the high reactivity of germanes, stannanes, and plumbanes. Silanes are believed to decompose by elimination of $:SiH_2$ by way of a transition state having a

[8] A. F. Wells, *Structural Inorganic Chemistry*, 5th ed., Oxford University Press, New York, 1984, pp. 1009–43.

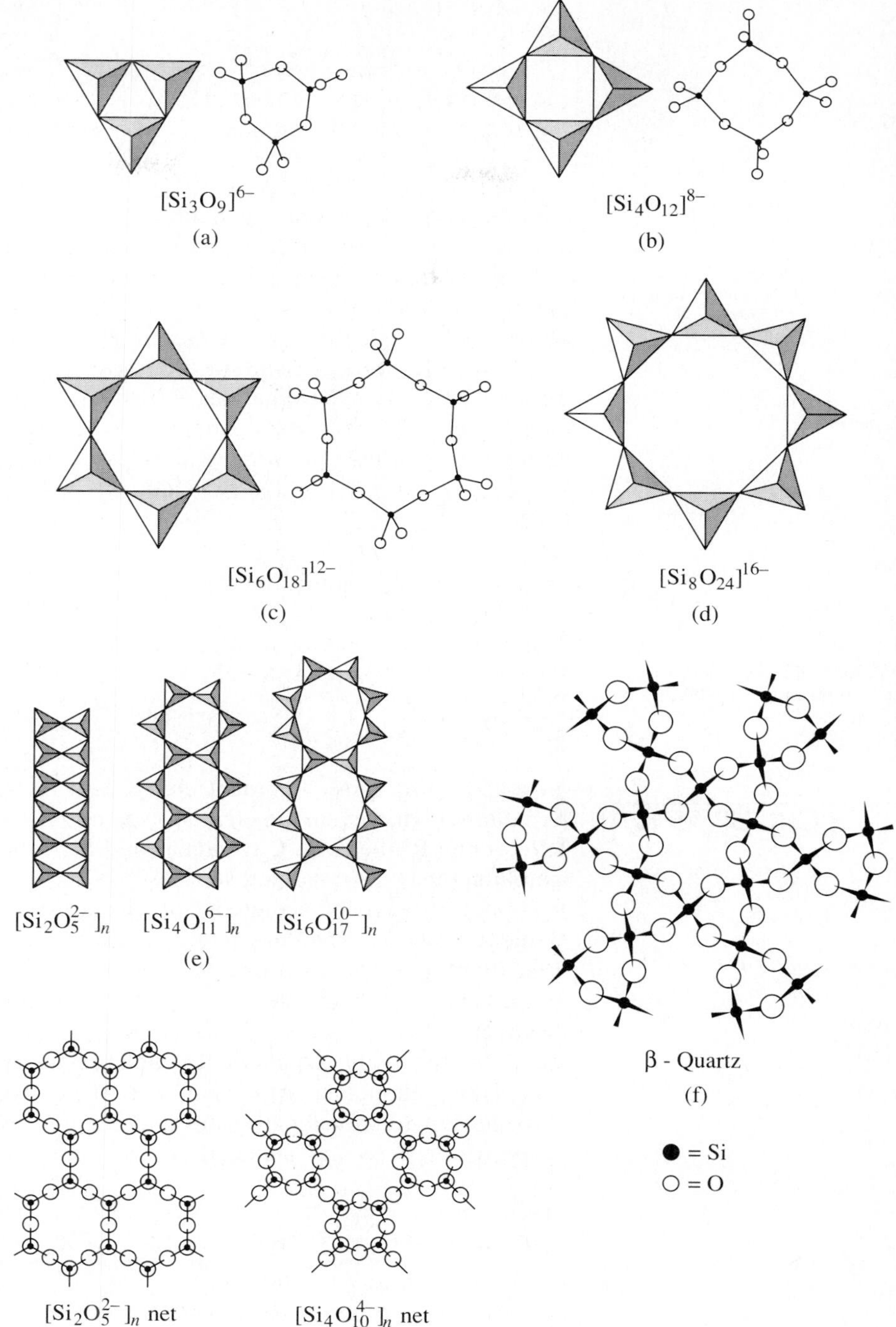

FIGURE 7-15 Examples of Silicate Structures. (Reproduced with permission (a, b, c, d, e) from N. N. Greenwood and A. Earnshaw, *Chemistry of the Elements,* Pergamon Press, Elmsford, N.Y., 1984, pp. 403, 405, copyright 1984, Pergamon Press PLC; and (f, g) from A. F. Wells, *Structural Inorganic Chemistry,* 5th ed., Oxford University Press, New York, 1984, pp. 1006, 1024.)

bridging hydrogen, as shown in Figure 7-16. This reaction, incidentally, can be used to prepare silicon of extremely high purity.

As has been mentioned, silicon has the diamond structure. Silicon carbide, SiC, occurs in numerous crystalline forms, some of them similar to the diamond. As carborundum, silicon carbide is widely used as an abrasive, with

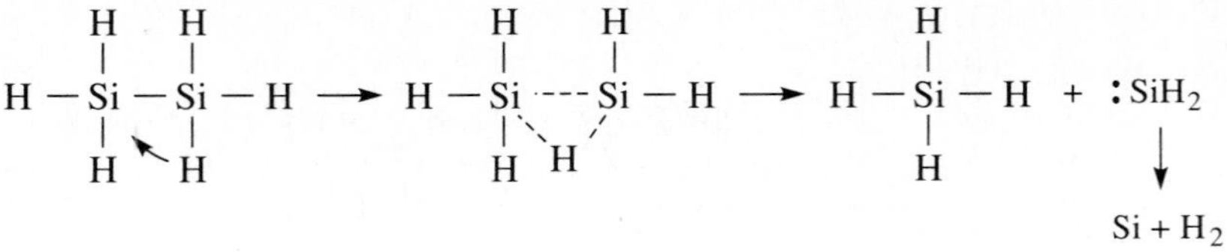

FIGURE 7-16 Decomposition of Silanes.

$$\mathrm{H_3Si{-}SiH_3} \longrightarrow \mathrm{H_3Si\cdots(H)\cdots SiH_2} \longrightarrow \mathrm{SiH_4} + :\mathrm{SiH_2}$$

$$:\mathrm{SiH_2} \longrightarrow \mathrm{Si} + \mathrm{H_2}$$

a hardness nearly as great as diamond and a low chemical reactivity. SiC has also recently gained interest as a high-temperature semiconductor.

The elements germanium, tin, and lead show increasing importance of the 2+ oxidation state, another example of the inert pair effect described in Section 7-5-2. All three, for example, show two sets of halides, of formula MX_4 and MX_2. For germanium, the most stable halides are of formula GeX_4, for lead PbX_2. For the dihalides, the metal exhibits a stereochemically active lone pair; this leads to bent geometry for the free molecules and to crystalline structures in which the lone pair is evident, as shown for $SnCl_2$ in Figure 7-17.

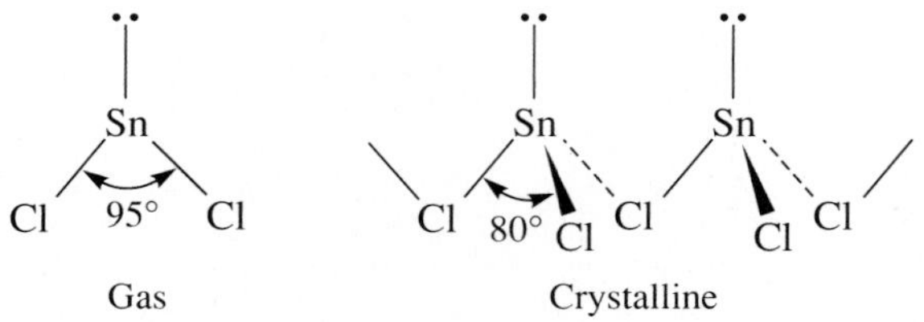

FIGURE 7-17 Structure of $SnCl_2$ in Gas and Crystalline Phases.

Gas: Sn, Cl, Cl, 95°. Crystalline: Sn, Sn, Cl, Cl, Cl, Cl, Cl, 80°.

7-7 GROUP 15 (VA)

Nitrogen is the most abundant component of the earth's atmosphere (78.1% by volume). However, the element was not successfully isolated from air until 1772, when Rutherford, Cavendish, and Scheele achieved the isolation nearly simultaneously by removing successively oxygen and carbon dioxide from air. Phosphorus was first isolated from urine by H. Brandt in 1669. Since the element glowed in the dark on exposure to air, it was named after the Greek *phos* (light) and *phoros* (bringing). Interestingly, the last three elements in this group had long been isolated by the time nitrogen and phosphorus were discovered. Their dates of discovery are lost in history, but all had been studied extensively, especially by alchemists, by the fifteenth century.

These elements again span the range from nonmetallic (nitrogen and phosphorus) to metallic (bismuth) behavior, with elements of intermediate properties in between. Selected physical properties are given in Table 7-9.

TABLE 7-9
Properties of the group 15 (VA) elements

Element	*Ionization energy (kJ mol⁻¹)*	*Electron affinity (kJ mol⁻¹)*	*Melting point (°C)*	*Boiling point (°C)*	*Electronegativity (Pauling)*
N	1402	−7	−210	−195.8	3.04
P	1012	72	44[a]	280.5	2.19
As	947	78	[b]	[b]	2.18
Sb	834	103	631	1587	2.05
Bi	703	91	271	1564	2.02

SOURCES: See Table 7-3.
NOTES: [a] α-P_4.
[b] Sublimes at 615°.

7-7-1 THE ELEMENTS

Nitrogen is a colorless diatomic gas. As discussed in Chapter 5, the dinitrogen molecule has a nitrogen–nitrogen triple bond of unusual stability (see problem 13 at the end of this chapter). In large part the stability of this bond is responsible for the low reactivity of this molecule (although it is by no means totally inert). Nitrogen is therefore suitable as an inert environment for many chemical studies on reactions that are either oxygen- or moisture-sensitive. Liquid nitrogen, at 77 K, is frequently used as a convenient, rather inexpensive coolant for studying low-temperature reactions, trapping of solvent vapors, and cooling superconducting magnets (actually, for preserving the liquid helium coolant, which boils at 4 K).

Phosphorus has numerous allotropes. The most common of these is white phosphorus, which exists in two modifications, α-P_4 (cubic) and β-P_4 (hexagonal). Condensation of phosphorus from the gas or liquid phases (both of which contain tetrahedral P_4 molecules) gives primarily the α form, which slowly converts to the β form at temperatures above $-76.9°C$. During slow air oxidation, α-P_4 emits a yellow-green light, an example of phosphorescence which has been known since antiquity (and is the source of the name of this element); to slow such oxidation, white phosphorus is commonly stored under water. White phosphorus was once used in matches; however, its extremely high toxicity has led to its replacement by other materials, especially P_4S_3 and red phosphorus, which are much less toxic.

Heating of white phosphorus in the absence of air gives red phosphorus, an amorphous material that exists in a variety of modifications. Still another allotrope, black phosphorus, can be obtained from white phosphorus by heating at very high temperatures. Black phosphorus converts to other forms at high pressure, each consisting of extended puckered layers of 3-coordinate phosphorus atoms. Examples of these structures are shown in Figure 7-18. The

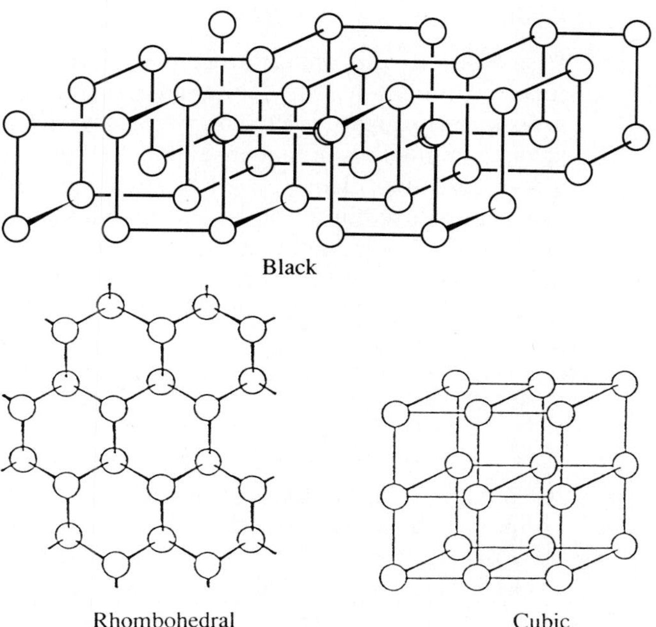

FIGURE 7-18 Allotropes of Phosphorus. (Reproduced with permission from N. N. Greenwood and A. Earnshaw, *Chemistry of the Elements*, Pergamon Press, Elmsford, N.Y., 1984, p. 558. Copyright 1984, Pergamon Press PLC.)

interested reader can find more detailed information on allotropes of phosphorus in other sources.[9]

As mentioned, phosphorus exists as tetrahedral P_4 molecules in the liquid and gas phases. At very high temperatures P_4 can dissociate: $P_4 \rightleftharpoons 2\ P{\equiv}P$; at approximately 1800°C, this dissociation reaches 50%.

Arsenic, antimony, and bismuth also exhibit a variety of allotropes. The most stable allotrope of arsenic is the gray (α) form, which is similar to the rhombohedral form of phosphorus. In the vapor phase, arsenic, like phosphorus, exists as tetrahedral As_4. Antimony and bismuth also have similar α forms. These three elements have a somewhat metallic appearance but are brittle and are only moderately good conductors; arsenic, for example, is the best conductor in this group but has an electrical resistivity nearly 20 times as great as copper. Bismuth is the heaviest element to have a stable, nonradioactive nucleus; polonium and all heavier elements are radioactive.

Anions

Nitrogen exists in two anionic forms, N^{3-} (nitride) and N_3^- (azide). Nitrides of primarily ionic character are formed by lithium and the group 2 (IIA) elements; numerous other nitrides having a greater degree of covalence are also known. In addition, N^{3-} is a strong π donor ligand toward transition metals (metal–ligand interactions are described in Chapter 8). Stable compounds containing the linear N_3^- ion include those of the group 1 and 2 (IA and IIA) metals. However, some other azides are explosive; $Pb(N_3)_2$, for example, is shock sensitive and used as a primer for explosives.

Although phosphides, arsenides, and analogous compounds are known with formulas that may suggest that they are ionic (for example, Na_3P, Ca_3As_2), such compounds are generally lustrous and have good thermal and electrical conductivity, properties more consistent with metallic than ionic bonding.

Hydrides

In addition to ammonia, nitrogen forms the hydrides N_2H_4 (hydrazine), N_2H_2 (diazene or diimide), and HN_3 (hydrazoic acid). Structures of these compounds are shown in Figure 7-19. The chemistry of ammonia and the ammonium ion is vast; ammonia is of immense industrial importance and is produced in larger molar quantities than any other chemical. More than 80% of the ammonia produced is used in fertilizers, with additional uses including the synthesis of explosives, the manufacture of synthetic fibers (such as rayon, nylon, and polyurethanes), and the synthesis of a wide variety of organic and inorganic compounds. As described in Chapter 6, liquid ammonia is used extensively as a nonaqueous ionizing solvent.

Oxidation of hydrazine is highly exothermic:

$$N_2H_4 + O_2 \longrightarrow N_2 + 2\ H_2O, \qquad \Delta H^\circ = -622\ \text{kJ mol}^{-1}$$

Advantage has been taken of this reaction in the major use of hydrazine (and its methyl derivatives), in rocket fuels. Hydrazine is also a convenient and versatile reducing agent, capable of being oxidized by a wide variety of oxidizing agents, in acidic (as the protonated hydrazonium ion, $N_2H_5^+$) and

[9] Ibid., pp. 838–40.

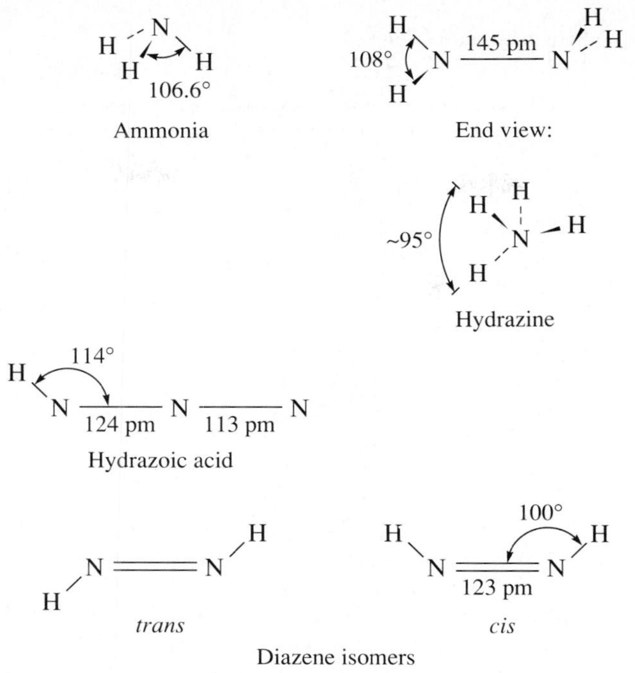

106.6°

Ammonia

108° 145 pm

End view:

~95°

Hydrazine

114°

124 pm 113 pm

Hydrazoic acid

100°

123 pm

trans *cis*

Diazene isomers

FIGURE 7-19 Nitrogen Hydrides. (Bond angles and distances are from A. F. Wells, *Structural Inorganic Chemistry*, 5th ed., Oxford University Press, New York, 1984.)

basic solutions. It may be oxidized by 1, 2, or 4 electrons, depending on the oxidizing agent:

	$\mathscr{E}°$ *(oxidation)*	*Examples of oxidizing agents*
$N_2H_5^+ \rightleftharpoons NH_4^+ + \frac{1}{2} N_2 + H^+ + e^-$	1.74 V	MnO_4^-, Ce^{4+}
$N_2H_5^+ \rightleftharpoons \frac{1}{2} NH_4^+ + \frac{1}{2} HN_3 + \frac{5}{2} H^+ + 2\,e^-$	−0.11 V	H_2O_2
$N_2H_5^+ \rightleftharpoons N_2 + 5\,H^+ + 4\,e^-$	0.23 V	I_2

Both the *cis* and *trans* isomers of diazene are known; they are unstable except at very low temperatures. The fluoro derivatives N_2F_2 are more stable and have been characterized structurally. Both isomers show N—N distances consistent with double bonds (*cis*: 120.9 pm; *trans*: 122.4 pm).

Phosphine, PH_3, is a highly poisonous gas. It does not engage significantly in hydrogen bonding or intermolecular attractions like those described for ammonia in Chapter 3; consequently, its melting and boiling points are much lower than those of ammonia (−133.5° and −87.5°C for PH_3 versus −77.8° and −34.5°C for NH_3). Phosphine derivatives of formula PR_3 (phosphines; R = H, alkyl, or aryl) and $P(OR)_3$ (phosphites) are important ligands that form numerous coordination compounds. Examples of phosphine compounds are discussed in Chapters 12 and 13. Arsines, AsR_3, and stibines, SbR_3, are also important ligands in coordination chemistry.

Nitrogen oxides and oxyions

Nitrogen oxides and ions containing nitrogen and oxygen are among the most frequently encountered species in inorganic chemistry. The most common of these are summarized in Table 7-10.

TABLE 7-10
Compounds and ions containing nitrogen and oxygen

Formula	*Name*	*Structure**	*Notes*
N_2O	Nitrous oxide	N=N=O	mp = −90.9°C; bp = −88.5°C
NO	Nitric oxide	 N≡O (115, dashed)	mp = −163.6°C; bp = −151.8°C; bond order approximately 2.5; paramagnetic
NO_2	Nitrogen dioxide	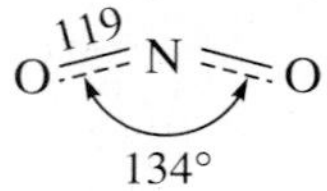 O–N–O (119; 134°)	Brown, paramagnetic gas; exists in equilibrium with N_2O_4: $2\ NO_2 \rightleftharpoons N_2O_4$
N_2O_3	Dinitrogen trioxide	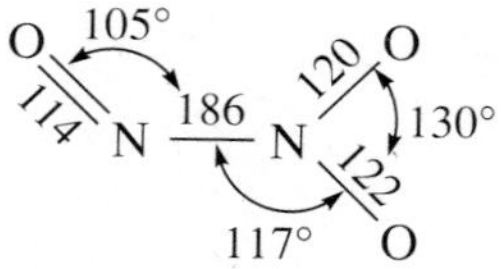 O=N–N(O)O (114, 186, 120, 122; 105°, 130°, 117°)	mp = −100.1°C; dissociates above melting point: $N_2O_3 \rightleftharpoons NO + NO_2$
N_2O_4	Dinitrogen tetroxide	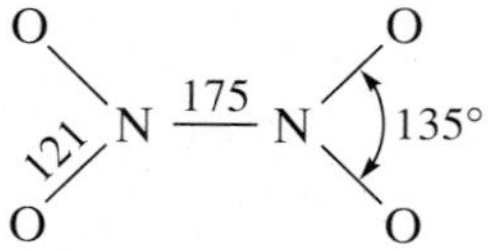 $O_2N–NO_2$ (121, 175; 135°)	mp = −11.2°C; bp = 21.15°C; dissociates into 2 NO_2 [ΔH (dissociation) = 57 kJ mol^{-1}]
N_2O_5	Dinitrogen pentoxide	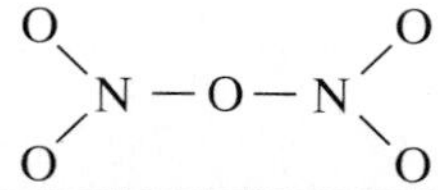 $O_2N–O–NO_2$	N—O—N bond may be bent; consists of NO_2^+ NO_3^- in the solid
NO^+	Nitrosonium or nitrosyl	 N≡O (106)	Isoelectronic with CO
NO_2^+	Nitronium or nitryl	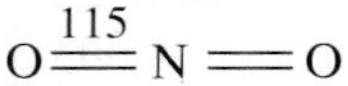 O=N=O (115)	Isoelectronic with CO_2
NO_2^-	Nitrite	 O–N–O (dashed)	N—O distance varies from 113 to 123 pm and bond angle varies from 116 to 132° depending on cation; versatile ligand (see Chapter 10)
NO_3^-	Nitrate	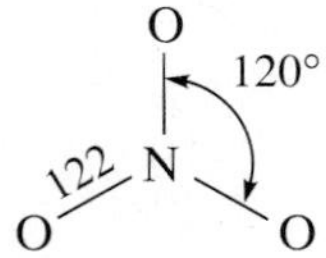 NO_3 (122; 120°)	Forms compounds with nearly all metals; as ligand, has a variety of coordination modes
$N_2O_2^{2-}$	Hyponitrite	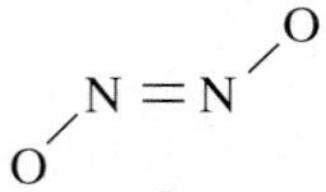 O–N=N–O	Useful reducing agent
NO_4^{3-}	Orthonitrate	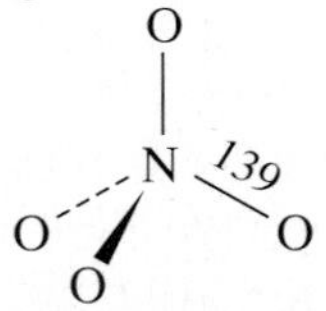 NO_4 (139)	Na and K salts known; decomposes in presence of H_2O and CO_2
HNO_2	Nitrous acid	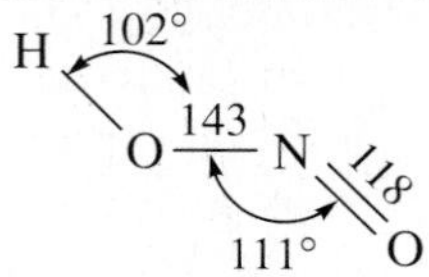 H–O–N=O (143, 118; 102°, 111°)	Weak acid (pK_a = 3.3 at 25°C); disproportionates: $3\ HNO_2 \rightleftharpoons H_3O^+ + NO_3^- + 2\ NO$ in aqueous solution
HNO_3	Nitric acid	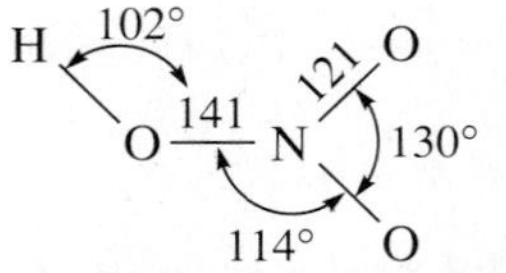 H–O–NO_2 (141, 121; 102°, 130°, 114°)	Strong acid in aqueous solution; concentrated aqueous solutions are strong oxidizing agents

SOURCE: Data from N. N. Greenwood and A. Earnshaw, *Chemistry of the Elements*, Pergamon Press, Elmsford, N.Y., 1984.
NOTE: * Distances in pm.

Nitrous oxide, N_2O, is commonly encountered as a mild dental anesthetic and propellant for aerosols; on atmospheric decomposition it yields the innocuous parent gases and therefore is an environmentally acceptable substitute for chlorofluorocarbons. Nitric oxide, NO, is an effective coordinating ligand; its function in this context is discussed in Chapter 12. The gases N_2O_4 and NO_2 form an interesting pair. At ordinary temperatures both exist in significant amounts in equilibrium:

$$N_2O_4\ (g) \rightleftharpoons 2\ NO_2\ (g), \qquad \Delta H^\circ = 57.20\ \text{kJ mol}^{-1}$$

The colorless, diamagnetic N_2O_4 has a weak N—N bond that can readily dissociate to give the brown, paramagnetic NO_2.

Nitric oxide is formed in the combustion of fossil fuels and is present in the exhausts of automobiles and power plants; it can also be formed from the action of lightning on atmospheric N_2 and O_2. In the atmosphere, NO is oxidized to NO_2 (which exists in equilibrium with N_2O_4). These gases, often collectively designated NO_x, contribute to the problem of acid rain, primarily because NO_2 reacts with atmospheric water to form nitric acid:

$$3\ NO_2 + H_2O \longrightarrow 2\ HNO_3 + NO$$

Nitrogen oxides are also believed to be instrumental in the destruction of the earth's ozone layer, as discussed in the following section.

Nitric acid is of immense industrial importance, especially in the synthesis of ammonium nitrate and other chemicals. Ammonium nitrate is used primarily as a fertilizer. In addition, it is thermally unstable and undergoes violently exothermic decomposition at elevated temperature:

$$2\ NH_4NO_3 \longrightarrow 2\ N_2 + O_2 + 4\ H_2O$$

Significant use is now made of ammonium nitrate as an explosive.

As mentioned in Chapter 6, nitric acid is also of interest as a nonaqueous solvent and undergoes the following autoionization:

$$2\ HNO_3 \rightleftharpoons H_2NO_3^+ + NO_3^- \rightleftharpoons H_2O + NO_2^+ + NO_3^-$$

7-8 GROUP 16 (VIA)

The first two elements in this group (which is occasionally designated the "chalcogen" group) are familiar as O_2, the colorless gas that comprises about 21% of the earth's atmosphere, and sulfur, a yellow solid of typical nonmetallic properties. The third element in this group, selenium, is perhaps not as well known, but is important in the xerography process, and selenides are used in the coloring of glasses. Although elemental selenium is highly poisonous, trace amounts of the element are essential for life. Tellurium is of less commercial interest but is used in small amounts in metal alloys, in tinting of glass, and in compounds used as catalysts in the rubber industry. Tellurium is also a photoconductor, a poor conductor ordinarily, but a good conductor when exposed to light. All isotopes of polonium, a metal, are radioactive. The highly exothermic radioactive decay of this element has made it a useful power source for satellites.

7-8-1 THE ELEMENTS

Sulfur, which occurs as the free element in numerous natural deposits, has been known since prehistoric times; it is the ''brimstone'' of the Bible. It was of considerable interest to the alchemists and, following the development of gunpowder (a mixture of sulfur, $NaNO_3$, and powdered charcoal) in the thirteenth century, to military leaders as well. Although elemental oxygen is ubiquitous in the earth's atmosphere, combined with other elements in the earth's crust (which contains 46% oxygen by mass), and in bodies of water, the pure element was not isolated and characterized until the 1770s, by C. W. Scheele and J. Priestley. Priestley's classic synthesis of oxygen by heating HgO with a magnifying glass was a landmark in the history of experimental chemistry. Selenium (1817) and tellurium (1782) were soon discovered and, because of their chemical similarities, named after the moon (Greek, *selene*) and earth (Latin, *tellus*). Polonium was discovered by Marie Curie in 1898; like radium, it was isolated in trace amounts from tons of uranium ore. Some important physical properties of these elements are summarized in Table 7-11.

TABLE 7-11
Properties of the group 16 (VIA) elements

Element	*Ionization energy ($kJ\ mol^{-1}$)*	*Electron affinity ($kJ\ mol^{-1}$)*	*Melting point (°C)*	*Boiling point (°C)*	*Electro-negativity (Pauling)*
O	1314	141	−218.8	−183.0	3.44
S	1000	200	112.8	444.7	2.58
Se	941	195	217	685	2.55
Te	869	190	452	990	2.10
Po	812	180*	250*	962	2.0*

SOURCES: See Table 7-3.
NOTES: * Approximate value.

Oxygen

Oxygen exists primarily in the diatomic form, O_2, but traces of ozone, O_3, are found in the upper atmosphere and in the vicinity of electrical discharges. O_2 is paramagnetic, O_3 diamagnetic. As discussed in Chapter 5, the paramagnetism of O_2 is the consequence of two electrons with parallel spin occupying π^* ($2p$) orbitals. In addition, two excited states of O_2 are known; these have π^* electrons of opposite spin and are higher in energy as a consequence of the effects of pairing energy and exchange energy (see Section 2-2-4):

	Relative Energy ($kJ\ mol^{-1}$)
⥮ __	157.85
↑ ↓	94.72
↑ ↑	0

The excited states of O_2 can be achieved when photons are absorbed in the liquid phase during molecular collisions; under these conditions a single photon can simultaneously excite two colliding molecules. This absorption occurs in the visible range, at 631 and 474 nm and gives rise to the blue color of the liquid.[10] The excited states are also important in many oxidation processes.

[10] E. A. Ogryzlo, *J. Chem. Educ.*, **1965**, *42*, 647.

O_2 is, of course, essential for respiration. The mechanism for oxygen transport to the cells via hemoglobin has received much attention; it is discussed briefly in Chapter 15.

Ozone absorbs ultraviolet radiation between 200 and 360 nm. It thus forms an indispensable shield in the upper atmosphere, protecting the earth's surface from most of the potentially hazardous effect of such high-energy electromagnetic radiation. Recently there has been increasing concern because atmospheric pollutants appear to be depleting the ozone layer, with the result that gaps in the layer have occurred over Antarctica. In the upper atmosphere, ozone is formed from O_2:

$$O_2 \xrightarrow{h\nu} 2\,O$$
$$O + O_2 \longrightarrow O_3$$

Absorption of ultraviolet radiation by O_3 causes it to decompose to O_2. In the upper atmosphere, therefore, a steady-state concentration of ozone is achieved, a concentration ordinarily sufficient to provide significant ultraviolet protection of the earth's surface. However, pollutants in the upper atmosphere such as nitrogen oxides (some of which occur in trace amounts naturally) from high-flying aircraft and chlorine atoms from photolytic decomposition of chlorofluorocarbons (from aerosols, refrigerants, and other sources) appear to be able to catalyze the decomposition of ozone. The overall processes governing the concentration of ozone in the atmosphere are extremely complex; the following reactions can be studied in the laboratory and are examples of the processes believed to be involved in the atmosphere:

$$NO_2 + O_3 \longrightarrow NO_3 + O_2$$
$$NO_3 \longrightarrow NO + O_2$$
$$NO + O_3 \longrightarrow NO_2 + O_2$$

Net: $$2\,O_3 \longrightarrow 3\,O_2$$

$$Cl + O_3 \longrightarrow ClO + O_2$$
$$ClO + O \longrightarrow Cl + O_2$$

Cl formed from photodecomposition of chlorofluorocarbons

Net: $$O_3 + O \longrightarrow 2\,O_2$$

Ozone is a more potent oxidizing agent than O_2; in acidic solution it is exceeded only by fluorine among the elements as an oxidizing agent.

Several diatomic and triatomic oxygen ions are known; these are summarized in Table 7-12.

Sulfur

More allotropes are known for sulfur than for any other element, with the most stable form at room temperature (orthorhombic, α-S_8) having eight sulfur atoms arranged in a puckered ring. Several of the most common sulfur allotropes are shown in Figure 7-20.[11]

Heating of sulfur results in very interesting changes in viscosity. At approximately 119°C, sulfur melts to give a yellow liquid, whose viscosity gradually decreases until approximately 155°C (Figure 7-21). However, further

[11] B. Meyer, *Chem. Rev.*, **1976**, *76*, 367.

TABLE 7-12
Neutral and ionic O_2 and O_3 species

Formula	*Name*	*O—O Distance (pm)*	*Notes*
O_2^+	Dioxygenyl	112.3	Bond order 2.5
O_2	Dioxygen	120.7	Coordinates to transition metals; singlet O_2 (excited state) important in photochemical reactions; oxidizing agent
O_2^-	Superoxide	128	Moderate reducing agent; most stable compounds: KO_2, RbO_2, CsO_2
O_2^{2-}	Peroxide	149	Forms ionic compounds with alkali metals, Ca, Sr, Ba; strong oxidizing agent
O_3	Ozone	127.8	Bond angle 116.8°; strong oxidizing agent; absorbs in UV range (220–290 nm)
O_3^-	Ozonide	134	Formed from reaction of O_3 with dry alkali metal hydroxides; decomposes to O_2^-

SOURCE: Data from N. N. Greenwood and A. Earnshaw, *Chemistry of the Elements*, Pergamon Press, Elmsford, N.Y., 1984.

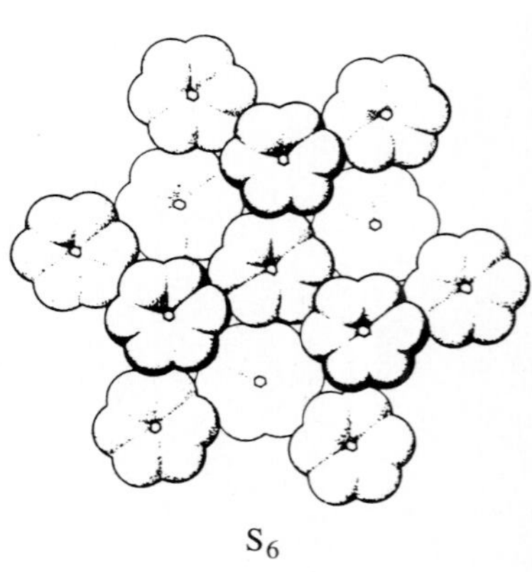

FIGURE 7-20 Allotropes of Sulfur. (Reproduced with permission from M. Schmidt and W. Siebert, "Sulphur," *Comprehensive Inorganic Chemistry,* Vol. 2, Pergamon Press, Elmsford, N.Y., 1973, pp. 804, 806. Copyright 1973, Pergamon Press PLC.)

heating causes the viscosity to increase very dramatically above 159°C until the liquid pours very sluggishly. Above about 200°, the viscosity again decreases, with the liquid eventually acquiring a more reddish hue at higher temperatures.[12]

[12] *The Sulphur Data Book*, W. N. Tuller, ed., McGraw-Hill, New York, 1954.

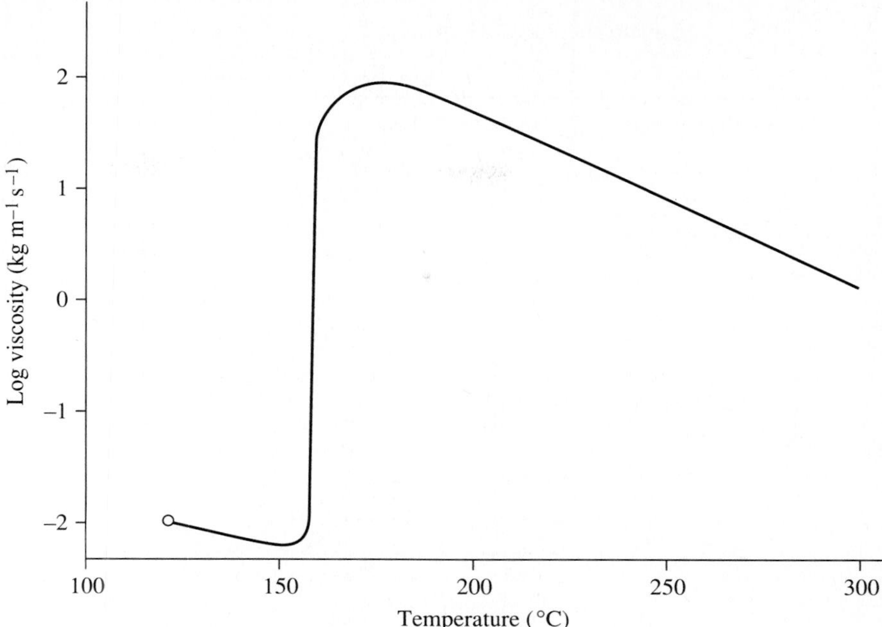

FIGURE 7-21 Viscosity of Sulfur.

The explanation of these changes in viscosity involves the tendency of S—S bonds to break and re-form at high temperatures. Above 159°C, the S_8 rings begin to open; the resulting S_8 chains can react with other S_8 rings to open them and form S_{16} chains, S_{24} chains, and so on:

$$S_8 \text{ (ring)} \longrightarrow \text{S–S–S–S–S–S–S–S (chain)} \longrightarrow S_{16} \longrightarrow S_{24} \longrightarrow \cdots$$

The longer the chains, the greater the viscosity (the more the chains can intertwine with each other). Large rings can also form by the linking of ends of chains. Chains exceeding 200,000 sulfur atoms are formed at the temperature of maximum viscosity, near 180°C. At higher temperatures, thermal breaking of sulfur chains occurs more rapidly than propagation of chains, and the average chain length decreases, accompanied by a decrease in viscosity. At very high temperatures, brightly colored species such as S_3 increase in abundance and give rise to the reddish coloration. Interestingly, hot, viscous sulfur, when poured into cold water, gives a rubbery solid that can be molded readily. This plastic material contains long sulfur chains that eventually convert to the yellow, crystalline α form, the most thermodynamically stable allotrope, which consists again of the S_8 rings.

Numerous compounds and ions containing sulfur and oxygen are known; many of these are important acids or conjugate bases (more sulfuric acid, for example, is produced than any other chemical). Some useful information about these compounds and ions is summarized in Table 7-13. The interested reader can find more detailed information in other sources.[13]

[13] N. N. Greenwood and A. Earnshaw, *Chemistry of the Elements*, Pergamon Press, Elmsford, N.Y., 1984, pp. 821–54.

TABLE 7-13
Molecules and ions containing sulfur and oxygen

Formula	*Name*	*Structure*	*Notes*
SO_2	Sulfur dioxide	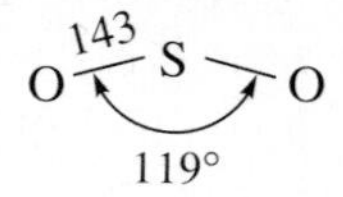 S–O 143; O–S–O 119°	mp = −75.5°C; bp = −10.0°C; colorless, choking gas; product of combustion of elemental sulfur
SO_3	Sulfur trioxide	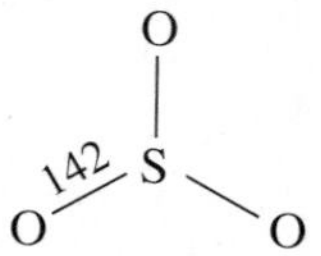 S–O 142	mp = 16.9°C; bp = 44.6°C; formed from oxidation of SO_2: $SO_2 + \frac{1}{2}O_2 \longrightarrow SO_3$; in equilibrium with trimer S_3O_9 in liquid and gas phases; reacts with water to form sulfuric acid
	Trimer:	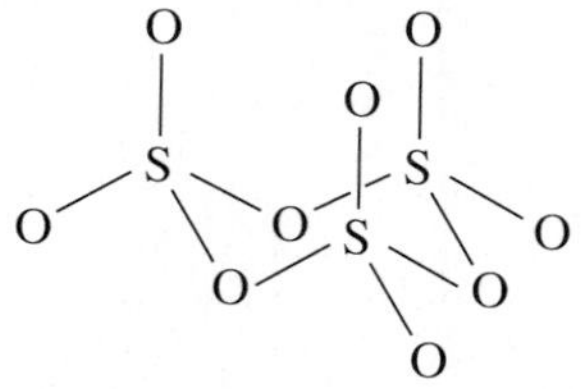	
SO_3^{2-}	Sulfite	 S–O 151; O–S–O 106°	Conjugate base of HSO_3^-, formed when SO_2 dissolves in water
SO_4^{2-}	Sulfate	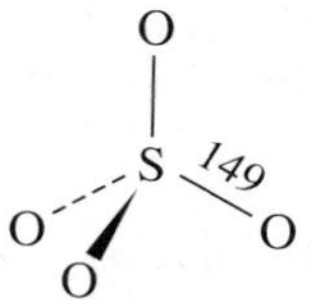 S–O 149	T_d symmetry; extremely common ion; used in gravimetric analysis
$S_2O_3^{2-}$	Thiosulfate	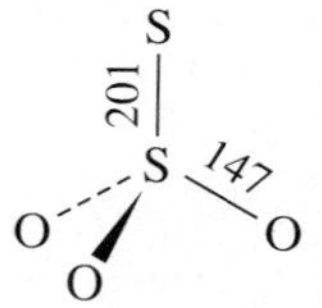 S–S 201; S–O 147	Moderate reducing agent; used in analytical determination of I_2: $I_2 + 2\ S_2O_3^{2-} \longrightarrow 2I^- + S_4O_6^{2-}$
$S_2O_4^{2-}$	Dithionite	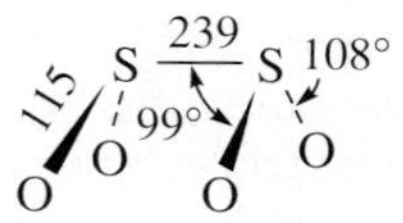 S–S 239; S–O 115; 99°; 108°	Very long S—S bond; dissociates into SO_2^-: $S_2O_4^{2-} \rightleftharpoons 2\ SO_2^-$; Zn and Na salts used as reducing agents
$S_2O_8^{2-}$	Peroxodisulfate	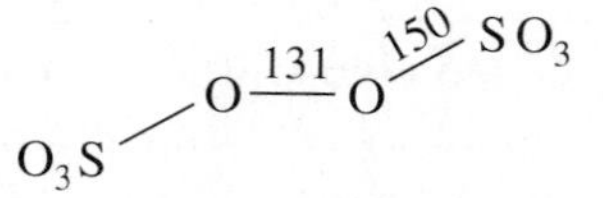 O_3S–O–O–SO_3; O–O 131; O–S 150	Useful oxidizing agent, readily reduced to sulfate: $S_2O_8^{2-} + 2e^- \rightleftharpoons 2\ SO_4^{2-}$, $\mathscr{E}° = 2.01$ V
H_2SO_4	Sulfuric acid	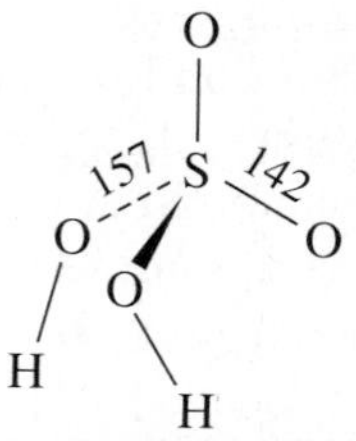 S–O(H) 157; S–O 142	C_2 symmetry; mp = 10.4°C; bp = ~300°C (dec); strong acid in aqueous solution; undergoes autoionization: $2\ H_2SO_4 \rightleftharpoons H_3SO_4^+ + HSO_4^-$, $pK = 3.57$ at 25°C

SOURCE: Data from N. N. Greenwood and A. Earnshaw, *Chemistry of the Elements*, Pergamon Press, Elmsford, N.Y., 1984.

7-9 GROUP 17 (VIIA): THE HALOGENS

7-9-1 THE ELEMENTS

Compounds containing the halogens (halogen = salt former) have been used since antiquity, with the first use probably that of rock or sea salt (primarily NaCl) as a food preservative. Isolation and characterization of the neutral elements, however, have occurred comparatively recently.[14] Chlorine was first recognized as a gas by J. B. van Helmont in approximately 1630 and first studied carefully by C. W. Scheele in the 1770s (hydrochloric acid, which was used in these early syntheses, had been prepared by the alchemists around the year 900). Iodine was next, obtained by Courtois in 1811 by subliming the product of the reaction of sulfuric acid with seaweed ash. A.-J. Balard obtained bromine in 1826 by reacting chlorine with $MgBr_2$, which was present in waters of salt marshes. Although hydrofluoric acid had been used to etch glass since the latter part of the seventeenth century, elemental fluorine was not isolated until 1886, when H. Moissan obtained a small amount of the very reactive gas on electrolysis of KHF_2 in anhydrous HF. Astatine, one of the last of the nontransuranium elements to be produced, was first synthesized in 1940 by D. R. Corson, K. R. Mackenzie, and E. Segre by bombardment of ^{209}Bi with alpha particles. All isotopes of astatine are radioactive (the longest-lived isotope has a half-life of 8.1 hours), and consequently the chemistry of this element has been studied only with the greatest difficulty.

All neutral halogens are diatomic and readily reduced to halide ions. All combine with hydrogen to form gases that, except for HF, are strong acids in aqueous solution. Some physical properties of the halogens are summarized in Table 7-14.

TABLE 7-14
Properties of the group 17 (VIIA) elements: the halogens

			Halogen molecules, X_2				
Element	*Ionization energy ($kJ\ mol^{-1}$)*	*Electron affinity ($kJ\ mol^{-1}$)*	*Melting point (°C)*	*Boiling point (°C)*	*Electronegativity (Pauling)*	*X—X Distance (pm)*	*ΔH of dissociation ($kJ\ mol^{-1}$)*
F	1681	328	−218.6	−188.1	3.98	143	158.8
Cl	1251	349	−101.0	−34.0	3.16	199	242.6
Br	1140	325	−7.25	59.5	2.96	228	192.8
I	1008	295	113.6[a]	185.2	2.66	266	151.1
At	930[b]	270[b]	302[b]		2.2		

SOURCE: See Table 7-3 for references. Ionization energy for At from J. Emsley, *The Elements*, Oxford University Press, New York, 1989.
NOTES: [a] Sublimes readily.
[b] Approximate value.

The halogens' chemistry is governed in large part by their tendency to acquire an electron to attain a noble gas electron configuration. Consequently, the halogens are excellent oxidizing agents, with F_2 the strongest oxidizing agent of all the elements. The tendency of the halogen atoms to attract electrons is also shown in their high electron affinities and electronegativities.

As mentioned, halogen molecules are diatomic. F_2 is extremely reactive and must be handled by special techniques; it is ordinarily prepared by electrolysis of molten fluorides such as KF. Cl_2 is a yellow gas; its odor is

[14] M. E. Weeks, *Discovery of the Elements*, 7th ed., Journal of Chemical Education, Easton, Pa., 1968, "The Halogen Family," pp. 701–49.

recognizable as the characteristic odor of chlorine bleach (an alkaline solution of the hypochlorite ion, ClO^-, which exists in equilibrium with small amounts of Cl_2). Br_2 is a dark red liquid that evaporates easily; it too is a strong oxidizing agent. I_2 is a black solid, readily sublimable at room temperature to give a purple vapor and, like the other halogens, highly soluble in nonpolar solvents. The color of iodine solutions varies significantly with the donor ability of the solvent as a consequence of charge transfer interactions, as described in Section 6-4-3. Iodine is also a moderately good oxidizing agent, but the weakest of the halogens. Because of its radioactivity, astatine has not been studied extensively; it would be interesting to be able to conveniently compare its properties and reactions with those of the other halogens.

The halogens are extremely interesting from the standpoint of trends in properties within the group. If we consider the halogens in order of increasing atomic number, several trends are immediately apparent from the data in Table 7-14. As the atomic number increases, the ability of the nucleus to attract the outermost electrons decreases; consequently, fluorine is the most electronegative and has the highest ionization energy, while astatine is lowest in both properties. With increasing size and mass in going down the periodic table, the London interactions between the diatomic molecules increase: F_2 and Cl_2 are gases, Br_2 a liquid, and I_2 a solid as a consequence of these interactions. The trends are not entirely predictable; however, fluorine and its compounds exhibit behavior that in some respects is substantially different than would be predicted by extrapolation of the characteristics of the other members of the group.

One of the most striking properties of F_2 is its remarkably low bond dissociation enthalpy, an extremely important factor in the high reactivity of this molecule. Extrapolation from the bond dissociation enthalpies of the other halogens would yield a value of approximately 290 kJ mol^{-1}, nearly double the actual value. Many suggestions have been made to account for this low value. It is likely that the weakness of the F—F bond is largely a consequence of repulsions between the nonbonding electron pairs.[15] The small size of the fluorine atom brings these pairs into close proximity when F—F bonds are formed; electrostatic repulsions between these pairs on neighboring atoms result in weaker bonding and an equilibrium bond distance significantly greater than would be expected in the absence of such repulsions. In orbital terms, the small size of the fluorine atoms leads to less overlap in the bonding molecular orbitals and more overlap of antibonding orbitals than would be expected by extrapolation from the other halogens.

For example, the covalent radius obtained for other compounds of fluorine is 64 pm; an F—F distance of 128 pm would therefore be expected in F_2. However, the actual distance is 143 pm. Oxygen and nitrogen share similar anomalies with fluorine: the O—O bonds in peroxides and the N—N bonds in hydrazines are longer than the sums of their covalent radii, and these bonds are weaker than the corresponding S—S and P—P bonds in the respective groups of these elements. In the case of oxygen and nitrogen, it is likely that the repulsion of electron pairs on neighboring atoms also plays a major role in the weakness of these bonds.[16]

[15] J. Berkowitz and A. C. Wahl, *Adv. Fluorine Chem.*, **1973**, *7*, 147.

[16] Anomalous properties of fluorine, oxygen, and nitrogen have been discussed by P. Politzer in *J. Am. Chem. Soc.*, **1969**, *91*, 6235 and *Inorg. Chem.*, **1977**, *16*, 3350.

Of the hydrohalic acids, HF is by far the weakest in aqueous solution (pK_a = 3.2 at 25°C); HCl, HBr, and HI are all strong acids. Although HF does react with water, strong hydrogen bonding occurs between F^- and the hydronium ion (F^---H^+--OH_2), to form the ion pair $H_3O^+F^-$, reducing the activity coefficient of H_3O^+. As the concentration of HF is increased, however, its tendency to form H_3O^+ increases as a result of further reaction of this ion pair with HF:

$$H_3O^+F^- + HF \rightleftharpoons H_3O^+ + HF_2^-$$

This view is supported by X-ray crystallographic studies of the ion pairs $H_3O^+F^-$ and $H_3O^+HF_2^-$.[17] (See also Chapter 6.)

Polyatomic Ions

In addition to the common monatomic halide ions, numerous polyatomic species, both cationic and anionic, have been prepared. Many readers will be familiar with the brown triiodide ion, I_3^-, formed from I_2 and I^-:

$$I_2 + I^- \rightleftharpoons I_3^-, \qquad K \approx 698 \text{ at } 25°C \text{ in aqueous solution}$$

Many other polyiodide ions have been characterized; in general, these may be viewed as aggregates of I_2 and I^- (sometimes I_3^-). Examples are shown in Figure 7-22.

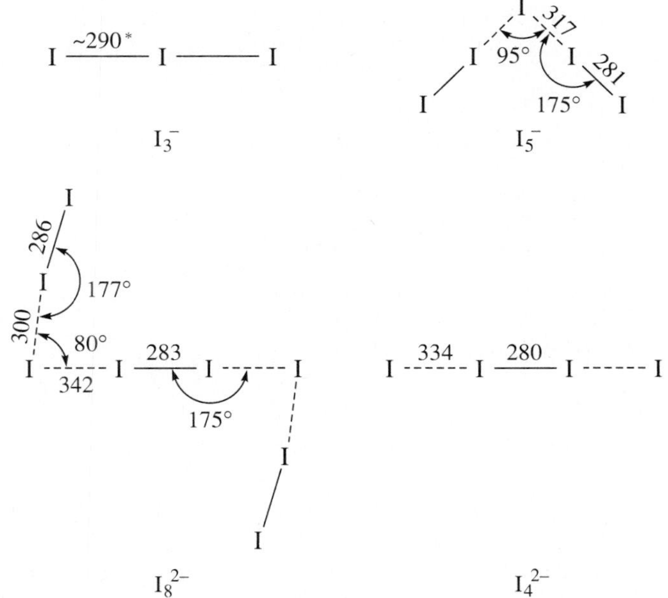

* Distances in triiodide vary depending on the cation. In some cases both I—I distances are identical, but in the majority of cases they are different. Differences in I—I distances as great as 33 pm have been reported.

FIGURE 7-22 Polyiodide Ions. [Bond angles and distances (in pm) are from A. F. Wells, *Structural Inorganic Chemistry*, 5th ed., Oxford University Press, New York, 1984, pp. 396–99.]

[17] D. Mootz, *Angew. Chem. Int. Ed. Engl.*, **1981**, *20*, 791.

The halogens Cl_2, Br_2, and I_2 can also be oxidized to cationic species. Examples include the diatomic ions Br_2^+ and I_2^+ (Cl_2^+ has been characterized in low-pressure discharge tubes but is much less stable), I_3^+, and I_5^+. I_2^+ dimerizes into I_4^{2+}:

$$2\,I_2^+ \rightleftharpoons I_4^{2+}$$

Interhalogens

Halogens form many compounds containing two or more different halogens. These may, like the halogens themselves, be diatomic (such as ClF) or polyatomic (such as ClF_3, BrF_5, or IF_7). In addition, polyatomic ions containing two or more halogens have been synthesized for many of the possible combinations. Selected neutral and ionic interhalogen species are listed in Table 7-15. The effect of size of the central atom can readily be seen, with iodine, for example, the only element able to have up to seven fluorine atoms in a neutral molecule, while chlorine and bromine have a maximum of five fluorines. The effect of size is also evident in the ions, with iodine the only halogen large enough to exhibit ions of formula XF_6^+ and XF_8^-.

Neutral interhalogens can be prepared in a variety of ways, including direct reaction of the elements (the favored product often depending on the ratio of halogens used) and reaction of halogens with metal halides or other halogenating agents. For example,

$$\begin{aligned} Cl_2 + F_2 &\longrightarrow 2\,ClF, \quad T = 225°C \\ I_2 + 5\,F_2 &\longrightarrow 2\,IF_5, \quad \text{room temperature} \\ I_2 + 3\,XeF_2 &\longrightarrow 2\,IF_3 + 3\,Xe, \quad T < -30°C \\ I_2 + AgF &\longrightarrow IF + AgI, \quad 0°C \end{aligned}$$

TABLE 7-15
Interhalogen species

Formal oxidation state of central atom	*Number of lone pairs on central atom*	*Compounds and ions*						
+7	0			IF_7				
				IF_6^+				
				IF_8^-				
+5	1	ClF_5	BrF_5	IF_5				
		ClF_4^+	BrF_4^+	IF_4^+				
			BrF_6^-	IF_6^-				
+3	2	ClF_3	BrF_3	IF_3		I_2Cl_6		
		ClF_2^+	BrF_2^+	IF_2^+		ICl_2^+	IBr_2^+	$IBrCl^+$
		ClF_4^-	BrF_4^-	IF_4^-		ICl_4^-		
+1	3	ClF	BrF	IF	BrCl	ICl	IBr	
		ClF_2^-	BrF_2^-	IF_2^-	$BrCl_2^-$	ICl_2^-	IBr_2^-	
					Br_2Cl^-	I_2Cl^-	I_2Br^-	$IBrCl^-$

Interhalogens can also serve as intermediates in the synthesis of other interhalogens:

$$ClF + F_2 \longrightarrow ClF_3, \quad T = 200° \text{ to } 300°C$$
$$ClF_3 + F_2 \longrightarrow ClF_5, \quad h\nu, \text{ room temperature}$$

Several interhalogens undergo autoionization in the liquid phase and have been studied as nonaqueous solvents. For example,

$$3\,IX \rightleftharpoons I_2X^+ + IX_2^-, \quad X = Cl, Br$$
$$2\,BrF_3 \rightleftharpoons BrF_2^+ + BrF_4^-$$
$$I_2Cl_6 \rightleftharpoons ICl_2^+ + ICl_4^-$$
$$2\,IF_5 \rightleftharpoons IF_4^+ + IF_6^-$$

Examples of acid–base reactions in autoionizing solvents were discussed in Chapter 6.

Pseudohalogens

Parallels have been observed between the chemistry of the halogens and a number of dimeric species. Dimeric molecules showing considerable similarity to the halogens are often called **pseudohalogens**. Some of the most important parallels in chemistry between the halogens and pseudohalogens include the following, illustrated for chlorine:

1. Neutral diatomic species — Cl_2
2. Ion of 1− charge — Cl^-
3. Formation of hydrohalic acids — HCl
4. Formation of interhalogen compounds — ICl, $BrCl$, ClF
5. Insolubility in water of salts with heavy metals such as Ag^+ and Pb^{2+} — $AgCl$, $PbCl_2$
6. Addition of halogen across multiple bonds — $Cl_2 + H_2C=CH_2 \longrightarrow H{-}CHCl{-}CHCl{-}H$ (i.e., $ClH_2C{-}CH_2Cl$)

For example, there are many similarities between the halogens and cyanogen, NCCN. The monoanion, CN^-, is, of course, well known; it combines with hydrogen to form the weak acid HCN and with Ag^+ and Pb^{2+} to form precipitates of low solubility in water. Interhalogen compounds FCN, ClCN, BrCN, and ICN are all known. Cyanogen, like the halogens, can add across double or triple carbon–carbon bonds. The pseudohalogen idea is a useful classification tool, although not many cases are known in which all the above characteristics are satisfied. Some examples of pseudohalogens are given in Table 7-16.[18]

[18] For additional examples of pseudohalogens, see J. Ellis, *J. Chem. Educ.*, **1976**, *53*, 2.

TABLE 7-16
Pseudohalogens

Characteristics	*Examples**		
Neutral dimeric species	Cl_2	$(CN)_2$	$[Co(CO)_4]_2$
Ion of 1− charge	Cl^-	CN^-	$[Co(CO)_4]^-$
Formation of hydrohalic acids	HCl (strong)	HCN (weak)	$HCo(CO)_4$ (strong)
Formation of interhalogen compounds	$Br_2 + Cl_2 \rightleftharpoons 2\,BrCl$	$Cl_2 + (CN)_2 \longrightarrow 2\,ClCN$	$[Co(CO)_4]_2 + I_2 \longrightarrow 2\,ICo(CO)_4$
Formation of heavy metal salts of low solubility	AgCl	AgCN	$AgCo(CO)_4$
Addition to unsaturated species	$Cl_2 + \text{>C=C<} \longrightarrow \text{−CCl−CCl−}$		$[Co(CO)_4]_2 + F_2C{=}CF_2 \longrightarrow (OC)_4Co{-}CF_2{-}CF_2{-}Co(CO)_4$

NOTE: * Metal carbonyl (CO) compounds will be discussed in Chapters 12 to 14.

7-10 GROUP 18 (VIIIA): THE NOBLE GASES

The elements in group 18 (VIIIA), long designated the inert or rare gases, no longer satisfy these early labels; they are now known to have an interesting, although somewhat limited, chemistry, and they are rather abundant. Helium, for example, is the second most abundant element in the universe, and argon is the third most abundant component of dry air, approximately 30 times as abundant by volume as carbon dioxide.

7-10-1 THE ELEMENTS

The first experimental evidence for the noble gases was obtained by Henry Cavendish in 1766. In a series of experiments on air, he was able to sequentially remove nitrogen (then known as "phlogisticated air"), oxygen ("dephlogisticated air"), and carbon dioxide ("fixed air") from air by chemical means, but a small residue, no more than one part in 120, resisted all attempts at reaction.[19] The nature of Cavendish's unreactive fraction of air remained a mystery for more than a century. This fraction was, of course, eventually shown to be a mixture of argon and other noble gases.[20]

During a solar eclipse in 1868, a new emission line, matching no known element, was found in the spectrum of the solar corona. J. N. Locklear and E. Frankland proposed the existence of a new element named, appropriately, helium (from the Greek for sun). The same spectral line was subsequently observed in the gases of Mt. Vesuvius.

In the early 1890s, Lord Rayleigh and William Ramsay observed a discrepancy in the apparent density of nitrogen isolated from air and from ammonia. The two researchers independently performed painstaking experiments to isolate and characterize what seemed either a new form of nitrogen (the formula N_3 was one suggestion) or a new element. Eventually the two worked cooperatively, with Ramsay apparently the first to suggest that the

[19] H. Cavendish, *Phil. Trans.*, **1785**, *75*, 372.

[20] Cavendish's experiments and other early developments in noble gas chemistry are described in E. N. Hiebert, "Historical Remarks on the Discovery of Argon: The First Noble Gas" in *Noble Gas Compounds*, H. H. Hyman, ed., University of Chicago Press, Chicago, 1963, pp. 3–20.

unknown gas might fit into the periodic table after the element fluorine. In 1895 they reported the details of their experiments and evidence for the element they had isolated, argon (Gr., no work or lazy).[21]

Within three years, Ramsay and M. W. Travers had isolated three additional elements by low-temperature distillation of liquid air, neon (Gr., new), krypton (Gr., concealed), and xenon (Gr., strange). The last of the noble gases, radon, was isolated as a nuclear decay product in 1902.

Helium is fairly rare on earth, but it is the second most abundant element in the universe (76% H, 23% He) and is a major component of stars. The other noble gases, with the exception of radon, are present in small amounts in air (see Table 7-17); they are commonly obtained by fractional distillation of liquid air. Helium is used as an inert atmosphere for arc welding, in weather and other balloons, and in gas mixtures used in deep-sea diving (where it gives a Donald Duck-like sound to voices). Recently, liquid helium (with a temperature of 4.2 K) has received increasing use as coolant for superconducting magnets in nuclear magnetic resonance instruments. Argon, the least expensive noble gas, is commonly used as an inert atmosphere for studying chemical reactions and for high-temperature metallurgical processes. It is also used for filling incandescent light bulbs. One useful property of the noble gases is that they emit light of vivid colors when an electrical discharge is passed through them, with neon's emission spectrum, for example, responsible for the bright orange-red of neon signs. Other noble gases are also used in discharge tubes, with the color dependent on the gases used. All isotopes of radon are radioactive; the longest-lived isotope, ^{222}Rn, has a half-life of only 3.825 days. Despite this short half-life, there has been recent concern over the level of radon in many homes. A potential cause of lung cancer, radon is formed from decay of trace amounts of uranium in certain rock formations and may enter homes through basement walls and floors.

Important properties of the noble gases are summarized in Table 7-17.

TABLE 7-17
Properties of the group 18 (VIIIA) elements: the noble gases

Element	*Ionization energy (kJ mol^{-1})*	*Melting point (°C)*	*Boiling point (°C)*	*Enthalpy of vaporization (kJ mol^{-1})*	*Abundance in dry air (% by volume)*
He	2372	*	−268.93	0.08	0.000524
Ne	2081	−248.61	−246.06	1.74	0.001818
Ar	1521	−189.37	−185.86	6.52	0.934
Kr	1351	−157.20	−153.35	9.05	0.000114
Xe	1170	−111.80	−108.13	12.65	0.0000087
Rn	1037	−71	−62	18.1	Trace

SOURCES: See Table 7-3.
NOTE: * Helium cannot be frozen at 1 atm pressure.

7-10-2 CHEMISTRY

For many years these elements were known as the inert gases because they were believed to be totally unreactive as a consequence of the very stable octet valence electron configurations of their atoms. Their chemistry was simple: they had none!

[21] Lord Rayleigh and W. Ramsay, *Phil. Trans., A.*, **1895**, *186*, 187.

The first chemical compounds containing noble gases were **clathrates**, cage compounds in which noble gas atoms could be trapped. Experiments begun in the late 1940s showed that when water or solutions containing hydroquinone (*p*-dihydroxybenzene, $HO—C_6H_4—OH$) were crystallized under pressures of certain gases, hydrogen-bonded lattices having rather large cavities could be formed, with gas molecules of suitable size trapped in the cavities. Clathrates containing the noble gases argon, krypton, and xenon, as well as those containing small molecules as SO_2, CH_4, and O_2, have been prepared. No clathrates have been found for helium and neon; these atoms are simply too small to be trapped.

Even though clathrates of three of the noble gases had been prepared, at the beginning of the 1960s no compounds containing covalently bonded noble gas atoms had been synthesized. Attempts had been made to react xenon with elemental fluorine, the most reactive of the elements, but without apparent success. In 1962, however, this situation changed dramatically. Neil Bartlett had observed that the compound PtF_6 changed color on exposure to air. With D. H. Lohmann, he demonstrated that PtF_6 was serving as a very strong oxidizing agent in this reaction and that the color change was due to the formation of $O_2^+[PtF_6]^-$.[22] Bartlett noted the similarity of the ionization energies of xenon (1169 kJ mol^{-1}) and O_2 (1175 kJ mol^{-1}) and repeated the experiment reacting Xe with PtF_6. He observed a color change from the deep red of PtF_6 to orange-yellow and reported the product as $Xe^+[PtF_6]^-$.[23] While the product of this reaction later proved to be a complex mixture of several xenon compounds (see below), these were the first covalently bonded noble gas compounds to be synthesized, and their discovery stimulated study of the chemistry of the noble gases in earnest. In a matter of months the compounds XeF_2 and XeF_4 had been characterized, and other noble gas compounds soon followed.[24]

Dozens of compounds of noble gas elements are now known, although the number remains modest in comparison with the other groups. The known stable compounds involve only the elements krypton, xenon, and radon (although argon forms some clathrates, no stable covalently bonded compounds of argon are known). Transient species containing helium and neon, as well as the other noble gases, have been observed using mass spectrometry. However, most of the stable noble gas compounds are those of xenon with the highly electronegative elements F, O, and Cl; a few compounds have also been reported with Xe—N, Xe—C, and even Xe—Cr bonds. Some of the compounds and ions of the noble gases are shown in Table 7-18.

Several of these compounds and ions have interesting structures which have provided tests for models of bonding. For example, the xenon fluorides' structures have been interpreted on the basis of the VSEPR model (Figure 7-23). XeF_2 and XeF_4 have structures entirely in accord with their VSEPR descriptions: XeF_2 is linear (three lone pairs on Xe) and XeF_4 planar (two lone pairs), as shown in Figure 7-22. XeF_6 and $[XeF_8]^{2-}$, on the other hand, are more difficult to interpret by VSEPR. Each has a single lone pair on the central xenon. The VSEPR model would predict this lone pair to occupy a definite position on the xenon, as do single lone pairs in such molecules as NH_3, SF_4, and IF_5. However, no definite location is found for the central lone pair of

[22] N. Bartlett and D. H. Lohmann, *Proc. Chem. Soc.*, **1962**, 115.

[23] N. Bartlett, *Proc. Chem. Soc.*, **1962**, 218.

[24] For a recent discussion of the development of the chemistry of xenon compounds, see P. Laszlo and G. L. Schrobilgen, *Angew. Chem. Int. Ed. Engl.*, **1988**, *27*, 479.

TABLE 7-18
Noble gas compounds and ions

Formal oxidation state of noble gas	*Number of lone pairs on central atom**	*Compounds and ions*			
+2	3	KrF^+	XeF^+		
		KrF_2	XeF_2		
+4	2		XeF_3^+		
			XeF_4	$XeOF_2$	
+6	1		XeF_5^+	$XeOF_4$	XeO_3
			XeF_6	XeO_2F_2	
			XeF_7^-	$KXeO_3F$	
			XeF_8^{2-}	$CsXeOF_5$	
+8	0			XeO_3F_2	XeO_4
					XeO_6^{4-}

NOTE: * In each case the noble gas is the central atom.

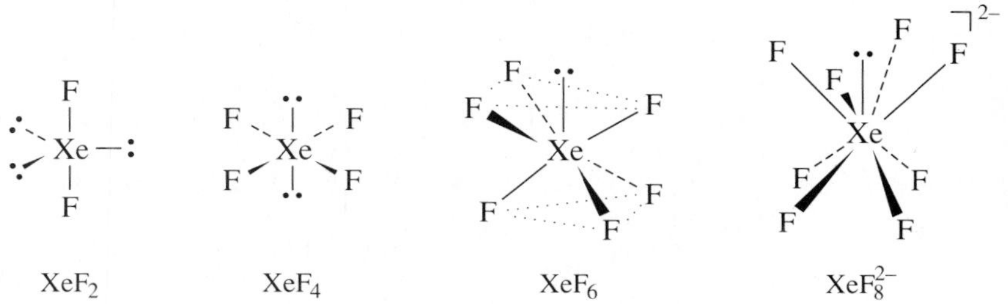

FIGURE 7-23 Structures of Xenon Fluorides.

XeF_6 or $[XeF_8]^{2-}$. One explanation is based on the degree of crowding around xenon. With a large number of fluorines attached to the central atom, repulsions between the electrons in the xenon–fluorine bonds are strong—too strong to enable a lone pair to itself occupy a well-defined position. The central lone pair does play a role, however. In XeF_6 the structure is not octahedral, but rather somewhat distorted as a consequence of the presence of the lone pair on xenon. Although the structure of XeF_6 in the gas phase has been very difficult to determine, spectroscopic evidence indicates that the lowest-energy form has C_{3v} symmetry, as shown in Figure 7-23. This is not a rigid structure, however; the molecule apparently undergoes rapid rearrangement from one C_{3v} structure to another (the lone pair appearing to move from the center of one face to another) by way of intermediates having C_{2v} or other symmetry.[25] Solid XeF_6 contains at least four phases, consisting of square pyramidal XeF_5^+ ions bridged by fluoride ions, as shown for one of the phases in Figure 7-24.[26]

The structure of XeF_8^{2-} is also distorted, but very slightly. As shown in Figure 7-23, XeF_8^{2-} is very nearly a square antiprism (D_{4d} symmetry), but one face is slightly larger than the opposite face (resulting in approximate C_{4v} symmetry).[27] While it is possible that this distortion is a consequence of the

[25] K. Seppelt and D. Lentz, "Novel Developments in Noble Gas Chemistry," *Progr. Inorg. Chem.*, Vol. 29, Wiley 1982, New York, pp. 172–80; E. A. V. Ebsworth, D. W. H. Rankin, and S. Cradock, *Structural Methods in Inorganic Chemistry*, Blackwell Scientific Publications, Oxford, England, 1987.

[26] R. D. Burbank and G. R. Jones, *J. Am. Chem. Soc.*, **1974**, *96*, 43.

[27] S. W. Peterson, J. H. Holloway, B. A. Coyle, and J. M. Williams, *Science*, **1971**, *173*, 1238.

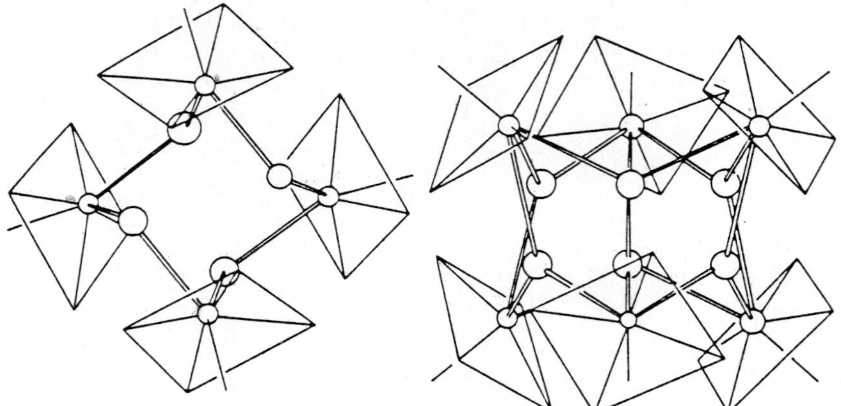

FIGURE 7-24 Xenon Hexafluoride (Crystalline Forms). (Reproduced with permission from R. D. Burbank and G. R. Jones, *J. Am. Chem. Soc.*, **1974**, *96*, 43. Copyright 1974 American Chemical Society.)

way in which these ions pack in the crystal, it is also possible that the distortion is caused by a lone pair exerting some influence on the size of the larger face.[28]

Positive ions containing xenon are also known. For example, Bartlett's original reaction of xenon with PtF_6 is now believed to proceed as follows:

$$Xe + 2\,PtF_6 \longrightarrow [XeF]^+\,[PtF_6]^- + PtF_5 \longrightarrow [XeF]^+\,[Pt_2F_{11}]^-$$

The ion XeF^+ does not ordinarily occur as a discrete ion but rather is attached covalently to a fluorine on the anion; an example, $[XeF]^+\,[RuF_6]^-$, is shown in Figure 7-25.[29]

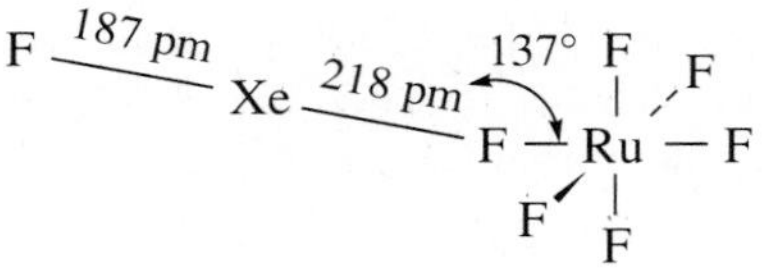

FIGURE 7-25 $[XeF]^+[RuF_6]^-$. (Data from N. Bartlett, M. Gennis, D. D. Gibler, B. K. Morrell, and A. Zalkin, *Inorg. Chem.*, **1973,** *12,* 1717.)

F —187 pm— Xe —218 pm— F — Ru(F)₅, 137°

Several reactions of the noble gas compounds are worth noting. Interest in using noble gas compounds as reagents in organic and inorganic synthesis has been stimulated in part because the byproduct of such reactions is often the (usually) chemically inert noble gas itself. The xenon fluorides XeF_2, XeF_4, and XeF_6 have been used as fluorinating agents for both organic and inorganic compounds. For example,

$$2\,SF_4 + XeF_4 \longrightarrow 2\,SF_6 + Xe$$
$$2\,C_6H_6 + XeF_2 \longrightarrow 2\,C_6H_5F + Xe + H_2$$

XeF_4 is also capable of selectively fluorinating aromatic positions in arenes such as toluene.

[28] The effect of lone pairs can be difficult to predict. For examples of sterically active and inactive lone pairs in ions of formula AX_6^{n-}, see K. O. Christe and W. Wilson, *Inorg. Chem.*, **1989**, *28*, 3275, and references therein.

[29] N. Bartlett, M. Gennis, D. D. Gibler, B. K. Morrell, and A. Zalkin, *Inorg. Chem.*, **1973**, *12*, 1717.

The oxides XeO_3 and XeO_4 are extremely explosive and must be handled under special precautions. XeO_3 is a powerful oxidizing agent in aqueous solution: the electrode potential of the half-reaction

$$XeO_3 + 6\,H^+ + 6e^- \longrightarrow Xe + 3\,H_2O$$

is 2.10 V. In basic solution, XeO_3 forms $HXeO_4^-$:

$$XeO_3 + OH^- \rightleftharpoons HXeO_4^-, \qquad K = 1.5 \times 10^{-3}$$

which subsequently disproportionates to form the perxenate ion, XeO_6^{4-}

$$2\,HXeO_4^- + 2\,OH^- \longrightarrow XeO_6^{4-} + Xe + O_2 + 2\,H_2O$$

The perxenate ion is an even more powerful oxidizing agent than XeO_3 and is capable of oxidizing Mn^{2+} to permanganate, MnO_4^-, in acidic solution.

The chemistry of krypton is much more limited, with less than a dozen compounds reported to date. The only neutral halide is KrF_2, and until the recent report of the synthesis of $F{-}Kr{-}N{\equiv}CH^+AsF_6^-$,[30] the only known bonds of krypton were with fluorine. The radioactivity of radon has made the study of its chemistry difficult; RnF_2 and a few other compounds have been observed through tracer studies.

GENERAL REFERENCES

More detailed descriptions of the chemistry of the main group elements can be found in N. N. Greenwood and A. Earnshaw, *Chemistry of the Elements*, Pergamon Press, Elmsford, N.Y., 1984, and F. A. Cotton and G. Wilkinson, *Advanced Inorganic Chemistry*, 5th ed., Wiley-Interscience, New York, 1988. A handy reference on the properties of the elements themselves, including many physical properties, is J. Emsley, *The Elements*, Oxford University Press, New York, 1989. For extensive structural information on inorganic compounds, see A. F. Wells, *Structural Inorganic Chemistry*, 5th ed., Oxford University Press, New York, 1984. Two useful references on the chemistry of nonmetals are P. Powell and P. Timms, *The Chemistry of the Nonmetals*, Chapman and Hall, London, 1974, and R. Steudel, *Chemistry of the Non-Metals*, Walter de Gruyter, Berlin, 1976 (English edition by F. C. Nachod and J. J. Zuckerman). The most complete reference on chemistry of the main group compounds through the early 1970s is the five-volume set *Comprehensive Inorganic Chemistry*, Pergamon Press, Elmsford, N.Y., 1973. We encourage the reader to consult these references to supplement the information in this chapter.

PROBLEMS

7-1 The ions H_2^+ and H_3^+ have been observed in gas discharges.

a. H_2^+ has been reported to have a bond distance of 106 pm and a bond dissociation enthalpy of 255 kJ mol^{-1}. Comparable values for the neutral molecule are 74.2 pm and 436 kJ mol^{-1}. Are these values for H_2^+ in agreement with the molecular orbital picture of this ion? Explain.

b. Assuming H_3^+ to be triangular (the believed geometry), describe the molecular orbitals of this ion and determine the expected H—H bond order.

7-2 The species He_2^+ and HeH^+ have been observed spectroscopically. Prepare molecular orbital diagrams for these two ions. What would you predict for the bond order of each?

[30] P. J. MacDougall, G. J. Schrobilgen, and R. F. W. Bader, *Inorg. Chem.*, **1989**, *28*, 763.

7-3 The equilibrium constant for the formation of the cryptand [Sr(cryptand[2.2.1])]$^{2+}$ is larger than the equilibrium constants for the analogous calcium and barium cryptands. Suggest an explanation. [See E. Kauffman, J.-M. Lehn, and J.-P. Sauvage, *Helv. Chim. Acta.*, **1976**, *59*, 1099.]

7-4 Gas-phase BeF_2 is monomeric and linear. Prepare a molecular orbital description of the bonding in BeF_2.

7-5 In the gas phase, $BeCl_2$ forms a dimer of structure: Describe the bonding of the chlorine bridges in this dimer in molecular orbital terms.

Cl — Be(μ-Cl)$_2$Be — Cl

7-6 BF can be obtained by reaction of BF_3 with boron at 1850°C and low pressure; BF is highly reactive but can be preserved at liquid nitrogen temperature (77 K). Prepare a molecular orbital diagram of BF. How would the molecular orbitals of BF differ from CO, with which BF is isoelectronic?

7-7 $Al_2(CH_3)_6$ is isostructural with diborane, B_2H_6. Describe the Al—C—Al bonding for the bridging methyl groups in $Al_2(CH_3)_6$ in molecular orbital terms.

7-8 Referring to the description of bonding in diborane in Figure 7-10:
a. Show that the representation Γ (sp^3) reduces to $A_g + B_{2g} + B_{1u} + B_{3u}$.
b. Show that the representation Γ ($1s$) reduces to $A_g + B_{3u}$.
c. Using the D_{2h} character table, verify that the sketches for the group orbitals match their respective symmetry designations (A_g, B_{2g}, B_{1u}, B_{3u}).

7-9 The compound $C(PPh_3)_2$ is bent at carbon; the P—C—P angle in one form of this compound has been reported as 130.1°. Account for the nonlinearity at carbon.

7-10 The C—C distances in carbides of formula MC_2 are in the range of 119 to 124 pm if M is a group 2 (IIA) metal or other metal commonly forming a 2+ ion but in the approximate range of 128 to 130 pm for group 3 (IIIB) metals, including the lanthanides. Why is the C—C distance greater for the carbides of the group 3 metals?

7-11 The half-life of ^{14}C is 5730 years. A sample taken for radiocarbon dating was found to contain 56% of its original ^{14}C. What was the age of the sample? (*Note:* Radioactive decay of ^{14}C follows first-order kinetics.)

7-12 Explain the increasing stability of the 2+ oxidation state for the group 14 (IVA) elements with increasing atomic number.

7-13 The reaction P_4 (g) $\rightleftharpoons$ 2 P_2 (g) has ΔH = 217 kJ mol^{-1}. If the bond energy of a single phosphorus–phosphorus bond is 200 kJ mol^{-1}, calculate the bond energy of the P≡P bond. Compare the value you obtain with the bond energy in N_2 (942 kJ mol^{-1}), and suggest an explanation for the difference in bond energies in P_2 and N_2.

7-14 The azide ion, N_3^-, is linear, with equal N—N bond distances.
a. Describe the pi molecular orbitals of azide.
b. Describe, in HOMO–LUMO terms, the reaction between azide and H^+ to form hydrazoic acid, HN_3.
c. The N—N bond distances in HN_3 are given in Figure 7-19. Explain why the terminal N—N distance is shorter than the central N—N distance in this molecule.

7-15 In aqueous solution, hydrazine is a weaker base than ammonia. Why? (pK_b values at 25°C: NH_3, 4.74; N_2H_4, 6.07.)

7-16 The bond angles for the hydrides of the group 15 (VA) elements are as follows: NH_3, 106.6°; PH_3, 93.6°; AsH_3, 91.8°; and SbH_3, 91.3°. Account for this trend.

7-17 Gas-phase measurements show that the nitric acid molecule is planar. Account for the planarity of this molecule.

7-18 With the exception of NO_4^{3-}, all the molecules and ions in Table 7-9 are planar. Assign their point groups.

7-19 The sulfur–sulfur distance in S_2, the major component of sulfur vapor above ~720°C, is 189 pm, significantly shorter than the sulfur–sulfur distance of 206 pm in S_8. Suggest an explanation for the shorter distance in S_2. (See C. L. Liao and C. Y. Ng, *J. Chem. Phys.*, **1986**, *84*, 778.)

7-20 Although sulfur forms the fluorides SF_2, SF_4, and SF_6, the only oxygen compound of similar formula is OF_2. Suggest why no stable compounds of formulas OF_4 and OF_6 are known.

7-21 Because of its high reactivity with most chemical reagents, F_2 is ordinarily synthesized electrochemically. However, the chemical synthesis of F_2 has recently been reported via the reaction

$$2\ K_2MnF_6 + 4\ SbF_5 \longrightarrow 4\ KSbF_6 + 2\ MnF_3 + F_2$$

This reaction can be viewed as a Lewis acid–base reaction. Explain. (See K. O. Christe, *Inorg. Chem.*, **1986**, *25*, 3721.)

7-22 The triiodide ion, I_3^-, is linear, while I_3^+ is bent. Explain.

7-23 While B_2H_6 has D_{2h} symmetry, I_2Cl_6 is planar. Account for the difference in the structures of these two molecules.

7-24 BrF_3 undergoes autodissociation according to the equilibrium:

$$2\ BrF_3 \leftrightarrows BrF_2^+ + BrF_4^-$$

Ionic fluorides such as KF behave as bases in BrF_3, while some covalent fluorides such as SbF_5 behave as acids. On the basis of the solvent system concept, write balanced chemical equations for these acid–base reactions of fluorides with BrF_3.

7-25 The diatomic cations Br_2^+ and I_2^+ are both known.

a. On the basis of the molecular orbital model, what would you predict for the bond orders of these ions? Would you predict these cations to have longer or shorter bonds than the corresponding neutral diatomic molecules?

b. Br_2^+ is red, I_2^+ bright blue. What electronic transition is probably responsible for absorption in these ions? Which ion has the more closely spaced HOMO and LUMO?

7-26 I_2^+ exists in equilibrium with its dimer I_4^{2+} in solution. I_2^+ is paramagnetic, the dimer is diamagnetic. Crystal structures of compounds containing I_4^{2+} have shown this ion to be planar and rectangular, with two short I—I distances (258 pm) and two longer distances (326 pm). Using molecular orbitals, propose an explanation for the interaction between two I_2^+ units to form I_4^{2+}.

7-27 Bartlett's original reaction of xenon with PtF_6 apparently yielded products other than the expected $Xe^+PtF_6^-$. However, when xenon and PtF_6 react in the presence of a large excess of sulfur hexafluoride, $Xe^+PtF_6^-$ is apparently formed. Suggest the function of SF_6 in this reaction. (See K. Seppelt and D. Lentz, "Novel Developments in Noble Gas Chemistry," *Progr. Inorg. Chem.*, Vol. 29, Wiley, New York, 1982, pp. 170–71.)

7-28 On the basis of VSEPR, predict the structures of $XeOF_2$, $XeOF_4$, XeO_2F_2, and XeO_3F_2. Assign the point group of each.

7-29 The sigma bonding in the linear molecule XeF_2 may be described as a 3-center, 4-electron bond. If the z axis is assigned as the internuclear axis, use the p_z orbitals on each atom to prepare a molecular orbital description of the sigma bonding in XeF_2.

7-30 The $OTeF_5$ group is able to stabilize compounds of xenon in formal oxidation states (IV) and (VI). On the basis of VSEPR, predict the structures of $Xe(OTeF_5)_4$ and $O{=}Xe(OTeF_5)_4$.

7-31 Write a balanced equation for the oxidation of Mn^{2+} to MnO_4^- by the perxenate ion in acidic solution (assume that neutral Xe is formed).

8

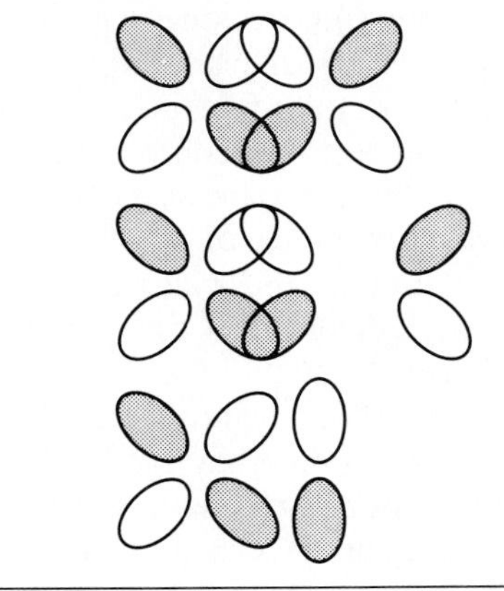

Coordination Chemistry I: Bonding

Coordination compounds, as the term is usually used in inorganic chemistry, include compounds composed of a metal atom or ion and one or more **ligands** (atoms, ions, or molecules) that can formally be thought of as donating electrons to the metal. This definition includes compounds with metal–carbon bonds; these are called **organometallic compounds,** and are described in Chapters 12 to 14. The name coordination compound comes from the coordinate covalent bond, which historically was considered to form by donation of a pair of electrons from one atom to another. Since these compounds are usually formed by donation of electron pairs of ligands to metals, the name is appropriate. Coordinate covalent bonds are identical to covalent bonds formally formed by combining one electron from each atom; only the formal electron counting distinguishes them. Coordination compounds are also acid–base adducts, as described in Chapter 6, and are frequently called **complexes** or, if charged, **complex ions.**

8-1 HISTORY[1]

Although the history of bonding and the interpretation of reactions of coordination compounds really begins with Alfred Werner (1866–1919), coordination compounds were known much earlier. Many coordination compounds have been used as pigments since antiquity. Examples still in use include Prussian

[1] G. B. Kauffman, "General Historical Survey to 1930," and J. C. Bailar, Jr., "Development of Coordination Chemistry since 1930," in G. Wilkinson, ed., *Comprehensive Coordination Chemistry,* Pergamon, Elmsford, N.Y., 1987, Vol. 1, pp. 1–30.

blue ($KFe[Fe(CN)_6]$), aureolin ($K_3[Co(NO_2)_6] \cdot 6\ H_2O$, yellow), and alizarin red dye (the calcium aluminum salt of 1,2-dihydroxy-9,10-anthraquinone). The striking colors of compounds such as these and their color changes on reaction were described in very early documents and provided impetus for further studies. The ion known today as tetramminecopper (II) (actually $[Cu(NH_3)_4(H_2O)_2]^{2+}$ in solution), which has a striking royal blue color, was certainly known in prehistoric times. With the gradual development of analytical methods, the formulas of many of these compounds became known late in the nineteenth century, and theories of structure and bonding became possible.

Inorganic chemists tried to use the advances in organic bonding theory and the simple ideas of ionic charges to explain bonding in coordination compounds, but found that the theories were inadequate. In a compound such as hexamminecobalt (III) chloride, $[Co(NH_3)_6]Cl_3$, the early bonding theories allowed only three atoms to be attached to the cobalt (because of its "valence" of 3). By analogy with ordinary salts, such as $FeCl_3$, the chlorides were assigned this role. This left the six ammonia molecules with no means of participating in bonding, and it was necessary to develop new ideas to explain the structure. One theory, proposed first by C. W. Blomstrand[2] (1826–1894) and developed further by S. M. Jørgensen[3] (1837–1914), was that the nitrogens could form chains much like those of carbon (and thus could have a valence of 5), and that chloride ions attached directly to cobalt were bonded more strongly than those bonded to nitrogen. Alfred Werner[4] (1866–1919) proposed instead that all six ammonias could bond directly to the cobalt ion. Werner allowed for a looser bonding of the chloride ions; we now consider them as independent ions. The series of compounds in Table 8-1 illustrates how both the chain theory and Werner's coordination theory predict the number of ions to be formed by a series of cobalt complexes. Blomstrand's theory allowed dissociation of chlorides attached to ammonia, but not of chlorides attached directly to cobalt. Werner's theory also included two kinds of chlorides, but the number attached to the cobalt (and therefore unavailable as ions) and the number of ammonia molecules totaled six. The other chlorides were considered less firmly bound and could therefore form ions in solution. We now consider them to be ions in the solid state as well.

Except for the last compound in the table, the predictions match, and the ionic behavior does not distinguish between them. Even with the last compound, problems with purity and conductance measurements left some ambiguity. The argument between Jørgensen and Werner continued for many years, each presenting data and explanations favoring his own position. This case illustrates some of the good features of such controversy; Werner was forced to develop his theory further and synthesize new compounds to test his ideas because Jørgensen defended the earlier theory so vigorously. Werner proposed an octahedral structure for compounds like those in Table 8-1. He prepared and characterized many isomers, including both green and violet forms of $[Co(H_2NC_2H_4NH_2)_2Cl_2]^+$. He claimed that these compounds had the

[2] C. W. Blomstrand, *Berichte*, **1871,** *4,* 40; translated by G. B. Kauffman, *Classics in Coordination Chemistry, Part 2,* Dover, New York, 1976, pp. 75–93.

[3] S. M. Jørgensen, *Zeit. Anorg. Chem.*, **1899,** *19,* 109; translated by G. B. Kauffman, op. cit., pp. 94–164.

[4] A. Werner, *Zeit. Anorg. Chem.*, **1893,** *3,* 267; *Berichte*, **1907,** *40,* 4817; **1911,** *44,* 1887; **1914,** *47,* 3087; A. Werner and A. Miolati, *Zeit. Phys. Chem.*, **1893,** *12,* 35; **1894,** *14,* 506; all translated by George B. Kauffman, *Classics in Coordination Chemistry, Part 1,* Dover, N.Y., 1968.

TABLE 8-1
Comparison of Blomstrand chain theory and Werner's coordination theory

Werner formula (modern form)	*Number of ions predicted*	*Blomstrand chain formula*	*Number of ions predicted*
$[Co(NH_3)_6]\mathit{Cl_3}$	4	NH_3—*Cl* Co—NH_3—NH_3—NH_3—NH_3—*Cl* NH_3—*Cl*	4
$[Co(NH_3)_5Cl]\mathit{Cl_2}$	3	NH_3—*Cl* Co—NH_3—NH_3—NH_3—NH_3—*Cl* Cl	3
$[Co(NH_3)_4Cl_2]\mathit{Cl}$	2	Cl Co—NH_3—NH_3—NH_3—NH_3—*Cl* Cl	2
$[Co(NH_3)_3Cl_3]$	0	Cl Co—NH_3—NH_3—NH_3—*Cl* Cl	2

NOTE: The italicized chlorides dissociate in solution, according to the two theories.

chlorides arranged ***trans*** (opposite each other) and ***cis*** (adjacent to each other), respectively, in an overall octahedral geometry, as in Figure 8-1. Jørgensen offered alternative isomeric structures, but finally conceded defeat in 1907, when Werner suceeded in synthesizing the green *trans* and the violet *cis* isomers of $[Co(NH_3)_4Cl_2]^+$, for which there were no counterparts in the chain theory.

However, even synthesis of this compound and the later discovery of optically active coordination compounds did not completely convince all chemists, although such compounds could not be explained directly by the chain theory. It was argued that Werner's optically active compounds still

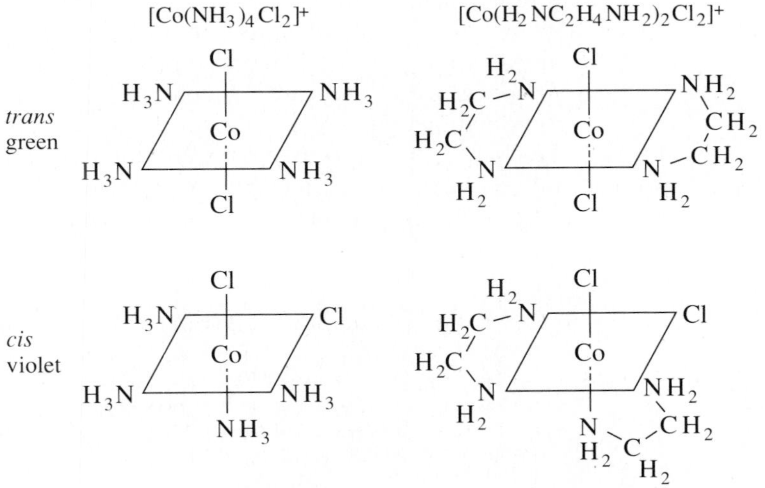

$[Co(NH_3)_4Cl_2]^+$ $[Co(H_2NC_2H_4NH_2)_2Cl_2]^+$

trans green

cis violet

FIGURE 8-1 *Cis* and *Trans* Isomers.

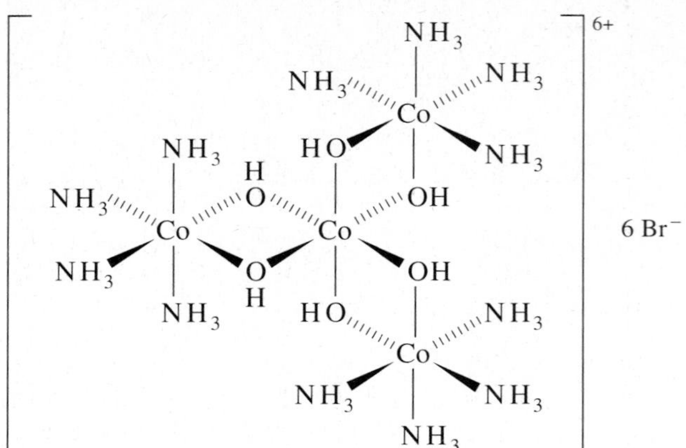

FIGURE 8-2 Werner's Totally Inorganic Optically Active Compound, $[Co(Co(NH_3)_4(OH)_2)_3]Br_6$.

contained carbon and that their chirality could be due to the carbon atoms. Finally, Werner resolved the compound $[Co(Co(NH_3)_4(OH)_2)_3]Br_6$ (Figure 8-2), initially prepared by Jørgensen, into its two optically active forms, using *d*- and *l*-α-bromocamphor-π-sulfonate as the resolving agents. With this final proof of optical activity without carbon, the validity of Werner's theory was finally accepted. Pauling[5] extended the theory in terms of hybrid orbitals, and more recent theories[6] have adapted arguments first used for electronic structures of ions in crystals to coordination compounds.

The Werner theory of coordination compounds was based on a group of compounds that is relatively slow to react in solution, and thus easier to study. For this reason, many of his examples were compounds of Co(III), Rh(III), Cr(III), Pt(II), and Pt(IV), which are kinetically inert, or slow to react. Examination of more reactive compounds over the years has confirmed their similarity to those originally studied, so we will include examples of both kinds in the descriptions that follow. Werner's theory required two kinds of bonding in the compound, a *primary* one in which the positive charge of the central metal ion is balanced by negative ions in the compound, and a *secondary* one in which molecules or ions (known collectively as **ligands**) are attached directly to the transition metal ion. The secondary bonded unit has been given many different names, such as the **complex ion** or the **coordination sphere,** and the formula is written with this part in brackets. Current practice considers this coordination sphere the more important, so the words primary and secondary no longer bear the same significance. In the examples in Table 8-1, the coordination sphere acts as a unit, while the ions outside the brackets balance the charge and are free ions in solution. Depending on the nature of the metal and the ligands, the metal can have from one up to at least 16 atoms attached to it, with 4 and 6 the most common numbers.[7] Additional water molecules

[5] L. Pauling, *J. Chem. Soc.*, **1948,** 1461; *The Nature of the Chemical Bond,* 3rd ed., Cornell University Press, Ithaca, N.Y., 1960, pp. 145–82.

[6] J. S. Griffith and L. E. Orgel, *Quart. Revs.*, **1957,** *XI,* 381.

[7] N. N. Greenwood and A. Earnshaw, *Chemistry of the Elements,* Pergamon, Elmsford, N.Y., 1984, p. 1077. The larger numbers depend on how the number of donors in organometallic compounds are counted; some would assign smaller coordination numbers because of the special nature of the organic ligands.

may be added to the coordination sphere when the compound is dissolved in water. We should include them specifically in the description of the compound, but in some cases they are omitted in order to concentrate on the other ligands. The discussion that follows concentrates on the coordination sphere; the other ions associated with it can frequently vary without changing the bonding.

Werner used compounds with four or six ligands in developing his theories, with the shapes of the coordination compounds established by synthesis of isomers. For example, he was able to synthesize only two isomers of the $[Co(NH_3)_4Cl_2]^+$ ion. The possible structures with six ligands are octahedral, trigonal prismatic, trigonal antiprismatic, and hexagonal (either planar or pyramidal). Since there are two possible isomers for the octahedral shape and three for each of the others, as shown in Figure 8-3, Werner claimed that the structure was octahedral. Such an argument cannot be conclusive, since a missing isomer may simply be difficult to synthesize or isolate. However, later experiments confirmed the octahedral shape, with *cis* and *trans* isomers as shown in Figure 8-3.

Werner's synthesis and separation of optical isomers proved the octahedral shape conclusively, since none of the other 6-coordinate geometries could have similar optical activity.

In a similar way, other experiments were consistent with square planar Pt(II) compounds, with the four ligands at the corners of a square. Only two isomers are found for $[Pt(NH_3)_2Cl_2]$. Although the two could have had different shapes (tetrahedral and square planar, for example), Werner assumed that they had the same overall shape; and since only one tetrahedral structure is possible for this compound, he argued that they must be square planar with *cis* and *trans* geometries. Again, his arguments were correct, although the evidence he presented could not be conclusive.

After Werner's evidence for the octahedral and square planar natures of many complexes, it was clear that any acceptable theory needed to account for bonds between ligands and metals and that the number of bonds required was more than commonly accepted at that time. Transition metal compounds with six ligands, for example, cannot fit the simple Lewis theory with 8 electrons around each atom, and even expanding the octet to 10 or 12 electrons does not work in cases such as $Fe(CN)_6^{4-}$, with a total of 18 electrons to accommodate. In fact the **18-electron rule** is sometimes useful in accounting for the bonding in a simple way; the total number of valence electrons around the central atom is counted, with 18 a common result. This approach is more often used in organometallic compounds; it is discussed in Chapter 12.

Pauling[8] used his **valence bond** approach to explain differences in magnetic behavior among coordination compounds by use of either 3*d* or 4*d* orbitals of the metal ion. Griffith and Orgel[9] developed and popularized the use of **ligand field theory,** derived from the **crystal field theory** of Bethe[10] and Van Vleck[11] on the behavior of metal ions in crystals and from the molecular orbital treatment of Van Vleck.[12] Several of these are described later, with emphasis on the ligand field theory.

[8] Pauling, op. cit.

[9] Griffith and Orgel, op. cit.; L. E. Orgel, *An Introduction to Transition-Metal Chemistry,* Methuen, London, 1960.

[10] H. Bethe, *Ann. Physik,* **1929,** *3,* 133.

[11] J. H. Van Vleck, *Phys. Rev.,* **1932,** *41,* 208.

[12] Ibid., p. 807.

cis- and *trans*- Tetramminedichlorocobalt (III), $[Co(NH_3)_4Cl_2]^+$

Hexagonal (three isomers)

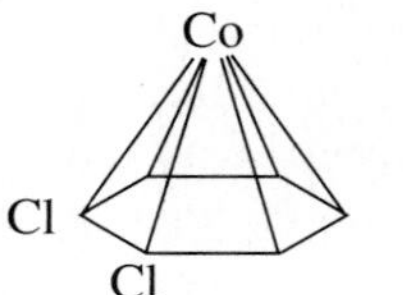

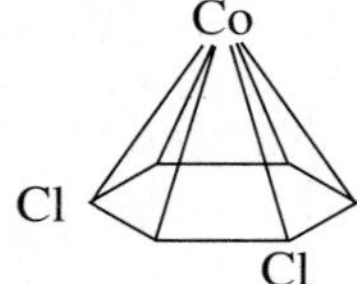

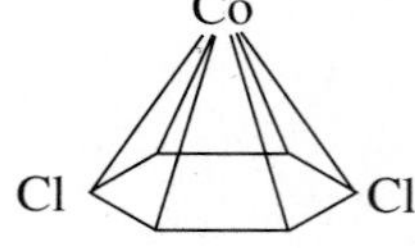

Hexagonal pyramidal (three isomers)

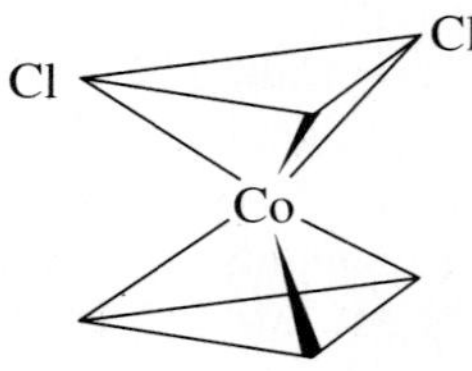

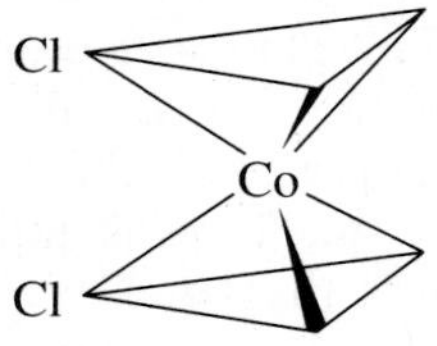

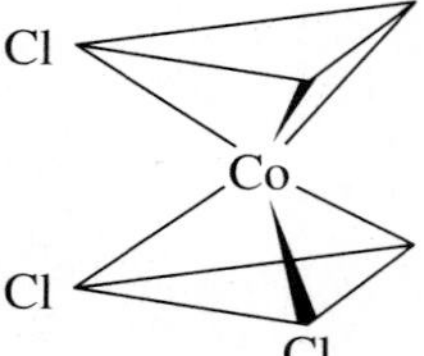

Trigonal prismatic (three isomers)

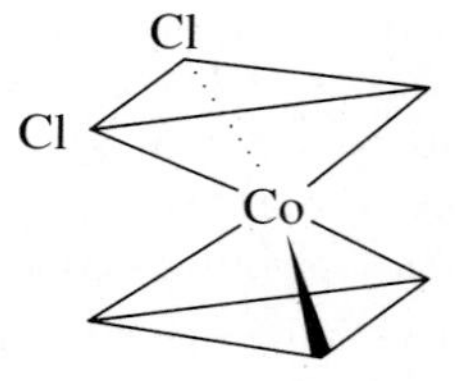

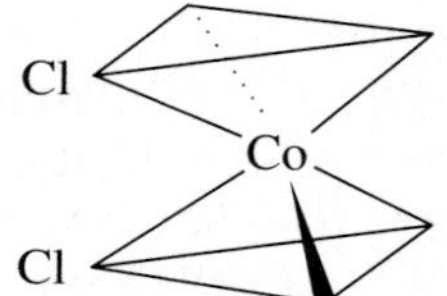

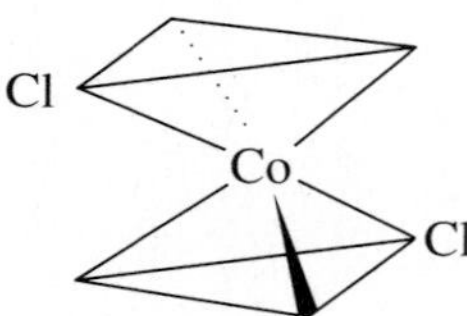

Trigonal antiprismatic (three isomers)

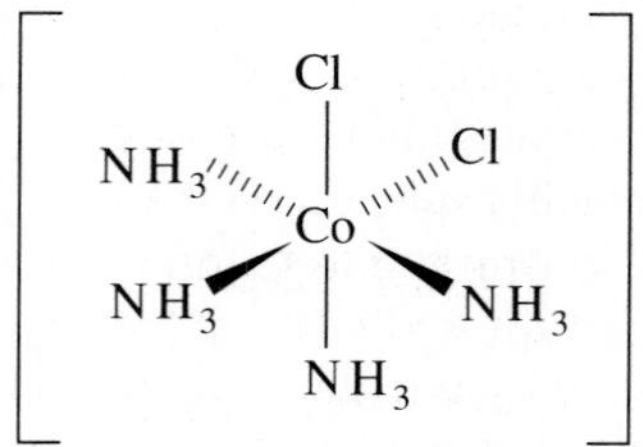

Octahedral (two isomers)

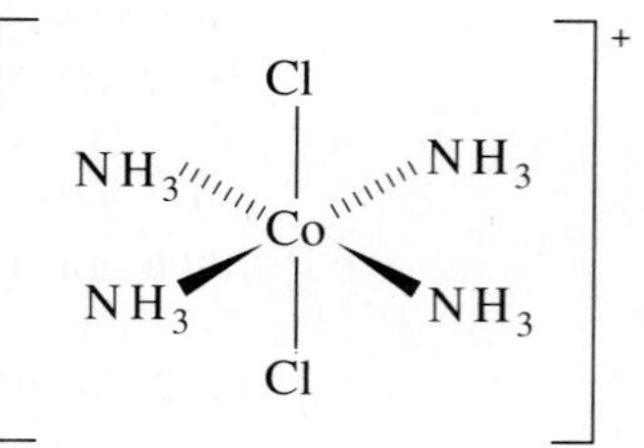

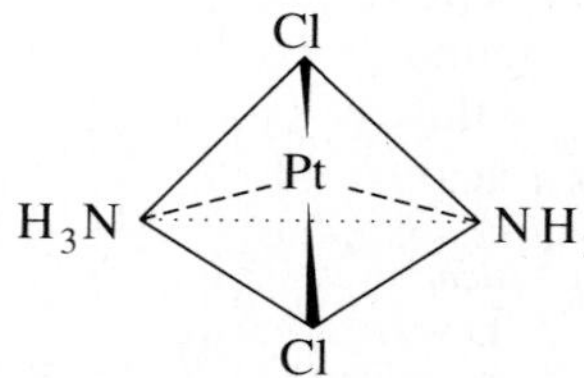

cis- and *trans*- Diamminedichloroplatinum (II), $[PtCl_2(NH_3)_2]$

Tetrahedral (one isomer)

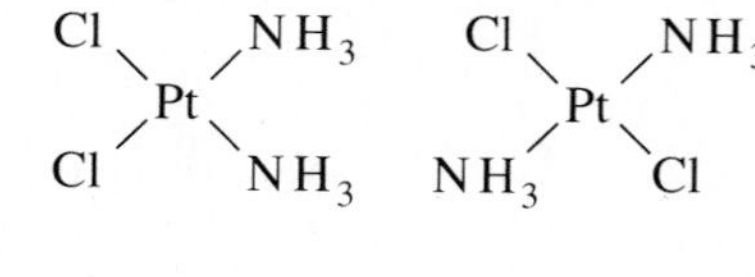

Square planar (two isomers)

FIGURE 8-3 Isomers for Different Geometries.

8-2 EXPERIMENTAL EVIDENCE FOR ELECTRONIC STRUCTURES

Any theory of bonding in coordination compounds must explain the experimental behavior of the compounds. Some of the methods most frequently used to study these compounds are described here. These, and others, have been used in the development of theories used to explain the electronic structure and bonding of coordination compounds.

8-2-1 THERMODYNAMIC DATA

Any explanation of bonding must include the energy of the compound as one of its primary goals. Experimentally, the energy is frequently not determined directly, but thermodynamic measurements of enthalpies and free energies of reaction are used to compare compounds.

Inorganic chemists, and coordination chemists in particular, frequently use **stability constants** (sometimes called **formation constants**). These are the equilibrium constants for formation of coordination compounds, usually measured in aqueous solution. Examples of the reactions and corresponding stability constant expressions include

$$Fe^{3+} + SCN^- = FeSCN^{2+} \qquad K_1 = \frac{[FeSCN^{2+}]}{[Fe^{3+}][SCN^-]}$$

$$Cu^{2+} + 4\,NH_3 = Cu(NH_3)_4^{2+}, \qquad K_4 = \frac{[Cu(NH_3)_4^{2+}]}{[Cu^{2+}][NH_3]^4}$$

As described in Chapter 6, enthalpies of reaction can be measured directly or the temperature dependence of equilibrium constants can be used to calculate enthalpies and entropies of reaction. Complicating factors such as solvent interaction with both reactants and products must be considered in any reactions in solution.

In practice, thermodynamic values rarely allow prediction of other properties of coordination compounds or absolute determination of structures or formulas. They are more valuable in considering relationships between similar compounds, such as a series of different metal ions all reacting with the same ligand or a series of different ligands reacting with the same metal ion. In such cases, correlation between thermodynamic properties and electronic structure can be made. An example is given in Section 8-4-2, where hydration enthalpies are correlated with electronic structure.

8-2-2 MAGNETIC SUSCEPTIBILITY[13]

Just as in the diatomic examples in Chapter 5, the magnetic properties of a compound can give indirect evidence of the orbital energy levels. Hund's rule requires the maximum number of unpaired electrons in energy levels with equal, or nearly equal, energies. Diamagnetic compounds, with all electrons paired, are slightly repelled by a magnetic field. When there are unpaired electrons, the compound is paramagnetic, and is attracted into a magnetic field. The measure of this magnetism is called the magnetic susceptibility. We will describe a modification of the Gouy method[14] for determining magnetic

[13] R. S. Drago, *Physical Methods in Chemistry,* 1977, Saunders, Philadelphia, pp. 411–31.

[14] B. Figgis and J. Lewis, *Techniques of Inorganic Chemistry,* ed. H. Jonassen and A. Weissberger, Vol. IV, Interscience, New York, 1965, p. 137.

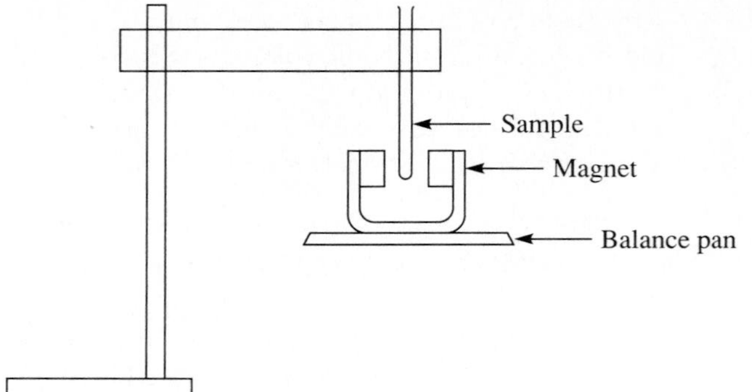

FIGURE 8-4 Gouy Magnetic Susceptibility Apparatus. (Adapted with permission from S. S. Eaton and G. R. Eaton, *J. Chem. Educ.*, **1979**, *56*, 170.)

susceptibility, which requires only an analytical balance and a small magnet as shown in Figure 8-4.[15] In this approach, the solid sample is placed in a small glass sample tube. A small high-field U-shaped magnet is weighed four times, (1) alone, (2) with the sample suspended between the poles of the magnet, (3) with a reference compound of known magnetic susceptibility suspended in the gap, and finally (4) with the empty sample tube suspended in the gap (to allow for correction for any magnetic effect in the sample tube). With a diamagnetic sample, the tube and magnet repel each other and the magnet appears slightly heavier. With a paramagnetic sample, the tube and magnet attract each other and the magnet appears slightly lighter. The measurement of the known compound provides a standard from which the mass susceptibility (susceptibility per gram) of the sample can be calculated and converted to the molar susceptibility. More precise measurements require temperature control and measurement at different magnetic field strengths to correct for possible impurities.

Magnetic susceptibility (χ) is commonly measured in units of cm^3/mole; the magnetic moment, μ, is defined as

$$\mu = 2.828(\chi_T)^{1/2} \quad \text{(in Bohr magnetons, } 9.27 \times 10^{-24} \text{ amperes meter}^2 \text{ or joules tesla}^{-1}\text{)}$$

The paramagnetism mentioned above arises because electrons behave as tiny magnets. Although there is no direct evidence for spinning movement by electrons, a charged particle spinning rapidly would generate a **spin magnetic moment,** and the popular term has therefore become **electron spin.** Electrons with $m_s = -\frac{1}{2}$ are said to have a negative spin; those with $m_s = +\frac{1}{2}$ have a positive spin. As described in Chapter 2, the total spin moment is characterized by the spin quantum number S. The orbital angular momentum, characterized by the quantum number L, results in an additional orbital magnetic moment. The combination of these two, added as vectors, is the total magnetic moment of the atom or molecule.

The equation for the magnetic moment is

$$\mu = g\sqrt{S(S+1) + \frac{1}{4}L(L+1)}$$

[15] S. S. Eaton and G. R. Eaton, *J. Chem. Educ.*, **1979**, *56*, 170.

where μ = magnetic moment

g = gyromagnetic ratio (conversion to magnetic moment)

S = spin quantum number

L = orbital quantum number

Although detailed determination of electronic structure requires consideration of the orbital moment, for most compounds of the first transition series, the spin-only moment is sufficient, as any orbital contribution is small. External fields from other atoms and ions may effectively quench the orbital moment in these compounds. For the heavier transition metals and the lanthanides, the orbital contribution is larger and must be taken into account. Since we are usually concerned primarily with the number of unpaired electrons in the compound, and the options are usually clear-cut and differ significantly, the errors introduced by considering only the spin moment are usually not large enough to cause difficulty. From this point, we will consider only the spin moment.

In Bohr magnetons, the gyromagnetic ratio, $g = 2.00023$, is frequently rounded to 2. If the orbital contribution is omitted, the equation for the spin-only moment μ_s becomes

$$\mu_s = 2\sqrt{S(S+1)} = \sqrt{4S(S+1)}$$

Since $S = \frac{1}{2}, 1, \frac{3}{2}, \ldots$, for 1, 2, 3, . . . , unpaired electrons, this equation can also be written

$$\mu_s = \sqrt{n(n+2)}$$

where n = number of unpaired electrons. This is the equation that is used most frequently. Table 8-2 shows the change in μ_s with n, along with some experimental moments.

EXERCISE 8-1

Show that $\sqrt{4S(S+1)}$ and $\sqrt{n(n+2)}$ are equivalent expressions.

TABLE 8-2
Calculated and experimental magnetic moments

Ion	*n*	*S*	*L*	μ_S	μ_{S+L}	*Observed*
V^{4+}	1	$\frac{1}{2}$	2	1.73	3.00	1.7–1.8
Cu^{2+}	1	$\frac{1}{2}$	2	1.73	3.00	1.7–2.2
V^{3+}	2	1	3	2.83	4.47	2.6–2.8
Ni^{2+}	2	1	3	2.83	4.47	2.8–4.0
Cr^{3+}	3	$\frac{3}{2}$	3	3.87	5.20	~3.8
Co^{2+}	3	$\frac{3}{2}$	3	3.87	5.20	4.1–5.2
Fe^{2+}	4	2	2	4.90	5.48	5.1–5.5
Co^{3+}	4	2	2	4.90	5.48	~5.4
Mn^{2+}	5	$\frac{5}{2}$	0	5.92	5.92	~5.9
Fe^{3+}	5	$\frac{5}{2}$	0	5.92	5.92	~5.9

SOURCE: Data from F. A. Cotton and G. Wilkinson, *Advanced Inorganic Chemistry*, 4th ed., Wiley, New York, 1980, pp. 627–28.

NOTE: All moments in Bohr magnetons.

There are several other ways to measure magnetic susceptibility, including nuclear magnetic resonance[16] and the Faraday method using an unsymmetrical magnetic field.[17]

8-2-3 ELECTRONIC SPECTRA

Direct evidence of orbital energy levels can be obtained from electronic spectra. The energy of the light absorbed as electrons are raised to higher levels is the difference in energy between the orbital energy levels. The observed spectra are frequently more complex than the simple energy diagrams used in this chapter seem to indicate; Chapter 9 gives a more complete picture of electronic spectra of coordination compounds. A very large amount of the information about bonding and electronic structures in complexes has come from the study of electronic spectra.

8-2-4 COORDINATION NUMBERS AND MOLECULAR SHAPES

Although a number of factors influence the number of ligands bonded to a metal and the shapes of the resulting molecules, in some cases we can determine which structure is favored by the electronic structure of the compound. For example, two 4-coordinate structures are possible, tetrahedral and square planar. Some metals, such as Pt(II), form almost exclusively square planar compounds, while others, such as Ni(II) and Cu(II), have both structures, depending on the ligands. Subtle differences in electronic structure, described later in this chapter, help to explain these differences. More discussion of these factors is given in Chapter 10, where we discuss shapes of coordination compounds, isomerism, and related topics.

8-3 THEORIES OF ELECTRONIC STRUCTURE

Terminology

Different names have been used for the theoretical approaches to the electronic structure of coordination compounds, depending on the preferences of the authors. The labels as we will use them are described here, in order of their historical development (and reverse order of discussion in this chapter):

Valence bond theory. A method that describes bonding using hybrid orbitals and electron pairs, as an extension of the electron-dot and hybrid orbital methods used for simpler molecules. Although the theory as originally proposed is seldom used today, the hybrid notation is still common in discussing bonding.

Crystal field theory. An electrostatic approach, used to describe the split in metal *d* orbital energies. It describes the electronic energy levels that determine the ultraviolet and visible spectra, but does not describe the bonding.

[16] D. F. Evans, *J. Chem. Soc.*, **1959,** 2003.

[17] L. N. Mulay and I. L. Mulay, *Anal. Chem.*, **1972,** *44,* 324R.

Ligand field theory. The molecular orbital approach to bonding, which uses some of the terminology of the crystal field theory but includes a more complete description of bonding as well as the electronic energy levels.

In the following pages, the ligand field theory and its use are described, including the method of angular overlap, which can be used to estimate the orbital energy levels. At the end of the chapter, the valence bond theory and the crystal field approach to spectra of coordination compounds are described briefly.

8-4 LIGAND FIELD THEORY

Before extending the molecular orbital theory to coordination compounds, a quick review of the treatment of simpler molecules is in order. The molecular orbitals of SF_6, the last example in Chapter 5, can be considered as a central sulfur 6+ ion accepting six pairs of electrons from the fluoride ions, acting as **σ donor ligands.** The results, shown in Figure 8-5, are essentially the same whether this molecule is treated beginning with atomic orbitals or with hybrid orbitals. The six ligand σ_p orbitals (p orbitals or hybrid orbitals with the same symmetry) match the symmetries of the $3s$, $3p$, $3d_{x^2-y^2}$, and $3d_{z^2}$ sulfur orbitals. The sulfur T_{2g} orbitals (d_{xy}, d_{xz}, and d_{yz}), do not have appropriate symmetry to interact with the ligands and are therefore nonbonding. When the 12 ligand electrons are filled in, they occupy six bonding orbitals (the six bonds in SF_6), and all the antibonding and nonbonding orbitals are empty. In hybridization terms, sulfur uses d^2sp^3 hybrids for bonding. The reducible representation and the necessary irreducible representations for this octahedral example are:

O_h	E	$8C_3$	$6C_2$	$6C_4$	$3C_2$	i	$6S_4$	$8S_6$	$3\sigma_h$	$6\sigma_d$	Sulfur orbitals
Γ	6	0	0	2	2	0	0	0	4	2	
A_{1g}	1	1	1	1	1	1	1	1	1	1	s
T_{1u}	3	0	−1	1	−1	−3	−1	0	1	1	(p_x, p_y, p_z)
E_g	2	−1	0	0	2	2	0	−1	2	0	$(d_{x^2-y^2}, d_{z^2})$
T_{2g}	3	0	1	−1	−1	3	−1	0	−1	1	(d_{xy}, d_{xz}, d_{yz})

A similar set of energy levels is common to all octahedral complexes. For the first-row transition metals, the principal orbitals involved are the $3d$, $4s$, and $4p$, with the $3d$ below the energy of the $4s$ and $4p$ orbitals. As a result, the orbitals with E_g symmetry form bonding and antibonding molecular orbitals and the T_{2g} orbitals are unchanged, all with energies between the bonding and antibonding orbitals formed from the s and p orbitals of the central atom, as shown in Figure 8-6. A similar ordering of the molecular orbitals appears in compounds of the heavier transition elements. In addition, the possibility of multiple bonding must also be considered when the d orbitals are close to the energies of the ligand orbitals. This adds some complexity to the description, since the other p orbitals of the ligands must be included. The $3d$ electrons of the transition metals also change the overall energy of the compounds, and we will consider them first.

In the discussion of ligand fields and the energy changes resulting from them, we must keep in mind that we are considering a relatively small part of the total bonding energy of the molecules. Electrons in bonding orbitals provide

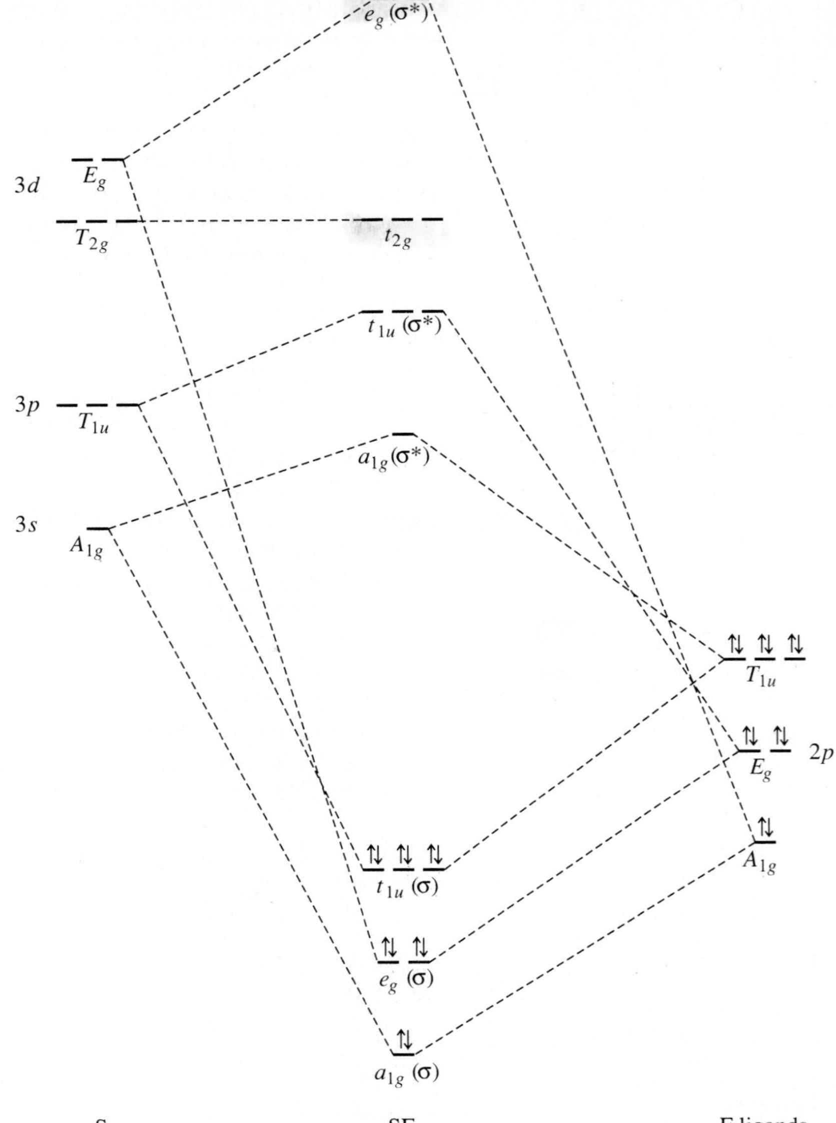

FIGURE 8-5 Molecular Orbitals for Octahedral SF_6. As in Chapter 5, the atomic orbital representations are labeled in capital letters and the molecular orbital representations are labeled in lowercase. (Adapted from T. A. Albright, J. K. Burdett, and M.-Y. Whangbo, *Orbital Interactions in Chemistry,* Wiley-Interscience, New York, 1985, pp. 259–60. Copyright © 1985, John Wiley & Sons, Inc. Reprinted by permission of John Wiley & Sons, Inc.)

the low potential energy that holds molecules together. Electrons in the higher levels affected by ligand field effects help determine the details of the structure, such as the shape, magnetic properties, and electronic spectrum.

8-4-1 ORBITAL SPLITTING AND ELECTRON SPIN

In octahedral coordination compounds, electrons from the ligands can be considered to fill all six bonding molecular orbitals, and any electrons from the metal ion occupy the nonbonding t_{2g} and the antibonding e_g orbitals. The split between these two sets of orbitals (t_{2g} and e_g) is called Δ_o (o for octahedral)

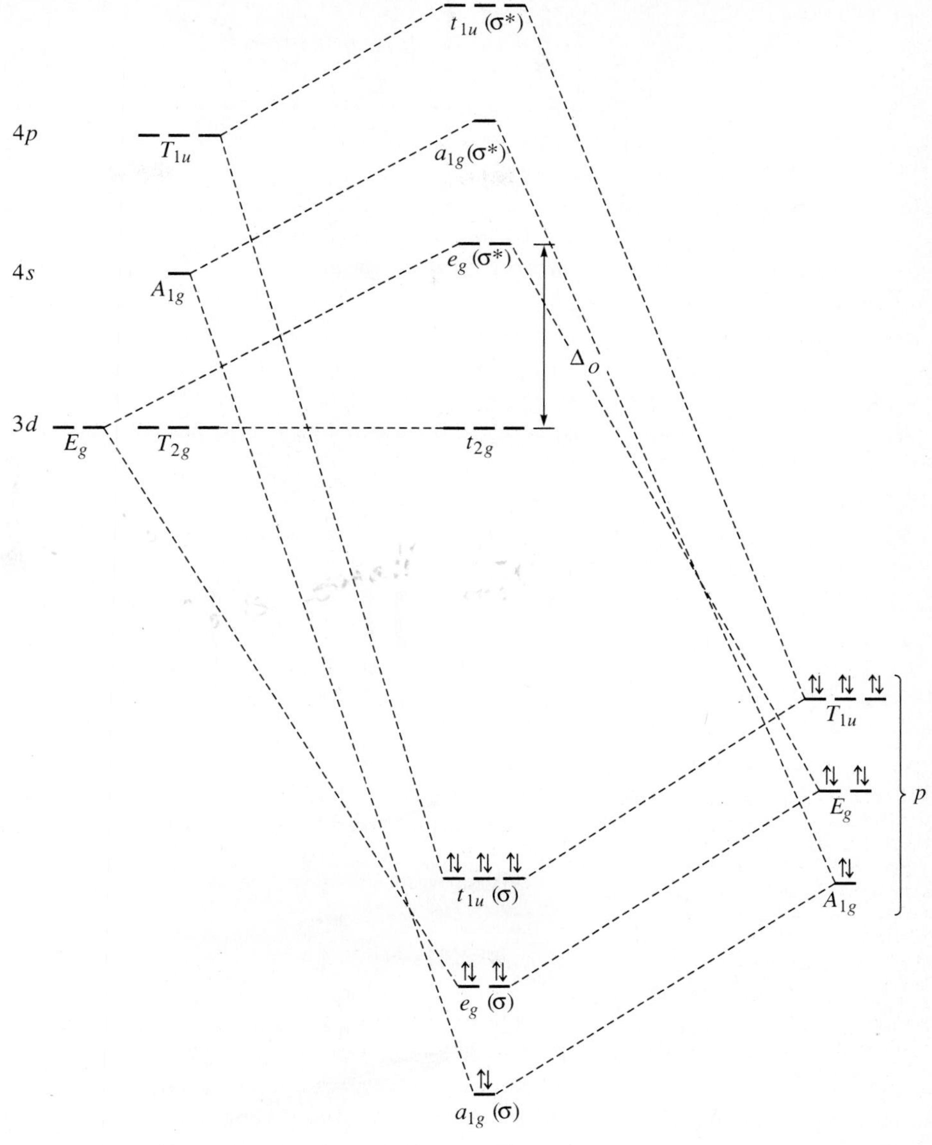

FIGURE 8-6 Molecular Orbitals for an Octahedral Transition Metal Complex. (Adapted from F. A. Cotton, *Chemical Applications of Group Theory*, 3rd ed., Wiley, New York, 1990, p. 232, omitting π orbitals. Copyright © 1990, John Wiley & Sons, Inc. Reprinted by permission of John Wiley & Sons, Inc.)

or $10Dq$. Ligands whose orbitals interact strongly with the metal orbitals are called **strong-field ligands.** With these, the split between the t_{2g} and e_g orbitals is large, and as a result Δ_o is large. Ligands with small interactions are called **weak-field ligands;** the split between the t_{2g} and e_g orbitals is smaller and Δ_o is small. For d^0 through d^3 and d^8 through d^{10} ions, there is only one possible electron configuration, so there is no difference in the net spin of the electrons for strong- and weak-field cases. On the other hand, the d^4 through d^7 ions exhibit **high-spin** and **low-spin** states, as shown in Table 8-3. Strong ligand fields lead to low-spin complexes; weak ligand fields lead to high-spin complexes.

TABLE 8-3
Spin states and ligand field strength

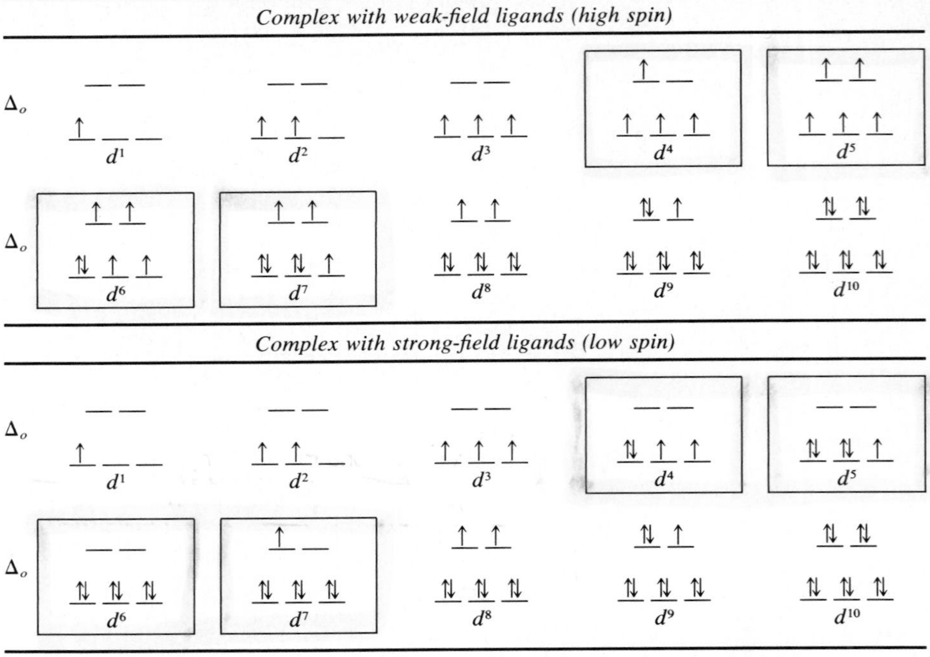

Complex with weak-field ligands (high spin)

	d^1	d^2	d^3	d^4	d^5
	— —	— —	— —	↑ —	↑ ↑
Δ_o					
	↑ — —	↑ ↑ —	↑ ↑ ↑	↑ ↑ ↑	↑ ↑ ↑

	d^6	d^7	d^8	d^9	d^{10}
	↑ ↑	↑ ↑	↑ ↑	⇅ ↑	⇅ ⇅
Δ_o					
	⇅ ↑ ↑	⇅ ⇅ ↑	⇅ ⇅ ⇅	⇅ ⇅ ⇅	⇅ ⇅ ⇅

Complex with strong-field ligands (low spin)

	d^1	d^2	d^3	d^4	d^5
	— —	— —	— —	— —	— —
Δ_o					
	↑ — —	↑ ↑ —	↑ ↑ ↑	⇅ ↑ ↑	⇅ ⇅ ↑

	d^6	d^7	d^8	d^9	d^{10}
	— —	↑ —	↑ ↑	⇅ ↑	⇅ ⇅
Δ_o					
	⇅ ⇅ ⇅	⇅ ⇅ ⇅	⇅ ⇅ ⇅	⇅ ⇅ ⇅	⇅ ⇅ ⇅

Terminology for these configurations is summarized as follows:

Ligand field	Δ_o	*Spin*
Strong	Large	Low
Weak	Small	High

The relationship between the energy level difference, the Coulomb energy, and the exchange energy (Δ_o, Π_c, and Π_e, respectively) determines where the electrons are located. As explained in Section 2-2-4, the pairing energy depends on the Coulomb energy of repulsion between two electrons in the same region of space, Π_c, and the purely quantum mechanical exchange energy, Π_e.[18]

For example, a d^5 ion could have five unpaired electrons, three in t_{2g} and two in e_g orbitals, as a **high-spin** case, or it could have only one unpaired electron, with all five electrons in the t_{2g} levels, as a **low-spin** case.

EXAMPLE

A d^6 ion has ten exchangeable pairs in a high spin complex and six in a low spin complex.

In the high spin complex, the electron spins are $\uparrow_1\downarrow_1\ \uparrow_2\ \uparrow_3\ \uparrow_4\ \uparrow_5$. The five ↑ electrons have exchangeable pairs 1-2, 1-3, 1-4, 1-5, 2-3, 2-4, 2-5, 3-4, 3-5, and 4-5, for a total of ten.

[18] Griffith and Orgel, op. cit.

In the low spin complex, the electron spins are $\uparrow_1 \downarrow_1\ \uparrow_2 \downarrow_2\ \uparrow_3 \downarrow_3$. Each set of three electrons with the same spin has exchangeable pairs 1-2, 1-3, and 2-3, for a total of six.

The difference between the high spin and low spin complexes is four exchangeable pairs.

EXERCISE 8-2

Find the number of exchangeable pairs for a d^5 ion, both as a high spin and as a low spin complex.

The loss of exchange energy on going from high spin to low spin is greater for d^5 than for d^6. The energy differences in Table 8-4 show this; the ratio of mean pairing energies is near 1.5.

TABLE 8-4
Orbital splitting (Δ_o) and mean pairing energy (Π) for aqueous ions (in cm^{-1})

	Ion	Δ_o	Π	*Ion*	Δ_o	Π
d^1				Ti^{3+}	20,300	
d^2				V^{3+}	18,000	
d^3	V^{2+}	11,800		Cr^{3+}	17,600	
d^4	Cr^{2+}	14,000	23,500	Mn^{3+}	21,000	28,000
d^5	Mn^{2+}	7,500	25,500	Fe^{3+}	14,000	30,000
d^6	Fe^{2+}	10,000	17,600	Co^{3+}	17,000–19,000	21,000
d^7	Co^{2+}	9,700	22,500	Ni^{3+}	—	27,000
d^8	Ni^{2+}	8,600				
d^9	Cu^{2+}	13,000				
d^{10}	Zn^{2+}	0				

SOURCE: Data from D. S. McClure, "The Effects of Inner-orbitals on Thermodynamic Properties," in T. M. Dunn, D. S. McClure, and R. G. Pearson, *Some Aspects of Crystal Field Theory*, Harper & Row, New York, 1965, p. 82. The value of Δ_o for Co^{3+} is from J. S. Griffith and L. E. Orgel, *Quart. Revs.*, **1957,** *XI*, 381.

Unlike the total pairing energy Π, Δ_o is strongly dependent on the ligands and on the metal. The values shown in Table 8-4 are for aqueous ions, with water as the relatively weak field ligand (with small Δ_o). In general, Δ_o for 3+ ions is larger than Δ_o for 2+ ions with the same number of electrons, and values for d^5 ions are smaller than for d^4 and d^6 ions. The number of unpaired electrons in the complex depends on the balance between the two terms. When $\Delta_o > \Pi$, there is a net loss in energy (increase in stability) on pairing electrons in the lower levels, and the low-spin configuration is more stable; when $\Delta_o < \Pi$, the total energy is lower with more unpaired electrons, and the high-spin configuration is more stable. In Table 8-4, only Co^{3+} has Δ_o near the size of Π, and it is the only aqua complex with low spin. All the other ions require a stronger field ligand than water for a low-spin configuration.

Another factor that influences electron configurations and the resulting spin is the position of the metal in the periodic table. Metals from the second and third transition series form low-spin complexes more readily than metals from the first transition series. This is a consequence of two cooperating effects, one the greater overlap between the larger $4d$ and $5d$ orbitals and the ligand orbitals, the other a decreased pairing energy due to the larger volume available for electrons in the $4d$ and $5d$ orbitals as compared to $3d$ orbitals.

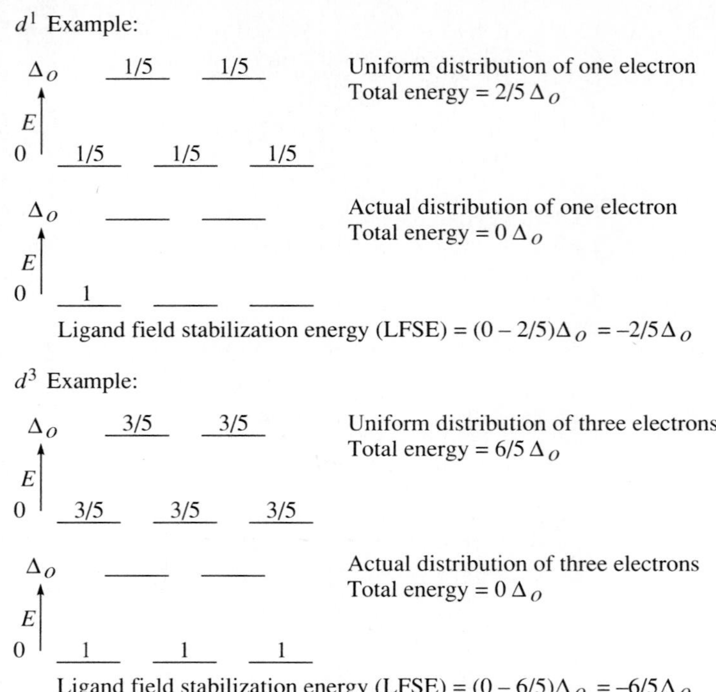

FIGURE 8-7 Calculation of LFSE.

8-4-2 LIGAND FIELD STABILIZATION ENERGY[19]

The difference between (1) the total energy of a coordination compound with the electron configuration resulting from ligand field splitting of the orbitals and (2) the total energy for the same compound with all the *d* orbitals equally populated is called the **ligand field stabilization energy,** or **LFSE.** As shown in Figure 8-7, if an electron were spread equally over all five of the *d* orbitals, we could say that we had $\frac{1}{5}$ electron in each orbital. For an octahedral complex, we can take the t_{2g} orbitals as zero in energy because they are unchanged on formation of a σ-bonded complex. The total energy of an evenly distributed 1-electron system would then be $\frac{2}{5}\Delta_o$, $\frac{1}{5}$ electron times Δ_o for each of the e_g levels. Since the actual configuration has the electron in one of the t_{2g} levels, the total energy is zero. The net difference is $-\frac{2}{5}\Delta_o$, which is the ligand field stabilization energy or LFSE. Similarly, for a d^3 complex, a uniform distribution of electrons would place $\frac{3}{5}$ electron in each level for a total energy of $\frac{6}{5}\Delta_0$ ($\frac{3}{5}$ electron in each of the e_g levels with an energy of Δ_o). The actual arrangement again has an energy of zero, so the LFSE is $-\frac{6}{5}\Delta_o$. Table 8-5 has the LFSE values for σ-bonded octahedral complexes with 1 through 10 electrons in both high- and low-spin arrangements. These values are commonly used as approximations even when significant π bonding is included.

The final columns in Table 8-5 show the difference in LFSE between low-spin and high-spin complexes with the same total number of *d* electrons and the associated pairing energies. For 1 to 3 and 8 to 10 electrons, there is no difference in the number of unpaired electrons or the LFSE. For 4 to 7 electrons, there is a significant difference in both.

[19] F. A. Cotton, *J. Chem. Educ.*, **1964,** *41,* 466.

TABLE 8-5
Ligand field stabilization energies

Number of d electrons	Weak-field arrangement t_{2g}			e_g		LFSE(Δ_o)	Coulomb energy	Exchange energy
1	↑					$-\frac{2}{5}$		
2	↑	↑				$-\frac{4}{5}$		Π_e
3	↑	↑	↑			$-\frac{6}{5}$		$3\Pi_e$
4	↑	↑	↑	↑		$-\frac{3}{5}$		$6\Pi_e$
5	↑	↑	↑	↑	↑	0		$10\Pi_e$
6	↑↓	↑	↑	↑	↑	$-\frac{2}{5}$	Π_c	$10\Pi_e$
7	↑↓	↑↓	↑	↑	↑	$-\frac{4}{5}$	$2\Pi_c$	$11\Pi_e$
8	↑↓	↑↓	↑↓	↑	↑	$-\frac{6}{5}$	$3\Pi_c$	$13\Pi_e$
9	↑↓	↑↓	↑↓	↑↓	↑	$-\frac{3}{5}$	$4\Pi_c$	$16\Pi_e$
10	↑↓	↑↓	↑↓	↑↓	↑↓	0	$5\Pi_c$	$20\Pi_e$

Number of d electrons	Strong-field arrangement t_{2g}			e_g		LFSE(Δ_o)	Coulomb energy	Exchange energy	Strong field − weak field
1	↑					$-\frac{2}{5}$			0
2	↑	↑				$-\frac{4}{5}$		Π_e	0
3	↑	↑	↑			$-\frac{6}{5}$		$3\Pi_e$	0
4	↑↓	↑	↑			$-\frac{8}{5}$	Π_c	$3\Pi_e$	$-\Delta_o + \Pi_c - 3\Pi_e$
5	↑↓	↑↓	↑			$-\frac{10}{5}$	$2\Pi_c$	$4\Pi_e$	$-2\Delta_o + 2\Pi_c - 6\Pi_e$
6	↑↓	↑↓	↑↓			$-\frac{12}{5}$	$3\Pi_c$	$6\Pi_e$	$-2\Delta_o + 2\Pi_c - 4\Pi_e$
7	↑↓	↑↓	↑↓	↑		$-\frac{9}{5}$	$3\Pi_c$	$9\Pi_e$	$-\Delta_o + \Pi_c - 2\Pi_e$
8	↑↓	↑↓	↑↓	↑	↑	$-\frac{6}{5}$	$3\Pi_c$	$13\Pi_e$	0
9	↑↓	↑↓	↑↓	↑↓	↑	$-\frac{3}{5}$	$4\Pi_c$	$16\Pi_e$	0
10	↑↓	↑↓	↑↓	↑↓	↑↓	0	$5\Pi_c$	$20\Pi_e$	0

NOTE: In addition to the LFSE, each pair formed has a positive Coulomb energy, Π_c, and each set of 2 electrons with the same spin has a negative exchange energy, Π_e. When $\Delta_o > \Pi_c - 3\Pi_e$ for d^4 or d^5 or when $\Delta_o > \Pi_c - 2\Pi_e$ for d^6 or d^7, the strong-field arrangement (low spin) is favored.

The most common example of LFSE in thermodynamic data appears in the exothermic enthalpy of hydration of bivalent ions of the first transition series, usually assumed to have six waters of hydration:

$$M^{2+}\ (g) + 6\ H_2O\ (l) = M(H_2O)_6^{2+}\ (aq)$$

Ions with spherical symmetry should have the magnitude of $-\Delta H$ increasing continuously across the transition series due to the decreasing radius of the ions with increasing nuclear charge and corresponding increase in electrostatic attraction for the ligands. In fact, the enthalpies show the characteristic double-hump shape shown in Figure 8-8, where $-\Delta H_{hyd}$ is plotted (larger values are more exothermic). The differences between the two curves are approximately equal to the LFSE values in Table 8-5 for high-spin complexes.[20] The total enthalpy change is the sum of the spherical contribution and the LFSE resulting from the nonuniform arrangement of the *d* orbitals and the water molecules. Similar effects appear in the enthalpies of formation of crystalline halides, where the metal ions are surrounded by negative ions and the orbital energy levels are split in a similar way.

[20] L. E. Orgel, *J. Chem. Soc.*, **1952,** 4756; P. George and D. S. McClure, *Prog. Inorg. Chem.*, **1959,** *1,* 381–463.

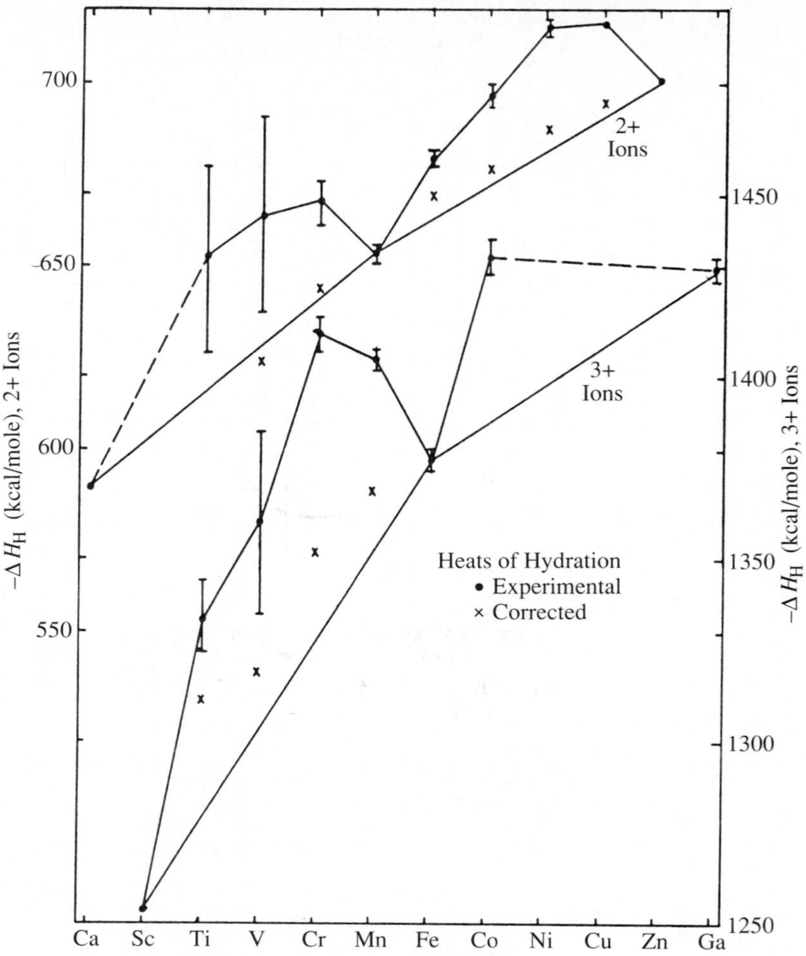

FIGURE 8-8 Heats of hydration of transition metal ions. Values of δH derived from spectroscopic Δ_o's are subtracted from each value of $-\Delta H_H$ and form the "corrected" curve. The straight lines Ca–Mn–Zn or Sc–Fe–Ga, are also shown. (Reproduced from P. George and D. S. McClure, *Prog. Inorg. Chem.*, **1959,** *1*, p. 418. Copyright © 1959, John Wiley & Sons, Inc. Reprinted by permission of John Wiley & Sons, Inc.)

8-4-3 PI BONDING

The description of LFSE and bonding in coordination compounds given up to this point has included only σ donor ligands. Addition of the other ligand orbitals allows the possibility of π bonding. This addition begins with consideration of the other p or π^* orbitals of the ligands (those that are not involved in σ bonding). These can be represented by the x and z axes in Figure 8-9. These axes (and their corresponding orbitals) must be taken as a single set of 12, because each axis can be converted into every other axis by one of the symmetry operations (C_4 or one of the σ). The reducible representation for these 12 orbitals can be found and reduced to its component irreducible representations by the methods presented in Chapter 4. The reducible representation has characters of zero for all the symmetry operations of O_h except E and C_2 ($= C_4^2$) and can be reduced to four components, as shown in Table 8-6:

$$\Gamma_\pi = T_{1g} + T_{2g} + T_{1u} + T_{2u}$$

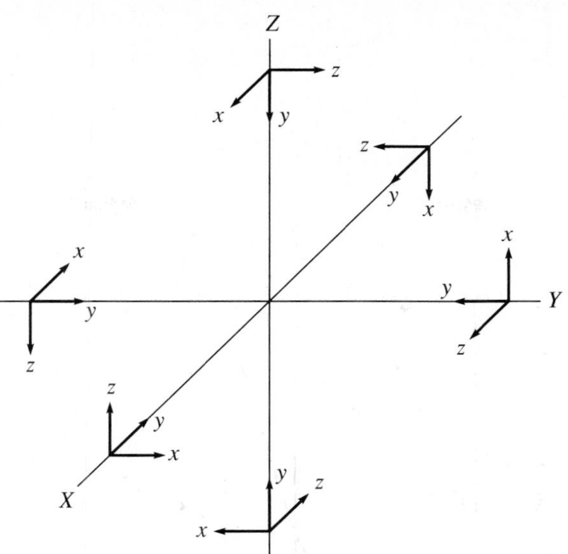

FIGURE 8-9 Coordinate System for Octahedral π Orbitals.

TABLE 8-6
Representations of octahedral π orbitals

O_h	E	$8C_3$	$6C_2$	$6C_4$	$3C_2(=C_4^2)$	i	$6S_4$	$8S_6$	$3\sigma_h$	$6\sigma_d$	
Γ_π	12	0	0	0	−4	0	0	0	0	0	
T_{1g}	3	0	−1	1	−1	3	1	0	−1	−1	
T_{2g}	3	0	1	−1	−1	3	−1	0	−1	1	(d_{xy}, d_{xz}, d_{yz})
T_{1u}	3	0	−1	1	−1	−3	−1	0	1	1	(p_x, p_y, p_z)
T_{2u}	3	0	1	−1	−1	−3	1	0	1	−1	

Of these four, T_{1g} and T_{2u} have no match in metal orbitals, T_{2g} matches the d_{xy}, d_{xz}, d_{yz} orbitals, and T_{1u} matches the p_x, p_y, p_z orbitals of the metal. Since the p orbitals of the metal are already used in σ bonding and will not overlap well with the ligand π orbitals because of the larger bond distances in coordination compounds, they are unlikely to be used also for π bonding. There are then three orbitals on the metal (d_{xy}, d_{xz}, d_{yz}) available for π bonds distributed over the six ligand–metal pairs. The t_{2g} orbitals of the metal, which are nonbonding in the σ-only orbital calculations shown in Figure 8-6, change because of the π interaction to produce a lower bonding set and a higher antibonding set.

Pi bonding in coordination compounds is possible when the ligand has p or π^* molecular orbitals available. Since the effects are smaller for occupied orbitals, we will first treat ligands with empty π^* orbitals, or **π acceptor ligands.**

The cyanide ion (Figure 8-10) provides an example. The molecular orbital picture of CN^- is intermediate between those of N_2 and CO given in Chapter 5, since the energy differences between C and N orbitals are significant but less than those between C and O orbitals. The HOMO for CN^- is a σ orbital with considerable bonding character and a concentration of electron density on the carbon. This is the donor orbital used by CN^- in forming σ orbitals in the complex. Above the HOMO, the LUMO orbitals of CN^- are two π^* orbitals that can be used for π bonding with the metal.

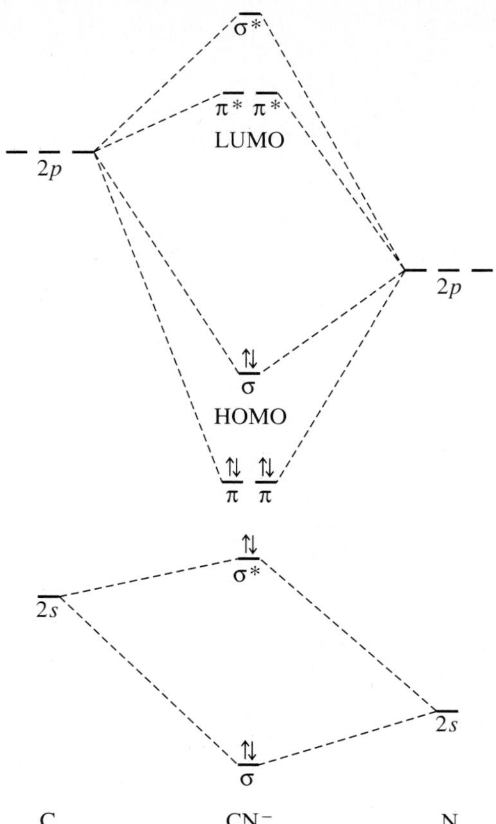

FIGURE 8-10 Cyanide Molecular Orbitals.

The ligand π^* orbitals have energies higher than, but near that of the metal t_{2g} (d_{xy}, d_{xz}, d_{yz}) orbitals, with which they overlap (Figure 8-11). As a result, they form molecular orbitals, with the bonding orbitals lower in energy than the initial metal t_{2g} orbitals. The corresponding antibonding orbitals are higher in energy than the e_g σ antibonding orbitals. Metal ion d electrons occupy the bonding orbitals (now the HOMO), resulting in a larger value for Δ_o, as shown in Figure 8-12. There can be significant energy stabilization from this added π bonding. This **metal to ligand (M $\longrightarrow$ L) π bonding** is also called **π back-bonding,** with electrons from d orbitals of the metal donated back to the ligands.

When the ligand has electrons in its p orbitals (as in F^- or Cl^-), the bonding molecular π orbitals will be occupied by these electrons, and there are two net results: the t_{2g} bonding orbitals strengthen the ligand–metal linkage slightly, and the corresponding t_{2g}^* levels are raised in energy and become antibonding; this reduces Δ_o, as in Figure 8-12. The metal ion d electrons are pushed into the higher orbital by the ligand electrons. This is described as **ligand to metal (L $\longrightarrow$ M) π bonding,** with the π electrons from the ligands being donated to the metal ion. Ligands participating in such interactions are called **π-donor ligands.** The decrease in the energy of the bonding orbitals is partly counterbalanced by the increase in the energy of the t_{2g}^* orbitals. In addition, the combined σ and π donations from the ligands give the metal more negative charge, which decreases attraction between the metal and the ligands and makes this kind of bonding less favorable.

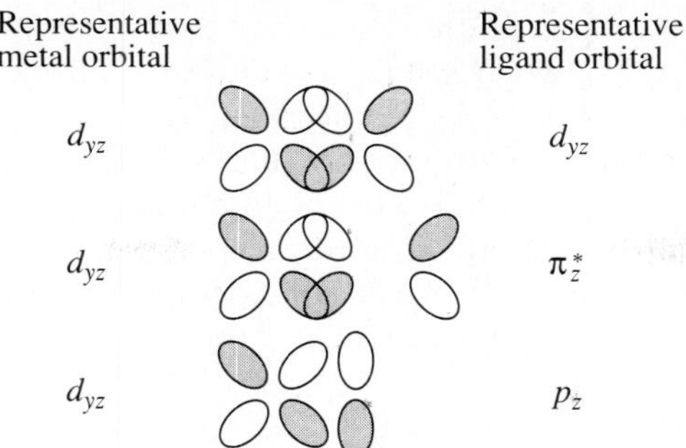

FIGURE 8-11 Overlap of d, π^*, and p Orbitals with Metal d Orbitals. Overlap is good with ligand d and π^* orbitals, poorer with ligand p orbitals.

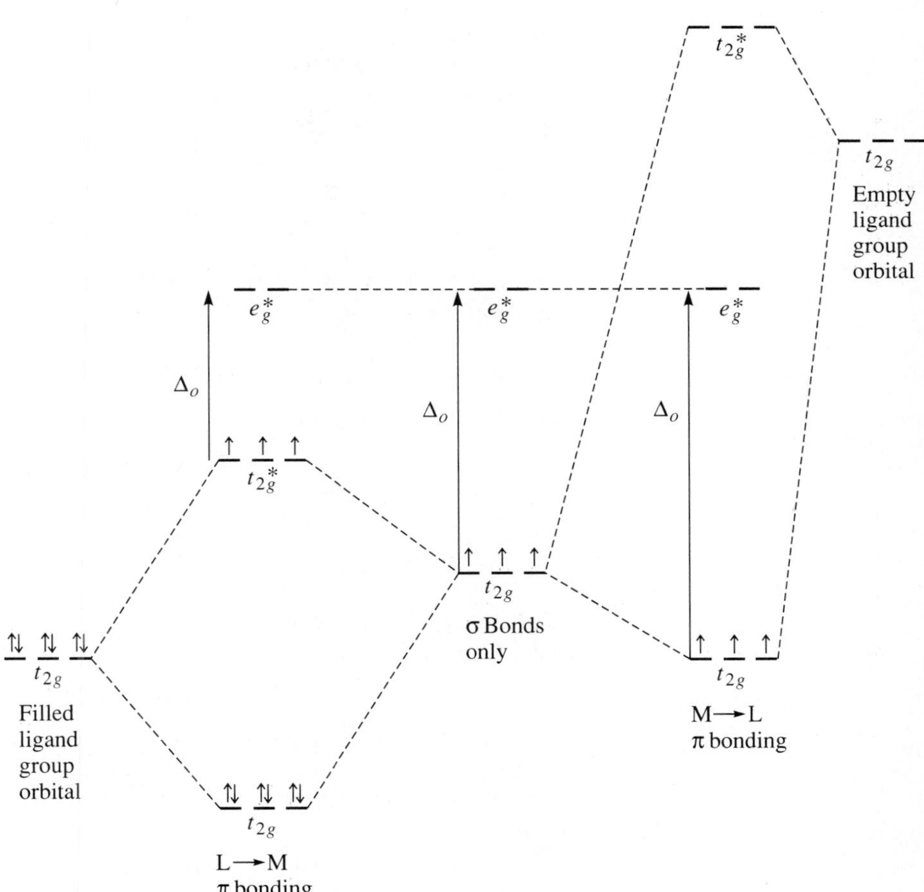

FIGURE 8-12 Effects of π Bonding on Δ_o (using a d^3 ion as example).

Overall, filled π^* or p orbitals on ligands (frequently with relatively low energy) result in L $\longrightarrow$ M π bonding and a smaller Δ_o for the overall complex. Empty, higher-energy π^* or d orbitals on the ligands result in M $\longrightarrow$ L π bonding and a larger Δ_o for the complex. Ligand-to-metal π bonding usually gives decreased stability for the complex; metal-to-ligand π bonding usually gives increased stability and favors low-spin configurations.

Part of the stabilizing effect of π back-bonding is a result of transfer of negative charge away from the metal ion. The positive ion accepts electrons from the ligands to form the σ bonds. The metal is then left with a large negative charge. When the π orbitals can be used to transfer part of this charge back to the ligands, the overall stability is improved. The π-acceptor ligands that can participate in π back-bonding are extremely important in organometallic chemistry and will be discussed further in Chapter 12.

Complexes with π bonding will have LFSE values modified by the changes in the t_{2g} levels described above. Many good π-acceptor ligands form complexes with large differences between t_{2g} and e_g levels. How much of this increase is caused by the π bonding and how much by the σ bonding strength of these ligands is uncertain, but the overall result is easily seen.

Square planar complexes

The same general approach works for any geometry, although some may be more complicated than others. Square planar complexes, with D_{4h} symmetry, provide one example. As before, the axes for the ligand atoms are chosen for convenience. The y axis of each ligand is directed toward the central atom, the x axis is in the plane of the molecule, and the z axis is perpendicular to the plane of the molecule, as shown in Figure 8-13. The p_y set is used in σ bonding. Unlike the octahedral case, there are two distinctly different sets of potential π-bonding orbitals, the parallel set ($p_{\parallel}$ or p_x, in the molecular plane) and the perpendicular set ($p_{\perp}$ or p_z, perpendicular to the plane). By taking each set in turn, we can use the techniques of Chapter 4 to find the representations that fit the different symmetries. Table 8-7 gives the results.

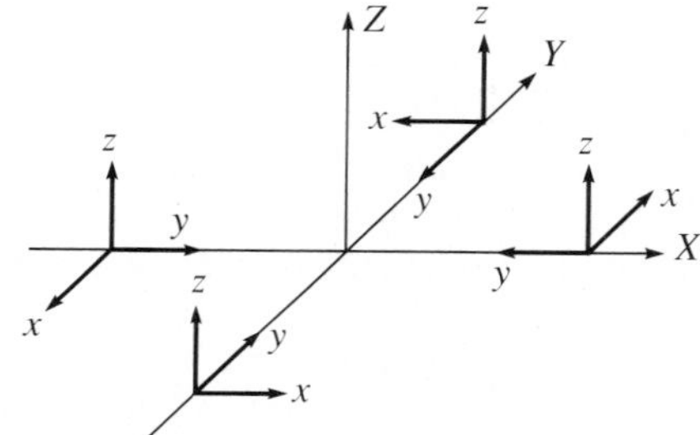

FIGURE 8-13 Coordinate System for Square Planar Orbitals.

> EXERCISE 8-3
> Derive the reducible representations for square planar bonding and then show that their component irreducible representations are those in Table 8-7.

The resulting σ-bonding orbitals for the first transition series are those with lobes in the x and y directions, $3d_{x^2-y^2}$, $4s$, $4p_x$, and $4p_y$ (in hybrid orbital terms, dsp^2).[21] Ignoring the other orbitals for the moment, we can construct the energy level diagram for the σ bonds, as in Figure 8-14. Comparing Figures 8-6 and 8-14, we see that changing from octahedral to square planar moves the d_{z^2} orbital from e_g to a_{1g} symmetry and splits the t_{2g} levels into degenerate b_{2g} and e_g levels. The difference between the degenerate b_{2g} and e_g levels and the a_{1g} level corresponds to Δ. In the more complete treatment that follows, the degeneracy is removed and several energy level differences appear.

[21] In principle, the d_{z^2} could be used in place of the s orbital to make the d^2p^2 hybrid. This is symmetrically equivalent to dsp^2, but less likely because of the small extension of d_{z^2} electron density in the xy plane (only the small collar is in position to overlap with the ligand orbitals).

TABLE 8-7
Representations and orbital symmetry for square planar complexes

D_{4h}	E	$2C_4$	C_2	$2C_2'$	$2C_2''$	i	$2S_4$	σ_h	$2\sigma_v$	$2\sigma_d$		
A_{1g}	1	1	1	1	1	1	1	1	1	1		x^2+y^2, z^2
A_{2g}	1	1	1	−1	−1	1	1	1	−1	−1	R_z	
B_{1g}	1	−1	1	1	−1	1	−1	1	1	−1		x^2-y^2
B_{2g}	1	−1	1	−1	1	1	−1	1	−1	1		xy
E_g	2	0	−2	0	0	2	0	−2	0	0	(R_x, R_y)	(xz, yz)
A_{1u}	1	1	1	1	1	−1	−1	−1	−1	−1		
A_{2u}	1	1	1	−1	−1	−1	−1	−1	1	1	z	
B_{1u}	1	−1	1	1	−1	−1	1	−1	−1	1		
B_{2u}	1	−1	1	−1	1	−1	1	−1	1	−1		
E_u	2	0	−2	0	0	−2	0	2	0	0	(x, y)	

D_{4h}	E	$2C_4$	C_2	$2C_2'$	$2C_2''$	i	$2S_4$	σ_h	$2\sigma_v$	$2\sigma_d$	
Γ_{p_x}	4	0	0	−2	0	0	0	4	−2	0	$p_\parallel$
Γ_{p_y}	4	0	0	2	0	0	0	4	2	0	p_σ
Γ_{p_z}	4	0	0	−2	0	0	0	−4	2	0	$p_\perp$

$\Gamma_{p_y} = A_{1g} + B_{1g} + E_u$ (σ) Matching orbitals on the central atom: $s, d_{x^2-y^2}, (p_x, p_y), d_{z^2}$

$\Gamma_{p_x} = A_{2g} + B_{2g} + E_u$ ($\parallel$) Matching orbitals on the central atom: $d_{xy}, (p_x, p_y)$

$\Gamma_{p_z} = A_{2u} + B_{2u} + E_g$ ($\perp$) Matching orbitals on the central atom: $p_z, (d_{xz}, d_{yz})$

When π interactions are also considered, the degeneracy of the b_{2g} and e_g levels is removed, and the diagram in Figure 8-15 results. Two kinds of π bonding become apparent, one parallel to the plane of the molecule using the ligand p_x and the metal d_{xy} orbitals, the other perpendicular to the plane of the molecule using the ligand p_z and metal d_{xz} and d_{yz} orbitals. The metal p orbitals have symmetries that would allow π bonding, but do not extend out far enough to overlap well. In addition, the metal p_x and p_y and the ligand p_y orbitals are used in σ bonding, forming the four bonding orbitals a_{1g}, e_u, and b_{2g}. Overall, there are 11 molecular orbitals remaining for the π electrons, three with bonding character, five nonbonding, and three antibonding. Monatomic ligands such as F^- or Cl^- each have four electrons in the p_x and p_y orbitals, for a total of 16 electrons, enough to fill the bonding and nonbonding orbitals. The important orbitals for detailed consideration are, in order of increasing energy, a_{1g}, e_g, b_{2g}, and b_{1g}, derived from the metal d_{z^2}, (d_{xz}, d_{yz}), d_{xy}, and $d_{x^2-y^2}$, respectively. They are also the orbitals into which any metal d electrons will go. The differences between them are labeled Δ_1, Δ_2, and Δ_3 from top to bottom. Because b_{2g} and e_g are π orbitals, their energies will change significantly with changes in the ligands.

Tetrahedral complexes

Hybrids for tetrahedral geometry can be described as either sp^3 or sd^3 and, in the case of coordination compounds, usually are a mixture of both. The σ-bonding orbitals are easily determined in the usual way, using the coordinate

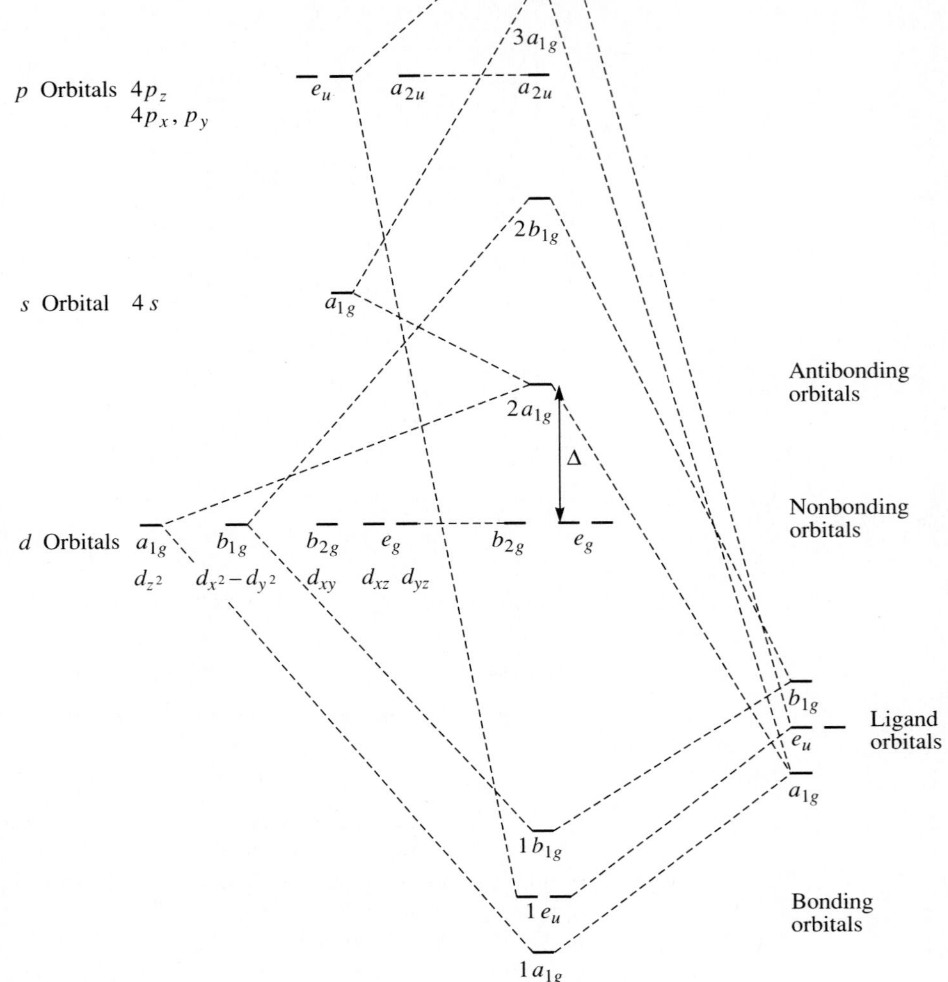

FIGURE 8-14 D_{4h} Molecular Orbitals (σ bonds only). (Adapted from T. A. Albright, J. K. Burdett, and M.-Y. Whangbo, *Orbital Interactions in Chemistry,* Wiley-Interscience, New York, 1985, p. 296. Copyright © 1985, John Wiley & Sons, Inc. Reprinted by permission of John Wiley & Sons, Inc.)

system illustrated in Figure 8-16. The reducible representation includes the A_1 and T_2 irreducible representations, allowing for four bonding MOs. The overall orbital energy level picture is inverted from the octahedral levels, with e the nonbonding and t_2 the bonding and antibonding levels. In addition, the split (now called Δ_t) is smaller than for octahedral geometry; the general result is $\Delta_t = \frac{4}{9}\Delta_o$.

The π orbitals are more difficult to see, but if the y axis of the ligand orbitals is chosen along the bond axis, and the x and z axes are arranged to allow the C_2 operation to work properly, the results in Table 8-8 are obtained. The reducible representation includes the E, T_1, and T_2 representations. The T_1 has no matching metal atom orbitals, E has d_{z^2} and $d_{x^2-y^2}$, and T_2 has d_{xy}, d_{xz}, and d_{yz}. The E and T_2 interactions lower the energy of the bonding orbitals and raise the corresponding antibonding orbitals. An additional complication appears when both bonding and antibonding π orbitals are available on the ligand, as is true for CO or CN^-. Figure 8-17 shows the special case of

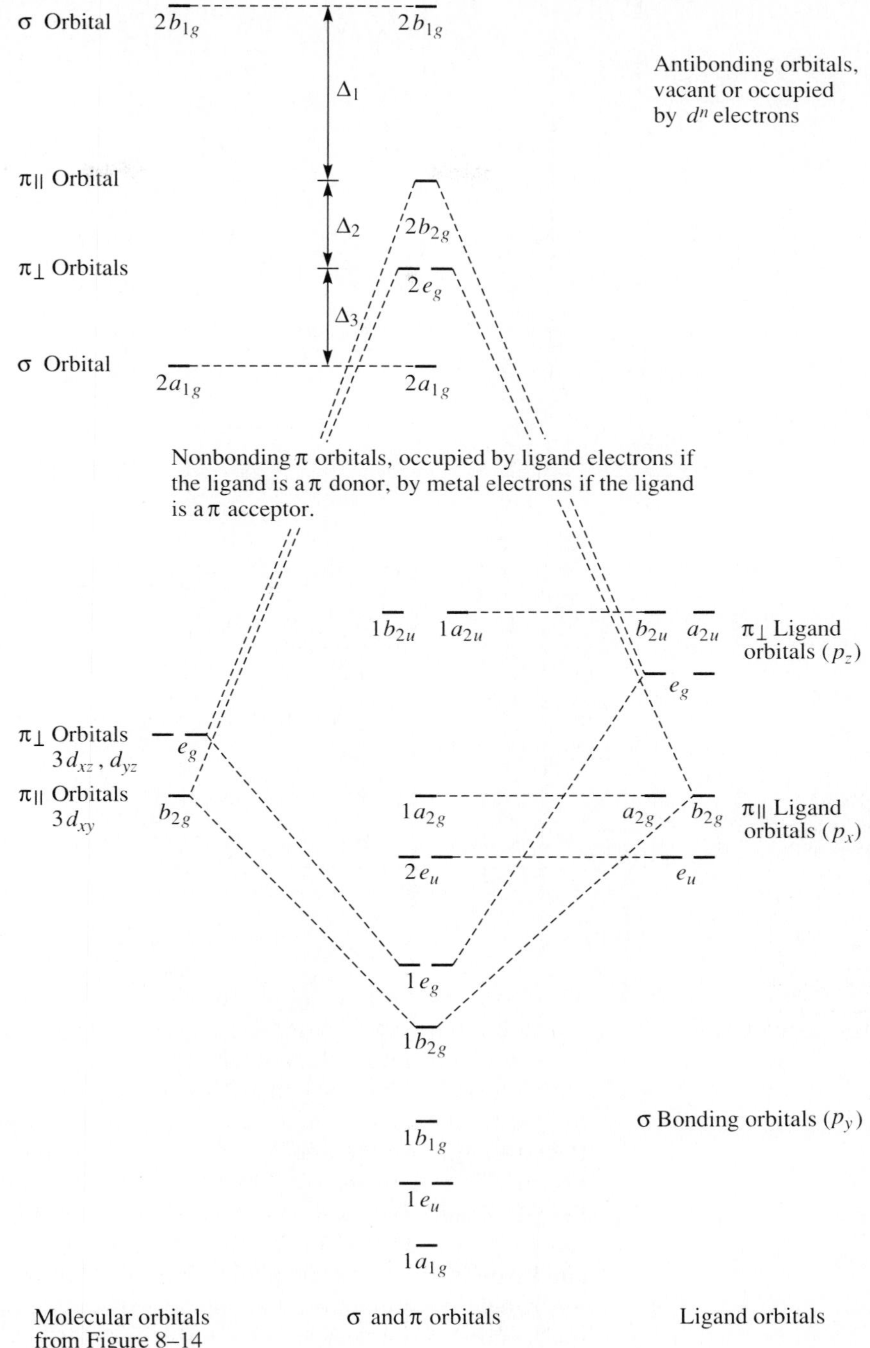

FIGURE 8-15 D_{4h} Molecular Orbitals (including π orbitals).

$Ni(CO)_4$, in which the interactions of the CO σ and π orbitals with the metal orbitals are probably small. Much of the bonding is from M ⟶ L π bonding. In cases where the *d* orbitals are not fully occupied, σ bonding is likely to be more important, with resulting shifts of the a_1 and t_2 orbitals to lower energies and the 4*s* and 4*p* orbitals to higher energies.

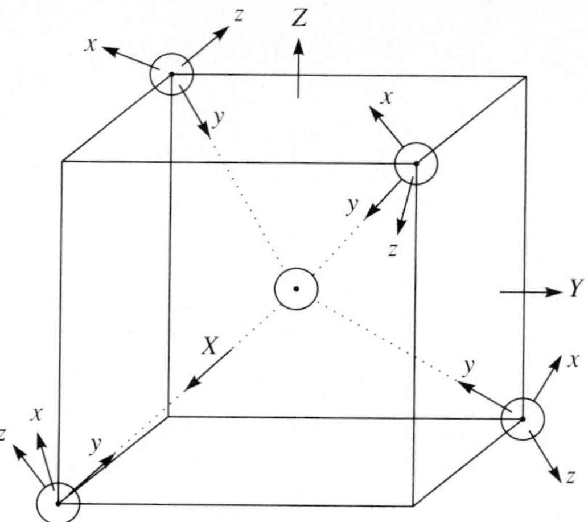

FIGURE 8-16 Coordinate System for Tetrahedral Orbitals.

TABLE 8-8
Representations of tetrahedral orbitals

T_d	E	$8C_3$	$3C_2$	$6S_4$	$6\sigma_d$		
A_1	1	1	1	1	1		$x^2 + y^2 + z^2$
A_2	1	−1	1	−1	−1		
E	2	−1	2	0	0		$(2z^2 - x^2 - y^2, x^2 - y^2)$
T_1	3	0	−1	1	−1	(R_x, R_y, R_z)	
T_2	3	0	−1	−1	1	(x, y, z)	(xy, yz, xz)
Γ_σ	4	1	0	0	2	$A_1 + T_2$	
Γ_π	8	−1	0	0	0	$E + T_1 + T_2$	

8-5 ANGULAR OVERLAP

The molecular orbital approach to energy levels in coordination compounds is more difficult to use when considering an assortment of ligands or structures with symmetry other than octahedral, square planar, or tetrahedral. A variation with the flexibility to deal with a variety of possible geometries and with a mixture of ligands is called the **angular overlap** model.[22] This approach estimates the strength of interaction between individual ligand orbitals and metal d orbitals, based on the overlap between them, and then combines these values for all ligands and all d orbitals for the complete picture. Both σ and π interactions are considered, and different coordination numbers and geometries can be treated. The term angular overlap is used because the amount of overlap is strongly dependent on the angles of the metal orbitals and the angle at which the ligand approaches.

In the angular overlap approach, the energy of a metal d orbital in a coordination compound is determined by summing the effects of each of the ligands on that orbital. Some ligands will have a strong effect on a particular d orbital, some a weaker effect, and some no effect at all, because of their angular dependence. Similarly, both σ and π interactions must be taken into

[22] E. Larsen and G. N. La Mar, *J. Chem. Educ.*, **1974,** *51,* 633. There are misprints on pages 635 and 636.

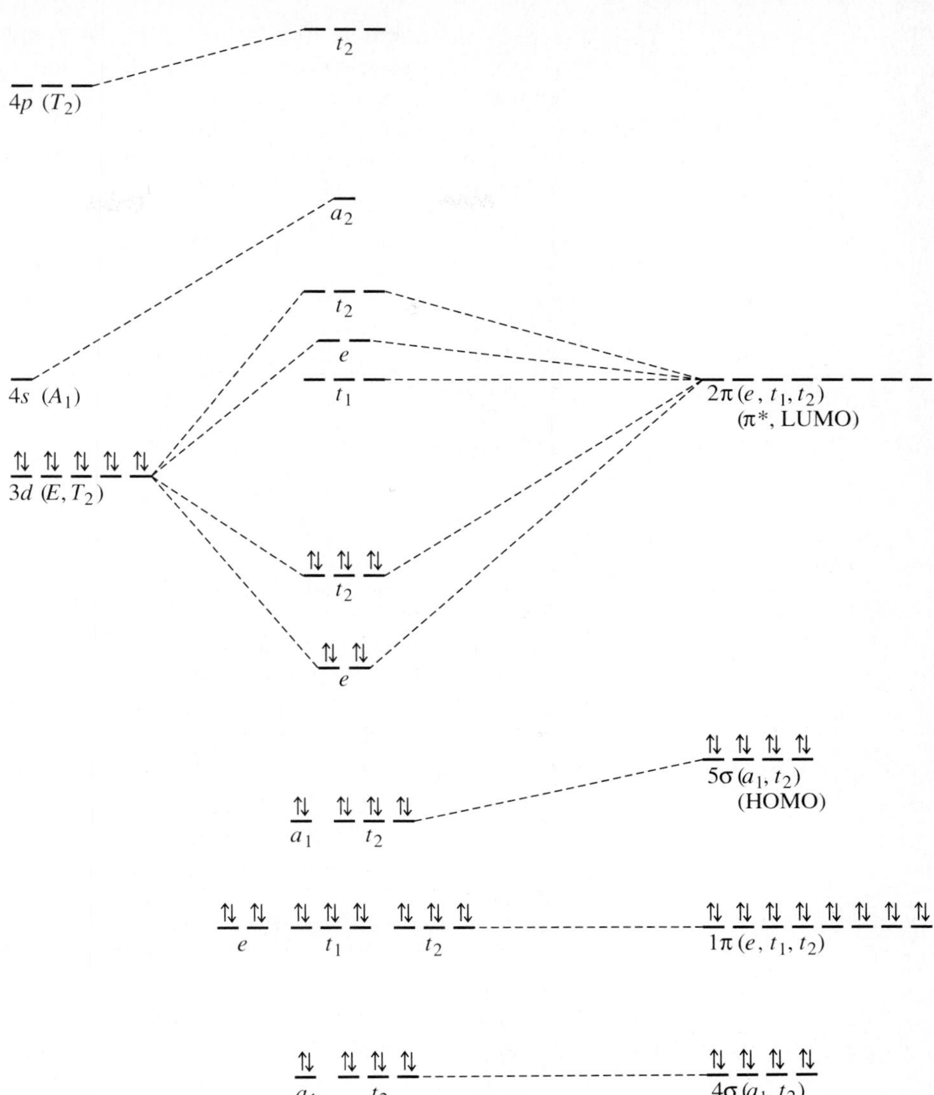

FIGURE 8-17 Tetrahedral Molecular Orbitals for $Ni(CO)_4$. The order of the orbitals is somewhat uncertain. C. W. Bauschlicher, Jr., and P. S. Bagus, *J. Chem. Phys.*, **1984,** *81,* 5889, argue that there is almost no σ bonding from the 4*s* and 4*p* orbitals of Ni and that the d^{10} configuration is the best starting place for the calculations. (See G. Cooper, K. H. Sze, and C. E. Brion, *J. Am. Chem. Soc.*, **1989,** *111,* 5051 for further references and an alternative diagram.)

account in order to determine the final energy of a particular orbital. By systematically considering each of the five *d* orbitals, we can use this approach to determine the overall energy pattern corresponding to the coordination geometry around the metal.

8-5-1 SIGMA DONOR INTERACTIONS

The strongest σ interaction is between a metal d_{z^2} orbital and a ligand *p* orbital (or a hybrid ligand orbital of the same symmetry). The strength of all other σ interactions is determined relative to the strength of this reference interaction. Interaction between these two orbitals results in a bonding orbital, which has a larger component of the ligand orbital, and an antibonding orbital, which is largely metal orbital in composition. Although the increase in energy of the antibonding orbital is larger than the decrease in energy of the bonding orbital, we will approximate the molecular orbital energies by an increase in the antibonding (mostly metal d) orbital of e_σ and a decrease in energy of the bonding (mostly ligand) orbital of e_σ:

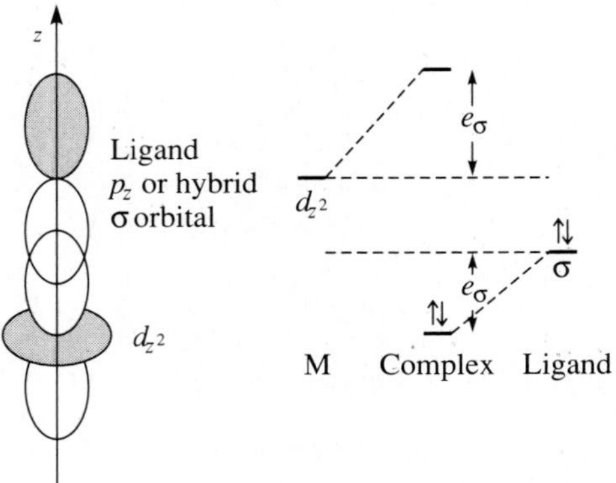

Similar changes in orbital energy result from other interactions between metal d orbitals and ligand orbitals, with the magnitude dependent on the ligand location and the specific d orbital being considered. Table 8-9 gives values of these energy changes for a variety of shapes. Calculation of the numbers in the table (all in e_σ units) is beyond the scope of this book, but the reader should be able to justify the numbers qualitatively by comparing the amount of overlap between the orbitals being considered.

TABLE 8-9
Angular overlap parameters: Sigma interactions

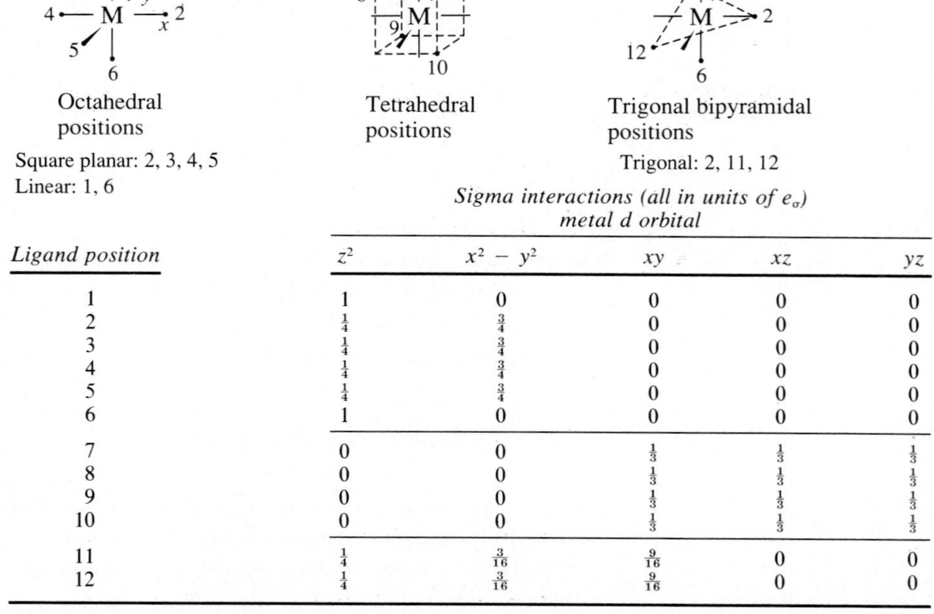

Sigma interactions (all in units of e_σ) metal d orbital

Ligand position	z^2	$x^2 - y^2$	xy	xz	yz
1	1	0	0	0	0
2	$\frac{1}{4}$	$\frac{3}{4}$	0	0	0
3	$\frac{1}{4}$	$\frac{3}{4}$	0	0	0
4	$\frac{1}{4}$	$\frac{3}{4}$	0	0	0
5	$\frac{1}{4}$	$\frac{3}{4}$	0	0	0
6	1	0	0	0	0
7	0	0	$\frac{1}{3}$	$\frac{1}{3}$	$\frac{1}{3}$
8	0	0	$\frac{1}{3}$	$\frac{1}{3}$	$\frac{1}{3}$
9	0	0	$\frac{1}{3}$	$\frac{1}{3}$	$\frac{1}{3}$
10	0	0	$\frac{1}{3}$	$\frac{1}{3}$	$\frac{1}{3}$
11	$\frac{1}{4}$	$\frac{3}{16}$	$\frac{9}{16}$	0	0
12	$\frac{1}{4}$	$\frac{3}{16}$	$\frac{9}{16}$	0	0

The angular overlap approach is best described by example. We will consider first the most common geometry for coordination compounds, octahedral.

EXAMPLE

$M(NH_3)_6^{n+}$

Examples of octahedral complexes with only σ interactions are the $M(NH_3)_6^{n+}$ ions. The ammonia ligands have no π orbitals available, either of donor or acceptor character, for bonding with the metal ion. The lone-pair orbital is mostly nitrogen p_z orbital in composition, and the other p orbitals are used in bonding to the hydrogens (see Figure 5-30).

In calculating these energies, the value for a given d orbital is the sum of the numbers for the appropriate ligands in the *vertical column* for that orbital in Table 8-9. The change in energy for a specific ligand orbital is the sum of the numbers for all d orbitals in the *horizontal row* for the required ligand position.

d_{z^2} orbital: The interaction is strongest with ligands in positions 1 and 6, along the z axis. Each interacts with the orbital to raise its energy by e_σ. The ligands in positions 2, 3, 4, and 5 interact more weakly with the d_{z^2} orbital, each raising the energy of the orbital by $\frac{1}{4}e_\sigma$. Overall, the energy of the d_{z^2} orbital is increased by the sum of all these interactions, for a total of $3e_\sigma$.

$d_{x^2-y^2}$ orbital: The ligands in positions 1 and 6 do not interact with this metal orbital. However, the ligands in positions 2, 3, 4, and 5 each interact to raise the energy of the metal orbital by $\frac{3}{4}e_\sigma$, for a total increase of $3e_\sigma$.

d_{xy}, d_{xz}, and d_{yz} orbitals: None of these orbitals interact in a sigma fashion with any of the ligand orbitals, so the energy of these metal orbitals remains unchanged.

The energy changes for the ligand orbitals are the same as those above for each interaction. The totals, however, are taken *across a row* of Table 8-9, including each d orbital.

Ligands in positions 1 and 6 interact strongly with d_{z^2} and are lowered by e_σ. They do not interact with the other d orbitals.

Ligands in positions 2, 3, 4, and 5 are lowered by $\frac{1}{4}e_\sigma$ by interaction with d_{z^2} and by $\frac{3}{4}e_\sigma$ by interaction with $d_{x^2-y^2}$, for a total of e_σ.

Overall, each ligand orbital is lowered by e_σ, the sum of the changes in energy for each position.

The resulting energy pattern is shown in Figure 8-18. This result is the same as the pattern obtained from the ligand field approach. Both describe how the metal complex is stabilized: as two of the d orbitals of the metal increase in energy and three remain unchanged, the six ligand orbitals fall in energy, and electron pairs in those orbitals are stabilized in the formation of ligand–metal bonds. The net stabilization is $12e_\sigma$ for the bonding pairs; any d electrons in the upper (e_g) level are destabilized by $3e_\sigma$.

The more complete MO picture includes use of the metal s and p orbitals in the formation of the bonding MOs and the four additional antibonding orbitals shown in Figure 8-6. The energy changes due to bonding orbital formation are similar for complexes of different shapes. There are no examples of complexes with electrons in the antibonding orbitals from s and p orbitals,

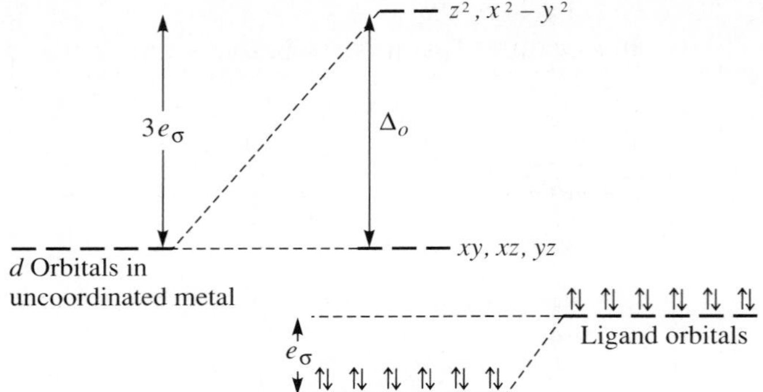

FIGURE 8-18 Energies of *d* Orbitals in Octahedral Complexes: Sigma Donor Ligands. Overall stabilization is $-12e_\sigma + 3me_\sigma$, where *m* is the number of electrons in the d_{z^2} and $d_{x^2-y^2}$ levels. $\Delta_o = 3e_\sigma$.

and these high-energy antibonding orbitals are not significant in describing the spectra of complexes, so we will not consider them further.

EXERCISE 8-4

Using the angular overlap model, determine the relative energies of *d* orbitals in a metal complex of formula ML_4 having tetrahedral geometry. Assume that the ligands are capable of σ interactions only.

This calculation shows that $\Delta_t = \frac{4}{9}\Delta_o$.

8-5-2 PI ACCEPTOR INTERACTIONS

Ligands such as CO, CN^-, and phosphines (of formula PR_3) are π acceptors, with empty orbitals that can interact with metal *d* orbitals in a π fashion. In the angular overlap model, the strongest π interaction is considered to be between a metal d_{xz} orbital and a ligand π^* orbital. Because the ligand π^* orbitals are higher in energy than the original metal *d* orbitals, the resulting bonding MOs are lower than the metal *d* orbitals (a difference of e_π), and the antibonding MO is higher in energy. The *d* electrons then occupy the bonding MO, with a net energy change of $-e_\pi$ for each electron.

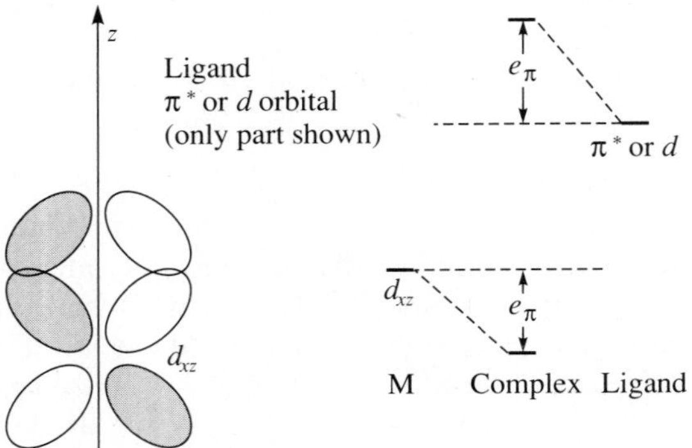

Because the overlap for these orbitals is smaller than the σ overlap described in the previous section, $e_\pi < e_\sigma$. The other π interactions are weaker than this reference interaction, with the magnitudes depending on the degree of overlap between the orbitals. Table 8-10 gives values for ligands at the same angles as in Table 8-9.

TABLE 8-10
Angular overlap parameters: Pi interactions

Ligand position	*Pi interactions (all in units of e_π)* metal d orbital				
	z^2	$x^2 - y^2$	xy	xz	yz
1	0	0	0	1	1
2	0	0	1	1	0
3	0	0	1	0	1
4	0	0	1	1	0
5	0	0	1	0	1
6	0	0	0	1	1
7	$\frac{2}{3}$	$\frac{2}{3}$	$\frac{2}{9}$	$\frac{2}{9}$	$\frac{2}{9}$
8	$\frac{2}{3}$	$\frac{2}{3}$	$\frac{2}{9}$	$\frac{2}{9}$	$\frac{2}{9}$
9	$\frac{2}{3}$	$\frac{2}{3}$	$\frac{2}{9}$	$\frac{2}{9}$	$\frac{2}{9}$
10	$\frac{2}{3}$	$\frac{2}{3}$	$\frac{2}{9}$	$\frac{2}{9}$	$\frac{2}{9}$
11	0	$\frac{3}{4}$	$\frac{1}{4}$	$\frac{1}{4}$	$\frac{3}{4}$
12	0	$\frac{3}{4}$	$\frac{1}{4}$	$\frac{1}{4}$	$\frac{3}{4}$

EXAMPLE

$M(CN)_6^{n-}$

The result of these interactions for $M(CN)_6^{n-}$ complexes is shown in Figure 8-19. The d_{xy}, d_{xz}, and d_{yz} orbitals are lowered by $4e_\pi$ each and the six ligand positions have an average increase in orbital energy of $2e_\pi$. The resulting ligand π^* orbitals have high energies and are not involved directly in the bonding. The new value of the $t_{2g} - e_g$ split is $\Delta_0 = 3e_\sigma + 4e_\pi$.

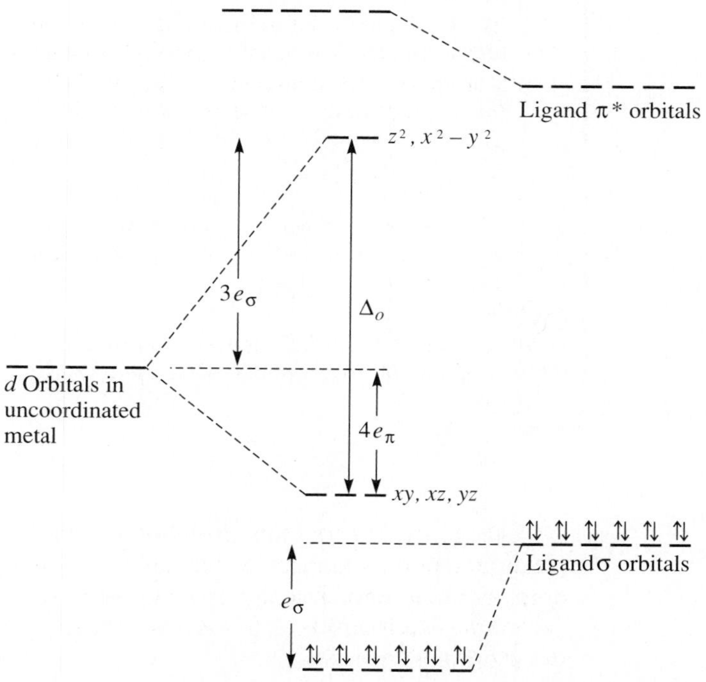

FIGURE 8-19 Energies of d Orbitals in Octahedral Complexes: Sigma Donor and Pi Acceptor Ligands. Overall stabilization is $-12e_\sigma - 4ne_\pi + 3me_\sigma$, where n is the number of electrons in the d_{xy}, d_{xz}, and d_{yz} levels and m is the number of electrons in the d_{z^2} and $d_{x^2-y^2}$ levels. $\Delta_o = 3e_\sigma + 4e_\pi$.

8-5-3 PI DONOR INTERACTIONS

The interactions between occupied ligand *d* or π^* orbitals and metal *d* orbitals are similar to those in the π acceptor case. In other words, the angular overlap model treats π donor ligands similarly to π acceptor ligands except that *for π donor ligands, the signs of the changes are reversed.*

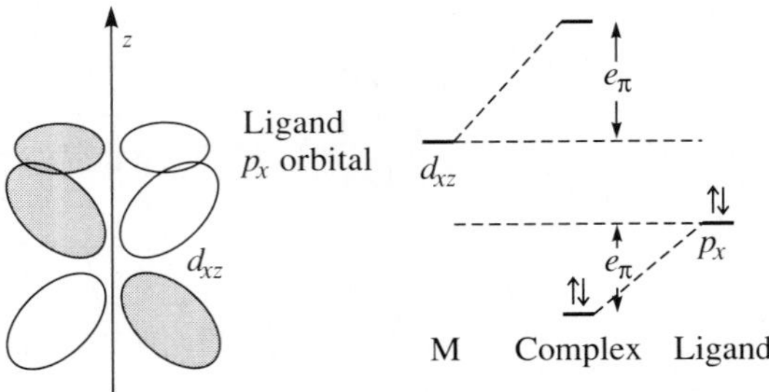

EXAMPLE

MX_6^{n-}

Halide ions donate electron density to a metal via p_y orbitals, a σ interaction; the ions also have p_x and p_z orbitals, which can interact with metal orbitals and donate additional electron density via π interactions. We will use MX_6^{n-} as our example, where X is a halide ion or other ligand that is both a σ and a π donor.

d_{z^2} and $d_{x^2-y^2}$ orbitals: Neither of these orbitals has the correct orientation for π interactions; therefore, the π orbitals have no effect on the energies of these *d* orbitals.

d_{xy}, d_{xz}, and d_{yz} orbitals: Each of these orbitals interacts in a π fashion with four of the ligands. For example, the d_{xy} orbital interacts with ligands in positions 2, 3, 4, and 5 with a strength of $1e_\pi$, resulting in a total increase of the energy of the d_{xy} orbital of $4e_\pi$ (the interaction with ligands at positions 1 and 6 is zero). The reader should verify by using Table 8-10 that the d_{xz} and d_{yz} orbitals are also raised in energy by $4e_\pi$.

The overall effect on the energies of the *d* orbitals of the metal, including both σ and π interactions, is shown diagrammatically in Figure 8-20.

EXERCISE 8-5

Using the angular overlap model, determine the splitting pattern of *d* orbitals for a tetrahedral complex of formula MX_4, where X is a ligand that can act as σ donor and π donor.

In general, in situations involving ligands that can behave as both π acceptors and π donors (such as CO and CN^-), the π acceptor nature predominates. While π donor ligands cause the value of Δ_o to decrease, π acceptor ligands cause Δ_o to increase. Pi acceptor ligands are better at splitting the *d* orbitals.

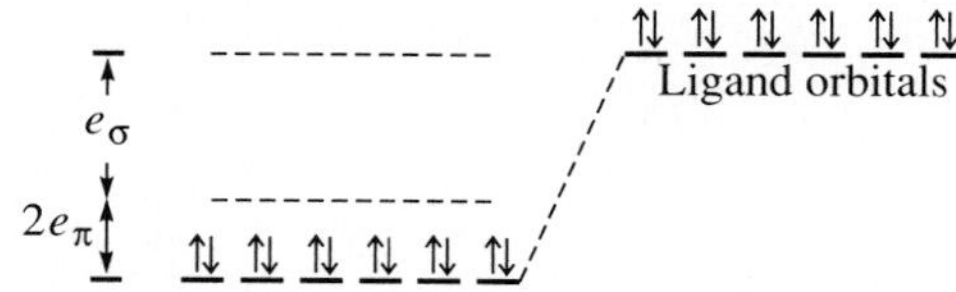

FIGURE 8-20 Energies of d Orbitals in Octahedral Complexes: Sigma Donor and Pi Donor Ligands. Overall stabilization is $-12e_\sigma - 24e_\pi + 4ne_\pi + 3me_\sigma$, where n is the number of electrons in the d_{xy}, d_{xz}, and d_{yz} levels and m is the number of electrons in the d_{z^2} and $d_{x^2-y^2}$ levels. $\Delta_o = 3e_\sigma - 4e_\pi$.

EXERCISE 8-6

Determine the energies of the d orbitals predicted by the angular overlap model for a square planar complex:

a. Considering σ interactions only.

b. Considering both σ and π (acceptor) interactions.

8-5-4 MAGNITUDES OF e_σ, e_π, AND Δ

Changing the ligand or the metal changes the magnitude of e_σ and e_π, with resulting changes in Δ and a possible change in the number of unpaired electrons. For example, water is a relatively weak field ligand. When combined with Co^{2+} in an octahedral geometry, the result is a high-spin complex with 3 unpaired electrons. Combined with Co^{3+}, water gives a low-spin complex with no unpaired electrons. The increase in charge on the metal changes Δ_o sufficiently to favor low spin, as in Figure 8-21.

Similar effects appear with different ligands. $Fe(H_2O)_6^{3+}$ is a high-spin species, while $Fe(CN)_6^{3-}$ is low spin. The change in ligand is enough to favor low spin and, in this case, the change in Δ_o is caused solely by the ligand. As described earlier, the balance between Δ, Π_c, and Π_e (the Coulomb and exchange energies) determines whether a specific compound will be high or low spin. Since Δ_t is small, low-spin tetrahedral complexes are unlikely; ligands with strong enough fields to give low-spin complexes are likely to form low-spin octahedral complexes instead.

Special cases

The angular overlap model has the capability to describe the electronic energy of complexes with different shapes or with combinations of different ligands. Although we will not consider cases of different ligands on the same metal here, it is possible to estimate approximately the magnitudes of e_σ and e_π with

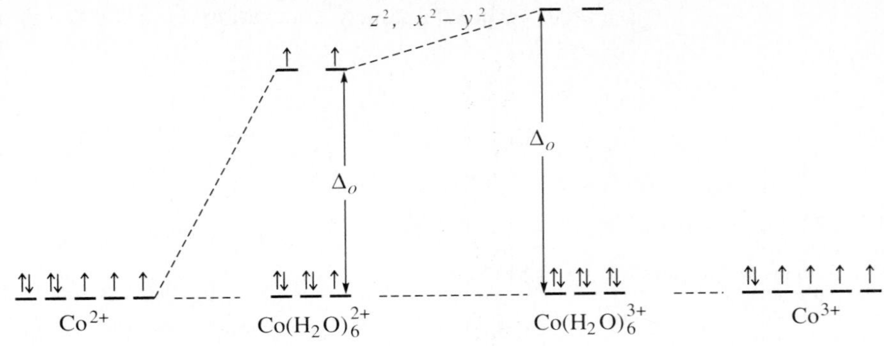

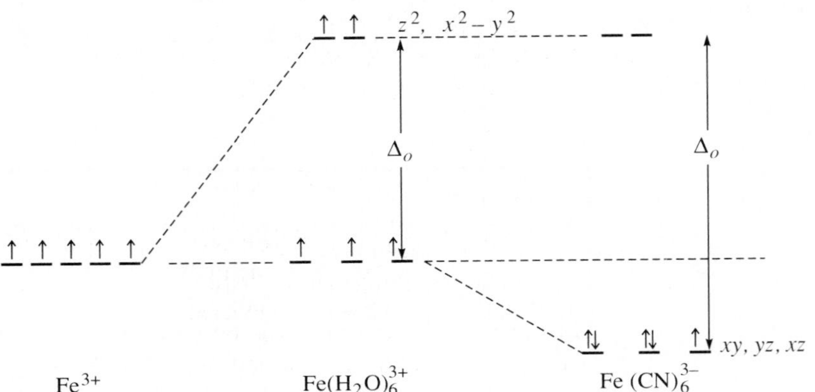

FIGURE 8-21 $Co(H_2O)_6^{2+}$, $Co(H_2O)_6^{3+}$, $Fe(H_2O)_6^{3+}$, $Fe(CN)_6^{3-}$, and Unpaired Electrons.

different ligands and to predict the effects on the electronic structure of complexes such as $Co(NH_3)_4Cl_2^+$. This compound, like all Co(III) complexes except CoF_6^{3-} and $Co(H_2O)_3F_3$, is low spin, so the magnetic properties do not depend on Δ_o. However, the magnitude of Δ_o does have a significant effect on the visible spectrum. Such questions are treated more thoroughly in Chapter 9. Angular overlap can be used to help compare the energies of different geometries, for example to predict whether a 4-coordinate complex is more likely to be tetrahedral or square planar. This application is discussed in Chapter 10. It is also possible to use the angular overlap model to calculate the energy change for reactions in which the transition state results in either a higher or lower coordination number. Examples of this use are given in Chapter 11.

8-5-5 TYPES OF LIGANDS AND THE SPECTROCHEMICAL SERIES

Ligands are frequently classified by their donor and acceptor capabilities. Some, like ammonia, are σ donors only, with no orbitals of appropriate symmetry for π bonding. Bonding by these ligands is relatively simple, using only the σ orbitals identified in Figure 8-6. The ligand field split, Δ, then depends on the relative energies of the metal ion and ligand orbitals and the degree of overlap. Ethylenediamine (en) has a stronger effect than ammonia among these ligands, generating a larger Δ. This is also the order of their basicity.

$$en > NH_3$$

The halide ions have ligand field strengths in the order

$$F^- > Cl^- > Br^- > I^-$$

which is also the order of basicity of these ligands. This is the expected order if the σ donating power of the ions depends on the electron density; the small fluoride has the highest electron density and the large iodide the smallest.

Ligands that have occupied *p* orbitals tend to donate these electrons to the metal along with the σ-bonding electrons and are classified as π donors. As shown earlier, this decreases Δ. Part of the low-field characteristics of the halide ions may be due to this effect. Other primarily σ donor ligands that can also act as π donors include H_2O, OH^-, and RCO_2^-. They fit into the series in the order

$$H_2O > F^- > RCO_2^- > OH^- > Cl^- > Br^- > I^-$$

with OH^- below H_2O in the series because it has more π-donating tendency.

When ligands have vacant π* or *d* orbitals, there is the possibility of π back-bonding, and the ligands are called π acceptors, as described earlier. The effect of this addition to the bonding scheme is to increase Δ. Ligands that do this very effectively include CN^- and CO, with many others also possible. A selected list of these ligands in order is

$$CO, CN^- > \text{phenanthroline (phen)} > NO_2^- > NCS^-$$

When the complete list of the ligands mentioned in this section is put together, thiocyanate turns out to have a smaller effect than ammonia. This list is called the **spectrochemical series** and runs roughly in order from strong π acceptor effect to strong π donor effect:

$$CO, CN^- > phen > NO_2^- > en > NH_3 > NCS^- > H_2O > F^- > RCO_2^- > OH^- > Cl^- > Br^- > I^-$$

Many of the large number of other ligands possible could be included in such a list, but the effects are changed by other circumstances (different metal ion, different charge on the metal, other ligands also present) and attempting to put a large number of ligands in such a list is not generally helpful.

8-6 THE JAHN–TELLER EFFECT

When there are unequal numbers of electrons in degenerate orbitals, the Jahn–Teller theorem[23] requires that the energies of the orbitals change to become nondegenerate. In other words, there cannot be unequal occupation of orbitals with identical energies because the energies, and the shape of the complex, will change slightly. In octahedral complexes, this effect is small when the t_{2g} orbitals are involved and larger when the uneven electron occupancy is in the e_g (d_{z^2} and $d_{x^2-y^2}$) orbitals. The e_g orbitals are directed toward octahedral

[23] H. A. Jahn and E. Teller, *Proc. Roy. Soc.*, **1937,** *A161,* 220.

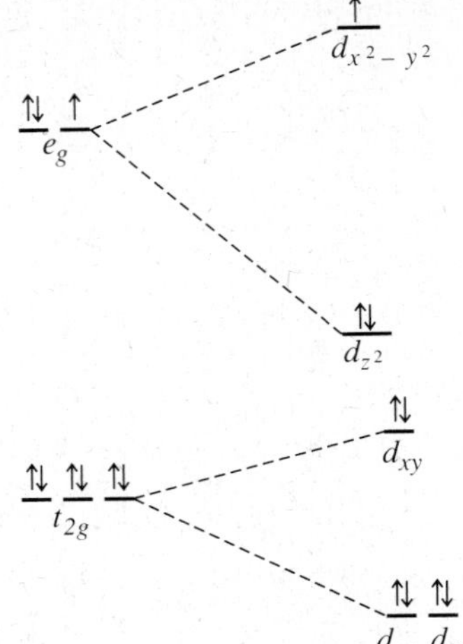

FIGURE 8-22 Jahn-Teller Effect on a d^9 Complex. Elongation along the z axis is coupled to a slight decrease in the other four bonding directions. The resulting split is larger for the e_g orbitals than for the t_{2g}. The energy differences are exaggerated in this figure.

ligands; distortion of the complex has a larger effect on these energy levels. The resulting distortion is most often an elongation along one axis as shown in Figure 8-22, but compression along one axis is also possible. The electron occupancies and expected Jahn–Teller effects are summarized in the following table:

Number of electrons	1	2	3	4	5	6	7	8	9	10
High-spin Jahn–Teller	w	w		s		w	w		s	
Low-spin Jahn–Teller	w	w		w	w		s		s	

w = weak Jahn–Teller effect expected (t_{2g} orbitals unevenly occupied)
s = strong Jahn–Teller effect expected (e_g orbitals unevenly occuped)
no entry = no Jahn–Teller effect expected

EXERCISE 8-7

Using the usual d orbital splitting diagrams, show that the Jahn–Teller effects in the table match the description in the preceding paragraph.

Examples of significant Jahn–Teller effects are found in compounds of Cr(II), high-spin Mn(III), and Cu(II). Ni(III) and low-spin Co(II) should also show this effect, but NiF_6^{3-} is the only known nearly octahedral example. It has a distorted structure consistent with the Jahn–Teller theorem.

Low-spin Cr(II) compounds are octahedral, with tetragonal distortion. They show two absorption bands, one in the visible and one in the near-infrared region, caused by this distortion. In a pure octahedral field, there should only be one d–d transition (see Chapter 9 for more details). Cr(II) also forms dimeric compounds with Cr—Cr bonds in many complexes. The acetate, $Cr_2(OAc)_4$, is an example in which the acetate ions bridge between the two chromiums, with significant Cr—Cr bonding resulting in a nearly diamagnetic compound.

Curiously, the $Mn(H_2O)_6^{3+}$ ion appears to form an undistorted octahedron in $CsMn(SO_4)_2 \cdot 12\,H_2O$, although other Mn(III) compounds show the expected distortion.[24,25]

[24] A. Avdeef, J. A. Costamagna, and J. P. Fackler, Jr., *Inorg. Chem.*, **1974,** *13,* 1854.

[25] J. P. Fackler, Jr. and A. Avdeef, *Inorg. Chem.*, **1974,** *13,* 1864.

Cu(II) forms the most common compounds with significant Jahn–Teller effects. In most cases, the distortion is an elongation of two bonds, but K_2CuF_4 forms a crystal with two shortened bonds in the octahedron. Elongation also plays a part in the change in equilibrium constants for complex formation. For example, $Cu(NH_3)_4^{2+}$ is readily formed in aqueous solution as a distorted octahedron with two water molecules at larger distances than the ammonias, but liquid ammonia is required for formation of hexammines. In many cases, Cu(II) complexes have square planar or nearly square planar geometry, with nearly tetrahedral shapes also possible. $CuCl_4^{2-}$, in particular, shows structures ranging from tetrahedral through square planar to distorted octahedral, depending on the cation present.[26]

8-7 VALENCE BOND THEORY

The valence bond theory proposed by Pauling uses the hybridization ideas presented in Chapter 5. For octahedral complexes, a d^2sp^3 hybrid of the metal orbitals is required. However, the d orbitals used by the first-row transition metals could be either $3d$ or $4d$. Pauling originally described the structures resulting from these as covalent and ionic, respectively. He later changed the terms to hyperligated and hypoligated, and they are also known as inner orbital (using $3d$) and outer orbital (using $4d$) complexes. The number of unpaired electrons, measured by the magnetic behavior of the compounds, determines which d orbitals are used.

Fe(III) has five unpaired electrons as an isolated ion, one in each of the $3d$ orbitals. In coordination compounds, it may have either one or five unpaired electrons. In complexes with one unpaired electron, the ligand electrons force the metal d electrons to pair up and leave two $3d$ orbitals available for hybridization and bonding. In complexes with 5 unpaired electrons, the ligands do not bond strongly enough to force pairing of the $3d$ electrons. Pauling proposed that the $4d$ orbitals could be used for bonding in such cases, with the arrangement of electrons shown in Figure 8-23.

When seven electrons must be provided for [as in Co(II)], there are either one or three unpaired electrons. In the **low-spin** case with one unpaired electron, the seventh electron must go into a higher orbital (unspecified by Pauling, but presumed to be $5s$).[27] In the **high-spin** case with three unpaired electrons, the $4d$ or outer orbital hybrid can be used for bonding, leaving the metal electrons in the $3d$ levels. Similar arrangements are necessary for eight or nine electrons

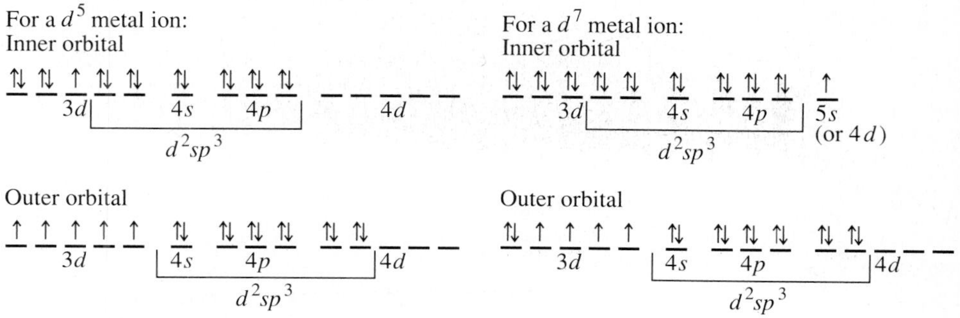

For a d^5 metal ion:
Inner orbital
⇅ ⇅ ↑ ⇅ ⇅ | ⇅ | ⇅ ⇅ ⇅ | _ _ _ _ _
$3d$ | $4s$ | $4p$ | $4d$
d^2sp^3

Outer orbital
↑ ↑ ↑ ↑ ↑ | ⇅ | ⇅ ⇅ ⇅ | ⇅ ⇅ _ _ _
$3d$ | $4s$ | $4p$ | $4d$
d^2sp^3

For a d^7 metal ion:
Inner orbital
⇅ ⇅ ⇅ ⇅ ⇅ | ⇅ | ⇅ ⇅ ⇅ | ↑
$3d$ | $4s$ | $4p$ | $5s$ (or $4d$)
d^2sp^3

Outer orbital
⇅ ↑ ↑ ↑ ↑ | ⇅ | ⇅ ⇅ ⇅ | ⇅ ⇅ _ _ _
$3d$ | $4s$ | $4p$ | $4d$
d^2sp^3

FIGURE 8-23 Inner and Outer Orbital Complexes.

[26] N. N. Greenwood and A. Earnshaw, *Chemistry of the Elements*, Pergamon Press, Elmsford, N.Y., 1984, pp. 1385–86.

[27] B. N. Figgis and R. S. Nyholm, *J. Chem. Soc.*, **1959**, 338; Griffith and Orgel, op. cit., p. 392.

[Ni(II) and Cu(II)], although they frequently change geometry to either tetrahedral or square planar structures.

The valence bond theory was of great importance in the development of bonding theory for coordination compounds, but it is rarely used today except in mention of the hybrid orbitals used in bonding. While it provided a set of orbitals for bonding, the use of the very high energy $4d$ orbitals seems unlikely, and the results do not lend themselves to a good explanation of the electronic spectra of complexes. Since much of our experimental data do come from spectra, this is a serious shortcoming.

8-8 CRYSTAL FIELD THEORY

As originally developed, the crystal field theory was used to describe the electronic structure of metal ions in crystals, where they might be surrounded by oxide ions or other anions that create an electrostatic field with symmetry dependent on the crystal structure. No attempt was made to deal with covalent bonding, since the ionic crystals did not require it.

When the d orbitals are placed in an octahedral field of ligand electron pairs, any electrons in them are repelled by the field. As a result, the $d_{x^2-y^2}$ and d_{z^2} orbitals, which are directed at the surrounding ligands, are raised in energy. The d_{xy}, d_{xz}, and d_{yz} orbitals, which are directed between the surrounding ions, are relatively unaffected by the field. This is the same t_{2g}–e_g split described in the molecular orbital treatment, and the resulting energy difference is identified as Δ_o or $10Dq$. This approach provides a simple means of identifying the d orbital splitting found in coordination compounds and can be extended to include more quantitative calculations. It requires extension to the more complete ligand field theory to include π bonding and more accurate calculations of the resulting energy levels.

The average energy of the five d orbitals is above that of the free ion orbitals, because the electrostatic field of the ligands raises their energy. The t_{2g} orbitals are $0.4\Delta_o$ (or $4Dq$) below and the e_g orbitals are $0.6\Delta_o$ (or $6Dq$) above this average energy, as shown in Figure 8-24. The three t_{2g} orbitals then have a total energy of $-0.4\Delta_o \times 3 = -1.2\Delta_o$, and the two e_g orbitals have a total energy of $+0.6\Delta_o \times 2 = +1.2\Delta_o$ compared to the average. Calculations of crystal field stabilization energies (CFSE) using these numbers and the actual distribution of electrons result in the same values as the ligand field stabilization energies (LFSE) calculations described earlier.

The chief drawbacks to the crystal field approach are in its concept of the repulsion of orbitals by the ligands and its lack of any explanation for bonding in coordination compounds. As we have seen in all our discussion of molecular orbitals, any combination of orbitals leads to both higher- and lower-energy molecular orbitals. The purely electrostatic approach does not allow

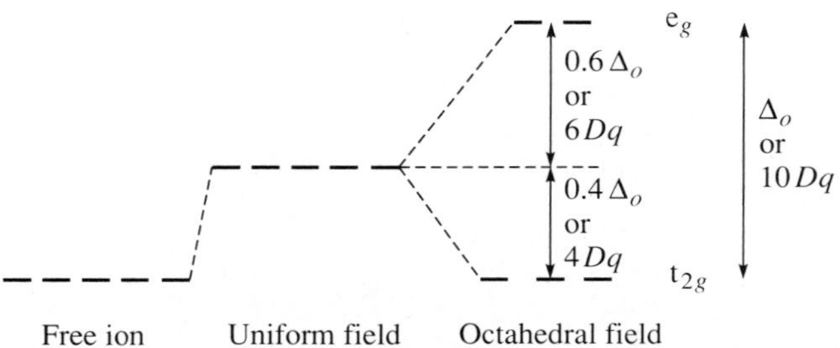

FIGURE 8-24 Crystal Field Splitting.

for the lower (bonding) molecular orbitals, and thus fails to provide a complete picture of the electronic structure.

All these theories can explain the experimentally determined number of unpaired electrons and offer some help in explaining electronic spectra. In this sense, they are all somewhat successful. However, current practice leans toward the molecular orbital, or ligand field, theory because it gives a much more complete picture of the bonding. We will use it for the remainder of this book.

8-9 GROUP THEORY AND NONOCTAHEDRAL COMPLEXES

Group theory can also be used to determine which d orbitals interact with ligand σ orbitals and to obtain a rough idea of the energies of the resulting molecular orbitals. The reducible representation for the ligand σ orbitals is determined and reduced to the irreducible representations as usual. The character table can then be used to determine which of the d orbitals match the representations. A qualitative estimate of the energies can usually be determined by examination of the shapes of the orbitals and their apparent overlap.

As an example, we will consider a trigonal bipyramidal complex. The point group is D_{3h}, and the reducible and irreducible representations are as follows:

D_{3h}	E	$2C_3$	$3C_2$	σ_h	$2S_3$	$3\sigma_v$	Orbitals
Γ	5	2	1	3	0	3	
A_1'	1	1	1	1	1	1	s
A_1'	1	1	1	1	1	1	d_{z^2}
A_2''	1	1	-1	-1	-1	1	p_z
E'	2	-1	0	2	-1	0	$(p_x, p_y), (d_{x^2-y^2}, d_{xy})$

These results show strong interaction with d_{z^2} and somewhat weaker interaction with $d_{x^2-y^2}$ and d_{xy} because their lobes are not directly aimed at the ligands. A partial energy level diagram is shown in Figure 8-25.

These results can be confirmed by the angular overlap method, described earlier in the chapter.

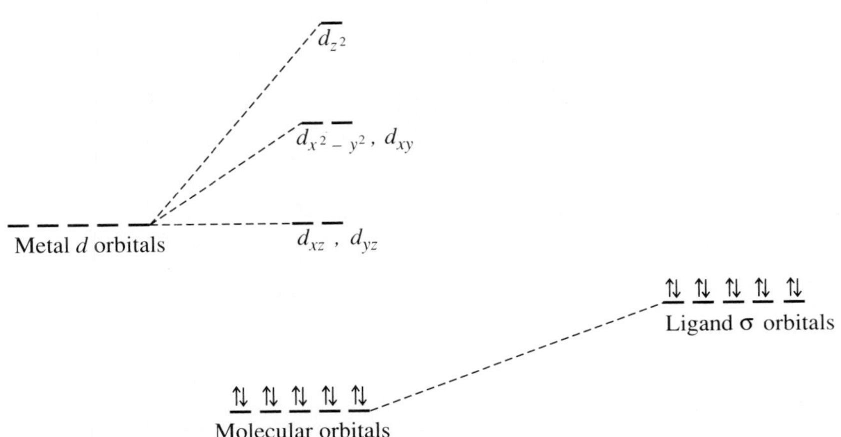

FIGURE 8-25 Trigonal Bipyramidal Energy Levels.

GENERAL REFERENCES

One of the best sources is *Comprehensive Coordination Chemistry,* G. Wilkinson, R. D. Gillard, and J. A. McCleverty, eds., Pergamon Press, Elmsford, N.Y., 1987. Volume 1, Theory and Background, and Volume 2, Ligands, are particularly useful. Others include the books cited in Chapter 4, which include chapters on coordination compounds. Some older, but still useful, sources are C. J. Ballhausen, *Introduction to Ligand Field Theory,* McGraw-Hill, New York, 1962; T. M. Dunn, D. S. McClure, and R. G. Pearson, *Crystal Field Theory,* Harper & Row, New York, 1965; and C. J. Ballhausen and H. B. Gray, *Molecular Orbital Theory,* Benjamin, New York, 1965.

PROBLEMS*

8-1 Predict the number of unpaired electrons for each of the following:

a. A tetrahedral d^6 ion
b. $Co(H_2O)_6^{2+}$
c. $Cr(H_2O)_6^{3+}$
d. A square planar d^7 ion
e. A coordination compound with a magnetic moment of 5.1 Bohr magnetons

8-2 Determine which one of the following is paramagnetic, explain your choice, and estimate its magnetic moment: $Fe(CN)_6^{4-}$, $Co(H_2O)_6^{3+}$, CoF_6^{3-}, or RhF_6^{3-}.

8-3 A compound with the empirical formula $Fe(H_2O)_4(CN)_2$ has a magnetic moment corresponding to $2\frac{2}{3}$ unpaired electrons per iron. How is this possible? (HINT: Two octahedral iron(II) species are involved, each containing a single type of ligand.)

8-4 Show graphically how you would expect ΔH for the reaction

$$M(H_2O)_6^{2+} + 6\ NH_3 \longrightarrow M(NH_3)_6^{2+} + 6\ H_2O$$

to vary for the first transition series (M = Sc through Zn).

8-5 The stepwise stability constants in aqueous solution at 25°C for the formation of the ions $M(H_2O)_4(en)^{2+}$, $M(H_2O)_2(en)_2^{2+}$ and $M(en)_3^{2+}$ for copper and nickel are given in the table. Why is there such a difference in the third values? (HINT: Consider the special nature of d^9 complexes.)

	$M(H_2O)_4(en)^{2+}$	$M(H_2O)_2(en)_2^{2+}$	$M(en)_3^{2+}$
Cu	3×10^{10}	1×10^{9}	0.1
Ni	1×10^{10}	1×10^{10}	1×10^{10}

8-6 Using the angular overlap model, determine the energies of the d orbitals of the metal for each of the following geometries, first for ligands that act as σ donors only, and second for ligands that act both as σ donors and as π acceptors.

a. Linear ML_2
b. Trigonal planar ML_3
c. Square pyramidal ML_5
d. Trigonal bipyramidal ML_5
e. Cubic ML_8 (HINT: A cube is two superimposed tetrahedra.)

* The problems designated by an asterisk will be considered in more detail in Chapter 10.

8-7 Consider a transition metal complex of formula ML_4L'. Using the angular overlap model and assuming trigonal bipyramidal geometry, determine the energies of the *d* orbitals:

a. Considering σ interactions only (assume L and L′ are similar in donor ability).

b. Considering L′ as a π acceptor as well. Consider L′ in both (1) an axial position and (2) an equatorial position.

c. Based on the preceding answers, would you expect π acceptor ligands to preferentially occupy axial or equatorial positions in 5-coordinate complexes? What other factors should be considered besides angular overlap?

8-8* On the basis of your answers to problems 6 and 7, which geometry, square pyramidal or trigonal bipyramidal, is predicted to be more likely for 5-coordinate complexes by the angular overlap model? Consider both σ donor and combined σ donor and π acceptor ligands.

8-9 What are the possible magnetic moments of Co(II) in tetrahedral, octahedral, and square planar complexes?

8-10 Kunze, Perry, and Wilson (*Inorg. Chem.*, **1977,** *16,* 594) prepared monothiocarbamate complexes of Fe(III), in which the ligand forms O—C—S—Fe rings. For the methyl and ethyl compounds, the magnetic moment, μ, is 5.7 to 5.8 at 300 K, changes to 4.70 to 5 at 150 K, and drops still farther to 3.6 to 4 at 78 K. The color changes from red to orange as the temperature is lowered. With larger R groups (propyl, piperidyl, pyrrolidyl), $\mu > 5.3$ at all temperatures and is greater than 6 in some. Explain these changes. Monothiocarbamate complexes have the structure

$$R_2N{-}C(S)(O)M$$

8-11 $Co(H_2O)_6^{3+}$ is a strong oxidizing agent that will oxidize water, while $Co(NH_3)_6^{3+}$ is stable in aqueous solution. Explain this difference. Table 8-4 gives data on the aqueous complex; Δ_o for the ammine complex is about 24,000 cm^{-1}. Both are low-spin complexes.

8-12 Find the number of unpaired electrons, magnetic moment, ground state term symbols, and ligand field stabilization energy for each of the following complexes:

$Co(CO)_4^-$ $\quad$ $Cr(CN)_6^{4-}$ $\quad$ $Fe(H_2O)_6^{3+}$ $\quad$ $Co(NO_2)_6^{4-}$

$Co(NH_3)_6^{3+}$ $\quad$ MnO_4^- $\quad$ $Cu(H_2O)_6^{2+}$

a. Why are two of these compounds tetrahedral and the rest octahedral?

b. Why is tetrahedral geometry more stable for Co(II) than for Ni(II)?

8-13 Explain the order of the magnitudes of the following Δ_o values for Cr(III) complexes in terms of the σ and π donor and acceptor properties of the ligands.

Ligand	F^-	Cl^-	H_2O	NH_3	*en*	CN^-
Δ_o (cm^{-1})	15,200	13,200	17,400	21,600	21,900	33,500

8-14 **a.** Explain the effect on the *d* orbital energies when an octahedral complex is compressed along the *z* axis.

b. Explain the effect on the *d* orbital energies when an octahedral complex is stretched along the *z* axis. In the limit, this results in a square planar complex.

8-15* The 2+ ions in the first transition series generally show a preference for octahedral geometry over tetrahedral geometry. Nevertheless, the number of tetrahedral complexes formed is in the order Co > Fe > Ni.

a. Calculate the ligand field stabilization energies for tetrahedral and octahedral symmetries for these ions. Do these numbers explain this order?

b. Does the angular overlap model offer any advantage in explaining this order?

8-16* Except in cases where ligand geometry requires it, square planar geometry is found only in d^7, d^8, and d^9 ions and with strong-field, π-acceptor ligands. Explain why these restrictions apply.

8-17 Oxygen is more electronegative than nitrogen; fluorine is more electronegative than the other halogens. Fluoride is a stronger field ligand than the other halides, but ammonia is a stronger field ligand than water. Why?

8-18 Use the group theory approach of Section 8-9 to prepare a molecular orbital diagram for a square pyramidal complex.

8-19 Solid CrF_3 contains a Cr(III) ion surrounded by six F^- ions in an octahedral geometry, all at distances of 190 pm. However, MnF_3 is in a distorted geometry, with Mn—F distances of 179, 191, and 209 pm. Suggest an explanation.

8-20 On the basis of molecular orbitals, explain why the Mn—O distance in MnO_4^{2-} is longer (by 3.9 pm) than in MnO_4^- (G. J. Palenik, *Inorg. Chem.*, **1967,** *6,* 503, 507).

8-21 Predict the magnetic moments (spin-only) of the following species:

a. $Cr(H_2O)_6^{2+}$ **b.** $Cr(CN)_6^{4-}$

c. $FeCl_4^-$ **d.** $Fe(CN)_6^{3-}$

e. $Ni(H_2O)_6^{2+}$ **f.** $Cu(en)_2(H_2O)_2^{2+}$

8-22 Use the angular overlap method to calculate the energies of both the ligand and metal orbitals for *trans*-$[Cr(NH_3)_4Cl_2]^+$, allowing for the fact that ammonia is a stronger σ donor ligand than chloride, but chloride is a stronger π donor. Use the 1 and 6 positions for the chloride ions.

9

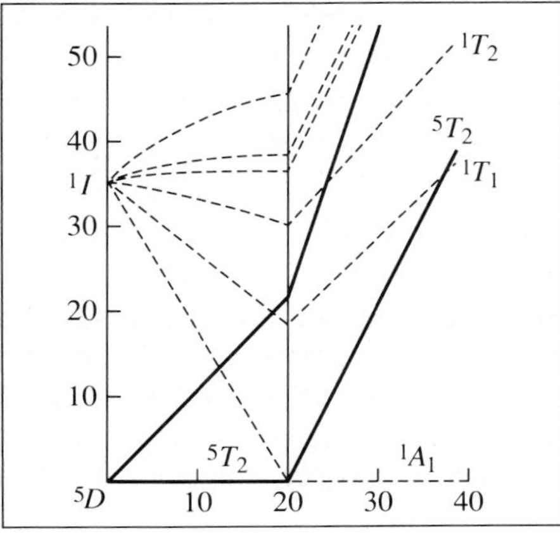

Coordination Chemistry II: Electronic Spectra

Perhaps the most striking aspect of many coordination compounds of transition metals is that they exhibit vivid colors. The dye Prussian blue, for example, has been used as a pigment for more than two centuries (and is still used in blueprints); it is a complicated coordination compound involving iron(II) and iron(III) coordinated octahedrally by cyanide. Many precious gems exhibit colors due to transition metal ions incorporated into their crystalline lattices. For example, emeralds are green as a consequence of incorporation of small amounts of chromium(III) into crystalline $Be_3Al_2Si_6O_{18}$; amethysts are violet as a result of the presence of small amounts of iron(II), iron(III), and titanium(IV) in an Al_2O_3 lattice; and rubies are red because of chromium(III), also in a lattice of Al_2O_3. The red color of blood is due to the red heme group, a coordination compound of iron present in hemoglobin. Most readers are probably familiar with blue $CuSO_4 \cdot 5\ H_2O$, a compound often used to demonstrate the growing of large, highly symmetric crystals.

It is desirable to understand just why so many coordination compounds are colored, in contrast to the majority of organic compounds, which are transparent, or nearly so, in the visible spectrum. In achieving such an understanding, it first will be necessary to review the concept of light absorption and how it is measured. The ultraviolet and visible spectra of coordination compounds of transition metals involve transitions between the *d* orbitals of the metals; we will need to look closely, therefore, at the energies of these orbitals (as discussed in Chapter 8) and at the possible ways in which electrons can be raised from lower to higher energy levels. In this connection, the energy levels of *d* electron configurations (as opposed to the energies of *individual* electrons) are somewhat more complicated than might be expected, and we will need to review how the quantum numbers of individual electrons can combine into term symbols that describe these configurations (as originally discussed in Chapter 2).

For many coordination compounds, the electronic absorption spectrum provides a convenient method for determining the magnitude of the effect of ligands on the *d* orbitals of the metal. While in principle we can study this effect for coordination compounds of any geometry, we will concentrate on the most common geometry, octahedral, and examine how the absorption spectrum can be used to determine the magnitude of the octahedral ligand field parameter Δ_o for a variety of complexes.

9-1 ABSORPTION OF LIGHT

In explaining the colors of coordination compounds, we are dealing with the phenomenon of *complementary colors*. If a compound absorbs light of one color, we see the complement of that color. For example, when white light (containing a broad spectrum of all visible wavelengths) passes through a substance that absorbs red light, the color observed is green; green is the complement of red, the color that predominates visually when red light is subtracted from white.

An example from coordination chemistry is the deep blue color of aqueous solutions of copper(II) compounds, containing the ion $[Cu(H_2O)_6]^{2+}$. The blue color is a consequence of absorption of light between approximately 600 and 1000 nm (maximum near 800 nm; see Figure 9-1) in the yellow to infrared region of the spectrum. The color observed, blue, is the complementary color of the light absorbed.

It is not always possible to make a simple prediction of color directly from the absorption spectrum, in large part because many coordination compounds contain two or more absorption bands of different energies and intensities. The net color observed is the color predominating after the various absorptions are removed from white light.

For reference, the approximate wavelengths and complementary colors to the principal colors of the visible spectrum are given in Table 9-1.

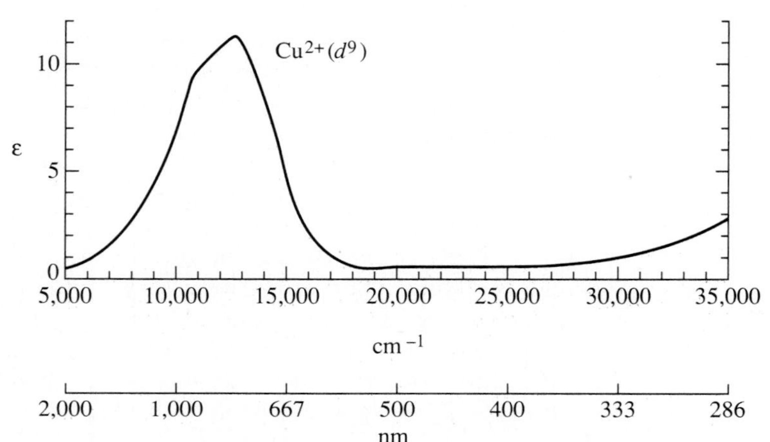

FIGURE 9-1 Absorption Spectrum of $[Cu(H_2O)_6]^{2+}$. (Reproduced with permission from B. N. Figgis, *Introduction to Ligand Fields*, Wiley-Interscience, New York, 1966, p. 221.)

TABLE 9-1
Visible light and complementary colors

Wavelength range (nm)	*Wave numbers (cm^{-1})*	*Color*	*Complementary color*
<400	>25,000	Ultraviolet	
400–450	22,200–25,000	Violet	Yellow
450–490	20,400–22,200	Blue	Orange
490–550	18,200–20,400	Green	Red
550–580	17,200–18,200	Yellow	Violet
580–650	15,400–17,200	Orange	Blue
650–700	14,300–15,400	Red	Green
>700	<14,300	Infrared	

9-1-1 BEER–LAMBERT ABSORPTION LAW

If light of intensity I_0 at a given wavelength passes through a solution containing a species that absorbs light, the light emerges with intensity I, which may be measured by a suitable detector (Figure 9-2). The Beer–Lambert law may be used to describe the absorption of light (ignoring scattering and reflection of light from cell surfaces) at a given wavelength by an absorbing species in solution:

$$\log \frac{I_0}{I} = A = \epsilon l c$$

where A = absorbance
ϵ = molar absorptivity (L mol^{-1} cm^{-1}) (also known as molar extinction coefficient)
l = path length through solution (cm)
c = concentration of absorbing species (mol L^{-1})

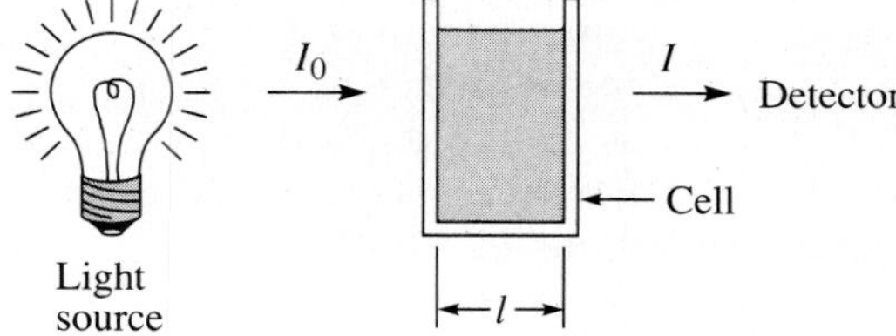

FIGURE 9-2 Absorption of Light by Solution.

Absorbance is a dimensionless quantity. An absorbance of 1.0 corresponds to 90% absorption at a given wavelength,[1] an absorbance of 2.0 corresponds to 99% absorption, and so on. The most common units of the other quantities in the Beer–Lambert law are shown in parentheses above.

Spectrophotometers commonly obtain spectra as plots of absorbance versus wavelength. The molar absorptivity is a characteristic of the species that is absorbing the light and is highly dependent on wavelength. A plot of molar absorptivity versus wavelength gives a spectrum that is characteristic of the molecule or ion in question, as in Figure 9-1. As we will see, this

[1] For absorbance = 1.0: $\log (I_0/I) = 1.0$. Therefore, $I_0/I = 10$, and $I = 0.10\ I_0 = 10\% \times I_0$. Ten percent of the light is transmitted, 90% is absorbed.

spectrum is a consequence of transitions between states of different energies and can provide valuable information about those states and, in turn, about the structure and bonding of the molecule or ion.

Although the quantity most commonly used to describe absorbed light is the wavelength, energy and frequency are also used. In addition, the wave number (the number of waves per centimeter), a quantity proportional to the energy, is frequently used, especially in reference to infrared light. For reference, the relations between these quantities are

$$E = h\nu = \frac{hc}{\lambda} = hc\frac{1}{\lambda} = hc\overline{\nu}$$

where

E = energy

h = Planck's constant = 6.626×10^{-34} J s

c = speed of light = 2.998×10^{8} m/s

ν = frequency (s^{-1})

λ = wavelength (often reported in nm)

$\frac{1}{\lambda} = \overline{\nu}$ = wave number (cm^{-1})

9-2 ELECTRONIC SPECTRA OF COORDINATION COMPOUNDS

Absorption of light results in the excitation of electrons from lower energy states to higher states; since such states are quantized, we observe absorption in bands (as in Figure 9-1), with the energy of each band corresponding to the difference in energy between the initial and final states. At this point, a review of term symbols used to describe these states is needed. If the brief discussion that follows is insufficient, review of Section 2-3 is recommended. An understanding of LS coupling, microstates, and free ion terms is essential background for the discussion of electronic spectra that follows.

In Chapter 2 a method for determining the microstates and free ion terms for electron configurations was presented. For example, a d^2 configuration gives rise to five free ion terms, ^3F, ^3P, ^1G, ^1D, and ^1S, with the ^3F term of lowest energy. Absorption spectra of coordination compounds in most cases involve the d orbitals of the metal, and it is consequently important to know the free ion terms for the possible d configurations. Determining the microstates and free ion terms for configurations of three or more electrons can be a tedious process. For reference, therefore, these are listed for the possible d electron configurations in Table 9-2.

TABLE 9-2
Free ion terms for d^n configurations

Configuration	*Free ion terms*						
d^1	^2D						
d^2		$^1S\ ^1D\ ^1G$	$^3P\ ^3F$				
d^3	^2D		$^4P\ ^4F$	$^2P\ ^2D\ ^2F\ ^2G\ ^2H$			
d^4	^5D	$^1S\ ^1D\ ^1G$	$^3P\ ^3F$	$^3P\ ^3D\ ^3F\ ^3G\ ^3H$	$^1S\ ^1D\ ^1F\ ^1G\ ^1I$		
d^5	^2D		$^4P\ ^4F$	$^2P\ ^2D\ ^2F\ ^2G\ ^2H$	$^2S\ ^2D\ ^2F\ ^2G\ ^2I$	^6S	$^4D\ ^4G$
d^6	Same as d^4						
d^7	Same as d^3						
d^8	Same as d^2						
d^9	Same as d^1						
d^{10}	^1S						

NOTE: For any configuration, the free ion terms are the sum of those listed; for example, for the d^2 configuration the free ion terms are $^1S + {}^1D + {}^1G + {}^3P + {}^3F$.

In the interpretation of spectra of coordination compounds, it is often important to identify the ground term. A quick and fairly simple way to do this is given here, using the examples of d^3 and d^4 configurations in octahedral symmetry.

1. Sketch the energy levels, showing the *d* electrons.	↑ ↑ ↑ (lower level); two empty upper levels
2. Spin multiplicity of lowest energy state = number of unpaired electrons + 1.[2]	Spin multiplicity = 3 + 1 = 4
3. Determine the maximum possible value of M_L (= sum of m_l values) for the configuration as shown. This determines the type of free ion term (*S*, *P*, *D*, etc.).	Maximum possible value of M_L for three electrons as shown: 2 + 1 + 0 = 3. Therefore, *F* term
4. Combine results of steps 2 and 3 to get ground term.	^4F

EXAMPLE

d^4 (low spin):

1. ⇅ ↑ ↑ (lower level); two empty upper levels

2. Spin multiplicity = 2 + 1 = 3

3. Highest possible value of M_L = 2 + 2 + 1 + 0 = 5; therefore, *H* term.

4. Ground term is therefore ^3H.

EXERCISE 9-1

Determine the ground terms for high-spin and low-spin d^6 configurations in O_h symmetry.

With this review of atomic states, we may now consider the electronic states of coordination compounds and how transitions between these states can give rise to the observed spectra. Before considering specific examples of spectra, however, we must also consider which types of transitions are most probable and therefore give rise to the most intense absorptions.

9-2-1 SELECTION RULES

The relative intensities of absorption bands are governed by a series of selection rules. On the basis of the symmetry and spin multiplicity of ground and excited electronic states, two of these rules may be stated as follows:[3,4]

[2] This is equivalent to spin multiplicity = 2*S* + 1, as shown in Chapter 2.

[3] B. N. Figgis, *Introduction to Ligand Fields*, Wiley-Interscience, New York, 1966, pp. 203–47.

[4] B. N. Figgis, "Ligand Field Theory," in *Comprehensive Coordination Chemistry*, Vol. 1, G. Wilkinson, R. D. Gillard, and J. A. McCleverty, eds., Pergamon, Elmsford, N.Y., 1987, pp. 243–46.

1. Transitions between states of the same parity (symmetry with respect to a center of inversion) are forbidden. This means that transitions between *d* orbitals are forbidden ($g \longrightarrow g$ transitions; *d* orbitals are symmetric to inversion), but transitions between *d* and *p* orbitals are allowed ($g \longrightarrow u$ transitions; *p* orbitals are antisymmetric to inversion). This is known as the **Laporte selection rule**.
2. Transitions between states of different spin multiplicities are forbidden. For example, transitions between $^4A_2$ and $^4T_1$ states are spin-allowed, but between $^4A_2$ and $^2A_2$ are spin-forbidden.

These rules would seem to rule out most electronic transitions for transition metal complexes. However, many such complexes are vividly colored, a consequence of various mechanisms by which these rules can be relaxed. Some of the most important of these mechanisms are:

1. The bonds in transition metal complexes are not rigid but undergo vibrations that may temporarily change the symmetry. Octahedral complexes, for example, vibrate in ways in which the center of symmetry is temporarily lost; this provides a way to relax the first selection rule. As a consequence, *d–d* transitions having molar absorptivities in the range of approximately 10 to 50 L mol^{-1} cm^{-1} commonly occur (and are often responsible for the bright colors of many of these complexes).
2. Tetrahedral complexes often absorb more strongly than octahedral complexes of the same metal in the same oxidation state. Metal–ligand sigma bonding in transition metal complexes of T_d symmetry can be described as involving a combination of sp^3 and sd^3 hybridization of the metal orbitals; both types of hybridization are consistent with the symmetry. The resulting mixing of *p* orbital character (of *u* symmetry) with *d* orbital character provides a second way of relaxing the first selection rule.
3. Spin–orbit coupling in some cases provides a mechanism of relaxation of the second selection rule, with the result that transitions may be observed from a ground state of one spin multiplicity to an excited state of different spin multiplicity. Such absorption bands are usually very weak, with typical molar absorptivities less than 1 L mol^{-1} cm^{-1}.

Examples of spectra illustrating the selection rules and the ways in which they may be relaxed are given in the following sections of this chapter. Our first example will be a metal complex having a d^2 configuration and octahedral geometry: $[V(H_2O)_6]^{3+}$.

In discussing spectra it will be particularly useful to be able to relate the electronic spectra of transition metal complexes to the ligand field splitting, Δ_o for octahedral complexes. To do this it will be necessary to introduce two special types of diagrams, **correlation diagrams** and **Tanabe–Sugano diagrams**.

9-2-2 CORRELATION DIAGRAMS

These diagrams make use of two extremes:

1. *Free ions* (no ligand field): In Chapter 2 the terms ^3F, ^3P, ^1G, ^1D, and ^1S were obtained for a d^2 configuration, with the ^3F term being of lowest energy. These terms describe the energy levels of a "free" d^2 ion (in our example, a V^{3+} ion), in the absence of any interactions with ligands. In

the correlation diagrams to be constructed, we will show these free ion terms on the far left.

2. *Strong ligand field:* There are three possible configurations for two d electrons in an octahedral ligand field (dots indicate a spin of either $+\frac{1}{2}$ or $-\frac{1}{2}$):

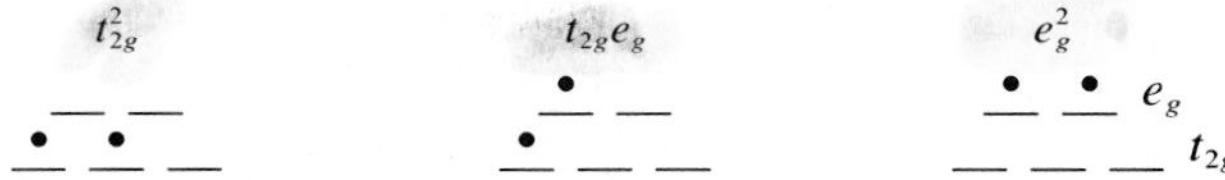

In our example, these would be the possible electron configurations of V^{3+} in an extremely strong ligand field (t_{2g}^2 would be the ground state; the others would be excited states). In correlation diagrams, we will show these states on the far right, as the strong field limit, that is, the situation in which the effect of the ligands is so strong as to completely override any effects that might arise from LS coupling.

In actual coordination compounds, the situation is intermediate between these extremes. The m_l and m_s values of the individual electrons couple to form, for d^2, the five terms ^3F, ^3P, ^1G, ^1D, and ^1S; these terms will represent five different atomic states having different energies. At the same time, effects of the ligands are important, too, and in our example tend to push the energy levels toward the t_{2g}^2, $t_{2g}e_g$, and e_g^2 configurations. In constructing a diagram to correlate the free ion and strong field limits, we will seek a way to deal with the in-between case in which both factors are important.

Some details of the method for achieving this are beyond the scope of this text; the interested reader should consult the literature[5] for details omitted here. The aspect of this problem that will interest us is that free ion terms (shown on the far left in the correlation diagrams) have symmetry characteristics that enable them to be reduced to their constituent irreducible representations (in our example, these will be irreducible representations in the O_h point group). These terms have the same symmetry as atomic orbitals of similar designation (for example, a P term will have the same symmetry as a p atomic orbital, a D term the same symmetry as a d orbital and so on). In an octahedral ligand field the free ion terms will be split into states corresponding to the irreducible representations as shown in Table 9-3.

TABLE 9-3
Splitting of free ion terms in octahedral symmetry

Term	*Irreducible representations*
S	A_{1g}
P	T_{1g}
D	$E_g + T_{2g}$
F	$A_{2g} + T_{1g} + T_{2g}$
G	$A_{1g} + E_g + T_{1g} + T_{2g}$
H	$E_g + 2T_{1g} + T_{2g}$
I	$A_{1g} + A_{2g} + E_g + T_{1g} + 2T_{2g}$

NOTE: Although representations based on atomic orbitals may have either g or u symmetry, the terms given here are for d orbitals and as a result have only g symmetry. See pp. 263–64 of footnote 5 for a discussion of these labels.

[5] F. A. Cotton, *Chemical Applications of Group Theory*, 3rd ed., Wiley-Interscience, New York, 1990, Chapter 9.

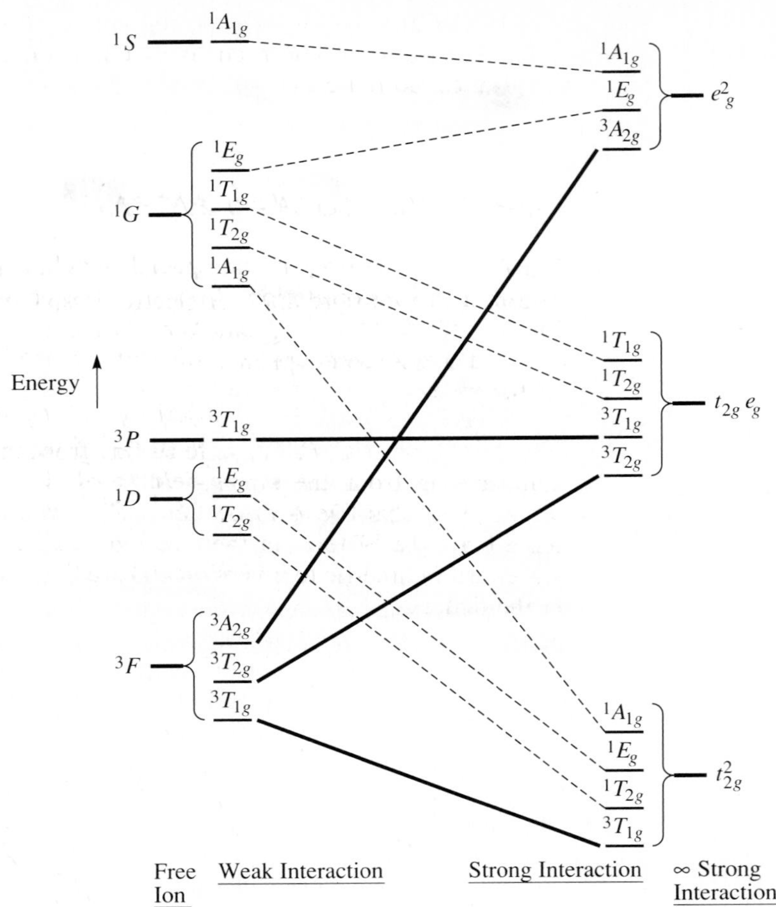

FIGURE 9-3 Correlation Diagram for d^2 in Octahedral Ligand Field.

Likewise, irreducible representations may be obtained for the strong-field limit configurations (in our example, t_{2g}^2, $t_g e_g$, and e_g^2). The irreducible representations for the two limiting situations *must* match; each irreducible representation for the free ion must match, or correlate with, a representation for the strong-field limit. A diagram showing these correlations is designated a correlation diagram; that for d^2 is shown in Figure 9-3.

Note especially the following characteristics of this correlation diagram:

1. The free ion states (terms) are shown on the far left.
2. The extremely strong field states are shown on the far right.
3. Both the free ion and strong-field states can be reduced to irreducible representations, as shown. Each free ion irreducible representation is matched with (correlates with) a strong-field irreducible representation having the same symmetry (same label). As mentioned in Section 9-2-1, transitions to excited states having the same spin multiplicity as the ground state are more likely than transitions to states of different spin multiplicity; consequently, states of the same spin multiplicity as the ground state are shown as heavy lines, and states having other spin multiplicities are shown as dashed lines.

In the correlation diagram the states are shown in order of energy. A noncrossing rule is observed; lines connecting states of the same symmetry designation do not cross.

9-2-3 TANABE–SUGANO DIAGRAMS

Tanabe–Sugano diagrams are special correlation diagrams that are particularly useful in the interpretation of electronic spectra of coordination compounds.[6] In Tanabe–Sugano diagrams the lowest energy state is plotted along the horizontal axis; consequently, the vertical distance above this axis is a measure of the energy above the ground state. For example, for the d^2 configuration the lowest energy state is described by the line in the correlation diagram (Figure 9-3) joining the $^3T_1$ state arising from the ^3F free ion term with the $^3T_1$ state arising from the strong-field term t_2^2. In the Tanabe–Sugano diagram (Figure 9-4) this line is made horizontal; it is labeled $^3T_1(F)$ and is shown to arise from the ^3F term in the free ion limit. In octahedral ligand fields these states are symmetric to inversion and are therefore designated with g subscripts in the figures.

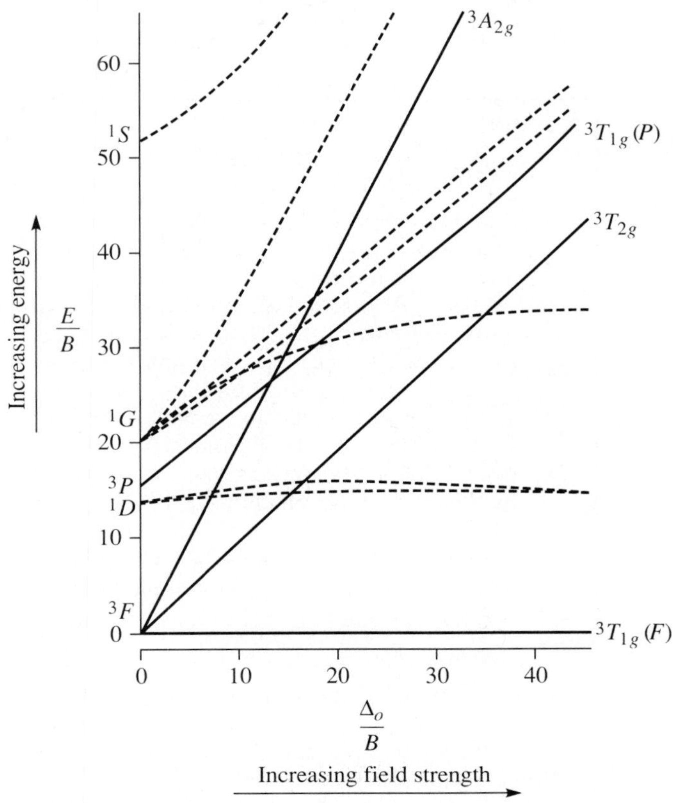

FIGURE 9-4 Tanabe–Sugano Diagram for d^2 in Octahedral Ligand Field.

[6] Y. Tanabe and S. Sugano, *J. Phys. Soc. Japan*, **1954**, *9*, 753, 766.

The Tanabe–Sugano diagram also shows excited states. In the d^2 diagram the excited states of same spin multiplicity as the ground state are the ${}^3T_{2g}$, ${}^3T_{1g}$ (P), and the ${}^3A_{2g}$. The reader should verify that these are the same triplet excited states shown in the d^2 correlation diagram. Excited states of other spin multiplicities are also shown, but, as we will see, they are generally not as important in the interpretation of spectra.

The quantities plotted in a Tanabe–Sugano diagram are as follows:

Horizontal axis: $\frac{\Delta_o}{B}$ where Δ_o is the octahedral ligand field splitting, described in Chapter 8.

B = **Racah parameter**, a measure of the repulsion between terms of the same multiplicity. For d^2, for example, the energy difference between ${}^3F$ and ${}^3P$ is $15B$. (For a discussion of Racah parameters, see page 232 of footnote 4.)

${}^3P$
↑
$15B$
↓
${}^3F$

Vertical axis: $\frac{E}{B}$ where E is the energy above the ground state.

As mentioned above, one of the most useful characteristics of Tanabe–Sugano diagrams is that *the ground electronic state is always plotted along the horizontal axis*; this makes it easy to determine values of E/B above the ground state.

EXAMPLE

$[V(H_2O)_6]^{3+}$ (d^2)

A good example of the utility of Tanabe–Sugano diagrams in explaining electronic spectra is provided by the d^2 complex $[V(H_2O)_6]^{3+}$. The ground state is ${}^3T_{1g}(F)$; under ordinary conditions this is the only electronic state that is appreciably occupied. Absorption of light should occur primarily to excited states also having a spin multiplicity of 3. There are three of these: ${}^3T_{2g}$, ${}^3T_{1g}(P)$, and ${}^3A_{2g}$. Therefore, three allowed transitions are expected, as shown in Figure 9-5. Consequently, we expect three absorption bands for $[V(H_2O)_6]^{3+}$, one corresponding to each allowed transition. Is this actually observed for $[V(H_2O)_6]^{3+}$? Two bands are readily observed at 17,800 and 25,700 cm^{-1}, as can be seen in Figure 9-6. A third band, at approximately 38,000 cm^{-1}, is apparently obscured in aqueous solution by charge transfer bands nearby[7] (charge transfer bands of coordination compounds are discussed later in this chapter). In the solid state, however, a band attributed to the ${}^3T_{1g} \longrightarrow {}^3A_{2g}$ transition is observed at 38,000 cm^{-1}. These bands match the transitions ν_1, ν_2, and ν_3 indicated on the Tanabe–Sugano diagram (Figure 9-5).

Other electron configurations

Tanabe–Sugano diagrams can be prepared for all possible d electron configurations. Diagrams for d^2 through d^8 are shown in Figure 9-7. The diagrams for d^4, d^5, d^6, and d^7 have apparent discontinuities, marked by vertical lines near

[7] The third band is in the ultraviolet and is off scale to the right in the spectrum shown. See Figgis, op. cit., p. 251.

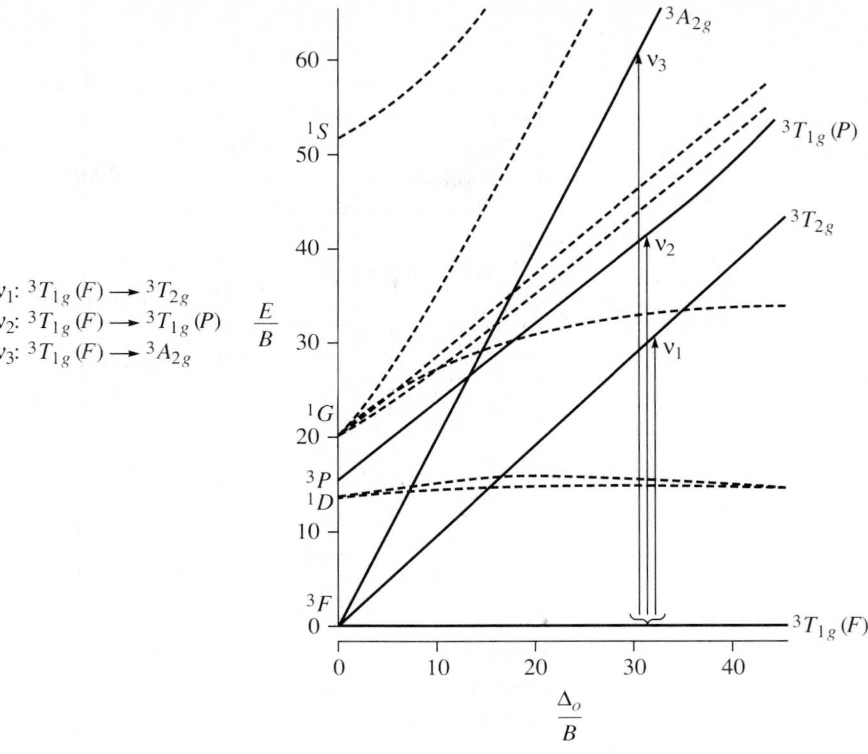

FIGURE 9-5 Spin-allowed Transitions for d^2 Configuration.

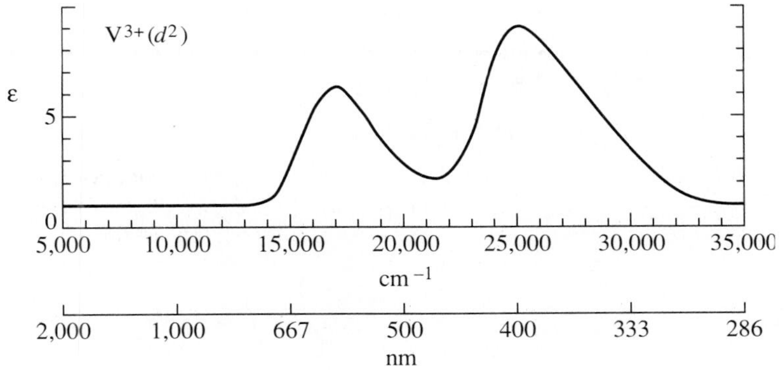

$V^{3+}(d^2)$

ε

5

0

5,000 10,000 15,000 20,000 25,000 30,000 35,000

cm^{-1}

2,000 1,000 667 500 400 333 286

nm

FIGURE 9-6 Absorption Spectrum of $[V(H_2O)_6]^{3+}$. (Reproduced with permission from B. N. Figgis, *Introduction to Ligand Fields*, Wiley-Interscience, New York, 1966, p. 221.)

the center. These are configurations for which both low and high spin are possible. For example, consider the configuration d^4.

High-spin (weak field) d^4 has 4 unpaired electrons, of parallel spin; such a configuration has a spin multiplicity of 5.	↑ __ ↑ ↑ ↑	$S = 4(\frac{1}{2}) = 2$ $M_s = 2S + 1 = 2(2) + 1 = 5$
Low-spin (strong field) d^4, on the other hand, has only 2 unpaired electrons and a spin multiplicity of 3.	__ __ ⇅ ↑ ↑	$S = 2(\frac{1}{2}) = 1$ $M_s = 2S + 1 = 2(1) + 1 = 3$

In the weak-field part of the Tanabe–Sugano diagram (left of $\Delta_o/B = 27$), the ground state is ${}^5E_g$, having the expected spin multiplicity of 5. On the right

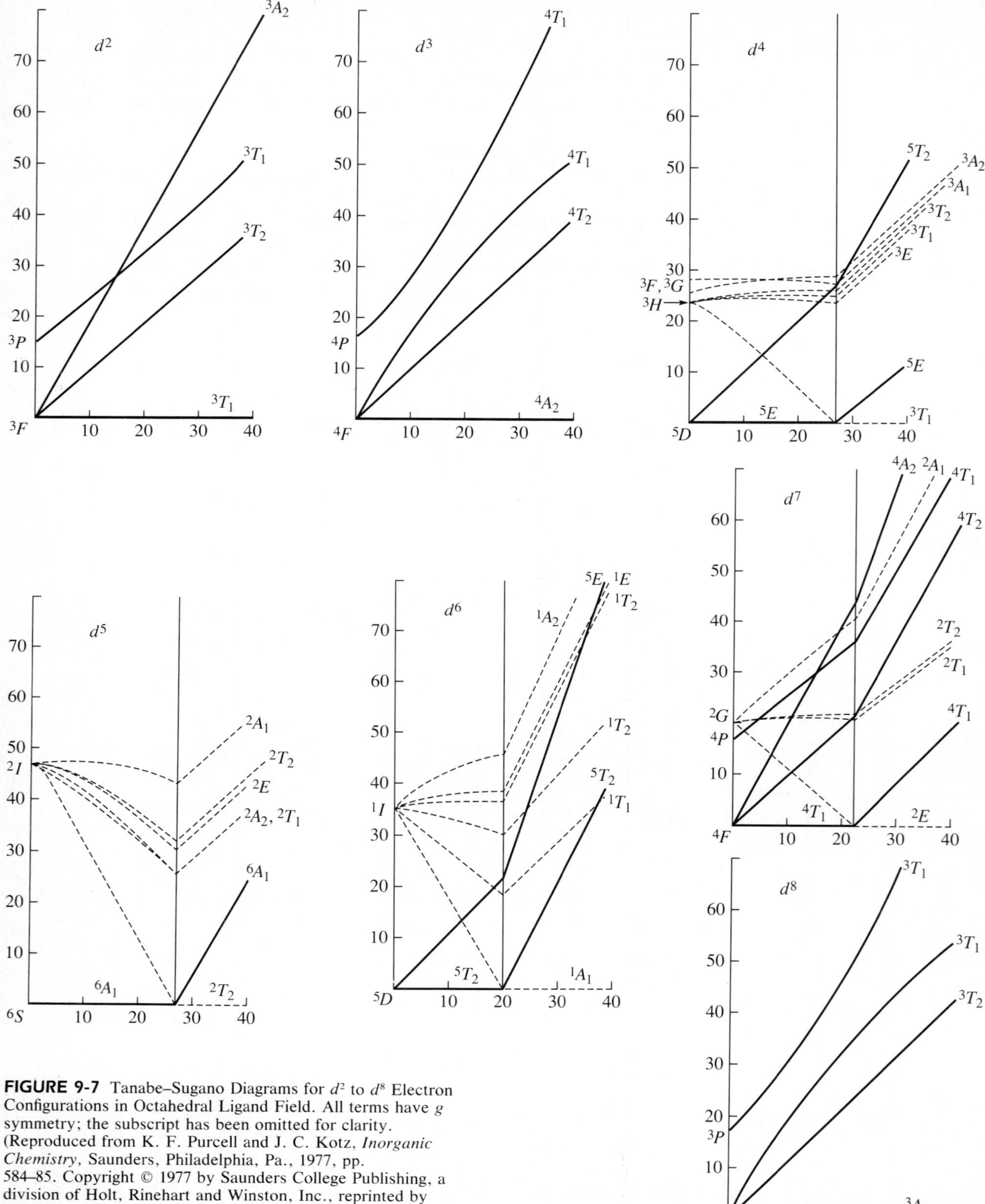

FIGURE 9-7 Tanabe–Sugano Diagrams for d^2 to d^8 Electron Configurations in Octahedral Ligand Field. All terms have *g* symmetry; the subscript has been omitted for clarity. (Reproduced from K. F. Purcell and J. C. Kotz, *Inorganic Chemistry*, Saunders, Philadelphia, Pa., 1977, pp. 584–85. Copyright © 1977 by Saunders College Publishing, a division of Holt, Rinehart and Winston, Inc., reprinted by permission of the publisher.)

(strong-field) side of the diagram the ground state is ${}^3T_{1g}$ (correlating with the ${}^3H$ term in the free ion limit), having the required spin multiplicity of 3. The vertical line is thus a dividing line between weak- and strong-field cases: high-spin (weak-field) complexes are to the left of this line, low-spin (strong-field) complexes to the right. At the dividing line, the ground state changes from ${}^5E_g$ to ${}^3T_{1g}$.

In Figure 9-8 are diagrams of spectra of first-row transition metal complexes of formula $[M(H_2O)_6]^{n+}$. It is an interesting exercise to compare the number of bands in these spectra with the number of bands expected from the respective Tanabe–Sugano diagrams. Note that in some cases absorption bands are off scale, farther into the ultraviolet than the spectral region shown.

In Figure 9-8, molar absorptivities (extinction coefficients) are shown on the vertical scale. The absorptivities for most bands are similar (1 to 20 L/mol

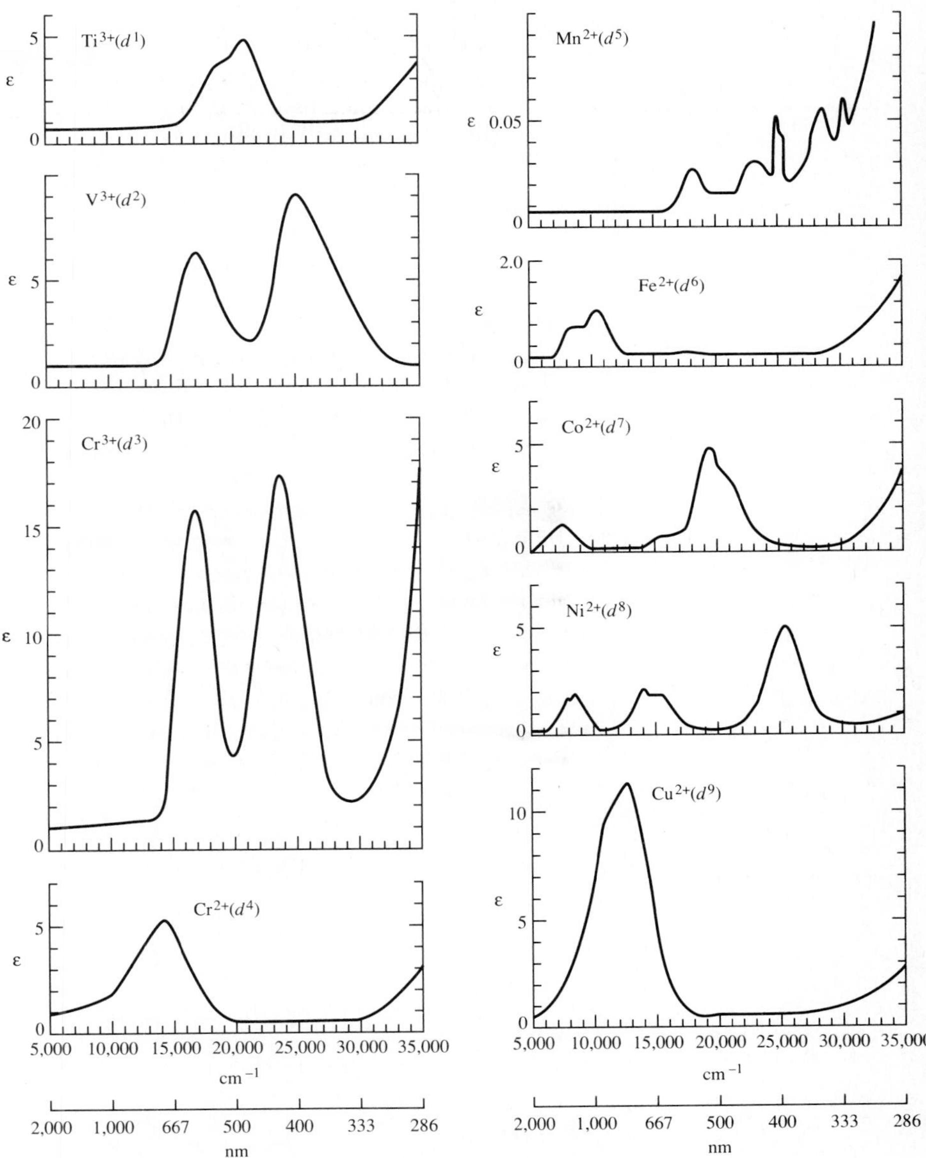

FIGURE 9-8 Electronic Spectra of First-row Transition Metal Complexes of Formula $[M(H_2O)_6]^{n+}$. (Reproduced with permission from B. N. Figgis, *Introduction to Ligand Fields*, Wiley-Interscience, New York, 1966, pp. 221, 224.)

cm) except for the spectrum of $[Mn(H_2O)_6]^{2+}$, which has much weaker bands. Solutions of $[Mn(H_2O)_6]^{2+}$ are an extremely pale pink, much more weakly colored than solutions of the other ions shown. Why is absorption by $[Mn(H_2O)_6]^{2+}$ so weak? To answer this question, it is necessary to examine the corresponding Tanabe–Sugano diagram, in this case for a d^5 configuration. We expect $[Mn(H_2O)_6]^{2+}$ to be a high-spin complex, since H_2O is a rather weak field ligand. The ground state for weak-field d^5 is the ${}^6A_{1g}$. There are *no* excited states of the same spin multiplicity (6), and consequently there can be no spin-allowed absorptions. That $[Mn(H_2O)_6]^{2+}$ is colored at all is a consequence of very weak forbidden transitions to excited states of spin multiplicity other than 6 (there are many such excited states, hence the rather complicated spectrum).

9-2-4 JAHN–TELLER DISTORTIONS AND SPECTRA

To this point we have not discussed the spectra of d^1 and d^9 complexes. By virtue of the simple d electron configurations for these cases, we might expect each to exhibit one absorption band corresponding to excitation of an electron from the t_{2g} to the e_g levels:

d^1 $\qquad\qquad$ d^9

However, this view must be at least a modest oversimplification, since examination of the spectra of $[Ti(H_2O)_6]^{3+}$ (d^1) and $[Cu(H_2O)_6]^{2+}$ (d^9) (see Figure 9-8) shows these coordination compounds to exhibit two closely overlapping absorption bands rather than a single band.

To account for the apparent splitting of bands in these examples, it is necessary to consider one additional aspect of electron configurations; that is, in some cases a configuration can lead to distortion of a molecule. In 1937, Jahn and Teller showed that nonlinear molecules having a degenerate electronic state should distort to lower the symmetry of the molecule and to reduce the degeneracy; this is commonly called the Jahn–Teller theorem, as discussed in Chapter 8.[8] For example, a d^9 metal in an octahedral complex has the electron configuration $t_{2g}^6e_g^3$; according to the Jahn–Teller theorem, such a complex should distort. If the distortion takes the form of an elongation along the z axis (the most common distortion observed experimentally), the t_{2g} and e_g orbitals are affected as shown:

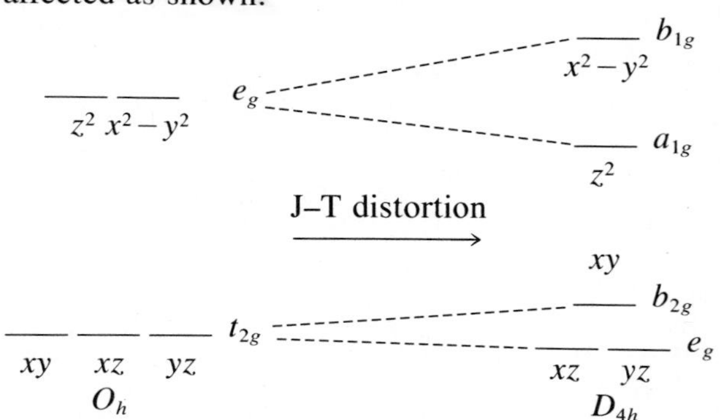

[8] I. B. Bersuker, *Coord. Chem. Rev.*, **1975**, *14*, 357.

Distortion from O_h to D_{4h} symmetry results in stabilization of the molecule: two e_g electrons are stabilized, while one is destabilized by the same amount. The t_{2g} split is much smaller than the e_g split because the t_{2g} orbitals are pointed between the ligands, while the e_g orbitals are pointed directly at the ligands.

When degenerate orbitals are asymmetrically occupied, Jahn–Teller distortions are likely. For example, the first two configurations below should give distortions, while the second two should not:

In practice, the only electron configurations for O_h symmetry that give rise to measurable Jahn–Teller distortions are those that have asymmetrically occupied e_g orbitals, such as the first configuration shown above. By far the most common distortion observed is elongation along the z axis. Although the Jahn–Teller theorem predicts that configurations having asymmetrically occupied t_{2g} orbitals, such as the second configuration above, should also be distorted, such distortions are too small to be measured in most cases. The structural consequences of Jahn–Teller distortions will be considered more fully in Chapter 10.

The Jahn–Teller effect on spectra can easily be seen from the example of $[Cu(H_2O)_6]^{2+}$, a d^9 complex. From the diagram at left showing the effect on d orbitals of distortion from O_h to D_{4h} geometry, we can see the additional splitting of orbitals accompanying the reduction of symmetry. In terms of *electron configurations*, the labels are reversed, as in Figure 9-9. The ^2D free-ion term is split into ^2E_g and $^2T_{2g}$ by a field of O_h symmetry and further split on distortion to D_{4h} symmetry. The labels of the states resulting from the free-ion term (Figure 9-9) are in reverse order to the labels on the orbitals; for example, the b_{1g} atomic orbital is of highest energy, whereas the B_{1g} state originating from the ^2D free-ion term is of lowest energy.[9]

For a d^9 configuration the ground state in octahedral symmetry is a ^2E_g term, the excited state a $^2T_{2g}$. On distortion to D_{4h} geometry, these terms split,

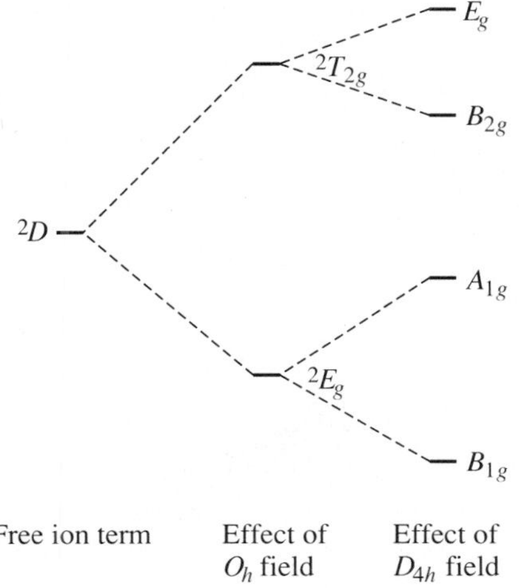

FIGURE 9-9 Splitting of Octahedral Ligand Field Terms on Jahn–Teller Distortion for d^9 Configuration.

[9] Figgis, op. cit., pp. 253, 255.

as shown in Figure 9-9. In an octahedral d^9 complex, we would expect excitation from the ^2E_g state to the $^2T_{2g}$ state and a single absorption band. Distortion of the complex to D_{4h} geometry splits the $^2T_{2g}$ level into two levels, the E_g and the B_{2g}. Excitation can now occur from the ground state (now the B_{1g} state) to either the E_g or the B_{2g}[10] (the splitting is exaggerated in Figure 9-9). If the distortion is strong enough, therefore, two separate absorption bands may be observed (more commonly, a broadened or narrowly split peak is observed, as in $[Cu(H_2O)_6]^{2+}$).

In predicting a spectrum for a d^1 complex, it is perhaps tempting to consider a simple excitation of a t_{2g} electron to an e_g orbital, with a single absorption band expected:

		$h\nu$ →	
— —	e_g		↑ —	e_g
↑ — —	t_{2g}		— — —	t_{2g}
Ground state ($^2T_{2g}$)			Excited state (^2E_g)	

However, examination of the spectrum of $[Ti(H_2O)_6]^{3+}$, an example of a d^1 complex, shows two apparently overlapping bands rather than a single band. How is this possible?

In this case the excited state can undergo Jahn–Teller distortion. As in the examples above, asymmetric occupation of the e_g orbitals can split these orbitals into two of slightly different energy (of A_{1g} and B_{1g} symmetry). Excitation can now occur from the t_{2g} level to either of these orbitals. Therefore, as in the case of the d^9 configuration, there are now two excited states of slightly different energy. The consequence may be a broadening of a spectrum into a two-humped peak, as in $[Ti(H_2O)_6]^{3+}$, or in some cases into two more clearly defined separate peaks.[11]

One additional point needs to be made in regard to Tanabe–Sugano diagrams. These diagrams, as shown in Figure 9-7, assume O_h symmetry in excited states as well as ground states. The consequence is that the diagrams are useful in predicting the general properties of spectra; in fact, many complexes do have sharply defined bands that fit the Tanabe–Sugano description well (see, for example, the d^2, d^3, and d^4 examples in Figure 9-8). However, distortions from pure octahedral symmetry are rather common, and the consequence can be the splitting of bands—or, in some cases of severe distortion, situations in which the bands are difficult to interpret. Additional examples of spectra showing the splitting of absorption bands can be seen in Figure 9-8.

EXERCISE 9-2

$[Fe(H_2O)_6]^{2+}$ has a two-humped absorption peak near 1000 nm. By using the appropriate Tanabe–Sugano diagram, account for the most likely origin of this absorption. Then account for the splitting of the absorption band.

[10] The $B_{1g} \longrightarrow A_{1g}$ transition is too low in energy to be observed in the visible spectrum.

[11] F. A. Cotton and G. Wilkinson, *Advanced Inorganic Chemistry*, 4th ed., Wiley-Interscience, New York, 1980, pp. 680–81.

9-2-5 APPLICATIONS OF TANABE–SUGANO DIAGRAMS: DETERMINING Δ_o FROM SPECTRA

It is desirable to be able to use absorption spectra for coordination compounds to determine the magnitude of the ligand field splitting, Δ_o in the case of octahedral complexes. It should be made clear from the outset that the accuracy with which Δ_o can be determined is to some extent limited by the mathematical tools brought to bear on the problem. Absorption spectra often have overlapping bands; to determine the positions of the bands accurately therefore requires an appropriate mathematical technique for reducing overlapping bands into their individual components. Such analysis is beyond the scope of this text. However, we can often obtain Δ_o values (and sometimes values of the Racah parameter B) of reasonable accuracy by simply using the positions of the absorption maxima taken directly from the spectra.

The ease with which Δ_o can be determined depends on the d electron configuration of the metal; in some cases Δ_o can be read easily from a spectrum, but in other cases a more complicated analysis is necessary. The following discussion proceeds from the simplest cases to the more complicated.

d^1, d^4 (high spin), d^6 (high spin), d^9

Each of these cases, as shown in Figure 9-10, corresponds to a simple excitation of an electron from t_{2g} to an e_g orbital, with the final (excited) electron configuration having the same spin multiplicity as the initial configuration. In each case there is a single excited state of the same spin multiplicity as the ground state. Consequently, there is a single spin-allowed absorption, with the energy of the absorbed light equal to Δ_o. Examples of such complexes include $[Ti(H_2O)_6]^{3+}$, $[Cr(H_2O)_6]^{2+}$, $[Fe(H_2O)_6]^{2+}$, and $[Cu(H_2O)_6]^{2+}$; note from Figure 9-8 that each of these complexes exhibits essentially a single absorption band (occasionally some splitting of bands due to Jahn–Teller distortions is observed, as discussed in Section 9-2-4).

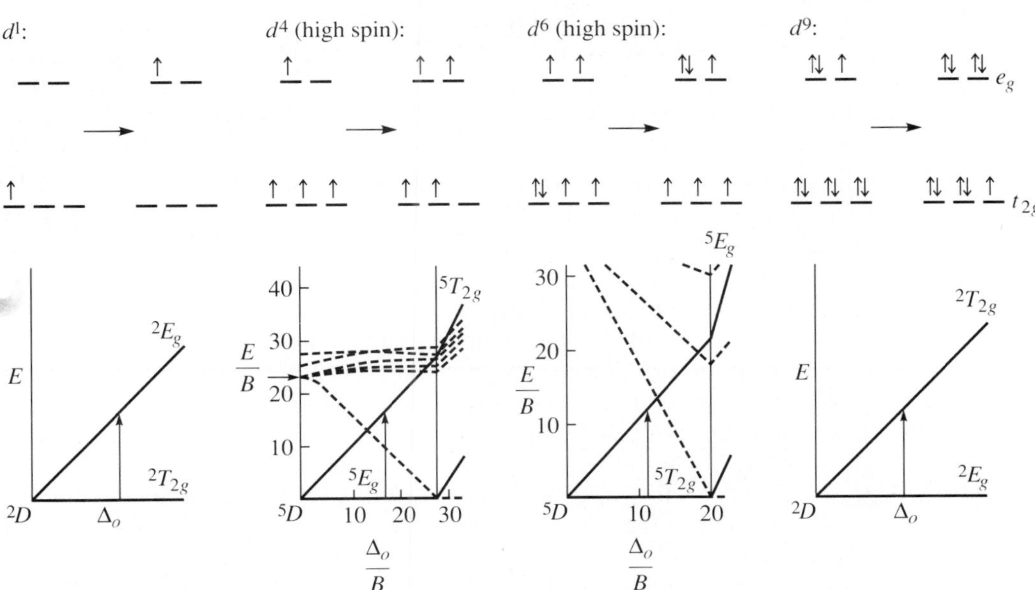

FIGURE 9-10 Determining Δ_o for d^1, d^4 (High Spin), d^6 (High Spin), and d^9 Configurations.

d^3, d^8

These electron configurations have a ground state F term. In an octahedral ligand field an F term splits into three terms, an A_{2g}, a T_{2g}, and a T_{1g}. As shown in Figure 9-11, the A_{2g} is of lowest energy for d^3 or d^8. For these configurations the difference in energy between the two lowest-energy terms, the A_{2g} and the T_{2g}, is equal to Δ_o. Therefore, to find Δ_o we need simply determine the energy of the lowest-energy transition in the absorption spectrum. Examples include $[Cr(H_2O)_6]^{3+}$ and $[Ni(H_2O)_6]^{2+}$. In each case the lowest energy band in the spectra of these complexes (Figure 9-8) is for the transition from the $^4A_{2g}$ ground state to the $^4T_{2g}$ excited state. The energies of these bands, approximately 17,500 and 8500 cm^{-1}, respectively, are the corresponding values of Δ_o.

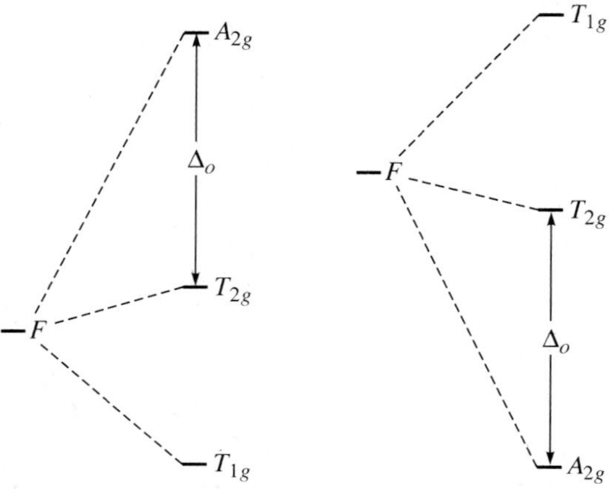

FIGURE 9-11 Splitting of F Terms in Octahedral Symmetry.

d^2, d^7 (high spin)

As in the case of d^3 and d^8, the ground free-ion terms for these two configurations are F terms. However, the determination of Δ_o is not as simple for d^2 and d^7. To explain this, it is necessary to take a close look at the Tanabe–Sugano diagrams in Figure 9-7. We will compare the d^3 and d^2 Tanabe–Sugano diagrams; the d^8 and d^7 (high spin) cases can be compared in a similar fashion [note the similarity of the d^3 and d^8 Tanabe–Sugano diagrams and of the d^2 and d^7 (high-spin region) diagrams].

As discussed above, in the d^3 case the ground state is a $^4A_{2g}$ state. There are three excited quartet states, $^4T_{2g}$, $^4T_{1g}$ (from the ^4F term) and $^4T_{1g}$ (from the ^4P term). Note the two states of the same symmetry ($^4T_{1g}$). An important property of such states is that *states of the same symmetry may mix*. The consequence of such mixing is that, as the ligand field is increased, the states appear to repel each other; the lines in the Tanabe–Sugano diagram curve away from each other. This effect can easily be seen in the Tanabe–Sugano diagram for d^3 (Figure 9-7). However, this causes no difficulty in obtaining Δ_o for a d^3 complex, since the lowest-energy transition ($^4A_{2g} \longrightarrow {}^4T_{2g}$) is not affected by such curvature. (The Tanabe–Sugano diagram shows that the energy of the $^4T_{2g}$ state varies linearly with the strength of the ligand field.)

The situation in the d^2 case is not quite as simple. For d^2 the free-ion ^3F term is also split into $^3T_{1g}$ + $^3T_{2g}$ + $^3A_{2g}$; these are the same states as obtained from d^3, *but in reverse order* (Figure 9-11). For d^2 the ground state is $^3T_{1g}$. This state can mix with the $^3T_{1g}$ state arising from the ^3P free-ion term, causing a slight curvature of $^3T_1(P)$ away from $^3T_1(F)$ in the Tanabe–Sugano diagram. This curvature can lead to some error in using the ground state to obtain values of Δ_o.

In the d^3 case considered on the previous page, it was stated that Δ_o is equal to the difference in energy between the A_{2g} and T_{2g} states. Therefore, for d^2, *if* we can identify the excited states for the absorption bands, we can determine Δ_o from

$$\Delta_o = \frac{\begin{array}{c}\text{energy of transition } {}^3T_{1g} \longrightarrow {}^3A_{2g} \\ -\text{energy of transition } {}^3T_{1g} \longrightarrow {}^3T_{2g}\end{array}}{\text{energy difference between } {}^3A_{2g} \text{ and } {}^3T_{2g} \quad \text{(see Figure 9-12)}}$$

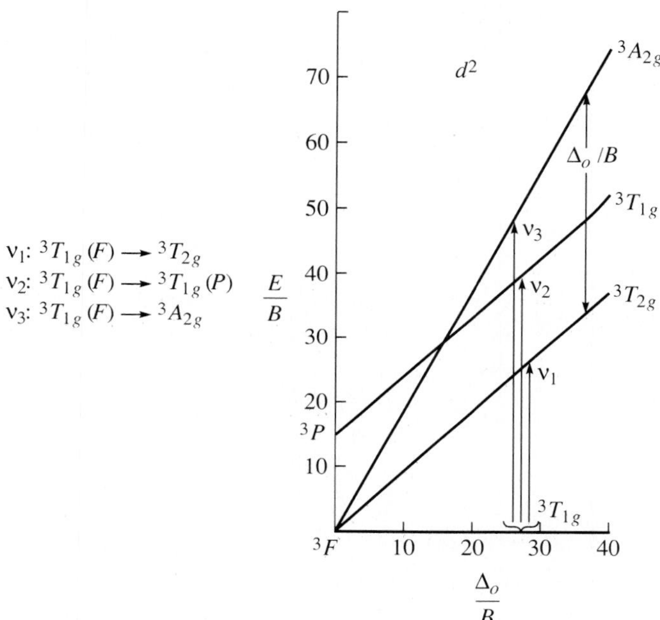

FIGURE 9-12 Spin-allowed Transitions for d^2 Configuration.

The difficulty with this approach is that the assignment of the absorption bands for a complex may be in question. From the Tanabe–Sugano diagram for d^2, we can see that while the lowest energy absorption band (to $^3T_{2g}$ state) is easily assigned, there are two possibilities for the next band: to $^3A_{2g}$ for very weak field ligands, to $^3T_{1g}(P)$ for stronger-field ligands. In addition, the second and third absorption bands may overlap, making it difficult to determine the exact positions of the bands (the positions of absorption maxima may be shifted if the bands overlap). In such cases a more complicated analysis, involving a calculation of the Racah parameter B, may be necessary. This procedure is best illustrated by example:

EXAMPLE

$[V(H_2O)_6]^{3+}$ has absorption bands at 17,800 and 25,700 cm^{-1}. Using the Tanabe–Sugano diagram for d^2, estimate values of Δ_o and B for this complex.

From the Tanabe–Sugano diagram there are three possible spin-allowed transitions (Figure 9-12):

$$^3T_{1g}(F) \longrightarrow {}^3T_{2g}, \qquad \nu_1 \text{ (lowest energy)}$$

$$^3T_{1g}(F) \longrightarrow {}^3T_{1g}(P), \qquad \nu_2$$

$$^3T_{1g}(F) \longrightarrow {}^3A_{2g}, \qquad \nu_3$$

When working with spectra, it is often useful to determine the ratio of energies of the absorption bands. In this example,

$$\frac{25{,}700 \text{ cm}^{-1}}{17{,}800 \text{ cm}^{-1}} = 1.44$$

The ratio of energy of the higher-energy transition (ν_2 or ν_3) to the lowest-energy transition (ν_1) must therefore be approximately 1.44. From the Tanabe–Sugano diagram, we can see that the ratio of ν_3 to ν_1 is approximately 2, regardless of the strength of the ligand field; we can therefore eliminate ν_3 as the possible transition occurring at 25,700 cm^{-1}. This means that the 25,700 cm^{-1} band must be ν_2, corresponding to $^3T_1(F) \longrightarrow {}^3T_1(P)$, and

$$1.44 = \frac{\nu_2}{\nu_1}$$

The ratio ν_2/ν_1 varies as a function of the strength of the ligand field. By taking values from Figure 9-12 and plotting the ratio ν_2/ν_1 versus Δ_o/B, we find that $\nu_2/\nu_1 = 1.44$ at approximately $\Delta_o/B = 31$ (see Figure 9-13).[12] At $\Delta_o/B = 31$,

$$\nu_2\text{:} \quad \frac{E}{B} = 42 \text{ (approximately)}; \quad B = \frac{E}{42} = \frac{25{,}700 \text{ cm}^{-1}}{42} = 610 \text{ cm}^{-1}$$

$$\nu_1\text{:} \quad \frac{E}{B} = 29 \text{ (approximately)}; \quad B = \frac{E}{29} = \frac{17{,}800 \text{ cm}^{-1}}{29} = 610 \text{ cm}^{-1}$$

Since $\dfrac{\Delta_o}{B} = 31$,

$$\Delta_o = 31 \times B = 31 \times 610 \text{ cm}^{-1} = 19{,}000 \text{ cm}^{-1}$$

It is usually necessary to follow this procedure for d^2 and d^7 complexes of octahedral geometry in order to obtain reasonably accurate values for Δ_o (and B).

EXERCISE 9-3

Use the Co(II) spectrum in Figure 9-8 and the Tanabe–Sugano diagrams of Figure 9-7 to find Δ_o and B. The broad band near 20,000 cm^{-1} can be considered to have the $^4T_1 \longrightarrow {}^4A_2$ transition in the low shoulder near 16,000 cm^{-1} and the $^4T_1 \longrightarrow {}^4T_1$ transition at the peak.

Other configurations: d^5 (high spin), d^4 to d^7 (low spin)

As has been mentioned previously, high-spin d^5 complexes have no excited states of the same spin multiplicity (6) as the ground state. The bands that are observed are therefore the consequence of spin-forbidden transitions and are typically very weak as, for example, in $[Mn(H_2O)_6]^{2+}$. The interested reader

[12] N. N. Greenwood and A. Earnshaw, *Chemistry of the Elements*, Pergamon, Elmsford, N.Y., 1984, p. 1161; B. N. Figgis, *Introduction to Ligand Fields*, Wiley-Interscience, New York, 1966, pp. 166–70.

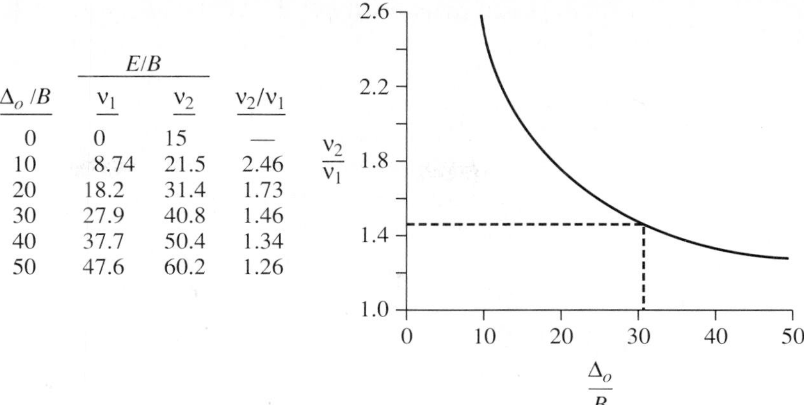

Δ_o/B	E/B ν_1	ν_2	ν_2/ν_1
0	0	15	—
10	8.74	21.5	2.46
20	18.2	31.4	1.73
30	27.9	40.8	1.46
40	37.7	50.4	1.34
50	47.6	60.2	1.26

FIGURE 9-13 Values of ν_2/ν_1 Ratio for d^2 Configuration.

is referred to the literature (see footnote 4) for an analysis of such spectra. In the case of low-spin d^4 to d^7 octahedral complexes, the analysis is typically rather complicated, since there are many excited states of the same spin multiplicity as the ground state (see right side of Tanabe–Sugano diagrams for d^4 to d^7, Figure 9-7). Again, the chemical literature provides examples and analyses of the spectra of such compounds, (see footnotes 3 and 4).

9-2-6 TETRAHEDRAL COMPLEXES

In general, tetrahedral complexes have more intense absorptions than octahedral complexes. This is a consequence of the first (Laporte) selection rule: transitions between *d* orbitals in a complex having a center of symmetry are forbidden. As a result, absorption bands for octahedral complexes are weak (small molar absorptivities); that they absorb at all is due to vibrational motions that act continually to distort molecules slightly from pure O_h symmetry. (This phenomenon is called **vibronic coupling**.)

In tetrahedral complexes the situation is different. The lack of a center of symmetry makes transitions between *d* orbitals more allowed; the consequence is that tetrahedral complexes often have much more intense absorption bands than octahedral complexes.[13]

As we have seen, the *d* orbitals for tetrahedral complexes are split in the opposite fashion to octahedral complexes:

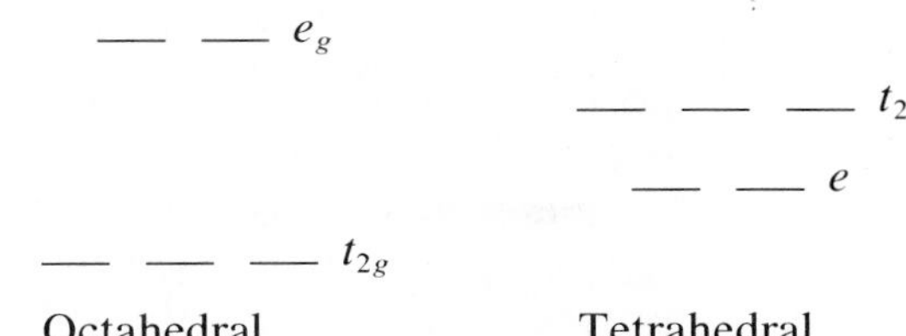

[13] Two types of hybrid orbitals are possible for a central atom of T_d symmetry: sd^3 and sp^3 (see Chapter 4). These types of hybrids may be viewed as mixing, to yield hybrid orbitals that contain some *p* character (note that *p* orbitals are not symmetric to inversion), as well as *d* character. The mixing in of *p* character can be viewed as making transitions between these orbitals more allowed. For a more thorough discussion of this phenomenon, see F. A. Cotton, *Chemical Applications of Group Theory*, 3rd ed., Wiley-Interscience, New York, 1990, pp. 295–96. Pages 289–97 of this reference also give a more detailed discussion of other selection rules.

A useful comparison can be drawn between these by using what is called the *hole formalism*. This can best be illustrated by example. Consider a d^1 configuration in an octahedral complex. The one electron occupies an orbital in a triply degenerate set (t_{2g}). Now consider a d^9 configuration in a tetrahedral complex. This configuration has a ''hole'' in a triply degenerate set of orbitals (t_2). It can be shown that, in terms of symmetry, the d^1 O_h configuration is analogous to the d^9 T_d configuration; the hole in d^9 results in the same symmetry as the single electron in d^1.

hole

↑↓ ↑↓ ↑

↑↓ ↑↓

↑

Octahedral Tetrahedral

In practical terms, this means that, for tetrahedral geometry, we can use the correlation diagram for the d^{10-n} configuration in octahedral geometry to describe the d^n configuration in tetrahedral geometry. Thus, for a d^2 tetrahedral case, we can use the d^8 octahedral correlation diagram; for the d^3 tetrahedral case, we can use the d^7 octahedral diagram; and so on. We can then identify the appropriate spin-allowed bands as in octahedral geometry, with allowed transitions occurring between the ground state and excited states of the same spin multiplicity.

Other geometries can of course also be considered according to the same approach as for octahedral and tetrahedral complexes. The interested reader is referred to the references in footnotes 3 and 5 for a discussion of different geometries.

9-2-7 CHARGE TRANSFER SPECTRA

Examples of charge transfer absorptions in solutions of halogens have been described in Chapter 6. In these cases a strong interaction between a donor solvent and a halogen molecule X_2 leads to the formation of a complex in which an excited state (primarily of X_2 character) can accept electrons from a HOMO (primarily of solvent character) on absorption of light of suitable energy:

$$X_2 \cdot \text{donor} \xrightarrow{h\nu} [\text{donor}^+][X_2^-]$$

The absorption band, known as a **charge transfer band**, can be very intense; it is responsible for the vivid colors of halogens in donor solvents.

It is extremely common for coordination compounds also to exhibit strong absorptions in the ultraviolet or, in some cases, in the visible portion of the spectrum. These absorptions may be much more intense than *d–d* transitions (which for octahedral complexes usually have ϵ values of 20 L mol^{-1} cm^{-1} or less); molar absorptivities of 50,000 L mol^{-1} cm^{-1} are not uncommon for these bands. Such absorptions may be considered charge transfer bands because they involve transfer of electrons from molecular orbitals that are primarily ligand in character to orbitals that are primarily metal in character (or vice versa). For example, consider an octahedral d^6

complex. The ligands donate electron pairs to the metal in a sigma fashion; as a consequence, these electron pairs are stabilized. Now the possibility exists that electrons can be excited, not only from the t_{2g} level to the e_g, but also from the sigma orbitals originating from the ligands to the e_g as in Figure 9-14. The latter excitation results in a charge transfer transition; it may be designated as CTTM (charge transfer to metal) or LMCT (ligand to metal charge transfer). This type of transition results in formal reduction of the metal. A CTTM excitation involving a cobalt(III) complex, for example, would exhibit an excited state of cobalt(II).

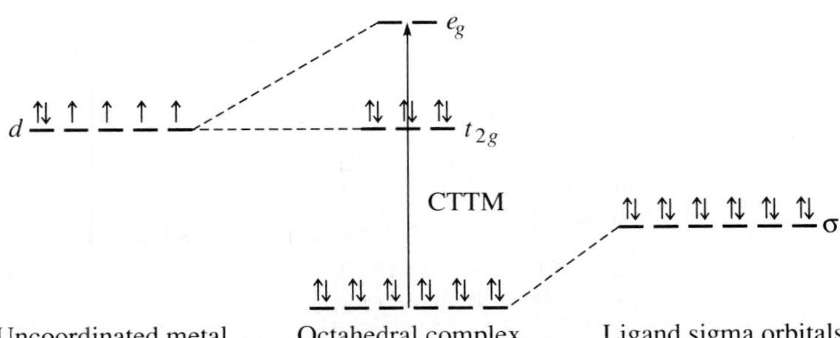

FIGURE 9-14 Charge Transfer to Metal.

Examples of charge transfer absorptions are numerous. The octahedral complexes $[IrBr_6]^{2-}$ (d^5) and $[IrBr_6]^{3-}$ (d^6) both show charge transfer bands. For $[IrBr_6]^{2-}$, two bands appear, near 600 and 270 nm, the former attributed to transitions to the t_{2g} levels, the latter to the e_g. In $[IrBr_6]^{3-}$, the t_{2g} levels are filled, and the only possible charge transfer to metal absorption is therefore to the e_g; consequently, no low-energy absorptions in the 600-nm range are observed, but strong absorption is seen near 250 nm, corresponding to charge transfer to e_g. A common example of tetrahedral geometry is the permanganate ion MnO_4^-, which is intensely purple because of a strong absorption involving charge transfer from orbitals derived primarily from the filled oxygen p orbitals to empty orbitals derived primarily from the manganese(VII).

Similarly, it is possible for there to be CTTL (charge transfer to ligand; also known as MLCT or metal to ligand charge transfer) transitions in coordination compounds having pi acceptor ligands. In these cases, empty π^* orbitals on the ligands become the acceptor orbitals on absorption of light. Figure 9-15 illustrates this phenomenon for a d^6 complex. Charge transfer to ligand results in oxidation of the metal; a CTTL excitation of an iron(III)

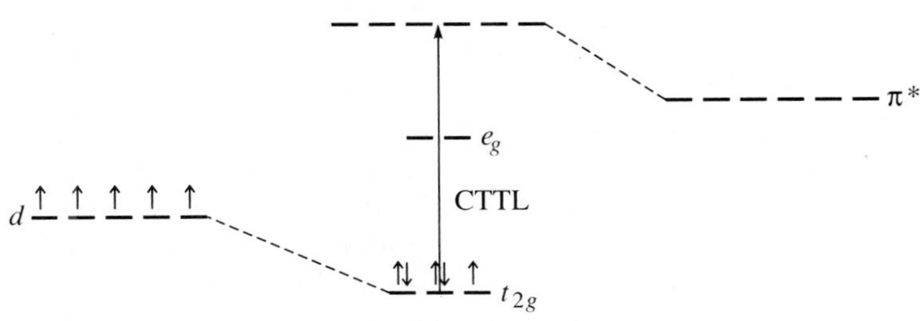

FIGURE 9-15 Charge Transfer to Ligand.

complex would give an iron(IV) excited state. Charge transfer to ligand most commonly occurs with ligands having empty π^* orbitals, such as CO, CN^-, bipyridine, and dithiocarbamate.

In complexes such as $Cr(CO)_6$, which have both sigma donor and pi acceptor orbitals, both types of charge transfer are possible. It is not always a simple matter to determine the type of charge transfer in a given coordination compound. Although $Cr(CO)_6$ is colorless, many ligands give highly colored complexes that have a series of overlapping absorption bands in the visible part of the spectrum as well as the ultraviolet. In such cases, the *d–d* transitions may be completely overwhelmed and essentially impossible to observe.

Finally, it is possible that the ligand itself may have a chromophore and that still another type of absorption band, an **intraligand band**, may be observed. These bands may sometimes be identified by comparing the spectra of complexes with the spectra of free ligands. However, coordination of a ligand to a metal may significantly alter the energies of the ligand orbitals, and such comparisons may be difficult, especially if charge transfer bands overlap the intraligand bands. Also, it should be pointed out that not all ligands exist in the free state; some ligands owe their very existence to the ability of metal atoms to stabilize molecules that are otherwise highly unstable. Examples of several such ligands will be discussed in later chapters.

GENERAL REFERENCES

Two references by B. N. Figgis, *Introduction to Ligand Fields*, Wiley-Interscience, New York, 1966, and "Ligand Field Theory," *Comprehensive Coordination Chemistry*, Vol. 1, G. Wilkinson, R. D. Gillard, and J. A. McCleverty, eds., Pergamon, Elmsford, N.Y., 1987, pp. 213–80, provide an extensive background in the theory of electronic spectra, with numerous examples. Important aspects of symmetry applied to this topic can be found in F. A. Cotton, *Chemical Applications of Group Theory*, 3rd ed., Wiley-Interscience, New York, 1990.

PROBLEMS

9-1 The most intense absorption band in the visible spectrum of $[Mn(H_2O)_6]^{2+}$ is at 24,900 cm^{-1} and has a molar absorptivity of 0.038 L mol^{-1} cm^{-1}. What concentration of $[Mn(H_2O)_6]^{2+}$ would be necessary to give an absorbance of 0.10 in a cell of path length 1.00 cm?

9-2 **a.** Determine the wavelength and frequency of 24,900 cm^{-1} light.
b. Determine the energy and frequency of 366 nm light.

9-3 Determine the ground terms for the following configurations:
a. d^8 (O_h symmetry)
b. High-spin and low-spin d^5 (O_h symmetry)
c. d^4 (T_d symmetry)
d. d^9 (D_{4h} symmetry, square planar)

9-4 The spectrum of $[Ni(H_2O)_6]^{2+}$ (Figure 9-8) shows three principal absorption bands, with two of the bands showing signs of further splitting. Referring to the Tanabe–Sugano diagram, estimate the value of Δ_o. Give a likely explanation for the further splitting of the spectrum.

9-5 From the following spectral data and using Tanabe–Sugano diagrams (Figure 9-7), calculate Δ_o for:
a. $[Cr(C_2O_4)_3]^{3-}$, which has absorption bands at 23,600 and 17,400 cm^{-1}. A third band occurs well into the ultraviolet.

b. $[Ti(NCS)_6]^{3-}$, which has an asymmetric, slightly split band at 18,400 cm^{-1}. (Also, suggest a reason for the splitting of this band.)

c. $[Ni(en)_3]^{2+}$, which has three absorption bands: 11,200, 18,350, and 29,000 cm^{-1}.

d. $[VF_6]^{3-}$, which has two absorption bands at 14,800 and 23,250 cm^{-1}, plus a third band in the ultraviolet. (Also calculate B for this ion.)

9-6 $[Co(H_2O)_6]^{2+}$ has three absorption bands at 8100, 16,000, and 19,400 cm^{-1}. Calculate Δ_o and B for this ion. (HINT: The graph in Figure 9-13 may be used for d^7 as well as d^2 complexes.)

9-7 The isoelectronic ions VO_4^{3-}, CrO_4^{2-}, and MnO_4^- all have intense charge transfer transitions. The wavelengths of these transitions increase in this series, with MnO_4^- having its charge transfer absorption at longest wavelength. Suggest a reason for this trend.

9-8 Of the first-row transition metal complexes of formula $[M(NH_3)_6]^{3+}$, which metals are predicted by the Jahn–Teller theorem to have distorted complexes?

9-9 MnO_4^- is a stronger oxidizing agent than ReO_4^-. Both ions have charge transfer bands; however, the charge transfer band for ReO_4^- is in the ultraviolet, while the corresponding band for MnO_4^- is responsible for its intensely purple color. Are the relative positions of the charge transfer absorptions consistent with the oxidizing abilities of these ions? Explain.

9-10 The complexes $[Co(NH_3)_5X]^{2+}$ (X = Cl, Br, I) have charge transfer to metal bands. Which of these complexes would you expect to have the lowest-energy charge transfer band? Why?

9-11 $[Fe(CN)_6]^{3-}$ exhibits two sets of charge transfer absorptions, one of lower intensity in the visible region of the spectrum and one of higher intensity in the ultraviolet. $[Fe(CN)_6]^{4-}$, however, shows only the high-intensity charge transfer in the ultraviolet. Explain.

9-12 The complexes $[Cr(O)Cl_5]^{2-}$ and $[Mo(O)Cl_5]^{2-}$ have C_{4v} symmetry.

a. Use the angular overlap approach (Chapter 8) to estimate the relative energies of the d orbitals in these complexes.

b. Using the C_{4v} character table, determine the symmetry labels (labels of irreducible representations) of these orbitals.

c. The $^2B_2 \longrightarrow {}^2E$ transition occurs at 12,900 cm^{-1} for $[Cr(O)Cl_5]^{2-}$ and at 14,400 cm^{-1} for $[Mo(O)Cl_5]^{2-}$. Account for the higher energy for this transition in the molybdenum complex. (See W. A. Nugent and J. M. Mayer, *Metal–Ligand Multiple Bonds*, Wiley, New York, 1988, pp. 33–35.)

9-13 For the isoelectronic series $[V(CO)_6]^-$, $Cr(CO)_6$, and $[Mn(CO)_6]^+$, would you expect the energy of metal to ligand charge transfer bands to increase or decrease with increasing charge on the complex? Why? (See K. Pierloot, J. Verhulst, P. Verbeke, and L. G. Vanquickenborne, *Inorg. Chem.*, **1989**, *28*, 3059.)

9-14 The compound *trans*-Fe(*o*-phen)$_2$(NCS)$_2$ has a magnetic moment of 0.65 Bohr magneton at 80 K, increasing with temperature to 5.2 Bohr magnetons at 300 K.

a. Assuming a spin-only magnetic moment, calculate the number of unpaired electrons at these two temperatures.

b. How can the increase in magnetic moment with temperature be explained? (HINT: There is also a significant change in the UV-visible spectrum with temperature.)

9-15 The absorption spectrum of the linear ion NiO_2^{2-} has bands attributed to *d–d* transitions at approximately 9000 and 16,000 cm^{-1}.

a. Using the angular overlap model (Chapter 8), predict the expected splitting pattern of the *d* orbitals of nickel in this ion.

b. Account for the two absorption bands.

c. Calculate the approximate value of e_σ and e_π.

(Reference: M. A. Hitchman, H. Stratemeier, and R. Hoppe, *Inorg. Chem.*, **1988**, *27*, 2506.)

10

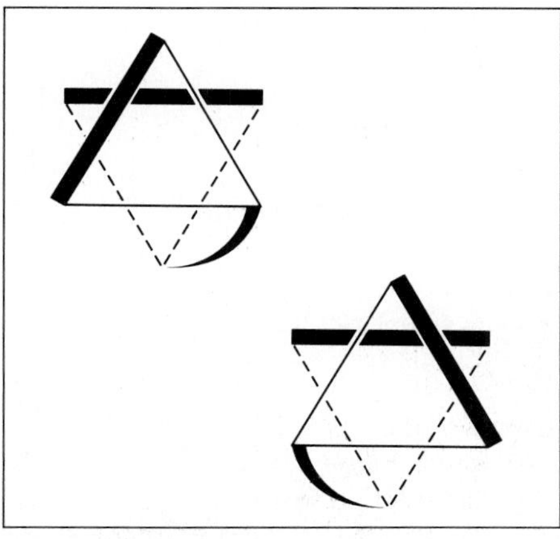

Coordination Chemistry III: Structures and Isomers

In the preceding two chapters, complexes with octahedral geometry were described most thoroughly. This is justified by the large number of octahedral compounds and their importance in the theories of coordination chemistry, but many other shapes and different coordination members are also found. This chapter describes a sampling of the different shapes of coordination compounds and explains some of the factors that influence shapes. Because of the complexity of the question, it is difficult to predict shapes with any confidence except when compounds of similar composition are already known. It is, however, possible to relate some structures to the individual factors that interact to produce them. Structures of organometallic compounds are even more difficult to predict, as will be seen in Chapters 12 through 14. This chapter also describes some of the isomers possible for these coordination compounds and some of the experimental methods used to study them.

10-1 NOMENCLATURE

As in any field of study, careful attention to nomenclature is required. The rules for names and formulas of coordination compounds are given here, with examples to show their use, but we need to be aware of changes in nomenclature with time. In many cases, the notation used by those who first prepared a compound is retained and enlarged on; in other cases, conflicting rules for names are proposed by different people and only after some time is a standard established. In such cases, the literature naturally includes papers using all the possible names, and sometimes careful research is necessary to interpret those that had relatively short lifetimes.

The following rules are the major ones required for the compounds in this text and the general literature. Reference to more complete sources may be needed for determining the names of other compounds.[1]

1. The positive ion (cation) comes first, followed by the negative ion (anion). This is the common order for simple salts as well.

Examples: diamminesilver(I) chloride, $[Ag(NH_3)_2]Cl$
potassium hexacyanoferrate(III), $K_3[Fe(CN)_6]$

2. Within the coordination sphere, the ligands are named before the metal, but in formulas the metal ion is written first. The inner coordination sphere is enclosed in square brackets in the formula.

Examples: tetraamminecopper(II) sulfate, $[Cu(NH_3)_4]SO_4$
hexaamminecobalt (III) chloride, $[Co(NH_3)_6]Cl_3$

3. Two systems exist for designating charge or oxidation number:
a. The Stock system puts the calculated oxidation number of the metal ion as a Roman numeral in parentheses after the name of the coordination sphere. This is the more common convention, although there are cases where it is difficult to assign oxidation numbers.
b. The Ewing–Bassett system puts the charge on the coordination sphere in parentheses after the name of the coordination sphere. This convention is used by *Chemical Abstracts* and offers an unambiguous identification of the species.

In either case, if the charge is negative, the suffix *-ate* is added to the name of the coordination sphere.

Examples: tetraammineplatinum(II) or tetraammineplatinum(2+), $[Pt(NH_3)_4]^{2+}$
tetrachloroplatinate(II) or tetrachloroplatinate(2−), $[PtCl_4]^{2-}$
hexachloroplatinate(IV) or hexachloroplatinate(2−), $[PtCl_6]^{2-}$

4. Ligands are named in alphabetical order (according to the name of the ligand, not the prefix), although exceptions to this rule are common. An earlier rule gave anionic ligands first, then neutral ligands, each listed alphabetically.

Examples: tetraamminedichlorocobalt(III), $[Co(NH_3)_4Cl_2]^+$
amminebromochloromethylamineplatinum(II),
$[Pt(NH_3)BrCl(CH_3NH_2)]$

5. Anionic ligands are given an *o* suffix. Neutral ligands retain their usual name. Coordinated water is called *aqua*.

Examples: chloro, Cl^-
bromo, Br^-
sulfato, SO_4^{2-}
methylamine, CH_3NH_2
ammine, NH_3 (the double m distinguishes NH_3 from alkyl amines)
aqua, H_2O

6. The number of ligands of one kind is given by the following prefixes. If the ligand name includes these prefixes or is complicated, it is set off in parentheses and the second set of prefixes is used.

[1] T. E. Sloan, "Nomenclature of Coordination Compounds," in G. Wilkinson, R. D. Gillard, and J. A. McCleverty, eds., *Comprehensive Coordination Chemistry,* Pergamon, Elmsford, N.Y., 1987, Vol. 1, pp. 109–34; IUPAC, *Nomenclature of Inorganic Chemistry: Recommendations 1990,* G. J. Leigh, ed., Blackwell Scientific Publications, Cambridge, Mass., 1990.

2	di	bis
3	tri	tris
4	tetra	tetrakis
5	penta	pentakis
6	hexa	hexakis
7	hepta	heptakis
8	octa	octakis
9	nona	nonakis
10	deca	decakis

Examples: Simple ligands are given on the previous page.
dichlorobis(ethylenediamine)cobalt(III), $[Co(NH_2CH_2CH_2NH_2)_2Cl_2]^+$
tris(bipyridine)iron(II), $[Fe(C_5H_4N\text{–}C_5H_4N)_3]^{2+}$

7. The prefixes *cis-* and *trans-* designate adjacent and opposite geometric locations. Examples are in Figure 10-1. Other prefixes are used as well and will be introduced as needed in the text.

Examples: *cis-* and *trans*-diamminedichloroplatinum(II), $[PtCl_2(NH_3)_2]$
cis- and *trans*-tetraamminedichlorocobalt(III), $[CoCl_2(NH_3)_4]^+$

8. Bridging ligands between two metal ions as in Figure 10-2 have the prefix μ-.

Examples: tris(tetraammine-μ-dihydroxocobalt)cobalt(6+), $[Co(Co(NH_3)_4(OH)_2)_3]^{6+}$
μ-amido-μ-hydroxobis(tetraamminecobalt)(4+), $[(NH_3)_4Co(OH)(NH_2)Co(NH_3)_4]^{4+}$

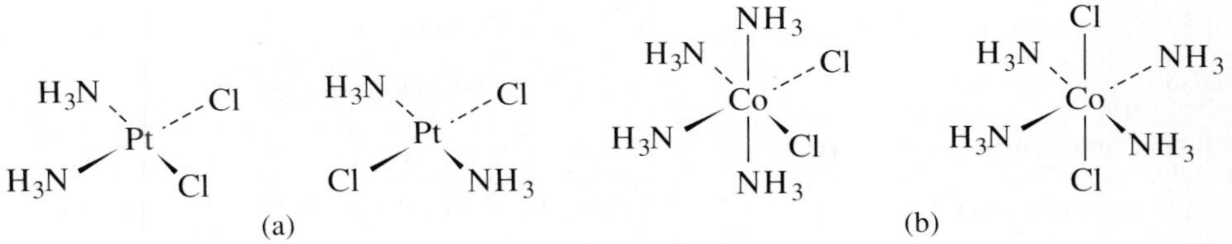

FIGURE 10-1 *Cis-* and *Trans-* Isomers. (a) *cis-* and *trans-* Dichlorodiammineplatinum(II), $[PtCl_2(NH_3)_2]$ (b) *cis-* and *trans-* dichlorotetraamminecobalt(III), $[CoCl_2(NH_3)_4]^+$

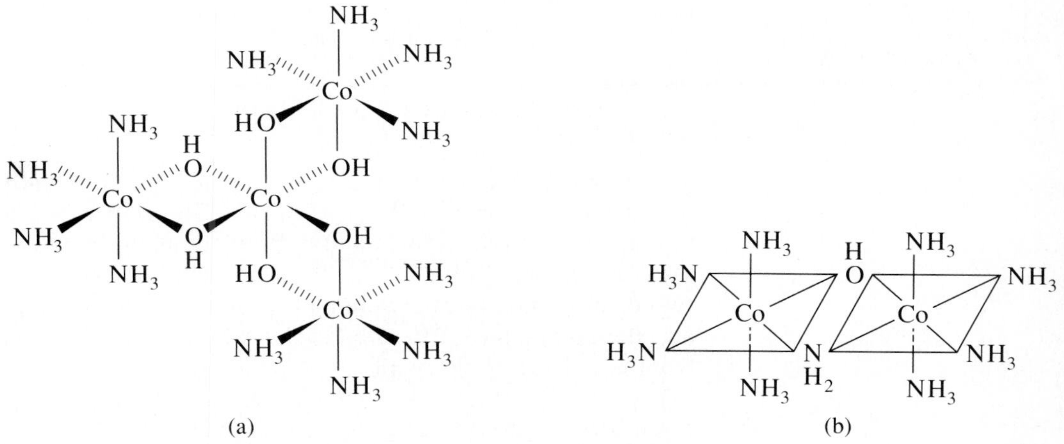

FIGURE 10-2 Bridging Ligands. (a) Tris(tetraammine-μ-dihydroxocobalt)cobalt(6+), $[Co(Co(NH_3)_4(OH)_2)_3]^{6+}$ (b) μ-Amido-μ-hydroxodi-(tetraamminecobalt)(4+), $[(NH_3)_4Co(OH)(NH_2)Co(NH_3)_4]^{4+}$

TABLE 10-1
Common ligands

Common name, formula, charge	*IUPAC name*	*Abbreviation*
fluoro, F^-	fluoro	F^-
chloro, Cl^-	chloro	Cl^-
bromo, Br^-	bromo	Br^-
iodo, I^-	iodo	I^-
cyano, CN^-	cyano	CN^-
thiocyano, SCN^-	thiocyanato-*S* (S-bonded)	SCN^-
isothiocyano, NCS^-	isothiocyanato-*N* (N-bonded)	NCS^-
hydroxo, OH^-	hydroxo	OH^-
aqua, H_2O	aqua	H_2O
carbonyl, CO	carbonyl	CO
thiocarbonyl, CS	thiocarbonyl	CS
nitrosyl, NO^+	nitrosyl	NO^+
nitro, NO_2^-	nitrito-*N* (N-bonded)	NO_2^-
nitrito, ONO^-	nitrito-*O* (O-bonded)	ONO^-
phosphine, PR_3	phosphane	PR_3
pyridine, C_5H_5N	pyridine	py
ammine, NH_3	ammine	NH_3
methylamine, CH_3NH_2	methylamine	$MeNH_2$
ethylenediamine, $NH_2CH_2CH_2NH_2$	1,2-ethanediamine	en
diethylenetriamine, $NH_2C_2H_4NHC_2H_4NH_2$	2,2′-diaminodiethylamine 1,4,7-triazaheptane	dien
triethylenetetramine, $NH_2C_2H_4NHC_2H_4NHC_2H_4NH_2$	1,4,7,10-tetraazadecane	trien
β, β′, β″-triaminotriethylamine, $N(C_2H_4NH_2)_3$	β, β′, β″-tris(2-aminoethyl)amine	tren
acetylacetonato, $CH_3COCH_2COCH_3$	2,4-pentanedione	acac
2,2′-bipyridine, $C_5H_4N–C_5H_4N$	2,2′-bipyridyl	bipy
1,10-phenanthroline, $C_{12}H_8N_2$	1,10-diaminophenanthrene	phen
dialkyldithiocarbamate, $S_2CNR_2^-$	dialkylcarbamodithioate	dtc
1,2-bis(diphenylphosphino)ethane, $PPh_2C_2H_4PPh_2$	1,2-ethanediyl-bis(diphenylphosphine)	dppe
o-phenylenebis(dimethylarsine), $C_6H_4(As(CH_3)_2)_2$	1,2-phenylenebis(dimethylarsine)	diars
dimethylglyoxime, $HON{=}C(CH_3){-}C(CH_3){=}NOH$	2,3-butanedione dioxime	DMG
ethylenediaminetetraacetate, $(^-OOCCH_2)_2NCH_2CH_2N(CH_2COO^-)_2$	(1,2-ethanediyldinitrilo)tetraacetate	EDTA

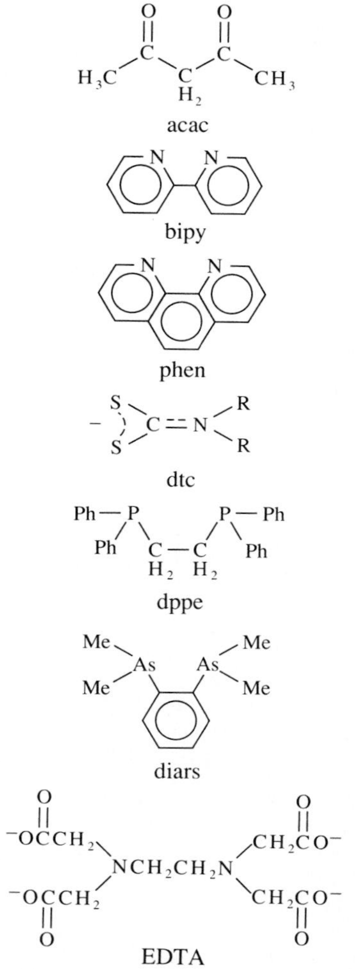

Structures

Organic (and some inorganic) ligands are frequently named with older trivial names rather than IUPAC names. The IUPAC names are more correct, but since the trivial names and abbreviations are commonly used, they must also be learned. Table 10-1 lists some of the common ligands. Ligands with two or more points of attachment to metal atoms are called **chelating ligands,** and the compounds are called **chelates** (pronounced key-late), a name derived from the Greek for claw of a crab. The ligands are described as **bidentate** for two points of attachment, with the prefixes **tri-, tetra-, penta-,** and **hexa-** for three through six bonding positions. **Chelate rings** may have any number of atoms; the most common contain five or six atoms, including the metal ion.

10-2 COMMON STRUCTURES

As an introduction to the structure of coordination compounds, we give a brief description of the common geometries seen in these compounds, along with examples of each. Explanations for some of the shapes are easy and follow the VSEPR approach presented in Chapter 3. Others do not seem to follow these rules and require more elaborate explanations.

Many factors interact to produce the overall shape of these compounds, often with one of the factors seeming most important for one compound and another factor taking the dominant role in another. For this reason, all predictions of shapes should be taken with a large dose of skepticism unless backed up by experimental evidence. To further confuse the observer, bond angles and distances are frequently distorted by the need for packing the compound into a crystal lattice (item 5 below). The resulting angles then fit none of the ideal cases, and choices between the various factors become difficult. Some of the factors involved in determining the structures of coordination compounds are:

1. VSEPR arguments, as used in the simpler cases of the representative elements.
2. Occupancy of *d* orbitals. LFSE and angular overlap energy calculations help explain this factor.
3. Steric interference by large ligands crowding each other around the central metal.
4. Other, more subtle interactions between the ligands.
5. Crystal packing effects. These include the effects resulting from the sizes of the ions and the overall shape of the coordination compounds. The regular shape of a compound may be distorted when it is packed into a crystalline lattice, and it is difficult to determine whether deviations from regular geometry are caused by effects within a given unit or by packing into a crystal.

In addition to these, there are subtle effects that influence isomer formation and bond formation. As in organic chemistry, exploitation of these effects and control of them by varying experimental conditions are part of the art and science of inorganic chemistry.

10-2-1 LOW COORDINATION NUMBERS (CN = 1, 2, AND 3)

Coordination number 1 is rare, except in ion pairs in the gas phase. Even species in aqueous solution that seem to be singly coordinated usually have water attached as well and have an overall coordination number higher than 1. Two organometallic compounds with coordination number 1 are the Cu(I) and Ag(I) complexes of 2,4,6-$Ph_3C_6H_2^-$ (5′-phenyl-*m*-terphenyl-2′-yl), shown in Figure 10-3, in which the very bulky ligand prevents any bridging between metals.[2] A transient species that seems to be singly coordinated is VO^{2+}.

Coordination number 2 is also rare. The best known example is $[Ag(NH_3)_2]^+$, the diamminesilver(I) ion. The silver 1+ ion is d^{10} (a filled, spherical subshell), so the only electrons to be considered in the VSEPR treatment are those forming the bonds with the ammonia ligands, and the structure is linear as

[2] R. Lingnau and J. Strähle, *Angew. Chem. Int. Ed. Engl.*, **1988,** *27,* 436.

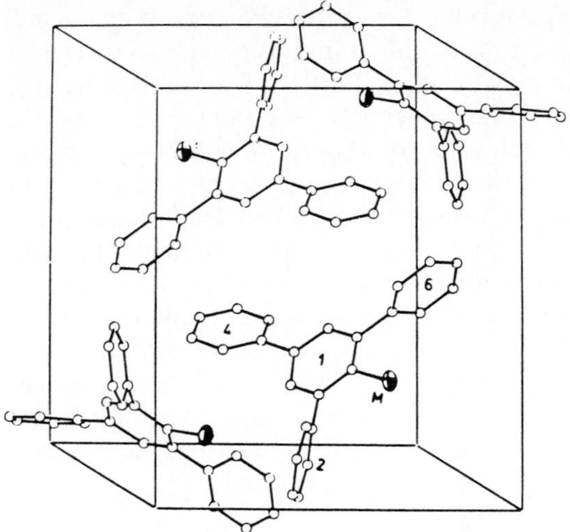

FIGURE 10-3 Coordination Number 1. 2,4,6-$Ph_3C_6H_{21}Cu$ or Ag. (Reproduced with permission from R. Lingnau and J. Strähle, *Angew. Chem. Int. Ed. Engl.*, **1988,** *27,* 436.)

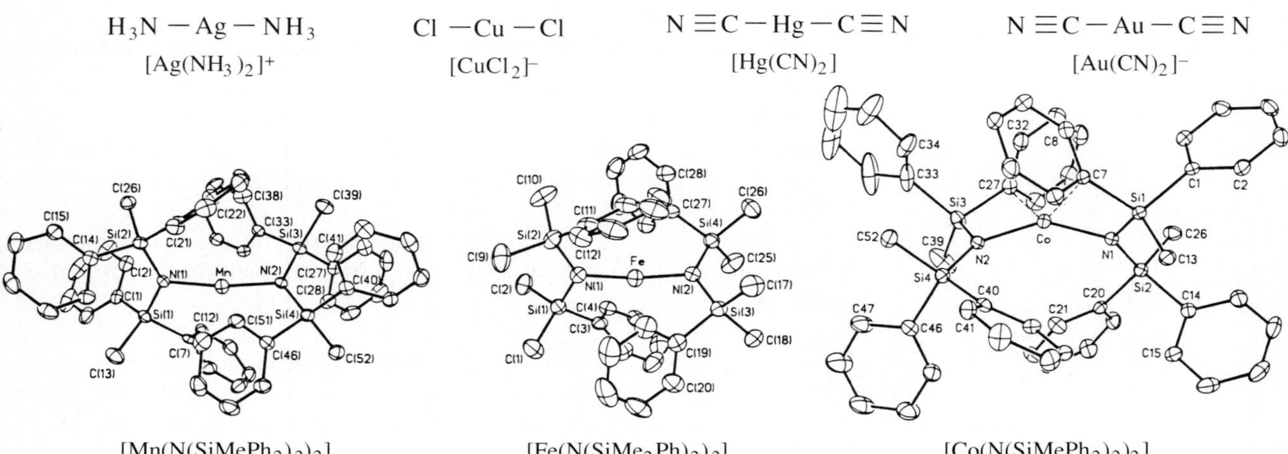

FIGURE 10-4 Compounds with Coordination Number 2. (Reproduced with permission from H. Chen, R. A. Bartlett, H. V. R. Dias, M. M. Olmstead, and P. P. Power, *J. Am. Chem. Soc.*, **1989,** *111,* 4338. Copyright 1989 American Chemical Society.)

expected for two bonding positions. The other examples shown in Figure 10-4 are also d^{10} and linear ($[CuCl_2]^-$, $[Hg(CN)_2]$, and $[Au(CN)_2]^-$), but others are d^5, d^6, or d^7 $[Mn(N(SiMePh_2)_2)_2]$, $[Fe(N(SiMe_2Ph)_2)_2]$, and $[Co(N(SiMePh_2)_2)_2]$).[3,4] All the ligands except CN^- and Cl^- are large, helping force a linear or near-linear arrangement in all but the cobalt species. The trimethylsilylamine cobalt compound may have a bent N—Co—N shape; the phenyl derivative has an N—Co—N angle of 147°.

Coordination number 3 also is more likely with d^{10} ions, with a planar trigonal structure the most common. Three-coordinate Au(I) and Cu(I) com-

[3] D. C. Bradley and K. J. Fisher, *J. Am. Chem. Soc.*, **1971,** *93,* 2058.

[4] H. Chen, R. A. Bartlett, H. V. R. Dias, M. M. Olmstead, and P. P. Power, *J. Am. Chem. Soc.*, **1989,** *111,* 4338.

pounds that are known include $[Au(PPh_3)_3]^+$, $[Au(PPh_3)_2Cl]$, and $[Cu(SPPh_3)_3]^+$.[5,6] Most 3-coordinate compounds seem to have a low coordination number because of ligand crowding. Ligands such as triphenylphosphine, PPh_3, and di(trimethylsilyl)amide, $N(SiMe_3)_2^-$, are bulky enough to prevent larger coordination numbers even when the electronic structure favors them. All the first-row transition metals except Mn(III) form such compounds, either with three identical ligands or two of one ligand and one of the other. These compounds are close to planar trigonal, although the VSEPR prediction for some is T-shaped. Others with three ligands are MnO_3^+, HgI_3^-, and the cyclic compound $Cu_3Cl_3(SPMe_3)_3$. Some of these compounds are shown in Figure 10-5.

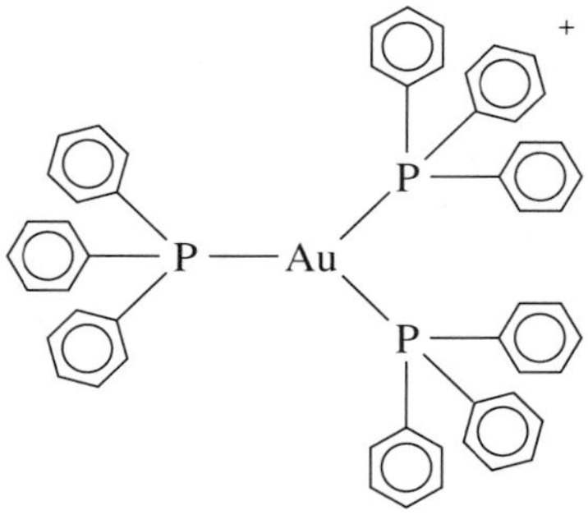

$[Au(PPh_3)_3]^+$
(a)

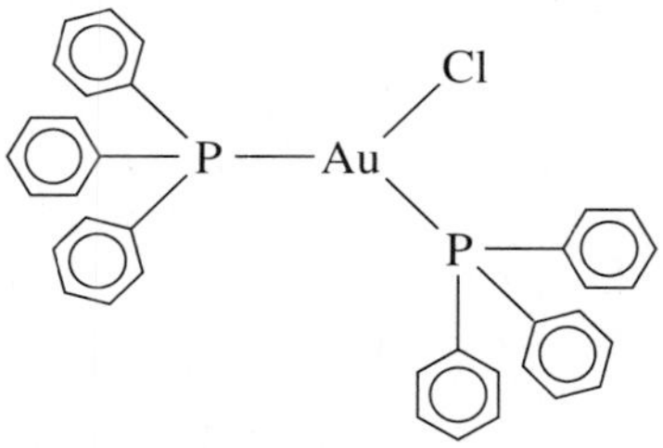

$[Au(PPh_3)_2Cl]$
(b)

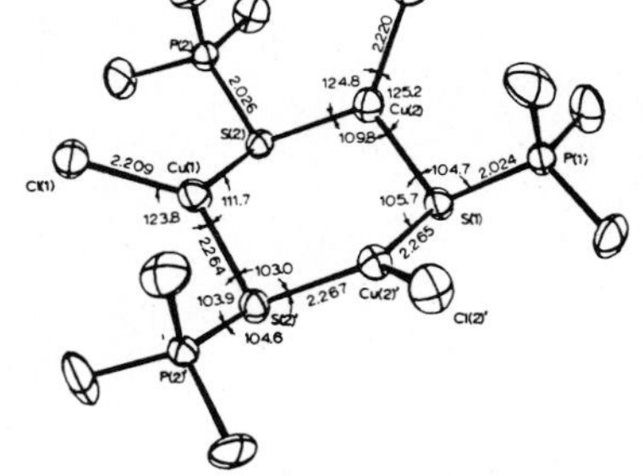

$[Cu(SP(CH_3)_3)Cl]_3$
(c)

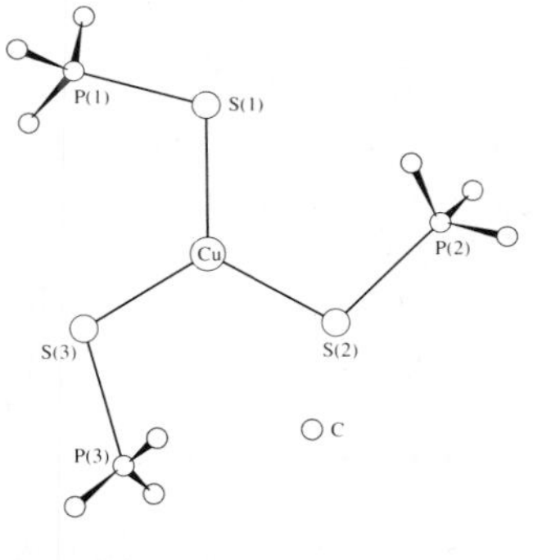

$[Cu(SP(CH_3)_3)_3]^+$
Tris(trimethylphosphinesulfide)copper(I)
(d)

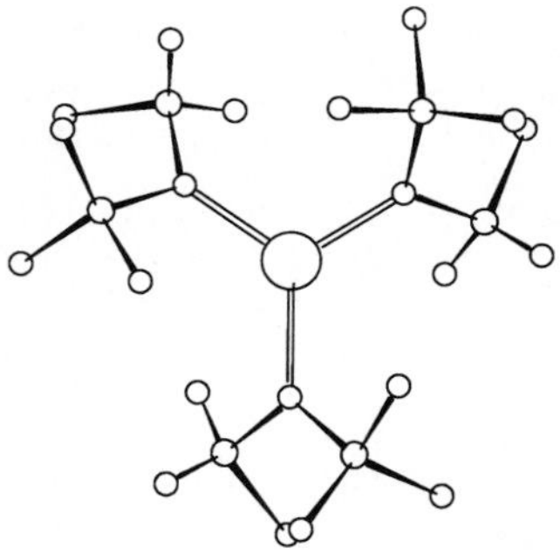

$[Fe(N(Si(CH_3)_3)_2)_3]$
Tris(hexamethyldisilylamine)iron(III)
(e)

FIGURE 10-5 Compounds with Coordination Number 3. ($[Cu(SP(CH_3)_3)Cl]_3$ figure reproduced with permission from J. A. Tiethof, J. K. Stalick, and D. W. Meek, *Inorg. Chem.*, **1973,** *12,* 1170. Copyright 1973 American Chemical Society. $[Cu(SP(CH_3)_3)_3]^+$ figure reproduced with permission from P. G. Eller and P. W. R. Corfield, *Chem. Commun.*, **1971,** 105. $[Fe(N(Si(CH_3)_3)_2)_3]$ reproduced with permission from D. C. Bradley, M. B. Hursthouse, and P. F. Rodesiler, *Chem. Commun.*, **1969,** 14.)

10-2-2 COORDINATION NUMBER 4[7]

Based on the fact that fewer bonds are formed, 4-coordinate compounds seem less likely than 6-coordinate compounds, but there are two common structures with four ligands, tetrahedral and square planar. Another structure, with four bonds and one lone pair, appears with compounds such as SF_4 and $TeCl_4$, described in Chapter 3. Crowding around small ions of high positive charge prevents octahedral shapes for ions such as Mn(VII) and Cr(VI), and large

[5] F. Klanberg, E. L. Muetterties, and L. J. Guggenberger, *Inorg. Chem.*, **1968,** *7,* 2273.
[6] N. C. Baenziger, K. M. Dittemore, and J. R. Doyle, *Inorg. Chem.*, **1974,** *13,* 805.
[7] M. C. Favas and D. L. Kepert, *Prog. Inorg. Chem.*, **1980,** *27,* 325.

ligands can prevent higher coordination for other ions. Many d^0 or d^{10} complexes have tetrahedral structures, such as BF_4^-, MnO_4^-, CrO_4^{2-}, $[Ni(CO)_4]$, and $[Cu(py)_4]^+$, with a few d^5, such as $MnCl_4^{2-}$. In such cases, the shape can be explained on the basis of VSEPR arguments, since the *d* orbital occupancy is spherically symmetrical. However, a number of tetrahedral Co(II) (d^7) species are also known, as well as some for other transition metal ions. Other tetrahedral complexes are $[Co(PF_3)_4]^-$, $TiCl_4$, $[NiCl_4]^{2-}$, and $[NiCl_2(PPh_3)_2]$. Tetrahedral structures are also found in the tetrahalide compounds of Cu(II). $Cs_2[CuCl_4]$ and similar salts with bromide or with tetramethylammonium ion are close to tetrahedral, as are the same ions in solution. Distortion of the tetrahedron as a result of Jahn–Teller effects makes two of the Cl—Cu—Cl bond angles near 102° and two near 125°. With other cations, $CuCl_4^{2-}$ can have other shapes, as discussed in Section 10-2-4. Examples of tetrahedral species are given in Figure 10-6.

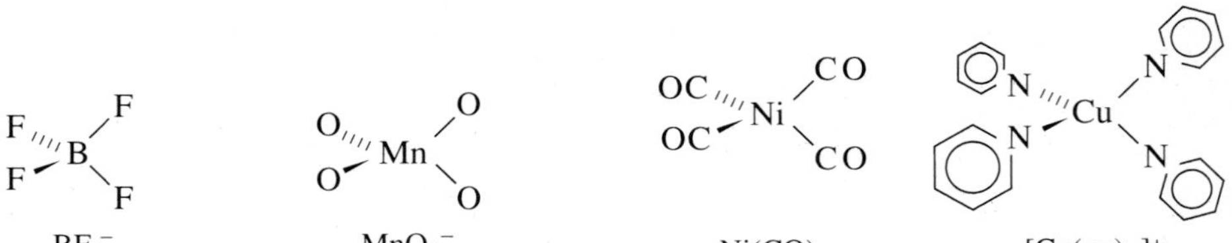

FIGURE 10-6 Compounds with Tetrahedral Geometry.

Square planar geometry is also possible for 4-coordinate species, with the same geometric requirements imposed by octahedral geometry (both require 90° angles between ligands). The only common square planar compounds whose structures are not imposed by a complex, planar ligand contain d^8 ions [Ni(II), Pd(II), Pt(II), for example]. Cu(II) also forms some square planar structures, but many copper compounds have distorted 6-coordinate structures between octahedral and square planar in shape. Ni(II) (d^8) and Cu(II) (d^9) compounds can have tetrahedral, square planar, or intermediate shapes, depending on both the ligand and the counterion in the crystal. Cases such as these indicate that the energy difference between the two structures is small, and crystal packing can have a large influence on the choice. Pd(II) and Pt(II) compounds are square planar, as are the d^8 compounds $[AgF_4]^-$, $[RhCl(PPh_3)_3]$, $[Ni(CN)_4]^{2-}$, and $[NiCl_2(PMe_3)_2]$. At least one compound, $[NiBr_2(P(C_6H_5)_2(CH_2C_6H_5))_2]$, has both square planar and tetrahedral isomers in the same crystal.[8] Some square planar compounds are shown in Figure 10-7.

10-2-3 COORDINATION NUMBER 5[9]

The structures possible for coordination number 5 are the trigonal bipyramid, the square pyramid, and the pentagonal plane (which is unknown, probably because of the crowding of the ligands required). The energy difference between the trigonal bipyramid and the square pyramid is very small. In fact, many molecules with five ligands either have structures between these two or

[8] B. T. Kilbourn, H. M. Powell, and J. A. C. Darbyshire, *Proc. Chem. Soc.*, **1963**, 207.

[9] R. R. Holmes, *Prog. Inorg. Chem.*, **1984**, *32*, 119; T. P. E. Auf der Heyde and H.-B. Bürgi, *Inorg. Chem.*, **1989**, *28*, 3960.

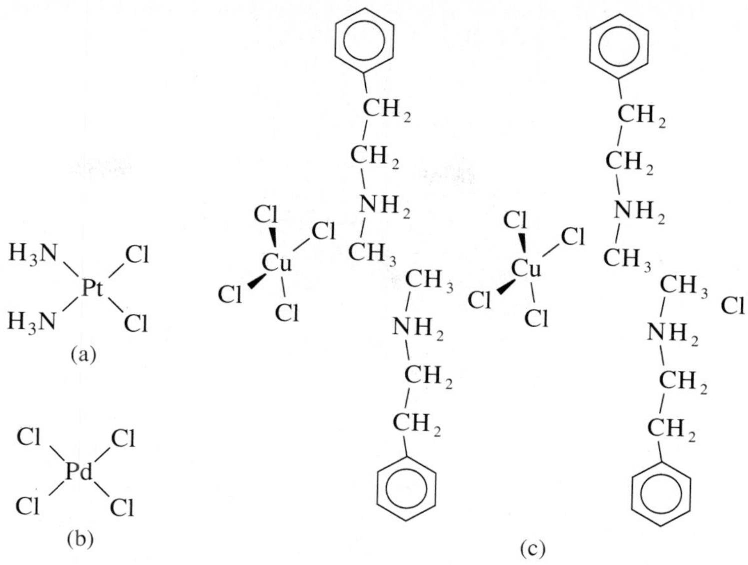

FIGURE 10-7 Compounds with Square Planar Geometry. (a) $PtCl_2(NH_3)_2$. (b) $[PdCl_4]^{2-}$. (c) N- Methylphenethylammonium tetrachlorocuprate(II) at 25°. At 70°, the $CuCl_4^-$ anion is nearly tetrahedral. (Adapted with permission from R. L. Harlow, W. J. Wells, III, G. W. Watt, and S. H. Simonsen, *Inorg. Chem.*, **1974,** *13,* 2106. Copyright 1974 American Chemical Society.)

can switch easily from one to the other in fluxional behavior. $Fe(CO)_5$ and PF_5 are examples; in both cases, nuclear magnetic resonance spectrometry (using ^{13}C and ^{19}F, respectively) shows only one peak, indicating that the atoms are identical on the NMR time scale. Since both the trigonal bipyramid and the square pyramid have ligands in two different environments, the experiment shows that the compounds switch from one structure to another rapidly or that they have a solution structure intermediate between the two. In the solid state, both are trigonal bipyramids. $[VO(acac)_2]$ is a square pyramid, with the doubly bonded oxygen in the apical site. Other 5-coordinate compounds are known for the full range of transition metals, including $[CuCl_5]^{3-}$ and $[FeCl(S_2C_2H_2)_2]$. Examples of five-coordinate compounds are shown in Figure 10-8.

10-2-4 COORDINATION NUMBER 6

Six is the most common coordination number. The most common structure is octahedral, with some trigonal prismatic structures. In a large number of octahedral and near-octahedral structures, no simple VSEPR argument for the shape is possible, but the metal is large enough to allow six ligands to fit around it. Such compounds exist for all the transition metals, with d^0 to d^{10} configurations.

Octahedral compounds have been used in many of the earlier illustrations in this chapter and others. Other octahedral complexes include tris(ethylenediamine)cobalt(III), $[Co(en)_3]^{3+}$, and hexanitritocobaltate(III), $[Co(NO_2)_6]^{3-}$, shown in Figure 10-9. Other examples will be introduced later, particularly in the discussion of isomerism.

For those compounds that are not regular octahedra, several kinds of distortion are possible. The first is elongation, leaving four short bonds in a square planar arrangement together with two longer bonds above and below the plane. Second is the reverse, a compression with two short bonds on top and bottom and four longer bonds in the plane. A trigonal elongation or

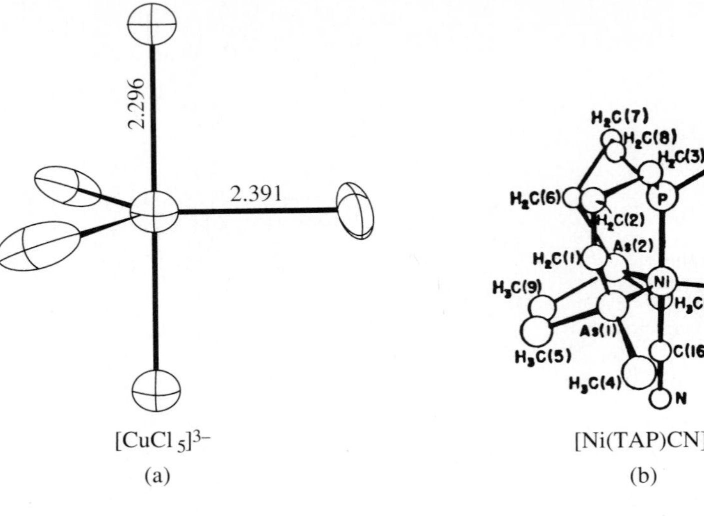

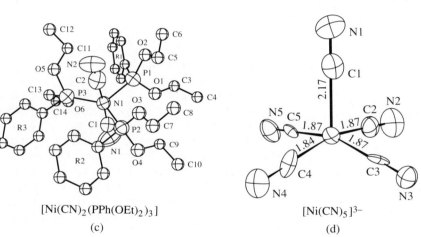

FIGURE 10-8 Compounds with Coordination Number 5. (a) ($[CuCl_5]^{3-}$ from $[Cr(NH_3)_6][CuCl_5]$, K. N. Raymond et al., *Inorg. Chem.*, © **1968,** *7,* 1111; (b) $[Ni(TAP)CN]^+$ from $[Ni(TAP)CN][ClO_4]$, D. L. Stevenson and L. F. Dahl, *J. Am. Chem. Soc.*, © **1967,** *89,* 3424; (c) $[Ni(CN)_2(PPh(OEt)_2)_3]$, J. K. Stalick and J. A. Ibers, *Inorg. Chem.*, © **1969,** *8,* 1084; (d) $[Ni(CN)_5]^{3-}$ from $[Cr(en)_3][Ni(CN)_5]$, K. N. Raymond, P. W. R. Corfield, and J. A. Ibers, *Inorg. Chem.*, © **1968,** *7,* 1362. All reproduced with permission of the American Chemical Society.)

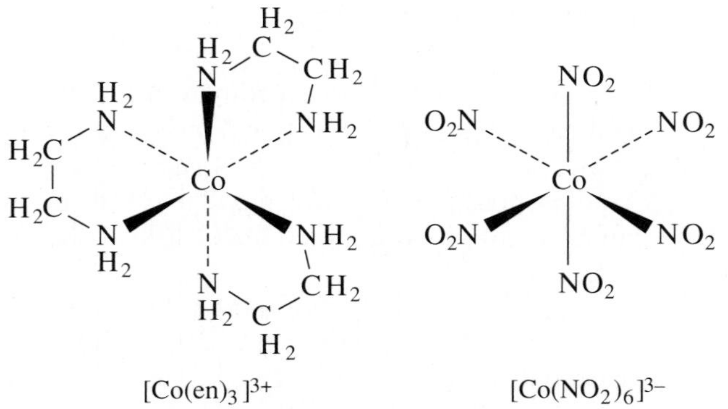

$[Ni(CN)_2(PPh(OEt)_2)_3]$

(c)

$[Ni(CN)_5]^{3-}$

(d)

FIGURE 10-9 Compounds with Octahedral Geometry.

$[Co(en)_3]^{3+}$ $[Co(NO_2)_6]^{3-}$

compression results in a trigonal antiprism when the angle between the top and bottom triangular faces is 60°, and a trigonal prism when the two triangular faces are eclipsed. Most prismatic compounds have three bidentate ligands (dithiolates or oxalates are common) linking the top and bottom triangular faces. Although similar in other ways, β-diketone complexes usually have near

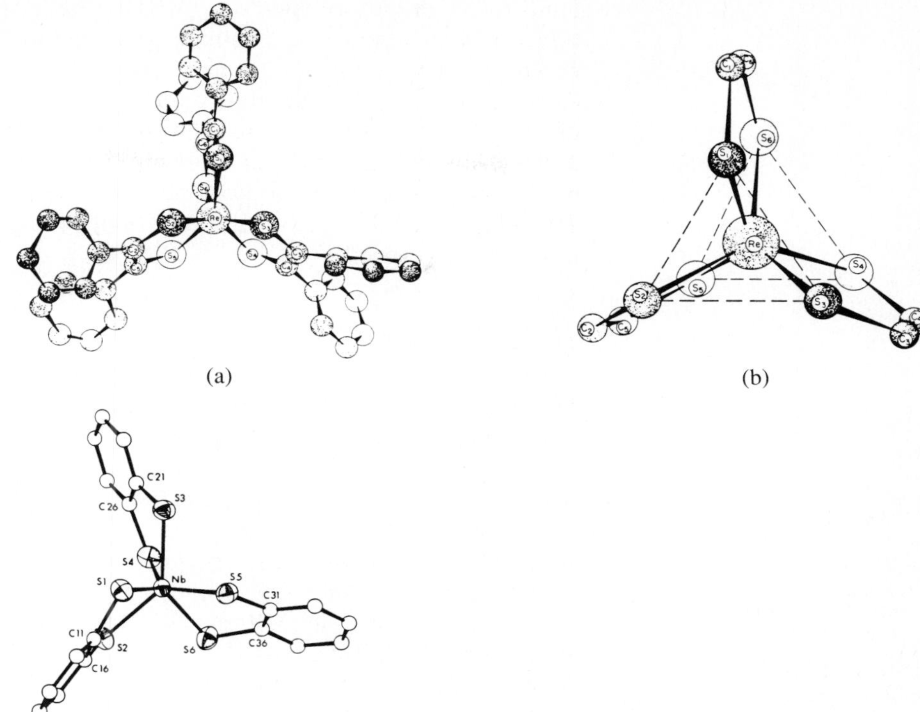

FIGURE 10-10 Compounds with Trigonal Prismatic Geometry. (a), (b) $Re(S_2C_2(C_6H_5)_2)_3$ (from R. Eisenberg and J. A. Ibers, *Inorg. Chem.*, © **1966,** *5,* 411). Part (b) is a perspective drawing of the coordination geometry excluding the phenyl rings. (c) Tris(benzene-1,2-dithiolato) niobate(V), $Nb(S_2C_6H_4)_3^-$, omitting the hydrogens (from M. Cowie and M. J. Bennett, *Inorg. Chem.*, © **1976,** *15,* 1589). All reproduced with permission of the American Chemical Society.

octahedral symmetry around the metal. Two trigonal prismatic structures are shown in Figure 10-10. The trigonal structures of compounds like these are explained as a result of π orbital interactions between the sulfur atoms in the trigonal faces.

A number of compounds that appear to be 4-coordinate are more accurately described as 6-coordinate. Although $(NH_4)_2[CuCl_4]$ is frequently cited as having a square planar $[CuCl_4]^{2-}$ ion, in the crystal the ions are packed so that two more chlorides are above and below the plane at considerably larger distances in a distorted octahedral structure. Similarly, $[Cu(NH_3)_4]SO_4 \cdot H_2O$ has the ammonias in a square planar arrangement, but each copper is also connected to a distant bridging water molecule above and below the plane.

When all the ligands are the same, these are also likely candidates for Jahn–Teller distortions. As described in Chapter 8, Jahn–Teller distortion is predicted to occur in situations in which degenerate sets of orbitals are asymmetrically occupied. In the case of octahedral geometry, the effect on molecular shape is greatest if the e_g orbitals (which point directly toward the ligands) are asymmetrically occupied; the effect, if any, of asymmetrically occupied t_{2g} orbitals (which point between the ligands) is generally too small to be observed. The configurations most likely to give rise to measurable Jahn–Teller distortions from octahedral geometry are therefore d^4 (high spin), d^7 (low spin), and d^9. Distortions for all three of these configurations have been observed, with the most common form of distortion involving elongation of bonds along one axis.

For example, the 2-pyridone complex $[Cu(C_5H_5NO)_6](ClO_4)_2$ has two *trans* Cu—O bonds elongated by 31% relative to the other four,[10] and the

[10] D. Taylor, *Aust. J. Chem.*, **1975,** *28,* 2615.

imidazole complex $[Cu(C_2H_4N_2)_6](NO_3)_2$ has a 28% elongation of two *trans* bonds.[11] By contrast, at temperatures below 3°C, $K_2Pb[Cu(NO_2)_6]$ has a *compression* of two *trans* bonds of approximately 5%,[12] but changes in the cations and higher temperatures shift the distortion to elongation. The choice of elongation versus compression appears to be largely dependent on how molecules pack into a crystal; in some cases, the deviations are too small to be observed or seem to be dynamic, dependent on vibrations of the molecule. The relative stabilities of a wide variety of examples of 6-coordinate geometries have been reviewed.[13]

10-2-5 COORDINATION NUMBER 7[14]

Three structures are possible for 7-coordinate compounds, the pentagonal bipyramid, capped trigonal prism, and capped octahedron. In the capped shapes, the seventh ligand is simply added to a face of the structure, with related adjustments in the other angles to allow it to fit. Although 7-coordination is not common, all three shapes are found experimentally, with the differences apparently due to different counterions and the steric requirements of the ligands (especially chelating ligands).

Examples include $[M(trenpy)]^{2+}$ (M = any of the metals from Mn to Zn, and trenpy = $(C_5H_4NCH{=}NCH_2CH_2)_3N$), in which the central nitrogen of the ligand caps a trigonal face of an octahedron), 2,13-dimethyl-3,6,9,12,18-pentaazabicyclo[12.3.1]-octadeca-1(18),2,12,14,16-pentaenebis(thiocyanato)iron(II), $[UO_2F_5]^{3-}$, and $[NbOF_6]^{3-}$, pentagonal bipyramids; $[NiF_7]^{2-}$ and $[NbF_7]^{2-}$, in both of which the seventh fluoride caps a rectangular face of a trigonal prism; and $[W(CO)_4Br_3]^-$, a monocapped octahedron. Some of these compounds are shown in Figure 10-11.

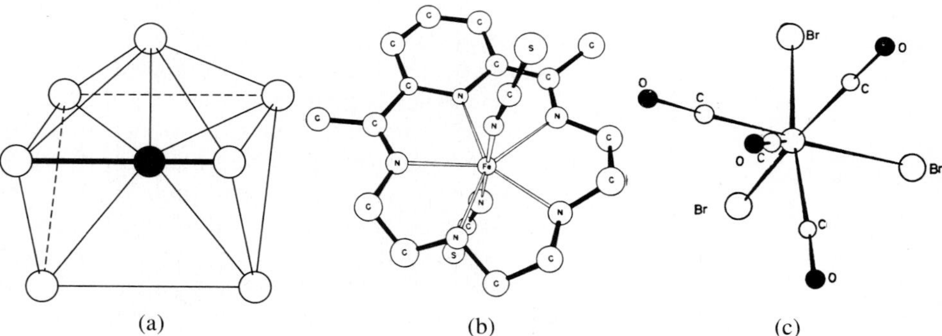

FIGURE 10-11 Compounds with Coordination Number 7. (a) Heptafluoroniobate(V), $[NbF_7]^{2-}$, a capped trigonal prism, (b) 2,13-dimethyl-3,6,9,12,18-pentaazabicyclo[12.3.1]-octadeca-1(18),2,12,14,16-pentaene complex of Fe(II) with two axial thiocyanates, a pentagonal bipyramid. (Reproduced with permission from E. Fleischer and S. Hawkinson, *J. Am. Chem. Soc.*, **1967,** *89,* 720. Copyright 1967 American Chemical Society.) (c) Tribromotetracarbonyltungstate(II) anion, $[W(CO)_4Br_3]^-$, a capped octahedron. The capping CO is directed toward the viewer. (Reproduced with permission from M. G. B. Drew and A. P. Wolters, *J. Chem. Soc. Chem. Comm.*, **1972,** 457.)

10-2-6 COORDINATION NUMBER 8[15]

Although the cube has 8-coordinate geometry, it exists only in simple ionic lattices like CsCl. The square antiprism and dodecahedron are common, and

[11] D. L. McFadden, A. T. McPhail, C. D. Garner, and F. E. Mabbs, *J. Chem. Soc., Dalton Trans.*, **1975,** 263.

[12] M. D. Joesten, S. Takagi, and P. G. Lenhert, *Inorg. Chem.*, **1977,** *16,* 2680.

[13] D. L. Kepert, "Coordination Numbers and Geometries," in G. Wilkinson, ed., *Comprehensive Coordination Chemistry,* Vol. 1, Pergamon, Elmsford, N.Y., 1987, pp. 49–68.

[14] D. L. Kepert, *Prog. Inorg. Chem.*, **1979,** *25,* 41.

[15] D. L. Kepert, *Prog. Inorg. Chem.*, **1978,** *24,* 179.

there are many 8-coordinate compounds. Because the central ion must be large in order to accommodate eight ligands, 8-coordination is rare among the first-row transition metals (although it is likely in $[Fe(edta)(H_2O)_2]^+$ in solution). Solid-state examples include $Na_7Zr_6F_{31}$, which has square antiprisms of ZrF_8 units, and $[Zr(acac)_2(NO_3)_2]$, a regular dodecahedron. $[AmCl_2(H_2O)_6]^+$ is a trigonal prism of water ligands with chloride caps on the trigonal faces. Two of these compounds are shown in Figure 10-12.

10-2-7 LARGER COORDINATION NUMBERS[16]

Coordination numbers are known up to 16, but most over 8 are special cases. Some examples are shown in Figure 10-13.

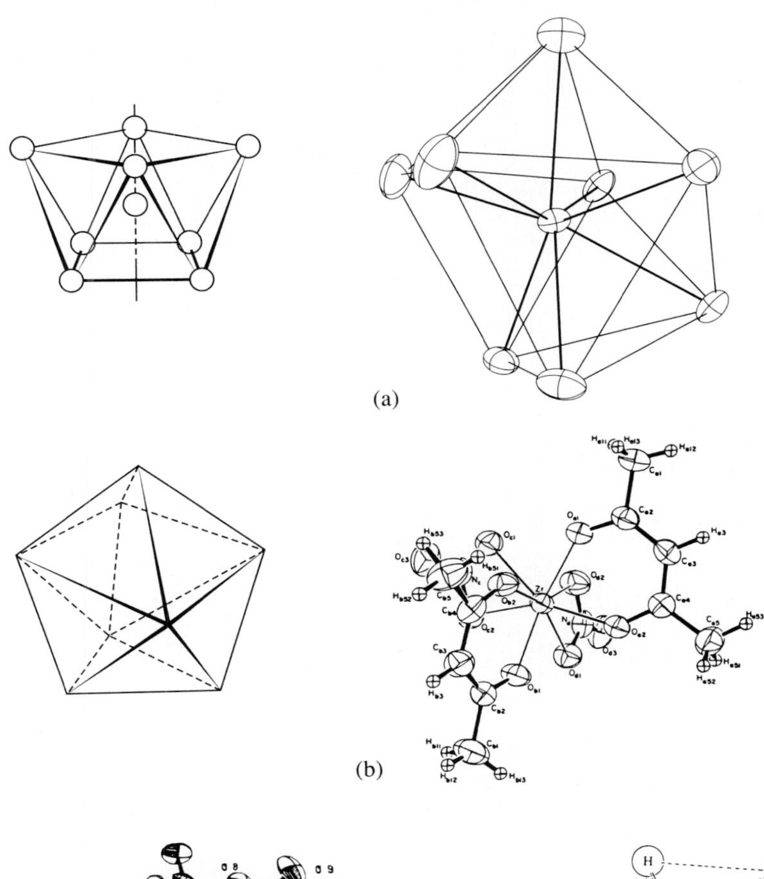

FIGURE 10-12 Compounds with Coordination Number 8. (a) $Na_7Zr_6F_{31}$, square antiprisms of ZrF_8. (Reproduced with permission from J. H. Burns et al., *Acta Cryst.*, **1968,** *B24,* 230.) (b) $[Zr(acac)_2(NO_3)_2]$, regular dodecahedron. (Reproduced with permission from V. W. Day and R. C. Fay, *J. Am. Chem. Soc.*, **1975,** *97,* 5136. Copyright 1975 American Chemical Society.)

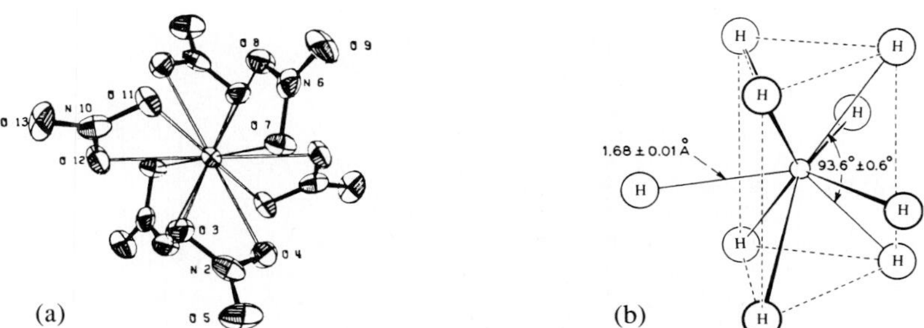

FIGURE 10-13 Compounds with Larger Coordination Numbers. (a) $[Ce(NO_3)_6]^{3-}$, with bidentate nitrates. (Reproduced with permission from T. A. Beinecke and J. Delgaudio, *Inorg. Chem.*, **1968,** *7,* 715. Copyright 1968 Americal Chemical Society.) (b) $[ReH_9]^{2-}$, tricapped trigonal prism. (Reproduced with permission from S. C. Abrahams, et al., *Inorg. Chem.*, **1964,** *3,* 558. Copyright 1964 American Chemical Society.)

[16] M. C. Favas and D. L. Kepert, *Prog. Inorg. Chem.*, **1981,** *28,* 309.

10-3 FOUR- AND SIX-COORDINATE PREFERENCES

The effect of d orbital occupancy on the shape of complexes can be considered mathematically using the techniques of Chapter 8. Since they are the most common shapes, we will consider the 4- and 6-coordinate species with tetrahedral, square planar, and octahedral shapes.

The ligand field stabilization energies for d^0 through d^{10} in Figure 10-14[17] show stabilization of square planar structures for any number of d electrons,

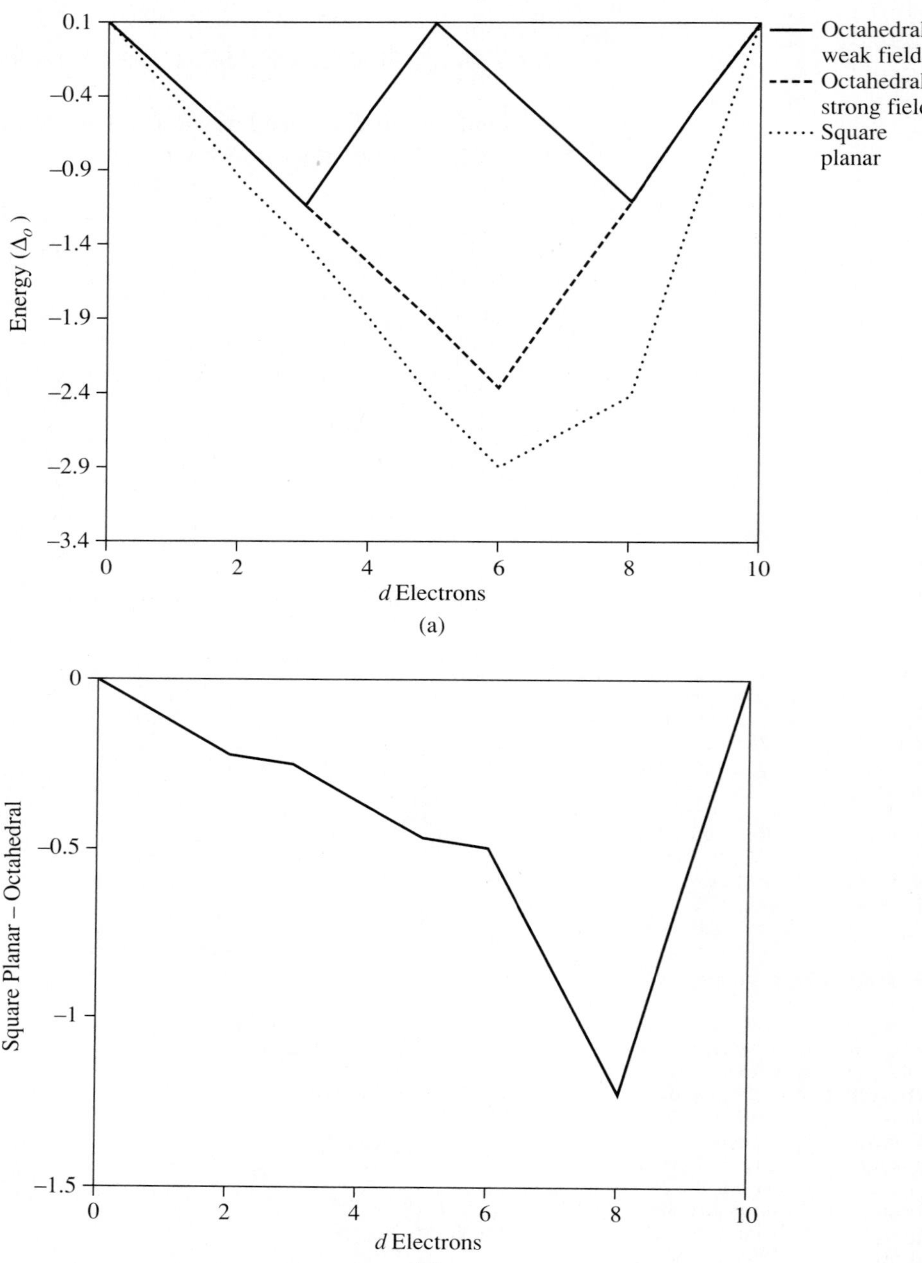

FIGURE 10-14 (a) Ligand Field Stabilization Energies of 4- and 6-Coordinate Complexes. (b) Difference between Square Planar and Octahedral LFSE (strong fields).

[17] The square planar values are calculated by the method of R. Krishnamurthy and W. B. Schaap, *J. Chem. Educ.*, **1969**, *46*, 799.

but particularly for d^8, and to a smaller extent for d^7 and d^9 species. This fits the observation that the largest number of square planar complexes have 8 d electrons. More complete angular overlap calculations for sigma interactions, in Figure 10-15, show that octahedral structures are more stable except in a moderately strong ligand field. The low-spin square planar structures are more stable than high-spin octahedral structures for d^5 through d^7 ions, with no difference for d^8 through d^{10}. As a result, square planar structures are possible for these species.

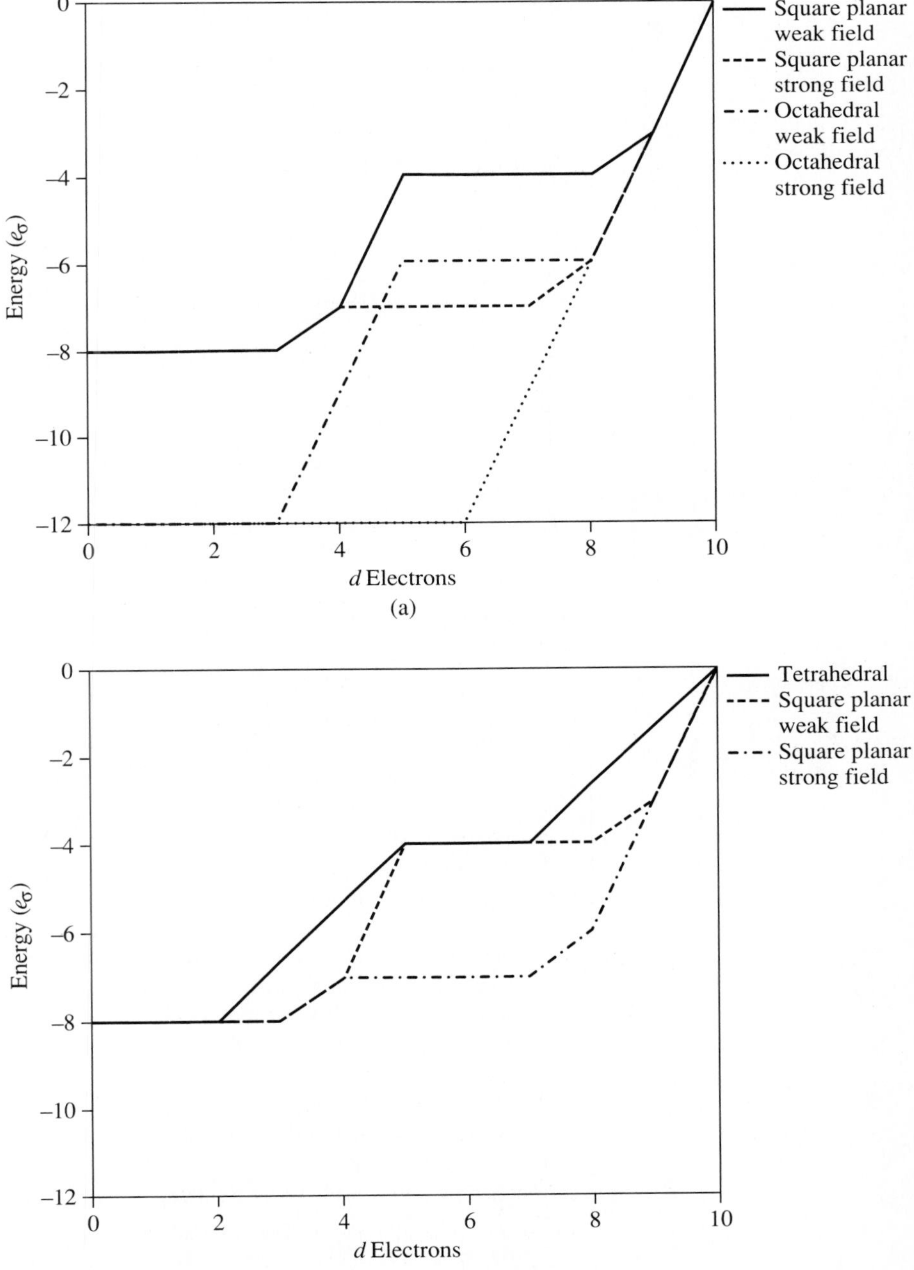

FIGURE 10-15 Angular Overlap Energies of 4- and 6-Coordinate Complexes. Only σ bonding is considered. (a) Octahedral and square planar geometries, in both weak and strong field cases. (b) Tetrahedral and square planar geometries, in both weak and strong field cases (there are no known low-spin tetrahedral complexes).

Angular overlap energy calculations also show that octahedral structures are more stable than tetrahedral for all but d^{10} ions, for which LFSE is zero for both, as shown in Figure 10-15.

The angular overlap model can be used to illustrate the effect of ligands in tetrahedral and square planar geometries on d orbital energy levels. From Table 8-5, the effect of four sigma donor ligands on the d orbitals of a central metal can be estimated for these two geometries, as shown in Figure 10-16.

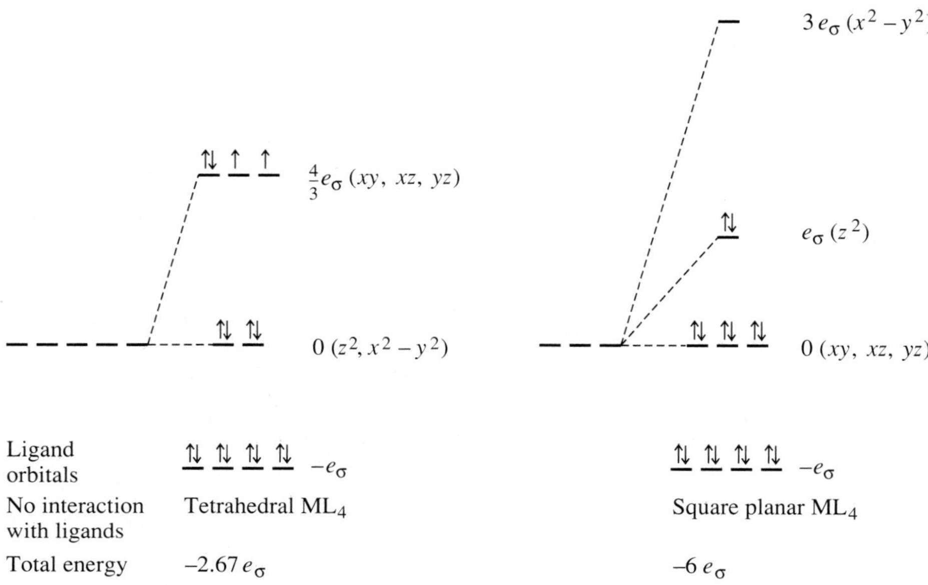

FIGURE 10-16 Tetrahedral and Square Planar Orbital Energies for d^8 Metals.

EXAMPLE

For a d^8 configuration, the energy of the d electrons (more correctly, electrons in molecular orbitals predominantly d in composition) in square planar geometry is lower by $3.33e_\sigma$ than the energy in tetrahedral geometry. Because of the large energy separation between the top two orbitals in square planar geometry, square planar d^8 complexes are generally low spin; tetrahedral d^8 complexes have only one possible electronic configuration, with two unpaired electrons.

If we take into account the effect of π acceptor ligands, this energy difference is enhanced by an additional $8.89e_\pi$, as shown in Figure 10-17. In many cases, the overall effect of both σ and π interactions is sufficient to cause square planar geometry for d^8 complexes. This advantage is enough to override the electron–electron repulsions of the metal–ligand bonds (VSEPR), which would be expected to favor tetrahedral geometry.

EXERCISE 10-1

For a d^9, 4-coordinate complex, use the angular overlap model to determine the energies of the d electrons for tetrahedral and square planar geometries:

a. For σ donor ligands only.
b. For σ donor and π acceptor ligands.

Does the d^9 configuration favor square planar geometry to a greater or lesser degree than d^8?

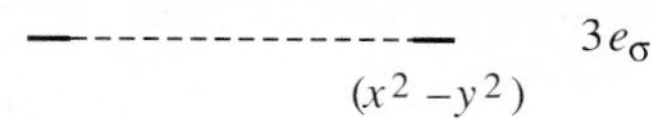

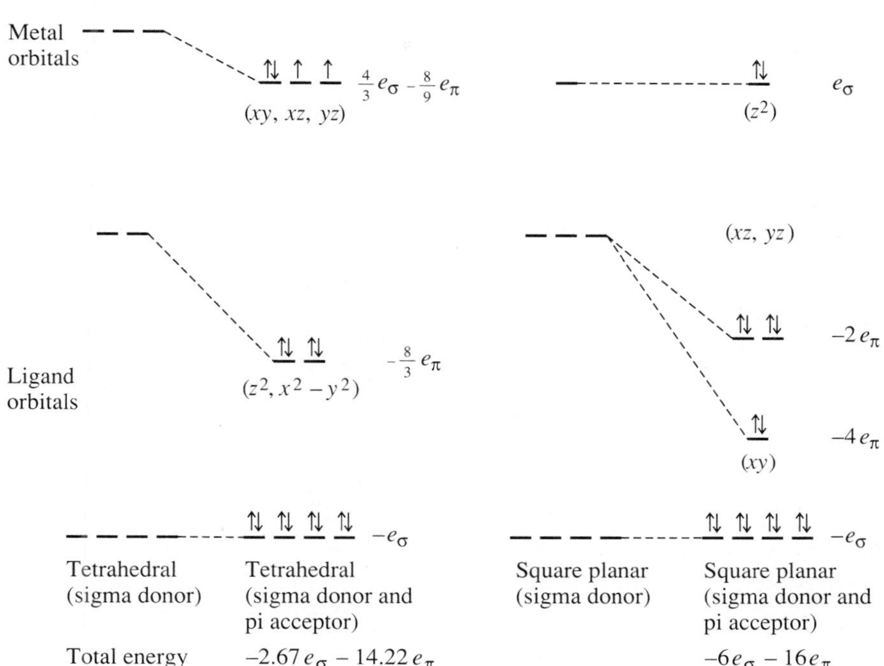

FIGURE 10-17 Addition of π-acceptor Interactions.

10-4 ISOMERISM

The larger variety of coordination numbers in these compounds as compared to organic compounds provides a comparable variety of isomers, even though we usually keep the ligand the same in considering isomers. For example, coordination compounds of the ligands 1-aminopropane and 2-aminopropane are isomers, but we do not include them in our discussion because they do not change the metal–ligand bonding. We will limit our discussion of isomers to those with the same ligands arranged in different geometries. Naturally, the number of possible isomers increases with coordination number. In the following examples, we also limit our discussion to the more common coordination numbers, primarily 4 and 6, but the reader should keep in mind the possibilities for isomerization in other cases as well.

Isomers in coordination chemistry include many types. **Hydrate** or **solvent isomers, ionization isomers,** and **coordination isomers** have different coordination species for the same overall formula. The names indicate whether solvent, anions, or other coordination compounds form the changeable part of the structure. The terms **linkage isomerism** or **ambidentate isomerism** are used for cases of bonding through different atoms of the same ligand. **Stereoisomers** are generally distinguished from other isomers, although they are just special cases of isomers with different shapes but with the same ligands. The diagram and examples that follow may help make the distinction clearer.

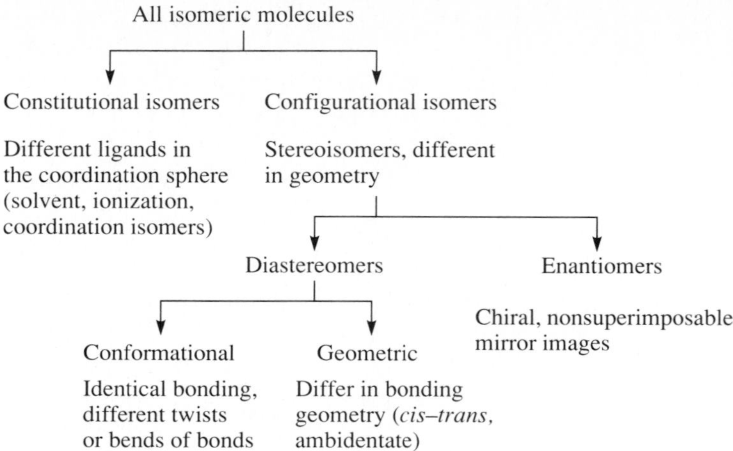

10-4-1 STEREOISOMERISM

Stereoisomers include *cis* and *trans* isomers, chiral isomers, compounds with different comformations of chelate rings, and other isomers that differ only in the geometry of attachment to the metal ion. As mentioned in Chapter 8, study of isomers provided much of the experimental evidence used by Werner to develop and defend his coordination theory. Similar study of new compounds is useful in establishing structures and reactions, even though development of experimental methods such as automated X-ray diffraction can shorten the process considerably.

10-4-2 FOUR-COORDINATE COMPOUNDS

Square planar compounds may have *cis* and *trans* isomers, but no chiral isomers are possible when the molecule has a mirror plane (as many square planar molecules do). In making decisions such as this, we usually ignore minor changes in the ligand such as rotation of substituent groups, conformational changes in ligand rings, and bending of bonds. Examples of square planar compounds that do have chiral isomers are (*meso*-stilbenediamine)(*iso*-butylenediamine)platinum(II) and palladium(II); they are shown in Figure 10-18. In this case, the geometry of the stilbenediamine ligand rules out the

FIGURE 10-18 Isomers of Square Planar Compounds. (a) *Cis* and *trans* isomers of $[Pt(NH_3)_2Cl_2]$. (b) Chiral isomers of (*meso*-stilbenediamine)(*iso*-butylenediamine)platinum(II) and palladium(II). (W. H. Mills and T. H. H. Quibell, *J. Chem. Soc.*, **1935,** *839;* A. G. Lidstone and W. H. Mills, *J. Chem. Soc.*, **1939,** 1754.)

mirror plane. If the compounds were tetrahedral, only one structure would be possible, with a mirror plane splitting the molecule between the two phenyl groups and between the two methyl groups.

Cis and *trans* isomers are common, with platinum(II) one of the most common metal ions used. Examples of $[Pt(NH_3)_2Cl_2]$ isomers are shown in Figure 10-18. The *cis* isomer is used in medicine as an antitumor agent (see Chapter 15). Chelate rings can require the *cis* structure, because the chelating ligand is too small to span the *trans* positions. The distance across the two *trans* positions is too large for all but very large ligands, and synthesis with such large rings is difficult.

10-4-3 CHIRALITY

Chiral molecules (named from the Greek for *hand*) have a degree of asymmetry that makes their mirror images nonsuperimposable. This condition can also be expressed in terms of symmetry elements, where a molecule can be chiral only if it has no rotation–reflection (S_n) axes. This means that chiral molecules either have no symmetry elements or have only axes of proper rotation (C_n). Tetrahedral molecules with four different ligands or with unsymmetrical chelating ligands can be chiral, as can octahedral molecules with bidentate or higher chelating ligands or with $[Ma_2b_2c_2]$, $[Mabc_2d_2]$, $[Mabcd_3]$, $[Mabcde_2]$, or [Mabcdef] structures (M = metal; a, b, c, d, e, f = monodentate ligands). Not all the isomers will be chiral, but the possibility must be considered for each.

The only isomers possible for tetrahedral compounds are chiral. All attempts to draw nonchiral isomers of tetrahedral compounds will fail because of the inherent symmetry of the tetrahedron.

10-4-4 SIX-COORDINATE COMPOUNDS

Compounds of the formula $ML_3L'_3$, where L and L′ are monodentate ligands, may have two isomeric forms called *fac*- and *mer*- (for facial and meridional). Similar isomers are possible with some chelating ligands. Examples with monodentate and tridentate ligands are shown in Figure 10-19.

Special nomenclature has been proposed for other isomers of a similar type. For example, triethylenetetramine compounds have three forms: α, with all three chelate rings in different planes; β, with two of the rings coplanar; and *trans*, with all three rings coplanar, as in Figure 10-20. Additional isomeric forms are possible, some of which will be discussed later in this chapter (both α and β have chiral isomers, and all three have additional isomers dependent on the conformations of the individual rings). Even when one multidentate ligand has a single geometry, other ligands may result in isomers. The β, β′, β″-triaminotriethylamine (tren) ligand bonds to four adjacent sites, but an asymmetric ligand like salicylate can then bond in the two ways shown in Figure 10-21.

Other isomers are possible when the number of different ligands is increased. There have been several schemes for calculating the maximum number of isomers for each case,[18] although omissions were difficult to avoid

[18] J. C. Bailar, Jr., *J. Chem. Educ.*, **1957**, *34*, 334; S. A. Meyper, *J. Chem. Educ.*, **1957**, *34*, 623.

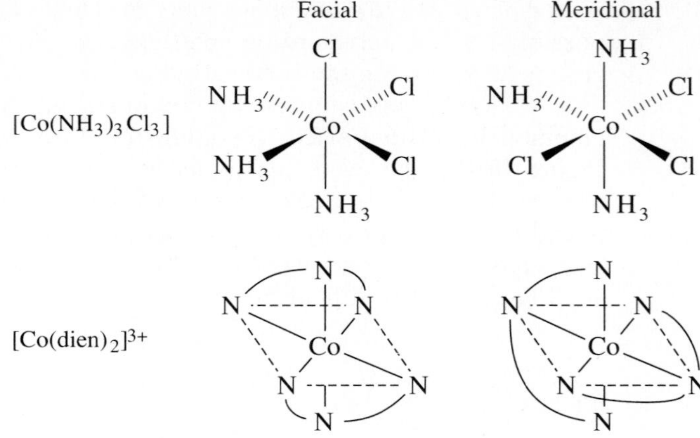

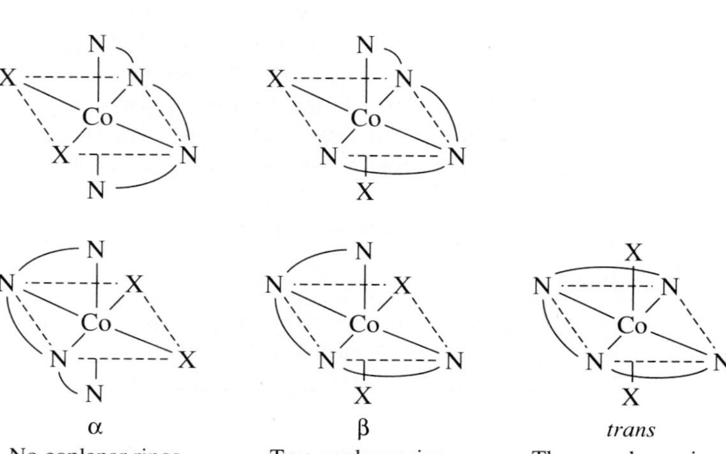

FIGURE 10-19 Facial and Meridional Isomers of $[Co(NH_3)_3Cl_3]$ and $[Co(dien)_2]^{3+}$.

FIGURE 10-20 Isomers of Triethylenetetraamine Compounds.

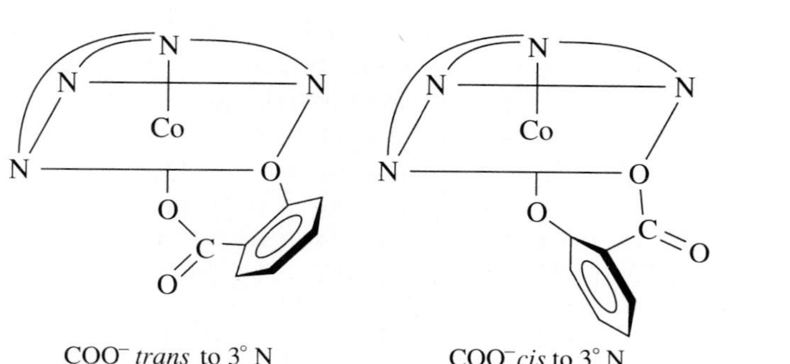

FIGURE 10-21 Isomers of $[Co(tren)(sal)]^+$.

until the advent of computer programs to assist in the process. One such program[19] begins with a single structure, generates all the others by switching ligands from one position to another, and then rotates the new form to all possible positions for comparison with the earlier structures. In the extreme case of six different ligands with an overall octahedral shape, the number of possible isomers is 30. Finding the number and identity of the isomers is

[19] W. E. Bennett, *Inorg. Chem.*, **1969,** *8,* 1325.

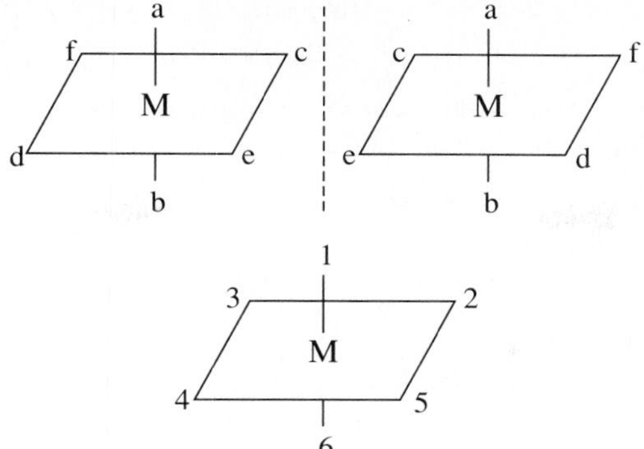

FIGURE 10-22 [M⟨ab⟩⟨cd⟩⟨ef⟩] Isomers and the Octahedral Numbering System.

primarily a matter of systematically listing the possible structures and then checking for identical species and chirality. Examples of the isomers for [Mabcdef] are given in Figure 10-22 and Table 10-2.

In these structures, the notation ⟨ab⟩ indicates that a and b are *trans* to each other, with M the metal ion and a, b, c, d, e, f monodentate ligands. The [M⟨ab⟩⟨cd⟩⟨ef⟩] isomers ([Pt⟨(py)(NH_3)⟩⟨(NO_2)(Cl)⟩⟨(Br)(I)⟩] is an example[20]) are shown in Figure 10-22. The six octahedral positions are commonly numbered as in Chapter 8, with positions 1 and 6 in axial positions and with 2 through 5, the other four, in counterclockwise order as viewed from the 1 position.

TABLE 10-2
[Mabcdef] isomers

	A	*B*	*C*
1	ab	ab	ab
	cd	ce	cf
	ef	df	de
2	ac	ac	ac
	bd	be	bf
	ef	df	de
3	ad	ad	ad
	bc	be	bf
	ef	cf	ce
4	ae	ae	ae
	bc	bf	bd
	df	cd	cf
5	af	af	af
	bc	bd	be
	de	ce	cd

Note: Each isomer (A1, A2, . . .) has the *trans* pairs listed. For example, A1 is shown in Figure 10-22. Each isomer also has a mirror image (enantiomer).

If the ligands are completely scrambled rather than limited to the *trans* pairs in the figure, there are 15 different diastereoisomers (different structures that are not mirror images of each other), each of which has an enantiomer (mirror image). The method suggested by Bailar uses a list of isomers. One *trans* pair, such as ⟨ab⟩, is held constant, the second pair has one component constant and the other is systematically changed, and the third pair is whatever

[20] L. N. Essen and A. D. Gel'man, *Zhur. Neorg. Khim.*, **1956**, *1*, 2475.

is left over. The second component of the first pair is changed and the process is continued. The result is Table 10-2.

EXAMPLE

The isomers of $Ma_2b_2c_2$ can be found by this method. The *trans* pairs of the isomers are:

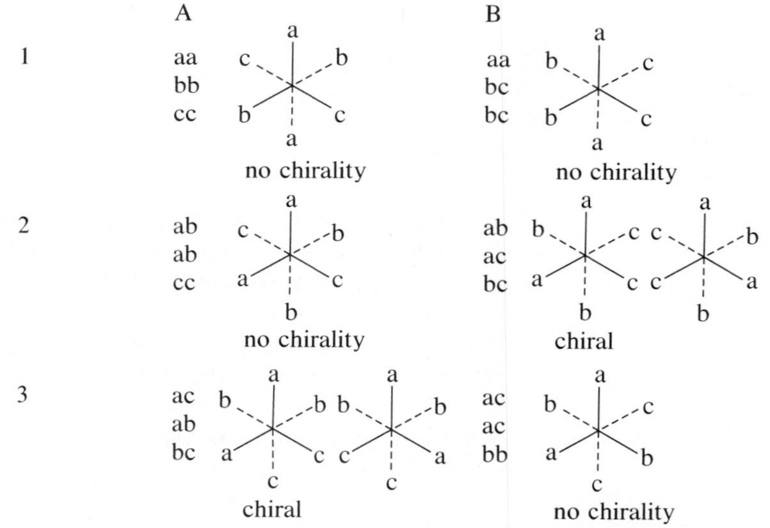

	A		B	
1	aa bb cc	no chirality	aa bc bc	no chirality
2	ab ab cc	no chirality	ab ac bc	chiral
3	ac ab bc	chiral	ac ac bb	no chirality

A3 and B2 are identical. Overall, there are four nonchiral isomers and one chiral pair, for a total of six.

EXERCISE 10-2

Find the number and identity of all the isomers of $[Ma_2b_2cd]$.

The same approach can be used for chelating ligands, with limits on the location of the ring. For example, a normal bidentate chelate ring cannot connect *trans* positions. After listing all the isomers without this restriction, those that are sterically impossible can be quickly eliminated and the others checked for duplicates and then for enantiomers. Table 10-3 lists the number of isomers and enantiomers for many general formulas, all calculated using a computer program similar to Bennett's.[21]

EXAMPLE

A methodical approach is important in finding isomers. For M(AA)(BB)cd, we first try c and d in *cis* positions:

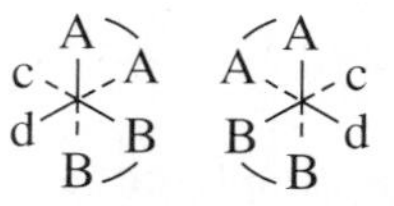

c opposite B
d opposite A

The mirror image is different, so there is a chiral pair.

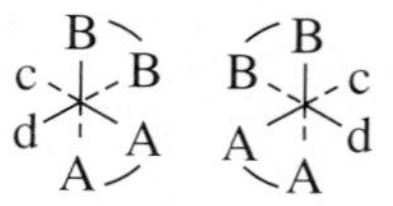

c opposite A
d opposite B

The mirror image is different, so there is a chiral pair.

[21] Bennett, op. cit.; B. A. Kennedy, D. A. MacQuarrie, and C. H. Brubaker, Jr., *Inorg. Chem.*, **1964,** *3*, 265.

Then trying c and d in *trans* positions:

$$\left(\begin{smallmatrix} & c & \\ A & \ast & B \\ A & & B \\ & d & \end{smallmatrix}\right) \quad \left(\begin{smallmatrix} & c & \\ B & \ast & A \\ B & & A \\ & d & \end{smallmatrix}\right)$$

The mirror image is identical, so there is only one isomer, for a total of five isomers.

EXERCISE 10-3

Find the number and identity of all isomers of [M(AA)bcde], where AA is a bidentate ligand with identical coordinating groups.

TABLE 10-3

Number of possible isomers for specific compounds

Formula	*Number of stereoisomers*	*Pairs of enantiomers*
Ma_6	1	0
Ma_5b	1	0
Ma_4b_2	2	0
Ma_3b_3	2	0
Ma_4bc	2	0
Ma_3bcd	5	1
Ma_2bcde	15	6
Mabcdef	30	15
$Ma_2b_2c_2$	6	1
Ma_2b_2cd	8	2
Ma_3b_2c	3	0
M(AA)(BC)de	10	5
M(AB)(AB)cd	11	5
M(AB)(CD)ef	20	10
$M(AB)_3$	4	2
M(ABA)cde	9	3
$M(ABC)_2$	11	5
M(ABBA)cd	7	3
M(ABCBA)d	7	3

NOTE: Capital letters represent chelating ligands; lowercase represent monodentate ligands.

10-4-5 COMBINATIONS OF CHELATE RINGS

Before discussing nomenclature rules for ring geometry, we need to establish clearly the idea of the handedness of propellers and helices. Consider first the propellers shown in Figure 10-23. The first is a left-handed propeller, which means that rotating it *counterclockwise* in air or water would move it away from the observer. The second, a right-handed propeller, moves away on *clockwise* rotation. The tips of the propeller blades describe left- and right-handed helices, respectively. With rare exceptions, the threads on screws and bolts are right-handed helices; a clockwise twist with a screwdriver or wrench drives them into a nut or piece of wood. The same clockwise motion drives a nut onto a stationary bolt. Another example of a helix is a coil spring, which can usually have either handedness without affecting its operation.

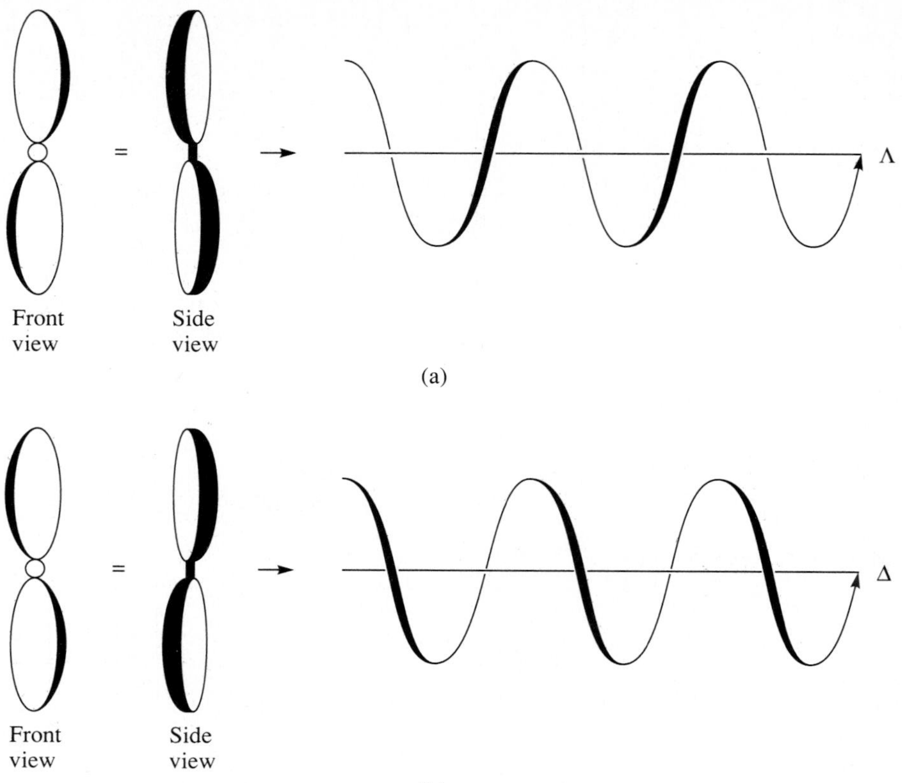

FIGURE 10-23 Right- and Left-handed Propellers. (a) Left-handed propeller and helix traced by tips of the blades. (b) Right-handed propeller and helix traced by tips of the blades.

Complexes with three rings, such as $[Co(en)_3]^{3+}$, can be treated like three-bladed propellers by looking at the molecule down a three-fold axis. Figure 10-24 shows a number of different ways to draw these structures, all equivalent. The clockwise (Δ) or counterclockwise (Λ) character can also be found by the procedure in the next paragraph.

Complexes with two or more nonadjacent chelate rings may have chiral character. Any two noncoplanar and nonadjacent chelate rings (not sharing a common atom bonded to the metal) can be used to determine the handedness. Rotate the molecule to place one ring at the back (away from the viewer) in a horizontal position. Imagine that the second ring was originally at the front, also in the horizontal plane. If it takes a counterclockwise (ccw) twist of the front ligand to place it as it is in the actual molecule, the rings have a Λ relationship. If it takes a clockwise (cw) twist to orient the front ligand properly, the rings have a Δ relationship. Figure 10-25 illustrates the process.

A molecule with more than one pair of rings may require more than one label, but is treated similarly. The handedness of each pair of skew rings is determined, and the final description then includes all the designations. For example, an EDTA complex has six points of attachment and five rings. All pairs of rings that are not coplanar and are not connected at the same atom are used in the description. The N—N ring is omitted because it is connected at the same atom with each of the other rings. Considering only the 4 O—N rings, there are three useful pairs (four pairs total without common atoms, one of which has coplanar rings). Each of the three pairs is used, as shown in Figure 10-26 with one isomer, where the rings are numbered arbitrarily R_1

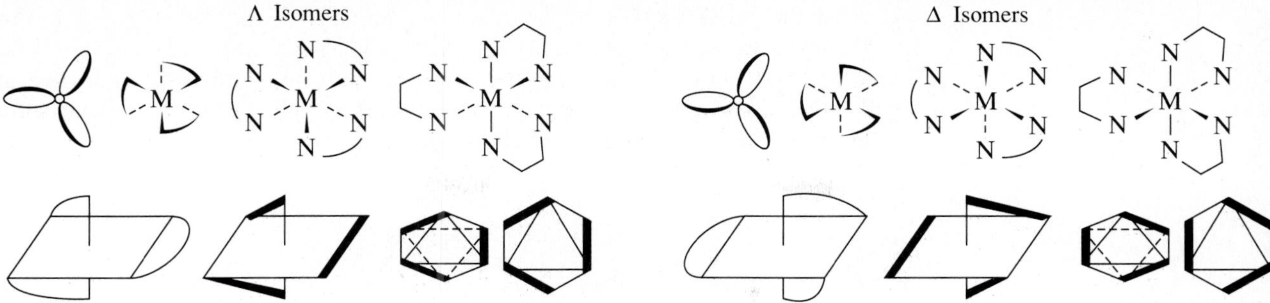

FIGURE 10-24 Right- and Left-handed Chelates.

FIGURE 10-25 Procedure for Determining Handedness.

1. Rotate the figure to place one ring horizontally across the back, at the top of one of the triangular faces.
2. Imagine the ring in the front triangular face as having originally been parallel to the ring at the back. Determine what rotation is required to obtain the actual configuration.
3. If the rotation from Step 2 is clockwise, the structure is designated delta (Δ). If the rotation is counterclockwise, the designation is lambda (Λ).

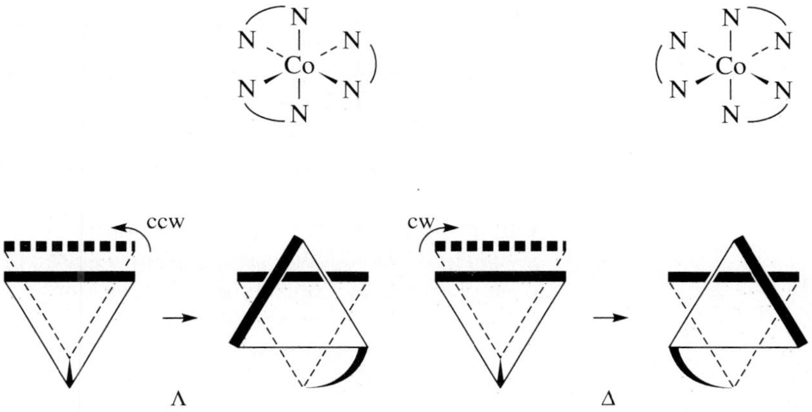

FIGURE 10-26 Labeling of Chiral Rings. The rings are numbered arbitrarily R_1 through R_5. The combination R_1–R_4 is Λ, R_1–R_5 is Δ, and R_2–R_5 is Λ. The notation for this structure is then Λ Δ Λ-(ethylenediaminetetraacetato)-cobaltate(III).

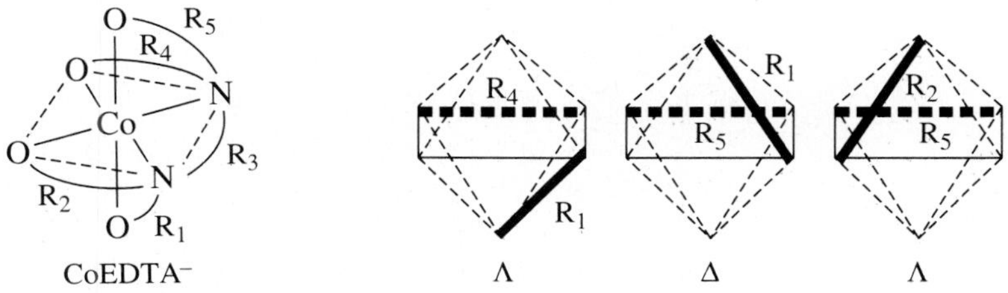

through R_5. The method described above shows the combination R_1–R_4 is Λ, R_1–R_5 is Δ, and R_2–R_5 is Λ. The notation for the compound given is then Λ Δ Λ-(ethylenediaminetetraacetato)cobaltate(III). The order of the designations is arbitrary and could as well be Λ Λ Δ or Δ Λ Λ.

10-4-6 LIGAND RING CONFORMATION

Because many chelate rings are not planar, they can have different conformations in different molecules, even in otherwise identical molecules. In some cases, these different conformations are also chiral. The notation used also requires using two lines to establish the handedness. The first line connects the atoms bonded to the metal. In the case of ethylenediamine, this line connects the two nitrogen atoms. The second line connects the two carbon atoms of the ethylenediamine, and the handedness of the two rings is found by the method described above for separate rings. A counterclockwise rotation of the second line is called λ and a clockwise rotation is called δ, as shown in Figure 10-27. Complete description of a complex then requires identification of the overall chirality and the chirality of each ring.

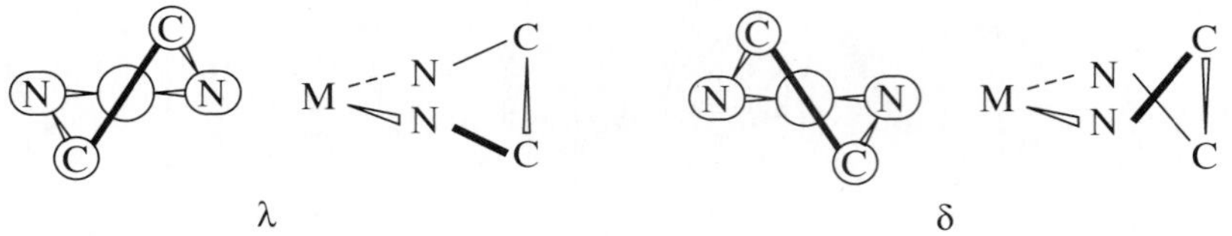

FIGURE 10-27 Chelate Ring Conformations.

Corey and Bailar[22] examined some examples and found the same steric interactions found in cyclohexane and other ring structures. For example, the Δ λλλ form of $[Co(en)_3]^{3+}$ was calculated to be 7.5 kJ/mol more stable than the Δ δδδ form because of interactions between protons on the nitrogens. For the Λ form, the δδδ ring conformations are more stable. Although there are examples where this preference is not followed, in general the experimental results confirm their calculations. In solution, the small difference in energy allows rapid interconversion of conformation between λ and δ, and the most abundant configuration for the Λ isomer is δδλ.[23]

An additional isomeric possibility arises because the symmetry of ligands can be changed by coordination. An example is a secondary amine in a ligand such as diethylenetriamine or triethylenetetramine. As a free ligand, inversion at the nitrogen is easy and only one isomer is possible. After coordination there may be additional chiral isomers. If there are chiral centers on the ligands, either inherent in their structure or created by coordination (as in some secondary amines), their structure must be described by the R and S notation familiar from organic chemistry.[24] These are illustrated in Figure 10-28.

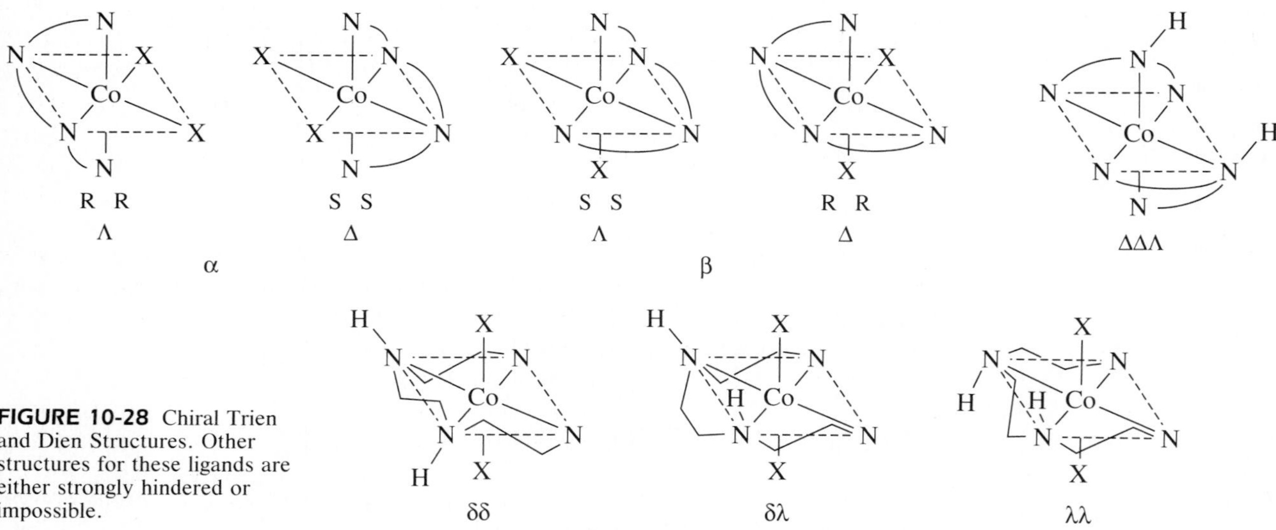

FIGURE 10-28 Chiral Trien and Dien Structures. Other structures for these ligands are either strongly hindered or impossible.

[22] E. J. Corey and J. C. Bailar, Jr., *J. Am. Chem. Soc.*, **1959,** *81,* 2620.

[23] J. K. Beattie, *Acc. Chem. Res.*, **1971,** *4,* 253.

[24] R. S. Cahn and C. K. Ingold, *J. Chem. Soc.*, **1951,** 612; R. S. Cahn, C. K. Ingold, and V. Prelog, *Experientia*, **1956,** *12,* 81.

10-4-7 EXPERIMENTAL SEPARATION AND IDENTIFICATION OF ISOMERS

Separation of geometric isomers frequently requires fractional crystallization with different counterions. Since different isomers will have slightly different shapes, the packing in crystals will depend on the fit of the ions and their overall solubility. One helpful idea, systematized by Basolo,[25] is that ionic compounds show the smallest solubility when the oppositely charged ions have the same magnitude of charge and their size is similar. For example, large cations of charge 2+ are best crystallized with large anions of charge 2−. While not a surefire method of separating isomers, comparisons such as this help in deciding what combinations to try.

Separation of chiral isomers requires chiral counterions. Cations are frequently resolved by using the anions *d*-tartrate, antimony *d*-tartrate, and α-bromocamphor-π-sulfonate; anionic complexes are resolved by the bases brucine or strychnine or by using resolved cationic complexes such as $[Rh(en)_3]^{3+}$.[26] In the case of compounds that racemize at appreciable rates, adding a chiral counterion may shift the equilibrium even if it does not precipitate one form. Apparently, interactions between the ions in solution are sufficient to stabilize one form over the other.[27]

The best method of identifying isomers, when crystallization allows it, is X-ray crystallography. Current methods allow rapid determination of the absolute configuration at costs that compare favorably with other more indirect methods, and in many cases new compounds are routinely examined this way.

Measurement of optical activity is a natural method for assigning absolute configuration to chiral isomers, but it usually requires more than simple determination of molar rotation at a single wavelength. Optical rotation changes markedly with the wavelength of the light used in the measurement and changes sign near absorption peaks. Many organic compounds have their largest rotation in the ultraviolet, and the old standard of molar rotation at the sodium D wavelength is a measurement of the tail of the much larger peak. Coordination compounds frequently have their major absorption (and therefore rotation) bands in the visible part of the spectrum, and it then becomes necessary to examine the rotation as a function of wavelength to determine the isomer present. Before the development of the X-ray methods now used, debates over assignments of configuration were common, since comparison of similar compounds could lead to contradictory assignments depending on which comparisons were made.

Chiral compounds rotate the plane of polarization of plane-polarized light (Figure 4-17). Measurement of this rotation as a function of energy is called **optical rotatory dispersion** (ORD). Such spectra result from differences in refractive index for left and right circularly polarized light and commonly show a positive value on one side of an absorption band and a negative value on the other, with a rapid change in rotation and sign near the absorption maximum. This is called the **Cotton effect,**[28] which is positive when the value

[25] F. Basolo, *Coord. Chem. Rev.*, **1968,** *3,* 213.

[26] R. D. Gillard, D. J. Shepherd, and D. A. Tarr, *J. Chem. Soc., Dalton,* **1976,** 594.

[27] F. P. Dwyer: $[Fe(phen)_3]^{2+}$ forms 100% (−) antimony tartrate solid. Werner: similar results with $[Cr(C_2O_4)]^{3-}$. Dwyer: $[Ni(phen)_3]^{2+}$ shows different rates of racemization depending on the anions or cations present.

[28] A. Cotton, *Ann. Chem. Phys.*, **1896,** *8,* 347.

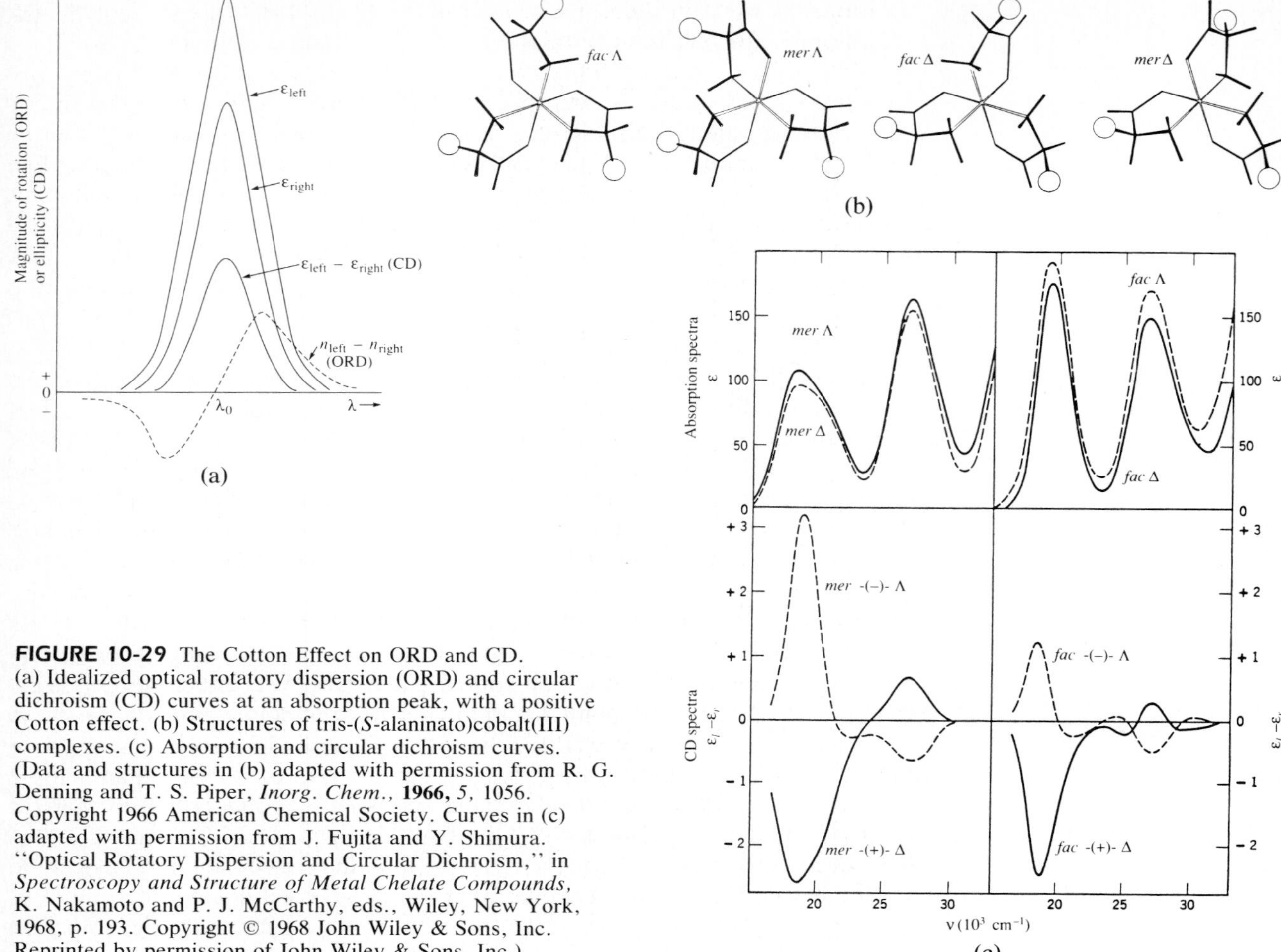

FIGURE 10-29 The Cotton Effect on ORD and CD. (a) Idealized optical rotatory dispersion (ORD) and circular dichroism (CD) curves at an absorption peak, with a positive Cotton effect. (b) Structures of tris-(*S*-alaninato)cobalt(III) complexes. (c) Absorption and circular dichroism curves. (Data and structures in (b) adapted with permission from R. G. Denning and T. S. Piper, *Inorg. Chem.*, **1966,** *5,* 1056. Copyright 1966 American Chemical Society. Curves in (c) adapted with permission from J. Fujita and Y. Shimura. "Optical Rotatory Dispersion and Circular Dichroism," in *Spectroscopy and Structure of Metal Chelate Compounds,* K. Nakamoto and P. J. McCarthy, eds., Wiley, New York, 1968, p. 193. Copyright © 1968 John Wiley & Sons, Inc. Reprinted by permission of John Wiley & Sons, Inc.)

is positive at low energy and becomes negative at higher energy, and negative for the reverse; this is illustrated in Figure 10-29. Unfortunately, the shape of the curves and the long tail makes resolution of the curves into individual transitions difficult and use of ORD for this purpose is not common.

Circular dichroism (CD) is easier to interpret and has become the method of choice. CD is the difference between the molar absorbances of left and right circularly polarized light ($\epsilon_l - \epsilon_r$). A positive Cotton effect shows a simple positive peak at the same energy as the absorption maximum, a negative Cotton effect a negative peak at the same position. Even with CD, spectra are not always easily interpreted because there may be overlapping bands of different signs.[29] Interpretation requires determination of the overall symmetry around the metal ion and assignment of absorption spectra to specific transitions (discussed in Chapter 9) in order to assign specific CD peaks to the appropriate

[29] R. D. Gillard, "Optical Rotatory Dispersion and Circular Dichroism," in H. A. O. Hill and P. Day, *Physical Methods in Advanced Inorganic Chemistry,* Wiley-Interscience, New York, 1968, pp. 183–85; C. J. Hawkins, *Absolute Configuration of Metal Complexes,* Wiley-Interscience, New York, 1971, p. 156.

transitions. Even then there are cases where the CD peaks do not match the absorption peaks and interpretation becomes much more difficult.

CD spectrometers have an optical system much like UV-visible spectrophotometers, with the addition of a crystal of ammonium phosphate mounted to allow imposition of a large electrostatic field on it. When the field is imposed, the crystal allows only circularly polarized light to pass through; changing the direction of the field rapidly provides alternating left and right circularly polarized light. The light received by the detector is compared electronically and presented as the difference between the absorbances.

10-4-8 HYDRATE ISOMERISM

Hydrate isomerism is not common, but deserves mention because it contributed to some of the confusion in describing coordination compounds before the Werner theory was generally accepted. It differs from other isomerism in having water as either a ligand or an added part of the crystal structure, as in the hydrates of sodium sulfate (Na_2SO_4, $Na_2SO_4 \cdot 7\ H_2O$, and $Na_2SO_4 \cdot 10\ H_2O$ are known). More strictly, it should be called solvent isomerism to allow for the possibility of ammonia or other ligands also used as solvents to participate in the structure, but many examples involve water.

The standard example is $CrCl_3 \cdot 6\ H_2O$, which can have three distinctly different crystalline compounds, now known as $[Cr(H_2O)_6]Cl_3$ (violet), $[CrCl(H_2O)_5]Cl_2 \cdot H_2O$ (blue-green), and $[CrCl_2(H_2O)_4]Cl \cdot 2\ H_2O$ (dark green). Existence of a fourth possibility, $[CrCl_3(H_2O)_3] \cdot 3\ H_2O$ (brown) is somewhat uncertain.[30] The three isomers can be separated by cation ion exchange from commercial $CrCl_3 \cdot 6\ H_2O$, in which the major component is $[CrCl_2(H_2O)_4]Cl \cdot 2\ H_2O$ in the *trans* configuration. Other examples are also known; a few are listed below.

$$[Co(NH_3)_4(H_2O)Cl]Cl_2 \quad \text{and} \quad [Co(NH_3)_4Cl_2]Cl \cdot H_2O$$

$$[Co(NH_3)_5(H_2O)](NO_3)_3 \quad \text{and} \quad [Co(NH_3)_5(NO_3)](NO_3)_2 \cdot H_2O$$

Ionization isomerism

Compounds with the same formula, but which give different ions in solution, exhibit ionization isomerization. The difference is in which ion is included as a ligand and which is present to balance the overall charge. Some examples are also hydrate isomers, such as the first one listed below.

$$[Co(NH_3)_4(H_2O)Cl]Br_2 \quad \text{and} \quad [Co(NH_3)_4Br_2]Cl \cdot H_2O$$

$$[Co(NH_3)_5SO_4]NO_3 \quad \text{and} \quad [Co(NH_3)_5NO_3]SO_4$$

$$[Co(NH_3)_4(NO_2)Cl]Cl \quad \text{and} \quad [Co(NH_3)_4Cl_2]NO_2$$

Many other examples, and even more possibilities, exist. Enthusiasm for preparing and characterizing such compounds is not great now, and new examples are more likely to be discovered as part of other studies.

[30] A. Recoura, *C. R. Acad. Sci.*, **1932,** *194,* 229; **1933,** *196,* 1853.

10-4-9 COORDINATION ISOMERISM

Examples of a complete series of coordination isomers require at least two metal ions and sometimes more. The total ratio of ligand to metal remains the same, but the ligands attached to a specific metal ion change. This is best described by example.

For the empirical formula $Pt(NH_3)_2Cl_2$, there are three possibilities:

$[Pt(NH_3)_2Cl_2]$	
$[Pt(NH_3)_3Cl][Pt(NH_3)Cl_3]$	(This compound apparently has not been reported, but the individual ions are known.)
$[Pt(NH_3)_4][PtCl_4]$	(Magnus's green salt, the first platinum ammine discovered, in 1828)

Other examples are possible with different metal ions and with different oxidation states:

$$[Co(en)_3][Cr(CN)_6] \quad \text{and} \quad [Cr(en)_3][Co(CN)_6]$$

$$\underset{\text{Pt(II)}}{[Pt(NH_3)_4]}\underset{\text{Pt(IV)}}{[PtCl_6]} \quad \text{and} \quad \underset{\text{Pt(IV)}}{[Pt(NH_3)_4Cl_2]}\underset{\text{Pt(II)}}{[PtCl_4]}$$

10-4-10 LINKAGE (AMBIDENTATE) ISOMERISM

Some ligands can bond to the metal through different atoms. The most common early examples were thiocyanate, SCN^-, and nitrite, NO_2^-. Class (a) metal ions (hard acids) tend to bond to the nitrogen of thiocyanate and class (b) metal ions (soft acids) bond through the sulfur, but the differences are small and the solvent used influences the bonding. Compounds of rhodium and iridium with the general formula $[M(PPh_3)_2(CO)(NCS)_2]$ form M—S bonds in solvents of large dielectric constant and M—N bonds in solvents of low dielectric constant.[31] There are also compounds such as isothiocyanatothiocyanato(1-diphenylphosphino-3-dimethylaminopropane)-palladium(II)[32] with both M—SCN (thiocyanato) and M—NCS (isothiocyanato), where the M—NCS is linear and the M—SCN is bent at the S atom. This bend means that the M—SCN isomer has a larger steric effect, particularly if it can rotate about the M—S bond. Examples of these isomers are shown in Figure 10-30.

The nitrite isomers of $[Co(NH_3)_5NO_2]^{2+}$ were studied by Jørgensen and Werner, who observed that there were two compounds of the same chemical formula but of different colors. A red form of low stability converted readily to a yellow form. The red form was thought to be the M—ONO nitrito isomer and the yellow form the M—NO_2 nitro isomer, based on comparison with compounds with similar colors. This conclusion was later confirmed, and kinetic[33] and ^{18}O labeling[34] experiments showed that conversion of one form

[31] J. L. Burmeister, R. L. Hassel, and R. J. Phelan, *Inorg. Chem.*, **1971,** *10,* 2032; J. E. Huheey and S. O. Grim, *Inorg. Nucl. Chem. Lett.*, **1974,** *10,* 973.

[32] D. W. Meek, P. E. Nicpon, and V. I. Meek, *J. Am. Chem. Soc.*, **1970,** *92,* 5351; G. R. Clark and G. J. Palenik, *Inorg. Chem.*, **1970,** *9,* 2754.

[33] B. Adell, *Z. Anorg. Chem.*, **1944,** *252,* 277.

[34] R. K. Murmann and H. Taube, *J. Am. Chem. Soc.*, **1956,** *78,* 4886.

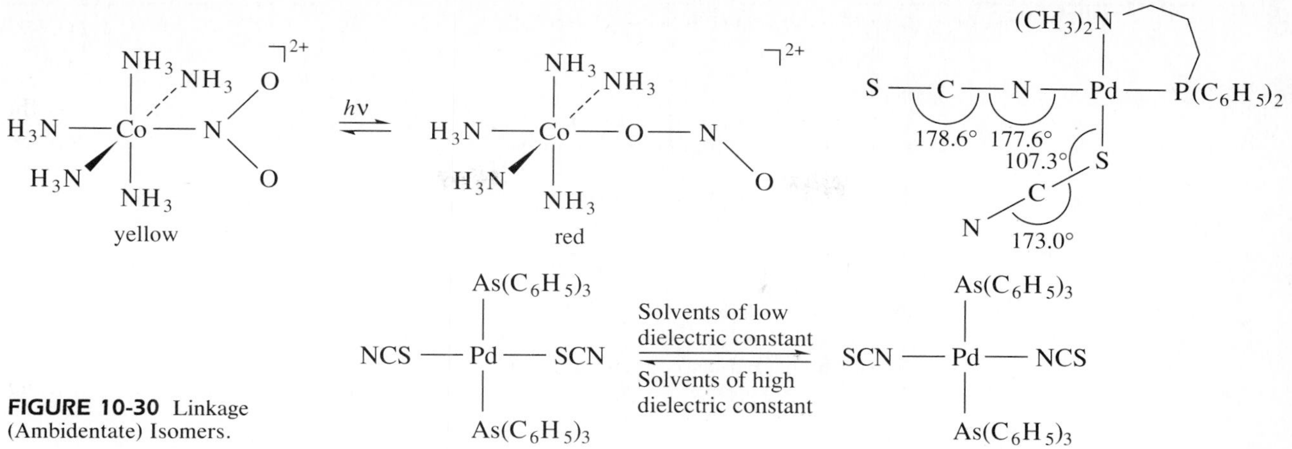

FIGURE 10-30 Linkage (Ambidentate) Isomers.

to the other is strictly intramolecular, not a result of dissociation of the NO_2^- ion followed by reattachment.

10-4-11 DISTORTIONAL ISOMERISM[35]

The final form of isomerism to be considered here is quite different than the others. The overall shapes of the molecules are the same, with the difference solely in the length of one or more bonds. A series of compounds of $Mo(O)Cl_2L_3$ was prepared by Butcher and Chatt[36] with blue and green isomeric forms. These were later found to differ in the length of the Mo—O bond; the blue forms have bond lengths of about 167 pm, and the green forms have bond lengths near 180 pm, with smaller differences in the bond distance to the ligand *trans* to the oxygen. This new form of isomerism was described as distortional isomerism.[37]

Other metals show the same phenomenon, including W, Ru, and Nb. Most of these isomers are converted into a single form in solution, but in at least one case, the hexafluorophosphate salt of $[LWOCl_2]^+$ (L = *N,N′,N″*-trimethyl-1,4,7,-triazacyclononane), the identity of the two isomers is retained for several days in acetonitrile solution.[38] The green form has a W—O distance of 189 pm, the blue form a W—O distance of 172 pm. Addition of water to the solution changes all the complex to the blue form. The causes and further effects of this type of isomerization are still under investigation; so far, no adequate explanation has been presented.

GENERAL REFERENCES

The best single reference for isomers and geometric structures is *Comprehensive Coordination Chemistry*, G. Wilkinson, R. D. Gillard, and J. A. McCleverty, eds., Pergamon Press, Elmsford, N.Y., 1987. The reviews cited in the individual sections are also very comprehensive.

[35] W. A. Nugent and J. M. Mayer, *Metal-ligand Multiple Bonds*, Wiley-Interscience, New York, 1988, pp. 152–54.

[36] A. V. Butcher and J. Chatt, *J. Chem. Soc. A*, **1970**, 2652.

[37] J. Chatt, L. Manojlovic-Muir, and K. W. Muir, *J. Chem. Soc., Chem. Comm.*, **1971**, 655.

[38] K. Wieghardt, G. Backes-Dahmann, B. Nuber, and J. Weiss, *Angew. Chem. Int. Ed. Engl.*, **1985,** *24*, 777.

PROBLEMS

10-1 Name (a) $[Fe(CN)_2(CH_3NC)_4]$, (b) $Rb[AgF_4]$, and (c) $[Ir(PPh_3)_2(CO)Cl]$(two isomers).

10-2 Give structures for:
a. Bis(en)Co(III)-μ-imido-μ-hydroxobis(en)Co(III) ion
b. Diaquadiiododinitrito Pd(IV)

10-3 Sketch structures of all isomers of $M(AB)_3$, and label them properly. AB is a bidentate unsymmetrical ligand.

10-4 Name (a) $[Co(N_3)(NH_3)_5]SO_4$, (b) $Na[AlCl_4]$, and (c) $[Co(en)_2CO_3]Cl$.

10-5 Give structures for:
a. Triammineaquadichlorocobalt(III) chloride
b. μ-Oxo-bis(pentaamminechromium(III) ion
c. Potassium diaquabis(oxalato)manganate(III)

10-6 Glycine has the structure NH_2CH_2COOH. It can lose a proton from the carboxyl group and form chelate rings bonded through both the N and one of the O atoms. Draw structures for all possible isomers of tris(glycinato)cobalt(III).

10-7 Two conceivable geometries of 3-coordinate compounds are trigonal planar (the most common geometry observed) and T-shaped. For both shapes, determine the energies of the *d* orbitals predicted by the angular overlap model:
a. For ligands that function as σ donors only.
b. For ligands that function as both σ donors and π acceptors.
c. What other factors are involved in determining the actual geometry?

10-8 Sketch all isomers of the following. Indicate clearly each pair of enantiomers.
a. $[Pt(NH_3)_3Cl_3]^+$
b. $[Co(NH_3)_2(H_2O)_2Cl_2]^+$
c. $[Co(NH_3)_2(H_2O)_2BrCl]^+$
d. $[Cr(H_2O)_3BrClI]$
e. $[Pt(en)_2Cl_2]^{2+}$
f. $[Cr(o\text{-phen})(NH_3)_2Cl_2]^+$
g. $[Pt(bipy)_2BrCl]^{2+}$
h. $[Fe(dtc)_3]$, dtc = $\left[S_2C{=}N(CH_3)(H)\right]^-$
i. $[Re(arphos)_2Br_2]$, arphos = $C_6H_4(As(CH_3)_2)(P(CH_3)_2)$
j. $[Re(dien)Br_2Cl]$

10-9 A molecule of formula $Cr(CO)_2(CN)_2Br_2$ has been synthesized. In the infrared spectrum, it shows two bands attributable to C—O stretching, but only one band attributable to C—N stretching. What is the most likely structure of this molecule? (See Section 4-4-2.)

10-10 Name all the compounds in problem 8 (omitting isomer designations).

10-11 Name all the compounds in problem 14 (omitting isomer designations).

10-12 Give chemical names for the following:
a. $[Cu(NH_3)_4]^{2+}$
b. $[PtCl_4]^{2-}$
c. $Fe(S_2CNMe_2)_3$
d. $[Mn(CN)_6]^{4-}$
e. $[ReH_9]^{2-}$
f. $[Ag(NH_3)_2][BF_4]$
g. $[Fe(CN)_2(CH_3NC)_4]$

h. $Rb[AgF_4]$
i. $[Co(en)_2CO_3]Br$
j. $[Co(N_3)(NH_3)_5]SO_4$

10-13 Give structural formulas, including all possible isomers, for the following:
a. Diamminebromochloroplatinum(II)
b. Diaquadiiododinitritopalladium(IV)
c. Tri-μ-carbonylbis(tricarbonyliron)(0)

10-14 Assign absolute configurations (Λ or Δ) to the following:

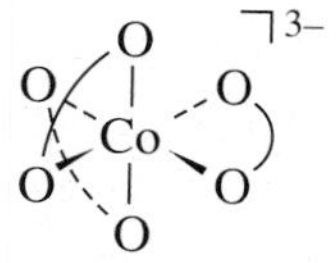

$[Co(O\frown O)_3]^{3-}$

O⌒O = oxalate

$Ru(CO)_2(S\frown S)_2$

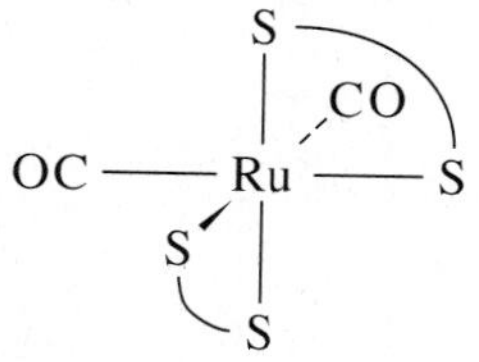

S⌒S = dimethyldithiocarbamate

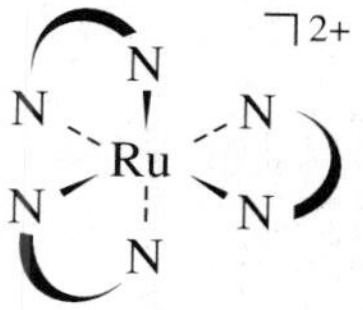

$[Ru(N\frown N)_3]^{2+}$

N⌒N = ethylenediamine

$NiCl_2(N\frown N)_2$

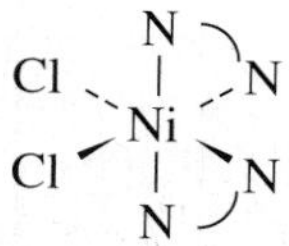

N⌒N = 2, 2'-bipyridine

10-15 Which of the following molecules are chiral?

a. Co(EDTA)

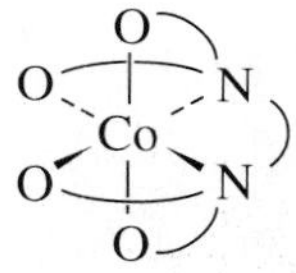

Ligand = EDTA

b. $Co(NH_3)_4(H_2N-CH_2-CH_2-NH_2)$

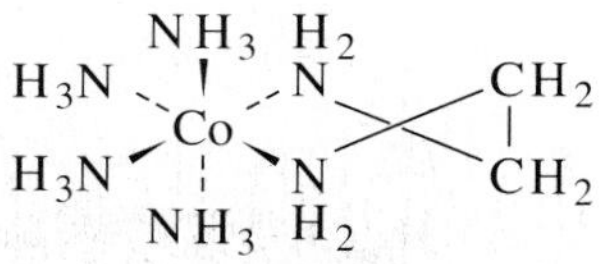

c. $RuCl_2(N-C-C-N)_2$

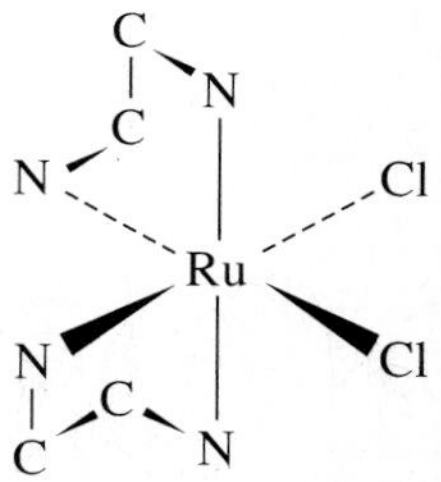

H's omitted for clarity

10-16 Give the symmetry designation (λ or δ) for the chelate rings in problem 15b and c.

10-17 $[Co(en)_2Cl(NO)]^+$ has a magnetic moment of approximately 3.9 Bohr magnetons.
a. How many unpaired electrons does this ion have (spin-only magnetic moment)?
b. What is the oxidation state of Co?
c. Draw all possible isomers of this ion.

10-18 Using the angular overlap method of Chapter 8, show that the ligand field stabilization energy favors square planar ML_4 over octahedral ML_6 for d^7, d^8, and d^9 metals.

11

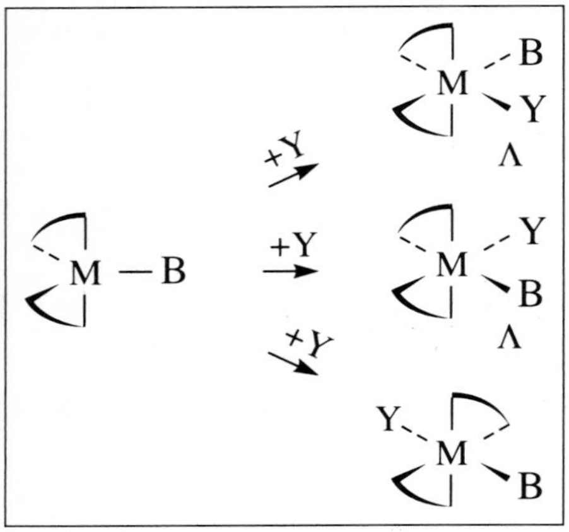

Coordination Chemistry IV: Reactions and Mechanisms

Reactions of coordination compounds share the usual characteristics of other reactions, whether of organic or inorganic molecules. An understanding of these reactions does not require totally new concepts, but does have some additional features because the molecules have more complex geometries and more possibilities for rearrangement, the metal atoms exhibit more variability in their reactions, and different factors influence the course of reactions. The types of reactions can be conveniently divided into substitution reactions at the metal center, reactions of the ligands that do not change the attachments to the metal center, and oxidation–reduction reactions. Reactions that include more elaborate rearrangements of ligand structures are more often observed in organometallic compounds; description of these reactions is given in Chapter 13.

11-1 HISTORY

Synthesis of coordination compounds has been a major part of chemistry since it began to become a science. Although the early chemists did not know the structures of the compounds they worked with, they did learn how to make many of them and described them according to the style of the time. The synthetic work done by Werner, Jørgensen, and others that established the current picture of coordination geometry began the systematic development of reactions for specific purposes. Many years of experimentation and consideration of possible reaction pathways have led to the ideas described in this chapter, and even now these ideas must be considered tentative and provisional in many cases. Through the rest of the chapter, keep in mind that all reactions ultimately must fit into the same framework of explanation, even though the current explanations may seem to be limited to a particular type of reaction or compound. This unification of reaction theory is still a goal of chemists, whether they work with organic, inorganic, coordination, organometallic, polymeric, solid-state, liquid, or gaseous compounds. The discovery of new reactions outruns the explanations, but correlation of these reactions with theoretical explanations gradually extends our knowledge. Although the ability to predict products and choose appropriate reaction conditions to obtain the desired products is still a matter of art as well as science, the list of known reactions is now long enough to provide considerable guidance.

11-2 SUBSTITUTION REACTIONS

11-2-1 INERT AND LABILE COMPOUNDS

Many synthetic reactions require substitution of one ligand for another; this is particularly true when the starting material is in aqueous solution, where the metal ion is likely to be $M(H_2O)_m^{n+}$. Some of the simpler reactions of this kind produce colored products that can be used to identify metal ions:

$$[Cu(H_2O)_6]^{2+} + 4\,NH_3 \rightleftharpoons [Cu(NH_3)_4(H_2O)_2]^{2+} + 4\,H_2O$$

blue — much more intense blue

$$[Fe(H_2O)_6]^{3+} + SCN^- \rightleftharpoons [Fe(H_2O)_5(SCN)]^{2+} + H_2O$$

colorless — red

These reactions, and others like them, are very fast and form species that are capable of undergoing a variety of reactions that are also very fast. Addition of HNO_3 (H^+), NaCl (Cl^-), H_3PO_4 (PO_4^{3-}), KSCN (SCN^-), and NaF (F^-) successively to a solution of $Fe(NO_3)_3 \cdot 9\,H_2O$ shows this very clearly. The initial solution is yellow due to the presence of $Fe(H_2O)_5(OH)^{2+}$ and other "hydrolyzed" species containing both water and hydroxide ion. Although the exact species formed in this series depend on solution concentrations, the products in the reactions given here are representative.

$$Fe(H_2O)_5(OH)^{2+} + H^+ \longrightarrow Fe(H_2O)_6^{3+}$$

yellow — very pale violet

$$Fe(H_2O)_6^{3+} + Cl^- \longrightarrow Fe(H_2O)_5(Cl)^{2+} + H_2O$$

yellow

$$Fe(H_2O)_5(Cl)^{2+} + PO_4^{3-} \longrightarrow Fe(H_2O)_5(PO_4) + Cl^-$$

colorless

$$Fe(H_2O)_5(PO_4) + SCN^- \longrightarrow Fe(H_2O)_5(SCN)^{2+} + PO_4^{3-}$$

red

$$Fe(H_2O)_5(SCN)^{2+} + F^- \longrightarrow Fe(H_2O)_5(F)^{2+} + SCN^-$$

colorless

Compounds like these that react rapidly are called **labile** (lā′-bil). In many cases, exchange of one ligand for another can take place in the time of mixing the solutions. Taube[1] has suggested a reaction half-life (the time of disappearance of half the initial compound) of one minute or less as the criterion for lability. Compounds that react more slowly are called **inert** or **robust** (a term used less often). An inert compound is not inert in the usual sense that no reaction can take place; it is simply slower to react. These kinetic terms must also be distinguished from the thermodynamic terms **stable** and **unstable**. A species like $Fe(H_2O)_5(F)^{2+}$ is very stable (has a large equilibrium constant for formation), but is also labile. On the other hand, hexaamminecobalt(III) is thermodynamically unstable in acid and can decompose to the equilibrium mixture on the right,

$$Co(NH_3)_6^{3+} + 6\ H_3O^+ \rightleftharpoons Co(H_2O)_6^{3+} + 6\ NH_4^+ \quad (\Delta G^\circ < 0)$$

but it reacts very slowly (has a very high activation energy) and is therefore called inert or robust. The possible confusion of terms is unfortunate, but no other terminology has gained general acceptance. One possibility is to call the compounds **substitutionally** or **kinetically labile** or **inert**, but these terms are not in general use at this time.

Werner studied cobalt(III), chromium(III), platinum(II), and platinum(IV) compounds because they are inert and can be more readily characterized than labile compounds. This tendency has continued, and much of the discussion in this chapter is based on inert compounds because they can be more easily crystallized from solution and their structures determined. Labile compounds have also been studied extensively, but their study frequently requires less direct techniques.

Although there are exceptions, general rules can be given for inert and labile electronic structures. Inert octahedral complexes are generally those with high ligand field stabilization energies (described in Chapter 8), specifically those with d^3 or low-spin d^4 through d^6 electronic structures. Complexes with d^8 configurations generally react somewhat faster, but slower than the d^7, d^9, or d^{10} compounds. With strong field ligands, d^8 atoms form square planar complexes, many of which are inert. Compounds with any other electronic structures tend to be labile.

11-2-2 MECHANISMS OF SUBSTITUTION

Langford and Gray[2] have described the range of possibilities for substitution reactions, listed in Table 11-1. At one extreme, the departing ligand leaves and a discernible intermediate with a lower coordination number is formed, a mechanism labeled D for **dissociation**. At the other extreme, the incoming ligand adds to the complex and a discernible intermediate with an increased coordination number is formed, a mechanism labeled A for **association**. Between the two extremes is **interchange**, I, in which the incoming ligand is presumed to assist in the reaction but no detectable intermediates appear. When the degree of assistance is small, and the reaction is primarily dissociation, it is called **dissociative interchange**, I_d. When the incoming ligand begins forming a bond to the central atom before the departing ligand bond is weakened appreciably, it is called **associative interchange**, I_a.

[1] H. Taube, *Chem. Rev.*, **1952**, *50*, 69.

[2] C. H. Langford and H. B. Gray, *Ligand Substitution Processes*, Benjamin, New York, 1966, pp. 7–11.

TABLE 11-1
Substitution mechanisms

Dissociative 5-coordinate transition state for octahedral reactant		*Associative* 7-coordinate transition state for octahedral reactant	
D	I_d	I_a	*A*
Dissociation	Dissociative interchange	Associative interchange	Association
Detectable intermediate	No detectable intermediate		Detectable intermediate
Alternative labels			
S_N1 lim (Limiting first order nucleophilic substitution)			S_N2 lim (Limiting second order nucleophilic substitution)

11-3 KINETIC CONSEQUENCES OF REACTION PATHWAYS

While the kinetic rate law is helpful in determining the mechanism of a reaction, it does not always provide sufficient information. In cases of ambiguity, other evidence must be used to find the mechanism. This chapter will describe a number of examples in which the rate law and other experimental evidence have been used to find the mechanism of a reaction. Our goal is to provide two related kinds of information: (1) the kind of information that is used to determine mechanisms, and (2) a selection of specific reactions for which the mechanisms seem to be fairly completely determined. The first is the more important, because it enables a chemist to examine data for other reactions critically and evaluate the proposed mechanisms. The second is also helpful, as it provides part of the collection of knowledge that is required for designing new syntheses. Each of the substitution mechanisms is described with its required rate law.

11-3-1 DISSOCIATION (*D*)

In a dissociative (*D*) reaction, loss of a ligand to form an intermediate with a lower coordination number is followed by addition of a new ligand to the intermediate.

$$ML_5X \underset{k_{-1}}{\overset{k_1}{\rightleftharpoons}} ML_5 + X$$

$$ML_5 + Y \xrightarrow{k_2} ML_5Y$$

[In the reactions of this chapter, X will indicate the ligand that is leaving the complex, Y the ligand that is entering, and L any ligands that are unchanged during the reaction. In cases of solvent exchange, all (X, Y, and L) may be chemically the same species, but in the more general case they may all be different. Charges will be omitted in the general case, but keep in mind that any of the species may be ions. The general examples will usually be 6-coordinate, but other coordination numbers could be chosen with no change in the arguments.]

The stationary state hypothesis, assuming a very small concentration of the intermediate ML_5, requires that the rates of formation and reaction of the intermediate must be equal. This in turn requires that the rate of change of $[ML_5]$ be zero during much of the reaction. Expressed as a rate equation,

$$\frac{d[ML_5]}{dt} = k_1[ML_5X] - k_{-1}[ML_5][X] - k_2[ML_5][Y] = 0$$

Solving for $[ML_5]$,

$$[ML_5] = \frac{k_1[ML_5X]}{k_{-1}[X] + k_2[Y]}$$

and substituting into the overall rate law,

$$\frac{d[ML_5Y]}{dt} = k_2[ML_5][Y]$$

leads to the rate law:

$$\frac{d[ML_5Y]}{dt} = \frac{k_2k_1[ML_5X][Y]}{k_{-1}[X] + k_2[Y]}$$

This mechanism requires that the intermediate, ML_5, be detectable during the reaction. Detection at the low concentrations expected is a very difficult experimental challenge, and there are very few clear-cut dissociative reactions. More often, the evidence is indirect, but no intermediate has been found. Such reactions are usually classified as following an interchange mechanism.

11-3-2 INTERCHANGE (*I*)

In an interchange (*I*) reaction, a rapid equilibrium between the incoming ligand and the 6-coordinate reactant forms an ion pair or outer-sphere transition species. This species, which is not described as having an increased coordination number and is usually not directly detectable, then reacts to form the product and release the initial ligand.

$$ML_5X + Y \underset{k_{-1}}{\overset{k_1}{\rightleftharpoons}} ML_5X \cdot Y$$

$$ML_5X \cdot Y \xrightarrow{k_2} ML_5Y + X$$

When $k_2 \ll k_{-1}$, the reverse reaction of step 1 is fast enough that the first step is independent of the second step, and the first step is an equilibrium with $K_1 = k_1/k_{-1}$.

Applying the stationary state hypothesis,

$$\frac{d[ML_5X \cdot Y]}{dt} = k_1[ML_5X][Y] - k_{-1}[ML_5X \cdot Y] - k_2[ML_5X \cdot Y] = 0$$

The concentration of the unstable transition species may be large enough to significantly change the concentration of the reactant. For this reason, we must solve for this species in terms of the total initial reactant concentrations of ML_5X and Y, which we will call $[M]_0$ and $[Y]_0$:

$$[M]_0 = [ML_5X] + [ML_5X \cdot Y]$$

assuming the concentration of final product, $[ML_5Y]$, is too small to be significant.

$$[Y]_0 \cong [Y]$$

assuming the fraction of Y in $ML_5X \cdot Y$ and the product ML_5Y is very small. In many experiments, $[Y]_0 > [M]_0$, making this even more likely.

From the stationary state equation above,

$$k_1([M]_0 - [ML_5X \cdot Y])[Y]_0 - k_{-1}[ML_5X \cdot Y] - k_2[ML_5X \cdot Y] = 0$$

The final rate equation then becomes

$$\frac{d[ML_5Y]}{dt} = k_2[ML_5X \cdot Y] = \frac{k_2K_1[M]_0[Y]_0}{1 + K_1[Y]_0 + (k_2/k_{-1})} \cong \frac{k_2K_1[M]_0[Y]_0}{1 + K_1[Y]_0}$$

where k_2/k_{-1} is very small and can be omitted because $k_2 << k_{-1}$ is required for the first step to be an equilibrium.

K_1 can be measured experimentally in some cases and estimated theoretically in others from calculation of the electrostatic energy of the interaction, with fair agreement in cases where both methods have been used.

Two variations on the interchange mechanism are I_d (dissociative interchange) and I_a (associative interchange). The difference between them is in the degree of bond formation in the first step of the mechanism. If bonding between the incoming ligand and the metal is more important, it is an I_a mechanism; if breaking the bond between the leaving ligand and the metal is more important, it is an I_d mechanism. The distinction between them is subtle and requires careful experimental design.

As can be seen from these equations, both *D* and *I* mechanisms have the same mathematical form for their rate laws:

$$\text{Rate} = \frac{k[M]_0[Y]}{1 + k'[Y]_0}$$

Both are second order at low [Y] and first order at high [Y], with the change from one to the other depending on the specific values of the rate constants.

11-3-3 ASSOCIATION (*A*)

In an associative reaction, the first step, forming an intermediate with an increased coordination number, is the rate-determining step. It is followed by a faster reaction in which the leaving ligand is lost:

$$ML_5X + Y \underset{k_{-1}}{\overset{k_1}{\rightleftharpoons}} ML_5XY$$

$$ML_5XY \xrightarrow{k_2} ML_5Y + X$$

The same stationary state approach used in the other rate laws results in the rate law:

$$\frac{d[ML_5Y]}{dt} = \frac{k_1k_2[ML_5X][Y]}{k_{-1} + k_2} = k[ML_5X][Y]$$

a second-order equation regardless of the concentration of Y.

EXERCISE 11-1

Show that the preceding equation is the result of the stationary state approach for an associative reaction.

In common with the dissociative mechanism, there are very few clear examples of associative mechanisms. Most reactions fit better between the two extremes, following associative or dissociative interchange mechanisms. The next section summarizes the evidence for the different mechanisms.

11-4 EXPERIMENTAL EVIDENCE IN OCTAHEDRAL SUBSTITUTION

11-4-1 DISSOCIATION

Most substitution reactions of octahedral complexes are believed to be dissociative, with the complex losing one ligand to become a 5-coordinate square pyramid in the transition state and the incoming ligand filling the vacant site to form the new octahedral product. Theoretical justification for the inert and labile classifications of Section 11-2-1 comes from ligand field theory, with calculation of the change in LFSE between the octahedral reactant and the presumed 5-coordinate transition state, either square pyramidal or trigonal bipyramidal in shape. Table 11-2 gives the **ligand field activation energy** (LFAE), calculated as the difference between the LFSE of the square pyramidal transition state and the LFSE of the octahedral reactant. LFAEs calculated for trigonal bipyramidal transition states are all less favorable. These calcula-

TABLE 11-2
Ligand field activation energies

For a square pyramidal transition state
LFAE = LFSE (sq. pyr.) − LFSE (oct.), all in Δ_o for σ donor only

	Strong fields			Weak fields		
System	*LFSE Oct.*	*LFSE Sq pyr.*	**LFAE**	*LFSE Oct.*	*LFSE Sq pyr.*	**LFAE**
d^0	0	0	**0**	0	0	**0**
d^1	−0.4	−0.457	**−0.057**	−0.4	−0.457	**−0.057**
d^2	−0.8	−0.914	**−0.114**	−0.8	−0.914	**−0.114**
d^3	−1.2	−1.0	**0.2**	−1.2	−1.0	**0.2**
d^4	−1.6	−1.457	**0.143**	−0.6	−0.914	**−0.314**
d^5	−2.0	−1.914	**0.086**	0	0	**0**
d^6	−2.4	−2.0	**0.4**	0.4	−0.457	**−0.057**
d^7	−1.8	−1.914	**−0.114**	−0.8	−0.914	**−0.114**
d^8	−1.2	−1.0	**0.2**	−1.2	−1.0	**0.2**
d^9	−0.6	−0.914	**−0.314**	−0.6	−0.914	**−0.314**
d^{10}	0	0	**0**	0	0	**0**

SOURCE: Adapted with permission from F. Basolo and R. G. Pearson, *Mechanisms of Inorganic Reactions*, 2nd ed., Wiley, New York, 1967, p. 146.

tions provide estimates of the energy necessary to form the transition state; large positive values are characteristic of slow reactions, and small or negative values are characteristic of fast reactions. Examination of these numbers shows that the activation energies of the square pyramidal transition state match the experimental facts (d^3, low-spin d^4 through d^6, and d^8 are inert) better than those for the trigonal bipyramidal transition state. Therefore, the calculation of LFAE supports a square pyramidal geometry (and a dissociative mechanism) for the transition state. However, all these numbers assume an idealized geometry not likely to be found in practice, and the LFAE is only one factor that must be considered in any reaction. Even for thermodynamically favorable reactions, a large activation energy means that reaction will be slow. For thermodynamically unfavorable reactions, even a fast reaction (with small activation energy) would be unlikely to occur.

Other metal ion factors that affect reaction rates of octahedral complexes include (relative rates for ligand exchange are indicated by the inequalities):

1. *Oxidation state of the central ion:* Central ions with higher oxidation states have slower ligand exchange rates.

$$[AlF_6]^{3-} > [SiF_6]^{2-} > [PF_6]^{-} > SF_6$$
$$3+ \qquad 4+ \qquad 5+ \qquad 6+$$

$$[Na(H_2O)_n]^{+} > [Mg(H_2O)_n]^{2+} > [Al(H_2O)_6]^{3+}$$
$$1+ \qquad 2+ \qquad 3+$$

2. *Ionic radius:* Smaller ions have slower exchange rates.

$$[Sr(H_2O)_6]^{2+} > [Ca(H_2O)_6]^{2+} > [Mg(H_2O)_6]^{2+}$$
$$112 \text{ pm} \qquad 99 \text{ pm} \qquad 66 \text{ pm}$$

Both of these effects can be attributed to a higher electrostatic attraction between the central atom and the attached ligands. A strong attraction between the two will slow the reaction, since reaction is presumed to require dissociation of a ligand from the complex. Figure 11-1 shows the rate of exchange of water molecules on aquated metal ions for many different ions, for which these effects can be seen. All the ions in the figure are labile, with half-lives for the aqua complexes shorter than one second; measurement of such fast reactions

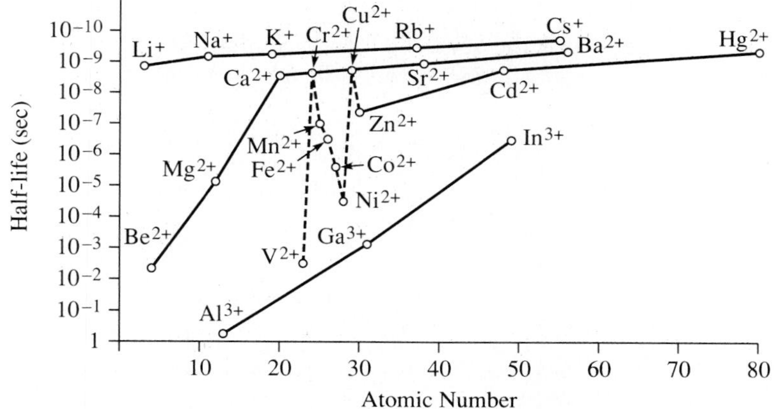

FIGURE 11-1 Rate of Water Exchange on Aquated Metal Ions. (Data from F. Basolo and R. C. Johnson, *Coordination Chemistry*, Science Reviews, 1986, p. 105.)

is done by indirect methods, particularly relaxation methods[3] (including temperature jump, pressure jump, and nuclear magnetic resonance).

The evidence for dissociative mechanisms can be grouped as follows:[4,5,6,7]

1. The rate of reaction changes only slightly with changes in the incoming ligand. In many cases, **aquation** (substitution by water) and **anation** (substitution by an anion) rates are comparable. If dissociation is the rate-determining reaction, the entering group should have no effect at all on the reaction rate. Incoming ligand effects indicate a probable A or I_a mechanism.
2. Increasing negative charge on the reactant decreases the rate of substitution. Larger electrostatic attraction between the positive metal ion and the negative ligand should slow the dissociation.
3. Steric crowding on the reactant increases the rate of ligand dissociation. When ligands on the reactant are crowded, loss of one of the ligands is made easier. (Organometallic examples of phosphine ligand dissociation are given in Chapter 13.) On the other hand, if the reaction has an A or I_a mechanism, steric crowding interferes with the incoming ligand and slows the reaction.
4. The rate of reaction correlates with the metal–ligand bond strength of the leaving group, in a linear free energy relationship (LFER, explained in the next section).
5. Activation energies and entropies are consistent with dissociation, although interpretation of these parameters is difficult. Another activation parameter now being measured by experiments at increased pressure is the volume of activation, which is the change in volume on forming the activated complex. Dissociative mechanisms generally result in positive values for ΔV_{act} because one species splits into two, and associative mechanisms result in negative ΔV_{act} values because two species combine into one, with a presumed total volume smaller than the reactants.

11-4-2 LINEAR FREE ENERGY RELATIONSHIPS (LFER)

Many kinetic effects can be related to thermodynamic effects by a **linear free energy relationship**.[8] Such effects are seen when, for example, the bond strength of a metal–ligand bond plays a major role in determining the dissociation rate of that ligand. When this is true, a plot of the logarithm of the rate constants for different leaving ligands versus the logarithm of the equilibrium constants for the same ligands in similar compounds is linear. The justification for this correlation is found in the Arrhenius equation for temperature dependence of

[3] F. Wilkinson, *Chemical Kinetics and Reactions Mechanisms,* Van Nostrand Reinhold, New York, 1980, pp. 83–91.

[4] F. Basolo and R. G. Pearson, *Mechanisms of Inorganic Reactions,* 2nd ed., Wiley, New York, 1967, pp. 158–70.

[5] R. G. Wilkins, *The Study of Kinetics and Mechanism of Reactions of Transition Metal Complexes,* Allyn and Bacon, Boston, 1974, pp. 185–96.

[6] J. D. Atwood, *Inorganic and Organometallic Reaction Mechanisms,* Brooks/Cole, Monterey, Calif., 1985, pp. 82–83.

[7] C. H. Langford and T. R. Stengle, *Ann. Rev. Phys. Chem.,* **1968**, *19*, 193.

[8] J. W. Moore and R. G. Pearson, *Kinetics and Mechanism,* 3rd ed., Wiley, 1981, pp. 357–63.

rate constants and the equation for temperature dependence of equilibrium constants. In logarithmic form, they are

$$\ln k = \ln A - \frac{E_a}{RT} \quad \text{and} \quad \ln K = \frac{-\Delta H^\circ}{RT} + \frac{\Delta S^\circ}{R}$$

If the preexponential factor A and the entropy ΔS° are nearly constant and the activation energy E_a depends on the enthalpy of reaction ΔH°, there will be a linear correlation between ln k and ln K. A straight line on such a log–log plot is indirect evidence for a strong influence of the thermodynamic parameter ΔH° on the activation energy of the reaction. Figure 11-2 shows an example from hydrolysis of $[Co(NH_3)_5X]^{2+}$. From this evidence, Langford argued that "The role of the X^- group in the transition state of acid hydrolysis is strongly similar to its role in the product: namely, that of a solvated anion,"[9] and additionally that water is at most weakly bound in the transition state. Another example from reactions of square planar platinum complexes is given in Section 11-6-2.

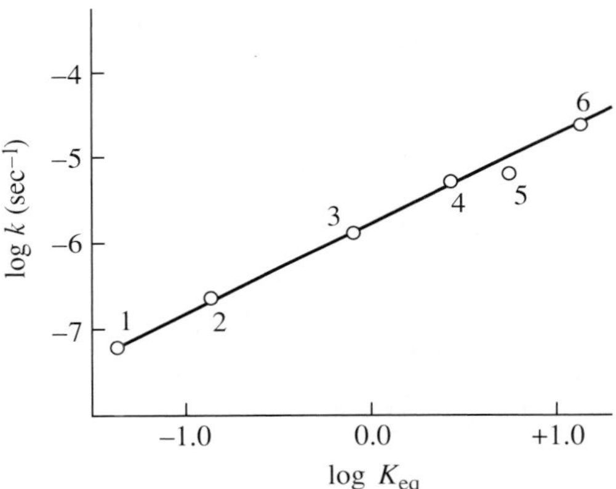

FIGURE 11-2 Linear Free Energy and $[Co(NH_3)_5X]^{2+}$ Hydrolysis. Plot of log rate constant versus log of equilibrium constant for the acid hydrolysis reaction of $[Co(NH_3)_5X]^{2+}$ ions. Measurements made at 25.0°C. Points are designated: 1, $X^- = F^-$; 2, $X^- = H_2PO_4^-$; 3, $X^- = Cl^-$; 4, $X^- = Br^-$; 5, $X^- = I^-$; and 6, $X^- = NO_3^-$. (Reproduced with permission from C. H. Langford, *Inorg. Chem.*, **1965**, *4*, 265. Copyright 1965 American Chemical Society. Data for F^- from S. C. Chan, *J. Chem. Soc.*, **1964**, 2375, and for I^- from R. G. Yalman, *Inorg. Chem.*, **1962**, *1*, 16. All other data from A. Haim and H. Taube, *Inorg. Chem.*, **1964**, *2*, 1199.)

Examples of the effect (or lack of effect) of incoming ligand are given in Tables 11-3 and 11-4. In Table 11-3, the data are for the first order region (large [L]). The rate constants are all relatively close to that for water exchange, as would be expected for a dissociative mechanism. Table 11-4 gives data for the second order region for anation of $Ni(H_2O)_6^{2+}$. The second order rate constant, k_0K_0, is the product of the ion pair equilibrium constant, K_0, and the rate constant k_0:

$$\text{Ni—OH}_2 + \text{L} \rightleftharpoons \text{Ni—OH}_2 \cdot \text{L} \qquad K_0$$
$$\text{Ni—OH}_2 \cdot \text{L} \longrightarrow \text{Ni—L} + \text{H}_2\text{O} \qquad k_0$$

K_0 is calculated from an electrostatic model that provides good agreement with the few cases where experimental evidence is also available. k_0 varies by a factor of 5 or less and is close to the rate constant for exchange of water. The close agreement for the wide variety of different ligands shows that the effect of the incoming ligand on the second step is minor, although the

[9] C. H. Langford, *Inorg. Chem.*, **1965**, *4*, 265.

TABLE 11-3
Limiting rate constants for anation or water exchange of $Co(NH_3)_5H_2O^{3+}$ at 45°

$$Co(NH_3)_5H_2O^{3+} + L^{m-} \longrightarrow Co(NH_3)_5L^{(3-m)+} + H_2O$$

L^{m-}	$10^6\ k_1(sec^{-1})$	$k_1/k_1(H_2O)$	Ref.
H_2O	100	1.0	a
N_3^-	100	1.0	b
SO_4^{2-}	24	0.24	c
Cl^-	21	0.21	d
NCS^-	16	0.16	d

SOURCES:
a. W. Schmidt and H. Taube, *Inorg. Chem.*, **1963**, *2*, 698.
b. H. R. Hunt and H. Taube, *J. Am. Chem. Soc.*, **1958**, *75*, 1463.
c. T. W. Swaddle and G. Guastalla, *Inorg. Chem.*, **1969**, *8*, 1604.
d. C. H. Langford and W. R. Muir, *J. Am. Chem. Soc.*, **1967**, *89*, 3141.

TABLE 11-4
Rate constants for substitution on $Ni(H_2O)_6^{2+}$

L	$10^3K_0k_0(M^{-1}\ sec^{-1})$	$K_0(M^{-1})$	$10^4k_0(sec^{-1})$
$CH_3PO_4^{2-}$	290	40	0.7
CH_3COO^-	100	3	3
NCS^-	6	1	0.6
F^-	8	1	0.8
HF	3	0.15	2
H_2O			3
NH_3	5	0.15	3
C_5H_5N, pyridine	~4	0.15	~3
$C_4H_4N_2$, pyrazine	2.8	0.15	2
$NH_2(CH_2)_2NMe_3^+$	0.4	0.02	2

SOURCES: Adapted with permission from R. G. Wilkins, *Acc. Chem. Res.*, **1970**, *3*, 408, except $C_4H_4N_2$ from J. M. Malin and R. E. Shepherd, *J. Inorg. Nucl. Chem.*, **1972**, *34*, 3203.

difference in ion pair formation is significant. Both of these reactions are consistent with D or I_d mechanisms, with ion pair formation likely as the first step in the nickel reactions.

11-4-3 ASSOCIATIVE MECHANISMS

Associative reactions are also possible in octahedral substitution, but are much less common.[10] Table 11-5 gives data for both dissociative and associative interchange for similar reactants. In the case of water substitution by several different anions in $Cr(NH_3)_5(H_2O)^{3+}$, the rate constants are quite similar (within a factor of 6), indicative of an I_d mechanism. On the other hand, the same ligands reacting with $Cr(H_2O)_6^{3+}$ show a large variation in rates (more than 2000-fold difference), indicative of an I_a mechanism. Data for similar Co(III) complexes are not conclusive, but their reactions generally have I_d mechanisms.

Reactions of Ru(III) compounds frequently have associative mechanisms, while those of Ru(II) compounds have dissociative mechanisms. Substitution reactions of $Ru(III)(EDTA)(H_2O)^-$ show a very large range of rate constants

[10] Atwood, op. cit., pp. 84–85.

TABLE 11-5
Effects of entering group and *cis*-ligands on rates

	Rate constants for anation	
Entering ligand	$Cr(H_2O)_6^{3+}$ $10^8k(M^{-1}\ s^{-1})$	$Cr(NH_3)_5H_2O^{3+}$ $10^4k(M^{-1}\ s^{-1})$
NCS^-	180	4.2
NO_3^-	73	—
Cl^-	2.9	0.7
Br^-	1.0	3.7
I^-	0.08	—
CF_3COO^-	—	1.4

SOURCES: Reproduced with permission from J. D. Atwood, *Inorganic and Organometallic Reaction Mechanisms*, Brooks/Cole, Monterey, Calif., 1985, p. 85. Data from D. Thusius, *Inorg. Chem.*, **1971**, *10*, 1106; T. Ramasami and A. G. Sykes, *J. Chem. Soc. Chem. Comm.*, **1976**, 378.

depending on the incoming ligand (Table 11-6), as required for an I_a mechanism, while those of Ru(II) are nearly the same for different ligands, as required for an I_d mechanism. The reasons for this difference are not certain. Both complexes have a free carboxylate (the EDTA is pentadentate, with the sixth position occupied by a water molecule). Hydrogen bonding between this free carboxylate and the bound water may distort the shape sufficiently in the Ru(III) complex to open a place for entry by the incoming ligand. While similar hydrogen bonding may be possible for the Ru(II) complex, the increased negative charge may reduce the Ru—H_2O bond strength enough to promote dissociation.

TABLE 11-6
Ruthenium substitution reactions

a. Rate constants for $Ru(III)(EDTA)(H_2O)^-$ substitution

Ligand	$k_1(M^{-1}\ s^{-1})$	$\Delta H^{\ddagger}(kJ\ mol^{-1})$	$\Delta S^{\ddagger}(J\ mol^{-1}\ K^{-1})$
Pyrazine	20,000 ± 1,000	5.7 ± 0.5	−20 ± 3
Isonicotinamide	8,300 ± 600	6.6 ± 0.5	−19 ± 3
Pyridine	6,300 ± 500		
Imidazole	1,860 ± 100		
SCN^-	270 ± 20	8.9 ± 0.5	−18 ± 3
CH_3CN	30 ± 7	8.3 ± 0.5	−24 ± 4

b. Rate constants for $Ru(II)(EDTA)(H_2O)^{2-}$ substitution

Ligand	$k_1(M^{-1}\ s^{-1})$
Isonicotinamide	30 ± 15
CH_3CN	13 ± 1
SCN^-	2.7 ± 0.2

SOURCE: Data from T. Matsubara and C. Creutz, *Inorg. Chem.*, **1979**, *18*, 1956.

11-4-4 THE CONJUGATE BASE MECHANISM

Other cases in which second-order kinetics seemed to require an associative mechanism have subsequently been found to have a **conjugate base mechanism**[11] (called S_N1CB, for substitution, nucleophilic, unimolecular, conjugate base in

[11] Wilkins, op. cit., pp. 207–10; Basolo and Pearson, op. cit., pp. 177–93.

Ingold's notation[12]). These reactions depend on amine, ammine, or aqua ligands that can lose protons to form amido or hydroxo species that are then more likely to dissociate:

$$(1) \quad Co(NH_3)_5X^{2+} + OH^- \rightleftharpoons Co(NH_3)_4(NH_2)X^+ + H_2O \quad \text{(equilibrium)}$$

$$(2) \quad Co(NH_3)_4(NH_2)X^+ \longrightarrow Co(NH_3)_4(NH_2)^{2+} + X^- \quad \text{(slow)}$$

$$(3) \quad Co(NH_3)_4(NH_2)^{2+} + H_2O \longrightarrow Co(NH_3)_5(OH)^{2+} \quad \text{(fast)}$$

Overall:

$$Co(NH_3)_5X^{2+} + OH^- \longrightarrow Co(NH_3)_5(OH)^{2+} + X^-$$

In the third step, addition of a ligand other than hydroxide is also possible; the rate constant is k_{OH} and the equilibrium constant is K_{OH} in basic solution.

Additional evidence for the mechanism above is provided by several related studies:

1. Base-catalyzed exchange of hydrogen from the amine groups takes place under the same conditions as these reactions.
2. The isotope ratio ($^{18}O/^{16}O$) in the product in ^{18}O-enriched water is the same as that in the water regardless of the leaving group ($X^- = Cl^-$, Br^-, NO_3^-). If an incoming water molecule had a large influence (an associative mechanism), a higher concentration of ^{18}O should be in the product, because the equilibrium constant $K = 1.040$ for the reaction

$$H_2{}^{16}O + {}^{18}OH^- \rightleftharpoons H_2{}^{18}O + {}^{16}OH^-$$

3. RNH_2 compounds (R = alkyl) react faster than NH_3 compounds, showing that steric crowding favors the 5-coordinate intermediate formed in step 2.
4. The rate constants and dissociation constants for these compounds form a linear free energy relationship (LFER), where a plot of $\ln k_{OH}$ versus $-\ln K_{OH}$ is linear.
5. When substituted amines are used, and there are no protons on the nitrogens available for ionization, the reaction is very slow or nonexistent.

Reactions with $[Co(tren)(NH_3)Cl]^{2+}$ isomers show that the position *trans* to the leaving group is the most likely deprotonation site for a conjugate base mechanism.[13] The reaction in Figure 11-3a is 10^4 times faster than that in Figure 11-3b. In addition, most of the product in both reactions is best explained by a trigonal bipyramidal intermediate or transition state with the deprotonated amine in the trigonal plane. The reaction in Figure 11-3a can form this state immediately; the reaction in Figure 11-3b requires rearrangement of an initial square pyramidal structure.

Explanations of the promotional effect of the amido group center on its basic strength, either as a σ donor or due to ligand-to-metal π interaction. The

[12] C. K. Ingold, *Structure and Mechanism in Organic Chemistry*, 2nd ed., Cornell University Press, Ithaca, N.Y., 1969, Chapters 5 and 7.

[13] D. A. Buckingham, P. J. Cressell, and A. M. Sargeson, *Inorg. Chem.*, **1975**, *14*, 1485.

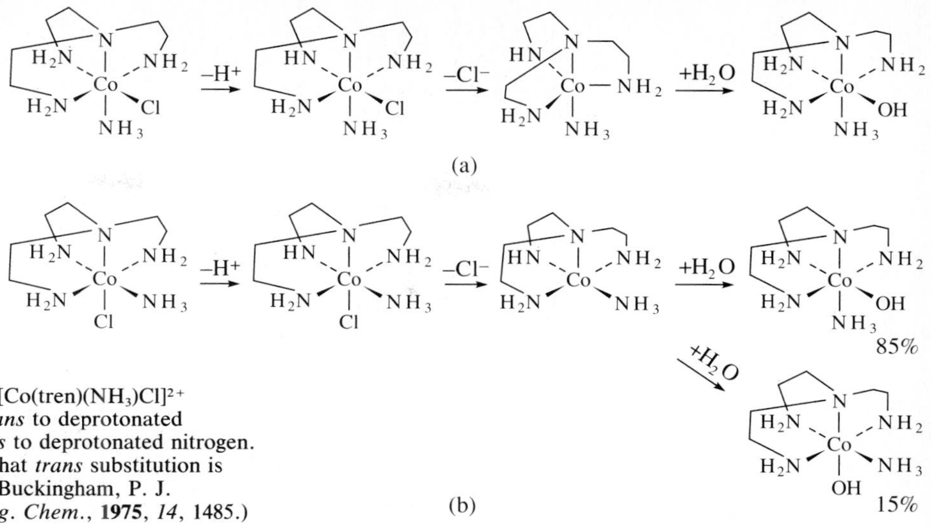

FIGURE 11-3 Base Hydrolysis of $[Co(tren)(NH_3)Cl]^{2+}$ Isomers. (a) Leaving group (Cl^-) *trans* to deprotonated nitrogen. (b) Leaving group (Cl^-) *cis* to deprotonated nitrogen. (a) is 10^4 faster than (b), indicating that *trans* substitution is strongly favored. (Data from D. A. Buckingham, P. J. Creswell, and A. M. Sargeson, *Inorg. Chem.*, **1975**, *14*, 1485.)

π interaction is most effective when the amido group is part of the trigonal plane in a trigonal bipyramidal geometry, but there is at least one case where this geometry is not necessarily achieved.[14]

11-4-5 THE CHELATE EFFECT

Substitution for a chelated ligand is generally a slower reaction than that for a similar monodentate ligand. Explanations for this effect center on two factors, the increased energy needed to remove the first bound atom and the probability of a reversal of this first step.[15]

The reaction must have two dissociation steps for a bidentate ligand, one for each bound atom (the addition of water in steps 2 and 4 is likely to be fast, because of its high concentration):

(1) $M(NH_2CH_2CH_2NH_2)$ (chelate) $\rightleftharpoons$ $M-NH_2-CH_2-CH_2-NH_2$ (one end free) (slow)

(2) $M-NH_2-CH_2-CH_2-NH_2 + H_2O \rightleftharpoons M(OH_2)(NH_2-CH_2-CH_2-NH_2)$ (fast)

(3) $M(OH_2)(NH_2-CH_2-CH_2-NH_2) \rightleftharpoons M-OH_2 + NH_2CH_2CH_2NH_2$ (slow)

(4) $M-OH_2 + H_2O \rightleftharpoons M(OH_2)_2$ (fast)

[14] D. A. Buckingham, P. A. Marzilli, and A. M. Sargeson, *Inorg. Chem.*, **1969**, *8*, 1595.

[15] D. W. Margerum, D. C. Weatherburn, and G. K. Pagenkopf, "Kinetics and Mechanisms of Complex Formation and Ligand Exchange," in A. E. Martell, ed., *Coordination Chemistry*, Vol. 2, American Chemical Society Monograph 174, Washington, D.C., 1978, pp. 1–220.

The first dissociation (1) is expected to be slower than a similar dissociation of ammonia because the ligand must bend and rotate to move the free amine group away from the metal. The second dissociation (3) is likely to be slow because the concentration of the intermediate is low and because the first dissociation can readily reverse. The uncoordinated nitrogen is held near the metal by the rest of the ligand, making reattachment more likely. Overall, this chelate effect reduces the rates of aquation reactions by factors from 20 to as much as 10^5.

There is also a thermodynamic chelate effect, in which the statistical difference between monodentate ligands and those with more points of attachment results in larger stability constants for the polydentate complexes.[16] As in the kinetic effect, the difference is in the attachment and dissociation of the second (and third or higher numbered) point of attachment for the ligand.

11-5 STEREOCHEMISTRY OF REACTIONS

While a common assumption is that reactions with dissociative mechanisms are more likely to result in random isomerization or racemization, and associative mechanisms are more likely to result in single-product reactions, the evidence is much less clear-cut. Dissociative mechanisms can lead to single-product reactions with either retention of configuration or a change of configuration, depending on the circumstances. For example,[17] base hydrolysis of Λ-*cis*-$[Co(en)_2Cl_2]^+$ in dilute (< 0.01 M) hydroxide yields Λ-*cis*-$[Co(en)_2(OH)_2]^+$, while in more concentrated (> 0.25 M) hydroxide it gives Δ-*cis*-$[Co(en)_2(OH)_2]^+$ (Figure 11-4). A conjugate base mechanism is expected in both cases, the hydroxide removing a proton from an ethylenediamine nitrogen, followed by loss of the chloride *trans* to the deprotonated nitrogen. In the more concentrated base, the higher concentration of ion pairs ($[Co(en)_2Cl_2]^+$–OH^-) is assumed to result in a water molecule (from the OH^- and the H^+ removed from ethylenediamine) positioned for easy addition with inversion of the chiral center.

A similar change in product, this time dependent on temperature, takes place in the substitution of ammonia for chloride in $[Co(en)_2Cl_2]^+$.[18] At low temperatures (-50°C, in liquid ammonia), there is inversion of configuration; at high temperatures (80°C, gaseous ammonia), there is retention. In both cases, there is also a large fraction of *trans* and racemic product. A similar result (inversion at low temperature, retention at high temperature), complicated by a change from α to β geometry, is found for the reactions of Λ α-$[Co(trien)Cl_2]^+$ with ammonia or ethylenediamine. The β isomer reacts with ethylenediamine with retention.[19]

Although not a complete explanation of these reactions, all the reported inversion reactions take place under conditions where a conjugate base mechanism is possible.[20] The orientation of the ligand entering the proposed trigonal bipyramidal intermediate then dictates the configuration of the product. In some cases, a preferred orientation of the other ligands may dictate the

[16] Basolo and Pearson, op. cit., pp. 27–31, 223–28; G. Schwarzenbach, *Helv. Chim. Acta,* **1952**, *35*, 2344.

[17] S. C. Chan and M. L. Tobe, *J. Chem. Soc.,* **1962**, 4531; L. J. Boucher, E. Kyuno, and J. C. Bailar, Jr., *J. Am. Chem. Soc.,* **1964**, *86*, 3656.

[18] J. C. Bailar, Jr., J. H. Haslam, and E. M. Jones, *J. Am. Chem. Soc.,* **1936**, *58*, 2226; R. D. Archer and J. C. Bailar, Jr., *J. Am. Chem. Soc.,* **1961**, *83*, 812.

[19] E. Kyuno and J. C. Bailar, Jr., *J. Am. Chem. Soc.,* **1966**, *88*, 1125.

[20] Basolo and Pearson, op. cit., p. 272.

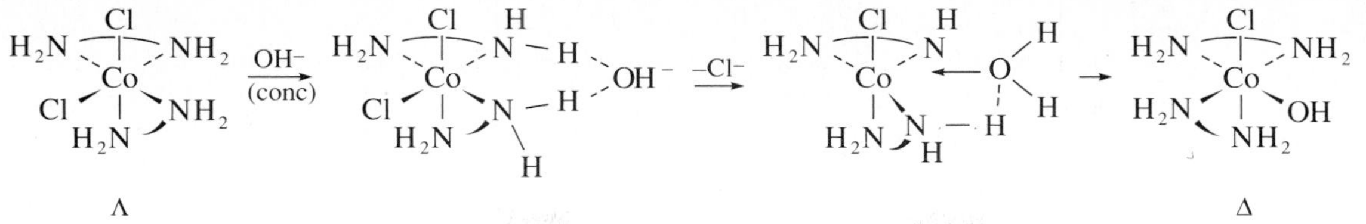

$$\underset{\Lambda}{\text{[Co(en)}_2\text{Cl}_2]^+} \xrightarrow[\text{(conc)}]{OH^-} \cdots \xrightarrow{-Cl^-} \cdots \rightarrow \underset{\Delta}{\text{[Co(en)}_2\text{(OH)Cl]}}$$

FIGURE 11-4 Mechanism of Inversion of Λ-*cis*-$[Co(en)_2Cl_2]^+$ in Concentrated Hydroxide.

product. For example, the β form of trien complexes is preferred over the α form; both are shown in Figure 10-20.

11-5-1 SUBSTITUTION IN *CIS* COMPLEXES

In many cases, substitution of Y for X in *cis*-$[M(LL)_2BX]$ (LL = a bidentate ligand, such as en) leads to *cis* products. If dissociation of X from the reactant leaves a square pyramidal intermediate that then adds the new ligand directly into the vacant site, the result is retention of the same configuration. If dissociation forms a trigonal bipyramid with B in the trigonal plane, there are three possible locations for addition of Y, all in the same trigonal plane. Two of these result in *cis* products, one in a *trans* product. Dissociation to form a trigonal bipyramid with B in an axial position allows two positions for attack by Y resulting in *cis* products (the third side of the triangle is blocked by an ethylenediamine ring). Statistically, this means that only one-sixth of the product from a trigonal bipyramidal intermediate should be *trans*. All these possibilities are shown in Figure 11-5. Examination of the *cis* products shows that one has an inverted chiral center and four have retained the original configuration.

Experimentally, aquation of *cis*-$[Co(en)_2BX]^{n+}$ in acid results in 100% *cis* isomer. Substitution reactions of optically active reactants in base show approximately a 2:1 preference for retention of configuration (Tables 11-7 and 11-8). The same reactions, with *trans* reactant, result in a mixture of isomers, the fractions of *cis* and *trans* depending on the retained ligand, B. The only definitive statement that can be made is that *cis* reactants retain their *cis* configuration while *trans* reactants give a mixture of *cis* and *trans* products. The *cis* complexes also usually react more rapidly than the *trans* isomers with the same ligands.

11-5-2 SUBSTITUTION IN *TRANS* COMPLEXES

Starting with a *trans* reactant with LL = en, a square pyramidal intermediate leads to a *trans* product; a trigonal bipyramidal intermediate with B in the trigonal plane leads to a mixture. The incoming ligand can enter along any of the three sides of the triangle, resulting in two *cis* possibilities and one *trans* possibility as shown in Figure 11-6. An intermediate with an axial B is less likely, because it requires more rearrangement of the ligands (a 90° change by one nitrogen and 30° changes by two others, in contrast to two 30° changes for the equatorial B). As a result, the statistical probability of a change from *trans* to *cis* is two-thirds for a trigonal bipyramidal intermediate. The chiral

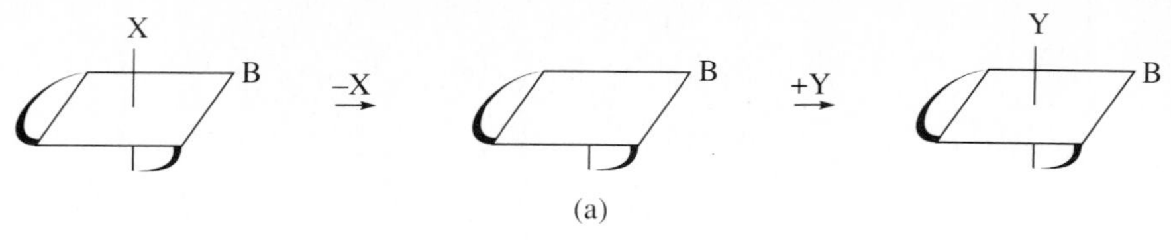

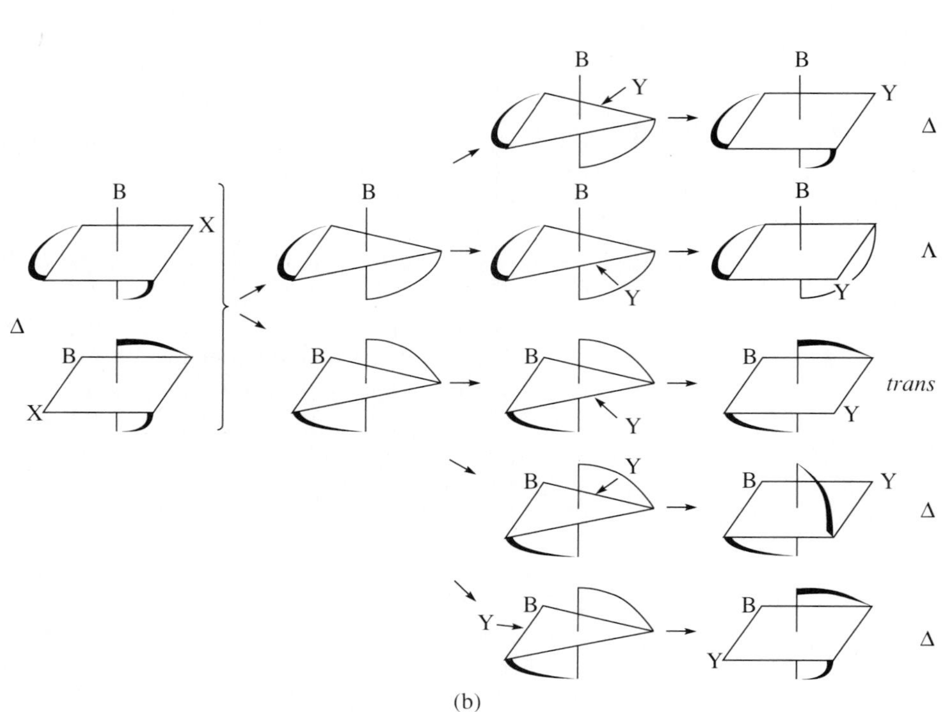

FIGURE 11-5 Dissociation Mechanism and Stereochemical Changes for *cis*-$[M(LL)_2BX]$. (a) Tetragonal pyramidal intermediate (retention of configuration). (b) Trigonal bipyramidal intermediate (three possible products).

TABLE 11-7
Stereochemistry of acid aquation
$Co(en)_2LX^{n+} + H_2O \longrightarrow Co(en)_2LH_2O^{1+n+} + X^-$

cis-L	*X*	*% cis in product*	*Ref.*	*trans-L*	*X*	*% cis in product*	*Ref.*
OH^-	Cl^-	100	a	OH^-	Cl^-	75	a
OH^-	Br^-	100	b	OH^-	Br^-	73	b
Br^-	Cl^-	100	b	Br^-	Cl^-	50	b
Cl^-	Cl^-	100	a	Br^-	Br^-	30	b
Cl^-	Br^-	100	b	Cl^-	Cl^-	35	a
N_3^-	Cl^-	100	c	Cl^-	Br^-	20	b
NCS^-	Cl^-	100	d	NCS^-	Cl^-	50–70	d
NCS^-	Br^-	100	d	NH_3	Cl^-	0	e
NO_2^-	Cl^-	100	f	NO_2^-	Cl^-	0	f

SOURCES: Table reproduced with permission from F. Basolo and R. G. Pearson, *Mechanisms of Inorganic Reactions*, 2nd ed., Wiley, New York, 1967, p. 257.
a. M. E. Baldwin, S. C. Chan, and M. L. Tobe, *J. Chem. Soc.*, **1961**, 4637.
b. S. C. Chan and M. L. Tobe, *J. Chem. Soc.*, **1963**, 5700.
c. P. J. Staples and M. L. Tobe, *J. Chem. Soc.*, **1960**, 4803, 4812; P. J. Stables, *J. Chem. Soc.*, **1963**, 3227; **1965**, 3300.
d. M. E. Baldwin and M. L. Tobe, *J. Chem. Soc.*, **1960**, 4275.
e. M. L. Tobe, *J. Chem. Soc.*, **1959**, 3776.
f. S. Asperger and C. K. Ingold, *J. Chem. Soc.*, **1956**, 2862.

TABLE 11-8
Stereochemistry of base substitution
$Co(en)_2LX^{n+} + OH^- \longrightarrow Co(en)_2LOH^{n+} + X^-$

		% cis product							
Δ–cis-L	*X*	*Δ*	*racemic**	*Λ*	*Ref.*	*trans-L*	*X*	*% cis product*	*Ref.*
OH^-	Cl^-	61		36	a	OH^-	Cl^-	94	a
OH^-	Br^-		96		a	OH^-	Br^-	90	a
Cl^-	Cl^-	21		16	a	Cl^-	Cl^-	5	a
Cl^-	Br^-		30		a	Cl^-	Br^-	5	a
Br^-	Cl^-		40		a	Br^-	Cl^-	0	a
N_3^-	Cl^-		51		b	N_3^-	Cl^-	13	b
NCS^-	Cl^-	56		24	b	NCS^-	Cl^-	76	d
NH_3	Br^-	59		26	c	NCS^-	Br^-	81	d
NH_3	Cl^-	60		24	c	NH_3	Cl^-	76	c
NO_2^-	Cl^-	46		20	e	NO_2^-	Cl^-	6	e

The total % *cis* product is the sum of Δ and Λ obtained from the Δ-*cis* starting material. The optically inactive *trans* isomer will of course yield *racemic cis*. % *trans* = 100% − % *cis*.

SOURCES: Table reproduced with permission from F. Basolo and R. G. Pearson, *Mechanisms of Inorganic Reactions*, 2nd ed., Wiley, New York, 1967, p. 262.
a. S. C. Chan and M. L. Tobe, *J. Chem. Soc.*, **1962**, 4531.
b. P. J. Staples and M. L. Tobe, *J. Chem. Soc.*, **1960**, 4803, 4812; P. J. Stables, *J. Chem. Soc.*, **1963**, 3227; **1965**, 3300.
c. R. S. Nyholm and M. L. Tobe, *J. Chem. Soc.*, **1956**, 1707.
d. C. K. Ingold, R. S. Nyholm, and M. L. Tobe, *J. Chem. Soc.*, **1956**, 1961.
e. S. Asperger and C. K. Ingold, *J. Chem. Soc.*, **1956**, 2862.
NOTE: * Racemic reactant, so the product is also racemic.

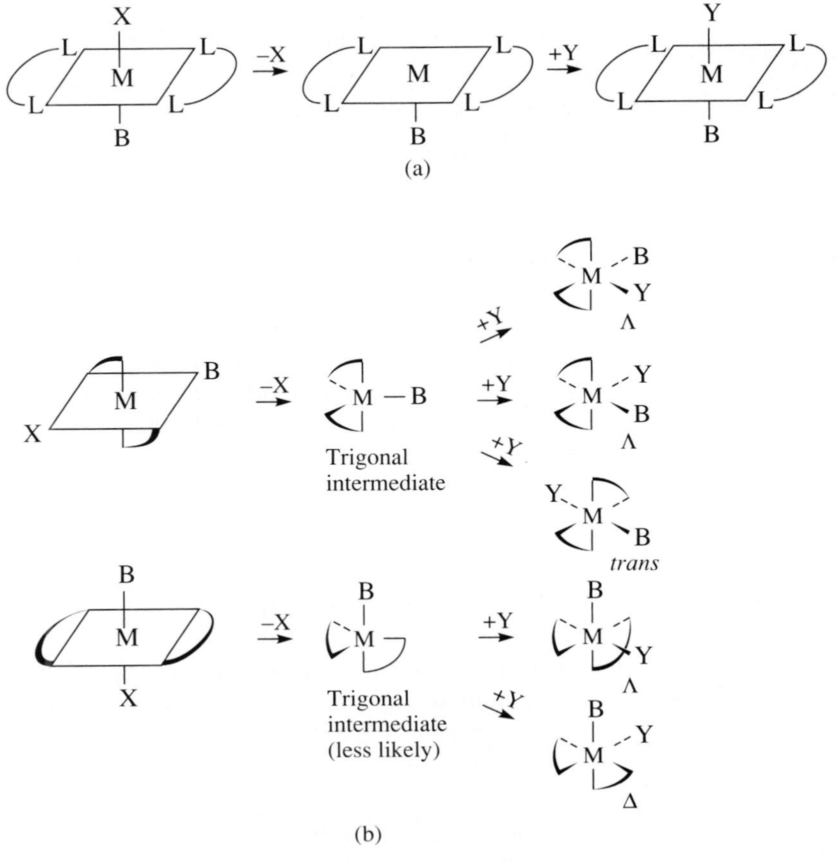

X L L M L L B −X L L M L L B +Y Y L L M L L B

(a)

B M X −X M — B Trigonal intermediate +Y +Y +Y B M Y Λ Y M B Λ Y M B *trans*

B M X −X B M Trigonal intermediate (less likely) +Y +Y B M Y Λ B M Y Δ

(b)

FIGURE 11-6 Dissociation Mechanism and Stereochemical Changes for *trans*-$[M(LL)_2BX]$ Substitution Reactions. (a) Tetragonal pyramidal intermediate (retention of configuration). (b) Trigonal bipyramidal intermediate (three possible products).

form of the product is determined by which square plane becomes trigonal in the intermediate; Δ and Λ products are equally likely.

In fact, experimental results indicate that the statistical distribution is seldom followed. Other factors are important enough to dictate, or at least influence, the outcome. Wilkins has collected data for *cis* ⟶ *trans* and *trans* ⟶ *cis* conversions, racemization, and water exchange for $[Co(en)_2(H_2O)X]^{n+}$, shown in Table 11-9. For X = Cl^-, SCN^-, and H_2O, racemization and *cis* ⟶ *trans* conversion are nearly equal in rate, making it likely that they have identical intermediates. For X = NH_3, racemization and *trans* ⟶ *cis* are much faster than *cis* ⟶ *trans*. In addition, the rate of water exchange is faster than isomerization for NH_3 and NO_2^-, suggesting different mechanisms for the two reactions. The differences in these rate constants show the difficulty in drawing simple conclusions for mechanisms or products. Predictions are always risky, but are best made from closely analogous compounds if necessary.

TABLE 11-9
Rate constants for reactions of $[Co(en)_2(H_2O)X]^{n+}$ at 25°C, 10^5k(sec^{-1})

X	*cis* ⟶ *trans*	*trans* ⟶ *cis*	*Racemization*	*H_2O exchange*
OH^-	200	300	—	160
Br^-	5.4	16.1	—	—
Cl^-	2.4	7.2	2.4	—
N_3^-	2.5	7.4	—	—
NCS^-	0.014	0.071	0.022	0.13
H_2O	0.012	0.68	~0.015	1.0
NH_3	<0.0001	0.002	0.003	0.10
NO_2^-	0.012	0.005	—	—

SOURCES: Adapted with permission from R. G. Wilkins, *The Study of Kinetics and Mechanism of Reactions of Transition Metal Complexes*, Allyn and Bacon, Boston, 1974, p. 344. Data from M. L. Tobe, in *Studies in Structures and Reactivity*, ed. J. H. Ridd, Methuen, London, 1966, and M. N. Hughes, *J. Chem. Soc., A*, **1969**, 1506.

11-5-3 OPTICALLY ACTIVE COMPOUNDS

Optically active compounds such as *cis*-$M(en)_2BX$ have similar possibilities for substitution reactions. As shown in Figure 11-7, products can retain the same configuration, convert to *trans* geometry, or be a racemic mixture. The geometry of the 5-coordinate intermediate or transition state does not allow a simple inversion reaction like that seen at a tetrahedral carbon atom. Among the compounds studied that retain their optical activity and geometry on hydrolysis are $M(en)_2Cl_2^+$ with M = Co, Rh, and Ru.[21]

11-5-4 ISOMERIZATION

Isomerization has been described above for a number of complexes with monodentate ligands or with two bidentate ligands. Similar reactions with three bidentate ligands or with more complex ligands can follow two kinds of mechanism. In some cases, one end of a chelate ring dissociates and the

[21] S. A. Johnson, F. Basolo, and R. G. Pearson, *J. Am. Chem. Soc.*, **1963**, *85*, 1741; J. A. Broomhead and L. Kane-Maguire, *Inorg. Chem.*, **1969**, *8*, 2124.

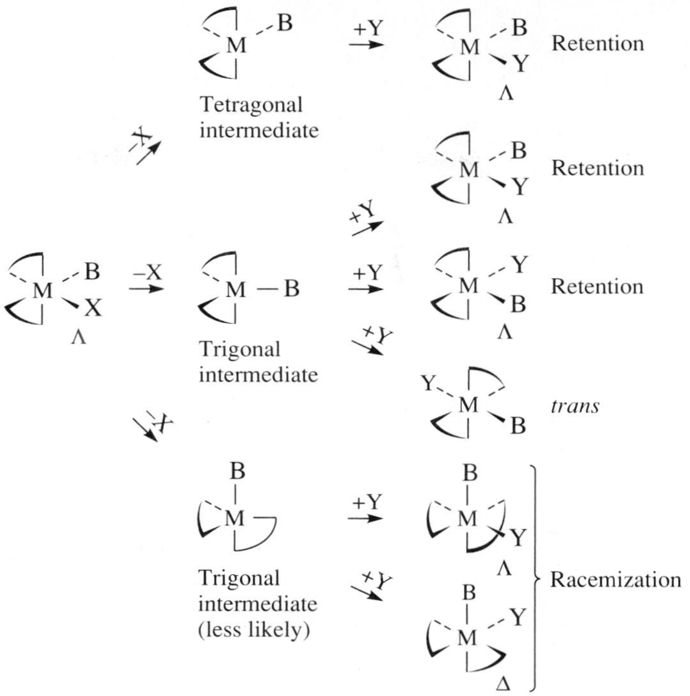

FIGURE 11-7 Dissociation Mechanism for *cis*-$[M(LL)_2BX]$ Substitution Reactions.

resulting 5-coordinate intermediate rearranges before reattachment of the loose end. This mechanism does not differ appreciably from the substitution reactions described above; the ligand that dissociates in the first step is the same one that adds in the final step, after rearrangement.

11-5-5 PSEUDOROTATION

Other isomerization mechanisms involving compounds containing chelating ligands are different kinds of twists. A number of twist mechanisms have been described, with different movements of the rings; those most commonly considered are shown in Figure 11-8.

The trigonal, or Bailar, twist (Figure 11-8a) requires twisting the two opposite trigonal faces through a trigonal prismatic transition state to the new structure. In the tetragonal twists, one chelate ring is held stationary while the other two are twisted to the new structure. The first one illustrated (Figure 11-8b) has a transition state with the stationary ring perpendicular to those being twisted. The second tetragonal twist (Figure 11-8c) requires twisting the two rings through a transition state with all three rings parallel. There have been attempts to determine which of these mechanisms is applicable, but the complexity of the reactions and the indirect means of measurement leave them subject to different interpretations. Nuclear magnetic resonance study of tris(trifluoroacetylacetonato) metal(III) chelates shows that a trigonal twist mechanism is not possible for M = Al, Ga, In, and *fac*-Cr, but leaves it a possibility for *fac*-Co.[22] The multiple-ring structure of *cis*-α-$[Co(trien)Cl_2]^+$ allows only a trigonal twist in its conversion to the β isomer, as shown in Figure 11-8d.

[22] R. C. Fay and T. S. Piper, *Inorg. Chem.*, **1964**, *3*, 348.

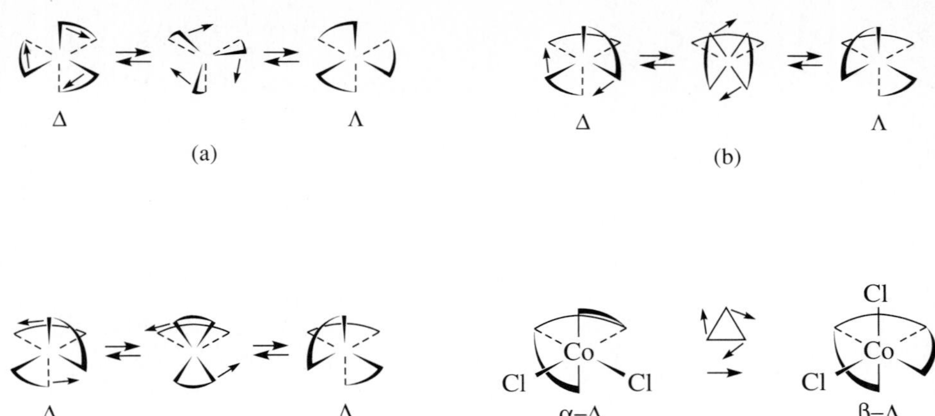

FIGURE 11-8 Twist Mechanisms for Isomerization of $M(LL)_3$ and $[Co(trien)Cl_2]^+$ Complexes. (a) Trigonal twist. The front triangular face rotates with respect to the back triangular face. (b) Twist with perpendicular rings. The back ring remains stationary as the front two rings rotate clockwise. (c) Twist with parallel rings. The back ring remains stationary as the front two rings rotate counterclockwise. (d) $[Co(trien)Cl_2]^+$ α–β isomerization. The connected rings limit this isomerization to a clockwise trigonal twist of the front triangular face.

11-6 SUBSTITUTION REACTIONS OF SQUARE PLANAR COMPLEXES

11-6-1 KINETICS OF SQUARE PLANAR SUBSTITUTIONS

Square planar substitution reactions frequently show two term rate laws, of the form

$$\text{Rate} = k_1\,[\text{Cplx}] + k_2\,[\text{Cplx}]\,[\text{Y}]$$

where [Cplx] = concentration of the complex and [Y] = concentration of the incoming ligand. Both pathways are considered to be associative, in spite of the difference in order. The k_2 term easily fits an associative mechanism in which the incoming ligand Y and the reacting complex form a 5-coordinate transition state. The accepted explanation for the k_1 term is a solvent-assisted reaction, solvent replacing X on the complex through a similar 5-coordinate transition state, and then itself being replaced by Y. The second step of this mechanism is presumed to be faster than the first (see Figure 11-9) and the concentration of solvent is large and unchanging, so the overall rate law for this path is first order in complex.

Since many of the reactions studied have been with platinum compounds, we will use the simplified reaction T—Pt—X + Y → T—Pt—Y + X, where

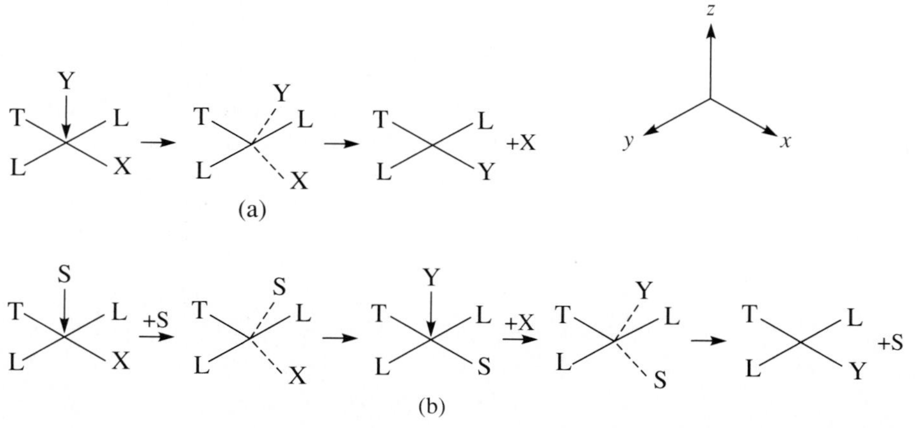

FIGURE 11-9 Interchange Mechanism in Square Planar Reactions. (a) Direct substitution by Y. (b) Solvent-assisted substitution.

T is the ligand *trans* to the departing ligand X, and Y is the incoming ligand. We will also designate the plane of the molecule the xy plane and the Pt axis through T—Pt—X the x axis. The other two ligands, L, are of lesser importance and will be ignored for the moment.

11-6-2 EVIDENCE FOR ASSOCIATIVE REACTIONS

It is generally accepted that reactions of square planar compounds are associative, although there is doubt about the degree of association, and they are classified as I_a. The reaction is shown in Figure 11-9. The incoming ligand approaches along the z axis. As it bonds to the Pt, the complex rearranges to approximate a trigonal bipyramid with Pt, T, X, and Y in the trigonal plane. As X leaves, Y moves into the plane of T, Pt, and the two L ligands. This same general description will fit whether the incoming ligand bonds strongly to Pt before the departing ligand bond is weakened appreciably (I_a) or the departing ligand bond is weakened considerably before the incoming ligand forms its bond (I_d).

The evidence for a 5-coordinate intermediate is very strong, including isolation of several 5-coordinate compounds with trigonal bipyramidal geometry [$Ni(CN)_5^{3-}$, $Pt(SnCl_3)_5^{3-}$, and similar compounds], although Basolo and Pearson argue that the transition state may well be 6-coordinate, with assistance from solvent.[23] The highest-energy transition state may be either during the formation of the intermediate or as the leaving ligand dissociates from the intermediate.

This mechanism explains naturally the effect of the incoming ligand. A strong Lewis base is likely to react readily, but the hard–soft nature of the base has an even larger effect. Pt(II) is generally a soft acid, so soft ligands react more readily with it. The order of ligand reactivity depends somewhat on the other ligands on the Pt, but the order for the reaction

$$trans\text{-}PtL_2Cl_2 + Y \longrightarrow trans\text{-}PtL_2ClY + Cl^-$$

for different Y in methanol was found to be (Table 11-10)[24]

$$PR_3 > CN^- > SCN^- > I^- > Br^- > N_3^- > NO_2^- > py > NH_3 \sim Cl^- > CH_3OH$$

A similar order, with some shuffling of the center of the list, is found for reactants with ligands other than chloride as T. The ratio of the rate constants for the extremes in the list is very large, with $k(PPh_3)/k(CH_3OH) = 9 \times 10^8$. Because T and Y have similar positions in the transition state, it is reasonable for them to have similar effects on the rate, and they do. Discussion of this ***trans* effect** is in the next section.

By the same argument, the leaving group X should also have a significant influence on the rate, and it does (Table 11-11).[25] The order of ligands is nearly the reverse of that given above, with hard ligands such as H_2O and NO_3^- leaving readily and quickly. Soft ligands with considerable π bonding, such as CN^- and NO_2^-, leave reluctantly; in the reaction

$$Pt(dien)X^+ + py \longrightarrow Pt(dien)(py)^{2+} + X^-$$

[23] Basolo and Pearson, op. cit., pp. 377–79, 395.

[24] U. Belluco, L. Cattalini, F. Basolo, R. G. Pearson, and A. Turco, *J. Am. Chem. Soc.*, **1965**, *87*, 241; R. G. Pearson, H. Sobel, and J. Songstad, *J. Am. Chem. Soc.*, **1968**, *90*, 319.

[25] R. G. Wilkins, op. cit., p. 231.

TABLE 11-10
Rate constants and LFER parameters for entering groups

	trans–$PtL_2Cl_2 + Y \longrightarrow$ *trans*–$PtL_2ClY + Cl^-$		
	$10^3k(M^{-1}\,s^{-1})$		
Y	L = py (s = 1)	L = PEt_3 (s = 1.43)	η_{Pt}
PPh_3	249,000		8.93
SCN^-	180	371	5.75
I^-	107	236	5.46
Br^-	3.7	0.93	4.18
N_3^-	1.55	0.2	3.58
NO_2^-	0.68	0.027	3.22
NH_3	0.47		3.07
Cl^-	0.45	0.029	3.04

SOURCES: Rate constants from U. Belluco, L. Cattalini, F. Basolo, R. G. Pearson, and A. Turco, *J. Am. Chem. Soc.*, **1965**, *87*, 241; PPh_3 and η_{Pt} data from R. G. Pearson, H. Sobel, and J. Songstad, *J. Am. Chem. Soc.*, **1968**, *90*, 319.

TABLE 11-11
Rate constants for leaving groups

	$Pt(dien)X^+ + py \longrightarrow Pt(dien)py^{2+} + X^-$ (Rate = $(k_1 + k_2[py])[Pt(dien)X^+]$)
X^-	$k_2(M^{-1}\,s^{-1})$
NO_3^-	very fast
Cl^-	5.3×10^{-3}
Br^-	3.5×10^{-3}
I^-	1.5×10^{-3}
N_3^-	1.3×10^{-4}
SCN^-	4.8×10^{-5}
NO_2^-	3.8×10^{-6}
CN^-	2.8×10^{-6}

SOURCE: Calculated from data in F. Basolo, H. B. Gray, and R. G. Pearson, *J. Am. Chem. Soc.*, **1960**, *82*, 4200.

there is an increase in the rate by a factor of 10^5 with H_2O as compared to $X^- = CN^-$ or NO_2^- as the leaving group. The bond-strengthening effect of the metal-to-ligand π bonding reduces the reactivity of these ligands significantly. In addition, π bonding to the leaving group uses the same orbitals as those bonding to the entering group in the trigonal plane. These two effects result in the slow displacement of π-bonding ligands when compared to ligands with only σ bonding or ligand-to-metal π bonding.

Good leaving groups (those that leave easily) show little discrimination between entering groups. Apparently, the ease of breaking the Pt—X bond takes precedence over the formation of the Pt—Y bond. On the other hand, for complexes with less reactive leaving groups, the other ligands have a significant role; the softer PEt_3 and $AsEt_3$ ligands show a large selective effect when compared to the harder dien or en ligands. The LFER equation [26] for this comparison is

$$\log k_Y = s\eta_{Pt} + \log k_S$$

[26] J. D. Atwood, *Inorganic and Organometallic Reaction Mechanisms*, Wadsworth, Belmont, Calif., 1985, pp. 60–63.

where k_Y = rate constant for reaction with Y

k_S = rate constant for reaction with solvent

s = **nucleophilic discrimination factor** (for the complex)

η_{Pt} = **nucleophilic reactivity constant** (for the entering ligand)

The parameter s is defined as 1 for *trans*-$[Pt(py)_2Cl_2]$, and has values from 0.44 for the hard $[Pt(dien)H_2O]^{2+}$ to 1.43 for the soft *trans*-$[Pt(PEt_3)_2Cl_2]$. Values of η_{Pt} are found by the equation $\eta_{Pt} = \log(k_Y/k_S)$, where k_Y and k_S refer to reactions with *trans*-$[Pt(py)_2Cl_2]$ in methanol at 30°C. Table 11-10 lists both these factors. For L = PEt_3, the change in rate constant is greater than for L = py due to the larger s value, and the parallel changes in the rate constants and η_{Pt} are obvious. Each of the parameters s and η_{Pt} may change by a factor of 3 from fast reactions to slow reactions, allowing for an overall ratio of 10^6 in the rates.

11-6-3 STEREOCHEMISTRY

The products of the reactions described above have the same configuration as the reactant, with direct replacement of the departing ligand by the new ligand. Because of the associative nature of the reactions, square planar substitution reactions result in retention of the initial geometry of the rest of the complex.

11-7 THE *TRANS* EFFECT

In 1926, Chernyaev[27] introduced the concept of the ***trans* effect** in platinum chemistry. In reactions of square planar Pt(II) compounds, ligands *trans* to chloride are more easily replaced than those *trans* to ligands such as ammonia; chloride is said to have a stronger *trans* effect than ammonia. When coupled with the fact that chloride itself is more easily replaced than ammonia, this *trans* effect allows formation of isomeric Pt compounds, as shown in the reactions of Figure 11-10. In reaction (a), after the first ammonia is replaced, the second replacement is *trans* to the first Cl^-. In reaction (b), the second replacement is *trans* to Cl^- (replacement of ammonia in the second reaction is possible, but leads to identical reactant and product). In (c) and (d), the greater lability of Cl^- relative to ammonia determines the geometry of the second product. The first steps in reactions (e) to (h) are the possible replacements, with nearly equal probabilities for replacement of ammonia or pyridine in any position. The second steps depend on the *trans* effect of Cl^-. By using reactions like these, it is possible to prepare specific isomers with different ligands. Chernyaev and his co-workers did much of this, preparing a wide variety of compounds and establishing the order of *trans* effect ligands:

$$CN^- \sim CO \sim C_2H_4 > PH_3 \sim SH_2 > NO_2^- > I^- > Br^- > Cl^- > NH_3 \sim py > OH^- > H_2O$$

EXERCISE 11-2

Predict the products of the reactions (there may be more than one product when there are conflicting preferences):

$$PtCl_4^{2-} + NO_2^- \longrightarrow (a) \qquad (a) + NH_3 \longrightarrow (b)$$

$$PtCl_3NH_3^- + NO_2^- \longrightarrow (c) \qquad (c) + NO_2^- \longrightarrow (d)$$

[27] I. I. Chernyaev, *Ann. Inst. Platine USSR*, **1926**, *4*, 261.

$PtCl(NH_3)_3^+ + NO_2^- \longrightarrow$ (e) $\quad$ (e) $+ NO_2^- \longrightarrow$ (f)

$PtCl_4^{2-} + I^- \longrightarrow$ (g) $\quad$ (g) $+ I^- \longrightarrow$ (h)

$PtI_4^{2-} + Cl^- \longrightarrow$ (i) $\quad$ (i) $+ Cl^- \longrightarrow$ (j)

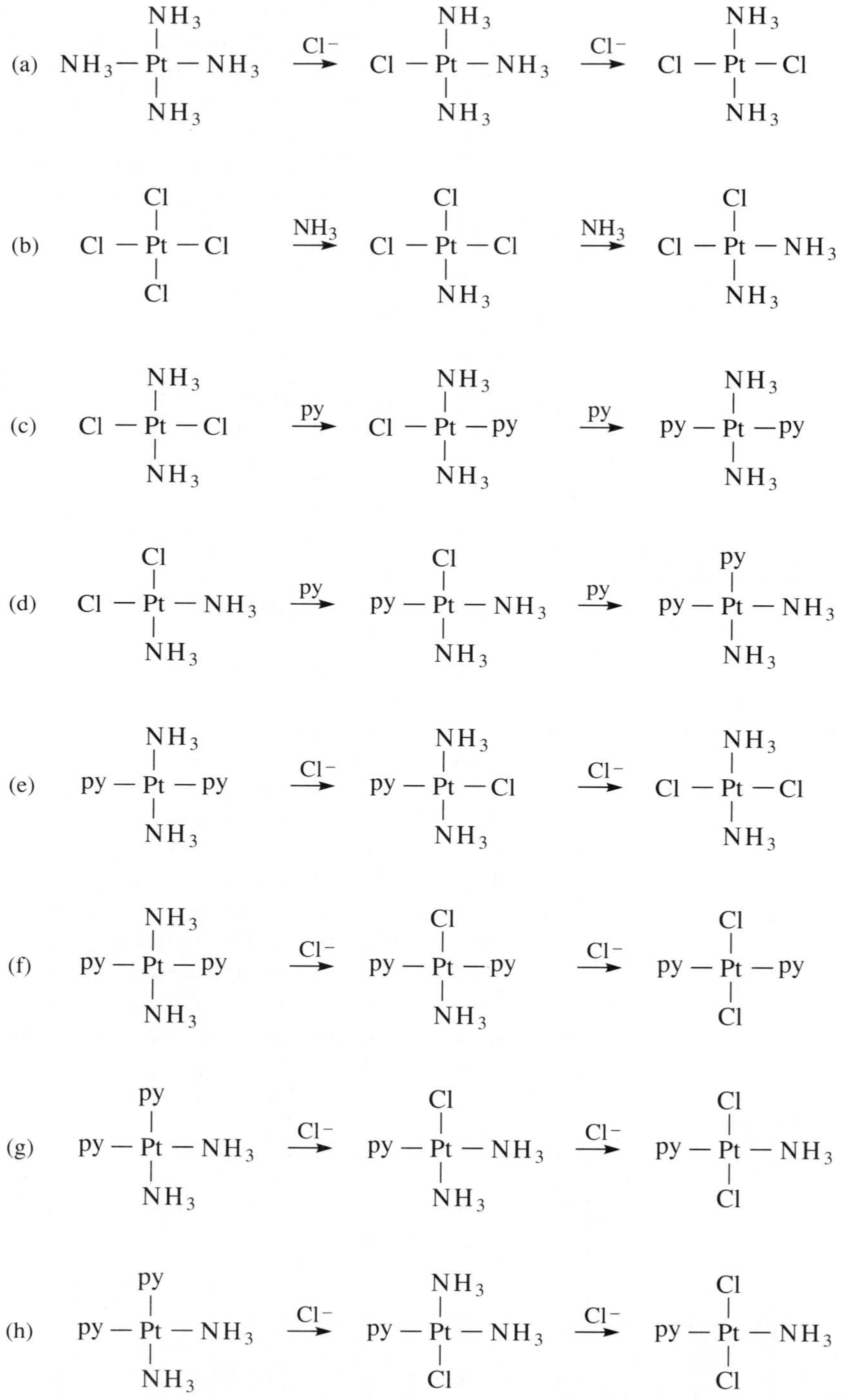

(a) $$\begin{array}{ccc} & NH_3 & \\ & | & \\ NH_3- & Pt & -NH_3 \\ & | & \\ & NH_3 & \end{array} \xrightarrow{Cl^-} \begin{array}{ccc} & NH_3 & \\ & | & \\ Cl- & Pt & -NH_3 \\ & | & \\ & NH_3 & \end{array} \xrightarrow{Cl^-} \begin{array}{ccc} & NH_3 & \\ & | & \\ Cl- & Pt & -Cl \\ & | & \\ & NH_3 & \end{array}$$

(b) $$\begin{array}{ccc} & Cl & \\ & | & \\ Cl- & Pt & -Cl \\ & | & \\ & Cl & \end{array} \xrightarrow{NH_3} \begin{array}{ccc} & Cl & \\ & | & \\ Cl- & Pt & -Cl \\ & | & \\ & NH_3 & \end{array} \xrightarrow{NH_3} \begin{array}{ccc} & Cl & \\ & | & \\ Cl- & Pt & -NH_3 \\ & | & \\ & NH_3 & \end{array}$$

(c) $$\begin{array}{ccc} & NH_3 & \\ & | & \\ Cl- & Pt & -Cl \\ & | & \\ & NH_3 & \end{array} \xrightarrow{py} \begin{array}{ccc} & NH_3 & \\ & | & \\ Cl- & Pt & -py \\ & | & \\ & NH_3 & \end{array} \xrightarrow{py} \begin{array}{ccc} & NH_3 & \\ & | & \\ py- & Pt & -py \\ & | & \\ & NH_3 & \end{array}$$

(d) $$\begin{array}{ccc} & Cl & \\ & | & \\ Cl- & Pt & -NH_3 \\ & | & \\ & NH_3 & \end{array} \xrightarrow{py} \begin{array}{ccc} & Cl & \\ & | & \\ py- & Pt & -NH_3 \\ & | & \\ & NH_3 & \end{array} \xrightarrow{py} \begin{array}{ccc} & py & \\ & | & \\ py- & Pt & -NH_3 \\ & | & \\ & NH_3 & \end{array}$$

(e) $$\begin{array}{ccc} & NH_3 & \\ & | & \\ py- & Pt & -py \\ & | & \\ & NH_3 & \end{array} \xrightarrow{Cl^-} \begin{array}{ccc} & NH_3 & \\ & | & \\ py- & Pt & -Cl \\ & | & \\ & NH_3 & \end{array} \xrightarrow{Cl^-} \begin{array}{ccc} & NH_3 & \\ & | & \\ Cl- & Pt & -Cl \\ & | & \\ & NH_3 & \end{array}$$

(f) $$\begin{array}{ccc} & NH_3 & \\ & | & \\ py- & Pt & -py \\ & | & \\ & NH_3 & \end{array} \xrightarrow{Cl^-} \begin{array}{ccc} & Cl & \\ & | & \\ py- & Pt & -py \\ & | & \\ & NH_3 & \end{array} \xrightarrow{Cl^-} \begin{array}{ccc} & Cl & \\ & | & \\ py- & Pt & -py \\ & | & \\ & Cl & \end{array}$$

(g) $$\begin{array}{ccc} & py & \\ & | & \\ py- & Pt & -NH_3 \\ & | & \\ & NH_3 & \end{array} \xrightarrow{Cl^-} \begin{array}{ccc} & Cl & \\ & | & \\ py- & Pt & -NH_3 \\ & | & \\ & NH_3 & \end{array} \xrightarrow{Cl^-} \begin{array}{ccc} & Cl & \\ & | & \\ py- & Pt & -NH_3 \\ & | & \\ & Cl & \end{array}$$

(h) $$\begin{array}{ccc} & py & \\ & | & \\ py- & Pt & -NH_3 \\ & | & \\ & NH_3 & \end{array} \xrightarrow{Cl^-} \begin{array}{ccc} & NH_3 & \\ & | & \\ py- & Pt & -NH_3 \\ & | & \\ & Cl & \end{array} \xrightarrow{Cl^-} \begin{array}{ccc} & Cl & \\ & | & \\ py- & Pt & -NH_3 \\ & | & \\ & Cl & \end{array}$$

FIGURE 11-10 *Trans* Effect in Pt(II) Reactions. Charges have been omitted for clarity.

11-7-1 EXPLANATIONS OF THE *TRANS* EFFECT[28]

Sigma bonding effects. Two factors dominate the explanations of the *trans* effect, weakening of the Pt—X bond and stabilization of the presumed 5-coordinate transition state. The energy relationships are given in Figure 11-11, with the activation energy the difference between the reactant ground state and the transition state.

The Pt—X bond is influenced by the Pt—T bond, since both use the Pt p_x and $d_{x^2-y^2}$ orbitals. When the Pt—T σ bond is strong, it uses a larger part of these orbitals and leaves less for the Pt—X bond. As a result, the Pt—X bond is weaker, and its ground state (sigma bonding orbital) is higher in energy, leading to a smaller activation energy for the breaking of this bond. This ground state effect is sometimes called the ***trans* influence**. This influence is a thermodynamic effect, separate from the *trans* effect, but contributing to the overall kinetic result. This part of the explanation predicts the order for the *trans* effect based on the relative sigma donor properties of the ligands:

$$H^- > PR_3 > SCN^- > I^-, CH_3^-, CO, CN^- > Br^- > Cl^- > NH_3 > OH^-$$

The order given here is not quite right for the *trans* effect, particularly for CO and CN^-, which have strong *trans* effects.

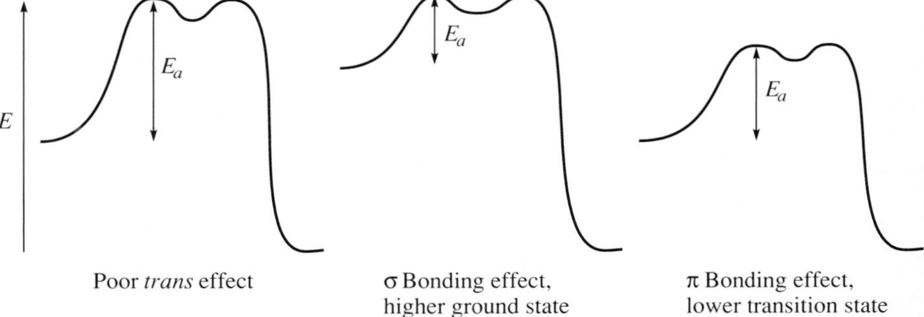

FIGURE 11-11 Activation Energy and the *Trans* Effect. The depth of the energy curve for the intermediate and the relative heights of the two maxima will vary with the specific reactants.

Pi bonding effects. The additional factor needed is π bonding in the Pt—T bond. When a ligand forms strong π-acceptor bonds with Pt, charge is removed from Pt, and the entrance of a ligand to form a 5-coordinate species is more likely. In addition to the charge effect, the $d_{x^2-y^2}$ orbital, which is involved in σ bonding in the square planar geometry, can contribute to π bonding in the trigonal bipyramidal transition state. Here the effect on the ground state of the reactant is small, but the energy of the transition state is lowered, again reducing the activation energy. The order of π-bonding ability of the ligands is

$$C_2H_4, CO > CN^- > NO_2^- > SCN^- > I^- > Br^- > Cl^- > NH_3 > OH^-$$

The expanded overall *trans* effect list is then the result of the combination of the two effects:

$$CO, CN^-, C_2H_4 > PR_3, H^- > CH_3^-, SC(NH_2)_2 > C_6H_5^- > NO_2^-, SCN^-, I^- > Br^- > Cl^- > py, NH_3, OH^-, H_2O$$

[28] Atwood, op. cit., p. 54; Basolo and Pearson, op. cit., pp. 369–75.

The order is not exactly as predicted by the σ and π effects due to subtle transition state differences. Ligands highest in the series are strong π acceptors, followed by strong σ donors. Ligands at the low end of the series have neither strong σ- nor π-bonding abilities. The *trans* effect can be very large; rates may differ as much as 10^6 between complexes with strong *trans* effect ligands and those with weak *trans* effect ligands.

11-8 OXIDATION–REDUCTION REACTIONS

Oxidation–reduction reactions of complexes have been studied by many different methods, including chemical analysis of the products, stopped-flow spectrophotometry, and the use of radioactive and stable isotope tracers. Henry Taube and his research group have been responsible for a large amount of the data, and his reviews cover the field.[29] One of the fundamental divisions in redox reactions is between those that depend on bonding through a common ligand (bridging or inner sphere reactions) and those where the contact is through the two coordination spheres, with no direct bonding between the two (nonbridging or outer-sphere reactions). Although it is not always easy to distinguish the two, the determination has been made for many reactions.

The rate of reaction for electron transfer depends on many factors, including the rate of substitution in the coordination sphere of the reactants, the match of energy levels of the two reactants, solvation of the two reactants, and the nature of the ligands.

11-8-1 INNER- AND OUTER-SPHERE REACTIONS

When the ligands of both reactants are tightly held and there is no change in the coordination sphere on reaction, the reaction proceeds by outer-sphere electron transfer. Examples of these reactions are given in Table 11-12 with their rate constants.

The rates show very large differences, depending on the details of the reactions. Characteristically, the rates depend on the ability of the electrons to tunnel through the ligands. This is a quantum mechanical property, by which electrons can pass through potential barriers that are too high to permit classical transfer. Ligands with π or p electrons or orbitals that can be used in bonding (as described in Chapter 8 for π donor and π acceptor ligands) provide better pathways for tunneling; those like NH_3, with no extra lone pairs and no low-lying antibonding orbitals, do not.

In outer-sphere reactions, where the ligands in the coordination sphere do not change, the primary change on electron transfer is a change in bond distance. A higher oxidation state on the metal leads to shorter σ bonds, with the extent of change depending on the electronic structure. The changes in bond distance are larger when e_g electrons are involved, as in the change from high-spin Co(II) ($t_{2g}^5 e_g^2$) to low-spin Co(III) (t_{2g}^6). Since the e_g orbitals are antibonding, removal of electrons from these orbitals results in a more stable compound and shorter bond distances. The combination of ligand-field energy change and stronger bonds to the ligands with stronger fields than water or fluoride make this reaction much easier. As a result, the electrode potential

[29] T. J. Meyer and H. Taube, "Electron Transfer Reactions," in *Comprehensive Coordination Chemistry,* Vol. 1, G. Wilkinson, R. D. Gillard, and J. A. McCleverty, eds., Pergamon, Elmsford, N.Y., 1987, pp. 331–84; H. Taube, *Electron Transfer Reactions of Complex Ions in Solution,* Academic, New York, 1970; *Adv. Inorg. Chem. Radiochem.,* **1959**, *1*, 1.

TABLE 11-12
Rate constants for outer-sphere electron transfer reactions (second order rate constants in $M^{-1}\ sec^{-1}$ at 25°C)

Oxidant	Reductants $[Cr(bipy)_3]^{2+}$	$[Ru(NH_3)_6]^{2+}$
$[Co(NH_3)_5(NH_3)]^{3+}$	6.9×10^2	1.1×10^{-2}
$[Co(NH_3)_5(F)]^{2+}$	1.8×10^3	
$[Co(NH_3)_5(OH)]^{2+}$	3×10^4	4×10^{-2}
$[Co(NH_3)_5(H_2O)]^{3+}$	5×10^4	3.0
$[Co(NH_3)_5(NO_3)]^{2+}$		3.4×10^1
$[Co(NH_3)_5(Cl)]^{2+}$	8×10^5	2.6×10^2
$[Co(NH_3)_5(Br)]^{2+}$	5×10^6	1.6×10^3
$[Co(NH_3)_5(I)]^{2+}$		6.7×10^3

SOURCES: $[Cr(bipy)_3]^{2+}$ data from J. P. Candlin, J. Halpern, and D. L. Trimm, *J. Am. Chem. Soc.*, **1964**, *86*, 1019. $[Ru(NH_3)_6]^{2+}$ data from J. F. Endicott and H. Taube, *J. Am. Chem. Soc.*, **1964**, *86*, 1686.

for reduction of the ammonia complex is much smaller than for reduction of aquated Co(III):

$$Co(NH_3)_6^{3+} + e^- \rightleftharpoons Co(NH_3)_6^{2+}, \qquad \mathscr{E}° = +0.108\ V$$

$$Co^{3+}(aq) + e^- \rightleftharpoons Co^{2+}(aq), \qquad \mathscr{E}° = +1.808\ V$$

Inner-sphere reactions also use the tunneling phenomenon, but in this case a single ligand is the conduit. The reactions proceed in three steps: (1) a substitution reaction that leaves the oxidant and reductant linked by the bridging ligand; (2) the actual transfer of the electron (frequently accompanied by transfer of the ligand); and (3) separation of the products:[30]

$$\underset{\text{Co(III) oxidant}}{[Co(NH_3)_5(Cl)]^{2+}} + \underset{\text{Cr(II) reductant}}{[Cr(H_2O)_6]^{2+}} \longrightarrow [\underset{\text{Co(III)}}{NH_3)_5Co}(Cl)\underset{\text{Cr(II)}}{Cr(H_2O)_5}]^{4+} + H_2O \qquad (1)$$

$$[(\underset{\text{Co(III)}}{NH_3)_5Co}(Cl)\underset{\text{Cr(II)}}{Cr(H_2O)_5}]^{4+} \longrightarrow [(\underset{\text{Co(II)}}{NH_3)_5Co}(Cl)\underset{\text{Cr(III)}}{Cr(H_2O)_5}]^{4+} \qquad (2)$$

$$[(NH_3)_5Co(Cl)Cr(H_2O)_5]^{4+} + H_2O \longrightarrow [(NH_3)_5Co(H_2O)]^{2+} + [(Cl)Cr(H_2O)_5]^{2+} \qquad (3)$$

In this case, these are followed by a reaction due to the labile nature of Co(II):

$$[(NH_3)_5Co(H_2O)]^{2+} + 5\,H_2O \longrightarrow [Co(H_2O)_6]^{2+} + 5\,NH_3$$

The transfer of chloride to the chromium in these reactions is easily followed because Cr(III) is substitutionally inert and the products can be separated by ion exchange techniques and their composition determined. When this is done, all the Cr(III) appears as $CrCl^{2+}$. The $[Cr(H_2O)_6]^{2+}$–$[Cr(H_2O)_5Cl]^{2+}$ reaction has also been studied, using radioactive ^{51}Cr as a tracer.[31] All the chloride in the product came from the reactant, with none entering from excess Cl^- in the solution. The rate of the reaction could also be determined by following the amount of radioactivity found in the $CrCl^{2+}$ at different times during the reaction.

[30] J. P. Candlin and J. Halpern, *Inorg. Chem.*, **1965**, *4*, 766.
[31] D. L. Ball and E. L. King, *J. Am. Chem. Soc.*, **1958**, *80*, 1091.

In many cases, the choice between inner-and outer-sphere mechanisms is difficult. In the examples of Table 11-12, the outer-sphere mechanism is required by the reducing agent. $Ru(NH_3)_6^{2+}$ is an inert species, and does not allow formation of bridging species fast enough for the rate constants observed. Although $Cr(bipy)_3^{2+}$ is labile, the parallels in the rate constants of the two species strongly suggest that its redox reactions are also outer sphere. In other cases, the oxidant may dictate an outer-sphere mechanism. In Table 11-13, $Co(NH_3)_6^{3+}$ and $Co(en)_3^{3+}$ have outer-sphere mechanisms because their ligands have no lone pairs with which to form bonds to the reductant. The other reactions are less certain, although Cr^{2+}(aq) is usually assumed to react by inner-sphere mechanisms in all cases where bridging is possible.

V^{2+}(aq) reactions seem to have a similar mechanism, although the decision is less clear for this reactant. The range of rate constants is smaller than that for Cr^{2+}, which seems to indicate that the ligands are less important and that an outer-sphere mechanism is likely. This is reinforced by comparison of the rate constants for V^{2+} and $Cr(bipy)_3^{2+}$. It may have different mechanisms for different oxidants, just as Cr^{2+} does.

Eu^{2+}(aq) is an unusual case. The rate constants do not parallel those of either the more common inner- or outer-sphere reactants, and the halide data are in reverse order from any others. The explanation offered for these rate constants is that the thermodynamic stability of the EuX^+ species helps drive the reaction faster for Cl^-, with slower rates and stabilities as we go down the series. Because of the smaller range of rate constants, Eu^{2+} reactions are usually classed as outer sphere.

TABLE 11-13
Rate constants for aquated reductants [$k(M^{-1}\ s^{-1})$]

	Cr^{2+}	Eu^{2+}	V^{2+}
$[Co(en)_3]^{3+}$	$\sim 2 \times 10^{-5}$	$\sim 5 \times 10^{-3}$	$\sim 2 \times 10^{-4}$
$[Co(NH_3)_6]^{3+}$	8.9×10^{-5}	2×10^{-2}	3.7×10^{-3}
$[Co(NH_3)_5(H_2O)]^{3+}$	5×10^{-1}	1.5×10^{-1}	$\sim 5 \times 10^{-1}$
$[Co(NH_3)_5(NO_3)]^{2+}$	$\sim 9 \times 10^{1}$	$\sim 1 \times 10^{2}$	
$[Co(NH_3)_5(Cl)]^{2+}$	6×10^{5}	3.9×10^{2}	~ 5
$[Co(NH_3)_5(Br)]^{2+}$	1.4×10^{6}	2.5×10^{2}	2.5×10^{1}
$[Co(NH_3)_5(I)]^{2+}$	3×10^{6}	1.2×10^{2}	1.2×10^{2}

SOURCES: Data from J. P. Candlin, J. Halpern, and D. L. Trimm, *J. Am. Chem. Soc.*, **1964**, *86*, 1019, except Cr^{2+} reactions with halide complexes from J. P. Candlin and J. Halpern, *Inorg. Chem.*, **1965**, *4*, 756 and $[Co(NH_3)_6]^{3+}$ Reactions with Cr^{2+} and V^{2+} from A. Zwickel and H. Taube, *J. Am. Chem. Soc.*, **1961**, *83*, 793.

TABLE 11-14
Rate constants for inner-sphere reactions with $[Co(CN)_5]^{3-}$

Oxidant	$k(M^{-1}\ s^{-1})$
$[Co(NH_3)_5(F)]^{2+}$	1.8×10^{3}
$[Co(NH_3)_5(OH)]^{2+}$	9.3×10^{4}
$[Co(NH_3)_5(NH_3)]^{3+}$	8×10^{4}*
$[Co(NH_3)_5(NCS)]^{2+}$	1.1×10^{6}
$[Co(NH_3)_5(N_3)]^{2+}$	1.6×10^{6}
$[Co(NH_3)_5(Cl)]^{2+}$	$\sim 5 \times 10^{7}$

SOURCE: Data from J. Candlin, J. Halpern, and S. Nakamura, *J. Am. Chem. Soc.*, **1963**, *85*, 2517.
NOTE: * Outer-sphere mechanism due to the oxidant. Complexes with other potential bridging groups (PO_4^{3-}, SO_4^{2-}, CO_3^{2-}, and several carboxylic acids) also react by an outer-sphere mechanism, with constants from 5×10^2 to 4×10^4.

Based on the products formed, $Co(CN)_5^{3-}$ reacts by an inner-sphere mechanism with many bridging ligands, with the rate constants given in Table 11-14. It also shows interesting behavior in reactions with thiocyanate or nitrite as bridging groups. With N-bonded $(NH_3)_5CoNCS^{2+}$, it reacts by bonding to the free S end of the ligand, since the cyanides soften the normally hard Co^{2+} ion. With S-bonded $(NH_3)_5CoSCN^{2+}$, it reacts initially by bonding to the free N end of the ligand, and then rearranges rapidly to the more stable S-bonded form. In a similar fashion, a transient O-bonded intermediate is detected in reactions of $(NH_3)_5Co(NO_2)^{2+}$ with $Co(CN)_5^{3-}$.[32]

Other reactions that follow an inner-sphere mechanism have been studied to determine which ligands bridge best. The overall rate of reaction usually depends on the first two steps above (substitution and transfer of electron), and in some cases it is possible to draw conclusions about the rates of the individual steps. For example, ligands that are reducible provide better pathways, and their complexes are more quickly reduced.[33] Benzoic acid is difficult to reduce, but 4-carboxy-N-methylpyridine is relatively easy to reduce. The $Co(NH_3)_5$(4-carboxy-N-methylpyridine) complex reacts ten times faster with Cr(II) than the corresponding benzoate complex, although both have similar structures and transition states (Table 11-15). For both ligands, the mechanism is inner-sphere, with transfer of the ligand to chromium, indicating that coordination to the Cr(II) is through the carbonyl oxygen. The substitution reactions should have similar rates, so the difference in overall rates is due to the transfer of electrons through the ligand. The data of Table 11-15 show these effects, and extend the data to glyoxylate and glycolate, which are still more easily reduced. The transfer of an electron through such ligands is very fast when compared to similar reactions with ligands which are not reducible.

TABLE 11-15
Ligand reducibility and electron transfer

Rate constants for the reaction
$[Co(NH_3)_5L]^{2+} + [Cr(H_2O)_6]^{2+} \longrightarrow Co^{2+} + 5NH_3 + [Cr(H_2O)_5L]^{2+} + H_2O$

L	$k_2(M^{-1}\ s^{-1})$	Comments
$C_6H_5C(=O)—O$	0.15	Benzoate is difficult to reduce
$CH_3NC_5H_5C(=O)—O$	1.3	N-Methyl-4-carboxy-pyridine is easily reduced
$CH_3C(=O)—O$	0.34	Acetic acid is difficult to reduce
$O{=}CHC(=O)—O$	3.1	Glyoxylate is moderately easy to reduce
$HOCH_2C(=O)—O$	7×10^3	Glycolate is very easy to reduce

SOURCE: H. Taube, *Electron Transfer Reactions of Complex Ions in Solution,* Academic Press, N.Y., 1970, pp. 64–66.

Remote attack on ligands with two potentially bonding groups is also found. Isonicotinamide bonded through the pyridine nitrogen can react with Cr^{2+} through the carbonyl oxygen on the other end of the molecule, transferring the ligand to the chromium and an electron through the ligand from the chromium to the other metal. The rate constants for different metals are shown in Table 11-16. The rate constants for the cobalt pentaammine and the chromium pentaaqua complexes are much closer than is usually found. The ratio is

[32] J. Halpern and S. Nakamura, *J. Am. Chem. Soc.,* **1965**, *87*, 3002; J. L. Burmeister, *Inorg. Chem.,* **1964**, *3*, 919.

[33] H. Taube, *Electron Transfer Reactions of Complex Ions in Solution,* Academic Press, New York, 1970, pp. 64–66; E. S. Gould and H. Taube, *J. Am. Chem. Soc.,* **1964**, *86*, 1318.

TABLE 11-16
Reductions of isonicotinamide (4-pyridine carboxylic acid amide) complexes by $[Cr(H_2O)_6]^{2+}$

Oxidant	$k_2(M^{-1}\,s^{-1})$
$[(NH_2\overset{O}{\overset{\parallel}{C}}—C_5H_5N)Co(NH_3)_5]^{3+}$	17.6
$[(NH_2\overset{O}{\overset{\parallel}{C}}—C_5H_5N)Cr(H_2O)_5]^{3+}$	1.8
$[(NH_2\overset{O}{\overset{\parallel}{C}}—C_5H_5N)Ru(NH_3)_5]^{3+}$	5×10^5

SOURCE: H. Taube, *Electron Transfer Reactions of Complex Ions in Solution,* Academic Press, New York, 1970, pp. 66–68.

frequently as large as 10^5, largely due to the greater oxidizing power of Co(III). In this case, the rate seems to depend more on the rate of electron transfer from Cr^{2+} to the bridging ligand, and the readily reducible isonicotinamide makes the two reactions more nearly equal in rate. The much faster rate found for the ruthenium pentaammine has been explained as the result of transfer of an electron through the π system of the ligand into the t_{2g} levels of Ru(III) (low-spin Ru(III) has a vacancy in the t_{2g} levels). A similar electron transfer to Co(III) or Cr(III) places the incoming electron in the e_g levels, which have σ symmetry.[34]

11-8-2 CONDITIONS FOR HIGH AND LOW OXIDATION NUMBERS

The overall stability of complexes with different charges on the metal ion depends on many factors, including LFSE, bonding energy of ligands, and redox properties of the ligands. When other factors are more or less equal, the hard and soft character of the ligands also has an effect. For example, all the very high oxidation numbers for the transition metals are found in combination with hard ligands, such as fluoride and oxide. Examples include MnO_4^-, CrO_4^{2-}, and FeO_4^{2-} with oxide and AgF_2, RuF_5, PtF_6, and OsF_6 with fluoride. At the other extreme, the lowest oxidation states are found with soft ligands, with carbon monoxide one of the most common. Zero is a common formal oxidation state for carbonyls; $V(CO)_6$, $Cr(CO)_6$, $Fe(CO)_5$, $Co_2(CO)_8$, and $Ni(CO)_4$ are examples. All these are stable enough for characterization in air, but some react slowly with air or decompose easily to the metal and CO. Their structure and reactions are explained further in Chapters 12 and 13.

Reactions of copper complexes show these ligand effects. Table 11-17 lists some of these reactions and their electrode potentials. As can be seen, complexing Cu(II) with the hard ligand ammonia stabilizes the higher oxidation number as compared to either Cu(I) or Cu(0). On the other hand, the soft ligand cyanide favors Cu(I), as do the halides. The halide cases are complicated by precipitation, but still show the effect and also show that the soft iodide ligand makes Cu(I) more stable than does the harder chloride.

In other cases, almost any ligand can serve to stabilize a particular species, and competing effects will have different results. Perhaps the most obvious example is the Co(III)–Co(II) couple, mentioned earlier. As the

[34] H. Taube and E. S. Gould, *Acc. Chem. Res.,* **1969**, *2*, 321.

TABLE 11-17
Electrode potentials of cobalt and copper species

Cu(II)–Cu(I) Reactions	$\mathscr{E}°(V)$
$Cu^{2+}(aq) + 2\,CN^- + e^- \rightleftharpoons [Cu(CN)_2]^-(aq)$	+1.103
$Cu^{2+}(aq) + I^- + e^- \rightleftharpoons CuI(s)$	+0.86
$Cu^{2+}(aq) + Cl^- + e^- \rightleftharpoons CuCl(s)$	+0.538
$Cu^{2+}(aq) + e^- \rightleftharpoons Cu^+(aq)$	+0.153
$[Cu(NH_3)_4]^{2+} + e^- \rightleftharpoons [Cu(NH_3)_2]^+ + 2\,NH_3$	−0.01
Cu(II)–Cu(O) Reactions	
$Cu^{2+}(aq) + 2e^- \rightleftharpoons Cu(s)$	+0.337
$[Cu(NH_3)_4]^{2+} + 2e^- \rightleftharpoons Cu(s) + 4\,NH_3$	−0.05
Co(III)–Co(II) Reactions	
$Co^{3+}(aq) + e^- \rightleftharpoons Co^{2+}(aq)$	+1.808
$[Co(NH_3)_6]^{3+} + e^- \rightleftharpoons [Co(NH_3)_6]^{2+}$	+0.108
$[Co(CN)_6]^{3-} + e^- \rightleftharpoons [Co(CN)_6]^{4-}$	−0.83

SOURCE: Data from T. Moeller, *Inorganic Chemistry: A Modern Introduction*, Wiley-Interscience, New York, 1982, pp. 580–81.

hydrated ion (or aqua complex), Co(III) is a very strong oxidizing agent, reacting readily with water to form oxygen and Co(II). However, when coordinated with any ligand other than water or fluoride, it is kinetically stable, and almost stable in the thermodynamic sense as well. Part of the explanation is that Δ_o is quite large with any ligand, leading to an easy change from the high-spin Co(II) configuration $t_{2g}^5\,e_g^2$ to the low-spin Co(III) configuration t_{2g}^6. This means that the reverse reduction is much less favorable, and the complex ions have little tendency to oxidize other species. The reduction potentials (Table 11-17) for Co(III) $\longrightarrow$ Co(II) with different ligands are in the order $H_2O > NH_3 > CN^-$, the order of increasing Δ_o and decreasing hardness. The increasing LFSE change is strong enough to overcome the usual effect of softer ligands stabilizing lower oxidation states.

11-9 REACTIONS OF COORDINATED LIGANDS

The reactions described up to this point are either substitution reactions or oxidation–reduction reactions. Other reactions are primarily those of the ligands; in these reactions, coordination to the metal changes the ligand properties sufficiently to change the rate of a reaction or to make possible a reaction that would otherwise not take place. Such reactions are important for many different kinds of compounds and many different circumstances. Chapter 13 describes such reactions for organometallic compounds, and Chapter 15 describes some reactions important in biochemistry. In this chapter, we describe only a few examples of these reactions; the interested reader can find many more examples in the references cited.

Organic chemists have long used inorganic compounds as reagents. For example, Lewis acids such as $AlCl_3$, $FeCl_3$, $SnCl_4$, $ZnCl_2$, and $SbCl_5$ are used in Friedel–Crafts electrophilic substitutions. The labile complexes formed by acyl or alkyl halides and these Lewis acids create positively charged carbon atoms that can react readily with aromatic compounds. The reactions are generally the same as without the metal salts, but their use speeds the reactions and makes them much more useful.

As usual, it is easier to study reactions of inert compounds, such as those of Co(III), Cr(III), Pt(II), and Pt(IV), where the products remain complexed to the metal and can be isolated for more complete study. However, useful catalysis requires that the products be easily separated from the catalyst, so relatively rapid dissociation from the metal is a desirable feature. While many of the reactions described here do not have this capability, those of biological significance do, and chemists studying ligand reactions for synthetic purposes try to incorporate it into their reactions.

11-9-1 HYDROLYSIS OF ESTERS, AMIDES, AND PEPTIDES

Amino acid esters, amides, and peptides can by hydrolyzed in basic solution, and the addition of many different metal ions speeds the reactions. Labile complexes of Cu(II), Co(II), Ni(II), Mn(II), Ca(II), and Mg(II), as well as other metal ions, promote the reactions. Whether the mechanism is through bidentate coordination of the α-amino group and the carbonyl, or only through the amine, is uncertain, but seems to depend on the relative concentrations. Since the reactions depend on complex formation and hydrolysis as separate steps, their temperature dependence is complex and interpretation of all the effects is difficult.[35]

Co(III) complexes also promote similar reactions. When four of the six octahedral positions are occupied by amine ligands and two *cis* positions are available for further reactions, it is possible to study not only the hydrolysis itself, but the steric preferences of the complexes. In general, these compounds catalyze the hydrolysis of N-terminal amino acids from peptides, and the amino acid that is removed remains as part of the complex. The reactions apparently proceed by coordination of the free amine to cobalt, followed either by coordination of the carbonyl to cobalt and subsequent reaction with OH^- or H_2O from the solution (path A in Figure 11-12) or reaction of the carbonyl carbon with coordinated hydroxide (path B).[36] As a result, the N-terminal amino acid is removed from the peptide and left as part of the cobalt complex in which the α-amino nitrogen and the carbonyl oxygen are bonded to the cobalt. Esters and amides are also hydrolyzed by the same mechanism, with the relative importance of the two pathways dependent on the specific compounds used.

Other compounds, such as phosphate esters, pyrophosphates, and amides of phosphoric acid, are hydrolyzed in similar reactions. Coordination may be through only one oxygen of these phosphate compounds, but the overall effect is similar.

11-9-2 TEMPLATE REACTIONS

Template reactions are those in which formation of a complex places the ligands in the correct geometry for reaction. One of the earliest was for formation of phthalocyanines, shown in Figure 11-13. Although the compounds

[35] M. M. Jones, *Ligand Reactivity and Catalysis,* Academic Press, New York, 1968, summarizes the arguments and mechanisms in Chapter III.

[36] J. P. Collman and D. A. Buckingham, *J. Am. Chem. Soc.*, **1963**, *85*, 3039; D. A. Buckingham, J. P. Collman, D. A. R. Hopper, and L. G. Marzelli, *J. Am. Chem. Soc.*, **1967**, *89*, 1082.

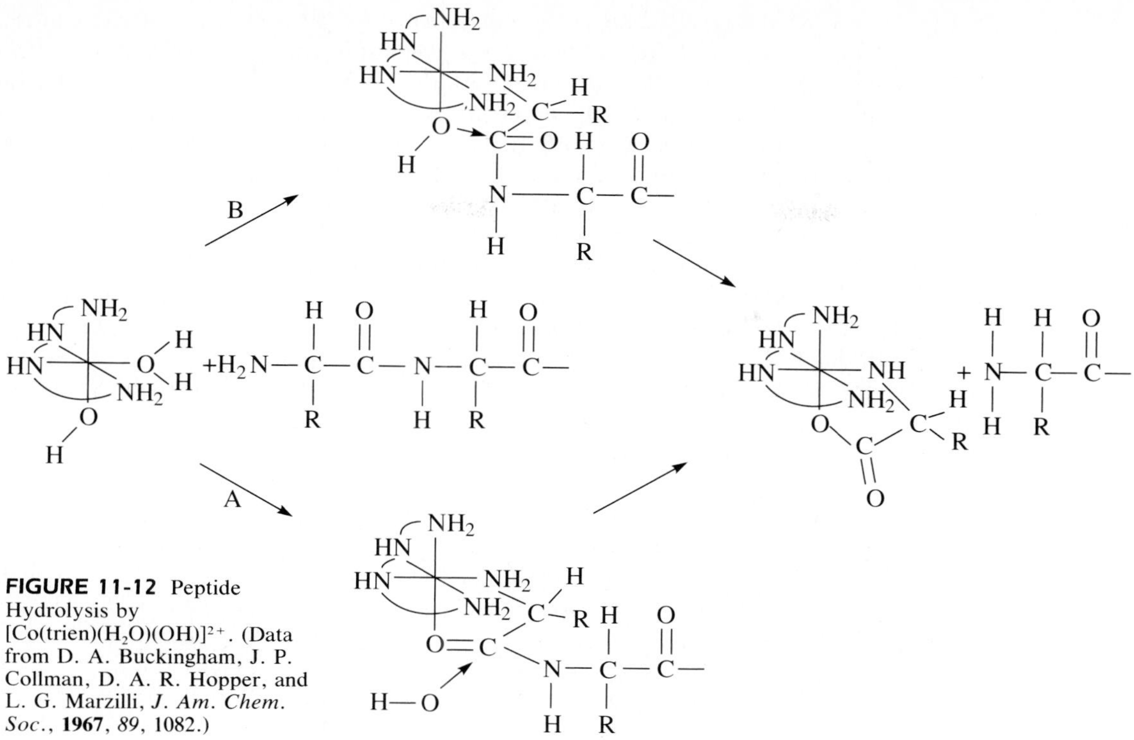

FIGURE 11-12 Peptide Hydrolysis by $[Co(trien)(H_2O)(OH)]^{2+}$. (Data from D. A. Buckingham, J. P. Collman, D. A. R. Hopper, and L. G. Marzilli, *J. Am. Chem. Soc.*, **1967**, *89*, 1082.)

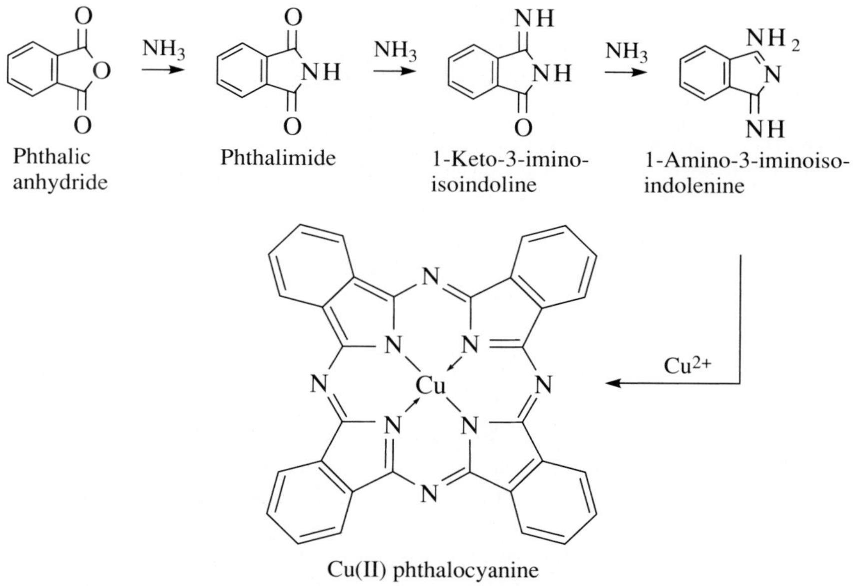

Phthalic anhydride → (NH_3) Phthalimide → (NH_3) 1-Keto-3-imino-isoindoline → (NH_3) 1-Amino-3-iminoiso-indolenine → (Cu^{2+}) Cu(II) phthalocyanine

FIGURE 11-13 Phthalocyanine Synthesis.

were known earlier, their study really began in 1928 after discovery of a dark blue impurity in phthalimide prepared by reaction of phthalic anhydride with ammonia in an enameled iron vessel. This impurity was later discovered to be the iron phthalocyanine complex, created from iron released into the mixture by a break in the enamel surface. A similar reaction takes place with copper, which forms more useful pigments. The intermediates shown in Figure 11-13 have been isolated. Phthalic acid and ammonia form first phthalimide, then 1-keto-3-iminoisoindoline, then 1-amino-3-iminoisoindolenine. The cyclization reaction then takes place, probably with the assistance of the metal ion, which holds the chelated reactants in position. This is confirmed by the lack of cyclization in the absence of the metals.[37] Other reagents can be used for this synthesis, but the essential feature of all these reactions is the formation of the cyclic compound by coordination to a metal ion.

More recently, similar reactions have been used extensively in the formation of macrocyclic compounds and others that form fewer coordinate bonds. Imine or Schiff base complexes have been extensively studied. In this case, the compounds can be formed without complexation, but the reaction is much faster in the presence of metal ions. An example is shown in Figure 11-14. In the absence of nickel, benzothiazoline is formed rather than the imine.

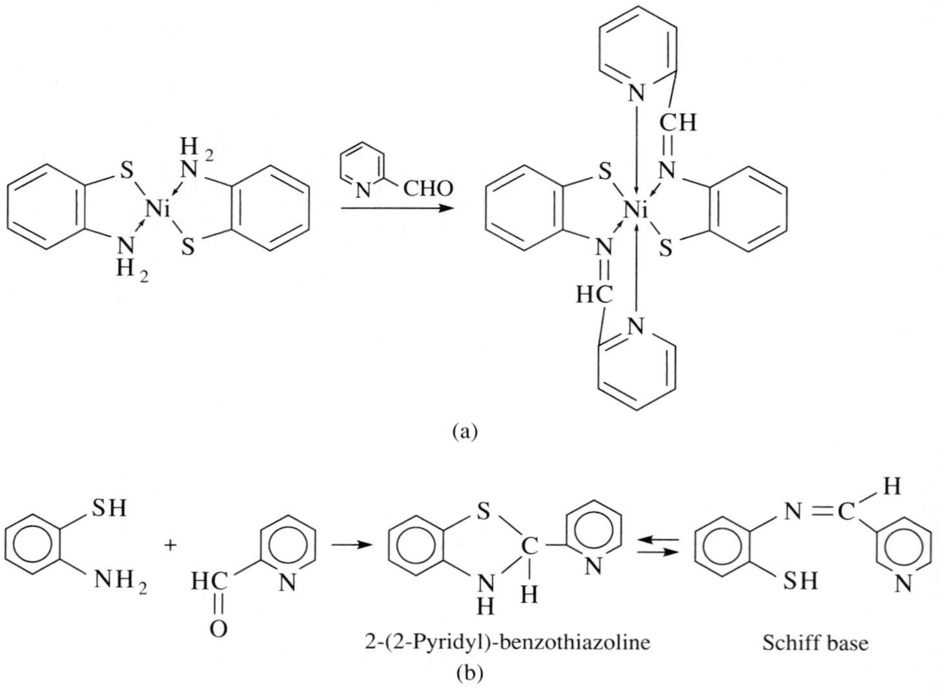

(a)

2-(2-Pyridyl)-benzothiazoline

Schiff base

(b)

FIGURE 11-14 Schiff Base Template Reaction. (a) The Ni(II)-*o*-aminothiophenol complex reacts with pyridine-1-carboxaldehyde to form the Schiff base complex. (b) In the absence of the metal ion, the product is benzthiazoline. (L. F. Lindoy and S. E. Livingstone, *Inorg. Chem.*, **1968**, *7*, 1149.)

A major feature of template reactions is geometric; formation of the complex brings the reactants into close proximity with the proper orientation for reaction. In addition, complexation may change the electronic structure sufficiently to promote the reaction. Both of these are common to all coordinated ligand reactions, but the geometric factor is more obvious in these; the final product has a structure determined by the coordination geometry. These reactions have been reviewed and a large number of reactions and products described.[38]

[37] R. Price, "Dyes and Pigments," in G. Wilkinson, R. D. Gillard, and J. A. McCleverty, *Comprehensive Coordination Chemistry*, Vol. 6, Pergamon, Elmsford, N.Y., 1987, pp. 88–89.

[38] D. St. C. Black, "Stoichiometric Reactions of Coordinated Ligands," in G. Wilkinson, R. D. Gillard, and J. A. McCleverty, op. cit., Vol. 6, pp. 155–226.

11-9-3 ELECTROPHILIC SUBSTITUTION

Acetylacetone complexes are known to undergo a wide variety of reactions that are at least superficially similar to aromatic electrophilic substitutions. Bromination, nitration, and similar reactions have been studied.[39] In all cases, coordination forces the ligand into an enol form and promotes reaction at the center carbon by preventing reaction at the oxygens and concentrating negative charge on the number three carbon. Figure 11-15 shows some of the reactions and a possible mechanism for the bromination reaction.

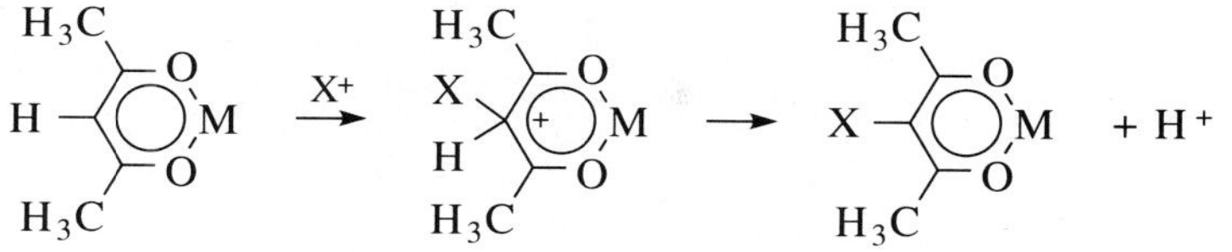

FIGURE 11-15 Electrophilic Substitution on Acetylacetone Complexes. X = Cl, Br, SCN, SAr, SCl, NO_2, CH_2Cl, $CH_2N(CH_3)_2$, COR, CHO.

GENERAL REFERENCES

The general principles of kinetics and mechanisms are described by J. W. Moore and R. G. Pearson, *Kinetics and Mechanism,* 3rd ed., Wiley-Interscience, New York, 1981 and F. Wilkinson, *Chemical Kinetics and Reaction Mechanisms,* Van Nostrand Reinhold, New York, 1980. The classic for coordination compounds is F. Basolo and R. G. Pearson, *Mechanisms of Inorganic Reactions,* 2nd ed., Wiley, New York, 1967. More recent books are by Jim D. Atwood, *Inorganic and Organometallic Reaction Mechanisms,* Brooks/Cole, Monterey, Calif., 1985, and D. Katakis and G. Gordon, *Mechanisms of Inorganic Reactions,* Wiley-Interscience, New York, 1987. The reviews in *Comprehensive Coordination Chemistry,* G. Wilkinson, R. D. Gillard, and J. A. McCleverty, eds., Pergamon, Elmsford, N.Y., 1987, provide a more comprehensive collection and discussion of the data. Volume 1, Theory and Background, covers substitution and redox reactions, while Volume 6, Applications, is particularly rich in data on ligand reactions.

PROBLEMS

11-1 The high-spin d^4 complex $[Cr(H_2O)_6]^{2+}$ is *labile,* but the low-spin d^4 complex ion $[Cr(CN)_6]^{4-}$ is *inert.* Explain.

11-2 Why is the existence of a series of entering groups with different rate constants evidence for an associative mechanism (A or I_a)?

11-3 Predict whether these complexes would be labile or inert, and explain your choices. The magnetic moment in Bohr magnetons is given after each complex.

Ammonium pentachlorooxochromate(V)	1.82 BM
Potassium hexaiodomanganate(IV)	3.82 BM
Potassium hexacyanoferrate(III)	2.40 BM
Hexaammineiron(II) chloride	5.45 BM

11-4 Consider the half-lives toward substitution of the pairs of complexes:

Half-lives less than 1 minute	*Half-lives greater than 1 day*
$Cr(CN)_6^{4-}$	$Cr(CN)_6^{3-}$
$Fe(H_2O)_6^{3+}$	$Fe(CN)_6^{4-}$
$Co(H_2O)_6^{2+}$	$Co(NH_3)_5(H_2O)^{3+}$ (H_2O exchange)

Interpret the differences in terms of the electronic structures.

[39] J. P. Collman, *Angew. Chem. Intern. Ed. English,* **1965**, *4*, 132.

11-5 The general rate law for substitution in square planar Pt(II) complexes is valid for the reaction

$$[Pt(NH_3)_4]^{2+} + Cl^- \longrightarrow [Pt(NH_3)_3Cl]^+ + NH_3$$

Design the experiments needed to verify this and to determine the rate constants. What experimental data are needed, and how are the data to be treated?

11-6 The graph shows plots of k_{obs} versus $[X^-]$ for the anation reactions *cis*-$[Co(en)_2(NO_2)(DMSO)]^{2+} + X^- \longrightarrow [Co(en)_2(NO_2)X]^+ + DMSO$. Δ, $X^- = NO_2^-$; ○, $X^- = Cl^-$: ●, $X^- = SCN^-$. The broken line shows the rate of the DMSO-exchange reaction. (Graph is reproduced with permission from W. R. Muir and C. H. Langford, *Inorg. Chem.*, **1968**, *7*, 1032. Copyright 1968 American Chemical Society.)

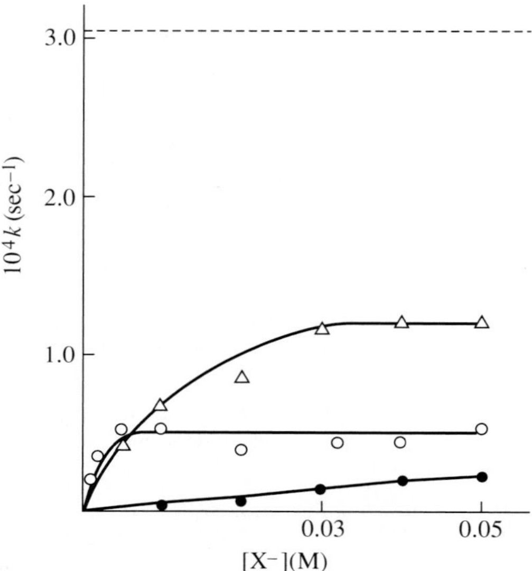

All three reactions are presumed to have the same mechanism.

a. Why is the DMSO exchange so much faster than the other reactions?
b. Why are the curves shaped as they are?
c. Explain what the limiting rate constants (at high concentration) are in terms of the rate laws for D and I_d mechanisms.
d. The limiting rate constants are $0.5 \times 10^{-4}\ s^{-1}$ and $1.2 \times 10^{-4}\ s^{-1}$ for Cl^- and NO_2^-, respectively. For SCN^-, the limiting rate constant can be estimated as $1 \times 10^{-4}\ s^{-1}$. Do these values constitute evidence for an I_d mechanism?

11-7 Account for the observation that two separate water exchange rates are found for $Cu(H_2O)_6^{2+}$ in aqueous solution.

11-8 Data for the reaction $Co(NO)(CO)_3 + As(C_6H_5)_3 \longrightarrow Co(NO)(CO)_2(As(C_6H_5)_3) + CO$ in toluene at 45°C is given in the following table. In all cases, the reaction is pseudo first order in $Co(NO)(CO)_3$. Find the rate constant(s) and discuss their probable significance (E. M. Thorsteinson and F. Basolo, *J. Am. Chem. Soc.*, **1966**, *88*, 3929).

$[As(C_6H_5)_3](M)$	$10^5 k_{obs}\ (s^{-1})$
0.014	2.3
0.098	3.9
0.525	12
1.02	23

11-9 The figure shows the log of the rate constant for substitution of CO on $Co(NO)(CO)_3$ by phosphorus and nitrogen ligands plotted against the half neutralization potential (ΔHNP) of the ligands. ΔHNP is a measure of the basicity of the compounds. Explain the linearity of such a plot and why there are two different lines. Incoming nucleophilic ligands: (1) $P(C_2H_5)_3$, (2) $P(n-C_4H_9)_3$, (3) $P(C_6H_5)(C_2H_5)_2$, (4) $P(C_6H_5)(C_2H_5)_2$, (5) $P(C_6H_5)_2(n-C_4H_9)$, (6) $P(p-CH_3C_6H_4)_3$, (7) $P(O-n-C_4H_9)_3$, (8) $P(C_6H_5)_3$, (9) $P(OCH_3)_3$, (10) $P(OCH_3)_2CH_3$, (11) $P(OC_6H_5)_3$, (12) 4-picoline, (13) pyridine, (14) 3-chloropyridine. (Graph is reproduced with permission from E. M. Thorsteinson and F. Basolo, *J. Am. Chem. Soc.*, **1966**, *88*, 3929. Copyright 1966 American Chemical Society.)

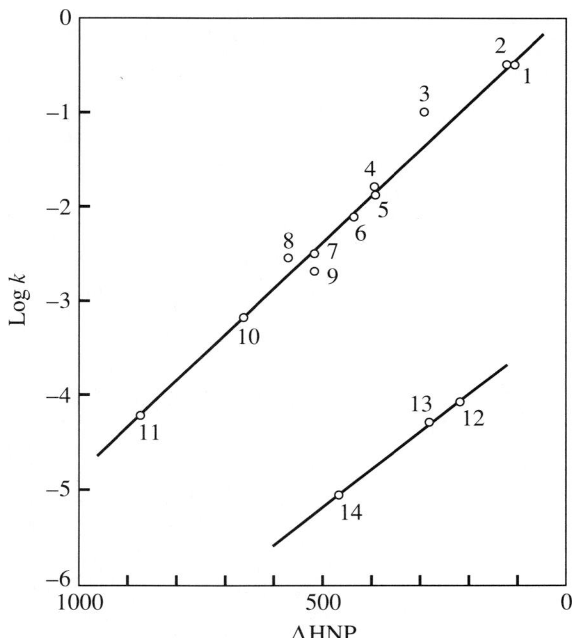

11-10 *Cis*-$PtCl_2(PEt_3)_2$ is stable in benzene solution. However, small amounts of free triethylphosphine catalyze establishment of an equilibrium with the *trans* isomer:

$$cis\text{-}PtCl_2(PEt_3)_2 \rightleftharpoons trans\text{-}PtCl_2(PEt_3)_2$$

For the conversion of *cis* to *trans* in benzene at 25°C, $\Delta H° = 10.3$ kJ mol^{-1} and $\Delta S° = 55.6$ J mol^{-1} K^{-1}.

a. Calculate the free energy change, $\Delta G°$, and the equilibrium constant for this isomerization.

b. Which isomer has the higher bond energy? Is this answer consistent with what you would expect on the basis of π bonding in the two isomers? Explain briefly.

c. Why is free triethylphosphine necessary to catalyze the isomerization?

11-11 The following table shows the effect of changing ligands on the dissociation rates of CO *cis* to those ligands. Explain the effect of these ligands on the rates of dissociation. Include the effect of these ligands on Cr—CO bonding and on the transition state (presumed to be square pyramidal) (J. D. Atwood and T. L. Brown, *J. Am. Chem. Soc.*, **1976**, *98*, 3160).

Compound	*k(s⁻¹) (CO dissociation)*
$Cr(CO)_6$	1×10^{-12}
$Cr(CO)_5(PPh_3)$	3.0×10^{-10}
$Cr(CO)_5I$	$< 10^{-5}$
$Cr(CO)_5Br$	2×10^{-5}
$Cr(CO)_5Cl$	1.5×10^{-4}

11-12 When the two isomers of $[Pt(NH_3)_2Cl_2)]$ react with thiourea [tu = $S{=}C(NH_2)_2$], one product is $[Pt(tu)_4]^{2+}$ and the other is $[Pt(NH_3)_2(tu)_2]^{2+}$. Identify the initial isomers and explain the results.

11-13 Predict the products (equimolar mixtures):
a. $[Pt(CO)Cl_3]^- + NH_3$
b. $[Pt(NH_3)Br_3]^- + NH_3$
c. $[(C_2H_4)PtCl_3]^- + NH_3$

11-14 The rate constant for electron exchange between $V^{2+}(aq)$ and $V^{3+}(aq)$ is observed to depend on the hydrogen ion concentration:

$$k_{obs} = a + \frac{b}{[H^+]}$$

Propose a mechanism, and express *a* and *b* in terms of the rate constants of the mechanism. [*Hint:* $V^{3+}(aq)$ hydrolyzes more easily than $V^{2+}(aq)$.]

11-15 Is the reaction $[Co(NH_3)_6]^{3+} + [Cr(H_2O)_6]^{2+}$ likely to proceed by an inner-sphere or outer-sphere mechanism? Explain your answer.

11-16 The rate constants for the exchange reaction

$$CrX^{2+} + {}^*Cr^{2+} \longrightarrow {}^*CrX^{2+} + Cr^{2+}$$

where *Cr is radioactive ^{51}Cr, are given in the table for reactions at 0°C and 1 *M* $HClO_4$. Explain the differences in the rate constants in terms of the probable mechanism of the reaction.

X⁻	*k(M⁻¹ s⁻¹)*	
F^-	1.2×10^{-3}	
Cl^-	11	
Br^-	60	
NCS^-	1.2×10^{-4}	(at 24°C)
N_3^-	>1.2	

12

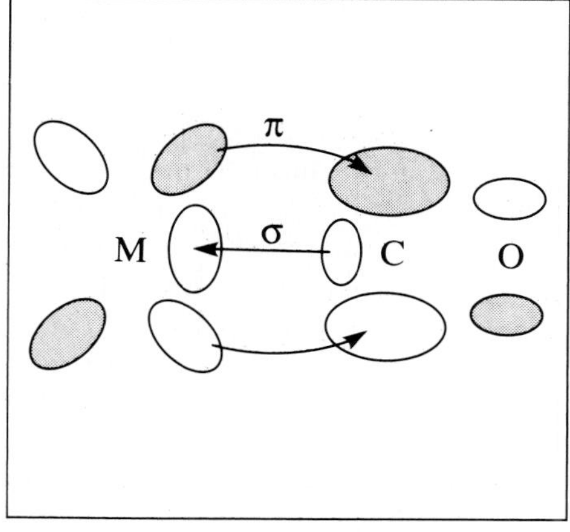

Organometallic Chemistry

Organometallic chemistry, the chemistry of compounds containing metal–carbon bonds, is one of the most interesting and certainly most rapidly growing areas of chemical research. It encompasses a wide variety of chemical compounds and their reactions, including compounds containing both sigma and pi bonds between metal atoms and carbon; many cluster compounds, containing one or more metal–metal bonds; molecules of structural types unusual or unknown in organic chemistry; and reactions that in some cases bear similarities to known organic reactions and in other cases are dramatically different. Aside from their intrinsically interesting nature, many organometallic compounds form useful catalysts and consequently are of significant industrial interest. In this chapter we describe a variety of types of organometallic compounds and present descriptions of organic ligands and how they bond to metals. Chapter 13 continues with an outline of major types of reactions of organometallic compounds and how these reactions may be combined into catalytic cycles. Chapter 14 discusses parallels that may be observed between organometallic chemistry and main group chemistry.

Certain organometallic compounds bear similarities to the types of coordination compounds already discussed in this text. $Cr(CO)_6$ and $[Ni(H_2O)_6]^{2+}$, for example, are both octahedral in structure. Both CO and H_2O are sigma donor ligands; in addition, CO is a strong pi acceptor. Other ligands that can exhibit both behaviors include CN^-, PPh_3, and SCN^-, as well as many organic ligands. The metal–ligand bonding and electronic spectra of compounds containing these ligands can be described using concepts discussed in Chapters 8 and 9 on coordination compounds. However, many organometallic molecules are strikingly different from any we have considered previously. Cyclic organic ligands containing delocalized pi systems can team up with metal atoms to form sandwich compounds, with a metal sandwiched between, for example, benzene or cyclopentadienyl rings. Sometimes rings incorporating other nonmetals can be included as well. Examples of these double- and multiple-decker sandwich compounds are shown in Figure 12-1.

A characteristic of metal atoms bonded to organic ligands, especially CO, is that they often exhibit the capacity to form covalent bonds to other metal atoms to form **cluster compounds** (some cluster compounds are also known that contain no organic ligands). These clusters may contain only two or three metal atoms or as many as several dozen; there is no limit to their size or variety. They may contain single, double, triple, or quadruple bonds between the metal atoms and may in some cases have ligands that bridge two or more of the metals. Examples of metal cluster compounds containing organic ligands are shown in Figure 12-2; clusters will be discussed further in Chapter 14.

Carbon itself may play quite a different role than commonly encountered in organic chemistry. Certain metal clusters encapsulate carbon atoms; the resulting molecules, called **carbide clusters,** in some cases contain carbon bonded to five, six, or more surrounding metals. The traditional notion of carbon forming bonds to, at most, four additional atoms must be reconsidered (a few examples of carbon bonded to more than four atoms are also known in organic chemistry). Two examples of carbide clusters are included in Figure 12-2.

Many other types of organometallic compounds have interesting structures and chemical properties. Figure 12-3 shows several additional examples of the variety of molecular structures encountered in this field.

Strictly speaking, the only compounds classified as organometallic are those that contain metal–carbon bonds, but in practice complexes containing several other ligands similar to CO in their bonding, such as NO and N_2, are frequently included. (Cyanide also forms complexes in a manner similar to

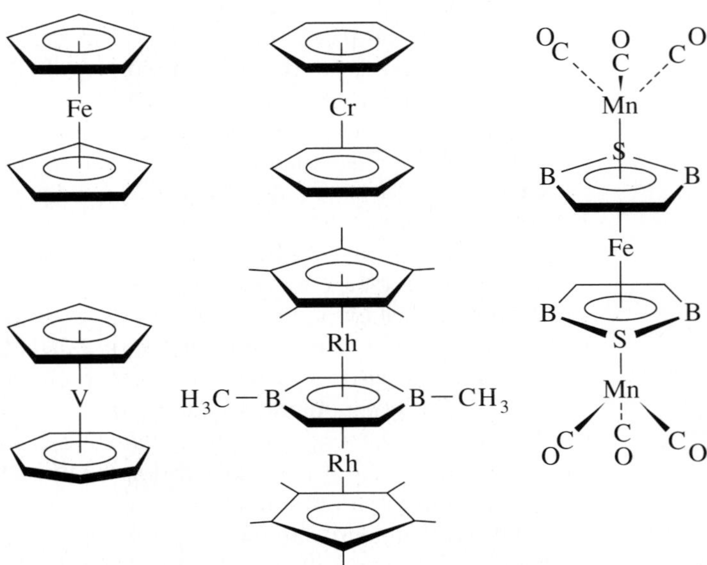

FIGURE 12-1 Examples of Sandwich Compounds.

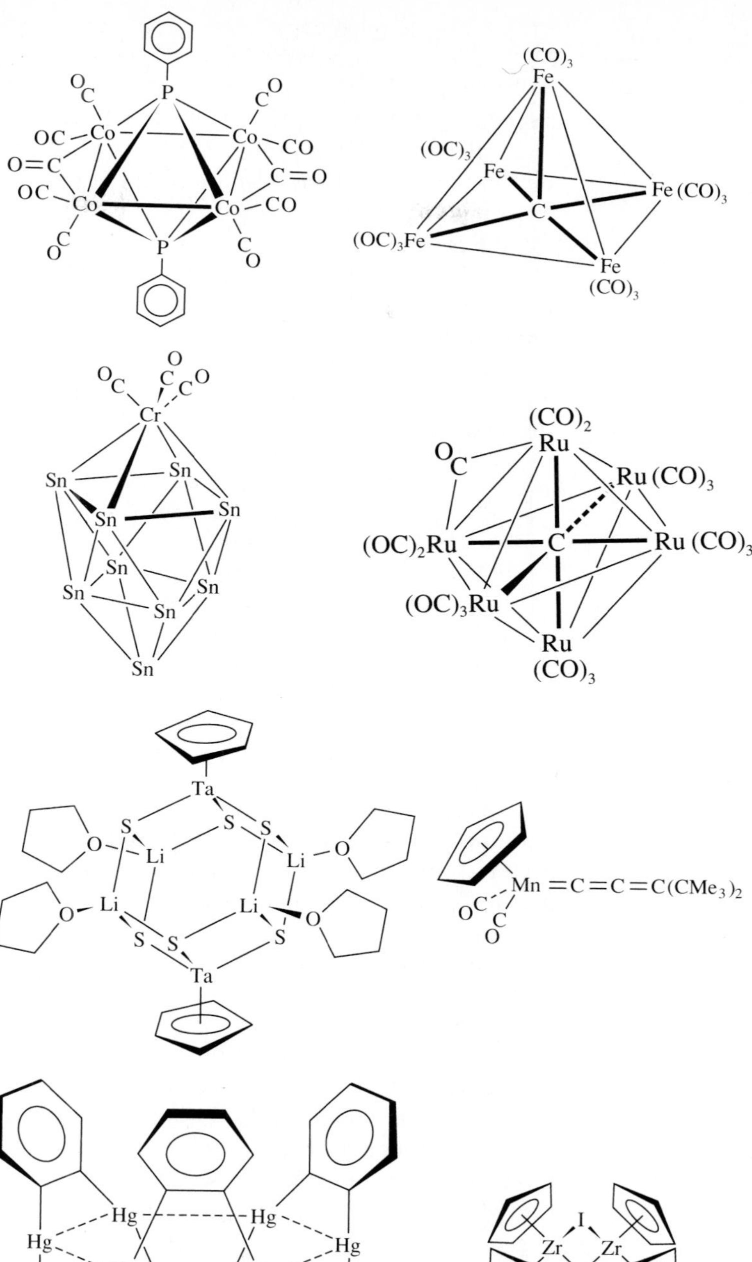

FIGURE 12-2 Examples of Cluster Compounds.

FIGURE 12-3 More Examples of Organometallic Compounds.

CO, but is usually considered a classical, nonorganic ligand.) Other pi-acceptor ligands, such as phosphines, often occur in organometallic complexes, and their chemistry may be studied in association with the chemistry of organic ligands. We will include examples of these and other nonorganic ligands as appropriate in our discussion of organometallic chemistry.

12-1 HISTORICAL BACKGROUND

The first organometallic compound to be reported was synthesized in 1827 by W. C. Zeise, who obtained yellow needlelike crystals after refluxing a mixture of $PtCl_4$ and $PtCl_2$ in ethanol, followed by addition of KCl solution.[1] Zeise correctly asserted that this yellow product (subsequently dubbed *Zeise's salt*) contained an ethylene group. This assertion was questioned by other chemists, most notably J. Liebig, and was not verified conclusively until experiments performed by K. Birnbaum in 1868. However, the structure of the compound proved extremely elusive and was not determined until more than 100 years later![2] Zeise's salt proved to be the first compound containing an organic molecule attached to a metal using the pi electrons of the former. It is an ionic compound of formula $K[Pt(C_2H_4)Cl_3] \cdot H_2O$; the structure of the anion, shown in Figure 12-4, is based on a square plane, with three chloro ligands occupying corners of the square and the ethylene occupying the fourth corner, but perpendicular to the plane.

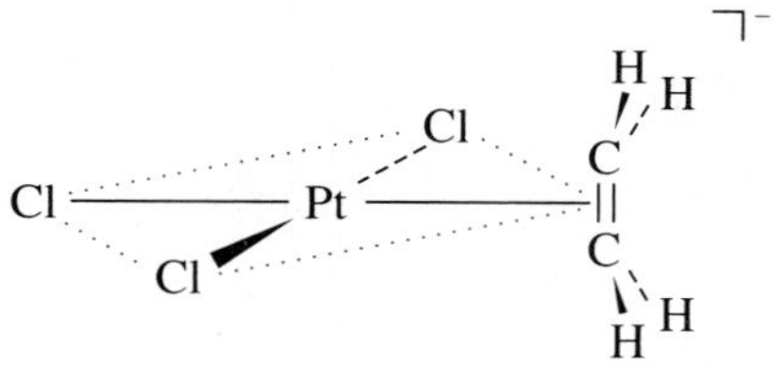

FIGURE 12-4 Anion of Zeise's Compound.

The first compound containing carbon monoxide as a ligand was another platinum chloride complex, reported in 1867. In 1890, Mond reported the preparation of $Ni(CO)_4$, a compound that became commercially useful for the purification of nickel. Other metal CO (carbonyl) complexes were soon obtained.

Reactions between magnesium and alkyl halides performed by Barbier in 1898 and 1899 and subsequently by Grignard led to the synthesis of alkyl magnesium complexes now known as Grignard reagents. These complexes, often complicated in structure, contain magnesium–carbon sigma bonds. Their synthetic utility was recognized early; by 1905 more than 200 research papers had appeared on the topic. Grignard reagents and other reagents containing metal–alkyl sigma bonds (such as organozinc and organocadmium reagents) have proved of immense importance in the development of organic chemistry.

From the discovery of Zeise's salt in 1827 to approximately 1950, organometallic chemistry developed rather slowly. Some organometallic compounds, such as Grignard reagents, found utility in organic synthesis, but there was little study of compounds containing metal–carbon bonds as a distinct research area. In 1951, in an attempt to synthesize fulvalene (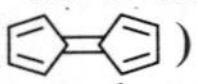) from cyclopentadienyl bromide, Kealy and Pauson reacted the Grignard reagent *cyclo*-C_5H_5MgBr with $FeCl_3$, using anhydrous diethyl ether as solvent.[3] This

[1] W. C. Zeise, *Annal. Physik Chemie,* **1831,** *21,* 497. A translation of excerpts from this paper can be found in *Classics in Coordination Chemistry,* Part 2, G. B. Kauffman, ed., Dover, New York, 1976, pp. 21–37.

[2] R. A. Love, T. F. Koetzle, G. J. B. Williams, L. C. Andrews, and R. Bau, *Inorg. Chem.,* **1975,** *14,* 2653.

[3] T. J. Kealy and P. L. Pauson, *Nature,* **1951,** *168,* 1039.

reaction did not yield the desired fulvalene but rather an orange solid of formula $(C_5H_5)_2Fe$, ferrocene:

$$cyclo\text{-}C_5H_5MgBr + FeCl_3 \longrightarrow (C_5H_5)_2Fe$$

The product was surprisingly stable; it could be sublimed in air without decomposition and was resistant to catalytic hydrogenation and Diels–Alder reactions. In 1956, X-ray diffraction showed the structure to consist of an iron atom sandwiched between two parallel C_5H_5 rings.[4] The details of the structure proved somewhat controversial, with the initial study indicating the rings to be in a staggered conformation (D_{5d} symmetry). Electron diffraction studies of gas-phase ferrocene, however, showed the rings to be eclipsed (D_{5h}), or very nearly so. More recent X-ray diffraction studies of solid ferrocene have identified several crystalline phases, with an eclipsed conformation at 98 K and with conformations having the rings slightly twisted (D_5) in higher-temperature crystalline modifications (Figure 12-5).[5]

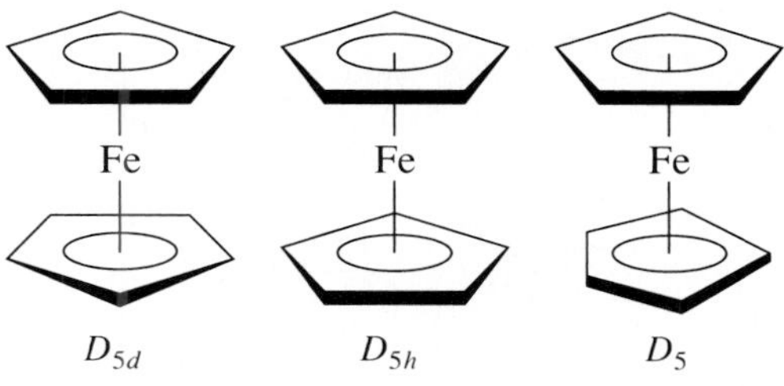

Staggered rings — Eclipsed rings — Skew rings

FIGURE 12-5 Conformations of Ferrocene.

The discovery of the prototype sandwich compound ferrocene rapidly led to the synthesis of other sandwich compounds, of other compounds containing metal atoms bonded to the C_5H_5 ring in a similar fashion, and to a vast array of other compounds containing other organic ligands. It is therefore often stated, and with justification, that the discovery of ferrocene began the era of modern organometallic chemistry, an area that has grown with increasing rapidity in the succeeding decades.

Finally, a discussion of the historical background of organometallic chemistry would be incomplete without mention of what surely qualifies as the oldest known organometallic compound, vitamin B_{12} coenzyme. This naturally occurring cobalt complex, whose structure is partially illustrated in Figure 12-6, contains a cobalt–carbon sigma bond. It is a cofactor in a number of enzymes that catalyze 1, 2 shifts in biochemical systems:

$$\begin{array}{ccc} \begin{array}{cc} R & H \\ | & | \\ -C- & C- \\ | & | \\ H & H \end{array} & \rightleftharpoons & \begin{array}{cc} H & R \\ | & | \\ -C- & C- \\ | & | \\ H & H \end{array} \end{array}$$

The chemistry of vitamin B_{12} is described briefly in Chapter 15.

[4] J. D. Dunitz, L. E. Orgel, and R. A. Rich, *Acta Crystallogr.*, **1956,** *9,* 373.

[5] E. A. V. Ebsworth, D. W. H. Rankin, and S. Cradock, *Structural Methods in Inorganic Chemistry,* Blackwell Scientific Publications, Oxford, England, 1987.

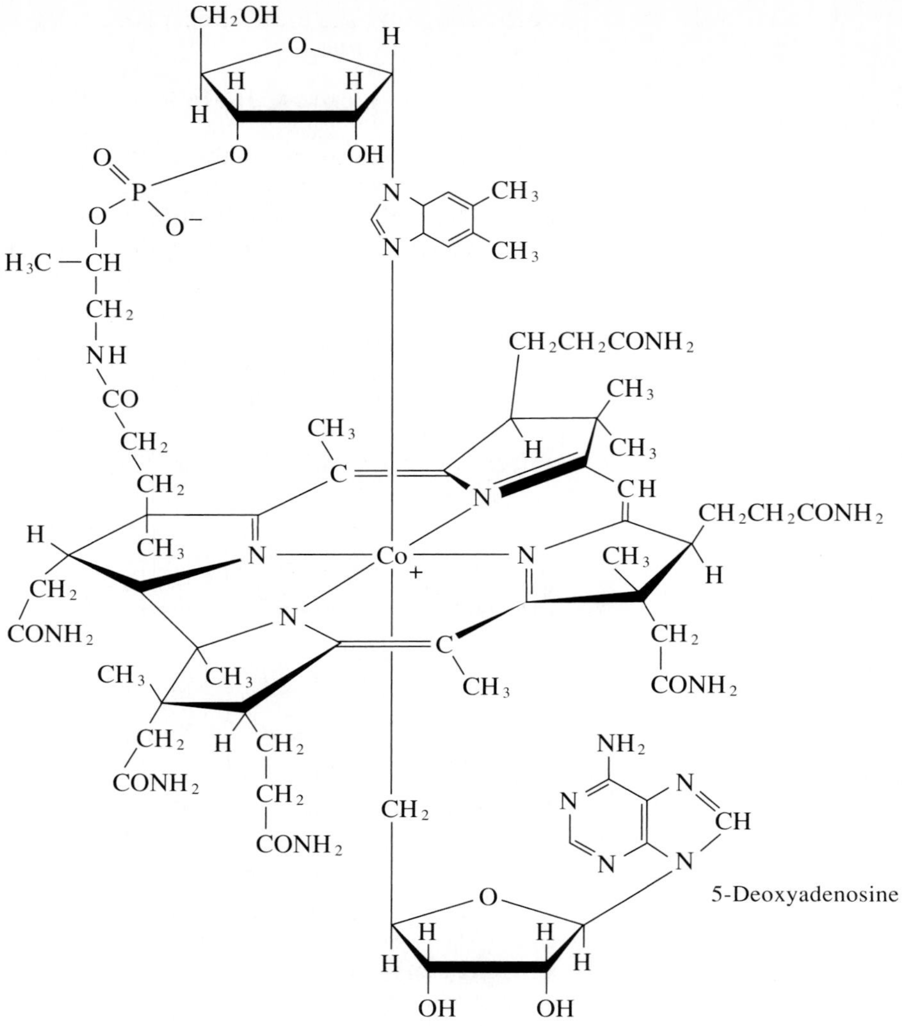

FIGURE 12-6 Vitamin B_{12} Coenzyme.

12-2 ORGANIC LIGANDS AND NOMENCLATURE

A certain amount of background information on organic ligands should prove useful in the discussion of organometallic compounds that will follow. Some of the most common organic ligands are shown in Figure 12-7. Special nomenclature has been devised to designate the manner in which these ligands bond to metal atoms; several of the ligands in Figure 12-7 may bond through different numbers of atoms, depending on the molecule in question. The number of atoms through which a ligand bonds is indicated by the Greek letter η (*eta*) followed by a superscript indicating the number of ligand atoms attached to the metal. For example, the cyclopentadienyl ligands in ferrocene bond through all five atoms; they are designated η^5-C_5H_5. The formula of ferrocene may therefore be written $(\eta^5\text{-}C_5H_5)_2Fe$ (in general we will write hydrocarbon ligands before the metal). In written or spoken form, however, the η^5-C_5H_5 ligand is designated the pentahapto cyclopentadienyl ligand. *Hapto* comes from the Greek word for *fasten*; pentahapto, therefore, means "fastened in five places." C_5H_5, probably the second most frequently encountered ligand in organometallic chemistry (after CO), most commonly bonds to metals

Ligand	*Name*
CO	Carbonyl
=C<	Carbene (alkylidene)
≡C—	Carbyne (alkylidyne)
	Cyclopropenyl (*cyclo*-C_3H_3)
	Cyclobutadiene (*cyclo*-C_4H_4)
	Cyclopentadienyl (*cyclo*-C_5H_5)(Cp)
	Benzene
	1,5-cyclooctadiene (1,5-COD) (1,3-cyclooctadiene complexes are also known)
$H_2C{=}CH_2$	Ethylene
$HC{\equiv}CH$	Acetylene
	π-Allyl (C_3H_5)
$—CR_3$	Alkyl
—C(=O)R	Acyl

FIGURE 12-7 Common Organic Ligands.

through five positions, but under certain circumstances may bond through only one or three positions. The corresponding formulas and names are designated according to this system as follows:

Number of bonding positions	*Formula*	*Name*	
1	η^1-C_5H_5	Monohaptocyclopentadienyl	M—
3	η^3-C_5H_5	Trihaptocyclopentadienyl	M—
5	η^5-C_5H_5	Pentahaptocyclopentadienyl	M—

For ligands having all carbons bonded to a metal, sometimes the superscript may be omitted. Ferrocene may therefore be written $(\eta\text{-}C_5H_5)_2Fe$ and dibenzenechromium $(\eta\text{-}C_6H_6)_2Cr$. Similarly, π with no superscript may occasionally be used to designate that all atoms in the pi system are bonded to the metal.

As in the case of other coordination compounds, bridging ligands, which are very common in organometallic chemistry, are designated by the prefix μ, followed by a subscript indicating the number of metal atoms bridged. Bridging carbonyl ligands, for example, are designated as follows:

Number of atoms bridged	*Formula*
None (terminal)	CO
2	μ_2-CO
3	μ_3-CO

12-3 THE 18-ELECTRON RULE[6]

In main group chemistry we have encountered the octet rule (Chapter 3) in which the electronic structures of many main group compounds could be rationalized on the basis of a valence shell requirement of 8 electrons. Similarly, in organometallic chemistry the electronic structures of many compounds are based on a total valence electron count of 18 on the central metal atom. As in the case of the octet rule, there are many exceptions to the 18-electron rule, but the rule nevertheless provides some useful guidelines to the chemistry of many organometallic complexes, especially those containing strong pi-acceptor ligands. In this section we will first examine how electrons are counted according to this rule and then consider the basis for its usefulness (and some of the reasons why it is not always useful).

12-3-1 COUNTING ELECTRONS

Several schemes exist for counting electrons in organometallic compounds. We will describe two of these using several examples. First, we give two examples that illustrate the special stability of 18-electron species:

EXAMPLES

$Cr(CO)_6$

A Cr atom has 6 electrons outside its noble gas core. Each CO is considered to act as a donor of 2 electrons (by both methods). The total electron count is therefore

Cr		6 electrons
6 (CO)	6 × 2 electrons	= 12 electrons
	Total =	18 electrons

$Cr(CO)_6$ is therefore considered an 18-electron complex. It is thermally stable; for example, it can be sublimed without decomposition. $Cr(CO)_5$, a 16-electron species, and $Cr(CO)_7$, a 20-electron species, are, on the other hand, of far lower stability and known only as transient species. Likewise, the 17-electron $[Cr(CO)_6]^+$ and 19-electron $[Cr(CO)_6]^-$ are of far lower stability than the neutral, 18-electron $Cr(CO)_6$. The bonding in $Cr(CO)_6$, which provides a rationale for the special stability of many 18-electron systems, is discussed in Section 12-3-2.

[6] Often called the effective atomic number (EAN) rule.

$(\eta^5\text{-}C_5H_5)Fe(CO)_2Cl$

Electrons in this complex may be counted in two ways:

Method A (Donor Pair Method)

This method considers ligands to donate electron pairs to the metal. To determine the total electron count, we must take into account the charge on each ligand and determine the formal oxidation state of the metal.

Pentahapto-C_5H_5 is considered by this method as $C_5H_5^-$, a donor of 3 electron pairs; it is a 6-electron donor. As usual, CO is counted as a 2-electron donor. Chloride is considered Cl^-, a donor of 2 electrons. Therefore, $(\eta^5\text{-}C_5H_5)Fe(CO)_2Cl$ is formally an iron(II) complex. Iron(II) has 6 electrons beyond its noble gas core. The electron count is therefore

Fe(II)	6 electrons
$\eta^5\text{-}C_5H_5^-$	6 electrons
2 (CO)	4 electrons
Cl^-	2 electrons
Total =	18 electrons

Method B (Neutral Ligand Method)

This method uses the number of electrons that would be donated by ligands if they were neutral. For simple inorganic ligands, this usually means that ligands are considered to donate the number of electrons equal to their negative charge as free ions (Cl is a 1-electron donor if singly bonded to a metal, O a 2-electron donor if doubly bonded, N a 3-electron donor if triply bonded, and so on).

To determine the total electron count by this method, we do not need to determine the oxidation state of the metal. For $(\eta^5\text{-}C_5H_5)Fe(CO)_2Cl$: An iron *atom* has 8 electrons beyond its noble gas core. $\eta^5\text{-}C_5H_5$ is now considered as if it were a neutral ligand (or radical), in which case it would contribute 5 electrons. CO is a 2-electron donor and Cl (counted as if it were a neutral species) a 1-electron donor. The electron count is

Fe atom	8 electrons
$\eta^5\text{-}C_5H_5$	5 electrons
2 (CO)	4 electrons
Cl	1 electron
Total =	18 electrons

Both methods give the same result: $(\eta^5\text{-}C_5H_5)Fe(CO)_2Cl$ is an 18-electron species.

Many organometallic complexes are charged species, and this charge must be included in determining the total electron count. The reader may wish to verify (by either method of electron counting) that $[Mn(CO)_6]^+$ and $[(\eta^5\text{-}C_5H_5)Fe(CO)_2]^-$ are both 18-electron ions. In addition, metal–metal bonds count as a single electron per metal. For example, in the dimeric complex $(CO)_5Mn\text{–}Mn(CO)_5$, the electron count per manganese atom is (by both methods)

Mn	7 electrons
5 (CO)	10 electrons
Mn–Mn bond	1 electron
Total =	18 electrons

The electron counts for common ligands according to both schemes are given in Table 12-1.

TABLE 12-1
Electron counting schemes for common ligands

Ligand	*Method A*	*Method B*
H	2 $(:H^-)$	1
Cl, Br, I	2 $(:\ddot{X}:^-)$	1
OH	2 $(:\ddot{O}:H^-)$	1
CN	2 $(:C{\equiv}N:^-)$	1
CH_3	2 $(:CH_3^-)$	1
NO (bent M—N—O)	2 $(:\dot{N}{=}\ddot{O}:^-)$	1
NO (linear M—N—O)	2 $(:N{\equiv}O:^+)$	3
CO, PR_3	2	2
NH_3, H_2O	2	2
=CRR′ (carbene)	2	2
$H_2C{=}CH_2$	2	2
=O, =S	4 $(:\ddot{O}:^{2-}, :\ddot{S}:^{2-})$	2
η^3-C_3H_5	2 $(C_3H_5^+)$	3
≡CR (carbyne)	3	3
≡N	6 (N^{3-})	3
Butadiene	4	4
η^5-C_5H_5	6 $(C_5H_5^-)$	5
η^6-C_6H_6	6	6
η^7-C_7H_7	6 $(C_7H_7^+)$	7

The electron counting method of choice is a matter of individual preference. Method A has the advantage of including the formal oxidation state of the metal, but may tend to overemphasize the ionic nature of some metal–ligand bonds. Method B is often quicker to use, especially for ligands having extended pi systems; for example, η^5 ligands have an electron count of 5, η^3 ligands an electron count of 3, and so on. Also, method B may be simpler in not requiring that the oxidation state of the metal be assigned. Other electron counting schemes have also been developed. It is generally best to select one method and to use it consistently.

Electron counting (by any method) does not imply anything about the degree of covalent or ionic bonding; it is strictly a bookkeeping procedure, as are the oxidation numbers that may be used in the counting. Physical measurements are necessary to provide evidence about the actual electron distribution in molecules.

In ligands such as CO that can interact with metal atoms in several ways, the number of electrons counted is usually based on sigma donation. For example, although CO is a pi acceptor and (weak) pi donor, its electron donating count of 2 is based on its sigma donor ability alone. However, the pi acceptor and donor abilities of ligands have significant effects on the degree to which the 18-electron rule is likely to be obeyed. Linear and cyclic organic pi systems interact with metals in more complicated ways, to be discussed later in this chapter.

EXAMPLES

Both methods of electron counting are illustrated for the following three complexes.

	Method A		*Method B*	
$ClMn(CO)_5$	Mn(I)	$6e^-$	Mn	$7e^-$
	Cl^-	$2e^-$	Cl	$1e^-$
	5 CO	$10e^-$	5 CO	$10e^-$
		$18e^-$		$18e^-$
$(\eta^5\text{-}C_5H_5)_2Fe$ (ferrocene)	Fe(II)	$6e^-$	Fe	$8e^-$
	2 $\eta^5\text{-}C_5H_5^-$	$12e^-$	2 $\eta^5\text{-}C_5H_5$	$10e^-$
		$18e^-$		$18e^-$
$[Re(CO)_5(PF_3)]^+$	Re(I)	$6e^-$	Re	$7e^-$
	5 CO	$10e^-$	5 CO	$10e^-$
	PF_3	$2e^-$	PF_3	$2e^-$
	+Charge	[a]	+Charge	$-1e^-$
		$18e^-$		$18e^-$

[a] Charge on ion is accounted for in assignment of oxidation state to Re.

EXERCISE 12-1

Determine the valence electron counts for the following:

a. $[Fe(CO)_4]^{2-}$
b. $[(\eta^5\text{-}C_5H_5)_2Co]^+$
c. $(\eta^3\text{-}C_5H_5)(\eta^5\text{-}C_5H_5)Fe(CO)$

EXERCISE 12-2

Identify the first-row transition metal for the following 18-electron species:

a. $[M(CO)_3(PPh_3)]^-$
b. $HM(CO)_5$
c. $(\eta^4\text{-}C_8H_8)M(CO)_3$
d. $[(\eta^5\text{-}C_5H_5)M(CO)_3]_2$ (assume a single M–M bond)

12-3-2 WHY 18 ELECTRONS?

An oversimplified rationale for the special significance of 18 electrons can be made by analogy with the octet rule in main group chemistry. If the octet represents a complete valence electron shell configuration (s^2p^6), then the number 18 can be considered to correspond to a filled valence shell for a transition metal ($s^2p^6d^{10}$). This analogy, while perhaps a useful way to relate the electron configurations to the idea of valence shells of electrons for atoms, does not provide an explanation for why so many complexes violate this rule. In particular, the valence shell rationale does not distinguish between types of ligands (sigma donors, pi acceptors, and pi donors); this distinction is an important consideration in determining which complexes obey and which violate the rule.

A good example of a complex that adheres to the 18-electron rule is $Cr(CO)_6$. The molecular orbitals of interest in this molecule are those that result primarily from interactions between the d orbitals of Cr and the σ-donor (HOMO) and π-acceptor orbitals (LUMO) of the six CO ligands. The molecular orbitals corresponding to these interactions are shown in Figure 12-8.

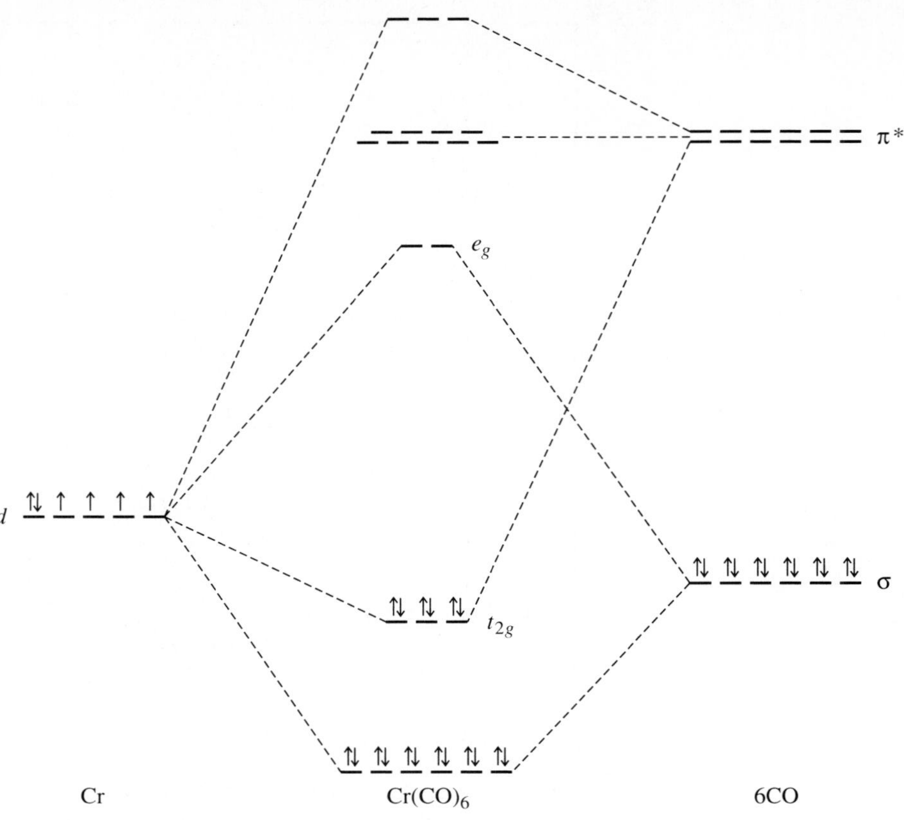

FIGURE 12-8 Molecular Orbitals of $Cr(CO)_6$.

* Only interactions between ligand orbitals and metal *d* orbitals are shown. A more complete molecular orbital diagram of octahedral ML_6, including symmetry labels of the orbitals, is given in Figure 8-6.

Chromium(0) has 6 electrons outside its noble gas core. Each CO contributes a pair of electrons to give a total electron count of 18. In the molecular orbital diagram, these 18 electrons appear as the 12 σ electrons (the sigma electrons of the CO ligands, stabilized by their interaction with the metal orbitals) and the 6 t_{2g} electrons. Addition of one or more electrons to $Cr(CO)_6$ would populate the e_g orbitals, which are antibonding; the consequence would be destabilization of the molecule. Removal of electrons from $Cr(CO)_6$ would depopulate the t_{2g} orbitals, which are slightly bonding in nature as a consequence of the strong π-acceptor ability of the CO ligands; a decrease in electron density in these somewhat bonding orbitals would also tend to destabilize the complex. The result is that the 18-electron configuration for this molecule is the most stable.

Continuing for the moment to consider 6-coordinate molecules of octahedral geometry, we can gain some insight as to when the 18-electron rule can be expected to be most valid. $Cr(CO)_6$ obeys the rule because of two factors: the strong σ-donor ability of CO raises the e_g orbitals in energy, making them considerably antibonding; and the strong π-acceptor ability of CO lowers the t_{2g} orbitals in energy, making them bonding. Ligands that are both strong sigma donors and pi acceptors should therefore be the most effective at forcing adherence to the 18-electron rule. Other ligands, including some organic ligands, do not have these features and consequently their compounds may or may not adhere to the rule.

Exceptions may be noted. $Zn(en)_3^{2+}$ is a 22-electron species; it has both the t_{2g} and e_g orbitals filled. While en (ethylenediamine) is a good σ donor, it is not as strong a donor as CO. As a result, the e_g electrons are not sufficiently antibonding to cause significant destabilization of the complex, and the 22-electron species, with 4 electrons in e_g orbitals, is stable. An example of a 12-electron species is TiF_6^{2-}. In this case the fluoride ligand is a π *donor* as well as a σ donor. The π-donor ability of F^- destabilizes the t_{2g} orbitals of the complex, making them slightly antibonding. The species TiF_6^{2-} has 12 electrons in the bonding σ orbitals and no electrons in the antibonding t_{2g} or e_g orbitals. These examples of exceptions to the 18-electron rule are shown schematically in Figures 12-9[7] and 8-12.

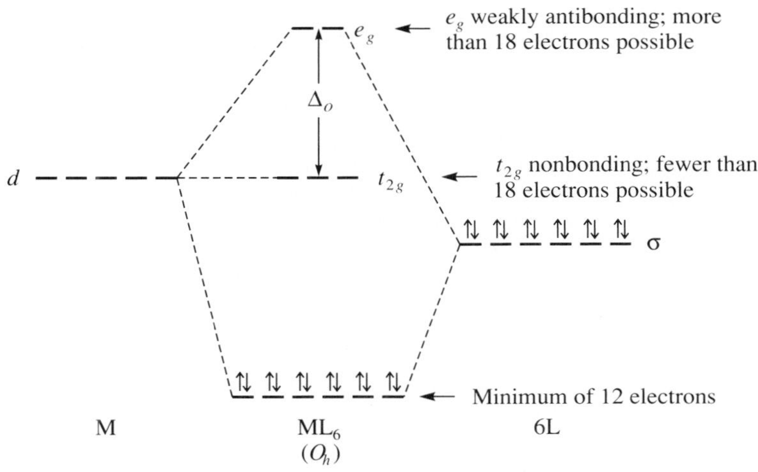

FIGURE 12-9 Exceptions to the 18-Electron Rule.

The same type of argument can be made for complexes of geometry other than octahedral; in most, but not all, cases, there is an 18-electron configuration of special stability for complexes of strongly π-accepting ligands. Examples include trigonal bipyramidal geometry (for example, $Fe(CO)_5$) and tetrahedral geometry (for example, $Ni(CO)_4$). The most common exception is square planar geometry, in which a 16-electron configuration may be the most stable, especially for complexes of d^8 metals.

12-3-3 SQUARE PLANAR COMPLEXES

Square planar complexes are extremely important. Examples include the d^8, 16-electron complexes shown in Figure 12-10. To understand why 16-electron square planar complexes might be especially stable, it is useful to examine the molecular orbitals of such a complex. An example of such a molecular orbital diagram for a square planar molecule of formula ML_4 (L = ligand that can function as both sigma donor and pi acceptor) is shown in Figure 12-11.

Four molecular orbitals of ML_4 are derived primarily from the σ-donor orbitals of the ligands; electrons occupying such orbitals would be bonding in nature. Three additional orbitals are slightly bonding (derived primarily from d_{xz}, d_{yz}, and d_{xy} orbitals of the metal) and one is essentially nonbonding (derived primarily from the d_{z^2} orbital of the metal). These bonding and nonbonding

[7] P. R. Mitchell and R. V. Parish, *J. Chem. Educ.*, **1969**, *46*, 811.

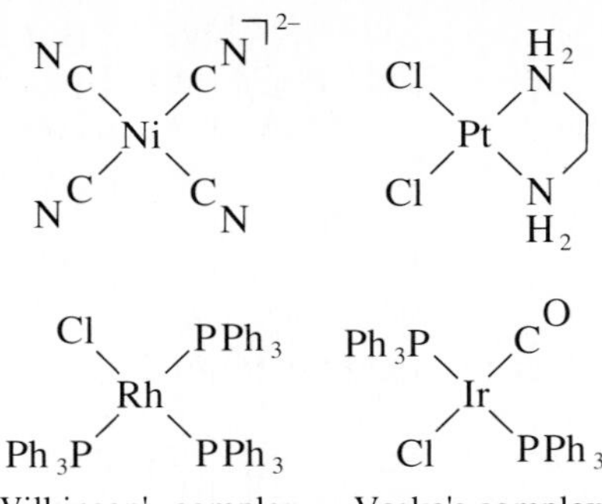

FIGURE 12-10 Examples of Square Planar d^8 Complexes.

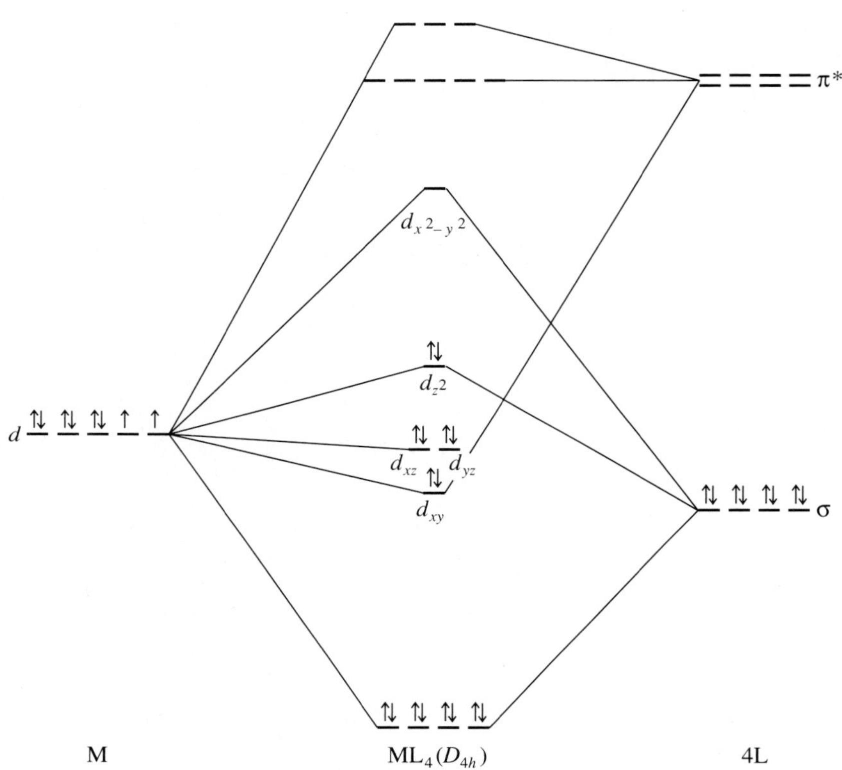

FIGURE 12-11 Molecular Orbitals for Square Planar Complex.

orbitals can be filled by 16 electrons. Additional electrons would occupy an antibonding orbital derived from the antibonding interaction of metal $d_{x^2-y^2}$ orbital with the sigma-donor orbitals of the ligands (the $d_{x^2-y^2}$ orbital points directly toward the ligands; its antibonding interaction is therefore the strongest). Consequently, for square planar complexes of ligands having both sigma-donor and pi-acceptor characteristics, a 16-electron configuration is more stable than an 18-electron configuration. Sixteen-electron square planar complexes may also be capable of accepting one or two ligands at the vacant coordination sites (along the z axis) and thereby achieving an 18-electron configuration. As will be shown in the next chapter, this is a common reaction of 16-electron square planar complexes.

EXERCISE 12-3

Verify that the complexes in Figure 12-10 are 16-electron species.

Sixteen-electron square planar species are most commonly encountered for d^8 metals, in particular for metals having formal oxidation states of 2+ (Ni^{2+}, Pd^{2+}, and Pt^{2+}) and 1+ (Rh^+, Ir^+). Such species may have important catalytic behavior, as discussed in the next chapter. Two examples of square planar d^8 complexes that are used as catalysts are Wilkinson's complex and Vaska's complex, shown in Figure 12-10; their chemistry is described in Chapter 13.

12-4 LIGANDS IN ORGANOMETALLIC CHEMISTRY

Hundreds of ligands are known to bond to metal atoms through carbon. Carbon monoxide forms a very large number of metal complexes and deserves special mention, along with several similar diatomic ligands. Many organic molecules containing linear or cyclic pi systems also form numerous organometallic compounds. Compounds containing such ligands will be discussed next, following a brief review of the pi systems in the ligands themselves. Finally, special attention will be paid to two types of organometallic compounds of recent interest: carbene complexes, containing metal–carbon double bonds, and carbyne complexes, containing metal–carbon triple bonds.

12-4-1 CARBONYL (CO) COMPLEXES

Carbon monoxide is the most common ligand in organometallic chemistry. It may serve as the only ligand in **binary carbonyls** such as $Ni(CO)_4$, $W(CO)_6$, and $Fe_2(CO)_9$ or more commonly, in combination with other ligands, both organic and inorganic. CO may bond to a single metal or it may serve as a bridge between two or more metals. In this section we will consider the bonding between metals and CO, the synthesis and some reactions of CO complexes, and examples of the various types of CO complexes formed.

Bonding

It is useful to review the bonding in CO. As described in Chapter 5 (Figure 5-14), the molecular orbital picture of CO is similar to that of N_2 (Figure 5-5); the shapes of the molecular orbitals derived primarily from the $2p$ atomic orbitals of these molecules are shown in Figure 12-12.

Two features of the molecular orbitals of CO deserve attention. First, the highest-energy occupied orbital (the HOMO) has its largest lobe on carbon. It is through this orbital, occupied by an electron pair, that CO exerts its σ-donor function, donating electron density directly toward an appropriate metal orbital (such as an unfilled p, d, or hybrid orbital). At the same time, CO has two empty π^* orbitals (the lowest unoccupied, or LUMO); these also have larger lobes on carbon than on oxygen. As a consequence of this localization of π^* orbitals on carbon, the carbon acts as the principal site of the π-acceptor function of the ligand; a metal atom having electrons in a d orbital of suitable symmetry can donate electron density to these π^* orbitals. These σ-donor and π-acceptor interactions are illustrated in Figure 12-13.

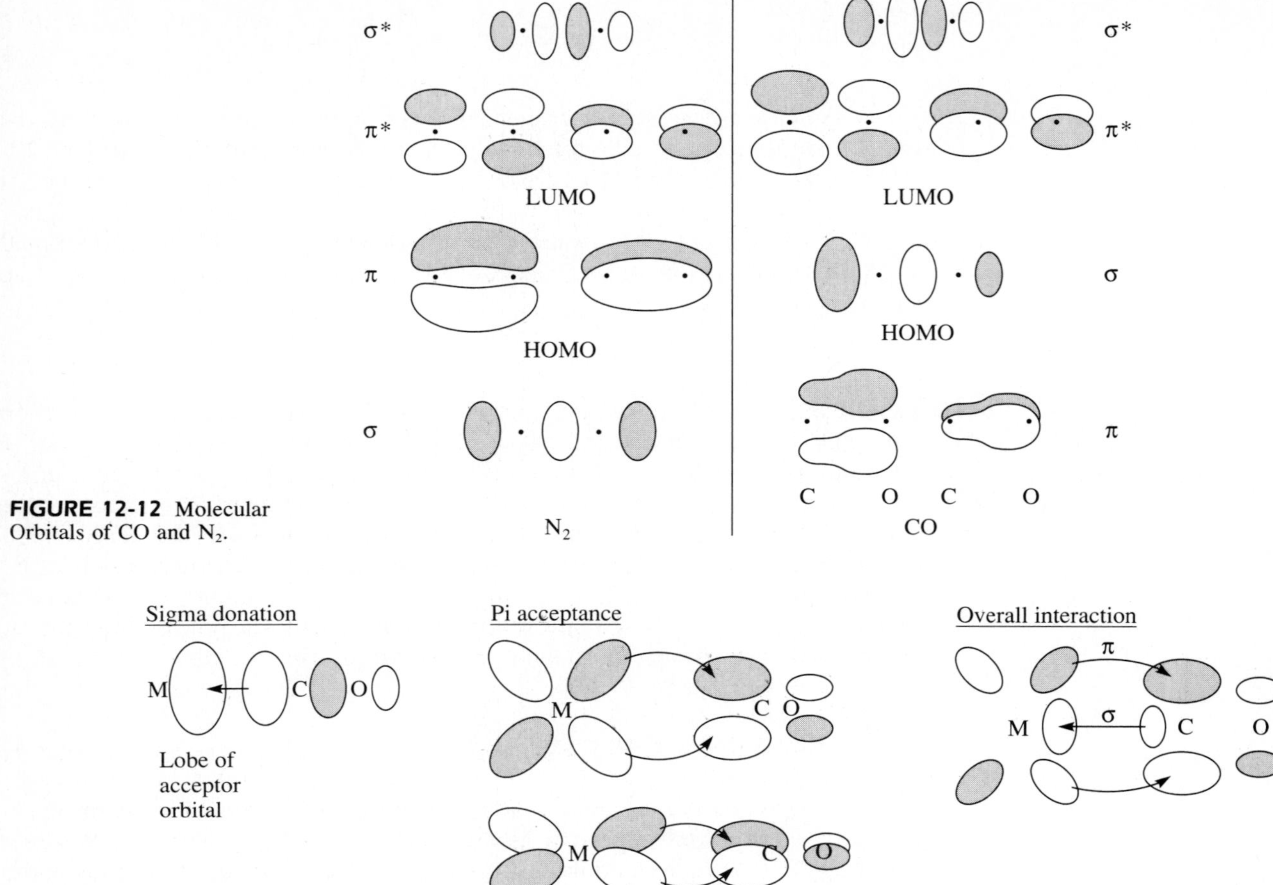

FIGURE 12-12 Molecular Orbitals of CO and N_2.

FIGURE 12-13 Sigma and Pi Interactions between CO and Metal Atom.

The overall effect is synergistic; CO can donate electron density via a σ orbital to a metal atom. The greater the electron density on the metal, the more effectively it is able to return electron density to the π* orbitals of CO. The net effect can be rather strong bonding between the metal and CO; however, as will be described later, the strength of this bonding is dependent on several factors, including the charge on the complex and the ligand environment of the metal.

EXERCISE 12-4

N_2 has molecular orbitals rather similar to those of CO, as shown in Figure 12-12. Would you expect N_2 to be a stronger or weaker pi acceptor than CO?

If this picture of bonding between CO and metal atoms is correct, it should be supported by experimental evidence. Two sources of such evidence are infrared spectroscopy and X-ray crystallography. First, any change in the bonding between carbon and oxygen should be reflected in the C—O stretching vibration as observed by IR. As in organic compounds, the C—O stretch in organometallic compounds is often very intense (stretching the C—O bond results in a substantial change in dipole moment), and its energy often provides valuable information about the molecular structure. Free carbon monoxide has a C—O stretch at 2143 cm^{-1}. $Cr(CO)_6$, on the other hand, has its C—O stretch at 2000 cm^{-1}. The lower energy for the stretching mode means that the C—O bond is weaker in $Cr(CO)_6$.

The energy necessary to stretch a bond is proportional to $\sqrt{\frac{k}{\mu}}$, where k = force constant and μ = reduced mass. For atoms of mass m_1 and m_2, the reduced mass is given by

$$\mu = \frac{m_1 m_2}{m_1 + m_2}$$

The stronger the bond between two atoms, the larger the force constant; consequently, the greater the energy necessary to stretch the bond and the higher the energy of the corresponding band (the higher the wave number, cm^{-1}) in the infrared spectrum. Similarly, the more massive the atoms involved in the bond, as reflected in a higher reduced mass, the less energy necessary to stretch the bond and the lower the energy of the absorption in the infrared spectrum.

Both σ donation (which donates electron density from a bonding orbital on CO) and π acceptance (which places electron density in C—O *anti*bonding orbitals) would be expected to weaken the C—O bond and to decrease the energy necessary to stretch that bond.

Additional evidence is provided by X-ray crystallography. In carbon monoxide the C—O distance has been measured at 112.8 pm. Weakening of the C—O bond by the factors described above would be expected to cause this distance to increase. Such an increase in bond length is found in complexes containing CO, with C—O distances approximately 115 pm for many carbonyls. While such measurements provide definitive measures of bond distances, in practice it is far more convenient to use infrared spectra to obtain data on the strength of C—O bonds.

The charge on a carbonyl complex is also reflected in its infrared spectrum. Three isoelectronic hexacarbonyls have the following C—O stretching bands[8] [compare with $\nu(CO)$ = 2143 cm^{-1} for free CO]:

Complex	*ν(CO), cm^{-1}*
$[V(CO)_6]^-$	1858
$Cr(CO)_6$	2000
$[Mn(CO)_6]^+$	2095

Of these three, $[V(CO)_6]^-$ has the metal with smallest nuclear charge; this means that vanadium has the weakest ability to attract electrons and the greatest tendency to "back" donate electron density to CO. The consequence is strong population of the π^* orbitals of CO and reduction of the strength of the C—O bond. In general, the more negative the charge on organometallic species, the greater the tendency of the metal to donate electrons to the π^* orbitals of CO and the lower the energy of the C—O stretching vibrations.

[8] K. Nakamoto, *Infrared and Raman Spectra of Inorganic and Coordination Compounds*, 4th ed., Wiley, New York, 1986, pp. 292–93.

Bridging modes of CO

Although CO is most commonly found as a terminal ligand attached to a single metal atom, many cases are known in which CO forms bridges between two or more metals. Many such bridging modes are known; the most common are shown in Table 12-2.

TABLE 12-2
Bridging modes of CO

Type of CO	*Approximate range for ν(CO) in neutral complexes (cm^{-1})*
Free CO	2143
Terminal M—CO	1850–2120
Symmetrical* μ_2-CO	
M–C(O)–M (C bridging two M)	1700–1860
Symmetrical* μ_3-CO	
M$_3$C–O (C bridging three M)	1600–1700

NOTE: * Asymmetrically bridging μ_2- and μ_3-CO are also known.

The bridging mode is strongly correlated with the position of the C—O stretching band. In cases where CO bridges two metal atoms, both metals can contribute electron density into π^* orbitals of CO to weaken the C—O bond and lower the energy of the stretch. Consequently, the C—O stretch for doubly bridging CO is at much lower energy than for terminal COs. An example is shown in Figure 12-14. Interaction of three metal atoms with a triply bridging CO further weakens the C—O bond; the infrared band for the C—O stretch is still lower than in the doubly bridging case. Ordinarily, terminal and bridging COs can be considered 2-electron donors, with the two donated electrons shared by the metals in the bridging cases.

A particularly interesting situation is that of nearly linear bridging carbonyls, such as in $[(\eta^5\text{-}C_5H_5)Mo(CO)_2]_2$. When a sample of $[(\eta^5\text{-}C_5H_5)Mo(CO)_3]_2$ is heated, some carbon monoxide is driven off; the product, $[(\eta^5\text{-}C_5H_5)Mo(CO)_2]_2$, reacts readily with CO to reverse this reaction:[9]

$$\underset{1960,\ 1915\ \text{cm}^{-1}}{[(\eta^5\text{-}C_5H_5)Mo(CO)_3]_2} \overset{\Delta}{\rightleftharpoons} \underset{1889,\ 1859\ \text{cm}^{-1}}{[(\eta^5\text{-}C_5H_5)Mo(CO)_2]_2} + 2\,CO$$

This reaction is accompanied by changes in the infrared spectrum in the CO region, as listed above. The Mo–Mo bonds also shorten by approximately 100

[9] D. S. Ginley and M. S. Wrighton, *J. Am. Chem. Soc.*, **1975,** *97,* 3533; R. J. Klingler, W. Butler, and M. D. Curtis, *J. Am. Chem. Soc.*, **1975,** *97,* 3535.

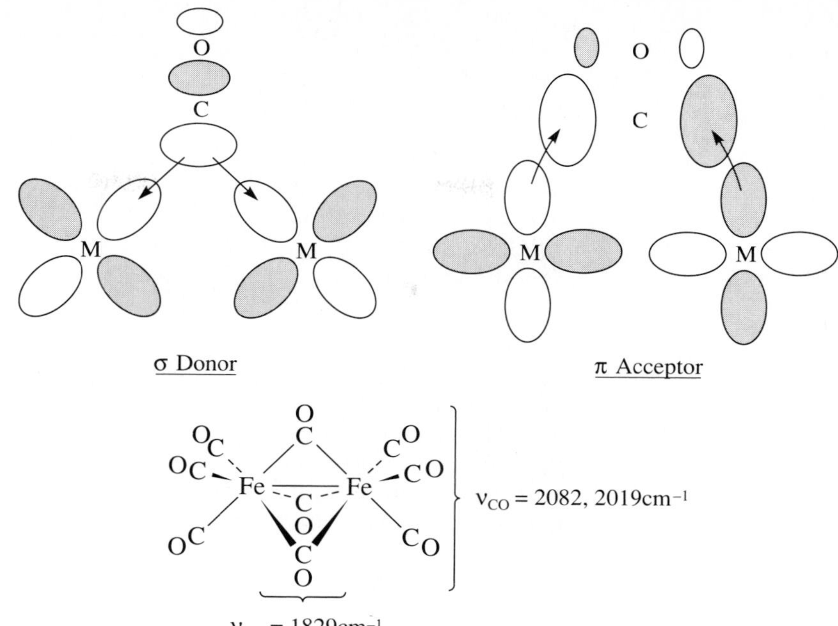

FIGURE 12-14 Bridging CO.

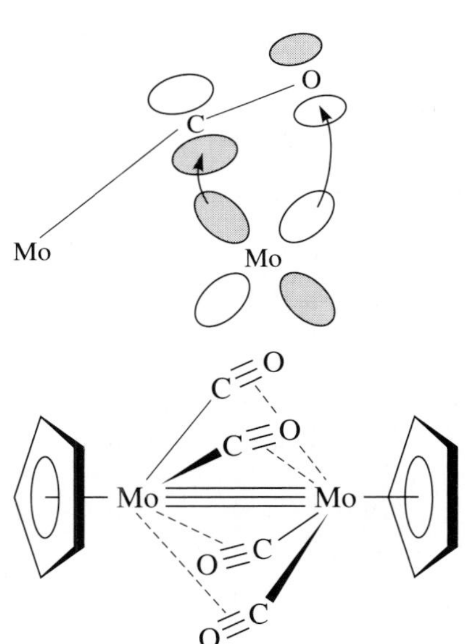

FIGURE 12-15 Bridging CO in $[(\eta^5\text{-}C_5H_5)Mo(CO)_2]_2$.

pm, consistent with an increase in the metal–metal bond order from 1 to 3. Although it was originally proposed that the "linear" CO ligands may donate some electron density to the neighboring metal from pi orbitals, subsequent calculations have indicated that a more important interaction is that of donation from a metal *d* orbital to the π^* orbital of CO, as shown in Figure 12-15.[10] Such donation weakens the carbon–oxygen bond in the ligand and results in the observed shift of the C—O stretching bands to lower energies.

[10] A. L. Sargent and M. B. Hall, *J. Am. Chem. Soc.*, **1989,** *111,* 1563 and references therein.

Additional information on infrared spectra of carbonyl complexes is included in Section 12-7 at the end of this chapter.

Binary carbonyl complexes

Binary carbonyls, containing only metal atoms and CO, are fairly numerous. Some representative binary carbonyl complexes are shown in Figure 12-16. Most of these complexes obey the 18-electron rule. The cluster compounds $Co_6(CO)_{16}$ and $Rh_6(CO)_{16}$ do not obey the rule, however; more detailed analysis of the bonding in cluster compounds is necessary to satisfactorily account for the electron counting in these and other cluster compounds. This question will be taken up in Chapter 14.

One other binary carbonyl does not obey the rule: the 17-electron $V(CO)_6$. This complex is one of a few cases in which strong pi-acceptor ligands do not succeed in requiring an 18-electron configuration. In $V(CO)_6$ the vanadium is apparently too small to permit a seventh coordination site; hence, no metal–metal bonded dimer (which would give an 18-electron configuration) is possible. However, $V(CO)_6$ is easily reduced to $[V(CO)_6]^-$, a well-studied 18-electron complex.

Mononuclear $[M(CO)_x]$

M = Ni, Pd

M = Fe, Ru, Os

M = V, Cr, Mo, W

Binuclear $[M_2(CO)_x]$

$Co_2(CO)_8$ (solution)

$Co_2(CO)_8$ (solid)

$Fe_2(CO)_9$

$M_2(CO)_{10}$
M = Mn, Tc, Re

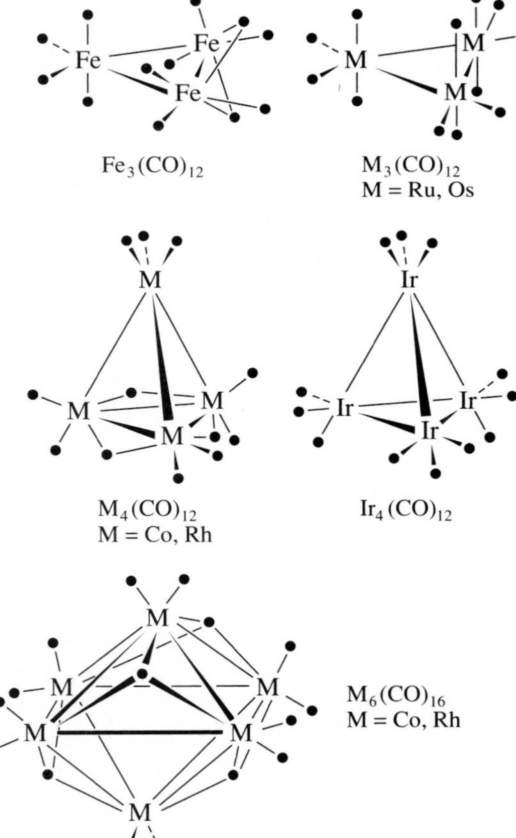

FIGURE 12-16 Binary Carbonyl Complexes.

EXERCISE 12-5

Verify the 18-electron rule for five of the binary carbonyls [other than $Co_6(CO)_{16}$ and $Rh_6(CO)_{16}$] shown in Figure 12-16.

An interesting feature of the structures of binary carbonyl complexes is that the tendency of CO to bridge transition metals tends to decrease in going down the periodic table. For example, in $Fe_2(CO)_9$ there are three bridging carbonyls, but in $Ru_2(CO)_9$ and $Os_2(CO)_9$ there is a single bridging CO. A possible explanation is that the orbitals of bridging CO are less able to interact effectively with transition metal atoms as the size of the metals increases.

Binary carbonyl complexes can be synthesized in many ways. Several of the most common methods are as follows:

1. Direct reaction of a transition metal with CO. The most facile of these reactions involves nickel, which reacts with CO at ambient temperature and 1 atm:

$$Ni + 4\,CO \longrightarrow Ni(CO)_4$$

$Ni(CO)_4$ is a volatile, extremely toxic liquid that must be handled with great caution. It was first observed by Mond, who found that CO reacted with nickel valves. The reverse reaction, involving thermal decomposition of $Ni(CO)_4$, can be used to prepare nickel of very high purity. Coupling of the forward and reverse reactions has been used commercially in the Mond process for obtaining purified nickel from ores. Other binary carbonyls can be obtained from direct reaction of metal powders with CO, but elevated temperatures and pressures are necessary.

2. Reductive carbonylation: reduction of a metal compound in the presence of CO and an appropriate reducing agent. Examples are

$$CrCl_3 + 6\,CO + Al \longrightarrow Cr(CO)_6 + AlCl_3$$

$$Re_2O_7 + 17\,CO \longrightarrow Re_2(CO)_{10} + 7\,CO_2$$

(CO acts as reducing agent; high temperature and pressure are required.)

3. Thermal or photochemical reaction of other binary carbonyls. Examples are

$$Fe(CO)_5 \xrightarrow{h\nu} Fe_2(CO)_9$$

$$Fe(CO)_5 \xrightarrow{\Delta} Fe_3(CO)_{12}$$

The most common reaction of carbonyl complexes is CO dissociation. This reaction, which may be initiated thermally or by absorption of ultraviolet light, characteristically involves loss of CO from an 18-electron complex to give a 16-electron intermediate, which may react in a variety of ways, depending on the nature of the complex and its environment. A common reaction is replacement of the lost CO by another ligand to form a new 18-electron species as product. For example,

$$Cr(CO)_6 + PPh_3 \xrightarrow[\text{or } h\nu]{\Delta} Cr(CO)_5(PPh_3) + CO$$

$$Re(CO)_5Br + en \xrightarrow{\Delta} fac\text{-}Re(CO)_3(en)Br + 2\,CO$$

This type of reaction therefore provides a pathway in which CO complexes can be used as precursors for a variety of complexes of other ligands. Additional aspects of CO dissociation reactions will be discussed in Chapter 13.

Oxygen-bonded carbonyls

This section would not be complete without mentioning one additional aspect of CO as a ligand: it can sometimes bond through oxygen as well as carbon. This phenomenon was first noted in the ability of the oxygen of a metal carbonyl complex to act as a donor toward Lewis acids such as $AlCl_3$, with the overall function of CO serving as a bridge between the two metals. Numerous examples are now known in which CO bonds through its oxygen to transition metal atoms, with the C—O—metal arrangement generally bent. Attachment of a Lewis acid to the oxygen results in significant weakening and lengthening of the C—O bond and a corresponding shift of the C—O stretching vibration to lower energy in the infrared. This shift to lower energy is typically between 100 and 200 cm^{-1}. Examples of O-bonded carbonyls (sometimes called isocarbonyls) are shown in Figure 12-17. The physical and chemical properties of oxygen-bonded carbonyls have been reviewed.[11]

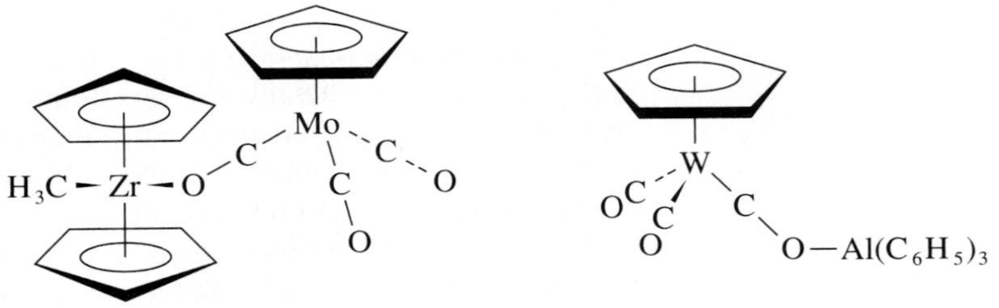

FIGURE 12-17 Oxygen-bonded Carbonyls. (a) B. Longato, B. D. Martin, J. R. Norton, and O. P. Anderson, *Inorg. Chem.*, **1985,** *24,* 1389. (b) J. M. Burlitch, M. E. Leonowicz, R. B. Petersen, and R. E. Hughes, *Inorg. Chem.*, **1979,** *18,* 1097.

$H_3C-Zr-O-C-Mo(CO)_2$ (a)

$W(CO)_2-C-O-Al(C_6H_5)_3$ (b)

12-4-2 LIGANDS SIMILAR TO CO

Several diatomic ligands similar to CO are worth brief mention. Two of these, CS (thiocarbonyl) and CSe (selenocarbonyl), are of interest in part for purposes of comparison with CO. In most cases, synthesis of CS and CSe complexes is somewhat more difficult than for analogous CO complexes, since CS and CSe do not exist as stable, free molecules and do not provide a ready ligand source.[12] Therefore, the comparatively small number of such complexes should not be viewed as an indication of their stability; the chemistry of CS and CSe complexes may eventually rival that of CO complexes in breadth and utility. Thiocarbonyl complexes are also of interest as possible intermediates in certain sulfur-transfer reactions in the removal of sulfur from natural fuels. In recent years the chemistry of complexes containing these ligands has developed more rapidly as avenues for their synthesis have been devised.

CS and CSe are similar to CO in their bonding modes; they behave as both σ donors and π acceptors and can bond to metals in terminal or bridging modes. Of these two ligands, CS has been studied more closely. It usually functions as a stronger σ donor and π acceptor than CO.[12,13]

[11] C. P. Horwitz and D. F. Shriver, *Adv. Organomet. Chem.*, **1984,** *23,* 219.

[12] E. K. Moltzen, K. J. Klabunde, and A. Senning, *Chem. Rev.*, **1988,** *88,* 391, provides a detailed review of CS chemistry.

[13] P. V. Broadhurst, *Polyhedron,* **1985,** *4,* 1801.

EXERCISE 12-6

Explain why the LUMO of CS might be expected to function as a slightly better pi acceptor than the LUMO of CO. (HINT: Consider the shape and energy of the LUMOs.)

Several other common ligands are isoelectronic with CO and, not surprisingly, exhibit structural and chemical parallels with CO. Two examples are CN^- and N_2. Cyanide is a stronger σ donor and a somewhat weaker π acceptor than CO; overall it is close to CO in the spectrochemical series. Unlike most organic ligands, which bond to metals in low formal oxidation states, cyanide bonds readily to metals having higher oxidation states. As a good σ donor, CN^- interacts strongly with positively charged metal ions; as a weaker π acceptor than CO (largely a consequence of the negative charge of CN^-), cyanide is not as able to stabilize metals in low oxidation states. Therefore, its compounds are often studied in the context of classical coordination chemistry rather than organometallic chemistry. Dinitrogen is a weaker donor and acceptor. However, N_2 complexes are of great interest, especially as possible intermediates in reactions that may simulate natural processes of nitrogen fixation.[14]

NO complexes

Although not an organic ligand, the NO (nitrosyl) ligand deserves discussion here because of its similarities to CO. Like CO, it is both a sigma donor and pi acceptor and can serve as a terminal or bridging ligand; useful information can be obtained about its compounds by analysis of its infrared spectra. Unlike CO, however, terminal NO has two common coordination modes, linear (like CO) and bent.

A formal analogy is often drawn between the linear bonding modes of both ligands; NO^+ is isoelectronic with CO; therefore, in its bonding to metals, linear NO is considered by electron counting scheme A as NO^+, a 2-electron donor. By the neutral ligand method (B), linear NO is counted a 3-electron donor (it has one more electron than the 2-electron donor CO).

The bent coordination mode of NO is often considered to arise formally from NO^-, with the bent geometry suggesting sp^2 hybridization at the nitrogen. By electron counting scheme A, therefore, bent NO is considered the 2-electron donor NO^-; by the neutral ligand model, it is considered a 1-electron donor. Useful information about the linear and bent bonding modes of NO is summarized in Figure 12-18. Numerous complexes containing each mode are known, and examples are also known in which both linear and bent NO occur in the same complex. While linear coordination usually gives rise to N—O stretching vibrations at higher energy than the bent mode, there is enough overlap in the ranges of these bands that infrared spectra alone may not be sufficient to distinguish between the two. Furthermore, the manner of packing in crystals may give rise to considerable bending of the metal—N—O bond from 180° in the linear coordination mode.

One compound containing only a metal and NO ligands is known, $Cr(NO)_4$, a tetrahedral molecule that is isoelectronic with $Ni(CO)_4$. Complexes containing bridging nitrosyl ligands are also known, with the bridging ligand generally considered formally a 3-electron donor.

[14] R. A. Henderson, G. J. Leigh, and C. J. Pickett, *Adv. Inorg. Chem. Radiochem.*, **1983,** *27,* 197.

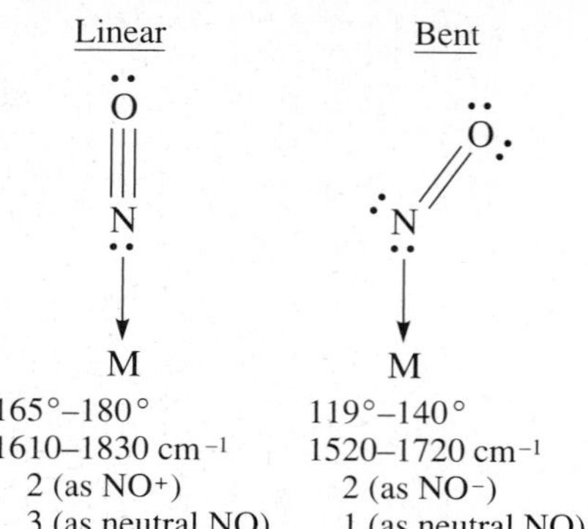

	Linear	Bent
M—N—O angle	165°–180°	119°–140°
ν (N-O) in neutral molecules	1610–1830 cm^{-1}	1520–1720 cm^{-1}
Electron donor count	2 (as NO^+) 3 (as neutral NO)	2 (as NO^-) 1 (as neutral NO)

FIGURE 12-18 Linear and Bent NO.

In recent years, several dozen compounds containing the isoelectronic NS (thionitrosyl) ligand have been synthesized. Infrared data have indicated that, like NO, NS can function in linear, bent, and bridging modes. In general, NS is similar to NO in its ability to act as a pi-acceptor ligand; the relative abilities of NO and NS to accept pi electrons depend on the electronic environment of the compounds being compared.[15]

12-4-3 DIHYDROGEN COMPLEXES[16]

Although complexes containing H_2 molecules coordinated to transition metals had been proposed for many years, and many complexes containing hydride ligands had been prepared, the first structural characterization of a dihydrogen complex did not occur until 1984.[17] Subsequently many H_2 complexes have been identified, and the chemistry of this ligand has developed rapidly.

The bonding between dihydrogen and a transition metal can be described as shown in Figure 12-19. The σ electrons in H_2 can be donated to a suitable empty orbital on the metal (such as a *d* or hybrid orbital), while the empty σ^* orbital of the ligand can accept electron density from occupied *d* orbitals of the metal. The result is an overall weakening and lengthening of the H—H bond in comparison with free H_2.

Dihydrogen complexes have frequently been suggested as possible intermediates in a variety of reactions of hydrogen at metal centers. Some of these reactions are steps in catalytic processes of significant commercial interest (see Chapter 13). As this ligand becomes more completely understood, the applications of its chemistry are likely to be extremely important.

12-4-4 LIGANDS HAVING EXTENDED PI SYSTEMS

While it is a relatively simple matter to describe pictorially how ligands such as CO and PPh_3 bond to metal atoms, it is a somewhat more involved process to explain bonding between metals and organic ligands having extended pi

[15] H. W. Roesky and K. K. Pandey, *Adv. Inorg. Chem. Radiochem.*, **1983,** *26,* 337.

[16] G. J. Kubas, *Comments Inorg. Chem.*, **1988,** *7,* 17; R. H. Crabtree, *Acc. Chem. Res.*, **1990,** *23,* 95; G. J. Kubas, *Acc. Chem. Res.*, **1988,** *21,* 120.

[17] G. J. Kubas, R. R. Ryan, B. I. Swanson, P. J. Vergamini, and H. J. Wasserman, *J. Am. Chem. Soc.*, **1984,** *106,* 451.

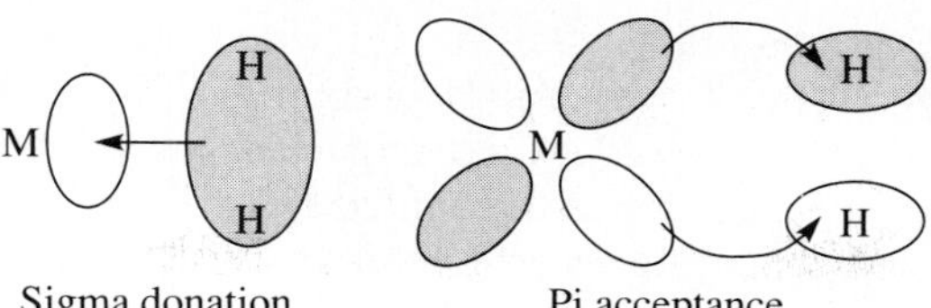

FIGURE 12-19 Bonding in H_2 Complexes.

systems. How, for example, are the C_5H_5 rings attached to Fe in ferrocene, and how can 1,3-butadiene bond to metals? To understand the bonding between metals and pi systems, it is necessary to consider first the pi bonding within the ligands themselves. In the following discussion we will first describe linear and then cyclic pi systems, after which we will consider the question of how molecules containing such systems can bond to metals.

Linear pi systems

The simplest case of an organic molecule having a linear pi system is ethylene, which has a single π bond resulting from the interactions of two $2p$ orbitals on its carbon atoms. Interactions of these p orbitals result in one bonding and one antibonding π orbital, as shown:

$H_2C{=}CH_2$ *p orbitals interacting*   *Relative energy*

π^*

$\uparrow\downarrow$ π

The antibonding orbital has a nodal plane perpendicular to the internuclear axis, while the bonding orbital has no such nodal plane.

Next is the three-atom pi system, the π-allyl radical, C_3H_5. In this case there are three $2p$ orbitals to be considered, one from each of the carbon atoms participating in the pi system. The possible interactions are as follows:

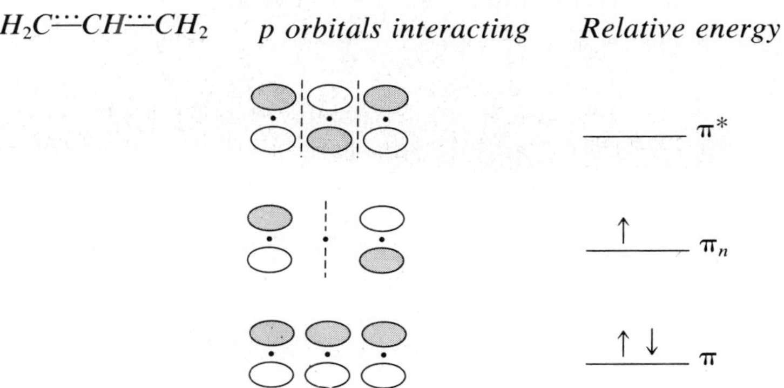

The lowest-energy π molecular orbital for this system has all three p orbitals interacting constructively, to give a bonding molecular orbital. Higher in energy is the nonbonding situation, in which a nodal plane bisects the molecule, cutting through the central carbon atom. In this case the *p* orbital on the central carbon does not participate in the molecular orbital (in general, nodal planes passing through the center of *p* orbitals and perpendicular to internuclear axes will cancel these orbitals from participation in the pi molecular orbitals). Highest in energy is the antibonding π^* orbital, in which there is an antibonding interaction between each neighboring pair of carbon *p* orbitals.

There is an increase in the number of nodes in going from lower- to higher-energy orbitals; for example, in the π-allyl system the number of nodes increases from zero to one to two from the lowest- to the highest-energy orbital.[18] This is a trend that will also appear in the examples that follow.

One additional example should suffice to illustrate this procedure. 1,3-Butadiene may exist in *cis* or *trans* forms. For our purposes it will be sufficient to treat both as linear systems; the nodal behavior of the molecular orbitals will be the same in each case as in a linear pi system of four atoms. As for ethylene and π-allyl, the 2*p* orbitals of the carbon atoms in the chain may interact in a variety of ways, with the lowest-energy π molecular orbital having all constructive interactions between neighboring *p* orbitals, and the energy of the other π orbitals increasing with the number of nodes between the atoms.

$H_2C{=}CH{-}CH{=}CH_2$ *p orbitals interacting* *Relative energy*

↑↓

↑↓

Similar patterns can be obtained for longer pi systems; two more examples are included in Figure 12-20.

Cyclic pi systems

The procedure for obtaining a pictorial representation of the orbitals of cyclic pi systems of hydrocarbons is similar to the procedure for the linear systems described above. The smallest such cyclic hydrocarbon is *cyclo*-C_3H_3. The lowest-energy π molecular orbital for this system is the one resulting from constructive interaction between each of the 2*p* orbitals in the ring:

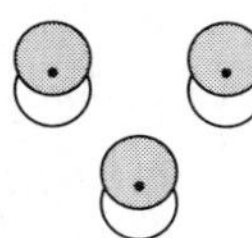

[18] The nodes are in addition to the nodal plane that is coplanar with the carbon chain, bisecting each *p* orbital participating in the pi system.

	p Orbitals interacting	Relative energy		p Orbitals interacting	Relative energy
C_5H_7			C_6H_8		

FIGURE 12-20 Pi Orbitals for Linear Systems.

Two additional π molecular orbitals are needed (since the number of molecular orbitals must equal the number of atomic orbitals used). Each of these has a single nodal plane that is perpendicular to the plane of the molecule and bisects the molecule; the nodes for these two molecular orbitals are perpendicular to each other:

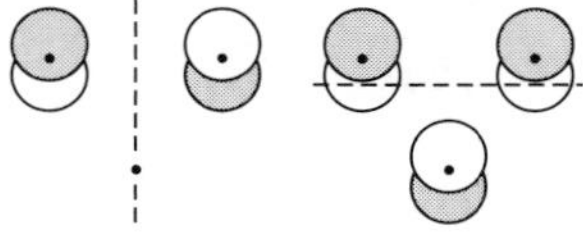

(Nodes are assigned in a procedure similar to the method used for the group orbitals in ammonia, as described in Chapter 5.) These molecular orbitals are of the same energy; in general, π molecular orbitals having the same number of nodes in cyclic pi systems of hydrocarbons are degenerate (have the same energy). The total π molecular orbital diagram for *cyclo*-C_3H_3 can therefore be summarized as follows:

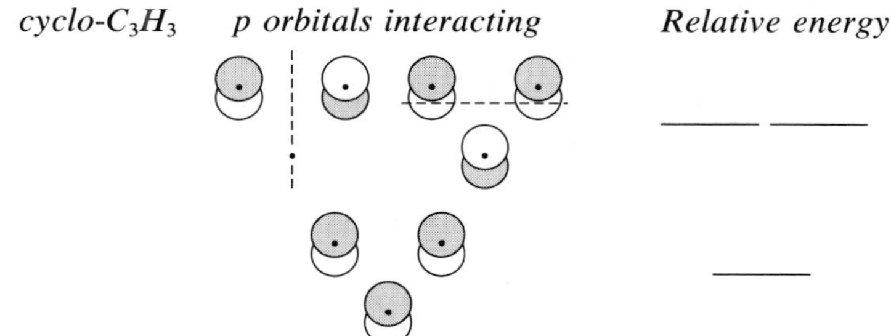

A simple way to determine the *p* orbital interactions and the relative energies of the cyclic pi systems having regular polyhedral geometry is to draw the polyhedron with one vertex pointed down; each vertex then corresponds to the relative energy of a molecular orbital. Furthermore, the number of nodal planes bisecting the molecule (and perpendicular to the plane of the molecule) increases as one goes to higher energy, with the bottom orbital having zero nodes, the next pair of orbitals a single node, and so on. For example, the next cyclic pi system, *cyclo*-C_4H_4 (cyclobutadiene), would be predicted by this scheme to have molecular orbitals as follows:[19]

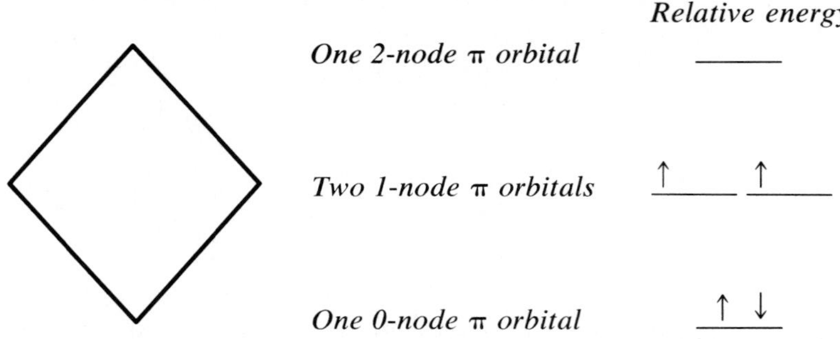

Similar results are obtained for other cyclic pi systems; two of these are shown in Figure 12-21. In these diagrams, nodal planes are disposed symmetrically. For example, in *cyclo*-C_4H_4 the single-node molecular orbitals bisect the molecule through opposite sides; the nodal planes are oriented at 90° angles to each other. The 2-node orbital for this molecule also has the nodal planes at 90° angles.

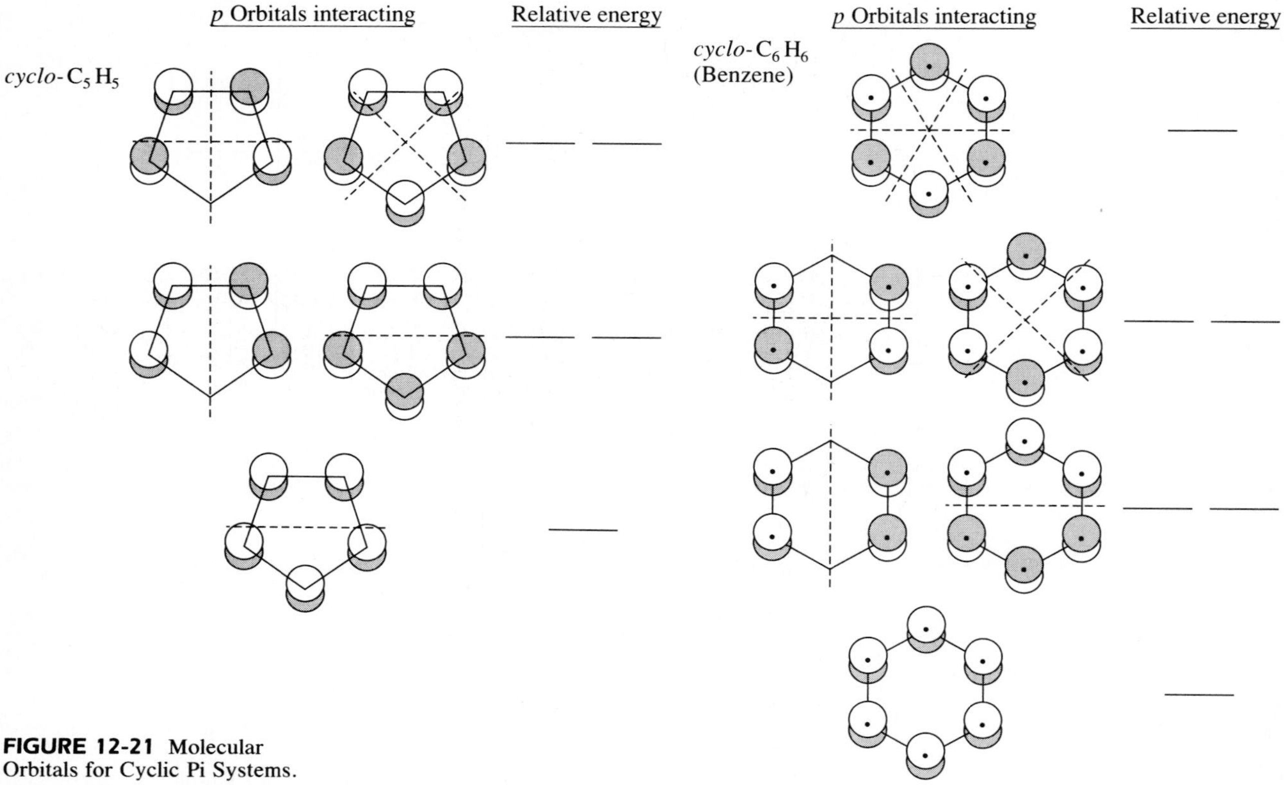

p Orbitals interacting | Relative energy

cyclo-C_5H_5

p Orbitals interacting | Relative energy

cyclo-C_6H_6 (Benzene)

FIGURE 12-21 Molecular Orbitals for Cyclic Pi Systems.

[19] This approach would predict a diradical for cyclobutadiene (one electron in each 1-node orbital). While cyclobutadiene itself is very reactive (P. Reeves, T. Devon, and R. Pettit, *J. Am. Chem. Soc.*, **1969,** *91,* 5890), complexes containing derivatives of this ligand are known.

This method may seem oversimplified, but the nodal behavior and relative energies are the same as obtained from molecular orbital calculations. The method for obtaining equations for the molecular orbitals and the equations themselves for cyclic hydrocarbons of formula C_nH_n (n = 3 to 8) are given by Cotton.[20]

Throughout this discussion we have shown not the actual shapes of the π molecular orbitals, but rather the p orbitals used. The nodal behavior of both sets (the π orbitals and the p orbitals used) is identical and therefore sufficient for the discussion of bonding with metals that follows. We have chosen for the sake of simplicity to use sketches of the p orbitals. Diagrams of many molecular orbitals for linear and cyclic pi systems can be found in the reference cited in footnote 21.

12-5 BONDING BETWEEN METAL ATOMS AND ORGANIC PI SYSTEMS

We are now ready to consider metal–ligand interactions involving such systems. We will begin with the simplest of the linear systems, ethylene, and conclude with the classic example of ferrocene.

12-5-1 LINEAR PI SYSTEMS

π-Ethylene complexes

Many complexes involve ethylene, C_2H_4, as a ligand, including the anion of Zeise's salt, $[Pt(\eta^2\text{-}C_2H_4)Cl_3]^-$, one of the earliest organometallic compounds. In such complexes, ethylene most commonly acts as a *sidebound* ligand with the following geometry with respect to the metal:

H H
C
Pt ‖
C
H H

Ethylene donates electron density to the metal in a sigma fashion, using its π bonding electron pair, as shown in Figure 12-22. At the same time, electron density can be donated back to the ligand in a pi fashion from a metal d orbital to the empty π^* orbital of the ligand. This is another example of the synergistic effect of sigma donation and pi acceptance encountered earlier with the CO ligand.

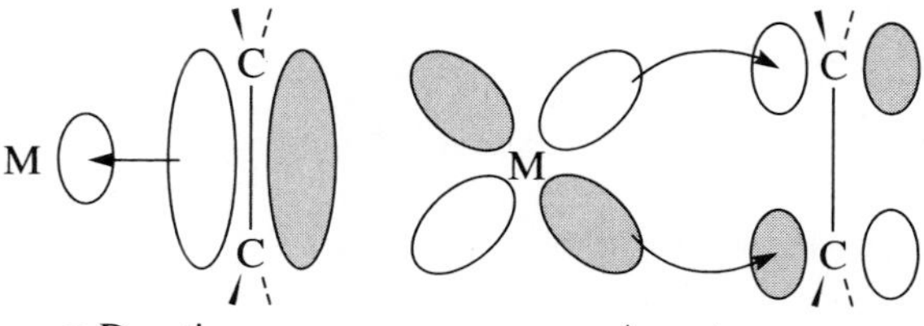

FIGURE 12-22 Bonding in Ethylene Complexes.

[20] F. A. Cotton, *Chemical Applications of Group Theory,* 3rd ed., Wiley-Interscience, New York, 1990, pp. 142–59.

[21] W. L. Jorgensen and L. Salem, *The Organic Chemist's Book of Orbitals,* Academic Press, New York, 1973.

If this picture of bonding in ethylene complexes is correct, it should be in agreement with the measured C—C distance. Free ethylene has a C—C distance of 133.7 pm, while the C—C distance in Zeise's salt is 137.5 pm. The lengthening of this bond can be explained by a combination of the two factors involved in the synergistic sigma donor–pi acceptor nature of the ligand: donation of electron density to the metal in a sigma fashion reduces the pi electron density within the ligand, hence weakening the C—C bond. Furthermore, the back-donation of electron density from the metal to the π^* orbital of the ligand also reduces the C—C bond strength by populating the antibonding orbital. The net effect weakens and hence lengthens the C—C bond in the C_2H_4 ligand.

π-Allyl complexes

The allyl group most commonly functions as a trihapto ligand, using delocalized pi orbitals as described previously, or as a monohapto ligand, primarily sigma bonded to a metal. Examples of these types of coordination are shown in Figure 12-23. Bonding between η^3-C_3H_5 and a metal atom is shown schematically in Figure 12-24. The lowest-energy pi orbital can donate electron density in a σ fashion to a suitable orbital on the metal. The next orbital, nonbonding in free allyl, can act as a donor or acceptor, depending on the electron distribution between the metal and the ligand. The highest-energy pi orbital acts as an acceptor; thus there can be synergistic sigma and pi interactions between allyl and the metal. In π-allyl complexes the plane of the ligand carbon atoms is often tilted somewhat with respect to the metal, with the consequence that the central carbon is slightly closer to the metal; this occurs to allow efficient overlap between metal and ligand orbitals. The C—C—C angle of the ligand is generally near 120°, consistent with sp^2 hybridization.

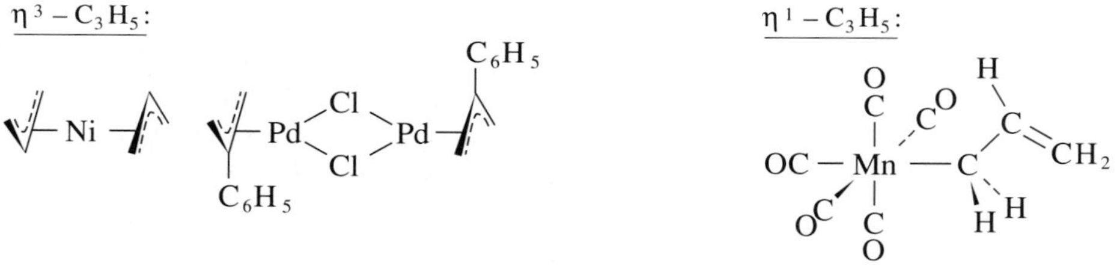

$\eta^3-C_3H_5$:

$\eta^1-C_3H_5$:

FIGURE 12-23 Examples of Allyl Complexes.

Allyl complexes (or complexes of substituted allyls) are intermediates in many reactions, some of which take advantage of the capacity of this ligand to function in both a η^3 and η^1 fashion. Loss of CO from carbonyl complexes containing η^1-allyl ligands often results in conversion of η^1- to η^3-allyl. For example,

$$[Mn(CO)_5]^- + C_3H_5Cl \longrightarrow (\eta^1\text{-}C_3H_5)Mn(CO)_5 + Cl^- \xrightarrow{\Delta \text{ or } h\nu} (\eta^3\text{-}C_3H_5)Mn(CO)_4 + CO$$

(All manganese-containing species in this sequence of reactions are 18-electron species.)

Other linear pi systems

Many other such systems are known; several examples of organic ligands having longer pi systems are shown in Figure 12-25. Butadiene and longer

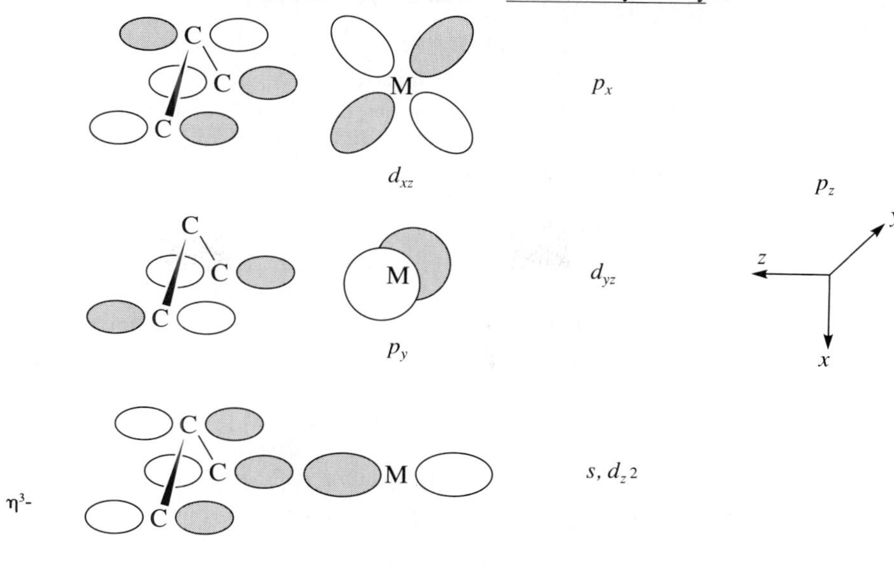

FIGURE 12-24 Bonding in η^3-Allyl Complexes.

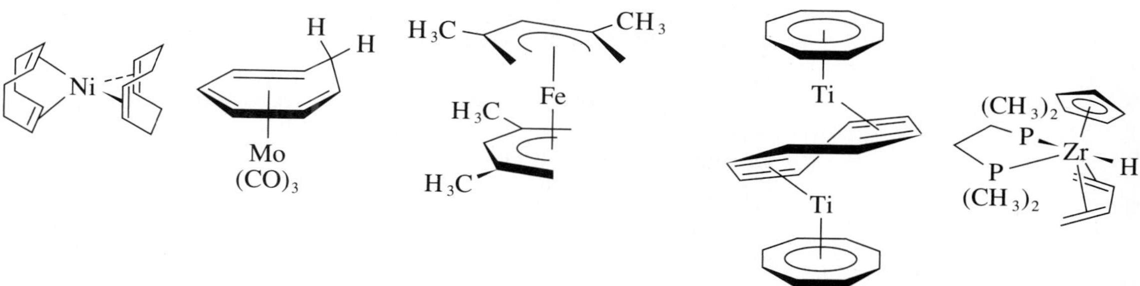

FIGURE 12-25 Examples of Molecules Containing Linear Pi Systems.

conjugated pi systems have the possibility of isomeric ligand forms (*cis* and *trans* for butadiene). Larger cyclic ligands may have a pi system extending through part of the ring. An example is cyclooctadiene (COD); the 1,3- isomer has a four-atom pi system comparable to butadiene; 1,5-cyclooctadiene has two isolated double bonds, one or both of which may interact with a metal in a manner similar to ethylene.

12-5-2 CYCLIC PI SYSTEMS

Cyclopentadienyl (Cp) complexes

The cyclopentadienyl group, C_5H_5, may bond to metals in a variety of ways, with many examples known of the η^1-, η^3-, and η^5- bonding modes. As described earlier in this chapter, the discovery of the first cyclopentadienyl complex, ferrocene, was a landmark in the development of organometallic chemistry and stimulated the search for other compounds containing pi-bonded organic ligands. Numerous substituted cyclopentadienyl ligands are also known, such as $C_5(CH_3)_5$ (often abbreviated Cp*) and $C_5(benzyl)_5$.

Ferrocene, $(\eta^5\text{-}C_5H_5)_2Fe$. Ferrocene is the prototype of a series of sandwich compounds, the **metallocenes,** with the formula $(C_5H_5)_2M$. The bonding in ferrocene can be viewed in two ways. One possibility is to consider it an iron(II) complex with two cyclopentadienide ($C_5H_5^-$) ions; another is to view it as iron(0) coordinated by two neutral C_5H_5 ligands. The actual bonding situation in ferrocene is, of course, much more complicated and requires an analysis of the various metal–ligand interactions in this molecule. As usual, we expect orbitals on the central Fe and on the two C_5H_5 rings to interact if they are of appropriate symmetry; furthermore, we expect interactions to be strongest if they are between orbitals of similar energy.

For the purpose of our analysis of this molecule, it will be useful to refer to Figure 12-21 for sketches of the molecular orbitals of a C_5H_5 ring; two of these rings will be arranged in a parallel fashion in ferrocene to "sandwich in" the metal atom. Our discussion will be based on the eclipsed D_{5h} conformation of ferrocene, the conformation consistent with gas-phase and low temperature data on this molecule.[22,23] The same approach could be taken using the staggered conformation and would yield a similar molecular orbital picture. Descriptions of the bonding in ferrocene based on D_{5d} symmetry are common in the chemical literature, since this was once believed to be the molecule's most stable conformation. The $C_5(CH_3)_5$ and $C_5(benzyl)_5$ analogues of ferrocene do have staggered, D_{5d} symmetry, as do several other metallocenes.[24]

In developing the group orbitals for a pair of C_5H_5 rings, we pair up molecular orbitals of the same energy and same number of nodes; for example, we pair the zero-node orbital of one ring with the zero-node orbital of the other.[25] We also must pair up the molecular orbitals in such a way that *the nodal planes are coincident*. Furthermore, in each pairing there are two possible orientations of the ring molecular orbitals: one in which lobes of like sign are pointed toward each other, and one in which lobes of opposite sign are pointed toward each other. For example, the zero-node orbitals of the C_5H_5 rings may be paired in the following two ways:

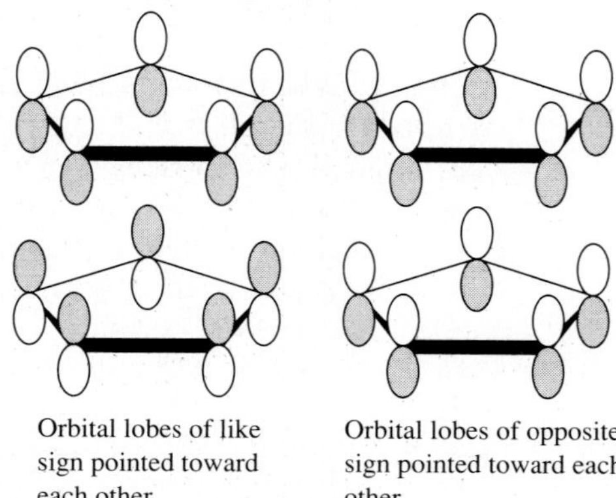

Orbital lobes of like sign pointed toward each other

Orbital lobes of opposite sign pointed toward each other

[22] A. Haaland and J. E. Nilsson, *Acta Chem. Scand.*, **1968,** *22,* 2653; see also A. Haaland, *Acc. Chem. Res.*, **1979,** *12,* 415.

[23] P. Seiler and J. Dunitz, *Acta Crystallogr., Sect. B,* **1982,** *38,* 1741.

[24] M. D. Rausch, W-M. Tsai, J. W. Chambers, R. D. Rogers, and H. G. Alt, *Organometallics,* **1989,** *8,* 816.

[25] Not counting the nodal planes that are coplanar with the C_5H_5 rings.

The ten group orbitals arising from the C_5H_5 ligands are shown in Figure 12-26.

2-Node group orbitals

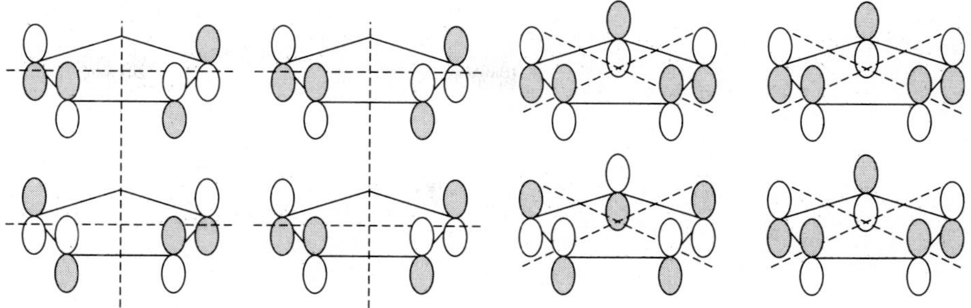

1-Node group orbitals

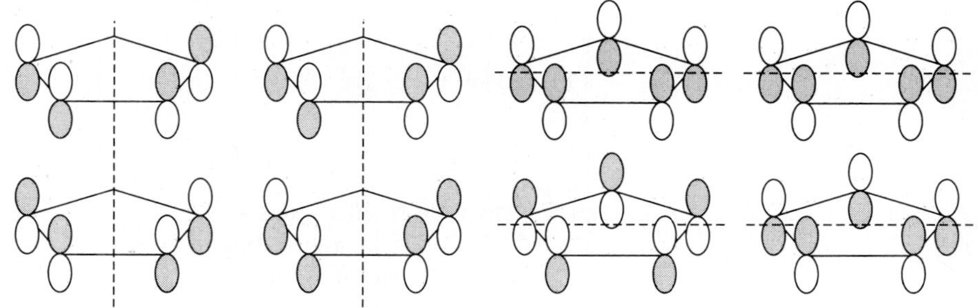

0-Node group orbitals

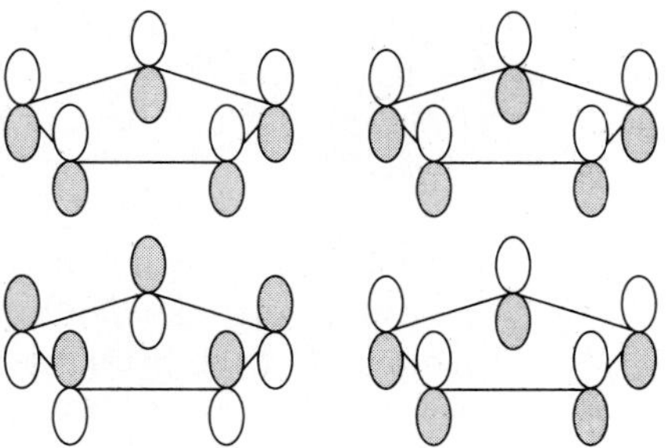

FIGURE 12-26 Group Orbitals for C_5H_5 Ligands of Ferrocene.

The process of developing the molecular orbital picture of ferrocene now becomes one of matching the group orbitals with the *s*, *p*, and *d* orbitals of appropriate symmetry on Fe.

EXERCISE 12-7

Determine which orbitals of Fe are appropriate for interaction with each of the group orbitals in Figure 12-26.

We will illustrate one of these interactions, between the d_{yz} orbital of Fe and its appropriate group orbital (one of the one-node group orbitals shown in

Figure 12-26). This interaction can occur in a bonding and an antibonding fashion:

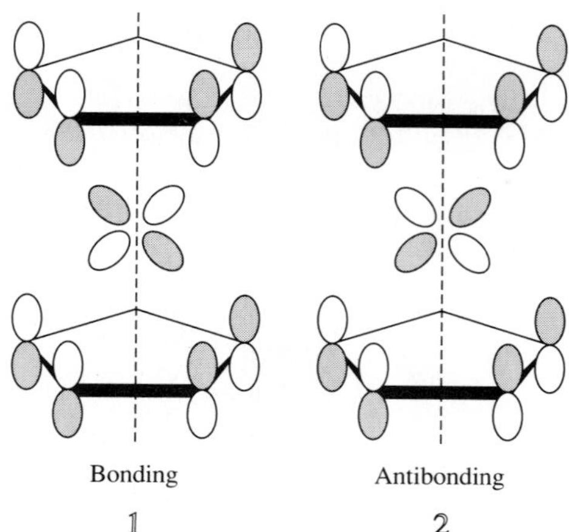

The complete energy-level diagram for the molecular orbitals of ferrocene is shown in Figure 12-27. The molecular orbital resulting from the d_{yz} bonding interaction, labeled 1 in the MO diagram, contains a pair of electrons. Its antibonding counterpart, 2, is empty. It is a useful exercise to match the other group orbitals from Figure 12-26 with the molecular orbitals in Figure 12-27 to verify the types of metal–ligand interactions that occur.

The orbitals of ferrocene that are of most interest are those having the greatest *d* orbital character; these are highlighted in the box in Figure 12-27. Two of these orbitals, having largely d_{xy} and $d_{x^2-y^2}$ character, are weakly bonding and are occupied by electron pairs; one, having largely d_{z^2} character, is essentially nonbonding and is also occupied by an electron pair; and two, having primarily d_{xz} and d_{yz} character, are empty. The relative energies of these orbitals and their *d* orbital–group orbital interactions are shown in Figure 12-28.[26,27]

The overall bonding in ferrocene can now be summarized. The occupied orbitals of the cyclopentadienyl ligands are stabilized by their interactions with iron [note the stabilization in energy of the group orbitals that form molecular orbitals that are primarily ligand in nature (the bonding interactions involving the 0-node and 1-node group orbitals)]. In addition, six electrons occupy molecular orbitals that are largely derived from iron *d* orbitals [as we would expect for iron(II)], but these molecular orbitals also have ligand character (the cone-shaped nodal surface of the d_{z^2} orbital points almost directly toward the lobes of its matching group orbital, so the resulting molecular orbital is essentially a nonbonding orbital localized on the iron). The molecular orbital picture in this case is consistent with the 18-electron rule.

Other metallocenes have similar structures but do not necessarily obey the rule. For example, cobaltocene and nickelocene are structurally similar

[26] The relative energies of the lowest three orbitals shown in Figure 12-28 have been a matter of controversy. Ultraviolet photoelectron spectroscopy indicates that the order is as shown, with the orbital having largely d_{z^2} character slightly higher in energy than the degenerate pair having substantial d_{xy} and $d_{x^2-y^2}$ character. This order may be reversed for some metallocenes. (See A. Haaland, *Acc. Chem. Res.*, **1979,** *12,* 415.)

[27] J. C. Giordan, J. H. Moore, and J. A. Tossell, *Acc. Chem. Res.*, **1986,** *19,* 281; E. Rühl and A. P. Hitchcock, *J. Am. Chem. Soc.*, **1989,** *111,* 5069.

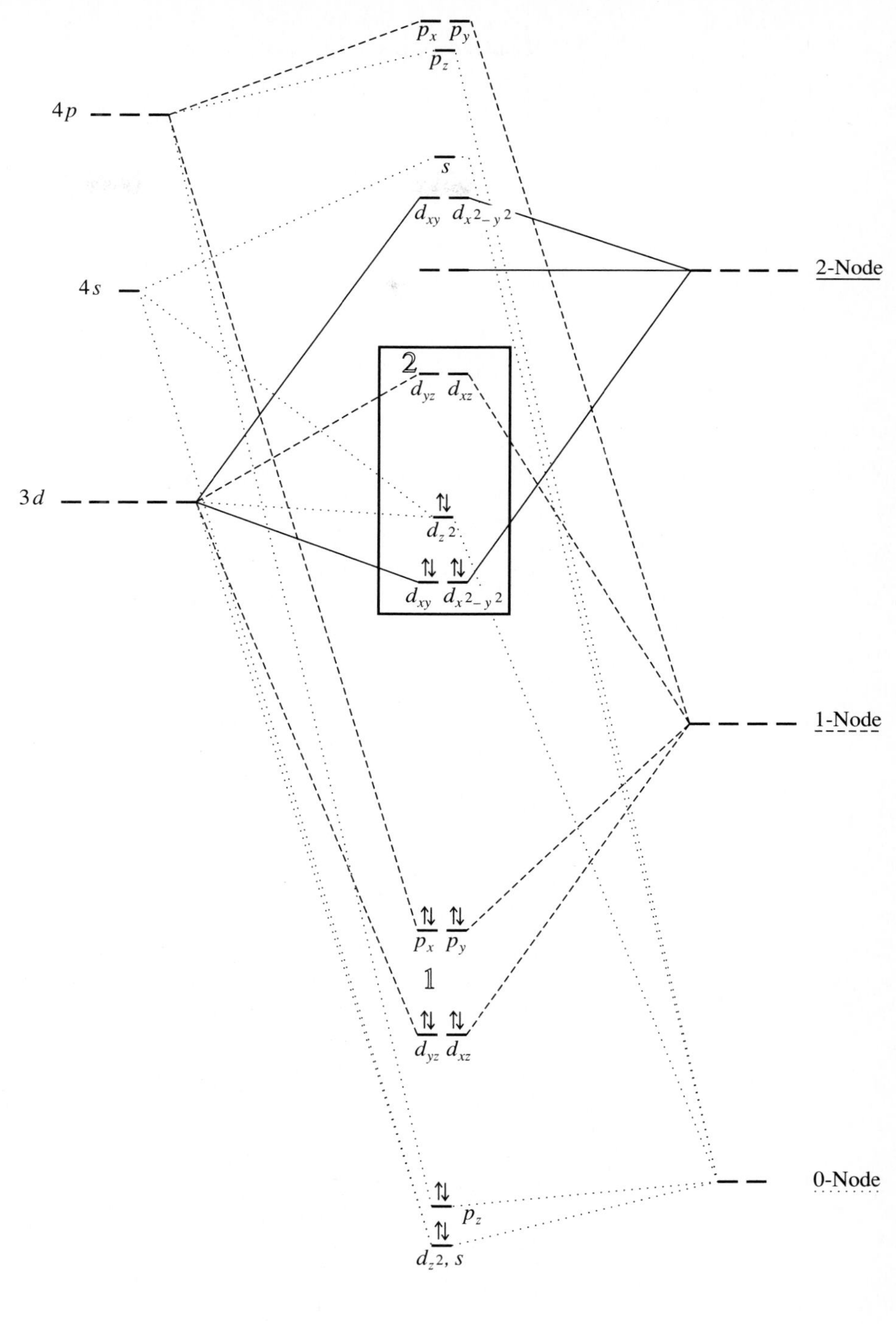

FIGURE 12-27 Molecular Orbitals of Ferrocene.

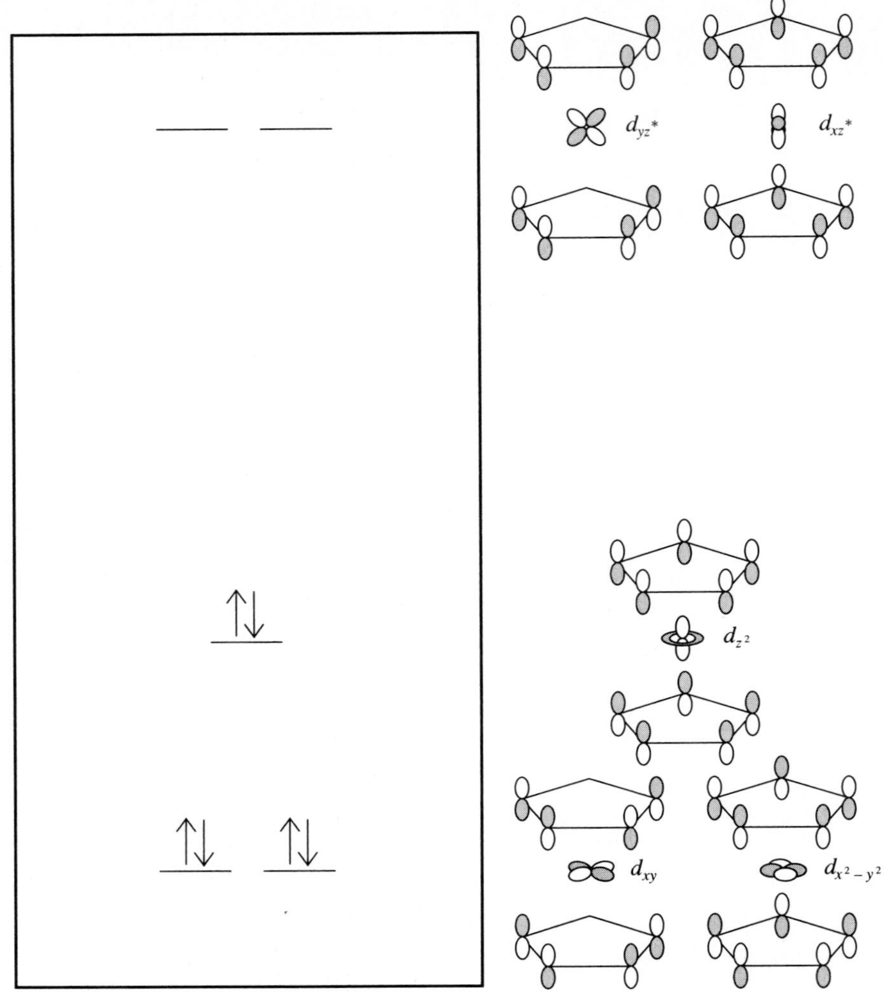

FIGURE 12-28 Molecular Orbitals of Ferrocene having Greatest *d* Character.

* For the molecular orbitals shown involving the d_{yz}, d_{xz}, and d_{z^2} orbitals, the *anti*bonding interactions are shown; the bonding interactions result in molecular orbitals of lower energy.

19- and 20-electron species. The extra electrons have important chemical and physical consequences, as can be seen from comparative data:

Complex	*Electron count*	*M—C distance (pm)*	*ΔH for M^{2+}—$C_5H_5^-$ dissociation (kJ/mol)*
$(\eta^5\text{-}C_5H_5)_2Fe$	18	206.4	1470
$(\eta^5\text{-}C_5H_5)_2Co$	19	211.9	1400
$(\eta^5\text{-}C_5H_5)_2Ni$	20	219.6	1320

Source: Data from A. Haaland, *Acc. Chem. Res.*, **1979,** *12, 415.*

The nineteenth and twentieth electrons of the metallocenes occupy slightly antibonding orbitals (largely d_{yz} and d_{xz} in character); as a consequence, the metal–ligand distance increases, and ΔH for metal–ligand dissociation decreases. Ferrocene itself shows much more chemical stability than cobaltocene and nickelocene; many of the chemical reactions of the latter are characterized by a tendency to yield 18-electron products. For example, ferrocene is unreactive toward iodine and rarely participates in reactions in which other ligands substitute for the cyclopentadienyl ligand. However, cobaltocene and nickelocene undergo the following reactions to give 18-electron products:

$$\underset{19e^-}{2\,(\eta^5\text{-}C_5H_5)_2Co} + I_2 \longrightarrow \underset{\substack{18e^- \\ \text{cobalticinium ion}}}{2\,[(\eta^5\text{-}C_5H_5)_2Co]^+} + 2\,I^-$$

$$\underset{20e^-}{(\eta^5\text{-}C_5H_5)_2Ni} + 4\,PF_3 \longrightarrow \underset{18e^-}{Ni(PF_3)_4} + \text{organic products}$$

Cobalticinium reacts with hydride to give a neutral, 18-electron sandwich compound in which one cyclopentadienyl ligand has been modified into η^4-C_5H_6, as shown in Figure 12-29.

Ferrocene, however, is by no means chemically inert. It undergoes a variety of reactions, including many on the cyclopentadienyl rings. A good example is that of electrophilic acyl substitution (Figure 12-30), a reaction parallelling that of benzene and its derivatives. In general, electrophilic aromatic substitution reactions are much more rapid for ferrocene than for benzene, an indication of greater concentration of electron density in the rings of the sandwich compound.

Complexes containing cyclopentadienyl and CO ligands

Not surprisingly, many complexes are known containing both Cp and CO ligands. These include "half-sandwich" compounds such as $(\eta^5\text{-}C_5H_5)Mn(CO)_3$ and dimeric and larger cluster molecules. Examples are shown in Figure 12-31. As for the binary CO complexes, complexes of the second- and third-row transition metals show a decreasing tendency of CO to act as a bridging ligand.

One of the most interesting of the complexes in Figure 12-31 is $(C_5H_5)_2Fe(CO)_2$. This compound contains both η^1- and η^5-C_5H_5 ligands (and

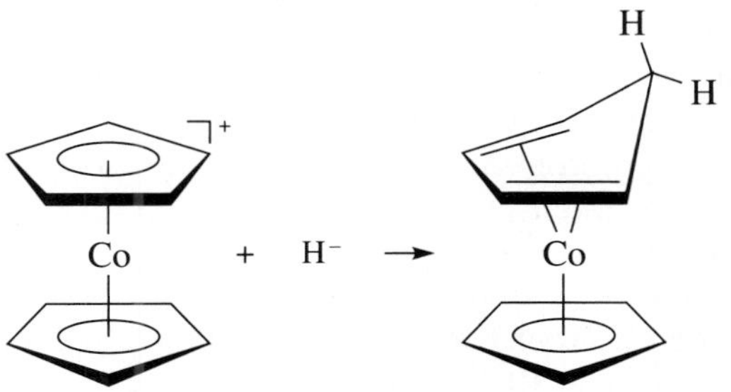

FIGURE 12-29 Reaction of Cobalticinium with Hydride.

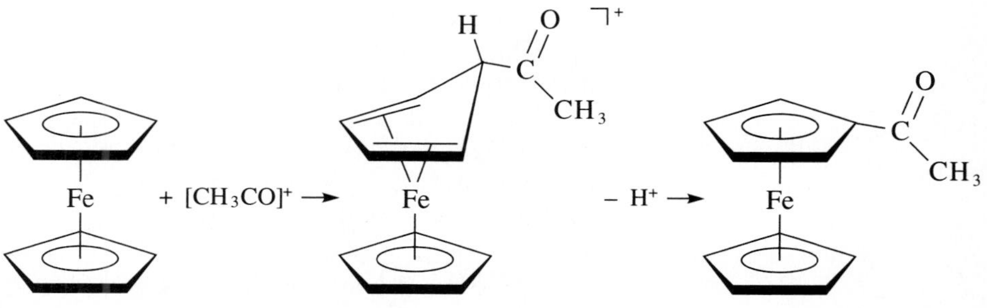

FIGURE 12-30 Electrophilic Acyl Substitution in Ferrocene.

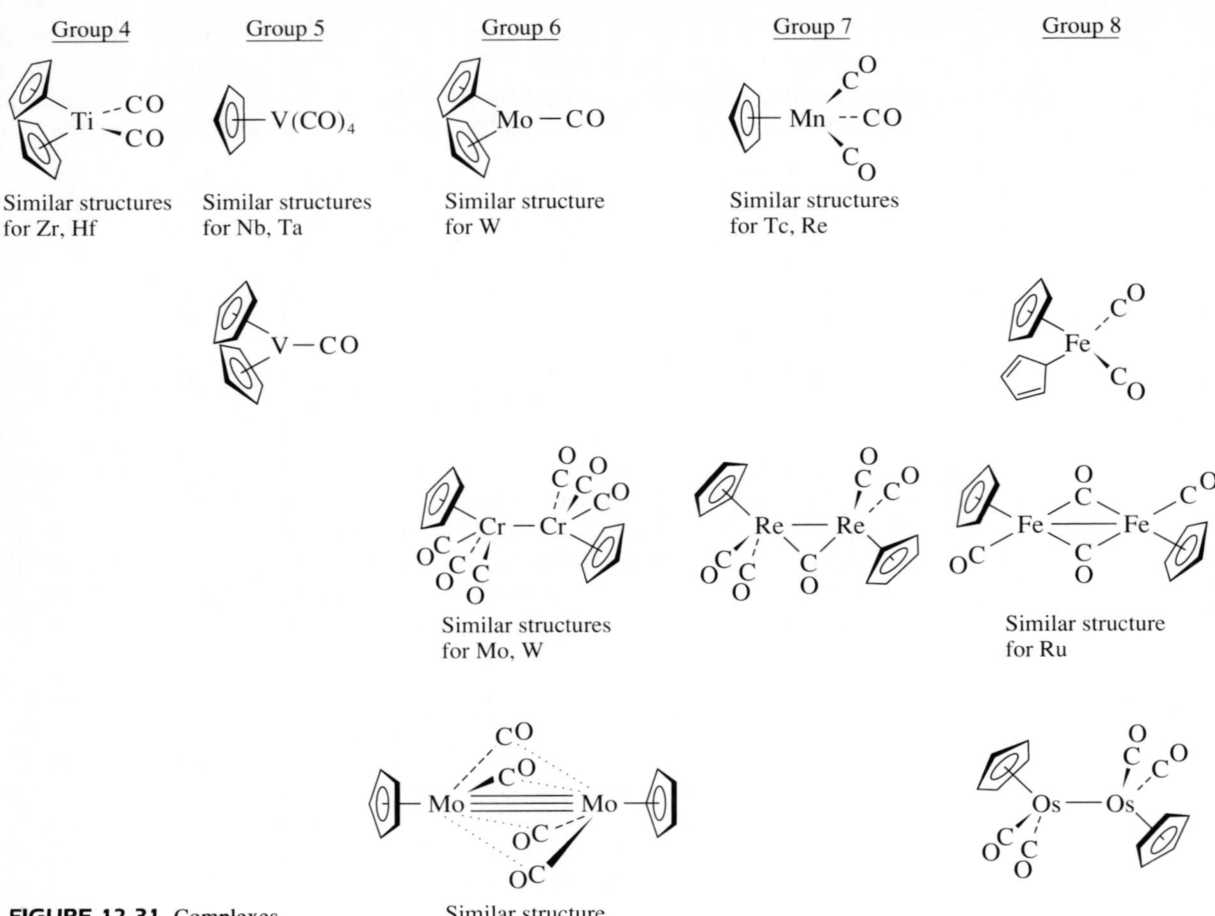

FIGURE 12-31 Complexes Containing C_5H_5 and CO.

consequently obeys the 18-electron rule). The ^1H NMR spectrum at 30°C shows two singlets of equal area. A singlet would be expected for the five equivalent protons of the η^5-C_5H_5 ring but is surprising for the η^1-C_5H_5 ring, since the protons are not all equivalent. A "ring whizzer" mechanism, Figure 12-32, has been proposed by which the five ring positions of the monohapto ring interchange via 1,2 shifts extremely rapidly, so rapidly that the NMR can see only the average signal.[28] At lower temperatures this process is slower, and the different resonances for the protons of η^1-C_5H_5 become apparent, as also shown in the figure. Additional applications of NMR to organometallic chemistry are discussed in Section 12-7-2.

Many other linear and cyclic pi ligands are known. Examples of complexes containing some of these ligands are shown in Figure 12-33. Depending on the ligand and the electron requirements of the metal (or metals), these ligands may be capable of bonding in a monohapto or polyhapto fashion, and they may bridge two or more metals. Particularly interesting are the cases in which cyclic ligands can bridge metals to give "triple-decker" and higher-order sandwich compounds (see Figure 12-1).

[28] C. H. Campbell and M. L. H. Green, *J. Chem. Soc. (A),* **1970,** 1318.

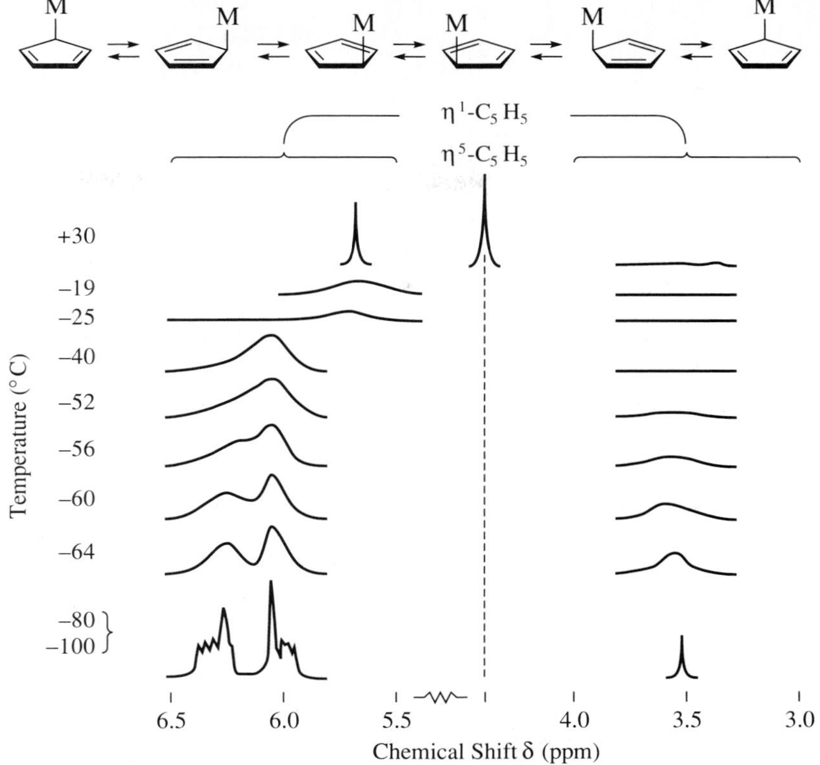

FIGURE 12-32 Ring Whizzer Mechanism and Variable Temperature NMR Spectra of $(C_5H_5)_2Fe(CO)_2$. (NMR spectra reproduced with permission from M. J. Bennett, Jr., F. A. Cotton, A. Davison, J. W. Faller, S. J. Lippard, and S. M. Morehouse, *J. Am. Chem. Soc.*, **1966,** *88,* 4371.)

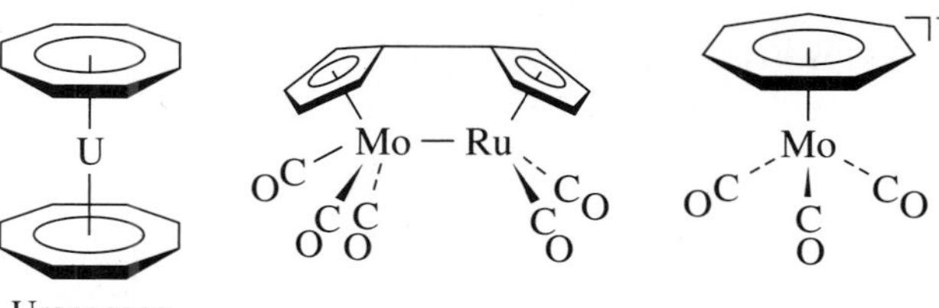

FIGURE 12-33 Examples of Molecules Containing Cyclic Pi Systems.

12-6 COMPLEXES CONTAINING METAL–CARBON SIGMA BONDS

12-6-1 ALKYL COMPLEXES

Some of the earliest known organometallic complexes were those having sigma bonds between main group metal atoms and alkyl groups. Examples include Grignard reagents, having magnesium–alkyl bonds, and alkyl complexes with alkali metals, such as methyllithium. The first stable transition metal alkyls were synthesized in the first decade of this century; many such complexes are now known. The metal–ligand bonding in these compounds may be viewed as primarily involving covalent sharing of electrons between the metal and the carbon in a sigma fashion. In terms of electron counting, the alkyl ligand may be considered the 2-electron donor $:R^-$ or the 1-electron donor $\cdot R$. Significant ionic contribution to the bonding may occur in complexes of highly electropositive elements, such as the alkali metals and alkaline earths.

Although many complexes contain alkyl ligands, complexes containing alkyl groups as the only ligands are relatively few and have a tendency to be kinetically unstable and difficult to isolate.[29] Examples include $Ti(CH_3)_4$, $W(CH_3)_6$, and $Cr[CH_2Si(CH_3)_3]_4$. This kinetic instability can be a very useful feature in catalytic processes, where the rapid reactivity of an intermediate can be crucial to the effectiveness of the overall process. Many alkyl complexes are important in catalytic processes; examples of reactions of these complexes will be considered in Chapter 13. An interesting and unusual use of alkyls is that diethylzinc has recently been selected by the Library of Congress for treating books (neutralizing the acid in the paper) for their long-term preservation.

12-6-2 CARBENE (ALKYLIDENE) COMPLEXES

Carbene complexes contain metal–carbon double bonds.[30] First synthesized in 1964 by E. O. Fischer,[31] carbene complexes are now known for the majority of transition metals and for a wide range of ligands, including the prototype carbene $:CH_2$. The majority of such complexes, including those first synthesized by Fischer, contain one or two highly electronegative *heteroatoms,* such as O, N, or S, directly bound to the carbene carbon and are sometimes designated Fischer carbenes. Complexes having only hydrogen or carbon bound to the carbene carbon have been studied extensively by R. R. Schrock's research group and by others; they are sometimes designated Schrock carbenes.[32] We will describe briefly the bonding in these complexes and give an example of the synthesis and properties of one Fischer carbene complex, $Cr(CO)_5[C(OCH_3)C_6H_5]$.

The formal double bond in carbene complexes may be compared with the double bond in alkenes; in the case of the carbene, the metal must use a *d* orbital (rather than a *p* orbital) in forming the pi bond with carbon, as illustrated in Figure 12-34.

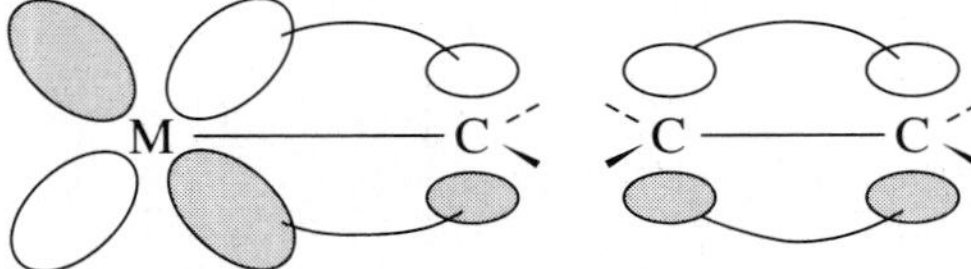

FIGURE 12-34 Pi Bonding in Carbene Complexes and in Alkenes.

Another aspect of bonding of importance to carbene complexes is that complexes having a highly electronegative atom, such as O, N, or S, attached to the carbene carbon tend to be more stable than complexes lacking such an atom. For example, $Cr(CO)_5[C(OCH_3)C_6H_5]$, with an oxygen on the carbene carbon, is much more stable than $Cr(CO)_5[C(H)C_6H_5]$. The stability of the complex is enhanced if the highly electronegative atom can participate in the pi bonding, with the result a delocalized, 3-atom pi system involving a *d* orbital on the metal and *p* orbitals on carbon and on the electronegative atom. Such a delocalized 3-atom system provides more stability to the bonding pi electron

[29] An interesting historical perspective on alkyl complexes is in G. Wilkinson, *Science,* **1974,** *185,* 109.

[30] IUPAC has recommended that "alkylidene" be used to describe all complexes containing metal–carbon double bonds and that carbene be restricted to free $:CR_2$. For a detailed description of the distinction between these two terms (and between carbyne and alkylidyne, described in Section 12-6-3), see W. A. Nugent and J. M. Mayer, *Metal–Ligand Multiple Bonds,* Wiley-Interscience, New York, 1988, pp. 11–16.

[31] E. O. Fischer and A. Maasböl, *Angew. Chem. Int. Ed. Engl.,* **1964,** *3,* 580.

[32] R. R. Schrock, *Acc. Chem. Res.,* **1979,** *12,* 98 and references therein.

pair than would a simple metal–carbon pi bond. An example of such a pi system is shown in Figure 12-35.

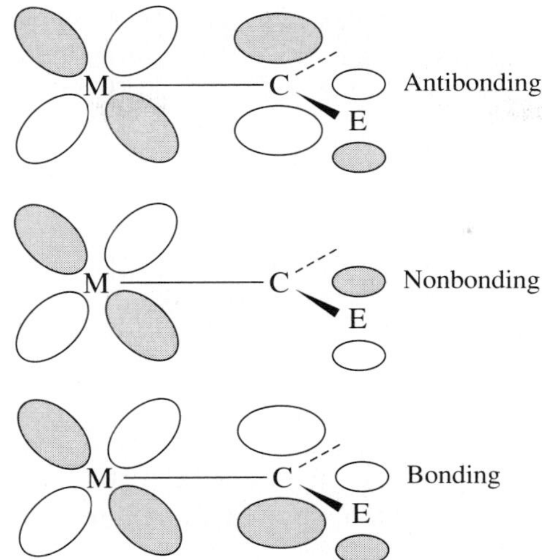

FIGURE 12-35 Delocalized Pi Bonding in Carbene Complexes. E designates a highly electronegative heteroatom such as O, N, or S.

The methoxycarbene complex, $Cr(CO)_5[C(OCH_3)C_6H_5]$, illustrates the bonding described above and some important related chemistry.[33] To synthesize this complex, we can begin with the hexacarbonyl, $Cr(CO)_6$. As in organic chemistry, highly nucleophilic reagents can attack the carbonyl carbon. For example, phenyllithium can react with $Cr(CO)_6$ to give the anion $[C_6H_5C(O)Cr(CO)_5]^-$, which has two important resonance structures, as shown:

$$Li^+{:}C_6H_5^- + O{\equiv}C{-}Cr(CO)_5 \longrightarrow \underbrace{C_6H_5{-}\overset{\overset{\displaystyle O}{\|}}{C}{-}Cr^-(CO)_5 \longleftrightarrow C_6H_5{-}\overset{\overset{\displaystyle O^-}{\|}}{C}{-}Cr(CO)_5}_{C_6H_5{-}\overset{\overset{\displaystyle O}{\|\,\diagdown^{-}}}{C}{\overset{\cdots}{=}}Cr(CO)_5} + Li^+$$

Alkylation by a source of CH_3^+ such as $[(CH_3)_3O][BF_4]$ or CH_3I gives the methoxycarbene complex:

$$C_6H_5{-}\overset{\overset{\displaystyle O}{|\,\diagdown^{-}}}{C}{\overset{\cdots}{=}}Cr(CO)_5 + [(CH_3)_3O][BF_4] \longrightarrow C_6H_5{-}\overset{\overset{\displaystyle OCH_3}{|}}{C}{=}Cr(CO)_5 + BF_4 + (CH_3)_2O$$

Evidence for double bonding between chromium and carbon is provided by X-ray crystallography, which measures this distance at 204 pm, compared with a typical Cr—C single bond distance of approximately 220 pm.

One very interesting aspect of this complex is that it exhibits a proton NMR spectrum that is temperature dependent. At room temperature, a single resonance is found for the methyl protons; however, as the temperature is lowered, this peak first broadens and then splits into two peaks. How can this behavior be explained?

A single proton resonance, corresponding to a single magnetic environment, is expected for the carbene complex as illustrated, with a double bond between chromium and carbon and a single bond (permitting rapid rotation about the bond) between carbon and oxygen. The room-temperature NMR is

[33] E. O. Fischer, *Adv. Organometallic Chem.*, **1976,** *14,* 1.

therefore as expected. However, the splitting of this peak at lower temperature into two peaks suggests two different proton environments.[34] Two environments are possible if there is hindered rotation about the C—O bond. A resonance structure for the complex can be drawn showing the possibility of some double bonding between C and O; were such double bonding significant, *cis* and *trans* isomers, as shown in Figure 12-36, might be observable at low temperatures.

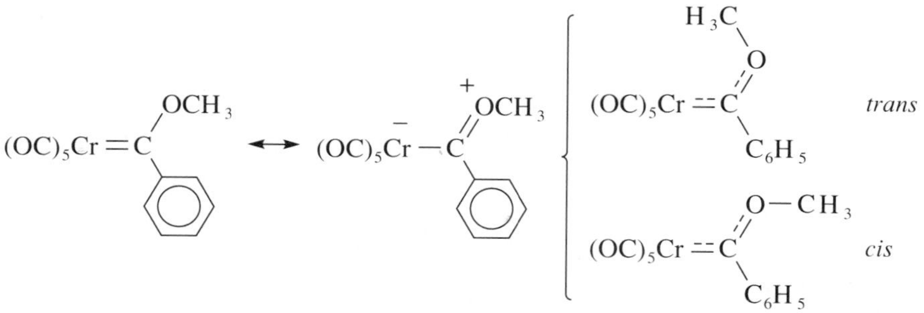

FIGURE 12-36 Resonance Structures and *cis* and *trans* Isomers for $Cr(CO)_5[C(OCH_3)C_6H_5]$.

Evidence for double-bond character in the C—O bond is also provided by crystal structure data, which show a C—O bond distance of 133 pm, compared with a typical C—O single-bond distance of 143 pm.[35] The double bonding between C and O, although weak (typical C═O bonds are much shorter, approximately 116 pm), is sufficient to slow down rotation about the bond so that at low temperatures proton NMR detects the *cis* and *trans* methyl protons separately. At higher temperature, there is sufficient energy to cause rapid rotation about the C—O bond so that the NMR sees only an average signal, which is observed as a single peak.

X-ray crystallographic data, as mentioned, show double-bond character in both the Cr—C and C—O bonds. This supports the statement made early in the discussion of carbene complexes that pi bonding in complexes of this type (containing a highly electronegative atom, in this case oxygen) may be considered delocalized over three atoms. While not absolutely essential for all carbene complexes (examples lacking such delocalization are known; however, these are generally of lower stability), the delocalization of pi electron density over three (or more) atoms provides an additional measure of stability to many of these compounds.[36]

Carbene complexes appear to be important intermediates in olefin metathesis reactions, which are of significant industrial interest; these reactions are discussed in Chapter 13.

12-6-3 CARBYNE (ALKYLIDYNE) COMPLEXES

Carbyne complexes have metal–carbon triple bonds; they are formally analogous to alkynes.[37] Many carbyne complexes are now known; examples of carbyne ligands include the following:

M≡C—R; R = aryl (first discovered), alkyl, H, $SiMe_3$, NEt_2, PMe_3, SPh, and Cl

[34] C. G. Kreiter and E. O. Fischer, *Angew. Chem. Int. Ed. Engl.*, **1969**, *8*, 761.

[35] O. S. Mills and A. D. Redhouse, *J. Chem. Soc. (A)*, **1968**, 642.

[36] *Transition Metal Carbene Complexes*, Verlag Chemie, Weinheim, West Germany, 1983, pp. 120–22.

[37] IUPAC has recommended that "alkylidyne" be used to designate complexes containing metal–carbon triple bonds.

Carbyne complexes were first synthesized fortuitously in 1973 as products of the reactions of carbene complexes with Lewis acids.[38] For example, the methoxycarbene complex $Cr(CO)_5[C(OCH_3)C_6H_5]$ was found to react with the Lewis acids BX_3 (X = Cl, Br, or I). First, the Lewis acid attacks the oxygen, the basic site on the carbene:

$$(CO)_5Cr{=}C\begin{matrix} OCH_3 \\ C_6H_5 \end{matrix} + BX_3 \longrightarrow [(CO)_5Cr{\equiv}C{-}C_6H_5]^+X^- + X_2BOCH_3$$

Subsequently, the intermediate loses CO, with the halogen coordinating in a position *trans* to the carbyne:

$$[(CO)_5Cr{\equiv}C-C_6H_5]^+X^- \longrightarrow X-Cr(CO)_4{\equiv}C-C_6H_5 + CO$$

The best evidence for the carbyne nature of the complex is provided by X-ray crystallography, which gives a Cr—C bond distance of 168 pm (for X = Cl), considerably shorter than the comparable distance for the parent carbene complex. The Cr≡C—C angle is, as expected, 180° for this complex; however, slight deviations from linearity are observed for many complexes in crystalline form, in part a consequence of the manner of packing in the crystal.

Bonding in carbyne complexes may be viewed as a combination of a σ bond plus two π bonds, as illustrated in Figure 12-37. The carbyne ligand has a lone pair of electrons in an *sp* hybrid on carbon; this lone pair can donate to a suitable hybrid orbital on Cr to form a sigma bond. In addition, the carbon has two *p* orbitals that can accept electron density from *d* orbitals on Cr to form pi bonds. Thus the overall function of the carbyne ligand is as both a σ donor and π acceptor. (For electron counting purposes, a $:CR^+$ ligand can be considered a 2-electron donor; it is usually more convenient to count neutral CR as a 3-electron donor.)

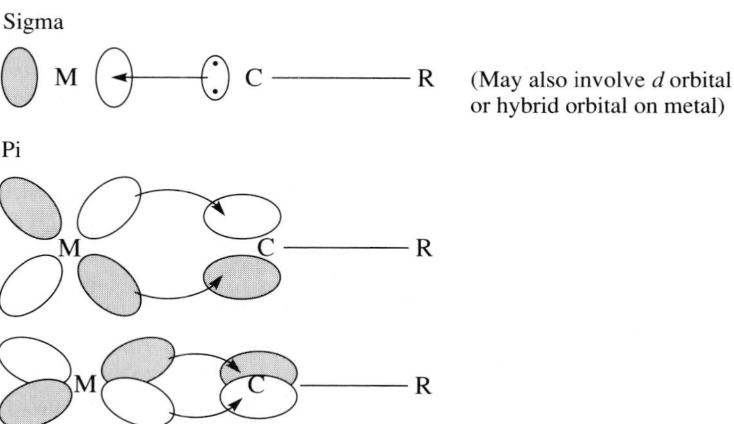

FIGURE 12-37 Bonding in Carbyne Complexes.

[38] E. O. Fischer, G. Kreis, C. G. Kreiter, J. Müller, G. Huttner, and H. Lorentz, *Angew. Chem., Int. Ed. Engl.*, **1973,** *12,* 564.

Carbyne complexes can be synthesized in a variety of ways in addition to Lewis acid attack on carbenes, as described above. Synthetic routes for carbynes and the reactions of these compounds have been reviewed.[39]

In some cases, molecules have been synthesized containing two or three of the types of ligands discussed in this section, alkyl, carbene, and carbyne. Such molecules provide an opportunity to make direct comparisons of lengths of metal–carbon single, double, and triple bonds, as shown in Figure 12-38.

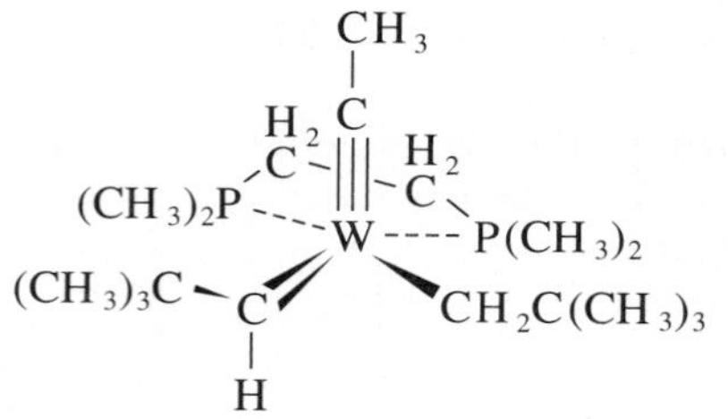

(a)

W—C	225.8 pm
W=C	194.2 pm
W≡C	178.5 pm

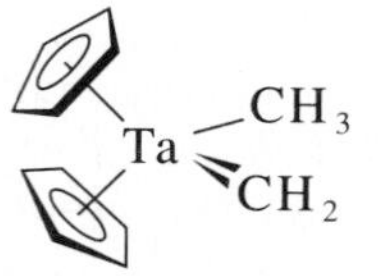

(b)

Ta—C	224.6 pm
Ta=C	202.6 pm

FIGURE 12-38 Complexes Containing Alkyl, Carbene, and Carbyne Ligands. (a) Data from M. R. Churchill and W. J. Youngs, *Inorg. Chem.*, **1979**, *18*, 2454. (b) Data from L. J. Guggenberger and R. R. Schrock, *J. Am. Chem. Soc.*, **1975**, *97*, 6578.

12-7 SPECTRAL ANALYSIS AND CHARACTERIZATION OF ORGANOMETALLIC COMPLEXES

One of the most challenging (and sometimes most frustrating!) aspects of organometallic research is the characterization of new reaction products. Assuming that specific products can be isolated (such as by chromatographic procedures, recrystallization, or other techniques), determining the structure may be an interesting challenge. Many complexes can be crystallized and characterized structurally by X-ray crystallography; however, not all organometallic complexes can be crystallized, and not all that crystallize do so in a manner lending themselves to structural solution by X-ray techniques. Furthermore, it is frequently desirable to be able to use more convenient, more rapid, and less expensive techniques if at all possible (although in some cases an X-ray structural determination is the only way to conclusively identify a compound—and may therefore be the most rapid and inexpensive technique). In the realm of organometallic chemistry, two techniques, infrared spectroscopy and nuclear magnetic resonance spectrometry, are often the most useful. In addition, mass spectrometry, elemental analysis, conductivity measurements, and other techniques may be valuable in characterizing products of organometallic reactions. We will consider primarily IR and NMR as techniques for characterization of organometallic compounds.

12-7-1 INFRARED SPECTRA

Infrared spectra can be useful in two respects. The number of infrared bands, as discussed briefly in Chapter 4, depends on molecular symmetry; conse-

[39] H. P. Kim and R. J. Angelici, *Adv. Organometallic Chem.*, **1987**, *27*, 51; H. Fischer, P. Hoffmann, F. R. Kreissl, R. R. Schrock, U. Schubert, and K. Weiss, *Carbyne Complexes*, VCH Publishers, Weinheim, West Germany, 1988.

quently, by determining the number of such bands for a particular ligand (such as CO), we may be able to decide among several alternative geometries for a compound or at least reduce the number of possibilities. In addition, the position of the IR band can indicate the function of a ligand (for example, terminal versus bridging modes) and, in the case of pi-acceptor ligands, can describe the electron environment of the metal.

Number of IR bands

In Chapter 4, a method was described for determining the number of IR-active stretching vibrations using molecular symmetry. The basis for this method is that vibrational modes, to be infrared active, must result in a change in the dipole moment of the molecule. In symmetry terms, the equivalent statement is that, to be IR active, vibrational modes must have irreducible representations of the same symmetry as the Cartesian coordinates x, y, or z (or a linear combination of these coordinates). The procedure developed in Chapter 4 is used in the following examples. It is suggested as an exercise that the reader verify some of these results using the method described in Chapter 4.

Our examples will be of carbonyl complexes. However, it should be kept in mind that identical reasoning applies to other linear monodentate ligands, such as CN^- and NO. We will begin by considering several simple cases.

Monocarbonyl complexes. These complexes have a single possible C—O stretching mode and consequently show a single band in the IR.

Dicarbonyl complexes. Two geometries, linear and bent, must be considered:

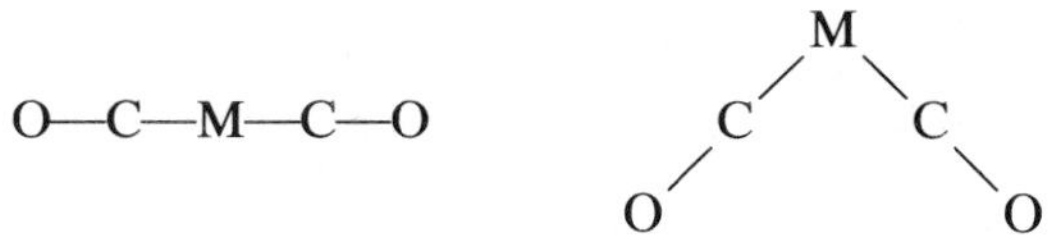

O—C—M—C—O

```
        M
       / \
      C   C
     /     \
    O       O
```

In the case of two CO ligands arranged linearly, only an asymmetric vibration of the ligands is IR active: a symmetric vibrational mode results in no change in dipole moment and hence is inactive. However, if two COs are oriented in a nonlinear fashion, both symmetric and asymmetric vibrations result in changes in dipole moment, and both are IR active:

Symmetric stretch

O←C—M—C→O

No change in dipole moment;
IR inactive

```
        M
       / \
      C   C
     ↙     ↘
    O       O
```

Change in dipole moment:
IR active

Asymmetric Stretch

O←C—M—C←O

Change in dipole moment;
IR active

```
        M
       / \
      C   C
     ↙     ↖
    O       O
```

Change in dipole moment;
IR active

Therefore, an infrared spectrum can be a convenient tool for determining the structure of molecules known to have exactly two CO ligands: a single band indicates linear orientation of the COs; two bands indicate nonlinear orientation.

For molecules containing exactly two CO ligands on the same metal atom, the relative intensities of the IR bands can be used to determine approximately the angle between the COs, using the equation

$$\frac{I_{\text{symmetric}}}{I_{\text{asymmetric}}} = \text{cotan}^2\left(\frac{\phi}{2}\right)$$

where the angle between the ligands is ϕ. For example, for two CO ligands at 90°, $\text{cotan}^2(45°) = 1$. For this angle, two IR bands of equal intensity would be observed. For an angle >90°, the ratio is less than 1; the IR band due to symmetric stretching would be less intense than the band due to asymmetric stretching. If $\phi < 90°$, the IR band for symmetric stretching would be the more intense. (For C—O stretching vibrations, the symmetric band occurs at higher energy than the corresponding asymmetric band.) In general, this calculation is approximate and requires integrated values of intensities of absorption bands (rather than the more easily determined intensity at the wavelength of maximum absorption).

Complexes containing three or more carbonyls. The predictions are not quite so simple. The exact number of carbonyl bands can be determined according to the symmetry approach of Chapter 4. For convenient reference, the numbers of bands expected for a variety of CO complexes are given in Table 12-3. In this table all ligands other than CO are assumed to be identical; if these ligands are not identical, the symmetry (point group) of the molecule will be different than shown, and the number of carbonyl bands may also be different.

Several additional points relating to the number of infrared bands are worth noting. First, while we can predict the number of infrared-active bands by the methods of group theory, fewer bands may sometimes be observed. In some cases, bands may overlap to such a degree as to be indistinguishable; alternatively, one or more bands may be of very low intensity and not readily observed. In some cases, isomers may be present in the same sample, and it may be difficult to sort out which IR absorptions belong to which compound.

In carbonyl complexes the number of C—O stretching bands cannot exceed the number of CO ligands. The alternative is possible in some cases (more CO groups than IR bands) when vibrational modes are not IR active (do not cause a change in dipole moment). Examples are in Table 12-3. Because of their symmetry, carbonyl complexes of T_d and O_h symmetry have a single carbonyl band in the infrared spectrum.

EXERCISE 12-8

The complex $Mo(CO)_3(NCC_2H_5)_3$ has the infrared spectrum shown. Is this complex more likely the *fac* or *mer* isomer?

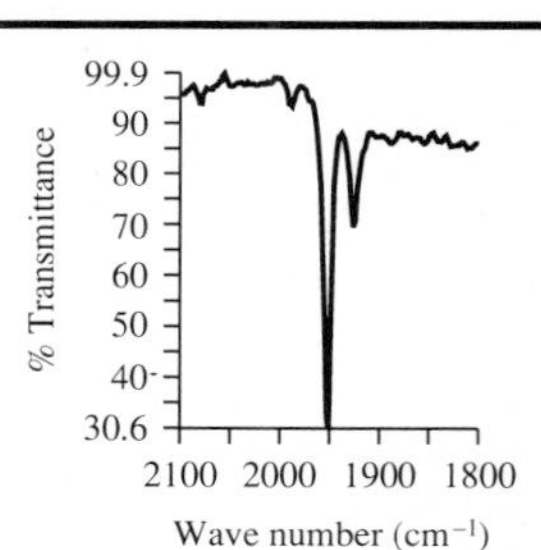

TABLE 12-3
Carbonyl stretching bands

Number of CO's		*Coordination number* 4	5	6
3		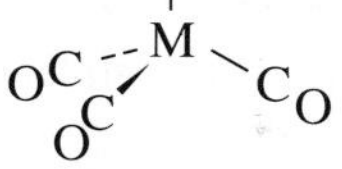	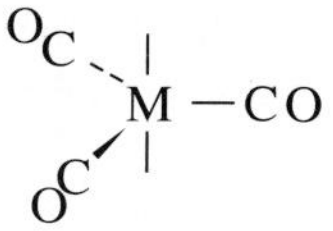	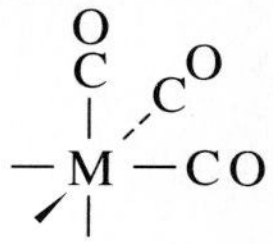
	IR bands:	2	1	2
			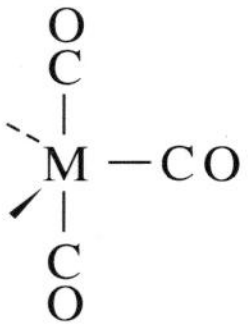	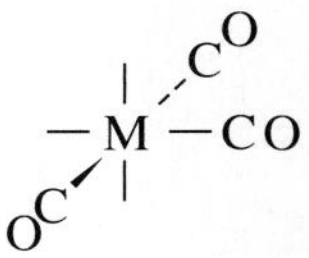
	IR bands:		3	3
			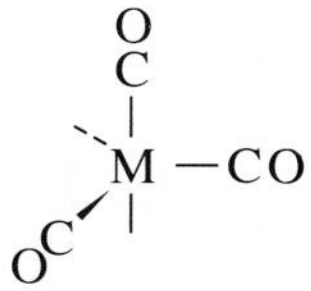	
	IR bands:		3	
4		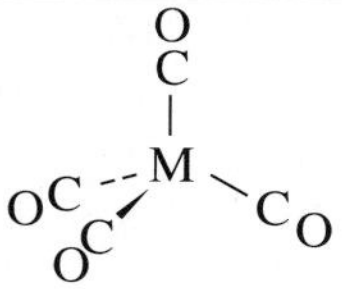	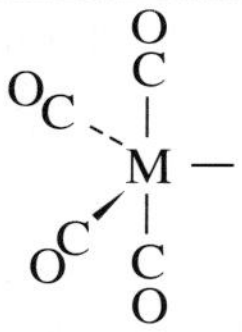	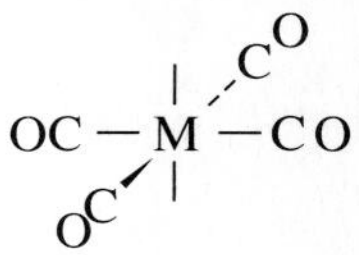
	IR bands:	1	4	1
			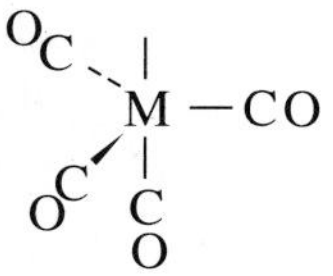	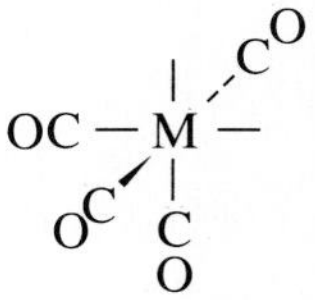
	IR bands:		3	4
5			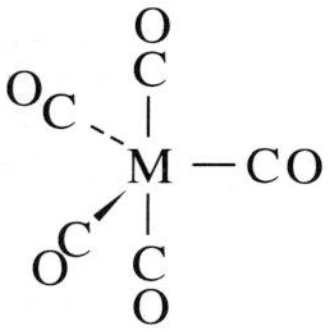	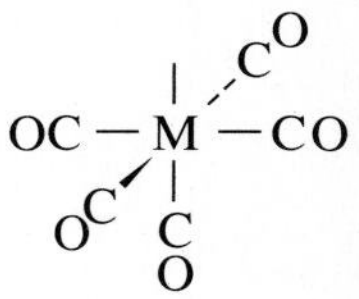
	IR bands:		2	3
6				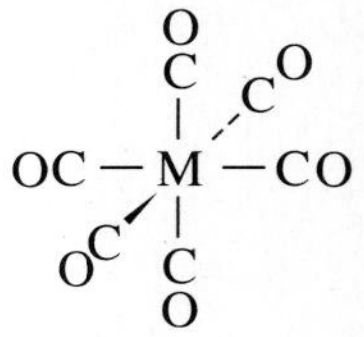
	IR bands:			1

Positions of infrared bands

We have already encountered in this chapter two examples in which the position of the carbonyl stretching band provides useful information. In the case of the isoelectronic species $[V(CO)_6]^-$, $Cr(CO)_6$, and $[Mn(CO)_6]^+$, an increase in negative charge on the complex causes a significant reduction in the energy of the C—O band as a consequence of additional π backbonding from the metal to the ligands. The bonding mode is also reflected in the infrared spectrum, with energy decreasing in the order

$$\text{Terminal CO} > \text{doubly bridging CO} > \text{triply bridging CO}$$

The positions of infrared bands are also a function of other ligands present. For example,[40]

Complex	*ν(CO), cm^{-1}*
fac-$Mo(CO)_3(PF_3)_3$	2090, 2055
fac-$Mo(CO)_3(PCl_3)_3$	2040, 1991
fac-$Mo(CO)_3(PClPh_2)_3$	1977, 1885
fac-$Mo(CO)_3(PMe_3)_3$	1945, 1854

Going down this series, the σ-donor ability of the phosphine ligands increases, and the π-acceptor ability decreases. PF_3 is the weakest of the donors (as a consequence of the highly electronegative fluorines) and the strongest of the acceptors. As a result, the molybdenum in $Mo(CO)_3(PMe_3)_3$ carries the greatest electron density; it is the most able to donate electron density to the π^* orbitals of the CO ligands. Consequently, the CO ligands in $Mo(CO)_3(PMe_3)_3$ have the weakest C—O bonds and the lowest-energy stretching bands. Many comparable series are known.

The important point is that the position of the carbonyl bands can provide important clues to the electronic environment of the metal. The greater the electron density on the metal (and the greater the negative charge), the greater the backbonding to CO and the lower the energy of the carbonyl stretching vibrations. Similar correlations between metal environment and infrared spectra can be drawn for a variety of other ligands, both organic and inorganic. NO, for example, has an infrared spectrum that is strongly correlated with environment in a manner similar to that of CO. In combination with information on the number of infrared bands, the positions of such bands for CO and other ligands can therefore be extremely useful in characterizing organometallic compounds.

12-7-2 NMR SPECTRA

Nuclear magnetic resonance is also an extremely valuable tool in characterizing organometallic complexes. The advent of high-field NMR instruments using superconducting magnets has in many ways revolutionized the study of these compounds. Convenient NMR spectra can now be taken using many metal nuclei, as well as the more traditional nuclei such as ^1H, ^{13}C, ^{19}F, and ^{31}P; the

[40] F. A. Cotton, *Inorg. Chem.*, **1964,** *3,* 702.

combined spectral data of several nuclei make it possible to identify many compounds by their NMR spectra alone.

As in organic chemistry, chemical shifts, splitting patterns, and coupling constants are useful in characterizing the environments of individual atoms in organometallic compounds. The reader may find it useful to review the basic theory of NMR as presented in an organic chemistry text. More advanced discussions of NMR, especially relating to ^{13}C, have been surveyed elsewhere.[41]

Carbon 13 NMR has become increasingly useful with the advent of modern instrumentation. Although the isotope ^{13}C has a low natural abundance (approximately 1.1%) and low sensitivity for the NMR experiment (about 1.6% the sensitivity of ^{1}H), Fourier transform techniques now make it possible to obtain useful ^{13}C spectra for most organometallic species of reasonable stability. Nevertheless, the time necessary to obtain a ^{13}C spectrum may still be an experimental difficulty for compounds present in very small amounts or of low solubility. Rapid reactions may also be inaccessible by this technique. Some useful features of ^{13}C spectra include:

1. An opportunity to observe organic ligands that do not contain hydrogen (such as CO and $F_3C—C{\equiv}C—CF_3$)
2. Direct observation of the carbon skeleton of organic ligands (often most useful when a spectrum is acquired with complete proton decoupling)
3. ^{13}C chemical shifts are more widely dispersed than ^{1}H shifts. This often makes it easy to distinguish between ligands in compounds containing several different organic ligands.

^{13}C NMR is also a valuable tool for observing intramolecular rearrangement processes.[42]

Chemical shifts for ^{13}C spectra of some categories of organometallic complexes are listed in Table 12-4.[43] Several features of these data are worth noting. Terminal carbonyl peaks are frequently in the range δ 195 to 225 ppm, a range sufficiently distinctive that the CO groups are usually easy to identify. The ^{13}C chemical shift is correlated with the strength of the C—O bond; the stronger the bond, the lower the chemical shift.[44] Bridging carbonyls have slightly greater chemical shifts than terminal carbonyls and consequently may lend themselves to easy identification (however, IR is usually a better tool than NMR for distinguishing between bridging and terminal COs). Cyclopentadienyl ligands have a fairly wide range of chemical shifts, with the value for ferrocene (68.2 ppm) at the low end for such values. Other organic ligands may also have fairly wide ranges in ^{13}C chemical shifts.

As in organic chemistry, integration of the various peaks can provide the ratio of atoms in different environments; it is usually accurate, for example, to assume that the area of a ^{1}H peak (or set of peaks) is proportional to the number of nuclei giving rise to that peak. However, for ^{13}C this approach is somewhat less reliable. Relaxation times of different carbon atoms in organometallic complexes vary widely; this may lead to inaccuracy in the correlation of peak area with number of atoms (the correlation between area and number

[41] B. E. Mann, *Adv. Organometallic Chem.*, **1974,** *12,* 135; P. W. Jolly and R. Mynott, *Adv. Organometallic Chem.*, **1981,** *19,* 257; E. Breitmaier and W. Voelter, *Carbon-13 NMR Spectroscopy,* VCH, New York, 1987.

[42] Breitmaier and Voelter, op. cit.

[43] Mann, op. cit., has extensive tables of chemical shifts for organic ligands.

[44] P. C. Lauterbur and R. B. King, *J. Am. Chem. Soc.*, **1965,** *87,* 3266.

TABLE 12-4
^{13}C Chemical shifts for organometallic compounds

Ligand	^{13}C *chemical shift (range)**			
M—CH_3	−28.9 to 23.5			
M=C<	190 to 400			
M≡C—	235 to 401			
M—CO	177 to 275			
Neutral binary CO	183 to 223			
M—(η^5-C_5H_5)	−790 to 1430			
Fe(η^5-C_5H_5)$_2$	69.2			
M—(η^3-C_3H_5)	*C_2*: 91 to 129	*C_1 and C_3*: 46 to 79		
M—C_6H_5	*M—C*: 130 to 193	*ortho*: 132 to 141	*meta*: 127 to 130	*para*: 121 to 131

NOTE: * Parts per million relative to $Si(CH_3)_4$.

of atoms is dependent on rapid relaxation). Addition of paramagnetic reagents may speed up relaxation and thereby improve the validity of integration data; one compound often used is $Cr(acac)_3$ [acac = acetylacetonate = $H_3CC(O)CHC(O)CH_3^-$].

12-7-3 EXAMPLES OF CHARACTERIZATION

In this chapter we have considered just a few types of reactions of organometallic compounds, principally the replacement of CO by other ligands and the reactions involved in syntheses of carbene and carbyne complexes. Additional types of reactions will be discussed in Chapter 13. We conclude this chapter with two examples of how spectral data may be used in the characterization of organometallic compounds. Further examples can be found in the problems at the end of this chapter and Chapter 13.

EXAMPLE

$$(H_3C)_2N-C(=S)-S-S-C(=S)-N(CH_3)_2$$

tds

$[(C_5H_5)Mo(CO)_3]_2$ reacts with tetramethylthiuramdisulfide (tds), in refluxing toluene to give a molybdenum-containing product having the following characteristics:

> ^1H NMR: Two singlets, at δ 5.48 (relative area = 5) and δ 3.18 (relative area = 6). (For comparison, $[(C_5H_5)Mo(CO)_3]_2$ has a single ^1H NMR peak at δ 5.30.)
>
> IR: Strong bands at 1950 and 1860 cm^{-1}.
>
> Mass spectrum: A pattern similar to the Mo isotope pattern with the most intense peak at *m/e* = 339. (The most abundant Mo isotope is ^{98}Mo.)

What is the most likely identity of this product?

Solution:

The ^1H NMR singlet at δ 5.48 suggests retention of the C_5H_5 ligand; the chemical shift is a close match for the starting material. The peak at δ 3.18 is most likely due to CH_3 groups originating from the tds. The 5:6 ratio suggests a 1:2 ratio of C_5H_5 ligands to CH_3 groups.

IR shows two bands in the carbonyl region, indicating at least two COs in the product.

The mass spectrum makes it possible to pin down the molecular formula. Subtracting from the total mass the molecular fragments believed to be present:

Total mass	339
−mass of Mo (from mass spectrum pattern)	−98
−mass of C_5H_5	−65
−mass of two CO's	−56
Remaining mass	120

120 is exactly half the mass of tds; it corresponds to the mass of $S_2CN(CH_3)_2$, the dimethyldithiocarbamate ligand. Therefore, the likely formula of the product is $(C_5H_5)Mo(CO)_2[S_2CN(CH_3)_2]$. This formula has the necessary 5:6 ratio of protons in two magnetic environments and should give rise to two C—O stretching vibrations (since the carbonyls would not be expected to be oriented at 180° angles with respect to each other in such a molecule).

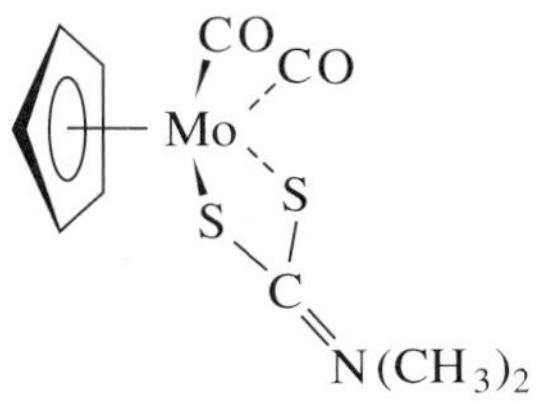

In practice, additional information is likely to be available to help characterize reaction products. For example, additional examination of the infrared spectrum in this case shows a moderately intense band at 1526 cm^{-1}, a common location for C—N stretching bands in dithiocarbamate complexes. Analysis of the fragmentation pattern of mass spectra may also provide useful information on molecular fragments.

EXAMPLE

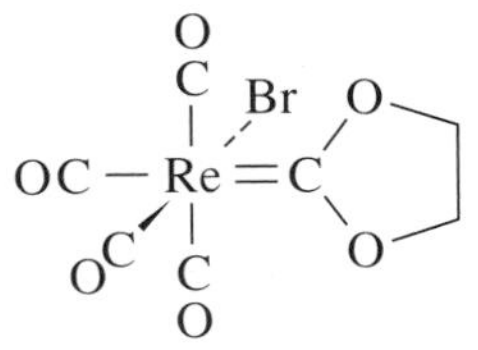

I

When a toluene solution containing **I** and excess triphenylphosphine is heated to reflux, first compound **II** is formed and then compound **III**. **II** has infrared bands at 2038, 1958, and 1906 cm^{-1}; **III**, at 1944 and 1860 cm^{-1}. ^1H and ^{13}C NMR data [δ values (relative area)]:

	I	II	III
^{1}H:	4.83 singlet	7.62, 7.41 multiplets (15)	7.70, 7.32 multiplets (15)
		4.19 multiplet (4)	3.39 singlet (2)
^{13}C:	224.31	231.02	237.19
	187.21	194.98	201.85
	185.39	189.92	193.83
	184.01	188.98	127.75–134.08 (several peaks)
	73.33	129.03–134.71 (several peaks)	68.80
		72.26	

Additional useful information: the ^{13}C signal of **I** at δ 224.31 is similar to the chemical shift of carbene carbons in similar compounds; the peaks between δ 184 and 188 correspond to carbonyls; and the peak at δ 73.33 is typical for CH_2CH_2 bridges in dioxycarbene complexes.

Identify **II** and **III**:

This is a good example of the utility of ^{13}C NMR. Both **II** and **III** have peaks with similar chemical shifts to the peak at δ 224.31 for **I**, suggesting that the carbene ligand is retained in the reaction. Similarly, **II** and **III** have peaks near δ 73.33, a further indication that the carbene ligand remains intact.

The ^{13}C peaks in the range δ 184 to 202 can be assigned to carbonyl groups. **II** and **III** show new peaks in the range δ 129 to 135. The most likely explanation is that the chemical reaction involves replacement of carbonyls by triphenylphosphines and that the new peaks in the 129 to 135 range are due to the phenyl carbons of the phosphines.

^{1}H NMR data are consistent with replacement of COs by phosphines. In both **II** and **III**, integration of the —CH_2CH_2— peaks (δ 4.19, 3.39, respectively) and the phenyl peaks (δ 7.32 to 7.70) gives the expected ratios for replacement of one and two COs.

Finally, infrared data are in agreement with the above conclusions. In **II** the three bands in the carbonyl region are consistent with the presence of three COs in either a *mer* or a *fac* arrangement.[45] In **III** the two C—O stretches correspond to two carbonyls *cis* to each other.

The chemical formulas of these products can now be written as follows:

II: $ReBr(CO)_3(\overline{COCH_2CH_2O})(PPh_3)$

III: *cis*-$ReBr(CO)_2(\overline{COCH_2CH_2O})(PPh_3)_2$

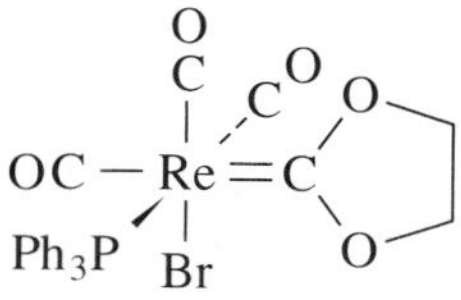

II

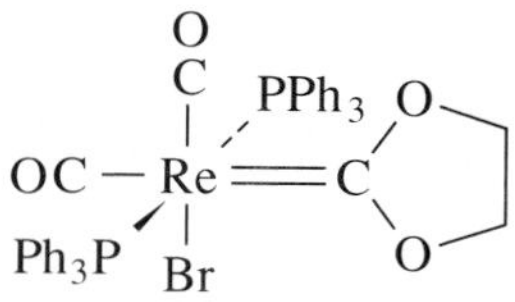

III

EXERCISE 12-9

Using the ^{13}C NMR data, determine if **II** is more likely the *fac* or *mer* isomer.[46]

[45] In an octahedral complex of formula *fac*-$ML_3(CO)_3$ (having C_{3v} symmetry) only two carbonyl stretching bands are expected if all ligands L are identical. However, in this case there are three different ligands in addition to CO, the point group is C_1, and three bands are expected.

[46] G. L. Miessler, S. Kim, R. A. Jacobson, and R. J. Angelici, *Inorg. Chem.*, **1987,** *26,* 1690.

GENERAL REFERENCES

Much information on organometallic compounds is included in the two general inorganic references, N. N. Greenwood and A. Earnshaw, *Chemistry of the Elements,* Pergamon, Elmsford, N.Y., 1984, and F. A. Cotton and G. Wilkinson, *Advanced Inorganic Chemistry,* 5th ed., Wiley-Interscience, New York, 1988. *Principles and Applications of Organotransition Metal Chemistry,* by J. P. Collman, L. S. Hegedus, J. R. Norton, and R. G. Finke, University Science Books, Mill Valley, Calif., 1987, provides extensive discussion, with numerous references, on many additional types of organometallic compounds in addition to those discussed in this chapter. The most comprehensive reference on organometallic chemistry is the nine-volume set *Comprehensive Organometallic Chemistry,* G. Wilkinson and F. G. A. Stone, eds., Pergamon Press, Elmsford, N.Y., 1982. Volume 9 of this set has an extensive listing of references of organometallic compounds that had been structurally characterized by X-ray, electron, or neutron diffraction through late 1980. A useful reference to literature sources on the synthesis, properties, and reactions of specific organometallic compounds is *Dictionary of Organometallic Compounds,* Chapman and Hall, London, 1984, to which supplementary volumes have also been published. The series *Advances in Organometallic Chemistry,* Academic Press, New York, provides valuable review articles on a variety of organometallic topics.

PROBLEMS

12-1 Which of the following obey the 18-electron rule?

a. $Fe(CO)_5$ **b.** $[Rh(bipy)_2Cl]^+$ **c.** $(\eta^5\text{-}Cp^*)Re({=}O)_3$; $Cp^* = C_5(CH_3)_5$
d. $Re(PPh_3)_2Cl_2N$ **e.** $Os(CO)({\equiv}CPh)(PPh_3)_2Cl$

12-2 Which of the following square planar complexes have 16-electron valence configurations?

a. $Ir(CO)Cl(PPh_3)_2$ **b.** $RhCl(PPh_3)_3$ **c.** $[Ni(CN)_4]^{2-}$ **d.** *cis*-$PtCl_2(NH_3)_2$

12-3 On the basis of the 18-electron rule, identify the first-row transition metal for each of the following:

a. $[M(CO)_7]^+$ **b.** $H_3CM(CO)_5$ **c.** $M(CO)_2(CS)(PPh_3)Br$
d. $[(\eta^3\text{-}C_3H_3)(\eta^5\text{-}C_5H_5)M(CO)]^-$ **e.** $(OC)_5M{=}C(OCH_3)(C_6H_5)$
f. $[(\eta^4\text{-}C_4H_4)(\eta^5\text{-}C_5H_5)M]^+$ **g.** $(\eta^3\text{-}C_3H_5)(\eta^5\text{-}C_5H_5)M(CH_3)(NO)$
h. $[M(CO)_4I(diphos)]^-$; diphos = 1,2-bis(diphenylphosphino)ethane

12-4 Determine the metal–metal bond order consistent with the 18-electron rule for the following:

a. $[(\eta^5\text{-}C_5H_5)Fe(CO)_2]_2$ **b.** $[(\eta^5\text{-}C_5H_5)Mo(CO)_2]_2^{2-}$

12-5 Identify the most likely second-row transition metal for each of the following:

a. $[M(CO)_3(NO)]^-$ **b.** $[M(PF_3)_2(NO)_2]^+$ (contains linear NO ligands)
c. $[M(CO)_4(\mu_2\text{-}H)]_3$ **d.** $M(CO)(PMe_3)_2Cl$ (square planar complex)

12-6 On the basis of the 18-electron rule, determine the expected charge on the following:

a. $[Co(CO)_3]^z$ **b.** $[Ni(CO)_3(NO)]^z$ (assume linear NO)
c. $[Ru(CO)_4(GeMe_3)]^z$ **d.** $[(\eta^3\text{-}C_3H_5)V(CNCH_3)_5]^z$ **e.** $[(\eta^5\text{-}C_5H_5)Fe(CO)_3]^z$
f. $[(\eta^5\text{-}C_5H_5)_3Ni_3(\mu_3\text{-}CO)_2]^z$

12-7 Determine the unknown quantity:

a. $[(\eta^5\text{-}C_5H_5)W(CO)_x]_2$ (has W—W single bond) **b.** $ReBr(CO)_x$(C(O)(O)) [cyclic carbene ligand with two O atoms]

c. $[(CO)_3Ni\text{–}Co(CO)_3]^z$ **d.** $[Ni(NO)_3(SiMe_3)]^z$ (assume linear NO ligands)
e. $[(\eta^5\text{-}C_5H_5)Mn(CO)_x]_2$ (has Mn=Mn bond)

12-8 Nickel tetracarbonyl, $Ni(CO)_4$, is an 18-electron species. Using a qualitative molecular orbital diagram, explain the stability of this 18 valence electron molecule. (Reference: G. Cooper, K. H. Sze, and C. E. Brion, *J. Am. Chem. Soc.*, **1989,** *111,* 5051.)

12-9 The Re—O stretching vibration in $Re(^{16}O)I(HC\equiv CH)_2$ is at 975 cm^{-1}. Predict the position of the Re—O stretching band in $Re(^{18}O)I(HC\equiv CH)_2$. (Reference: J. M. Mayer, D. L. Thorn, and T. H. Tulip, *J. Am. Chem. Soc.*, **1985,** *107,* 7454.)

12-10 The compound $W(O)Cl_2(CO)(PMePh_2)_2$ has $\nu(CO)$ at 2006 cm^{-1}. Would you predict $\nu(CO)$ for $W(S)Cl_2(CO)(PMePh_2)_2$ to be at higher or lower energy? Explain briefly. (Reference: J. C. Bryan, S. J. Geib, A. L. Rheingold, and J. M. Mayer, *J. Am. Chem. Soc.*, **1987,** *109,* 2826.)

12-11 The vanadium–carbon distance in $V(CO)_6$ is 200 pm, but only 193 pm in $[V(CO)_6]^-$. Explain.

12-12 Describe, using sketches, how the following ligands can act as both sigma donors and pi acceptors:
a. CN^- **b.** $P(CH_3)_3$ **c.** SCN^-

12-13 **a.** Account for the following trend in infrared frequencies:

$[Cr(CN)_5(NO)]^{4-}$, $\nu(NO) = 1515\ cm^{-1}$

$[Mn(CN)_5(NO)]^{3-}$, $\nu(NO) = 1725\ cm^{-1}$

$[Fe(CN)_5(NO)]^{2-}$, $\nu(NO) = 1939\ cm^{-1}$

b. The ion $[RuCl(NO)_2(PPh_3)_2]^+$ has N—O stretching bands at 1687 and 1845 cm^{-1}. The C—O stretching bands of dicarbonyl complexes typically are much closer in energy. Explain.

12-14 Sketch the π molecular orbitals for the following:
a. CO_2 **b.** 1,3,5-hexatriene **c.** Cyclobutadiene, C_4H_4 **d.** *Cyclo*-C_7H_7

12-15 For the hypothetical molecule $(\eta^4\text{-}C_4H_4)Mo(CO)_4$:
a. Assuming C_{4v} geometry, predict the number of infrared-active C—O bands.
b. Sketch the π molecular orbitals of cyclobutadiene. For each, indicate which *s, p,* and *d* orbitals of Mo are of suitable symmetry for interaction. (Suggestion: Assign the *z* axis to be collinear with the C_4 axis.)

12-16 Using the D_{5h} character table in Appendix C:
a. Assign symmetry labels (labels of irreducible representations) for the group orbitals shown in Figure 12-25.
b. Assign symmetry labels for the atomic orbitals of Fe in a D_{5h} environment.
c. Verify that the orbital interactions for ferrocene shown in Figure 12-26 are between atomic orbitals of Fe and group orbitals of matching symmetry.

12-17 Dibenzenechromium, $(\eta^6\text{-}C_6H_6)_2Cr$, is a sandwich compound having two parallel benzene rings in an eclipsed conformation. For this molecule:
a. Sketch the π orbitals of benzene.
b. Sketch the group orbitals, using the π orbitals of the two benzene rings.

c. For each of the (12) group orbitals, identify the Cr orbital(s) of suitable symmetry for interaction.
d. Sketch an energy-level diagram of the molecular orbitals.

12-18 Predict the number of infrared-active C—O stretching vibrations for (η^6-C_6H_6)W(CO)$_3$, assuming C_{3v} geometry.

12-19 Refluxing W(CO)$_6$ in butyronitrile (C_3H_7CN) gives, in succession, products in which one, two, and three carbonyls have been replaced by coordinating butyronitrile. The following carbonyl stretching bands (in cm^{-1}) are observed:

W(CO)$_5$(NCC$_3$H$_7$),	2077, 1975, 1938
W(CO)$_4$(NCC$_3$H$_7$)$_2$,	2107, 1898, 1842
W(CO)$_3$(NCC$_3$H$_7$)$_3$,	1910, 1792

a. On the basis of the number of infrared bands, determine the most likely isomers of W(CO)$_4$(NCC$_3$H$_7$)$_2$ and W(CO)$_3$(NCC$_3$H$_7$)$_3$ formed in this reaction.
b. Account for the trend in the position of the C—O bands as CO ligands are replaced by butyronitrile.
(Reference: G. J. Kubas, *Inorg. Chem.*, **1983,** *22,* 692.)

12-20 Samples of Fe(CO)(PF$_3$)$_4$ show *two* carbonyl stretching bands, at 2038 and 2009 cm^{-1}.
a. How is it possible for this compound to exhibit two carbonyl bands?
b. Fe(CO)$_5$ has carbonyl bands at 2025 and 2000 cm^{-1}. Would you place PF$_3$ above or below CO in the spectrochemical series? Explain briefly.
(Reference: H. Mahnke, R. J. Clark, R. Rosanske, and R. K. Sheline, *J. Chem. Phys.*, **1974,** *60,* 2997.)

12-21 Account for the observation that [Co(CO)$_3$(PPh$_3$)$_2$]$^+$ has only a single carbonyl stretching frequency.

12-22 The following carbonyl bands have been reported:

(η^5-C$_5$H$_5$)Re(CO)$_3$	2024, 1937 cm^{-1}
(η^5-C$_5$H$_5$)Re(CO)$_2$(CSe)	2005, 1946 cm^{-1}

On the basis of this information, which ligand, CO or CSe, is the better pi acceptor? Explain briefly. (Reference: I. S. Butler, D. Cozak, and S. Stobart, *Inorg. Chem.*, **1977,** *16,* 1779.)

12-23 One of the first thionitrosyl complexes to be reported was (η^5-C$_5$H$_5$)Cr(CO)$_2$(NS) (T. J. Greenhough, B. W. S. Kolthammer, P. Legzdins, and J. Trotter, *J. Chem. Soc. Chem. Commun.*, **1978,** 1036). This compound has carbonyl bands at 1962 and 2033 cm^{-1}. The corresponding bands for (η^5-C$_5$H$_5$)Cr(CO)$_2$(NO) are at 1955 and 2028 cm^{-1}. On the IR evidence, is NS behaving as a stronger or weaker pi acceptor in these compounds? Explain briefly.

12-24 Predict the products of the following reactions:
a. Mo(CO)$_6$ + Ph$_2$P—CH$_2$—PPh$_2$ $\xrightarrow{\Delta}$
b. (η^5-C$_5$H$_5$)(η^1-C$_5$H$_5$)Fe(CO)$_2$ $\xrightarrow{h\nu}$
c. (η^5-C$_5$Me$_5$)Rh(CO)$_2$ $\xrightarrow{\Delta}$(dimeric product, contains one CO per metal)
d. V(CO)$_6$ + NO $\longrightarrow$
e. W(CO)$_5$(C$_6$H$_5$)(OC$_2$H$_5$) + BF$_3$ $\longrightarrow$
f. [(η^5-C$_5$H$_5$)Fe(CO)$_2$]$_2$ + Al(C$_2$H$_5$)$_3$ $\longrightarrow$

12-25 Complexes of formula $Rh(CO)(phosphine)_2Cl$ have the C—O stretching bands shown below. Match the infrared bands with the appropriate phosphine.

Phosphines: $P(p\text{-}C_6H_4F)_3$, $P(p\text{-}C_6H_4Me)_3$, $P(t\text{-}C_4H_9)_3$, $P(C_6F_5)_3$

$\nu(CO)$, cm^{-1}: 1923, 1965, 1984, 2004

12-26 For each of the following sets, which complex would be expected to have the highest C—O stretching frequency?
a. $Fe(CO)_5$ $Fe(CO)_4(PF_3)$ $Fe(CO)_4(PCl_3)$ $Fe(CO)_4(PMe_3)$
b. $[Re(CO)_6]^+$ $W(CO)_6$ $[Ta(CO)_6]^-$
c. $Mo(CO)_3(PCl_3)_3$ $Mo(CO)_3(PCl_2Ph)_3$ $Mo(CO)_3(PPh_3)_3$ $Mo(CO)_3py_3$ (py = pyridine)

12-27 Arrange the following complexes in order of the expected frequency of their $\nu(CO)$ bands (Reference: M. F. Ernst and D. M. Roddick, *Inorg. Chem.*, **1989**, *28*, 1624).

$Mo(CO)_4(F_2PCH_2CH_2PF_2)$

$Mo(CO)_4[(C_6F_5)_2PCH_2CH_2P(C_6F_5)_2]$

$Mo(CO)_4(Et_2PCH_2CH_2PEt_2)$, (Et = C_2H_5)

$Mo(CO)_4(Ph_2PCH_2CH_2PPh_2)$, (Ph = C_6H_5)

$Mo(CO)_4[(C_2F_5)_2PCH_2CH_2P(C_2F_5)_2]$

12-28 Free N_2 has a stretching vibration (not observable by IR; why?) at 2331 cm^{-1}. Would you expect the stretching vibration for coordinated N_2 to be at higher or lower energy? Explain briefly.

12-29 The ^{1}H NMR spectrum of the carbene complex shown below shows two peaks of equal intensity at 40°C. However, at −40°C the NMR shows four peaks, two of a lower intensity and two of a higher intensity. The solution may be warmed and cooled repeatedly without changing the NMR properties at these temperatures. Account for this NMR behavior.

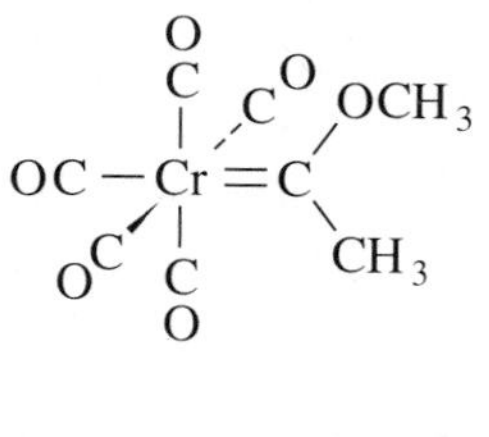

```
       O
       C  CO   OCH3
       | /     /
OC — Cr = C
     / |       \
   OC  C        CH3
       O
```

12-30 The ^{1}H NMR spectrum of $(C_5H_5)_2Fe(CO)_2$ shows only a single peak at room temperature but gives four resonances of relative intensity 5:2:2:1 at low temperatures. Explain. (Reference: C. H. Campbell and M. L. H. Green, *J. Chem. Soc., (A)*, **1970**, 1318.)

12-31 Of the compounds $Cr(CO)_5(PF_3)$ and $Cr(CO)_5(PCl_3)$, which would you expect to have:
a. The shorter C—O bonds?
b. The higher-energy Cr—C stretching bands in the infrared spectrum?

12-32 Select the best choice for each of the following:

a. Higher N—O stretching frequency:

$[Fe(NO)(mnt)_2]^-$

$[Fe(NO)(mnt)_2]^{2-}$

mnt^{2-} = (S)(NC)C=C(S)(CN) $]^{2-}$

b. Longest N—N bond:

N_2

$(CO)_5Cr:N\equiv N$

$(CO)_5Cr:N\equiv N:Cr(CO)_5$

c. Shorter Ta—C distance in $(\eta^5\text{-}C_5H_5)_2Ta(CH_2)(CH_3)$:

Ta—CH_2

Ta—CH_3

d. Shortest Cr—C distance:

$Cr(CO)_6$

Cr—CO in *trans*-$Cr(CO)_4I(CCH_3)$

Cr—CCH_3 in *trans*- $Cr(CO)_4I(CCH_3)$

e. Lowest C—O stretching frequency:

$Ni(CO)_4$

$[Co(CO)_4]^-$

$[Fe(CO)_4]^{2-}$

12-33 In this chapter the assertion was made that highly symmetric binary carbonyls of T_d and O_h symmetry should show only a single C—O stretching band in the infrared. Check this assertion by analyzing the C—O vibrations of $Ni(CO)_4$ and $Cr(CO)_6$ by the symmetry method described in Chapter 4.

12-34 $Mn_2(CO)_{10}$ and $Re_2(CO)_{10}$ have D_{4d} symmetry. How many infrared-active carbonyl stretching bands would you predict for these compounds?

12-35 A solution of blue $Mo(CO)_2(PEt_3)_2Br_2$ was treated with a tenfold excess of 2-butyne to give **X**, a dark green product. **X** had bands in the ^1H NMR at δ 0.90 (relative area = 3), 1.63 (2), and 3.16 (1). The peak at 3.16 was a singlet at room temperature but split into two peaks at temperatures below −20°C. ^{31}P NMR showed only a single resonance. Infrared showed a single strong band at 1950 cm^{-1}. Molecular weight determinations suggest that **X** has a molecular weight of 580 ± 15. Suggest a structure for **X**, and account for as much of the data as possible. (Reference: P. B. Winston, S. J. Nieter Burgmayer, and J. L. Templeton, *Organometallics*, **1983,** *2,* 167.)

12-36 Photolysis at −78° of $[(\eta^5\text{-}C_5H_5)Fe(CO)_2]_2$ results in loss of a colorless gas and formation of an iron-containing product having a single carbonyl band at 1785 cm^{-1} and containing 14.7% oxygen by weight. Suggest a structure for the product.

12-37 Nickel carbonyl reacts with cyclopentadiene to produce a red, diamagnetic compound of formula $NiC_{10}H_{12}$. The ^1H NMR spectrum of this compound shows four different types of hydrogen; integration gives relative areas of

5:4:2:1, with the most intense peak in the aromatic region. Suggest a structure for $NiC_{10}H_{12}$ that is consistent with this NMR spectrum.

12-38 The carbonyl carbon–molybdenum–carbon angle in $Cp(CO)_2Mo[\mu\text{-}S_2C_2(CF_3)_2]_2MoCp$ (Cp = $\eta^5\text{-}C_5H_5$) is 76.05°. Calculate the ratio of intensities $I_{symmetric}/I_{asymmetric}$ expected for the C—O stretching bands of this compound. (Reference: K. Roesselet, K. E. Doan, S. D. Johnson, P. Nicholls, G. L. Miessler, R. Kroeker, and S. H. Wheeler, *Organometallics,* **1987,** *6,* 480.)

13

Organometallic Reactions and Catalysis

Organometallic compounds undergo a rich, almost bewildering, variety of reactions, comparable in diversity to the reactions of organic molecules. These may involve loss or gain of ligands (or both), molecular rearrangement, formation or breaking of metal–metal bonds, or reactions at the ligands themselves. Often reaction mechanisms involve multiple steps, and frequently reactions yield not one, but a variety of products. Sequences of reactions may be combined into catalytic cycles that may be useful, in some cases commercially. In this chapter we will not attempt to cover all possible types of organometallic reactions but will concentrate on those that have proved most common and useful, particularly for syntheses and catalytic processes. We will discuss organometallic reactions according to the following outline:

I. Reactions Involving Gain or Loss of Ligands
- **A.** Ligand dissociation and substitution
- **B.** Oxidative addition
- **C.** Reductive elimination
- **D.** Nucleophilic displacement

II. Reactions Involving Modification of Ligands
- **A.** Insertion
- **B.** Carbonyl insertion (alkyl migration)
- **C.** Hydride elimination
- **D.** Abstraction

13-1 REACTIONS INVOLVING GAIN OR LOSS OF LIGANDS

Some of the most important reactions of organometallic compounds involve a change in coordination number of the metal by a gain or loss of ligands. If the formal oxidation state of the metal is retained, these reactions are considered addition or dissociation reactions; if the formal oxidation state is changed, they are termed oxidative additions or reductive eliminations.

Type of reaction	*Change in coordination number*	*Change in formal oxidation state of metal*
Addition	Increase	None
Dissociation	Decrease	None
Oxidative addition	Increase	Increase
Reductive elimination	Decrease	Decrease

In classifying these reactions, it will frequently be necessary to determine formal oxidation states of the metals in organometallic compounds. In general, method A (the donor pair method) described in Chapter 12 can be used in assigning oxidation states. Examples will be given later in this chapter in the discussion of oxidative addition reactions.

We will first consider ligand dissociation reactions. When coupled with addition reactions, dissociation reactions can be very useful synthetically, providing an avenue to replace ligands such as carbon monoxide and phosphines by other ligands.

13-1-1 LIGAND DISSOCIATION AND SUBSTITUTION

CO dissociation

Chapter 12 gave a brief introduction to carbonyl dissociation reactions, in which CO may be lost thermally or photochemically. Such a reaction may result in rearrangement of the molecule or replacement of CO by another ligand:

$$(\eta^5\text{-}C_5H_5)(OC)_3Mo{-}Mo(CO)_3(\eta^5\text{-}C_5H_5) \xrightarrow{\Delta} (\eta^5\text{-}C_5H_5)(OC)_2Mo{\equiv}Mo(CO)_2(\eta^5\text{-}C_5H_5) + 2CO$$

$$Fe(CO)_5 + P(CH_3)_3 \xrightarrow{\Delta} Fe(CO)_4(P(CH_3)_3) + CO$$

The second type of reaction, involving ligand replacement, is an important way to introduce new ligands into complexes and deserves further discussion.

Most thermal reactions involving replacement of CO by another ligand L have rates that are independent of concentration of L; they are first order with respect to the metal complex. This behavior is consistent with a **dissociative** mechanism involving slow loss of CO, followed by rapid reaction with L:

$Ni(CO)_4 \xrightarrow{k_1} Ni(CO)_3 + CO$ (slow), loss of CO from an 18-electron complex

$Ni(CO)_3 + L \xrightarrow{k_2} Ni(CO)_3L$ (fast), addition of L to a 16-electron intermediate

Loss of CO from the stable, 18-electron $Ni(CO)_4$ is slow relative to addition of L to the more reactive, 16-electron $Ni(CO)_3$. Consequently, the first step is rate limiting and this mechanism has the rate law

$$\text{Rate} = k_1[Ni(CO)_4]$$

Some reactions show more complicated kinetics. For example, study of the reaction

$$Mo(CO)_6 + L \xrightarrow{\Delta} Mo(CO)_5L + CO \qquad (L = \text{phosphine})$$

has shown that for some phosphine ligands the rate law has the form

$$\text{Rate} = k_1[Mo(CO)_6] + k_2[Mo(CO)_6][L]$$

The two terms in the rate law imply parallel pathways for the formation of $Mo(CO)_5L$. The first term is again consistent with a dissociative mechanism:

$$Mo(CO)_6 \xrightarrow{k_1} Mo(CO)_5 + CO \quad (\text{slow})$$

$$Mo(CO)_5 + L \longrightarrow Mo(CO)_5L \quad (\text{fast})$$

$$\text{Rate}_1 = k_1[Mo(CO)_6]$$

The second term in the rate law is consistent with an **associative** process involving a bimolecular reaction of $Mo(CO)_6$ and L to form a transition state, which then loses CO:

$$Mo(CO)_6 + L \xrightarrow{k_2} [Mo(CO)_6\text{---}L], \qquad \text{association of } Mo(CO)_6 \text{ and L}$$

$$[Mo(CO)_6\text{---}L] \longrightarrow Mo(CO)_5L + CO, \qquad \text{loss of CO from transition state}$$

Formation of the transition state is the rate-limiting step in this mechanism; the rate law for this pathway is therefore

$$\text{Rate}_2 = k_2[Mo(CO)_6][L]$$

There is also strong evidence that solvent is involved in the first-order mechanism for replacement of CO; however, the rate law obtained in this case is the same as shown above.[1]

Because of the two pathways, the overall rate of formation of $Mo(CO)_5L$ is the sum of the rates of the unimolecular and bimolecular mechanisms, $\text{rate}_1 + \text{rate}_2$.

While most CO substitution reactions proceed primarily by a dissociative mechanism, an associative path is more likely for complexes of large metals (providing favorable sites for incoming ligands to attack) and for reactions involving highly nucleophilic ligands.

Dissociation of phosphine

Carbon monoxide is by no means the only ligand that can undergo dissociation from metal complexes. Many other ligands can dissociate, with the ease of

[1] W. D. Covey and T. L. Brown, *Inorg. Chem.*, **1973,** *12,* 2820.

dissociation a function of the strength of metal–ligand bonding and, in some cases, the degree of crowding of ligands around the metal. These steric effects have been investigated for a variety of ligands, especially phosphines and similar ligands.

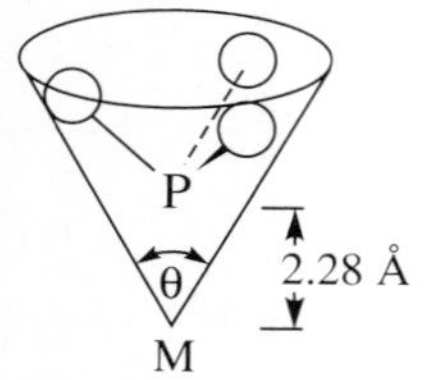

FIGURE 13-1 Ligand Cone Angle.

To describe steric effects, Tolman has defined the **cone angle** as the apex angle θ of a cone that encompasses the van der Waals radii of the outermost atoms of a ligand, as shown in Figure 13-1. Values of cone angles of selected ligands are given in Table 13-1.[2]

As might be expected, the presence of bulky ligands, having large cone angles, can lead to more rapid ligand dissociation as a consequence of crowding around the metal. For example, the rate of the reaction

$$cis\text{-}Mo(CO)_4L_2 + CO \longrightarrow Mo(CO)_5L + L \quad (L = \text{phosphine or phosphite})$$

which is first order in *cis*-$Mo(CO)_4L_2$, increases with increasing ligand bulk, as shown in Figure 13-2; the larger the cone angle, the more rapidly the phosphine or phosphite is lost.[3]

TABLE 13-1
Ligand cone angles

Ligand	θ	*Ligand*	θ
PH_3	87°	$P(CH_3)(C_6H_5)_2$	136°
PF_3	104°	$P(CF_3)_3$	137°
$P(OCH_3)_3$	107°	$P(C_6H_5)_3$	145°
$P(OC_2H_5)_3$	109°	$P(cyclo\text{-}C_6H_{11})_3$	170°
$P(CH_3)_3$	118°	$P(t\text{-}C_4H_9)_3$	182°
PCl_3	124°	$P(C_6F_5)_3$	184°
PBr_3	131°	$P(o\text{-}C_6H_4CH_3)_3$	194°
$P(C_2H_5)_3$	132°		

Numerous other examples of the effect of ligand bulk on the dissociation of ligands have been reported in the chemical literature.[4] For many dissociation reactions, the effect of ligand crowding may be more important than electronic effects in determining reaction rates.

13-1-2 OXIDATIVE ADDITION REACTIONS

These reactions, as the name suggests, involve an increase in both the *formal* oxidation state and the coordination number of the metal. Oxidative addition reactions are among the most important of organometallic reactions and are essential steps in many catalytic processes; the reverse type of reaction, designated reductive elimination, is also very important.

For example, heating $Fe(CO)_5$ in the presence of I_2 leads to formation of *cis*-$I_2Fe(CO)_4$. The reaction has two steps:

$$Fe(CO)_5 \xrightarrow{\Delta} Fe(CO)_4 \xrightarrow{I_2} cis\text{-}I_2Fe(CO)_4$$

[2] C. A. Tolman, *J. Am. Chem. Soc.*, **1970,** *92,* 2953; *Chem. Rev.*, **1977,** *77,* 313.

[3] D. J. Darensbourg and A. H. Graves, *Inorg. Chem.*, **1979,** *18,* 1257.

[4] For example, M. J. Wovkulich and J. D. Atwood, *Organometallics,* **1982,** *1,* 1316; J. D. Atwood, M. J. Wovkulich, and D. C. Sonnenberger, *Acc. Chem. Res.*, **1983,** *16,* 350.

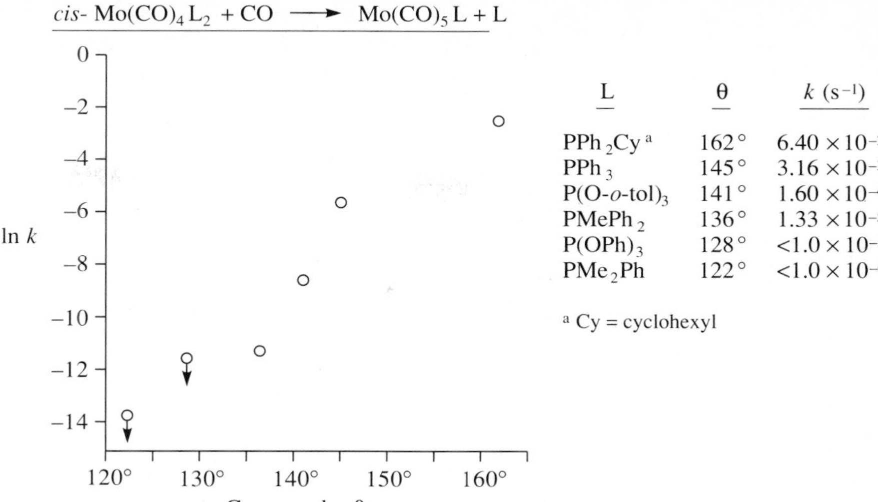

L	θ	k (s⁻¹)
PPh_2Cy[a]	162°	6.40×10^{-2}
PPh_3	145°	3.16×10^{-3}
$P(O\text{-}o\text{-}tol)_3$	141°	1.60×10^{-4}
$PMePh_2$	136°	1.33×10^{-5}
$P(OPh)_3$	128°	$<1.0 \times 10^{-5}$
PMe_2Ph	122°	$<1.0 \times 10^{-6}$

[a] Cy = cyclohexyl

FIGURE 13-2 Reaction Rate versus Cone Angle for Phosphine Dissociation.

The first step involves dissociation of CO to give a 4-coordinate iron(0) intermediate. In the second step, iron is formally oxidized to iron (II) and the coordination number expanded by addition of two iodo ligands. This second step is an example of oxidative addition.

It may be useful at this point to review briefly the assignment of oxidation states. Coordinated ligands are generally assigned the charges of the free ligand (zero for neutral ligands such as CO, 1− for Cl^-, CN^-, and so on). Hydrogen atom ligands and organic radicals are treated as anions:

H^-	CH_3^-	$C_6H_5^-$	$C_5H_5^-$
hydride	methyl	phenyl	η^5-C_5H_5

(The assigned charges on these ligands may have little chemical significance. For example, in methyl complexes the carbon–metal bond is largely covalent, and such complexes should not be viewed as containing the free ion CH_3^-. The assignment of these charges is a formalism, another electron counting scheme.)

Oxidative addition reactions of square planar d^8 complexes are of special chemical significance, and we will therefore use one such complex, *trans*-$Ir(CO)Cl(PEt_3)_2$, to illustrate these reactions (Figure 13-3). In each of the examples shown, the formal oxidation state of iridium increases from (I) to (III), and its coordination number increases from 4 to 6. The new ligands may add in a *cis* or *trans* fashion, with their orientation a function of the mechanistic pathway involved. An important feature of such reactions is that, in the expansion of the coordination number of the metal, the newly added ligands are brought into close proximity to the original ligands; this may enable chemical reactions to occur between ligands. Such reactions, encountered frequently in the mechanisms of catalytic cycles involving organometallic compounds, will be discussed later in this chapter.

Cyclometallations

These are reactions that incorporate metals into organic rings. The most common of these are *orthometallations*, oxidative additions in which the *ortho*

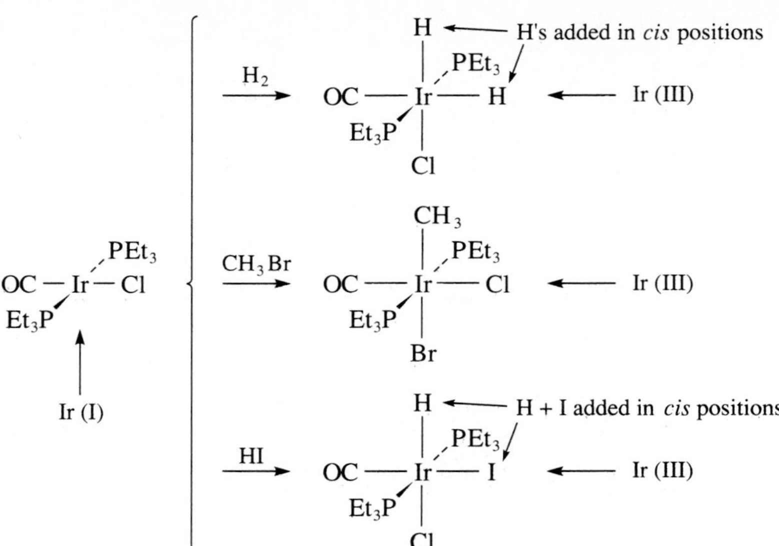

FIGURE 13-3 Examples of Oxidative Addition Reactions.

position of an aromatic ring on a ligand becomes attached to the metal. The first example in Figure 13-4 is an oxidative addition in which an *ortho* carbon and the hydrogen originally in the *ortho* position add to iridium. Not all cyclometallation reactions are oxidative additions; the second example in Figure 13-4 shows a cyclometallation that is not an oxidative addition overall (although one step in the mechanism may be oxidative addition).

13-1-3 REDUCTIVE ELIMINATION

The following equilibrium has been studied:

$$(\eta^5\text{-}C_5H_5)_2TaH + H_2 \rightleftharpoons (\eta^5\text{-}C_5H_5)_2TaH_3$$

The forward reaction involves formal oxidation of the metal, accompanied by an increase in coordination number; it is an oxidative addition. The reverse reaction is labeled reductive elimination; it involves a decrease in both oxidation number and coordination number. Reductive elimination is simply the reverse of oxidative addition.

Reductive elimination reactions often involve elimination of molecules such as:

$$\text{R—H,} \quad \text{R—R}', \quad \text{R—X,} \quad \text{H—H} \qquad (\text{R, R}' = \text{alkyl, aryl; X} = \text{halogen})$$

The products eliminated by these reactions may be important and useful organic compounds (R—H, R—R′, R—X). In some cases the organic fragments (R, R′) undergo rearrangement or other reactions while coordinated to the metal. Examples of this phenomenon will be shown in the discussion of catalysis later in this chapter.

13-1-4 NUCLEOPHILIC DISPLACEMENT REACTIONS

Ligand displacement reactions may be described as nucleophilic substitutions, involving incoming ligands as nucleophiles. Organometallic complexes, espe-

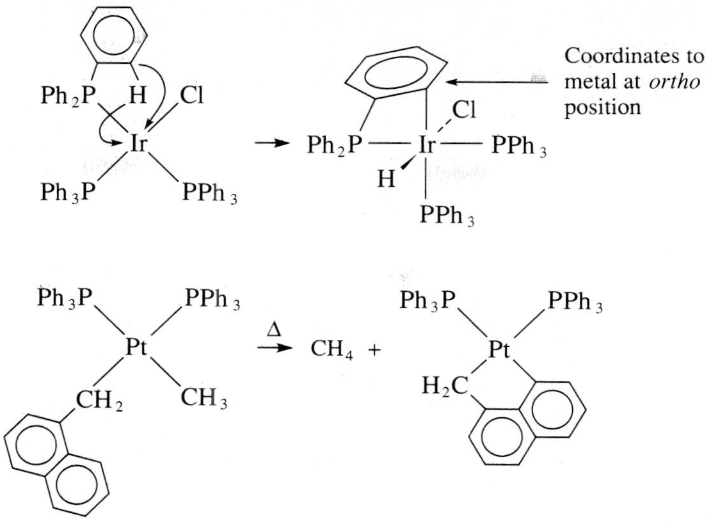

Coordinates to metal at *ortho* position

FIGURE 13-4 Cyclometallation Reactions.

cially those carrying negative charges, may themselves behave as nucleophiles in displacement reactions. For example, the anion $[(\eta^5\text{-}C_5H_5)Mo(CO)_3]^-$ can displace iodide from methyl iodide:

$$[(\eta^5\text{-}C_5H_5)Mo(CO)_3]^- + CH_3I \longrightarrow (\eta^5\text{-}C_5H_5)(CH_3)Mo(CO)_3 + I^-$$

An extremely useful organometallic nucleophile is $[Fe(CO)_4]^{2-}$. Cooke and Collman developed the synthesis for the parent compound of this nucleophile, $Na_2Fe(CO)_4$, by reacting sodium with $Fe(CO)_5$ in dioxane:[5]

$$2\,Na + Fe(CO)_5 \xrightarrow[100°C]{\text{dioxane}} Na_2Fe(CO)_4 \cdot 1.5\,\text{dioxane} + CO$$

The product of this reaction is extremely useful in the synthesis of a variety of organic compounds. For example, nucleophilic attack of $[Fe(CO)_4]^{2-}$ on an organic halide RX yields $[RFe(CO)_4]^-$, which can subsequently be converted to alkanes, ketones, carboxylic acids, aldehydes, acid halides, or other organic products. These reactions are outlined in Figure 13-5; note that $[RFe(CO)_4]^-$ undergoes other types of reactions in addition to nucleophilic displacements, as shown for some examples in the figure. Additional details of these reactions can be found in the literature.[6]

Another useful anionic nucleophile is $[Co(CO)_4]^-$, whose chemistry has been developed by Heck.[7] A rather mild nucleophile, $[Co(CO)_4]^-$, can be synthesized by reduction of $Co_2(CO)_8$ by sodium; it reacts with organic halides to generate alkyl complexes:

$$[Co(CO)_4]^- + RX \longrightarrow RCo(CO)_4 + X^-$$

[5] M. P. Cooke, *J. Am. Chem. Soc.*, **1970,** *92,* 6080; J. P. Collman, *Acc. Chem. Res.*, **1975,** *8,* 342; R. G. Finke and T. N. Sorrell, *Org. Synth.*, **1979,** *59,* 102.

[6] J. P. Collman, R. G. Finke, J. N. Cawse, and J. I. Brauman, *J. Am. Chem. Soc.*, **1977,** *99,* 2515; *J. Am. Chem. Soc.*, **1978,** *100,* 4766.

[7] R. F. Heck, in I. Wender and P. Pino, eds., *Organic Synthesis via Metal Carbonyls,* Vol. 1, Wiley, New York, 1968, pp. 373–404.

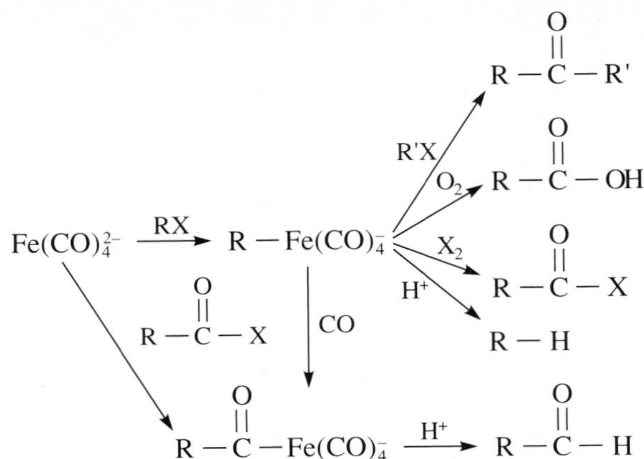

FIGURE 13-5 Synthetic Pathways Using $Fe(CO)_4^{2-}$.

The alkyl complex reacts with carbon monoxide to *apparently* insert CO into the cobalt–alkyl bond (insertion reactions will be discussed later in this chapter) to give an acyl complex (containing a —C(=O)R ligand):

$$RCo(CO)_4 + CO \longrightarrow R\overset{O}{\overset{\|}{C}}Co(CO)_4$$

The acyl complex can then react with alcohols to generate esters:

$$R\overset{O}{\overset{\|}{C}}Co(CO)_4 + R'OH \longrightarrow R\overset{O}{\overset{\|}{C}}OR' + HCo(CO)_4$$

Reaction of $HCo(CO)_4$, a strong acid, with base can regenerate the $[Co(CO)_4]^-$ and, hence, make the overall process catalytic.

Many other nucleophilic anionic organometallic complexes have been studied. Parallels between these anions and anions of main group elements will be discussed in Chapter 14.

13-2 REACTIONS INVOLVING MODIFICATION OF LIGANDS

Many cases are known in which a ligand or molecular fragment appears to insert into a metal–ligand bond. While some of these reactions are believed to occur by direct, single-step insertion, many insertion reactions are much more complicated and do not involve a direct insertion step at all. The most studied of these reactions are the carbonyl insertions; these will be discussed following a brief introduction to some common insertion reactions.

13-2-1 INSERTION REACTIONS

Many reactions occur with formal insertion of a molecule or molecular fragment into a metal–ligand bond. Examples of insertion reactions are shown in Figure 13-6. The reactions in Figure 13-6 may be designated 1,1 insertions, indicating that both bonds to the inserted molecule are made to the same atom in that molecule (for example, in the second reaction both the Mn and CH_3 are bonded to the sulfur of the inserted SO_2).

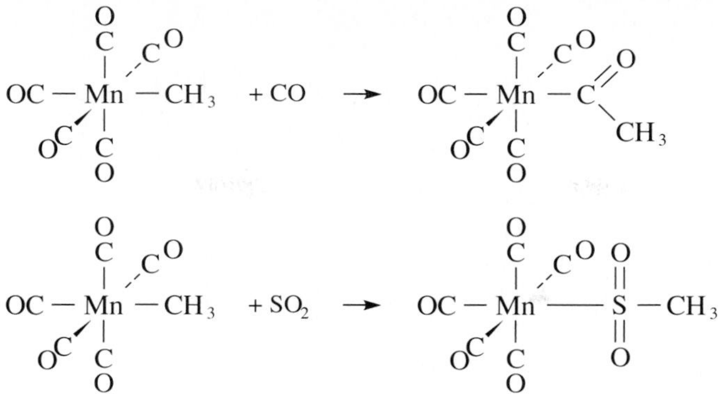

$$\mathrm{(CO)_5Mn{-}CH_3} + \mathrm{CO} \rightarrow \mathrm{(CO)_5Mn{-}C({=}O)CH_3}$$

$$\mathrm{(CO)_5Mn{-}CH_3} + \mathrm{SO_2} \rightarrow \mathrm{(CO)_5Mn{-}S(O)_2{-}CH_3}$$

FIGURE 13-6 Examples of Insertion Reactions.

The first reaction might on first view be expected to occur by a mechanism involving direct insertion of an incoming ligand into the metal–ligand bond. However, as we will see, studies have shown that the CO inserted into the Mn—CH_3 bond must come from within the molecule. This is an example of the most common "insertion" mechanism: not direct insertion of an entering ligand, but rather a more complicated pathway involving *intra*molecular ligand migration prior to addition of the incoming ligand.

1,2 insertions give products in which bonds to the inserted molecule are made to adjacent atoms in that molecule. For example, in the reaction of $HCo(CO)_4$ with tetrafluoroethylene, as shown in Figure 13-7, the product has the $Co(CO)_4$ group attached to one carbon, H attached to the neighboring carbon. 1,2 insertion reactions should look familiar; they are analogous in form to additions across multiple bonds in organic chemistry (although the mechanisms are not necessarily similar).

13-2-2 CARBONYL INSERTION (ALKYL MIGRATION) REACTIONS

Perhaps the most commonly encountered insertion reaction is the carbonyl insertion, in which the CO ligand is apparently inserted into a metal–ligand bond. The insertion of CO into a metal–carbon bond in alkyl complexes is of particular interest (such a reaction would lengthen a carbon chain and might therefore be useful in organic synthesis), and its mechanism deserves careful consideration.

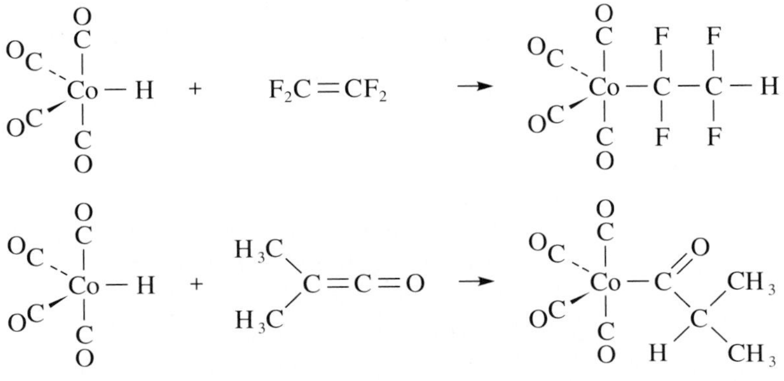

$$\mathrm{(CO)_4Co{-}H} + \mathrm{F_2C{=}CF_2} \rightarrow \mathrm{(CO)_4Co{-}CF_2{-}CF_2{-}H}$$

$$\mathrm{(CO)_4Co{-}H} + \mathrm{(H_3C)_2C{=}C{=}O} \rightarrow \mathrm{(CO)_4Co{-}C({=}O){-}CH(CH_3)_2}$$

FIGURE 13-7 Examples of 1,2 Insertion Reactions.

The reaction of $CH_3Mn(CO)_5$ with CO has the following stoichiometry:

$$H_3C{-}Mn(CO)_5 + CO \longrightarrow H_3C{-}\overset{\overset{\displaystyle O}{\|}}{C}{-}Mn(CO)_5$$

From the net equation, we might expect that the CO inserts directly into the Mn—CH_3 bond; were such the case, the label "CO insertion" would be entirely appropriate for this reaction. However, other mechanisms are possible that would give the overall reaction stoichiometry while involving steps other than insertion of an incoming CO. Three plausible mechanisms have been suggested for this reaction:

Mechanism 1: CO insertion

Direct insertion of CO into metal–carbon bond.

Mechanism 2: CO migration

Migration of CO to give *intramolecular* CO insertion. This would give rise to a 5-coordinate intermediate, with a vacant site available for attachment of an incoming CO.

Mechanism 3: Alkyl migration

In this case the alkyl group would migrate, rather than the CO, and attach itself to a CO *cis* to the alkyl. This would also give a 5-coordinate intermediate with a vacant site available for an incoming CO.

These mechanisms are described schematically in Figure 13-8. In both mechanisms 2 and 3, the intramolecular migration is considered to occur to the migrating group's nearest neighbors, located in *cis* positions.

Experimental evidence that may be applied to evaluation of these mechanisms includes:[8]

1. Reaction of $CH_3Mn(CO)_5$ with ^{13}CO gives a product with the labeled CO in carbonyl ligands only; *none* is found in the acyl position.
2. The reverse reaction

$$H_3C{-}\overset{\overset{\displaystyle O}{\|}}{C}{-}Mn(CO)_5 \longrightarrow H_3C{-}Mn(CO)_5 + CO$$

   (which occurs readily on heating $CH_3C({=}O)Mn(CO)_5$), when carried out with ^{13}C in the acyl position, yields product $CH_3Mn(CO)_5$ with the labeled CO entirely *cis* to CH_3. No labeled CO is lost in this reaction.
3. The reverse reaction, when carried out with ^{13}C in a carbonyl ligand *cis* to the acyl group, gives a product that has a 2:1 ratio of *cis* to *trans* product (*cis* and *trans* referring to the position of labeled CO relative to CH_3 in the product). Some labeled CO is also lost in this reaction.

The mechanisms can now be evaluated on the basis of these data. First, mechanism 1 is definitely ruled out by the first experiment. Direct insertion of

[8] T. C. Flood, J. E. Jensen, and J. A. Statler, *J. Am. Chem. Soc.*, **1981,** *103,* 4410, and references cited therein.

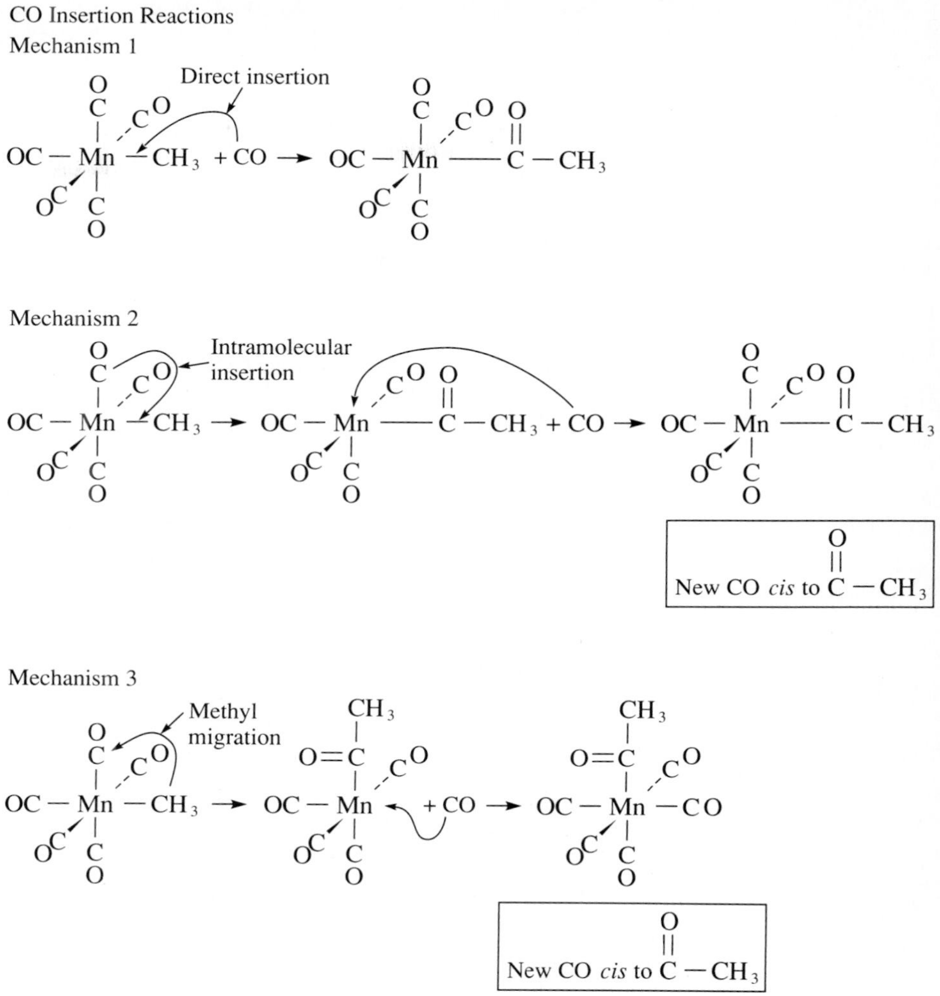

CO Insertion Reactions

Mechanism 1

Direct insertion

$$(OC)_5Mn{-}CH_3 + CO \rightarrow (OC)_5Mn{-}C(=O){-}CH_3$$

Mechanism 2

Intramolecular insertion

$$(OC)_5Mn{-}CH_3 \rightarrow (OC)_4Mn{-}C(=O){-}CH_3 + CO \rightarrow (OC)_5Mn{-}C(=O){-}CH_3$$

New CO *cis* to $C(=O){-}CH_3$

Mechanism 3

Methyl migration

$$(OC)_5Mn{-}CH_3 \rightarrow (OC)_4Mn{-}C(=O)CH_3 + CO \rightarrow (OC)_4Mn(CO){-}C(=O)CH_3$$

New CO *cis* to $C(=O){-}CH_3$

FIGURE 13-8 Possible Mechanisms for CO Insertion Reactions.

^{13}C must result in ^{13}C in the acyl ligand; since none is found, the mechanism cannot be a direct insertion. Mechanisms 2 and 3, on the other hand, are both compatible with the results of this experiment.

The principle of microscopic reversibility requires that any reversible reaction must have identical pathways for the forward and reverse reactions, simply proceeding in opposite directions. (This principle is similar to the idea that the lowest pathway over a mountain chain must be the same regardless of the direction of travel.) If the forward reaction is carbonyl migration (mechanism 2), the reverse reaction must proceed by loss of CO from the acyl compound, followed by migration of CO from the acyl ligand to the empty site. Since this migration is unlikely to occur to a *trans* position, all the product should be *cis*. If the mechanism is alkyl migration, the reverse reaction must proceed by loss of CO from the acyl compound, followed by migration of the alkyl portion of the acyl ligand to the vacant site. Again, all the product should be *cis*. Both mechanisms 2 and 3 would transfer labeled CO in the acyl group to a *cis* position and are therefore consistent with the experimental data for the second experiment (Figure 13-9).

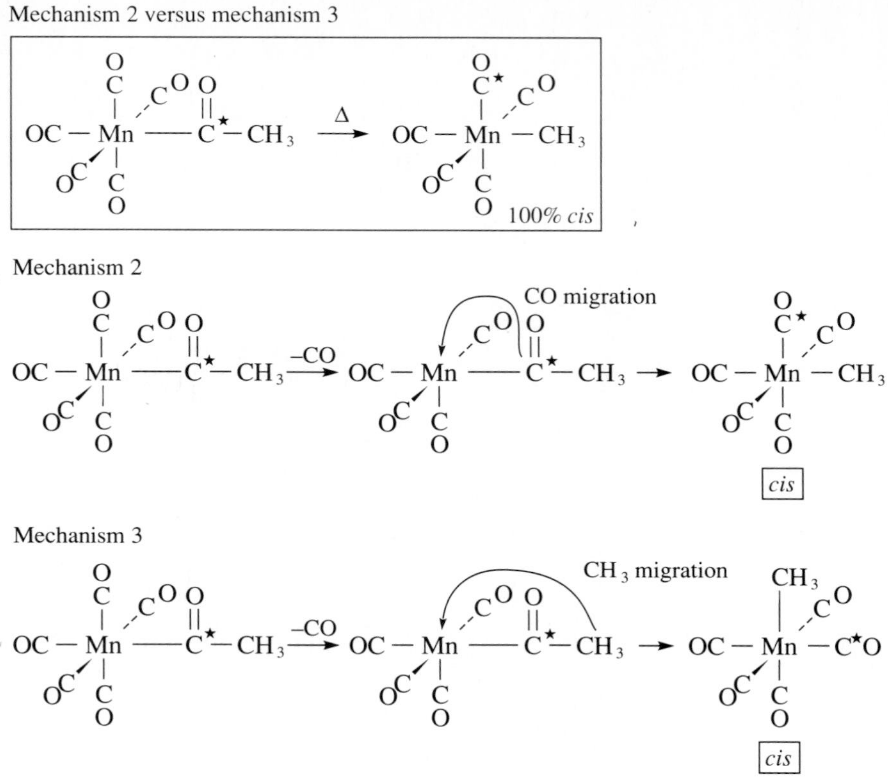

Mechanism 2 versus mechanism 3

Mechanism 2

Mechanism 3

FIGURE 13-9 Mechanisms of Reverse Reactions for CO Migration and Alkyl Insertion (1). ★ indicates location of ^{13}C.

EXERCISE 13-1

Show that heating of $H_3C—{}^{13}C(=O)—Mn(CO)_5$ would not be expected to give the *cis* product by mechanism 1.

The third experiment allows a choice between mechanisms 2 and 3. The CO migration of mechanism 2, with ^{13}CO *cis* to the acyl ligand, requires loss of CO from the acyl ligand to the vacant site. As a result, 25% of the product should have no ^{13}CO label, and 75% should have the labeled CO *cis* to the alkyl, as shown in Figure 13-10. On the other hand, alkyl migration (mechanism 3) should yield 25% with no label, 50% with the label *cis* to the alkyl, and 25% with the label *trans* to the alkyl. Since this is the ratio of *cis* to *trans* found in the experiment, the evidence supports mechanism 3, which is the accepted pathway for this reaction.

The result is that a reaction that initially appears to involve CO insertion, and is often so designated, does not involve CO insertion at all! It is not uncommon for reactions on close study to differ substantially from how they might at first appear; the "carbonyl insertion" reaction may in fact be substantially more complicated than described here. In this reaction, as in all chemical reactions, it is extremely important for chemists to be willing to undertake mechanistic studies and to keep an open mind on possible alternative mechanisms. No mechanism can be proved; it is always possible to suggest alternatives consistent with the known data.

Mechanism 2

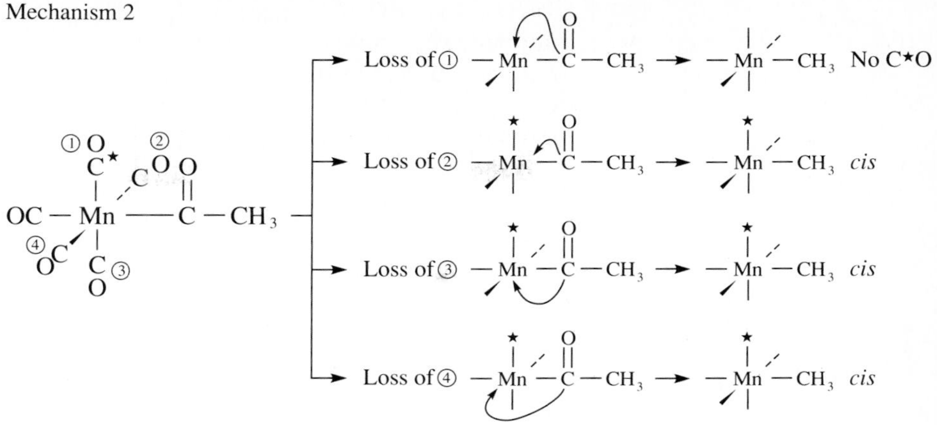

Loss of ① → No C*O; Loss of ② → cis; Loss of ③ → cis; Loss of ④ → cis

Mechanism 3

Loss of ① → No C*O; Loss of ② → cis; Loss of ③ → trans; Loss of ④ → cis

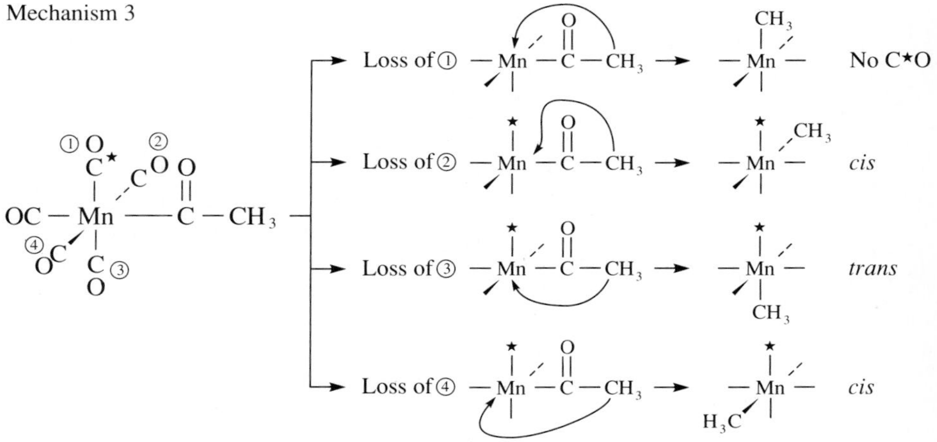

FIGURE 13-10 Mechanisms of Reverse Reactions for CO Migration and Alkyl Insertion (2). ★ indicates location of ^{13}C.

One final point about the mechanism of these reactions should be made. In the discussion of mechanisms 2 and 3 above, it was assumed that the intermediate was a square pyramid and that no rearrangement to other geometries (such as trigonal bipyramidal) occurred. Other labeling studies, involving reactions of labeled $CH_3Mn(CO)_5$ with phosphines, have supported a square pyramidal intermediate.[9]

EXERCISE 13-2

Predict the product distribution for the reaction of *cis*-$CH_3Mn(CO)_4(^{13}CO)$ with $PR_3(R = C_2H_5)$. (You may check your answer by consulting footnote 8.)

13-2-3 HYDRIDE ELIMINATION REACTIONS

Hydride elimination reactions are characterized by transfer of a hydrogen atom from a ligand to a metal. Effectively, this may be considered an oxidative addition, with both the coordination number and the formal oxidation state of the metal being increased (the hydrogen transferred is formally considered as hydride, H^-). The most common type is β *elimination,* with a proton in a beta

[9] Flood, Jensen, and Statler, op. cit.

position on an alkyl ligand being transferred to the metal by way of an intermediate in which the metal, the alpha and beta carbons, and the hydride are coplanar. An example is shown in Figure 13-11. Beta eliminations, as will be seen later in this chapter, are important in many catalytic processes involving organometallic complexes.

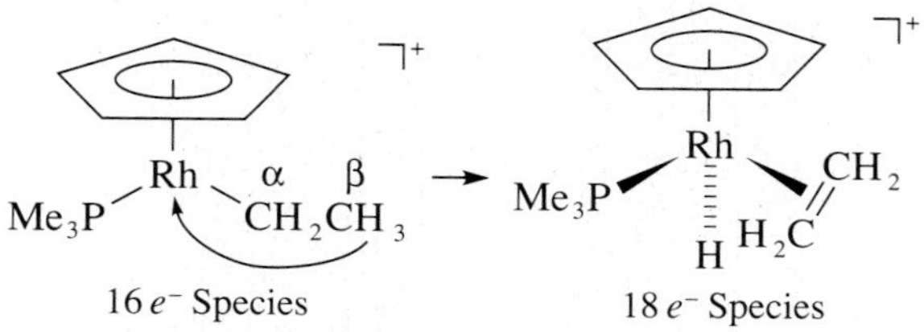

FIGURE 13-11 Beta Elimination.

Several general comments can be made about beta-elimination reactions. First, since only those complexes that have β hydrogens can undergo these reactions, alkyl complexes that lack β hydrogens tend to be more stable thermally than those that have such hydrogens (although the former may undergo other types of reactions). Furthermore, coordinatively saturated complexes (complexes in which all coordination sites are filled) containing β hydrogens are in general more thermally stable than complexes having empty coordination sites; the β-elimination mechanism requires transfer of a hydrogen to an empty coordination site. Finally, other types of elimination reactions are also known (such as elimination of hydrogen from α and γ positions); the interested reader is referred to other sources for examples of such reactions.[10]

13-2-4 ABSTRACTION

Abstraction reactions are elimination reactions in which the coordination number of the metal does not change. In general, they involve removal of a substituent from a ligand, often by the action of an external reagent, such as a Lewis acid. Two types of abstractions, α and β *abstractions*, are illustrated in Figure 13-12; they involve, respectively, removal of substituents from the

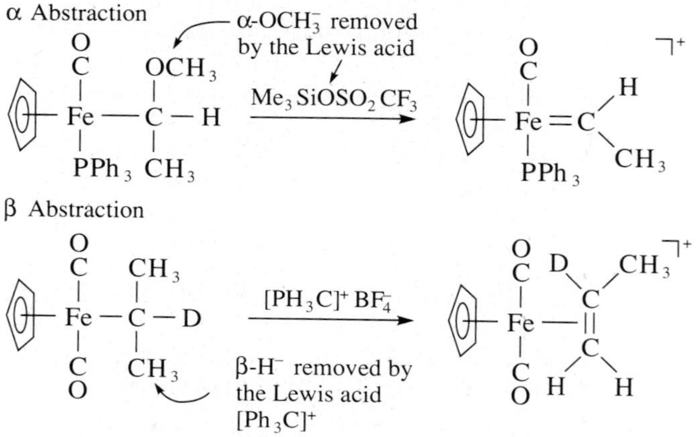

FIGURE 13-12 Abstraction Reactions.

[10] J. D. Fellmann, R. R. Schrock, and D. D. Traficante, *Organometallics*, **1982,** *1*, 481; J. P. Collman, L. S. Hegedus, J. R. Norton, and R. G. Finke, *Principles and Applications of Organotransition Metal Chemistry*, University Science Books, Mill Valley, Calif., 1987, and references therein.

alpha and beta positions (with respect to the metal) of coordinating ligands. α-Abstraction has been encountered previously, in the synthesis of carbene complexes discussed in Section 12-6-2.

13-3 ORGANOMETALLIC CATALYSTS

In addition to having an intrinsic interest for chemists, organometallic reactions are also of great interest industrially, especially in the development of catalysts for reactions of commercial importance. The commercial interest in catalysis has been spurred by the fundamental problem of how to convert relatively inexpensive feedstocks (such as coal, petroleum, and water) into molecules of greater commercial value. This frequently involves, as part of the industrial process, conversion of simple molecules into molecules of more complexity (such as ethylene into acetaldehyde, methanol into acetic acid, or organic monomers into polymers), conversion of one molecule into another of the same type (one alkene into another), or a selective reaction at a particular molecular site (replacement of hydrogen by deuterium, selective hydrogenation of a specific double bond). Historically, many catalysts have been **heterogeneous** in nature; that is, solid materials having catalytically active sites on their surface, with only the surface in contact with the reactants. **Homogeneous** catalysts, soluble in the reaction medium, are molecular species that are easier to study and modify for specific applications than are heterogeneous catalysts. Appropriate design of catalyst molecules may provide high selectivity in the processes catalyzed; it is not surprising that development of highly selective homogeneous catalysts has been of considerable industrial interest. Not every catalytic cycle, however, is efficient or profitable enough to be commercially feasible.

In the examples of catalysis that follow, the reader will find it useful to identify the catalysts, the species regenerated in each complete reaction cycle. In addition, the individual steps in these cycles will provide examples of the various types of organometallic reactions introduced earlier in this chapter. In each case, the proposed mechanisms presented in this section are subject to modification as additional research data are obtained on these processes.

13-3-1 CATALYTIC DEUTERATION[11]

If deuterium gas (D_2) is bubbled through a benzene solution of $(\eta^5\text{-}C_5H_5)_2TaH_3$ at elevated temperature, the hydrogen atoms of benzene are replaced by deuterium; eventually, perdeuterobenzene, C_6D_6, can be obtained (for use, for example, as an NMR solvent). Replacement of H by D occurs in the series of steps outlined in Figure 13-13. The initial step in this process is loss of H_2 (formally, reductive elimination) from the 18-electron $(\eta^5\text{-}C_5H_5)_2TaH_3$ to give the 16-electron $(\eta^5\text{-}C_5H_5)_2TaH$. $(\eta^5\text{-}C_5H_5)_2TaH$ can then react with benzene in the second step (oxidative addition) to give an 18-electron species containing a phenyl group sigma bonded to the metal. This species can undergo a second loss of H_2 to give another 16-electron species ($(\eta^5\text{-}C_5H_5)_2Ta\text{-}C_6H_5$). $(\eta^5\text{-}C_5H_5)_2Ta\text{-}C_6H_5$ subsequently adds D_2 (another oxidative addition) to form an 18-electron species (step 4), which in the last step, eliminates C_6H_5D. Repetition of this sequence in the presence of excess D_2 eventually leads to C_6D_6. In each subsequent cycle the catalytic species $(\eta^5\text{-}C_5H_5)_2TaD$ is regenerated.

[11] J. W. Lauher and R. Hoffmann, *J. Am. Chem. Soc.*, **1976,** *98,* 1729, and references therein.

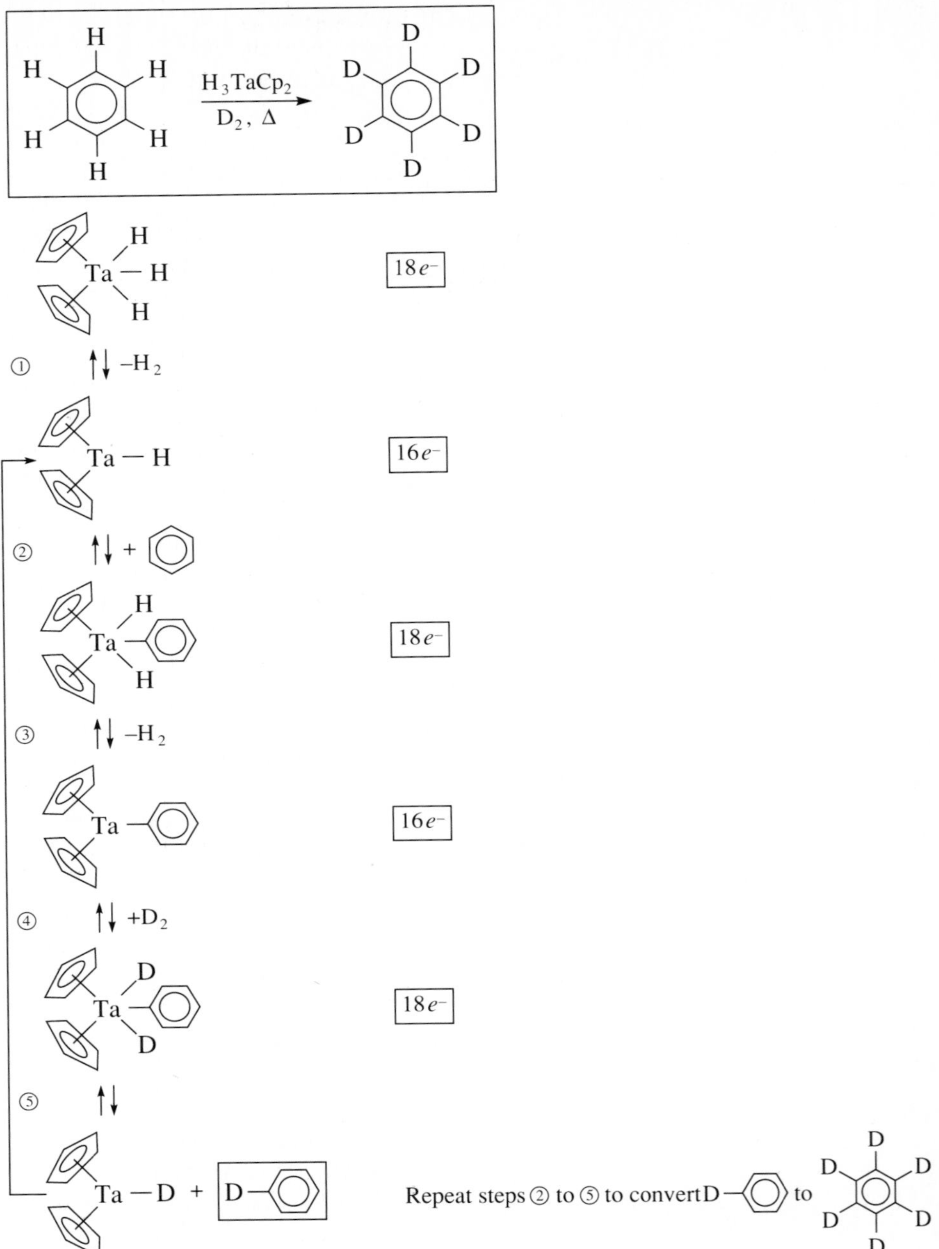

FIGURE 13-13 Catalytic Deuteration.

13-3-2 HYDROFORMYLATION

The hydroformylation, or oxo, process is a commercially useful process for converting terminal alkenes into a variety of other organic products, especially those having their carbon chain increased in length by one. One of these processes, the conversion of an alkene of formula $R_2C{=}CH_2$ into an aldehyde $R_2CH{-}CH_2{-}CHO$, is illustrated in Figure 13-14.[12]

[12] R. F. Heck, and D. S. Breslow, *J. Am. Chem. Soc.*, **1961,** *83,* 4023. See also Collman, Hegedus, Norton, and Finke, op. cit., pp. 621–36.

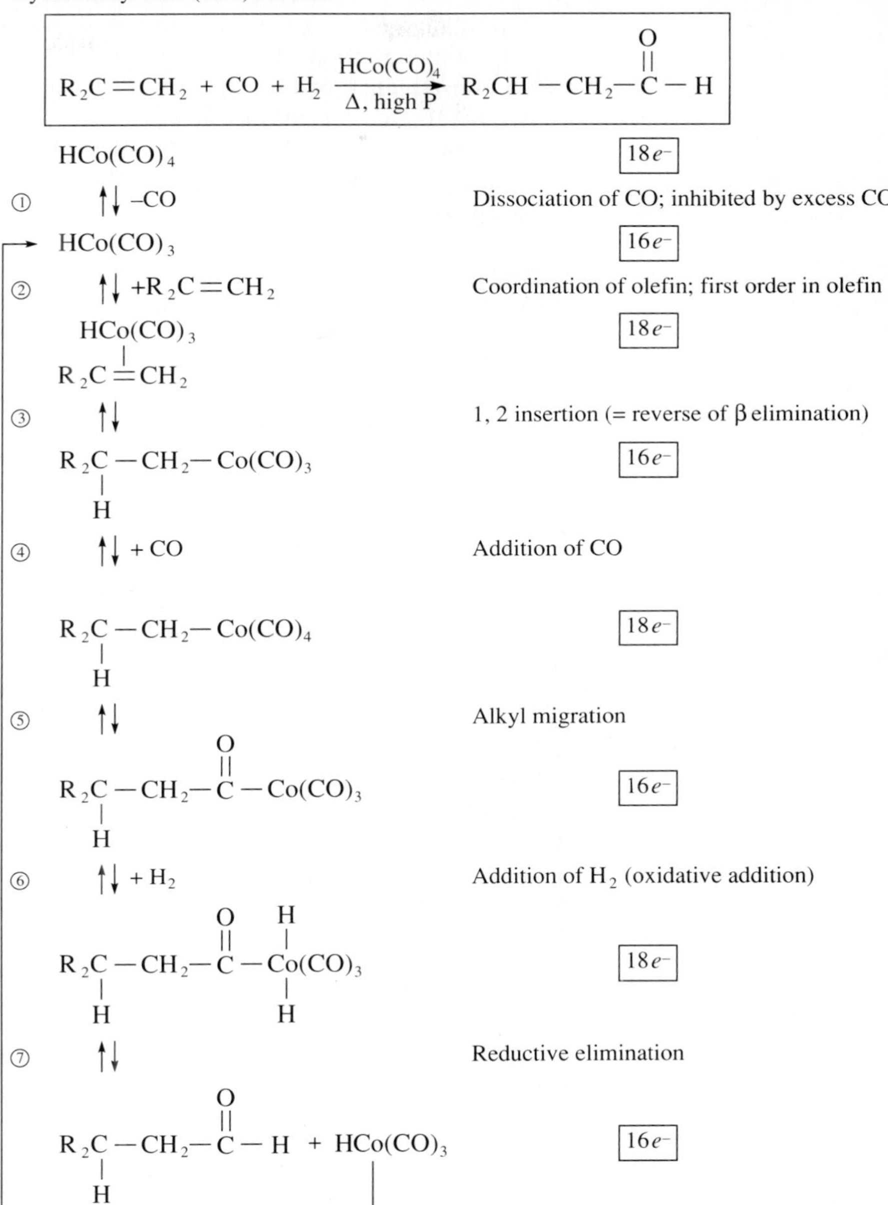

FIGURE 13-14 Hydroformylation Process.

Each step of the hydroformylation cycle may be categorized according to its characteristic type of organometallic reaction, as indicated in the figure. The $HCo(CO)_4$ is derived from hydrogenation of $Co_2(CO)_8$:

$$H_2 + Co_2(CO)_8 \longrightarrow 2\ HCo(CO)_4$$

The cobalt-containing intermediates in this cycle alternate between 18- and 16-electron species. The 18-electron species react to reduce their electron count by two (by ligand dissociation, 1,2-insertion of coordinated alkene, alkyl migration, and reductive elimination), while the 16-electron species are capable of an increase in their formal electron count (by coordination of alkene or CO or by oxidative addition). Such a pattern is commonly encountered in catalytic cycles involving organometallic complexes, with the catalytic activity in large part a consequence of the capacity of the metal to react by way of a variety of 18- and 16-electron intermediates.

A few aspects of the steps of the hydroformylation process are worth examining. The first step, involving dissociation of CO from $HCo(CO)_4$, is inhibited by high CO pressure, yet the fourth step requires CO; thus careful control of this pressure is necessary for optimum yields and rates. The second step is first order in alkene; it is the rate-determining step. In step 4 the product is formed preferentially with a CH_2 group rather than a CR_2 group bonded to the metal; this preference for the CH_2 end of the alkene bonding to metal is favored by bulky R groups. Step 6 involves addition of H_2 (oxidative addition); however, high H_2 pressure can lead to addition of H_2 to the 16-electron intermediate from step 3, which would then eliminate an alkane:

$$R_2CH{-}CH_2{-}Co(CO)_3 + H_2 \longrightarrow R_2CH{-}CH_2{-}Co(H)_2(CO)_3, \qquad \text{oxidative addition}$$

$$R_2CH{-}CH_2{-}Co(H)_2(CO)_3 \longrightarrow R_2CH{-}CH_3 + HCo(CO)_4, \qquad \text{reductive elimination}$$

Again, careful control of the experimental conditions is necessary to maximize yield of the desired products. The actual catalytic species in this mechanism is the 16-electron $HCo(CO)_3$.

Industrially, the main application of hydroformylation is in the production of butanal from propene ($R_2C{=}CH_2$ = $CH_3CH{=}CH_2$ in Figure 13-14). Subsequent hydrogenation gives butanol, which is an important industrial solvent. Other aldehydes are also produced industrially by hydroformylation, using either cobalt catalysts like the one described above or rhodium-based catalysts. Replacement of one or more of the carbonyls by phosphines can in some cases improve the selectivity of these processes and enable reactions to be catalyzed under milder conditions. An example is shown later in Section 13-3-5.

EXERCISE 13-3

Show how $(CH_3)_2CH{-}CH_2{-}CHO$ can be prepared from $(CH_3)_2C{=}CH_2$ by the hydroformylation process.

13-3-3 MONSANTO ACETIC ACID PROCESS

A process that has been used with great commercial success by Monsanto since 1971 has been the synthesis of acetic acid from methanol and CO. The mechanism of this process is very complex; an outline that has been proposed is shown in Figure 13-15.[13] As for the hydroformylation process, the individual steps of this mechanism are the characteristic types of organometallic reactions described earlier in this chapter; the intermediates are 18- or 16-electron species having the capacity to lose or gain, respectively, two electrons. (Solvent molecules may occupy empty coordination sites in the 4- and 5- coordinate 16-electron intermediates.) The fourth step is especially interesting: it is believed

[13] D. Forster, *Adv. Organomet. Chem.*, **1979,** *17,* 255.

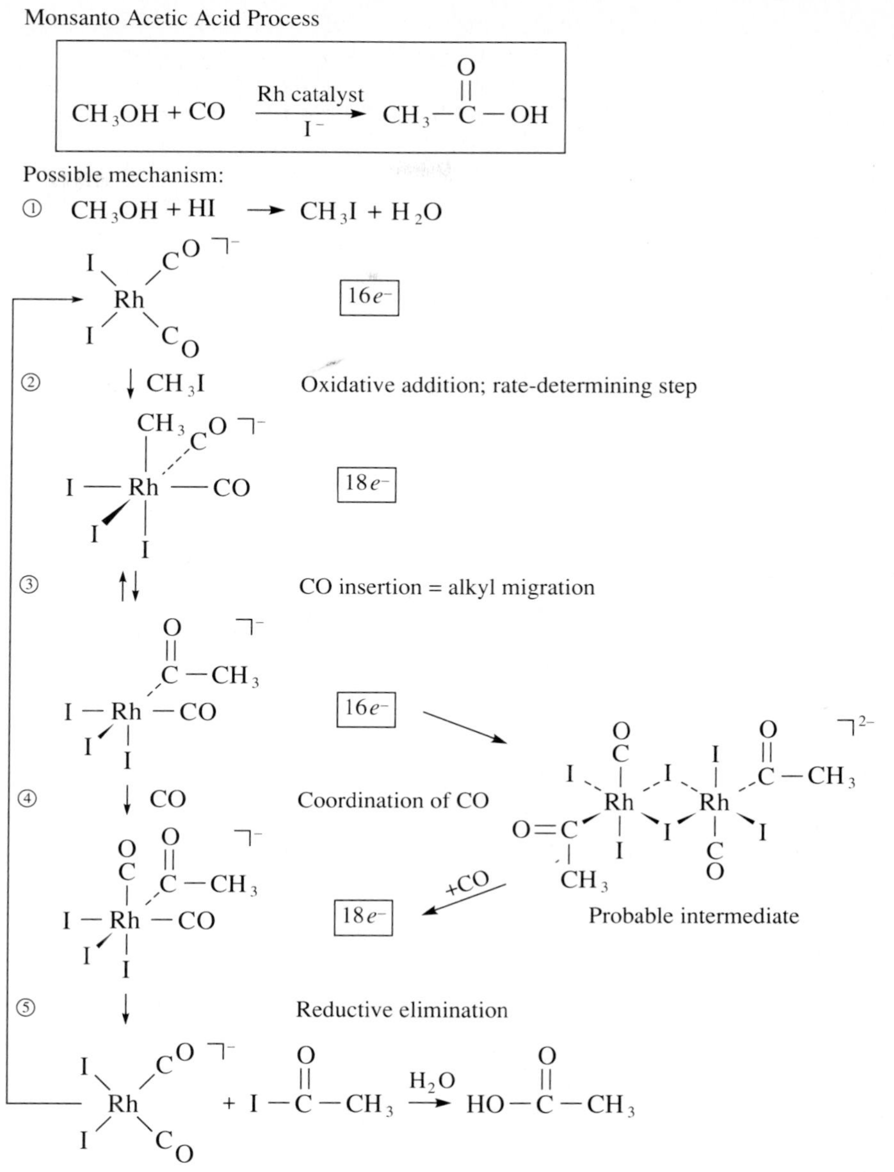

Monsanto Acetic Acid Process

$$CH_3OH + CO \xrightarrow[I^-]{\text{Rh catalyst}} CH_3{-}C({=}O){-}OH$$

Possible mechanism:

① $CH_3OH + HI \longrightarrow CH_3I + H_2O$

$[RhI_2(CO)_2]^-$ 16e^-

② $\downarrow CH_3I$ Oxidative addition; rate-determining step

$[RhI_3(CH_3)(CO)_2]^-$ 18e^-

③ ⇅ CO insertion = alkyl migration

$[RhI_3(CO)(COCH_3)]^-$ 16e^- → Probable intermediate: $[Rh_2I_6(CO)_2(COCH_3)_2]^{2-}$ (iodide-bridged dimer)

④ $\downarrow$ CO Coordination of CO ($+CO$ from probable intermediate)

$[RhI_3(CO)_2(COCH_3)]^-$ 18e^-

⑤ $\downarrow$ Reductive elimination

$$[RhI_2(CO)_2]^- + I{-}C({=}O){-}CH_3 \xrightarrow{H_2O} HO{-}C({=}O){-}CH_3$$

FIGURE 13-15 Monsanto Acetic Acid Process.

that two $[RhI_3(CO)(COCH_3)]^-$ units dimerize by way of iodide bridges to give the intermediate shown; this intermediate is then able to add CO to form an 18-electron species.

The final step involving rhodium is reductive elimination of $IC({=}O)CH_3$. Acetic acid is formed by hydrolysis of this compound. The catalytic species, $[Rh(CO)_2I_2]^-$, (likely containing solvent in the empty coordination sites) is regenerated, as shown in the figure.

13-3-4 WACKER (SMIDT) PROCESS

The Wacker process, used to synthesize acetaldehyde from ethylene, involves a catalytic cycle using $PdCl_4^{2-}$. A brief outline of a cycle that has been proposed for this process is shown in Figure 13-16. An important feature of this process is that it uses the ability of palladium to form complexes with the reactant ethylene, with the important chemistry of ethylene occurring while it is attached to the metal. In other words, the palladium modifies the chemical behavior of ethylene to enable reactions to occur that would not be possible for free ethylene. Incidentally, the first ethylene complex with palladium in Figure 13-16 is isoelectronic with Zeise's complex, $[PtCl_3(\eta^2\text{-}H_2C{=}CH_2)]^-$.

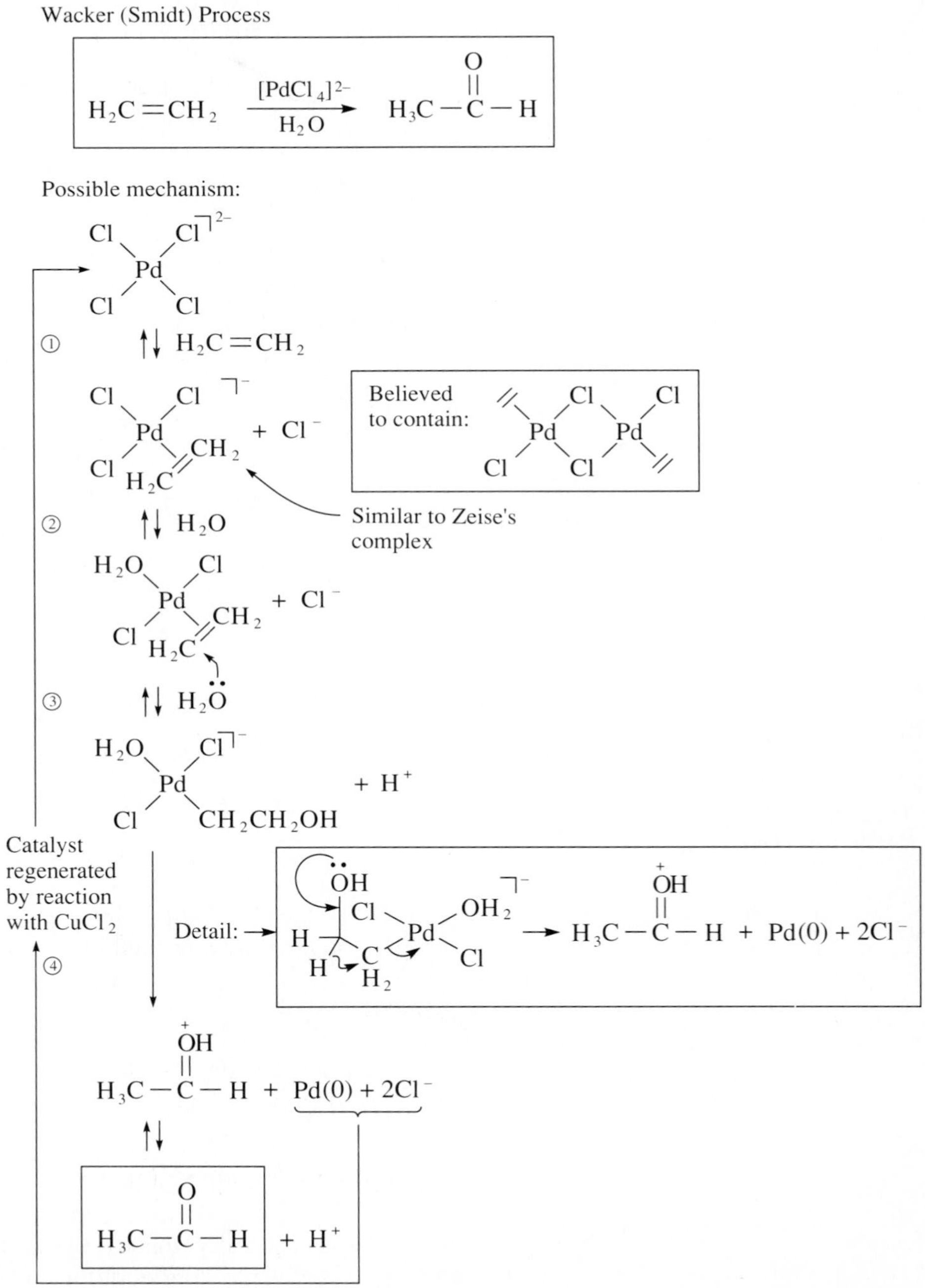

Wacker (Smidt) Process

$$H_2C{=}CH_2 \xrightarrow[H_2O]{[PdCl_4]^{2-}} H_3C{-}C({=}O){-}H$$

Possible mechanism:

$[PdCl_4]^{2-}$

① $\uparrow\downarrow$ $H_2C{=}CH_2$

$[PdCl_3(\eta^2\text{-}H_2C{=}CH_2)]^-$ + Cl^-

Similar to Zeise's complex

Believed to contain: $[(\eta^2\text{-}C_2H_4)PdCl(\mu\text{-}Cl)_2PdCl(\eta^2\text{-}C_2H_4)]$

② $\uparrow\downarrow$ H_2O

$PdCl_2(H_2O)(\eta^2\text{-}H_2C{=}CH_2)$ + Cl^-

③ $\uparrow\downarrow$ $H_2\ddot{O}$

$[PdCl_2(H_2O)(CH_2CH_2OH)]^-$ + H^+

Detail: $[PdCl_2(OH_2)(CH_2CH_2OH)]^- \rightarrow H_3C{-}C({=}\overset{+}{O}H){-}H + Pd(0) + 2Cl^-$

④

$H_3C{-}C({=}\overset{+}{O}H){-}H + Pd(0) + 2Cl^-$

$\uparrow\downarrow$

$H_3C{-}C({=}O){-}H$ + H^+

Catalyst regenerated by reaction with $CuCl_2$

FIGURE 13-16 Wacker (Smidt) Process.

13-3-5 HYDROGENATION BY WILKINSON'S CATALYST

Wilkinson's catalyst, $RhCl(PPh_3)_3$, is not itself an organometallic compound but participates in the same kinds of reactions as expected for 4-coordinate organometallic compounds (for example, many reactions bear similarities to those of Vaska's catalyst, *trans*-$IrCl(CO)(PPh_3)_2$). $RhCl(PPh_3)_3$ participates in a wide variety of catalytic and noncatalytic processes. The bulky phosphine ligands play an important role in making the complex a selective one, for example in limiting coordination of Rh to positions on alkenes having low steric hindrance. An example, involving catalytic hydrogenation of an alkene, is shown in Figure 13-17.[14] The first two steps in this process give the catalytic species $RhCl(H)_2(PPh_3)_2$, which has a vacant coordination site. A C═C double bond can coordinate to this site, gain the two hydrogens coordinated to Rh, and subsequently leave, *if* the double bond is not sterically hindered. In molecules containing several double bonds, the least hindered double bonds are reduced; the most hindered positions are not able to coordinate effectively to Rh (largely due to the presence of the bulky phosphines) and hence do not react. Consequently, Wilkinson's catalyst is useful for selective hydrogenations of C═C bonds that are not sterically hindered. Examples are shown in Figure 13-18. Since the selectivity of Wilkinson's catalyst is largely a consequence of the bulky triphenylphosphine ligands, the selectivity can be "fine tuned" somewhat by using phosphines having different cone angles than PPh_3.

Wilkinson's catalyst and similar compounds having different phosphine ligands are useful in a variety of other catalytic cycles. The complex $HRh(CO)_2(PPh_3)_2$ can be used to catalyze hydroformylation; a proposed mechanism is shown in Figure 13-19.

13-3-6 OLEFIN METATHESIS[15]

Olefin metathesis involves the formal exchange of methylene (:CH_2) fragments between alkenes. For example, metathesis between molecules of formula $H_2C{=}CH_2$ and $HRC{=}CHR$ would form two molecules of $H_2C{=}CHR$:

$$\begin{matrix} H_2C & & HRC \\ \| & & \| \\ H_2C & & HRC \end{matrix} \quad \rightleftharpoons \quad \begin{matrix} H_2C{=}CHR \\ \\ H_2C{=}CHR \end{matrix}$$

This reaction, which is reversible and can be catalyzed by a variety of organometallic complexes, has been the subject of much investigation and controversy. These reactions are now believed to proceed by way of formation of carbene (alkylidene) complexes, which can then react with alkenes via metallacyclobutane intermediates.

The first step is apparently the formation of a carbene complex; although carbene complexes have not been isolated for most known metathesis processes, such complexes have been shown in many cases to catalyze metathesis. A terminal alkene (containing $=CH_2$) can then add across the metal–carbon

[14] B. R. James, *Adv. Organomet. Chem.*, **1979,** *17,* 319; see also Collman, Hegedus, Norton, and Finke, op. cit., pp. 531–35, and references cited therein.

[15] K. J. Ivin, *Olefin Metathesis,* Academic Press, New York, 1983.

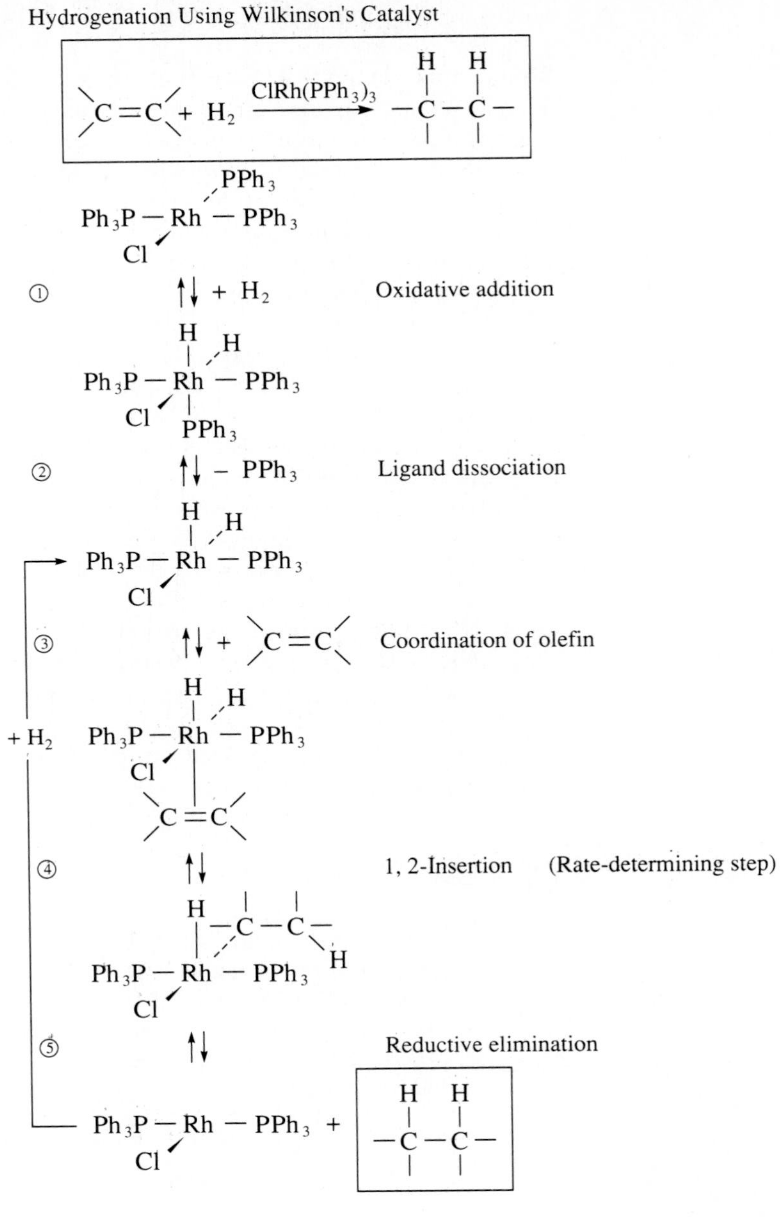

Hydrogenation Using Wilkinson's Catalyst

$$>C=C< + H_2 \xrightarrow{ClRh(PPh_3)_3} -\overset{H}{\underset{|}{C}}-\overset{H}{\underset{|}{C}}-$$

$ClRh(PPh_3)_3$

① $\uparrow\downarrow$ + H_2 — Oxidative addition

$H_2RhCl(PPh_3)_3$

② $\uparrow\downarrow$ − PPh_3 — Ligand dissociation

$H_2RhCl(PPh_3)_2$

③ $\uparrow\downarrow$ + $>C=C<$ — Coordination of olefin

$H_2RhCl(PPh_3)_2(>C=C<)$

④ $\uparrow\downarrow$ — 1, 2-Insertion (Rate-determining step)

$HRhCl(PPh_3)_2(-\overset{|}{C}-\overset{|}{C}-H)$

⑤ $\uparrow\downarrow$ — Reductive elimination

$ClRh(PPh_3)_2$ + $-\overset{H}{\underset{|}{C}}-\overset{H}{\underset{|}{C}}-$

+ H_2 (returns to step ③)

FIGURE 13-17 Catalytic Hydrogenation Involving Wilkinson's Catalyst.

$\xrightarrow[ClRu(PPh_3)_3]{H_2}$

$\xrightarrow[ClRu(PPh_3)_3]{H_2}$

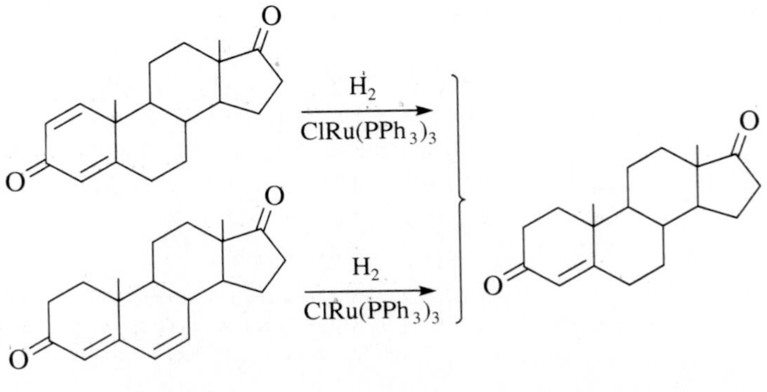

FIGURE 13-18 Selective Hydrogenation by Wilkinson's Catalyst.

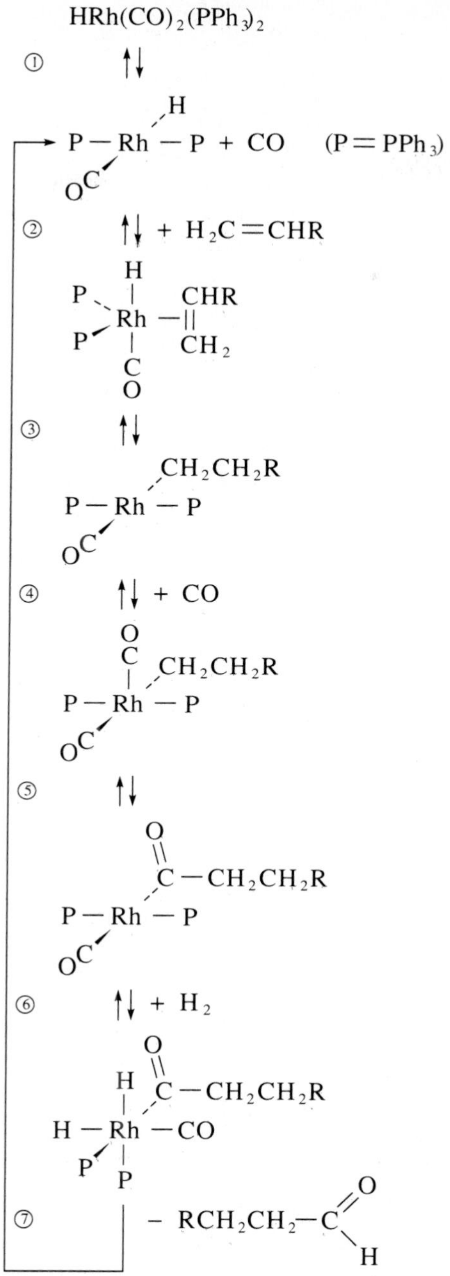

$HRh(CO)_2(PPh_3)_2$

① ⇅

$P{-}Rh(H)(CO){-}P + CO \quad (P{=}PPh_3)$

② ⇅ $+ H_2C{=}CHR$

$(P)_2Rh(H)(CO)(\eta^2\text{-}CH_2{=}CHR)$

③ ⇅

$P{-}Rh(CH_2CH_2R)(CO){-}P$

④ ⇅ $+ CO$

$P{-}Rh(CO)_2(CH_2CH_2R){-}P$

⑤ ⇅

$P{-}Rh(CO)(C({=}O){-}CH_2CH_2R){-}P$

⑥ ⇅ $+ H_2$

$H{-}Rh(H)(P)_2(CO)(C({=}O){-}CH_2CH_2R)$

⑦ $- RCH_2CH_2{-}C({=}O)H$

FIGURE 13-19 Hydroformylation Using $HRh(CO)_2(PPh_3)_2$. (Based on information from C. K. Brown and G. Wilkinson, *J. Chem. Soc. A*, **1970**, 2753.)

double bond to form a metallacyclobutane intermediate (Figure 13-20), which subsequently reacts to form the new alkene (which has gained the CH_2) and a metal complex that has a now modified carbene ligand. Although strong evidence has been gathered in support of this mechanistic pathway, the complete mechanism still is not well understood, and olefin metathesis remains a very active field of research.

Olefin metathesis serves as an important route for converting internal alkenes to terminal alkenes, which are widely used in the manufacture of detergents, perfumes, and other products.

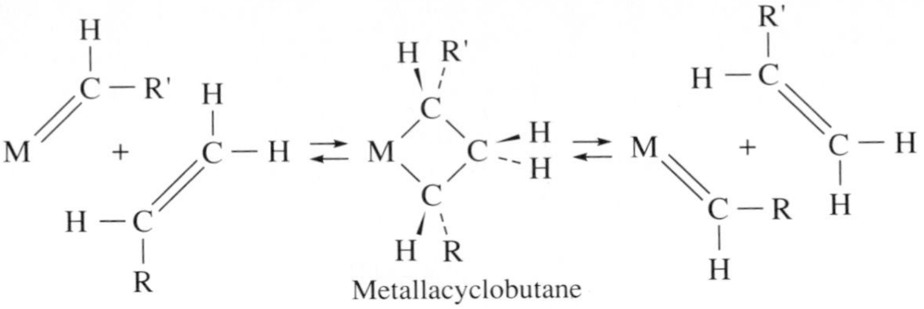

FIGURE 13-20 Olefin Metathesis.

Metallacyclobutane

An interesting variation on olefin metathesis is the use of carbene complexes to catalyze alkene polymerization, also via a metallacyclobutane intermediate. An example is the ring-opening polymerization of norbornene using W(CH—*t*-Bu)(OCH$_2$—*t*-Bu)$_2$Br$_2$ as catalyst in the presence of $GaBr_3$, as shown in Figure 13-21.[16] Proton and ^{13}C NMR spectra are consistent with the proposed structure of the metallacyclobutane, as well as the polymer growing off the carbene carbon.

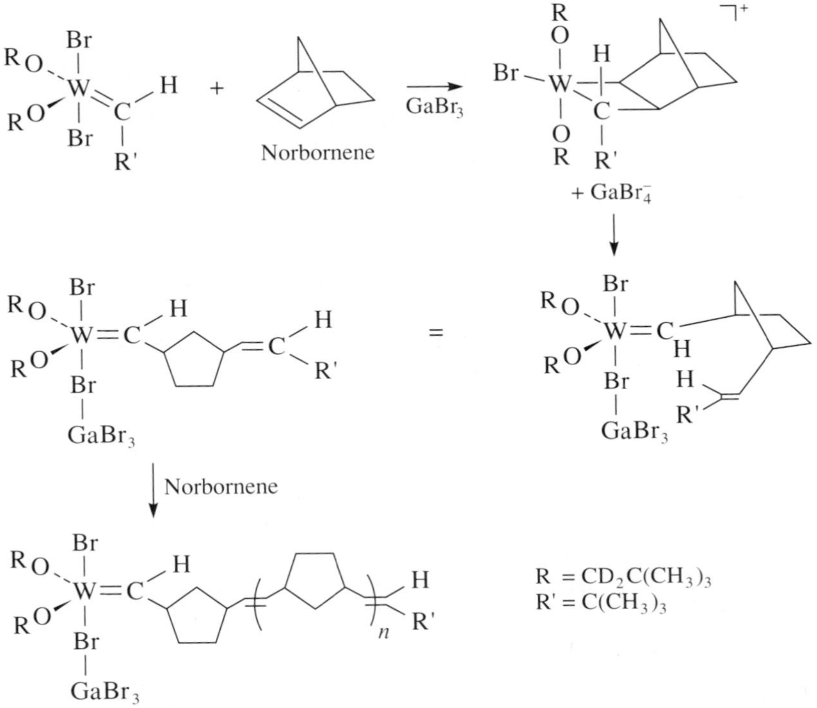

Norbornene

$GaBr_3$

$+ GaBr_4^-$

Norbornene

$R = CD_2C(CH_3)_3$
$R' = C(CH_3)_3$

FIGURE 13-21 Polymerization of Norbornene Using Carbene Catalyst.

Alkynes can also undergo metathesis reactions catalyzed by transition metal carbyne complexes. The intermediates in these reactions are believed to be metallacyclobutadiene species, formed from addition of an alkyne across a metal–carbon triple bond of the carbyne (Figure 13-22). The structures of a variety of metallacyclobutadiene complexes have been determined, and some have been shown to catalyze alkyne metathesis.[17]

[16] J. Kress, J. A. Osborn, R. M. E. Greene, K. J. Ivin, and J. J. Rooney, *J. Am. Chem. Soc.*, **1987,** *109,* 899.

[17] W. A. Nugent and J. M. Mayer, *Metal–Ligand Multiple Bonds,* Wiley-Interscience, New York, 1988, p. 311, and references therein.

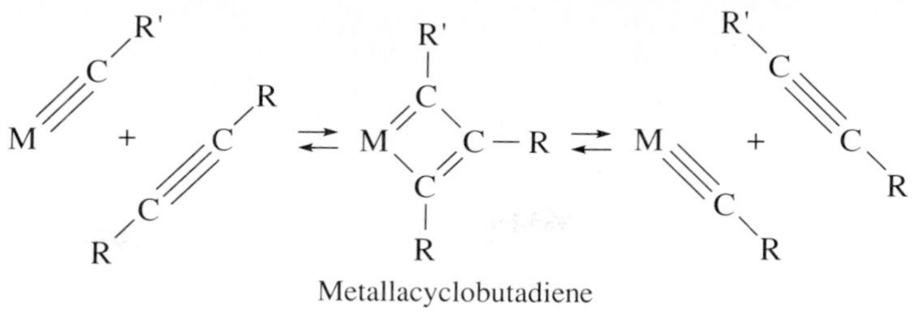

$$\underset{M}{}\!\equiv\! C\!-\!R' + R\!-\!C\!\equiv\! C\!-\!R \rightleftharpoons \text{M}\!\!\begin{array}{c}C(R')\\ \diagup\!\!\!\!=\;\;\diagdown\\ \diagdown\;\;=\!\!\!\!\diagup\\ C(R)\end{array}\!\!C\!-\!R \rightleftharpoons M\!\equiv\! C\!-\!R + R'\!-\!C\!\equiv\! C\!-\!R$$

Metallacyclobutadiene

FIGURE 13-22 Alkyne Metathesis.

13-4 HETEROGENEOUS CATALYSTS

In addition to the homogeneous catalytic processes described above, heterogeneous processes, involving solid catalytic species, are very important, although the exact nature of the reactions occurring on the surface of the catalyst may be extremely difficult to ascertain. Of the 20 organic chemicals produced in greatest quantities in the United States, 15 are produced commercially by processes that involve metal catalysts; most of these processes involve heterogeneous catalysis (Table 13-2).[18] In many cases the methods of preparing the catalysts and information on their function are proprietary, the product of very substantial corporate investment. Nevertheless, it is important to mention several of these processes as important practical applications of organometallic reactions.

13-4-1 ZIEGLER–NATTA POLYMERIZATIONS

In 1955 Ziegler and co-workers reported that solutions of $TiCl_4$ in hydrocarbon solvents in the presence of $Al(C_2H_5)_3$ gave heterogeneous systems capable of polymerizing ethylene.[19] Subsequently, many other heterogeneous processes have been developed for polymerizing alkenes using aluminum alkyls in combination with transition metal complexes. An outline of a possible mechanism for the Ziegler–Natta process proposed by Cossee and Arlman is given in Figure 13-23.[20]

First, reaction of $TiCl_4$ with aluminum alkyl gives $TiCl_3$, which on further reaction with the aluminum alkyl gives a titanium alkyl complex, as shown in the figure. Ethylene (or propylene) can then insert into the titanium–carbon bond, forming a longer alkyl. This alkyl is further susceptible to insertion of ethylene to lengthen the chain. Although the mechanism of the Ziegler–Natta process has proved extremely difficult to understand, direct insertions of multiply bonded organics into titanium–carbon bonds have been demonstrated, supporting the Cossee–Arlman mechanism.[21]

However, an alternative mechanism, involving a metallacyclobutane intermediate, has also been proposed.[22] This mechanism, also shown in Figure 13-23, involves initial formation of alkylidene from a metal alkyl complex, followed by addition of ethylene to give the metallacyclobutane, which then

[18] R. Chang and W. Tikkanen, *The Top Fifty Industrial Chemicals,* Random House, New York, 1988.

[19] K. Ziegler, E. Holzkamp, H. Breil, and H. Martin, *Angew. Chem.,* **1955,** *67,* 541.

[20] P. Cossee, *J. Catal.,* **1964,** *3,* 80; E. J. Arlman, *J. Catal.,* **1964,** *3,* 89; E. J. Arlman and P. Cossee, *J. Catal,* **1964,** *3,* 99.

[21] J. J. Eisch, A. M. Piotrowski, S. K. Brownstein, E. J. Gabe, and F. L. Lee, *J. Am. Chem. Soc.,* **1985,** *107,* 7219.

[22] K. J. Ivin, J. J. Rooney, C. D. Stewart, M. L. H. Green, and R. Mahtab, *J. Chem. Soc., Chem. Commun.,* **1978,** 604.

TABLE 13-2
Leading organic compounds and metal catalysts

Rank	Compound	U.S. production, 1988 (× 10^9 kg)	Metal-containing catalysts used
1	Ethylene	16.58	
2	Propylene	9.06	$TiCl_3$ or $TiCl_4$ + AlR_3, (R = alkyl) (Ziegler–Natta)
3	Urea	7.15	
4	Ethylene dichloride	6.19	$FeCl_3$, $AlCl_3$
5	Benzene	5.37	Pt on Al_2O_3 support
6	Xylene (all isomers)	5.23	Pt + Re
7	Ethylbenzene	4.51	$AlCl_3$
8	Terephthalic acid	4.35	Co, Mn compounds
9	Vinyl chloride	4.11	$CuCl_2$ on KCl support
10	Styrene	3.90	ZnO, Cr_2O_3
11	Methanol	3.33	ZnO + other metal oxides
12	Formaldehyde	3.05	Cu, Ag
13	Toluene	2.93	Pt on Al_2O_3 support
14	Ethylene oxide	2.44	Ag
15	Ethylene glycol	2.22	Rh complexes
16	Cumene	2.18	
17	Methyl-*t*-butyl ether	2.12	
18	Phenol	1.60	
19	Butadiene	1.45	Fe_2O_3, other metal oxides
20	Acetic acid	1.43	Mn acetate, Rh organometallic complexes

SOURCE: *Chem. Eng. News*, June 19, 1989, p. 39.

yields product having ethylene inserted into the original metal–carbon bond. It appears that both mechanisms may, under appropriate circumstances, catalyze ethylene polymerization; future studies are necessary to determine more precisely the mechanism of the Ziegler–Natta process.

13-4-2 WATER GAS REACTION

This reaction occurs at elevated temperatures and pressures between water (steam) and natural sources of carbon, such as coal or coke:

$$H_2O + C \longrightarrow H_2 + CO$$

The products of this reaction, an equimolar mixture of H_2 and CO (called "synthesis gas" or "syn gas"; some CO_2 may be produced as a byproduct), can be used in conjunction with various metallic heterogeneous catalysts in the synthesis of a variety of useful organic products. For example, the **Fischer–Tropsch process,** developed by German chemists in the early 1900s, uses transition metal catalysts to prepare hydrocarbons, alcohols, and other products from synthesis gas. For example,

$$H_2 + CO \longrightarrow \text{alkanes}, \quad \text{Co catalyst}$$

$$3\,H_2 + CO \longrightarrow CH_4 + H_2O, \quad \text{Ni catalyst}$$

$$2\,H_2 + CO \longrightarrow CH_3OH, \quad \text{Co or Zn/Cu catalyst}$$

The catalysts in these processes may be metals on alumina (Al_2O_3) or other supports or metal oxides.

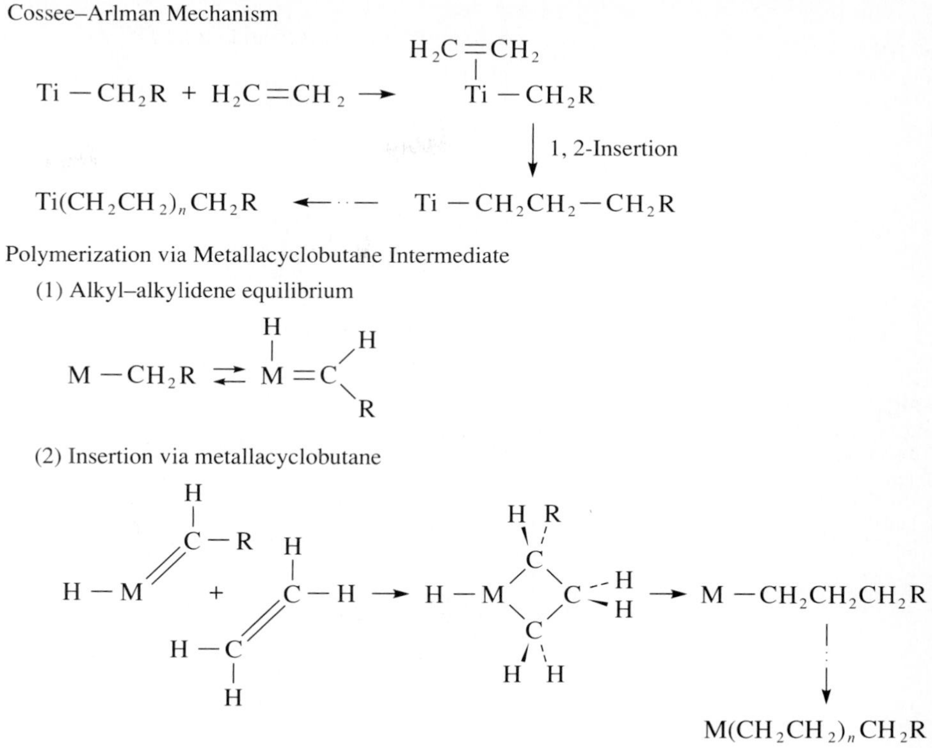

Cossee–Arlman Mechanism

$$Ti-CH_2R + H_2C{=}CH_2 \longrightarrow \underset{Ti-CH_2R}{\overset{H_2C{=}CH_2}{|}}$$

$$\downarrow \text{1, 2-Insertion}$$

$$Ti(CH_2CH_2)_nCH_2R \longleftarrow\cdots\; Ti-CH_2CH_2-CH_2R$$

Polymerization via Metallacyclobutane Intermediate

(1) Alkyl–alkylidene equilibrium

$$M-CH_2R \rightleftharpoons \overset{H}{\overset{|}{M}}{=}C\begin{matrix}H\\R\end{matrix}$$

(2) Insertion via metallacyclobutane

$$H-M{=}\overset{H}{\overset{|}{C}}-R \;+\; \overset{H}{\overset{|}{C}}(-H){=}\underset{H}{\underset{|}{C}}(H-) \longrightarrow H-M\begin{matrix}C(H)(R)\\ \quad\; C(H)(H)\\ C(H)(H)\end{matrix} \longrightarrow M-CH_2CH_2CH_2R \;\cdots\!\!\longrightarrow M(CH_2CH_2)_nCH_2R$$

FIGURE 13-23 Ziegler–Natta Polymerization.

Most of these processes have been conducted under heterogeneous conditions. However, recently there has been considerable interest in developing homogeneous systems to catalyze the Fischer–Tropsch conversion.

These processes for obtaining synthetic fuels were used by a number of countries during World War II. They are, however, uneconomical in most cases, since hydrogen and carbon monoxide in sufficient quantities must be obtained from coal or petroleum sources. Currently, South Africa, which has large coal reserves, makes the greatest use of Fischer–Tropsch reactions in the synthesis of fuels in its Sasol plants.

In **steam reforming,** natural gas (consisting chiefly of methane) is mixed with steam at high temperatures and pressures over a heterogeneous catalyst to generate carbon monoxide and hydrogen:

$$CH_4 + H_2O \longrightarrow CO + 3\,H_2, \qquad \text{Ni catalyst, 700–1000°C}$$

(Other alkanes also react with steam to give mixtures of CO and H_2.) Steam reforming is the principal industrial source of hydrogen gas. Additional hydrogen can be produced by recycling the CO to react further with steam in the **water gas shift reaction:**

$$CO + H_2O \longrightarrow CO_2 + H_2, \qquad \text{Fe/Cr or Zn/Cu catalyst, 400°C}$$

This reaction is favored thermodynamically: at 400°C, $\Delta G° = -14.0$ kJ/mol. Removal of CO_2 from the product can yield hydrogen of greater than 99% purity. This reaction has been studied extensively with the objective of being able to catalyze formation of H_2 *homo*geneously. An example is shown in

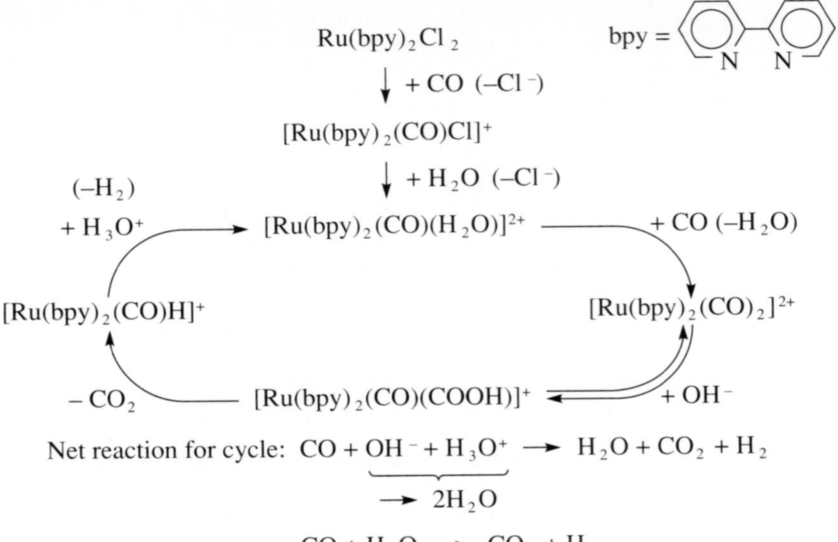

$$\text{Net reaction for cycle: } CO + \underbrace{OH^- + H_3O^+}_{\rightarrow\ 2H_2O} \rightarrow H_2O + CO_2 + H_2$$

$$CO + H_2O \rightarrow CO_2 + H_2$$

FIGURE 13-24 Homogeneous Catalysis of Water Gas Shift Reaction. (Adapted with permission from H. Ishida, K. Tanaka, M. Morimoto, and T. Tanaka, *Organometallics,* **1986,** *5,* 724.)

Figure 13-24.[23] However, these processes have not yet proved efficient enough for commercial use.

In general, these processes, when performed using heterogeneous catalysts, require significantly elevated temperatures and pressures. Consequently, as in the case of the water gas shift reaction, there is high interest in developing homogeneous catalysts that can perform the same functions but under much milder conditions.

GENERAL REFERENCES

Principles and Applications of Organotransition Metal Chemistry, by J. P. Collman, L. S. Hegedus, J. R. Norton, and R. G. Finke, University Science Books, Mill Valley, Calif., 1987, provides a detailed discussion, with numerous references, of the reactions and catalytic processes described in this chapter, as well as a variety of other types of organometallic reactions. In addition to providing extensive information on structural and bonding properties of organometallic compounds, *Comprehensive Organometallic Chemistry,* G. Wilkinson, F. G. A. Stone, and W. Abel, eds. Pergamon Press, Elmsford, N.Y., 1982, gives the most comprehensive information on organometallic reactions, with numerous references to the original literature. A recent article, ''Homogeneous, Heterogeneous, and Enzymatic Catalysis,'' by S. T. Oyama and G. A. Somorjai, *J. Chem. Educ.,* **1988,** *65,* 765, gives examples of the types and amounts of catalysts used in a variety of industrial processes. The other references listed at the end of Chapter 12 are also useful in connection with this chapter.

PROBLEMS

13-1 Predict the transition metal-containing products of the following reactions:

a. $[Mn(CO)_5]^- + H_2C{=}CH{-}CH_2Cl \longrightarrow$ initial product $\xrightarrow{-CO}$ final product

b. *trans*-$Ir(CO)Cl(PPh_3)_2 + CH_3I \longrightarrow$

c. $Ir(PPh_3)_3Cl \xrightarrow{\Delta}$

d. $(\eta^5\text{-}C_5H_5)Fe(CO)_2(CH_3) + PPh_3 \longrightarrow$

e. $(\eta^5\text{-}C_5H_5)Mo(CO)_3(C(=O)CH_3) \xrightarrow{\Delta}$

[23] J. P. Collin, R. Ruppert, and J. P. Sauvage, *Nouv. J. Chim.,* **1985,** *9,* 395.

f. $H_3C{-}Mn(CO)_5 + SO_2 \longrightarrow$ (no gases are evolved)
g. $H_3C{-}Mn(CO)_5 + P(CH_3)(C_6H_5)_2 \longrightarrow$ (No gases are evolved)
h. $H_3CCH{=}CH(CH_2)_3{-}C({=}O)Co(CO)_4 \xrightarrow{\Delta}$ (colorless gas is evolved)
i. $CHCl_3$ + excess $[Co(CO)_4]^- \longrightarrow$
j. $[Mn(CO)_5]^- + (\eta^5\text{-}C_5H_5)Fe(CO)_2Br \longrightarrow$
k. *trans*-$Ir(CO)Cl(PPh_3)_2 + H_2 \longrightarrow$
l. $W(CO)_6 + C_6H_5Li \longrightarrow$
m. *cis*-$Re(CH_3)(PEt_3)(CO)_4 + {}^{13}CO \longrightarrow$
n. *fac*-$Mn(CO)_3(CH_3)(PMe_3)_2 + {}^{13}CO \longrightarrow$
o. *cis*-$Mn(CO)_4({}^{13}CO)(COCH_3) \xrightarrow{\Delta}$ (show all expected products and percent of each)
p. $C_6H_5CH_2{-}Mn(CO)_5 \xrightarrow{h\nu} CO +$
q. $[V(CO)_6]^- + NO^+ \longrightarrow$
r. $Cr(CO)_6 + Na/NH_3 \longrightarrow$
s. $Fe(CO)_5 + NaC_5H_5 \longrightarrow$
t. $H_3C{-}Rh(PPh_3)_3 \xrightarrow{\Delta} +$

13-2 Heating $[(C_5H_5)Fe(CO)_3]^+$ with NaH in solution gives **A**, which has empirical formula $C_7H_6O_2Fe$. **A** reacts rapidly at room temperature to eliminate a colorless gas **B**, forming a purple-brown solid **C** having empirical formula $C_7H_5O_2Fe$. Treatment of **C** with iodine generates a brown solid **D** of empirical formula $C_7H_5O_2FeI$, which on treatment with TlC_5H_5 gives a solid **E** of formula $C_{12}H_{10}O_2Fe$. **E**, on heating, gives off a colorless gas, leaving an orange solid **F** of formula $C_{10}H_{10}Fe$. Propose structural formulas for **A** through **F**.

13-3 An acyl metal carbonyl ($R{-}\overset{\overset{O}{\|}}{C}{-}M(CO)_x$) is much easier to protonate than either a metal carbonyl or an organic ketone, such as acetone. Suggest an explanation.

13-4 $Na[(\eta^5\text{-}C_5H_5)Fe(CO)_2]$ reacts with $ClCH_2CH_2SCH_3$ to give **A**, a monomeric and diamagnetic substance of stoichiometry $C_{10}H_{12}FeO_2S$ having two strong IR bands at 1980 and 1940 cm^{-1}. Heating of **A** gives **B**, a monomeric, diamagnetic substance having strong IR bands at 1920 and 1630 cm^{-1}. Identify **A** and **B**.

13-5 The reaction of $V(CO)_5(NO)$ with $P(OCH_3)_3$ to give $V(CO)_4[P(OCH_3)_3](NO)$ has the rate law

$$\frac{-d[V(CO)_5(NO)]}{dt} = k_1[V(CO)_5(NO)] + k_2[P(OCH_3)_3][V(CO)_5(NO)]$$

a. Suggest mechanisms for this reaction consistent with the rate law.
b. One possible mechanism consistent with the last term in the rate law includes a transition state of formula $V(CO)_5[P(OCH_3)_3](NO)$. Would this necessarily be a 20-electron species? Explain.

13-6 The rate law for the reaction $H_2 + Co_2(CO)_8 \longrightarrow 2\ HCo(CO)_4$ is

$$\text{rate} = \frac{k[Co_2(CO)_8][H_2]}{[CO]}$$

Propose a mechanism consistent with this rate law.

13-7 Which of the following complexes would you expect to react most rapidly with CO? Least rapidly? Briefly explain your choices. (Reference: M. J. Wovkulich and J. D. Atwood, *Organometallics*, **1982**, *1*, 1316.)
a. $Cr(CO)_4(PPh_3)_2$
b. $Cr(CO)_4(PPh_3)(PBu_3)$, Bu = *n*-butyl
c. $Cr(CO)_4(PPh_3)[P(OMe)_3]$
d. $Cr(CO)_4(PPh_3)[P(OPh)_3]$

13-8 The equilibrium constants for the ligand dissociation reaction $NiL_4 \rightleftharpoons NiL_3 + L$ have been determined for a variety of phosphines (Reference: C. A. Tolman, W. C. Seidel, and L. W. Gosser, *J. Am. Chem. Soc.*, **1974,** *96,* 53). For L = PMe_3, PEt_3, $PMePh_2$, and PPh_3, arrange these equilibria in order of the expected magnitudes of their equilibrium constants (from largest K to smallest).

13-9 The complex shown below loses carbon monoxide on heating. Would you expect this carbon monoxide to be ^{12}CO, ^{13}CO, or a mixture of both? Why?

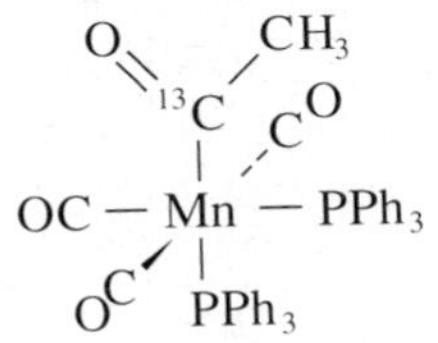

```
 O     CH3
  \\  /
   13C   CO
     |  /
OC — Mn — PPh3
    /  |
  OC   PPh3
```

13-10 **a.** Predict the products of the following reaction, showing clearly the structure of each:

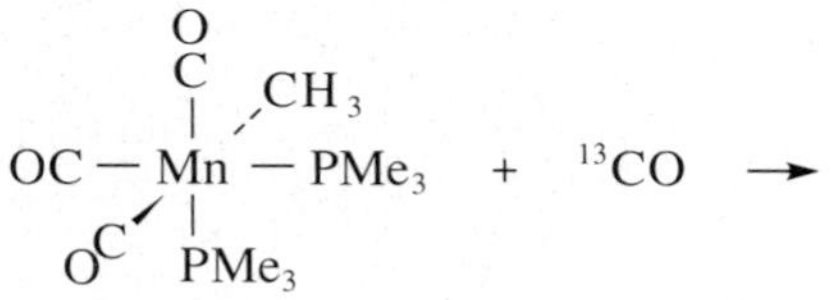

```
       O
       C
       |   CH3
       |  /
OC — Mn — PMe3   +   13CO   →
     /  |
   OC   PMe3
```

b. Each product of this reaction has a new, rather strong IR band that is distinctly different in energy from any bands in the reactants. Account for this band, and predict its approximate location (in cm^{-1}) in the IR spectrum.

13-11 Give structural formulas for **A** through **D**:

$$(C_5H_5)_2Fe_2(CO)_4 \xrightarrow{\text{Na/Hg}} \mathbf{A} \xrightarrow{\text{Br}_2} \mathbf{B} \xrightarrow{\text{LiAlH}_4} \mathbf{C}$$

$$\nu_{CO} = 1961,\ 1942,\ 1790\ cm^{-1}$$

A has strong IR bands at 1880 and 1830 cm^{-1}; **C** has a ^1H NMR spectrum consisting of two singlets of relative intensity 1:5 a approximately δ −12 ppm and δ 5 ppm, respectively. (HINT: Metal hydrides often have protons with negative chemical shifts.)

$$\mathbf{C} \xrightarrow{\text{PhNa}} \mathbf{A} + \mathbf{D} \quad \text{(a hydrocarbon)}$$

13-12 $Re(CO)_5Br$ reacts with the ion $Br—CH_2CH_2—O^-$ to give compound **Y** + Br^-.

a. What is the most likely site of attack of this ion on $Re(CO)_5Br$? [HINT: Consider the hardness (Chapter 6) of the Lewis base.]

b. Using the following information, propose a structural formula for **Y** and account for each of the following:

Y obeys the 18-electron rule.

No gas is evolved in the reaction.

^{13}C NMR indicates that there are five distinct magnetic environments for carbon in **Y**.

Addition of a solution of Ag^+ to a solution of **Y** gives a white precipitate.

(Reference: M. M. Singh and R. J. Angelici, *Inorg. Chem.*, **1984,** *23,* 2699.)

13-13 The carbene complex **I** shown below undergoes the following reactions. Propose structural formulas for the reaction products.

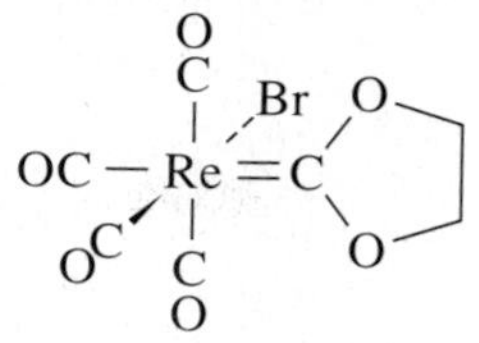

I

a. When a toluene solution containing **I** and excess triphenylphosphine is heated to reflux, first compound **II** is formed and then compound **III**. **II** has infrared bands at 2038, 1958, and 1906 cm^{-1}, **III** at 1944 and 1860 cm^{-1}. ^1H NMR data δ values (relative area) are:

II: 7.62, 7.41 multiplets (15)
4.19 multiplet (4)

III: 7.70, 7.32 multiplets (15)
3.39 singlet (2)

b. When a solution of **I** in toluene is heated to reflux with 1,1-bis(diphenylphosphino)methane, a colorless product **IV** is formed that has the following properties:

IR: 2036, 1959, 1914 cm^{-1}
Elemental analysis (accurate to ±0.3%): 35.87% C, 2.73% H

c. **I** reacts rapidly with the dimethyldithiocarbamate ion, $S_2CN(CH_3)_2^-$, in solution to form $Re(CO)_5Br$ + **V**, a product that does not contain a metal atom. This product has no infrared bands between 1700 and 2300 cm^{-1}. However, it does show moderately intense bands at 1500 and 977 cm^{-1}. The ^1H NMR spectrum of **V** shows bands at δ 3.91 (triplet), 3.60 (triplet), 3.57 (singlet), and 3.41 (singlet) (Reference: G. L. Miessler, S. Kim, R. A. Jacobson, and R. J. Angelici, *Inorg. Chem.*, **1987**, *26*, 1690).

13-14 The complex **I** in Problem 13 can be synthesized from $Re(CO)_5Br$ and 2-bromoethanol in ethylene oxide solution with solid NaBr present. Suggest a mechanism for the formation of the carbene ligand.

13-15 The anion $[Mn(CO)_5]^-$ reacts with 1,3-dibromopropane to form $Br—CH_2CH_2CH_2—Mn(CO)_5$. However, the reaction does not stop here; the product reacts with additional $[Mn(CO)_5]^-$ to yield a carbene complex. Propose a structure for this complex, and suggest a mechanism for its formation.

13-16 Show how transition metal complexes could be used to effect the following syntheses:

a. Acetaldehyde from ethylene
b. $CH_3CH_2COOCH_3$ from CH_3CH_2Cl
c. $CH_3CH_2CH_2CH_2CHO$ from $CH_3CH_2CH{=}CH_2$
d. $PhCH_2CH_2CH_2CHO$ from an alkene (Ph = phenyl)
e.

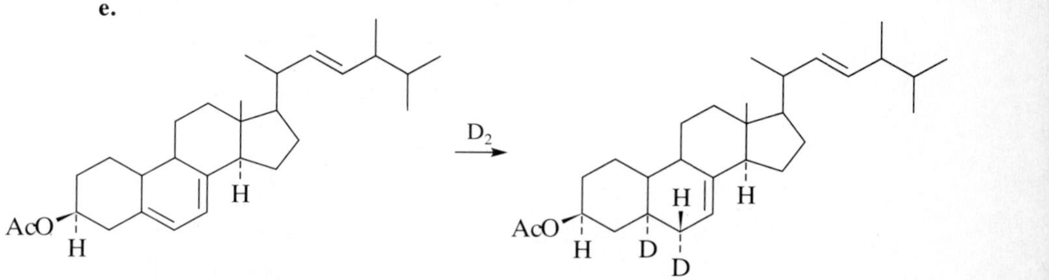

$\xrightarrow{D_2}$

f. $C_6D_5CH_3$ from $C_6H_5CH_3$.

13-17 The complex $Rh(H)(CO)_2(PPh_3)_2$ can be used in the catalytic synthesis of *n*-pentanal from an alkene having one less carbon. Propose a mechanism for this process. Give an appropriate designation for each type of reaction step (such as oxidative addition or alkyl migration) and identify the catalytic species.

13-18 It is possible by using an appropriate transition metal catalyst to synthesize the following aldehyde from an appropriate 5-carbon alkene:

$$H_3C{-}CH_2{-}\underset{}{\overset{\displaystyle CH_3}{\overset{|}{C}}}H{-}CH_2{-}\overset{\displaystyle O}{\overset{\|}{C}}{-}H$$

Show how this synthesis could be effected catalytically. Identify the catalytic species.

13-19 At low temperature and pressure, a gas-phase reaction can occur between iron atoms and toluene. The product, a rather unstable sandwich compound, reacts with ethylene to give compound **X**. Compound **X** decomposes at room temperature to liberate ethylene; at $-20°C$ it reacts with $P(OCH_3)_3$ to give Fe(toluene)$[P(OCH_3)_3]_2$. Suggest a structure for compound **X**. (Reference: U. Zenneck and W. Frank, *Angew. Chem. Int. Ed. Engl.*, **1986,** *25,* 831.)

13-20 The reaction of $RhCl_3 \cdot 3\ H_2O$ with tri-*o*-tolylphosphine in ethanol at 25°C gives a blue-green complex **I** ($C_{42}H_{42}P_2Cl_2Rh$) that has ν (Rh—Cl) at 351 cm^{-1} and μ_{eff} = 2.3 BM. At higher temperature, a diamagnetic yellow complex **II** that has an Rh:Cl ratio of 1:1 is formed that has an intense band near 920 cm^{-1}. Addition of NaSCN to **II** replaces Cl with SCN to give a product **III** having the following ^1H NMR spectrum:

Chemical shift	*Relative area*	*Type*
6.9–7.5	12	Aromatic
3.50	1	Doublet of 1:2:1 triplets
2.84	3	Singlet
2.40	3	Singlet

Treatment of **II** with NaCN gives a phosphine ligand **IV** of empirical formula $C_{21}H_{19}P$ and a molecular weight of 604. **IV** has an absorption band at 965 cm^{-1} and the following ^1H NMR spectrum:

Chemical shift	*Relative area*	*Type*
7.64	1	Singlet
6.9–7.5	12	Aromatic
2.37	6	Singlet

Determine the structural formulas of compounds **I** through **IV** and account for as much of the data as possible. (Reference: M. A. Bennett and P. A. Longstaff, *J. Am. Chem. Soc.*, **1969,** *91,* 6266.)

14

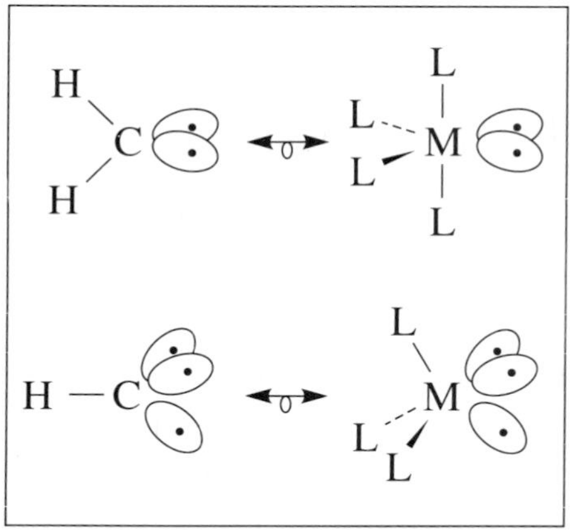

Parallels Between Main Group and Organometallic Chemistry

It is common to treat organic and inorganic chemistry as separate topics and, within inorganic chemistry, to consider separately the chemistry of main group compounds and organometallic compounds, as we have generally done so far in this text. However, valuable insights can be gained by examining parallels between these different classifications of compounds. Such an examination may lead to a more thorough understanding of the different types of compounds being compared and may suggest new chemical compounds or new types of reactions. The objective of this chapter is to consider several of these parallels, especially between main group and organometallic compounds.

14-1 MAIN GROUP PARALLELS WITH BINARY CARBONYL COMPLEXES

Several comparisons within main group chemistry have already been discussed in earlier chapters. These included the similarities (and differences) between borazine and benzene, the relative instability of silanes in comparison with alkanes, and differences in bonding in homonuclear and heteronuclear diatomic species (such as the isoelectronic N_2 and CO). In general, these parallels have centered around isoelectronic species. Similarities also occur between main group and transition metal species that are *electronically equivalent,* species that require the same number of electrons to achieve a filled valence configuration.[1] For example, a halogen atom, one electron short of a valence shell octet, may be considered electronically equivalent to $Mn(CO)_5$, a 17-electron

[1] J. E. Ellis, *J. Chem. Ed.*, **1976,** *53,* 2.

species one electron short of an 18-electron configuration. In this section we will discuss briefly some parallels between main group atoms and ions and electronically equivalent binary carbonyl complexes.

Much chemistry of main group and metal carbonyl species can be rationalized from the way in which these species can achieve closed shell (octet or 18-electron) configurations. These methods of achieving more stable configurations will be illustrated for the following electronically equivalent species:

Electrons short of filled shell	*Examples of electronically equivalent species*	
	Main group	*Metal carbonyl*
1	Cl, Br, I	$Mn(CO)_5$, $Co(CO)_4$
2	S	$Fe(CO)_4$, $Os(CO)_4$
3	P	$Co(CO)_3$, $Ir(CO)_3$

Halogen atoms, one electron short of a valence shell octet, exhibit chemical similarities with 17-electron organometallic species; some of the most striking are the parallels between halogen atoms and $Co(CO)_4$, as summarized in Table 14-1. Both can reach filled shell electron configurations by acquiring an electron or by dimerization. The neutral dimers are capable of adding across multiple carbon–carbon bonds and can undergo disproportionation by Lewis bases. Anions of both electronically equivalent species have a 1− charge and can combine with H^+ to form acids: both HX (X = Cl, Br, or I) and $HCo(CO)_4$ are strong acids in aqueous solution. Both types of anions form precipitates with heavy metal ions such as Ag^+ in aqueous solution. The parallels between 7-electron halogen atoms and 17-electron binary carbonyl species are sufficiently strong to justify extending the label *pseudohalogen* (Chapter 7) to these carbonyls.

Similarly, six-electron main group species show chemical similarities with 16-electron organometallic species. As for the halogens and 17-electron organo-

TABLE 14-1
Parallels between Cl and $Co(CO)_4$

Characteristic	*Examples*	*Examples*
Ion of 1− charge	Cl^-	$[Co(CO)_4]^-$
Neutral dimeric species	Cl_2	$[Co(CO)_4]_2$
Hydrohalic acid	HCl (strong acid in aqueous solution)	$HCo(CO)_4$ (strong acid in aqueous solution)*
Formation of interhalogen compounds	$Br_2 + Cl_2 \rightleftharpoons 2\ BrCl$	$I_2 + [Co(CO)_4]_2 \longrightarrow 2\ ICo(CO)_4$ (unstable)
Formation of heavy metal salts of low solubility in water	AgCl	$AgCo(CO)_4$
Addition to unsaturated species	$Cl_2 + H_2C{=}CH_2 \longrightarrow H{-}CCl(H){-}CCl(H){-}H$	$[Co(CO)_4]_2 + F_2C{=}CF_2 \longrightarrow (CO)_4Co{-}CF_2{-}CF_2{-}Co(CO)_4$
Disproportionation by Lewis bases	$Cl_2 + N(CH_3)_3 \longrightarrow [ClN(CH_3)_3]Cl$	$[Co(CO)_4]_2 + C_5H_{10}NH$ (piperidine) $\longrightarrow [(CO)_4Co(C_5H_{10}NH][Co(CO)_4]$

NOTE: * However, $HCo(CO)_4$ is only slightly soluble in water.

TABLE 14-2
Parallels between sulfur and $Fe(CO)_4$

Characteristic	*Examples*	
Ion of 2− charge	S^{2-}	$[Fe(CO)_4]^{2-}$
Neutral compound	S_8	$Fe_2(CO)_9$, $Fe_3(CO)_{12}$
Hydride	H_2S: $pK_1 = 7.24$* $pK_2 = 14.92$	$H_2Fe(CO)_4$: $pK_1 = 4.44$* $pK_2 = 14$
Phosphine adduct	Ph_3PS	$Ph_3PFe(CO)_4$
Polymeric mercury compound	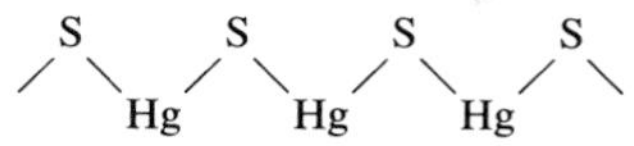 —S—Hg—S—Hg—S—Hg—S—	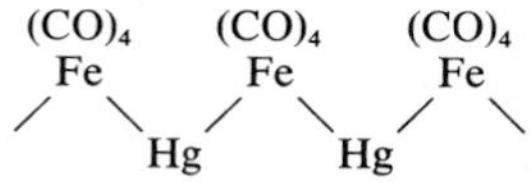 —$Fe(CO)_4$—Hg—$Fe(CO)_4$—Hg—$Fe(CO)_4$—
Compound with ethylene	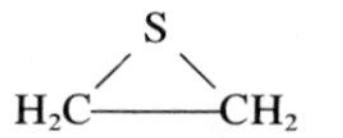 S bridging H_2C—CH_2 (ethylene sulfide)	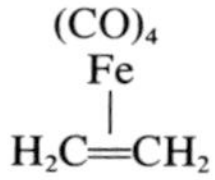 $(CO)_4Fe$ bonded to $H_2C{=}CH_2$ (π complex)

NOTE: * pK values in aqueous solution at 25°C.

metallic complexes, many of these similarities can be accounted for on the basis of ways in which the species can acquire or share electrons to achieve filled shell configurations. Some similarities between sulfur and the electronically equivalent $Fe(CO)_4$ are listed in Table 14-2.

The concept of electronically equivalent groups can also be extended to 5-electron main group elements [group 15(VA)] and 15-electron organometallic species. For example, phosphorus and $Ir(CO)_3$ both form tetrahedral tetramers, as shown in Figure 14-1. The 15-electron $Co(CO)_3$, which is isoelectronic with $Ir(CO)_3$, can replace one or more phosphorus atoms in the P_4 tetrahedron, as also shown in this figure.

The parallels between electronically equivalent main group and organometallic species are interesting and summarize a considerable amount of their chemistry. The limitations of these parallels should also be recognized, however. For example, main group compounds having expanded octets may not have organometallic analogues; organometallic analogues of such compounds as IF_7 and XeF_6 are not known. Organometallic complexes of ligands significantly weaker than CO in the spectrochemical series may not follow the 18-electron rule and may consequently behave quite differently than electronically equivalent main group species. In addition, the reaction chemistry of organometallic compounds may be very different than main group chemistry. For example, loss of ligands such as CO is far more common in organometallic chemistry than in main group chemistry. Therefore, as in any scheme based on as simple a framework as electron counting, the concept of electronically

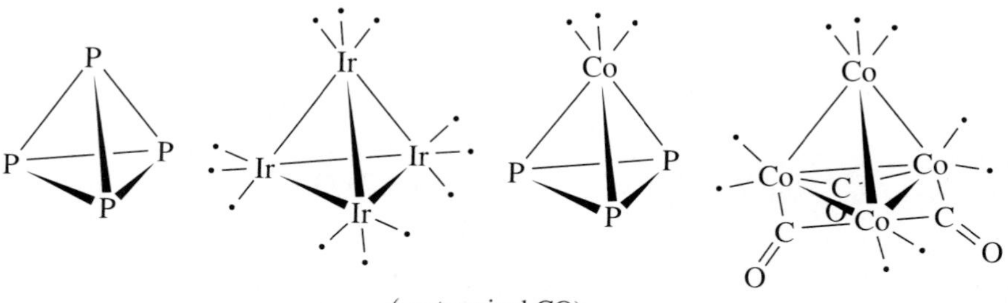

FIGURE 14-1 P_4, $[Ir(CO)_3]_4$, $P_3[Co(CO)_3]$, and $Co_4(CO)_{12}$.

equivalent groups, while useful, has its limitations. It serves as valuable background, however, for a potentially more versatile way to seek parallels between main group and organometallic chemistry, the concept of isolobal groups.

14-2 THE ISOLOBAL ANALOGY

An important contribution to the understanding of parallels between organic and inorganic chemistry has been the concept of isolobal molecular fragments, described most elaborately by Roald Hoffmann in his 1982 Nobel lecture.[2] Hoffmann defined molecular fragments to be isolobal

> if the number, symmetry properties, approximate energy, and shape of the frontier orbitals and the number of electrons in them are similar—not identical, but similar.

To illustrate this definition, we will find it useful to compare fragments of methane with fragments of an octahedrally coordinated transition metal complex ML_6. For simplicity, we will consider only sigma bonding between the metal and the ligands in this complex. The fragments to be discussed are shown in Figure 14-2

The parent compounds have filled valence shell electron configurations, an octet for CH_4, 18 electrons for ML_6 (an example of such a compound is $Cr(CO)_6$). Methane may be considered to use sp^3 hybrid orbitals in bonding, with eight electrons occupying bonding pairs formed from interactions between

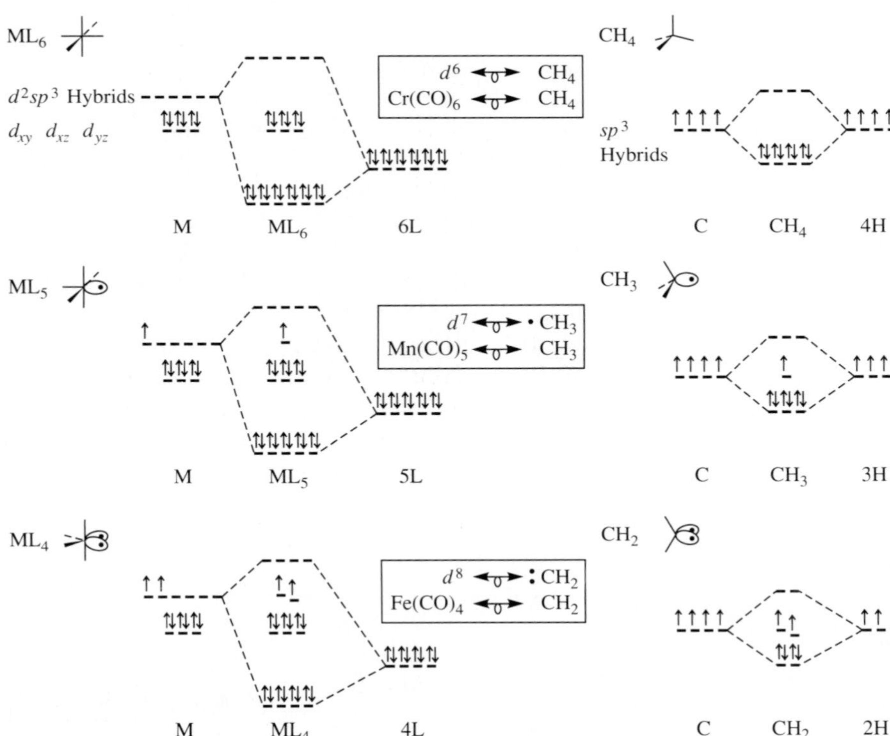

FIGURE 14-2 Orbitals of Octahedral and Tetrahedral Fragments.

[2] R. Hoffmann, *Angew. Chem. Int. Ed. Engl.*, **1982,** *21,* 711; see also H-J. Krause, *Z. Chem.*, **1988,** *28,* 129.

the hybrids and 1s orbitals on hydrogen. The metal in ML_6, by similar reasoning, uses d^2sp^3 hybrids in bonding to the ligands, with 12 electrons occupying bonding orbitals and six essentially nonbonding electrons occupying d_{xy}, d_{xz}, and d_{yz} orbitals. (The model can be refined further to include pi interactions between d_{xy}, d_{xz}, and d_{yz} orbitals with ligands having suitable donor and/or acceptor orbitals.)

Molecular fragments containing fewer ligands than the parent polyhedra can now be described; for the purpose of the analogy, these fragments will be assumed to preserve the geometry of the remaining ligands.

In the 7-electron fragment CH_3, three of the sp^3 orbitals of carbon are involved in sigma bonding with the hydrogens. The fourth hybrid is singly occupied and at higher energy than the sigma-bonding pairs of CH_3, as shown in Figure 14-2. This situation is similar to the 17-electron fragment $Mn(CO)_5$. The sigma interactions between the ligands and Mn in this fragment may be considered to involve five of the metal's d^2sp^3 hybrid orbitals. The sixth hybrid is singly occupied and at higher energy than the five sigma-bonding orbitals.

As Figure 14-2 shows, each of these fragments has a single electron in a hybrid orbital at the vacant site of the parent polyhedron. These orbitals are sufficiently similar to meet Hoffmann's isolobal definition. Using Hoffmann's symbol $\leftarrow\!\!\circ\!\!\rightarrow$ to designate groups as isolobal, we may write

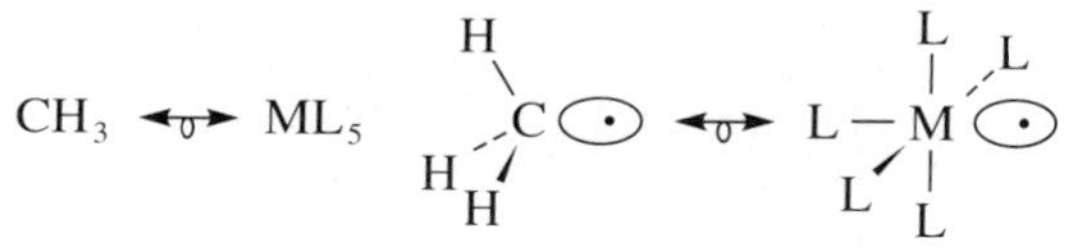

$$CH_3 \leftarrow\!\!\circ\!\!\rightarrow ML_5$$

Similarly, 6-electron CH_2 and 16-electron ML_4 are isolobal. Each of these fragments represents the parent polyhedron, with single electrons occupying hybrid orbitals at otherwise vacant sites and having two electrons less than the filled shell octet or 18-electron configurations. Absence of a third ligand similarly gives a pair of isolobal fragments, CH and ML_3.

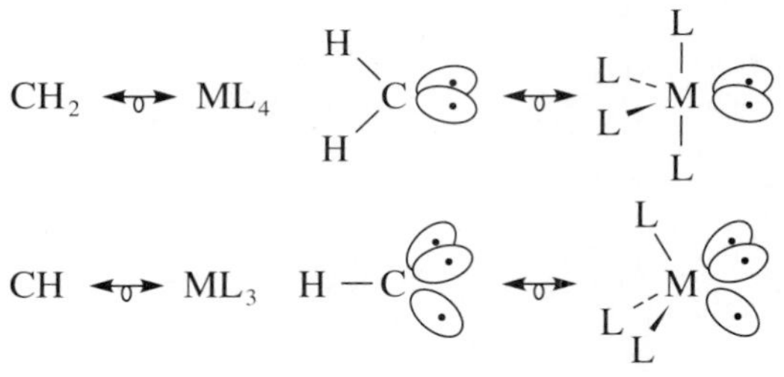

$$CH_2 \leftarrow\!\!\circ\!\!\rightarrow ML_4$$

$$CH \leftarrow\!\!\circ\!\!\rightarrow ML_3$$

To summarize:

	Organic	Inorganic	Organo-metallic example	Vertices missing from parent polyhedron	Electrons short of filled shell
Parent	CH_4	ML_6	$Cr(CO)_6$	0	0
Fragments	CH_3	ML_5	$Mn(CO)_5$	1	1
	CH_2	ML_4	$Fe(CO)_4$	2	2
	CH	ML_3	$Co(CO)_3$	3	3

These fragments can be combined into molecules. For example, two CH_3 fragments form ethane, and two $Mn(CO)_5$ fragments form the dimeric

$(OC)_5Mn—Mn(CO)_5$. Furthermore, these organic and organometallic fragments can be combined into $H_3C—Mn(CO)_5$, which is also a known compound.

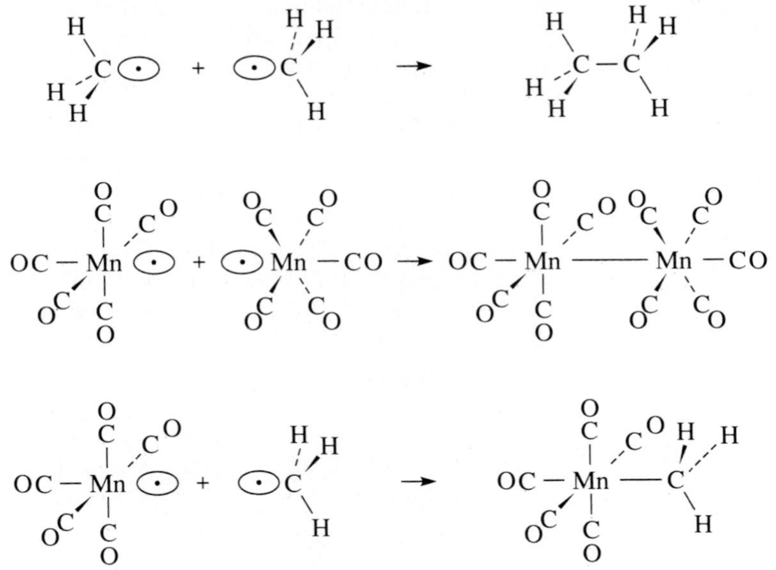

The organic and organometallic parallels are not always this complete. For example, while two 6-electron CH_2 fragments form ethylene, $H_2C{=}CH_2$, the dimer of the isolobal $Fe(CO)_4$ is not nearly as stable; it is known as a transient species obtained photochemically from $Fe_2(CO)_9$.[3] However, both CH_2 and $Fe(CO)_4$ form three-membered rings, cyclopropane and $Fe_3(CO)_{12}$. Although cyclopropane is a trimer of three CH_2 fragments, $Fe_3(CO)_{12}$ has two bridging carbonyls and is therefore not a perfect trimer of $Fe(CO)_4$. The isoelectronic $Os_3(CO)_{12}$, on the other hand, is a trimeric combination of three $Os(CO)_4$ fragments, which are isolobal with both $Fe(CO)_4$ and CH_2 and can correctly be described as $[Os(CO)_4]_3$.

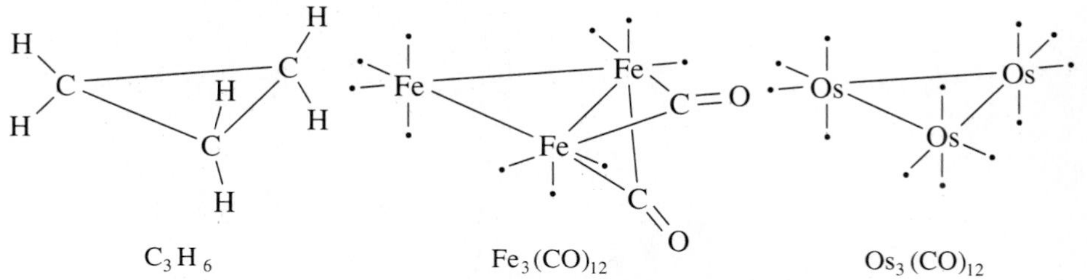

C_3H_6 $Fe_3(CO)_{12}$ $Os_3(CO)_{12}$

(• = terminal carbonyl)

As mentioned previously, $Ir(CO)_3$, a 15-electron fragment, forms $[Ir(CO)_3]_4$, which has T_d symmetry. The isoelectronic complex $Co_4(CO)_{12}$ has a nearly tetrahedral array of cobalt atoms but has three bridging carbonyls and hence C_{3v} symmetry. Compounds are also known having a central tetrahedral structure, with one or more $Co(CO)_3$ fragments [which are isolobal and isoelectronic with $Ir(CO)_3$] replaced by the isolobal CR fragment. This is similar to the replacement of phosphorus atoms in the P_4 tetrahedron by $Co(CO)_3$ fragments; P may also be described as isolobal with CR.

[3] M. Poliakoff and J. J. Turner, *J. Chem. Soc. (A)*, **1971**, 2403.

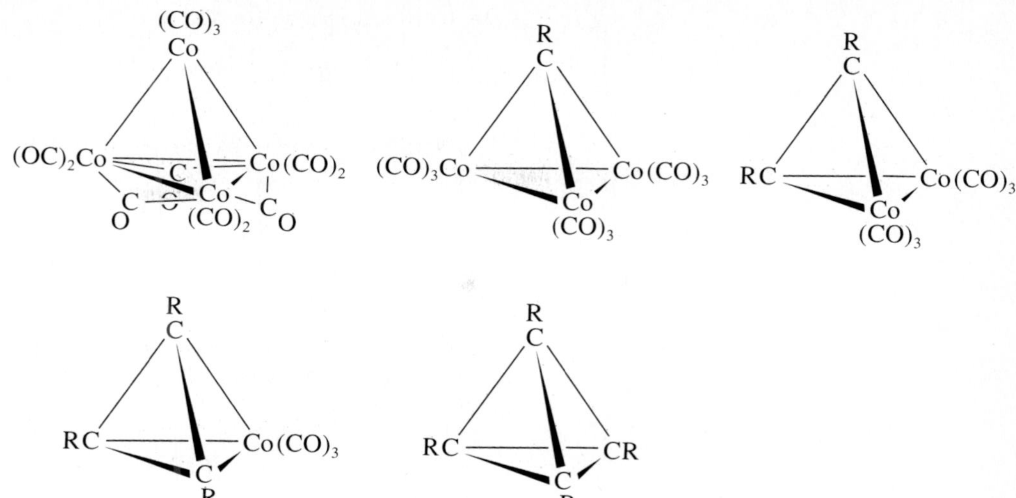

14-2-1 EXTENSIONS OF THE ANALOGY

The concept of isolobal fragments can be extended beyond the examples given so far to include charged species, a variety of ligands other than CO, and organometallic fragments based on structures other than octahedral. Some of the ways of extending the isolobal parallels can be summarized as follows:

1. The isolobal definition may be extended to isoelectronic fragments having the same coordination number. For example,

$$\text{Since } Mn(CO)_5 \underset{\text{o}}{\longleftrightarrow} CH_3, \qquad \begin{matrix} Re(CO)_5 \\ [Fe(CO)_5]^+ \\ [Cr(CO)_5]^- \end{matrix} \underset{\text{o}}{\longleftrightarrow} CH_3$$

2. Gain or loss of electrons from two isolobal fragments yields isolobal fragments. For example,

$$\text{Since } \underset{\text{(17-electron fragment)}}{Mn(CO)_5} \underset{\text{o}}{\longleftrightarrow} \underset{\text{(7-electron fragment)}}{CH_3}, \qquad \underset{\text{(16-electron fragments)}}{\begin{matrix} [Mn(CO)_5]^+ \\ Cr(CO)_5 \\ Mo(CO)_5 \end{matrix}} \underset{\text{o}}{\longleftrightarrow} \underset{\text{(6-electron fragment)}}{CH_3^+}$$

$$\underset{\text{(18-electron fragments)}}{\begin{matrix} [Mn(CO)_5]^- \\ Fe(CO)_5 \\ Ru(CO)_5 \end{matrix}} \underset{\text{o}}{\longleftrightarrow} \underset{\text{(8-electron fragment)}}{CH_3^-}$$

3. Other 2-electron donors are treated similarly to CO:

$$Mn(CO)_5 \leftarrow\!\!\!\circ\!\!\!\rightarrow Mn(PR_3)_5 \leftarrow\!\!\!\circ\!\!\!\rightarrow [MnCl_5]^{5-} \leftarrow\!\!\!\circ\!\!\!\rightarrow Mn(NCR)_5 \leftarrow\!\!\!\circ\!\!\!\rightarrow CH_3$$

4. η^5-C_5H_5 is considered to occupy three coordination sites and to be a 6-electron donor:

$$(\eta^5\text{-}C_5H_5)Fe(CO)_2 \leftarrow\!\!\!\circ\!\!\!\rightarrow [Fe(CO)_5]^+ \leftarrow\!\!\!\circ\!\!\!\rightarrow Mn(CO)_5 \text{ (17-electron fragments)} \leftarrow\!\!\!\circ\!\!\!\rightarrow CH_3$$

$$(\eta^5\text{-}C_5H_5)Mn(CO)_2 \leftarrow\!\!\!\circ\!\!\!\rightarrow [Mn(CO)_5]^+ \leftarrow\!\!\!\circ\!\!\!\rightarrow Cr(CO)_5 \text{ (16-electron fragments)} \leftarrow\!\!\!\circ\!\!\!\rightarrow CH_2$$

5. Fragments of formula ML_n (where M has a d^x configuration) are isolobal with fragments of formula ML_{n-2} (where M has a d^{x+2} configuration and L = 2-electron donor):

$$\underset{d^6}{Cr(CO)_5} \leftarrow\!\!\!\circ\!\!\!\rightarrow \underset{d^8}{Fe(CO)_3} \leftarrow\!\!\!\circ\!\!\!\rightarrow \underset{d^8}{[PtCl_3]^-}$$

$$\underset{d^8}{Fe(CO)_4} \leftarrow\!\!\!\circ\!\!\!\rightarrow \underset{d^{10}}{Ni(PR_3)_2} \leftarrow\!\!\!\circ\!\!\!\rightarrow \underset{d^{10}}{Pt(PR_3)_2}$$

The fifth of these extensions of the isolobal analogy is less obvious than the others and deserves explanation. We will consider two examples, the parallels between d^6 ML_5 and d^8 ML_3 fragments, and the parallels between d^8 ML_4 and d^{10} ML_2 fragments. The ML_3 and ML_2 groups may be considered fragments of a square planar ML_4 molecule, as shown in Figure 14-3. They will be compared with the fragments of an octahedral ML_6 molecule shown in Figure 14-2.

A d^8 ML_3 fragment has an empty lobe of a nonbonding hybrid orbital as its LUMO. This is comparable to the LUMO of a d^6 fragment of an octahedron; such a fragment (for example, $Cr(CO)_5$) would have one less electron than shown for $Mn(CO)_5$ in Figure 14-2. A d^8 fragment such as $[PtCl_3]^-$ or $Fe(CO)_3$ would therefore be isolobal with $Cr(CO)_5$ and other ML_5 fragments provided the empty lobe in each case had suitable energy.[4]

A d^{10} ML_2 fragment such as $Ni(PR_3)_2$ would have two valence electrons more than the example of $PtCl_2$ shown in Figure 14-3. These electrons would occupy two hybrid orbitals not used in the sigma bonding between nickel and the ligands. This situation is very comparable to the $Fe(CO)_4$ fragment shown in Figure 14-2; in each case two singly occupied lobes are exhibited.

Examples of isolobal fragments containing CO and η^5-C_5H_5 ligands are given in Table 14-3.

EXERCISE 14-1

For the following, propose examples of isolobal organometallic fragments other than those given above and in Table 14-3:

a. A fragment isolobal with CH_2^+
b. A fragment isolobal with CH^-
c. Three fragments isolobal with CH_3

[4] The highest occupied orbitals of ML_3 have similar, but not equal, energies. For a detailed analysis of the energies and symmetries of ML_5, ML_3, and other fragments, see M. Elian and R. Hoffmann, *Inorg. Chem.*, **1975,** *14,* 1058, and T. A. Albright, R. Hoffmann, J. C. Thibeault, and D. L. Thorn, *J. Am. Chem. Soc.*, **1979,** *101,* 3801.

EXERCISE 14-2

Find organic fragments isolobal with each of the following:

a. $Ni(\eta^5\text{-}C_5H_5)$

b. $Cr(CO)_2(\eta^6\text{-}C_6H_6)$

c. $[Fe(CO)_2(PPh_3)]^-$

Analogies are by no means limited to organometallic fragments of octahedra; similar arguments can be used to derive fragments of different

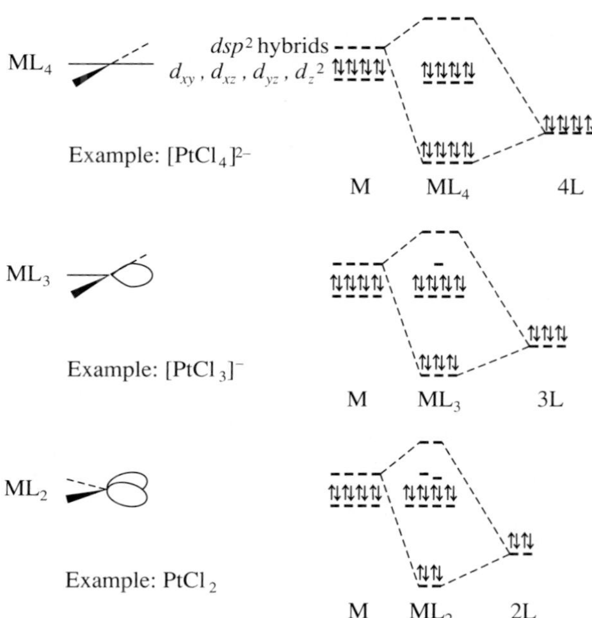

FIGURE 14-3 Orbitals of Square Planar Fragments.

TABLE 14-3
Examples of isolobal fragments

Neutral hydrocarbon	CH_4	CH_3	CH_2	CH	C
Isolobal organometallic fragments	$Cr(CO)_6$	$Mn(CO)_5$	$Fe(CO)_4$	$Co(CO)_3$	$Ni(CO)_2$
	$[Mn(CO)_6]^+$	$[Fe(CO)_5]^+$	$[Co(CO)_4]^+$	$[Ni(CO)_3]^+$	$[Cu(CO)_2]^+$
	$CpMn(CO)_3$	$CpFe(CO)_2$	$CpCo(CO)$	CpNi	
Anionic hydrocarbon fragments obtained by loss of H^+	CH_3^-	CH_2^-	CH^-		
Isolobal organometallic fragments	$Fe(CO)_5$	$Co(CO)_4$	$Ni(CO)_3$		
Cationic hydrocarbon fragments obtained by gain of H^+		CH_4^+	CH_3^+	CH_2^+	CH^+
Isolobal organometallic fragments		$V(CO)_6$	$Cr(CO)_5$	$Mn(CO)_4$	$Fe(CO)_3$

polyhedra. For example, $Co(CO)_4$, a 17-electron fragment of a trigonal bipyramid, is isolobal with $Mn(CO)_5$, a 17-electron fragment of an octahedron:

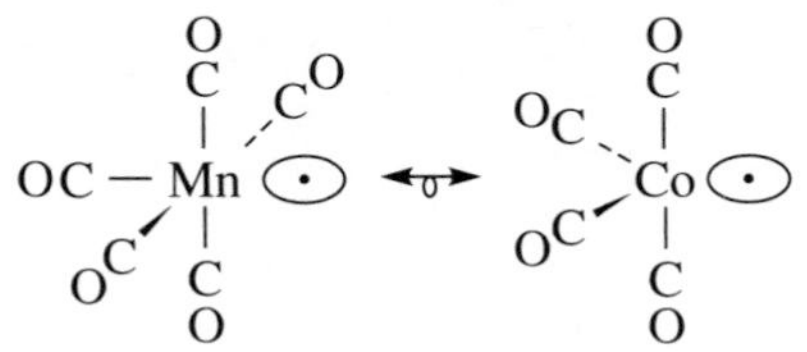

$$(OC)_5Mn\cdot \longleftrightarrow \cdot Co(CO)_4$$

Examples of electron configurations of isolobal fragments of polyhedra having five through nine vertices are given in Table 14-4.

TABLE 14-4
Isolobal relationships for fragments of polyhedra

Organic fragment	*Coordination number of transition metal for parent polyhedron*					*Valence electrons of fragment*
	5	*6*	*7*	*8*	*9*	
CH_3	d^9-ML_4	d^7-ML_5	d^5-ML_6	d^3-ML_7	d^1-ML_8	17
CH_2	d^{10}-ML_3	d^8-ML_4	d^6-ML_5	d^4-ML_6	d^2-ML_7	16
CH		d^9-ML_3	d^7-ML_4	d^5-ML_5	d^3-ML_6	15

Again, credit is due Hoffmann for his systematic extension of the definition of isolobal groups to include the types of examples cited here. The interested reader is strongly encouraged to refer to his Nobel lecture for further information on how the isolobal analogy can be extended to include other ligands and geometries.

14-2-2 EXAMPLES OF APPLICATIONS OF THE ANALOGY

The isolobal analogy can be extended to any molecular fragment having frontier orbitals of suitable size, shape, symmetry, and energy. For example, $Au(PPh_3)$, a 13-electron fragment, has a single electron in a hybrid orbital pointing away from the phosphine.[5] This electron is in an orbital of similar symmetry but of somewhat higher energy than the singly occupied hybrid in the $Mn(CO)_5$ fragment.

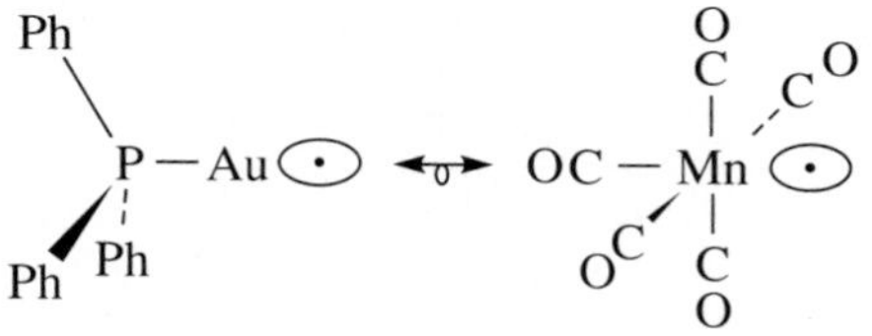

$$Ph_3P{-}Au\cdot \longleftrightarrow OC{-}Mn(CO)_4\cdot$$

Nevertheless, $Au(PPh_3)$ can combine with the isolobal $Mn(CO)_5$ and CH_3 to form $(OC)_5Mn{-}Au(PPh_3)$ and $H_3C{-}Au(PPh_3)$.

Even a hydrogen atom, with a single electron in its 1*s* orbital, can in some cases be viewed as a fragment isolobal with such species as CH_3,

[5] D. G. Evans and D. M. P. Mingos, *J. Organomet. Chem.*, **1982,** *232,* 171.

$Mn(CO)_5$, and $Au(PPh_3)$. Hydrides of the first two are well known, and $Au(PPh_3)$ and H in some cases show surprisingly similar behavior, such as in their ability to bridge the triosmium clusters shown below.[6,7]

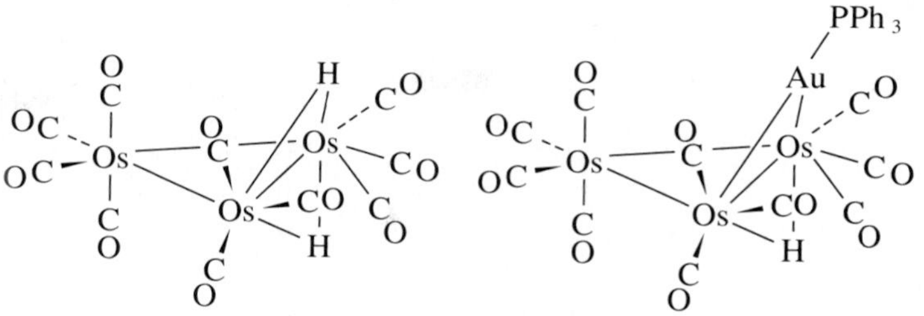

Potentially the greatest practical use of isolobal analogies is in the suggested syntheses of new compounds. For example, CH_2 is isolobal with 16-electron $Cu(\eta^5\text{-}C_5Me_5)$ (extension 4 of the analogy, as described previously) and 14-electron PtL_2 (L = PR_3, CO; extension 5). Recognition of these fragments as isolobal has been exploited in syntheses of new organometallic compounds composed of fragments isolobal with fragments of known compounds.[8] Some of the compounds obtained in these studies are:

Previously known compounds	*New compounds composed of isolobal fragments*
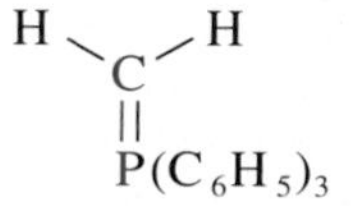 $H_2C{=}P(C_6H_5)_3$	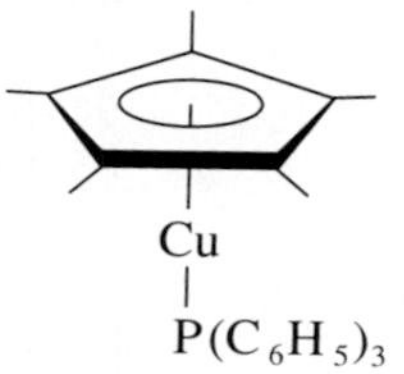 $(\eta^5\text{-}C_5Me_5)Cu{-}P(C_6H_5)_3$
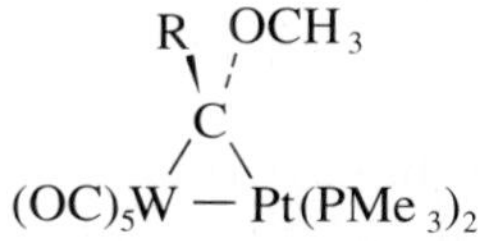 $(OC)_5W{-}Pt(PMe_3)_2$ bridged by $C(R)(OCH_3)$	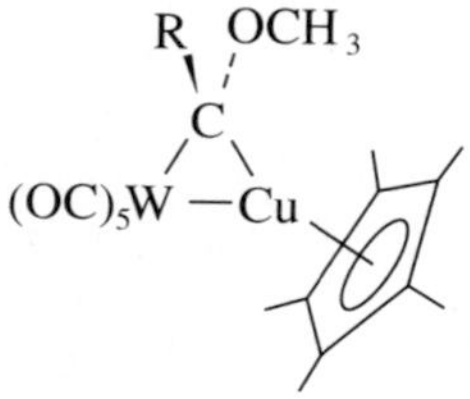 $(OC)_5W{-}Cu(C_5Me_5)$ bridged by $C(R)(OCH_3)$
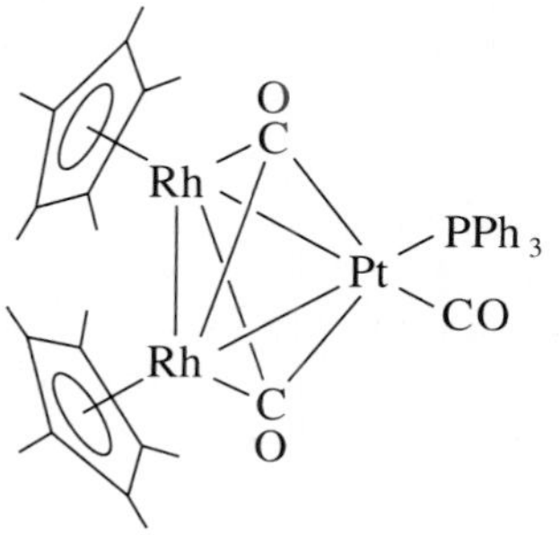 $Rh_2(C_5Me_5)_2(\mu\text{-}CO)_2Pt(PPh_3)(CO)$	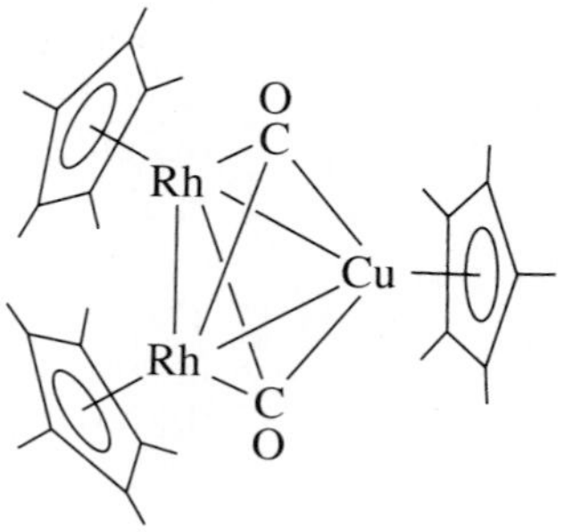 $Rh_2(C_5Me_5)_2(\mu\text{-}CO)_2Cu(C_5Me_5)$

[6] A. G. Orpen, A. V. Rivera, E. G. Bryan, D. Pippard, G. M. Sheldrick, and K. D. Rouse, *J. Chem. Soc. Chem. Commun.*, **1978**, 723.

[7] B. F. G. Johnson, D. A. Kaner, J. Lewis, and P. R. Raithby, *J. Organomet. Chem.*, **1981**, *215*, C33.

[8] G. A. Carriedo, J. A. K. Howard, and F. G. A. Stone, *J. Organomet. Chem.*, **1983**, *250*, C28.

14-3 CLUSTER COMPOUNDS

Examples of cluster compounds have been given in previous sections of this chapter and in several earlier chapters in this text. Transition metal cluster chemistry has developed rapidly in recent years. Beginning with simple dimeric molecules such as $Co_2(CO)_8$ and $Fe_2(CO)_9$,[9] chemists have developed syntheses of far more complex clusters, some having interesting and unusual structures and chemical properties. Large clusters have been studied with the objective of developing catalysts that may duplicate or improve on the properties of heterogeneous catalysts; the surface of a large cluster may in these cases mimic the behavior of the surface of a solid catalyst.

Before turning our attention in more detail to transition metal clusters, we will find it useful to consider compounds of boron, which has an extremely detailed cluster chemistry. As mentioned in Chapter 7, boron forms numerous hydrides (boranes) of interesting structure. Some of these compounds exhibit similarities in their bonding and structures to transition metal clusters.

14-3-1 BORANES

There are a great many neutral and ionic species composed of boron and hydrogen, far too numerous to describe in this text; structures of some have been shown in Chapter 7. For the purposes of illustrating parallels between these species and transition metal clusters, we will first consider one category of boranes, *closo* (cagelike) boranes that have the formula $B_nH_n^{2-}$. These boranes consist of closed polyhedra having n corners, with each polyhedron having all triangular faces (triangulated polyhedra). Each corner is occupied by a BH group; unlike diborane (B_2H_6; see Chapter 7) and some other boranes, there are no bridging hydrogens in *closo* boranes.

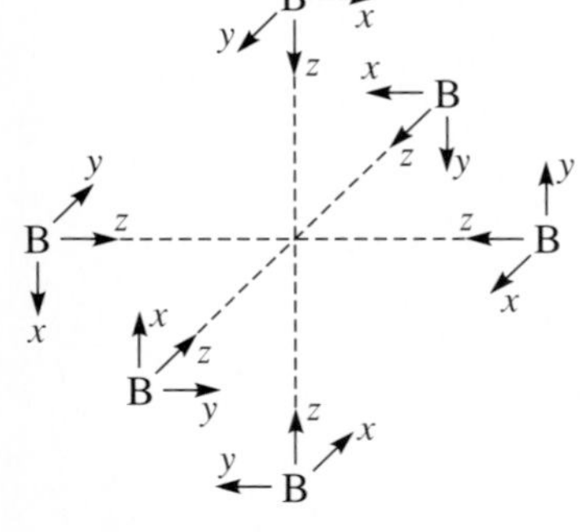

FIGURE 14-4 Coordinate System for Bonding in $B_6H_6^{2-}$.

Molecular orbital calculations have shown that *closo* boranes have $2n + 1$ bonding molecular orbitals, including n B—H sigma-bonding orbitals and $n + 1$ bonding orbitals in the central core (described as **framework** or **skeletal** bonding orbitals).[10] A useful example is $B_6H_6^{2-}$, which has O_h symmetry. In this ion, each boron has four valence orbitals that can participate in bonding, giving a total of 24 boron valence orbitals for the cluster. These orbitals can be classified into two sets. If the z axis of each boron atom is chosen to point toward the center of the octahedron (see Figure 14-4), the p_z and s orbitals are a set of suitable symmetry to bond with the hydrogen atoms. A second set of orbitals, consisting of the p_x and p_y orbitals of the borons, is then available for boron–boron bonding.

The p_z and s orbitals of the borons collectively have the same symmetry, which reduces to the irreducible representations $A_{1g} + E_g + T_{1u}$ (an analysis of the orbitals in terms of symmetry is left as an exercise in problem 10) and therefore may be considered to combine in the formation of sp hybrid orbitals. These hybrid orbitals on each boron point out toward the hydrogen atoms and in toward the center of the cluster. Six of the hybrids form bonds with the $1s$ orbitals of the hydrogens. The remaining hybrids and the unhybridized $2p$ orbitals of the borons remain to participate in bonding within the B_6 core. Seven orbital combinations lead to bonding interactions; these are shown in Figure 14-5. Constructive overlap of all six hybrid orbitals at the center of the octahedron yields a framework bonding orbital of A_{1g} symmetry; as its symmetry

[9] Some chemists define clusters as having at least three metal atoms.

[10] K. Wade, *Electron Deficient Compounds*, Thomas Nelson, London, 1971.

label indicates, this orbital is completely symmetric with respect to all symmetry operations of the O_h point group. Additional bonding interactions are of two types: overlap of two *sp* hybrid orbitals with parallel *p* orbitals on the remaining four boron atoms (three such interactions, collectively of T_{1u} symmetry) and overlap of *p* orbitals on four boron atoms within the same plane (three interactions, T_{2g} symmetry). The remaining orbital interactions lead to non-bonding or antibonding molecular orbitals. To summarize:

From the 24 valence atomic orbitals of boron are formed:

13 bonding orbitals ($= 2n + 1$), consisting of:

- 7 framework molecular orbitals ($= n + 1$), consisting of:
  - 1 bonding orbital (A_{1g}) from overlap of *sp* hybrid orbitals
  - 6 bonding orbitals from overlap of *p* orbitals of boron with *sp* hybrid orbitals (T_{1u}) or with other boron *p* orbitals (T_{2g})
- 6 boron–hydrogen bonding orbitals ($= n$)

11 nonbonding or antibonding orbitals

Similar descriptions of bonding can be derived for other *closo* boranes. In each case, one particularly useful similarity can be found: there is one more

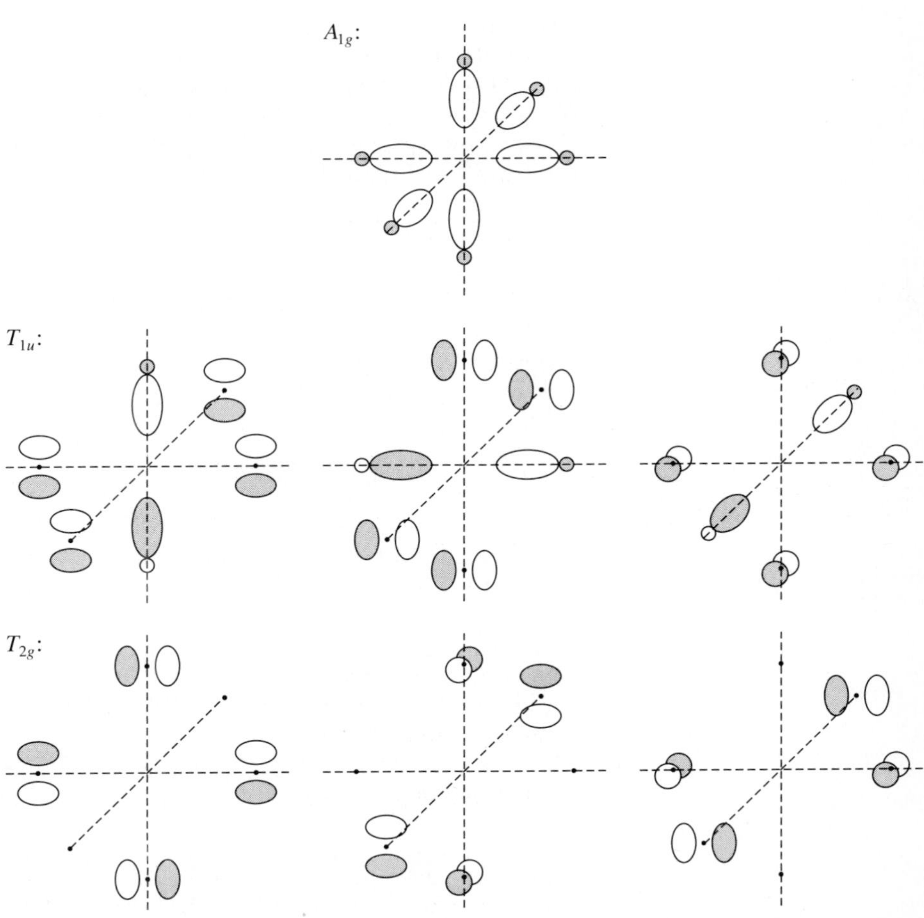

FIGURE 14-5 Bonding in $B_6H_6^{2-}$.

framework bonding pair than the number of corners in the polyhedron. The extra framework bonding pair is in a totally symmetric orbital (like the A_{1g} orbital in $B_6H_6^{2-}$) resulting from overlap of atomic (or hybrid) orbitals at the center of the polyhedron. In addition, there is a significant gap in energy between the highest bonding orbital (HOMO) and the lowest nonbonding orbital (LUMO).[11] The numbers of bonding pairs for common geometries are shown in Table 14-5.

TABLE 14-5
Bonding pairs for *closo* boranes

Formula	*Total valence electron pairs*	*Framework bonding pairs* A_1 *symmetry**	*Framework bonding pairs* *Other symmetry*	*B—H bonding pairs*
$B_6H_6^{2-}$	13	1	6	6
$B_7H_7^{2-}$	15	1	7	7
$B_8H_8^{2-}$	17	1	8	8
$B_nH_n^{2-}$	$2n + 1$	1	n	n

NOTE: * Symmetry designation depends on point group (such as A_{1g} for O_h symmetry).

Together the *closo* structures make up only a very small fraction of all known borane species. Additional structural types can be obtained by removing one or more corners from the *closo* framework. Removal of one corner yields a **nido** (*nestlike*) structure, removal of two corners an **arachno** (*spiderweblike*) structure, and removal of three corners a **hypho** (*netlike*) structure. Examples of three related *closo, nido,* and *arachno* borane structures are shown in Figure 14-6, and the structures for these boranes having 6 to 12 boron atoms are shown in Figure 14-7.

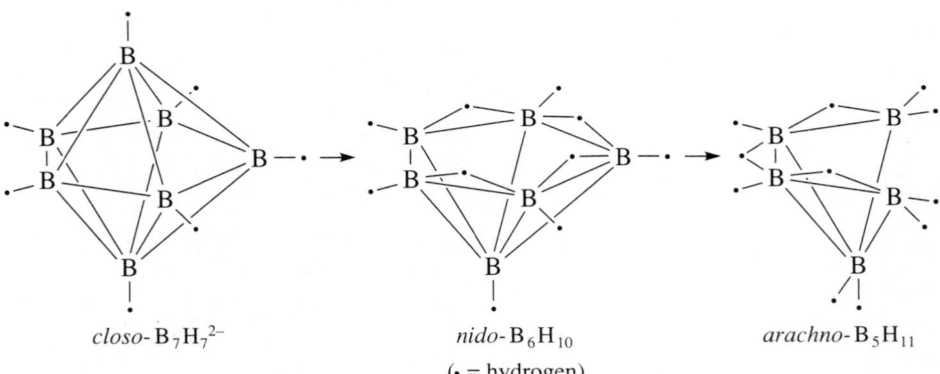

FIGURE 14-6 *Closo, nido,* and *arachno* Borane Structures.

The classification of structural types can often be done more conveniently on the basis of valence electron counts. Various schemes for relating electron counts to structures have been proposed, with most proposals based on a set of rules formulated by Wade in 1971.[12] The classification scheme based on these rules in summarized in Table 14-6.

[11] K. Wade, "Some Bonding Considerations," in *Transition Metal Clusters,* B. F. G. Johnson, ed., Wiley, New York, 1980, p. 217.

[12] K. Wade, *Adv. Inorg. Chem. Radiochem.*, **1976,** *18,* 1.

TABLE 14-6
Classification of cluster structures

Structure type	*Corners occupied*	*Pairs of framework bonding electrons*	*Empty corners*
closo-	n corners of n-cornered polyhedron	$n + 1$	0
nido-	$n - 1$ corners of n-cornered polyhedron	$n + 1$	1
arachno-	$n - 2$ corners of n-cornered polyhedron	$n + 1$	2
hypho-	$n - 3$ corners of n-cornered polyhedron	$n + 1$	3

In addition, it is sometimes useful to relate the total valence electron count in boranes to the structural type. In *closo* boranes the total number of valence electron pairs is equal to the sum of the number of vertices in the polyhedron (at each boron one pair is involved in boron–hydrogen bonding) and the number of skeletal bond pairs. For example, in $B_6H_6^{2-}$ there are 26 valence electrons, or 13 pairs (= $2n + 1$, as mentioned previously). Six of these pairs are involved in bonding to the hydrogens (one per boron), and seven pairs are involved in framework bonding. The polyhedron of the *closo* structure is the parent polyhedron for the other structural types.

Boranes can conveniently be classified by considering

closo boranes to have the formula $B_nH_n^{2-}$;
nido boranes to be derived from $B_nH_n^{4-}$ ions;
arachno boranes to be derived from $B_nH_n^{6-}$ ions; and
hypho boranes to be derived from $B_nH_n^{8-}$ ions.

The formulas of boranes can be related to these formulas by formally subtracting H^+ ions. For example, to classify $B_9H_{14}^-$ we can formally consider it to be derived from $B_9H_9^{6-}$:

$$B_9H_{14}^- - 5\ H^+ = B_9H_9^{6-}$$

The appropriate classification for this borane is therefore *arachno*. Table 14-7 summarizes electron counting and classifications for several examples of boranes.

EXAMPLES

Classify the following boranes by structural type:

$B_{10}H_{14}$

$B_{10}H_{14} - 4\ H^+ = B_{10}H_{10}^{4-}$ The classification is *nido*.

$B_2H_7^-$

$B_2H_7^- - 5\ H^+ = B_2H_2^{6-}$ The classification is *arachno*.

B_8H_{16}

$B_8H_{16} - 8\ H^+ = B_8H_8^{8-}$ The classification is *hypho*.

EXERCISE 14-3
Classify the following boranes by structural type:
a. $B_{11}H_{13}^{2-}$ b. $B_5H_8^-$ c. $B_7H_7^{2-}$ d. $B_{10}H_{18}$

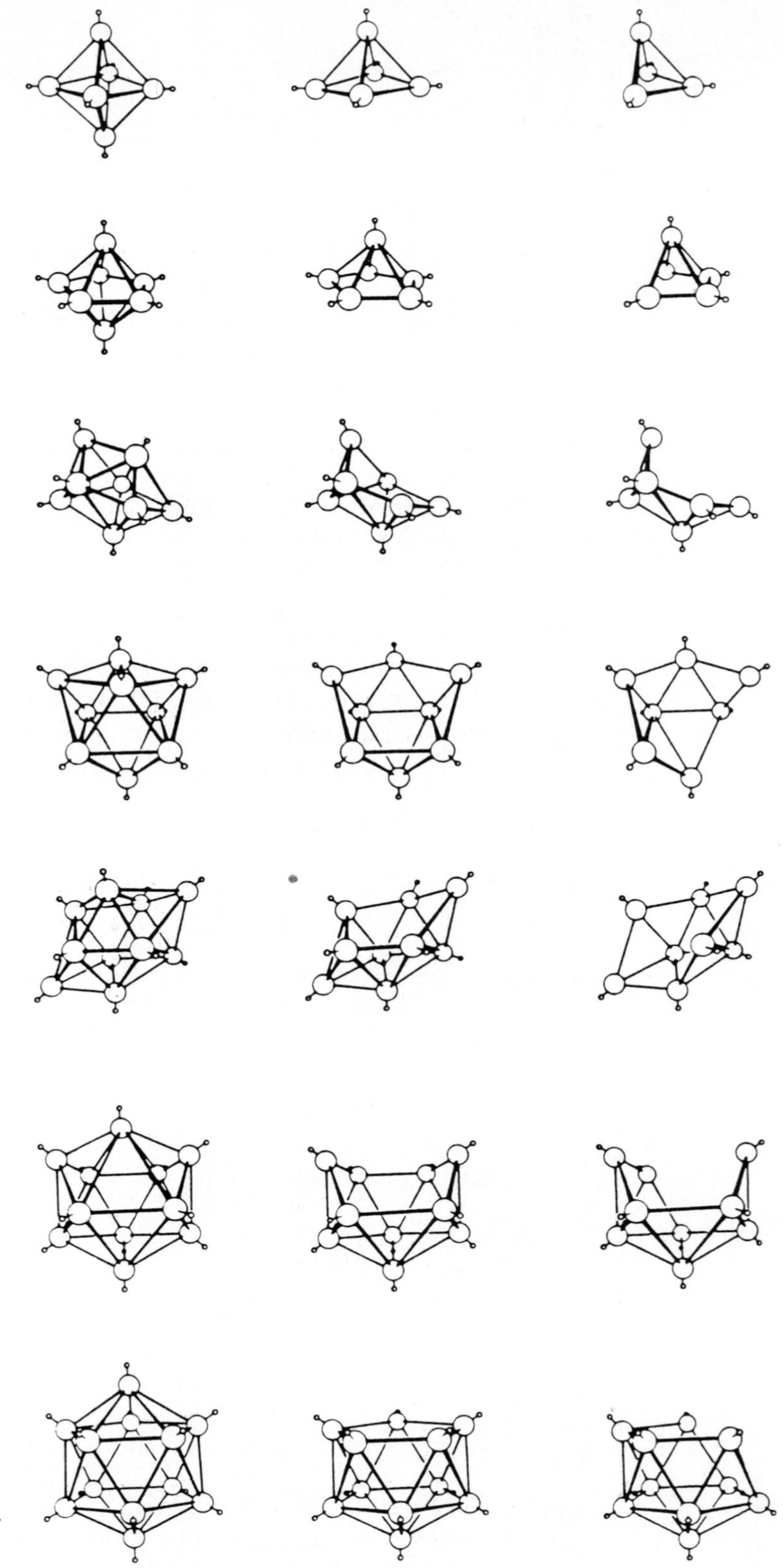

FIGURE 14-7 Structures of *closo, nido,* and *arachno* Boranes. (Reproduced and adapted with permission from R. W. Rudolph, *Acc. Chem. Res.*, **1976,** *9,* 446. Copyright 1976 American Chemical Society.)

TABLE 14-7
Examples of electron counting in boranes

Vertices in parent polyhedron	*Classification*	*Boron atoms in cluster*	*Valence electrons*	*Framework electron pairs*	*Examples*	*Formally derived from*
6	*closo*	6	26	7	$B_6H_6^{2-}$	$B_6H_6^{2-}$
	nido	5	24	7	B_5H_9	$B_5H_5^{4-}$
	arachno	4	22	7	B_4H_{10}	$B_4H_4^{6-}$
7	*closo*	7	30	8	$B_7H_7^{2-}$	$B_7H_7^{2-}$
	nido	6	28	8	B_6H_{10}	$B_6H_6^{4-}$
	arachno	5	26	8	B_5H_{11}	$B_5H_5^{6-}$
12	*closo*	12	50	13	$B_{12}H_{12}^{2-}$	$B_{12}H_{12}^{2-}$
	nido	11	48	13	$B_{11}H_{13}^{2-}$	$B_{11}H_{11}^{4-}$
	arachno	10	46	13	$B_{10}H_{15}^{-}$	$B_{10}H_{10}^{6-}$

14-3-2 HETEROBORANES

The electron counting schemes can be extended to isoelectronic species such as the **carboranes** (also known as **carbaboranes**). The CH^+ unit is isoelectronic with BH; many compounds are known in which one or more BH groups have been replaced by CH^+ (or by C, which has the same number of electrons as BH). For example, replacement of two BH groups in *closo*-$B_6H_6^{2-}$ yields *closo*-$C_2B_4H_6$, a neutral compound. *Closo, nido,* and *arachno* carboranes are all known, most commonly containing two carbon atoms; examples are shown in Figure 14-8.

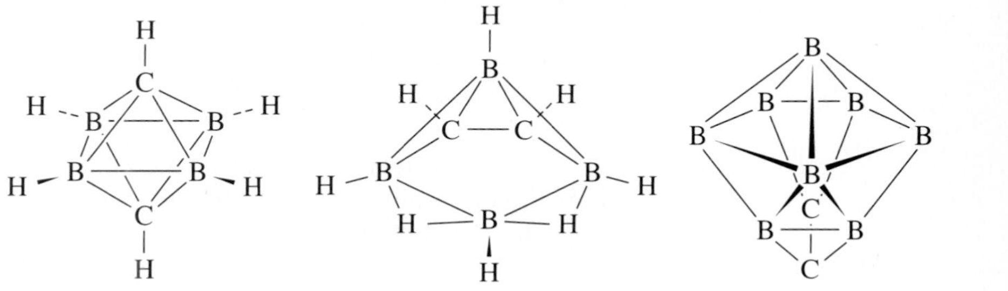

$C_2B_4H_6$

closo

$C_2B_4H_8$

nido

$C_2B_8H_{10}^{4-}$
(has one terminal H on each B and C atom)
arachno

FIGURE 14-8 Examples of Carboranes.

Examples of chemical formulas corresponding to these designations are:

Type	*Borane*	*Example*	*Carborane*	*Example*
Closo	$B_nH_n^{2-}$	$B_{12}H_{12}^{2-}$	$C_2B_{n-2}H_n$	$C_2B_{10}H_{12}$
Nido	$B_nH^*_{n+4}$	$B_{10}H_{14}$	$C_2B_{n-2}H_{n+2}$	$C_2B_8H_{12}$
Arachno	$B_nH^*_{n+6}$	B_9H_{15}	$C_2B_{n-2}H_{n+4}$	$C_2B_7H_{13}$

NOTE: * *Nido* boranes may also have the formulas $B_nH_{n+3}^{-}$ and $B_nH_{n+2}^{2-}$; *arachno* boranes may also have the formulas $B_nH_{n+5}^{-}$ and $B_nH_{n+4}^{2-}$.

Carboranes may be classified by structural type using the same method as described previously for boranes. Since a carbon atom has the same number of valence electrons as a boron atom plus a hydrogen atom, formally each C should be converted to BH in the classification scheme. For example, for a carborane having the formula $C_2B_8H_{10}$:

$$C_2B_8H_{10} \longrightarrow B_{10}H_{12}$$

$$B_{10}H_{12} - 2\,H^+ \quad = \quad B_{10}H_{10}^{2-}$$

The classification of the carborane $C_2B_8H_{10}$ is therefore *closo*.

EXAMPLES

Classify the following carboranes by structural type:

$\mathbf{C_2B_9H_{12}^-}$

$C_2B_9H_{12}^- \longrightarrow B_{11}H_{14}^-$

$B_{11}H_{14}^- - 3\,H^+ = B_{11}H_{11}^{4-}$ The classification is *nido*,

$\mathbf{C_2B_7H_{13}}$

$C_2B_7H_{13} \longrightarrow B_9H_{15}$

$B_9H_{15} - 6\,H^+ = B_9H_9^{6-}$ The classification is *arachno*.

$\mathbf{C_4B_2H_6}$

$C_4B_2H_6 \longrightarrow B_6H_{10}$

$B_6H_{10} - 4\,H^+ = B_6H_6^{4-}$ The classification is *nido*.

EXERCISE 14-4

Classify the following carboranes by structural type:

a. $C_3B_3H_7$ b. $C_2B_5H_7$ c. $C_2B_7H_{12}^-$

Many derivatives of boranes containing other main group atoms (designated heteroatoms) are also known. These heteroboranes may be classified by formally converting the heteroatom to a BH_x group having the same number of valence electrons, then proceeding as in previous examples. For some of the most common heteroatoms, the substitutions are:

Heteroatom	*Replace with*
C, Si, Ge, Sn	BH
N, P, As	BH_2
S, Se	BH_3

EXAMPLES

Classify the following heteroboranes by structural type.

$\mathbf{SB_9H_{11}}$

$SB_9H_{11} \longrightarrow B_{10}H_{14}$

$B_{10}H_{14} - 4\,H^+ = B_{10}H_{10}^{4-}$ The classification is *nido*.

$CPB_{10}H_{11}$

$CPB_{10}H_{11} \longrightarrow PB_{11}H_{12} \longrightarrow B_{12}H_{14}$

$B_{12}H_{14} - 2\,H^+ = B_{12}H_{12}^{2-}$ The classification is *closo*.

EXERCISE 14-5

Classify the following heteroboranes by structural type.

a. SB_9H_9 b. $GeC_2B_9H_{11}$ c. $SB_9H_{12}^-$

While it may not be surprising that the same set of electron counting rules can be used to describe satisfactorily such similar compounds as boranes and carboranes, it is of interest to examine how far the comparison can be extended. Can Wade's rules, for example, be used effectively on compounds containing metals bonded to boranes or carboranes? Can the rules be extended even further to describe the bonding in polyhedral metal clusters?

14-3-3 METALLABORANES AND METALLACARBORANES

The CH group of a carborane is isolobal with 15-electron fragments of an octahedron such as $Co(CO)_3$. Similarly, BH, which has four valence electrons, is isolobal with 14-electron fragments such as $Fe(CO)_3$ and $Co(\eta^5\text{-}C_5H_5)$. These organometallic fragments have been found in substituted boranes and carboranes in which the organometallic fragments substitute for the isolobal main group fragments. For example, the organometallic derivatives of B_5H_9 shown in Figure 14-9 have been synthesized. Theoretical calculations on the iron derivatives have supported the view that $Fe(CO)_3$ in these compounds is bonding in a fashion isolobal with BH.[13] In both fragments the orbitals involved in framework bonding within the cluster are similar (Figure 14-10). In BH the orbitals participating in framework bonding are an sp_z hybrid pointing toward the center of the polyhedron (similar to the orbitals participating in bonding of A_{1g} symmetry in $B_6H_6^{2-}$; Figure 14-5) and p_x and p_y orbitals tangential to the

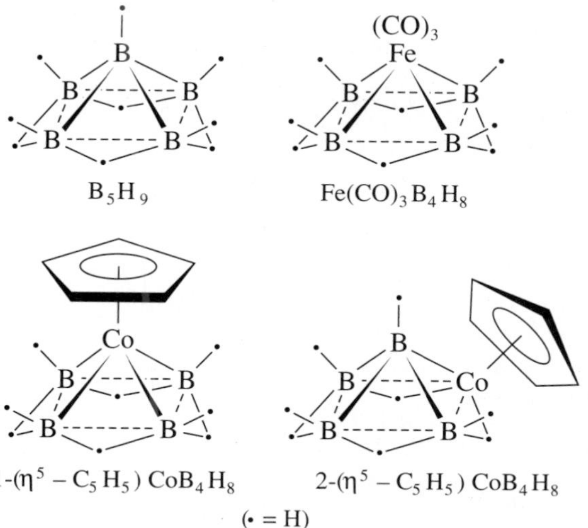

FIGURE 14-9 Organometallic Derivatives of B_5H_9.

[13] R. L. DeKock and T. P. Fehlner, *Polyhedron,* **1982,** *1,* 521.

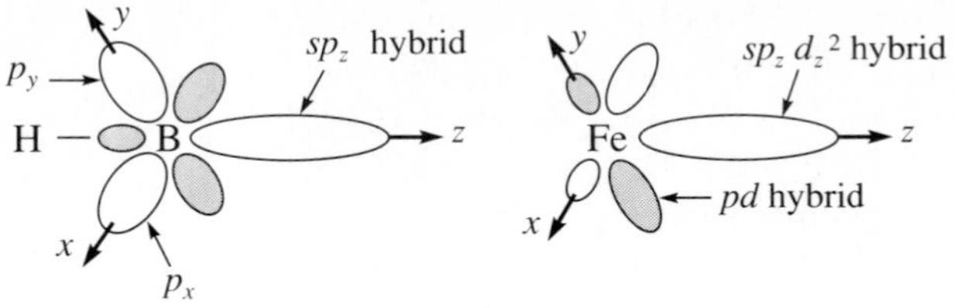

FIGURE 14-10 Orbitals of Isolobal Fragments BH and $Fe(CO)_3$.

surface of the cluster. In $Fe(CO)_3$ an $sp_zd_{z^2}$ hybrid points toward the center, and *pd* hybrid orbitals are oriented tangentially to the cluster surface.

Examples of metallaboranes and metallacarboranes are numerous. Selected examples with *closo* structures are given in Table 14-8.

TABLE 14-8
Metallaboranes and metallacarboranes with *closo* structures

Number of skeletal atoms	*Shape*	*Examples*	
6	Octahedron	 $B_4H_6(CoCp)_2$	$C_2B_3H_5Fe(CO)_3$
7	Pentagonal bipyramid	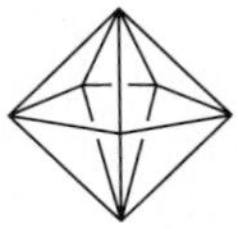 $C_2B_4H_6Ni(PPh_3)_2$	$C_2B_3H_5(CoCp)_2$
8	Dodecahedron	 $C_2B_4H_4[(CH_3)_2Sn]CoCp$	
9	Capped square antriprism	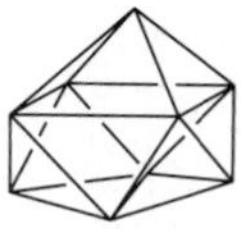 $C_2B_6H_8Pt(PMe_3)_2$	$C_2B_5H_7(CoCp)_2$
10	Bicapped square antiprism	 $[B_9H_9NiCp]^-$	$CB_7H_8(CoCp)(NiCp)$
11	Octadecahedron	 $[CB_9H_{10}CoCp]^-$	$C_2B_8H_{10}IrH(PPh_3)_2$
12	Icosahedron	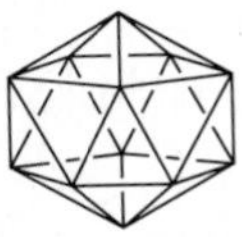 $C_2B_7H_9(CoCp)_3$	$C_2B_9H_{11}Ru(CO)_3$

Anionic boranes and carboranes can also act as ligands toward metals in a manner resembling that of cyclic organic ligands. For example, *nido* carboranes of formula $C_2B_9H_{11}^{2-}$ have *p* orbital lobes pointing toward the "missing" site of the icosahedron (remember that the *nido* structure corresponds to a *closo* structure, in this case the 12-vertex icosahedron, with one vertex missing). This arrangement of *p* orbitals can be compared with the *p* orbitals of the cyclopentadienyl ring, as shown in Figure 14-11.

nido-7, 8-$C_2B_9H_{11}^{2-}$ $(\eta^5\text{-}C_5H_5)^-$ *nido*-7, 9-$C_2B_9H_{11}^{2-}$

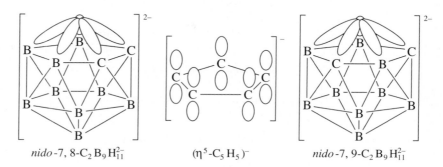

FIGURE 14-11 Comparison of $C_2B_9H_{11}^{2-}$ with $C_5H_5^-$. (Adapted with permission from N. N. Greenwood and A. Earnshaw, *Chemistry of the Elements*, Pergamon, Oxford, 1984, p. 210. Copyright 1984, Pergamon Press PLC.)

Although the comparison between these ligands is not exact, the similarity is sufficient that $C_2B_9H_{11}^{2-}$ can bond to iron in the formation of a carborane analogue of ferrocene, $[Fe(\eta^5\text{-}C_2B_9H_{11})_2]^{2-}$. A mixed ligand sandwich compound containing one carborane and one cyclopentadienyl ligand, $[Fe(\eta^5\text{-}C_2B_9H_{11})(\eta^5\text{-}C_5H_5)]$, has also been made (Figure 14-12). Numerous other examples of boranes and carboranes serving as ligands to transition metals are also known.[14]

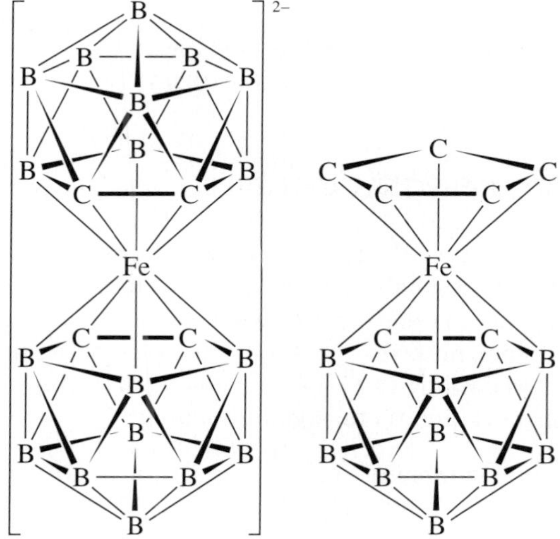

FIGURE 14-12 Carborane Analogues of Ferrocene. (Adapted with permission from N. N. Greenwood and A. Earnshaw, *Chemistry of the Elements*, Pergamon, Oxford, 1984, pp, 211, 212. Copyright 1984, Pergamon Press PLC.)

Metallaboranes and metallacarboranes can be classified structurally by using a procedure similar to the method described previously for boranes and their main group derivatives. To classify borane derivatives with transition metal-containing fragments, it is convenient to determine how many electrons the metal-containing fragment needs to satisfy the requirements of the 18-electron rule. This fragment can be considered equivalent to a BH_x fragment

[14] K. P. Callahan and M. F. Hawthorne, *Adv. Organomet. Chem.*, **1976**, *14*, 145.

needing the same number of electrons to satisfy the octet rule. For example, a 14-electron fragment such as $Co(\eta^5\text{-}C_5H_5)$ is four electrons short of 18; this fragment may be considered the equivalent of the 4-electron fragment BH, which is four electrons short of an octet. Examples of organometallic fragments and their corresponding BH_x fragments:

Valence electrons in organometallic fragment	*Example*	*Replace with*
13	$Mn(CO)_3$	B
14	CoCp	BH
15	$Co(CO)_3$	BH_2
16	$Fe(CO)_4$	BH_3

EXAMPLES

Classify the following metallaboranes by structural type:

$B_4H_6(CoCp)_2$

$B_4H_6(CoCp)_2 \longrightarrow B_4H_6(BH)_2 = B_6H_8$

$B_6H_8 - 2\,H^+ = B_6H_6^{2-}$ The classification is *closo*.

$B_3H_7[Fe(CO)_3]_2$

$B_3H_7[Fe(CO)_3]_2 \longrightarrow B_3H_7[BH]_2 = B_5H_9$

$B_5H_9 - 4\,H^+ = B_5H_5^{4-}$ The classification is *nido*.

EXERCISE 14-6

Classify the following by structural type:

a. $C_2B_7H_9(CoCp)_3$ b. $C_2B_4H_6Ni(PPh_3)_2$

14-3-4 CARBONYL CLUSTERS

The structures of several carbonyl cluster compounds were shown in Chapter 12. Many carbonyl clusters have structures similar to boranes; it is therefore of interest to determine to what extent the approach used to describe bonding in boranes may also be applicable to bonding in carbonyl clusters.

According to Wade, the valence electrons in a cluster can be assigned to framework and metal–ligand bonding:[15]

$$\begin{pmatrix}\text{total number of}\\\text{valence electrons}\\\text{in cluster}\end{pmatrix} = \begin{pmatrix}\text{number of electrons}\\\text{involved in framework}\\\text{bonding}\end{pmatrix} + \begin{pmatrix}\text{number of electrons}\\\text{involved in metal–}\\\text{ligand bonding}\end{pmatrix}$$

As we have seen previously, the number of electrons involved in framework bonding in boranes is related to the classification of the structure as *closo, nido, arachno,* or *hypho*. Rearranging this equation gives

$$\begin{pmatrix}\text{number of electrons}\\\text{involved in framework}\\\text{bonding}\end{pmatrix} = \begin{pmatrix}\text{total number of}\\\text{valence electrons}\\\text{in cluster}\end{pmatrix} - \begin{pmatrix}\text{number of electrons}\\\text{involved in metal–}\\\text{ligand bonding}\end{pmatrix}$$

[15] K. Wade, *Adv. Inorg. Chem. Radiochem.*, **1980,** *18,* 1.

For a borane, one electron pair is assigned to one boron–hydrogen bond on each boron. The remaining valence electron pairs are regarded as framework bonding pairs.[16] For a transition metal carbonyl complex, on the other hand, Wade suggests that six electron pairs per metal are either involved in metal–carbonyl bonding (to all carbonyls on a metal) or are nonbonding and therefore unavailable for participation in framework bonding. The result is that there is a net difference of five electron pairs, or ten electrons, per framework atom in comparing boranes with transition metal carbonyl clusters. A metal carbonyl analogue of *closo*-$B_6H_6^{2-}$, which has 26 valence electrons, would therefore need a total of 86 valence electrons to adopt a *closo* structure. An 86-electron cluster that satisfies this requirement is $Co_6(CO)_{16}$. Like $B_6H_6^{2-}$, $Co_6(CO)_{16}$ has an octahedral framework. As in the case of boranes, *nido* structures correspond to *closo* geometries from which one vertex is empty, *arachno* structures lack two vertices, and so on.

A simpler way to compare electron counts in boranes and transition metal clusters is to consider the different numbers of valence orbitals available to the framework atoms. Transition metals, with nine valence orbitals (one *s*, three *p*, and five *d* orbitals), have five more orbitals available for bonding than boron, which has only four valence orbitals; these five extra orbitals, when filled as a consequence of bonding within the framework and with surrounding ligands, give an increased electron count of 10 electrons per framework atom. Consequently, a useful rule of thumb is to increase the electron requirement of the cluster by 10 per framework atom when replacing a boron with a transition metal atom. In the example cited above, replacing the six borons in *closo*-$B_6H_6^{2-}$ with six cobalts should therefore increase the electron count from 26 to 86 for a comparable *closo* cobalt cluster. $Co_6(CO)_{16}$, an 86-electron cluster, meets this requirement.

The valence electron counts corresponding to the various structural classifications for main group and transition metal clusters are summarized in Table 14-9.[17]

TABLE 14-9
Electron counting in main group and transition metal clusters

Structure type	*Main group cluster*	*Transition metal cluster*
closo-	$4n + 2$	$14n + 2$
nido-	$4n + 4$	$14n + 4$
arachno-	$4n + 6$	$14n + 6$
hypho-	$4n + 8$	$14n + 8$

Examples of *closo, nido,* and *arachno* borane and transition metal clusters are given in Table 14-10. Transition metal clusters formally containing seven metal–metal framework bonding pairs are among the most common; examples illustrating the structural diversity of these clusters are given in Table 14-11 and Figure 14-13.

The predictions of structures of transition metal carbonyl complexes using Wade's rules are often, but not always, accurate. For example, the

[16] For structures involving bridging hydrogen atoms, the bridging hydrogens are considered to be involved in framework bonding.

[17] D. M. P. Mingos, *Acc. Chem. Res.*, **1984,** *17,* 311.

TABLE 14-10
Closo, nido, and *arachno* borane and transition metal clusters

Atoms in cluster	Vertices in parent polyhedron	Skeletal electron pairs	Valence electrons (boranes) Closo	Nido	Arachno	Example	Valence electrons (transition metal clusters) Closo	Nido	Arachno	Example
4	4	5	18				58			
	5	6		20		$B_4H_7^-$		60		$Co_4(CO)_{12}$
	6	7			22	B_4H_{10}			62	$[Fe_4C(CO)_{12}]^{2-}$
5	5	6	22			$C_2B_3H_5$	72			$Os_5(CO)_{16}$
	6	7		24		B_5H_9		74		$Os_5C(CO)_{15}$
	7	8			26	B_5H_{11}			76	$[Ni_5(CO)_{12}]^{2-}$
6	6	7	26			$B_6H_6^{2-}$	86			$Co_6(CO)_{16}$
	7	8		28		B_6H_{10}		88		$Os_6(CO)_{17}[P(OMe)_3]_3$
	8	9			30	B_6H_{12}			90	

TABLE 14-11
Clusters that formally contain seven metal–metal skeletal bond pairs

Number of skeletal atoms	Cluster type	Shape	Examples
7	Capped *closo*[a]	Capped octahedron	$[Rh_7(CO)_{16}]^{3-}$ $Os_7(CO)_{21}$
6	*Closo*	Octahedron	$Rh_6(CO)_{16}$ $Ru_6C(CO)_{17}$
6	Capped *nido*[a]	Capped square pyramid	$H_2Os_6(CO)_{18}$
5	*Nido*	Square pyramid	$Ru_5C(CO)_{15}$
4	*Arachno*	Butterfly	$[Fe_4(CO)_{13}H]^{-b}$

SOURCE: K. Wade, "Some Bonding Considerations," in *Transition Metal Clusters,* B. F. G. Johnson, ed., Wiley, 1980, p. 232.
NOTES: [a] A capped *closo* cluster has a valence electron count equivalent to neutral B_nH_n. A capped *nido* cluster has the same electron count as a *closo* cluster.
[b] This complex has an electron count matching a *nido* structure, but it adopts the butterfly structure expected for *arachno.* This is one of many examples in which the structure of metal clusters is not predicted accurately by Wade's rules. Limitations of Wade's rules are discussed in R. N. Grimes, "Metallacarboranes and Metallaboranes" in *Comprehensive Organometallic Chemistry,* Vol. 1, G. Wilkinson, F. G. A. Stone, and W. Abel, eds., Pergamon Press, Elmsford, N.Y., 1982, p. 473.

clusters $M_4(CO)_{12}$ (M = Co, Rh, Ir) have 60 valence electrons and are predicted to be *nido* complexes ($14n + 4$ valence electrons). A *nido* structure would correspond to a trigonal bipyramid (the parent structure) with one position vacant. X-ray crystallographic studies, however, have shown these complexes to have tetrahedral metal cores.

14-3-5 CARBIDE CLUSTERS

In recent years, many compounds have been synthesized, often fortuitously, in which one or more atoms have been partially or completely encapsulated

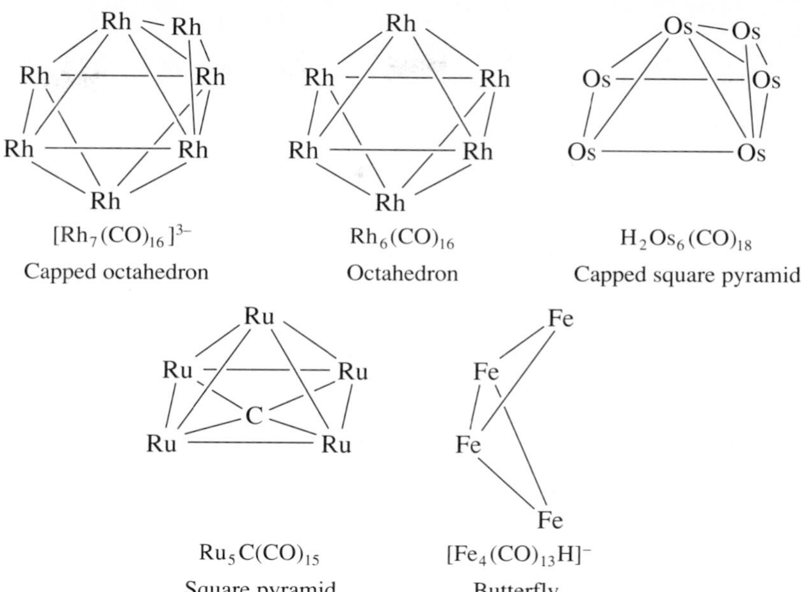

FIGURE 14-13 Metal Cores for Clusters Containing Seven Skeletal Bond Pairs.

within metal clusters. The most common of these cases have been the carbide clusters, with carbon exhibiting coordination numbers and geometries not found in classical organic structures. Examples of these unusual coordination geometries are shown in Figure 14-14. Additional examples of carbide clusters are shown in Figure 1-6. Encapsulated atoms contribute their valence electrons to the total electron count. For example, carbon contributes its four valence electrons in $Ru_6C(CO)_{17}$ to give a total of 86 electrons, corresponding to a *closo* electron count (Table 14-10).

How can carbon, with only four valence orbitals, form bonds to more than four surrounding transition metal atoms? $Ru_6C(CO)_{17}$, with a central core of O_h symmetry, is a useful example. The $2s$ orbital of carbon has A_{1g} symmetry and the $2p$ orbitals have T_{1u} symmetry in the O_h point group. The octahedral Ru_6 core has skeletal bonding orbitals of the same symmetry as in $B_6H_6^{2-}$ described earlier in this chapter (see Figure 14-5): a centrally directed A_{1g} group orbital and two sets of orbitals oriented tangentially to the core, of T_{1u} and T_{2g} symmetry. There are therefore two ways in which the symmetry match is correct for interactions between the carbon and the Ru_6 core, the interactions of A_{1g} and T_{1u} symmetry shown in Figure 14-15 (the T_{2g} orbitals participate in Ru–Ru bonding but not in bonding with the central carbon). The net result is formation of four C–Ru bonding orbitals, occupied by electron pairs in the cluster, and four unoccupied antibonding orbitals.[18]

14-3-6 METAL–METAL BONDS IN CLUSTER COMPOUNDS

For nearly a century, compounds containing two or more metal atoms have been known. The first of these compounds to be correctly identified, by

[18] G. A. Olah, G. K. S. Prakash, R. E. Williams, L. D. Field, and K. Wade, *Hypercarbon Chemistry,* Wiley, New York, 1987, pp. 123–33.

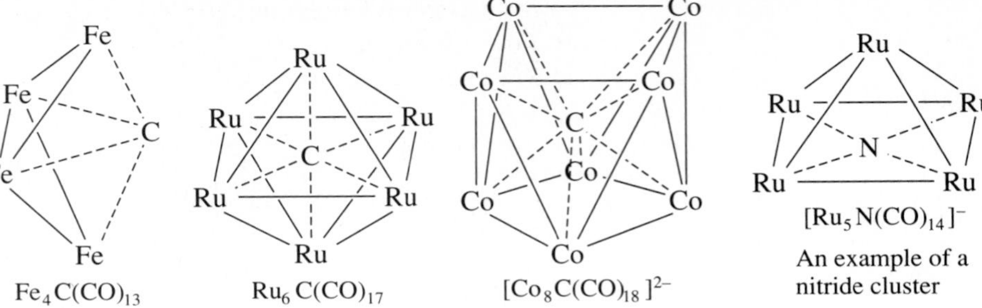

FIGURE 14-14 Carbide Clusters. CO ligands have been omitted for clarity.

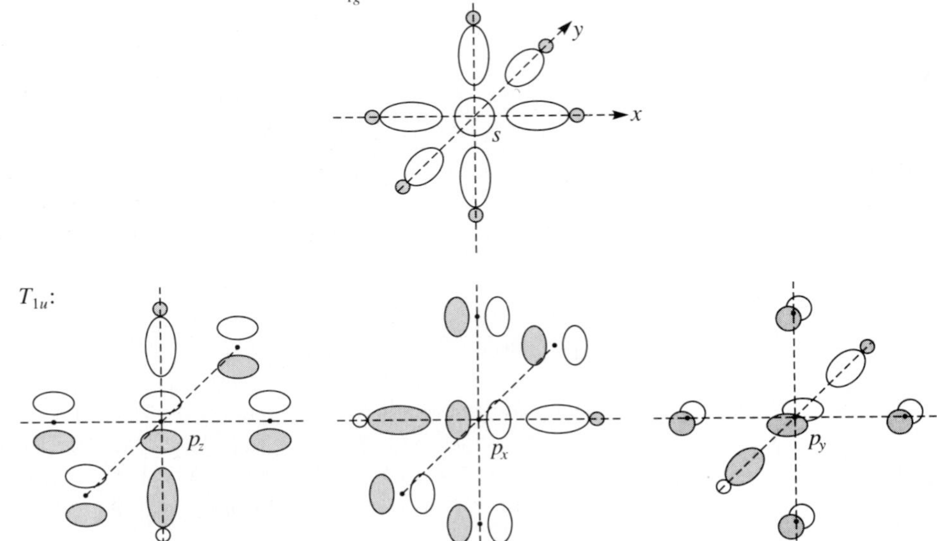

FIGURE 14-15 Bonding Interactions Between Central Carbon and Octahedral Ru_6. Orbitals shown for Ru are derived from *s* and *p* orbitals. *d* Orbitals also have suitable symmetry to interact with A_{1g} and T_{1u} orbitals of carbon.

Werner, were held together by bridging ligands shared by the metals involved; X-ray crytallographic studies eventually showed that the metal atoms were too far apart to be likely participants in direct metal–metal orbital interactions.

Not until 1935 did X-ray crystallography demonstrate direct metal–metal bonding. In that year, C. Brosset reported the structure of $K_3W_2Cl_9$, which contained the $W_2Cl_9^{3-}$ ion. In this ion the tungsten–tungsten distance (240 pm) was found to be substantially shorter than the interatomic distance in tungsten metal (275 pm):

$[Cl_3W(\mu\text{-}Cl)_3WCl_3]^{3-}$

The close proximity of the metals in this ion raised for the first time the serious possibility of direct bonding interactions between metal orbitals. However, little attention was paid to this interesting question for many years, even though several additional compounds having very short metal–metal distances were synthesized.

The modern development of the chemistry of metal–metal bonded species was spurred by the crystal structures of $Re_3Cl_{12}^{3-}$ and $Re_2Cl_8^{2-}$.[19] $Re_3Cl_{12}^{3-}$, originally believed to be monomeric $ReCl_4^-$, was shown in 1963 to be a trimeric, cyclic ion having very short rhenium-rhenium distances (248 pm). In the following year, in the course of a study on the synthesis of trirhenium complexes, the dimeric $Re_2Cl_8^{2-}$ was synthesized; this ion had a remarkably short metal–metal distance (224 pm) and was the first complex found to have a quadruple bond.

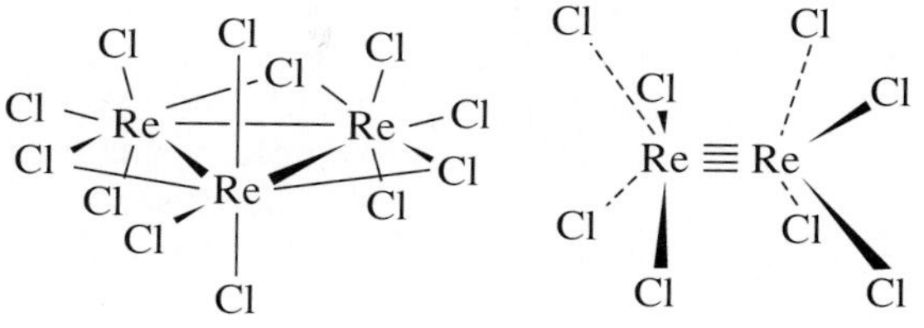

$Re_3Cl_{12}^{3-}$ $Re_2Cl_8^{2-}$

During the succeeding quarter of a century, many thousands of cluster compounds of transition metals have been synthesized, including hundreds containing quadruple bonds. We need therefore to consider briefly how metal atoms can bond to each other and, in particular, how quadruple bonds between metals are possible.

14-3-7 METAL–METAL BONDS

Transition metals may form single, double, triple, or quadruple bonds (or bonds of fractional order) with other metal atoms. Examples are shown in Figure 14-16.

How are quadruple bonds possible? In main group chemistry, atomic orbitals in general can interact in a sigma or pi fashion, with the highest possible bond order of 3 a combination of a sigma bond and two pi bonds. When two transition metal atoms interact, the most important interactions are between their outermost *d* orbitals. These *d* orbitals can combine to form not

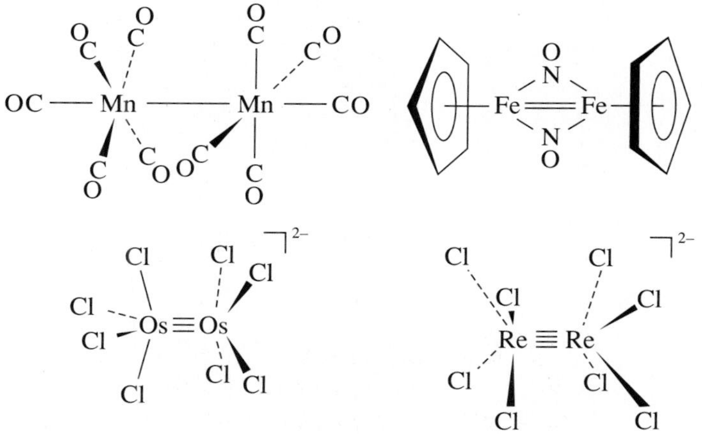

FIGURE 14-16 Examples of Transition Metal Single and Multiple Bonds.

[19] F. A. Cotton, *Chem. Soc. Rev.*, **1975**, *4*, 27.

only σ and π orbitals, but δ orbitals, as shown in Figure 14-17. If the z axis is chosen as the internuclear axis, the strongest interaction (involving greatest overlap) is the sigma interaction between the d_{z^2} orbitals. Next in effectiveness of overlap are the d_{xz} and d_{yz} orbitals, which form pi orbitals as a result of interactions in two regions in space. The last, and weakest, of these interactions are between the d_{xy} and $d_{x^2-y^2}$ orbitals; these orbitals interact in four regions in the formation of δ molecular orbitals.

The relative energies of the resulting molecular orbitals are shown schematically in Figure 14-18. In the absence of ligands, an M_2 fragment would have five bonding orbitals resulting from *d-d* interactions, with molecular orbitals increasing in energy in the order σ, π, δ, δ^*, π^*, σ^*, as shown. In $Re_2Cl_8^{2-}$, our example of quadruple bonding, the configuration is eclipsed (D_{4h} symmetry). For convenience, we can choose the Re—Cl bonds to be oriented in the xz and yz planes. The ligand orbitals interact most strongly with the metal orbitals pointing toward them, the dsp^2 hybrid orbitals, which include the $d_{x^2-y^2}$ atomic orbital. The $d_{x^2-y^2}$ orbital is therefore unavailable to participate in metal–metal bonding. The consequence of these interactions is that new molecular orbitals are formed as shown on the right side of Figure 14-18. The relative energies of these orbitals depend on the strength of the metal–ligand interactions and therefore vary for different complexes.

In $Re_2Cl_8^{2-}$, each rhenium is formally Re(III) and has four d electrons. If the eight d electrons for this ion are placed into the four lowest energy orbitals shown in Figure 14-18 (not including the low energy orbital arising from the $d_{x^2-y^2}$ interactions, occupied by ligand electrons) the total bond order is 4, corresponding to (in increasing energy) a sigma bond, two pi bonds, and a

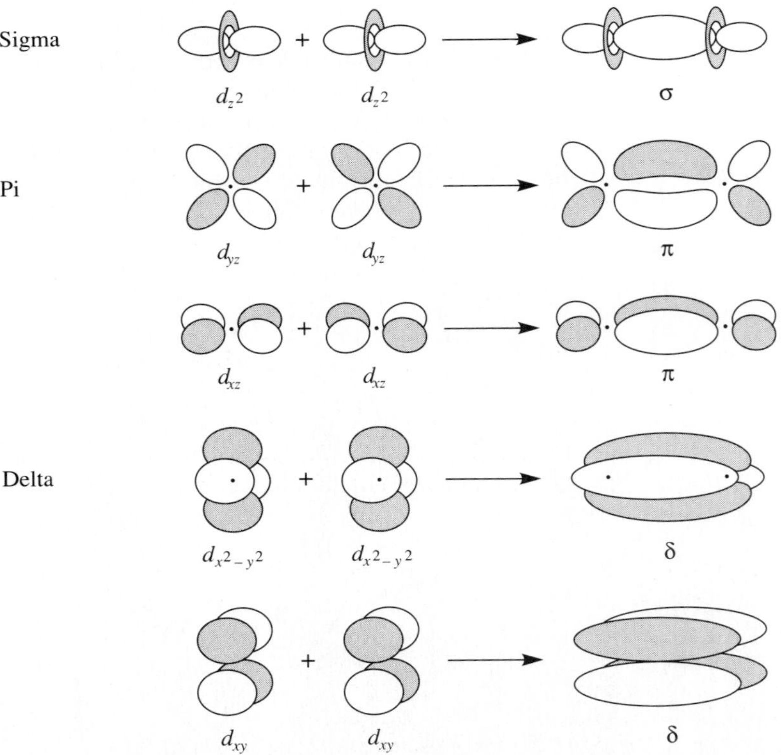

FIGURE 14-17 Bonding Interactions between Metal *d* Orbitals.

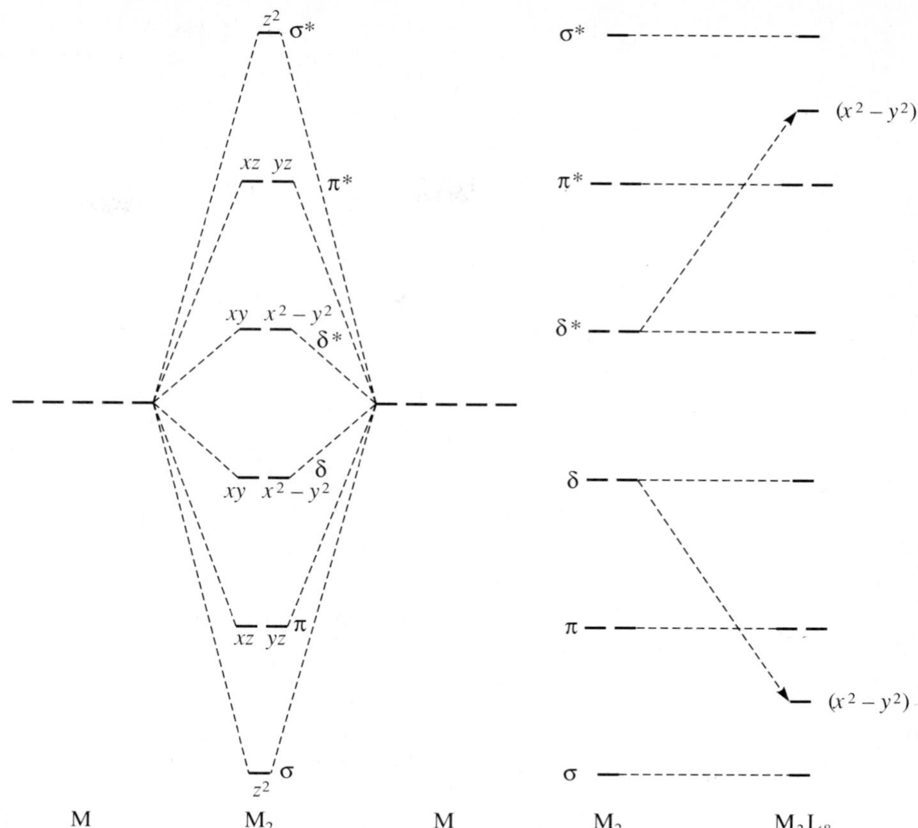

FIGURE 14-18 Relative Energies of Orbitals Formed from *d* Orbital Interactions.

delta bond. The delta bond is weakest in energy; however, it is strong enough to maintain this ion in its eclipsed conformation.

The weakness of the delta bond is illustrated by the small separation in energy of the δ and δ* orbitals. This energy difference typically corresponds to the energy of visible light, with the consequence that most quadruply bonded complexes are vividly colored. For example, $Re_2Cl_8^{2-}$ is royal blue and $Mo_2Cl_8^{4-}$ is bright red. By comparison, main group compounds having filled π and empty π* orbitals are often colorless (for example, N_2 and CO), since the energy difference between them is in the ultraviolet part of the spectrum.

Additional electrons populate δ* orbitals and reduce the bond order. For example, $Os_2Cl_8^{2-}$, an osmium(III) species with a total of ten *d* electrons, has a triple bond. The delta bond order in this ion is zero; in the absence of such a bond, the eclipsed geometry as found in quadruply bonded complexes such as $Re_2Cl_8^{2-}$ is absent. X-ray crystallographic analysis has shown $Os_2Cl_8^{2-}$ to be very nearly staggered (D_{4d} geometry), as would be expected from VSEPR considerations.

Cl Cl Cl
Cl Os≡Os
Cl
Cl Cl Cl

$Os_2Cl_8^{2-}$

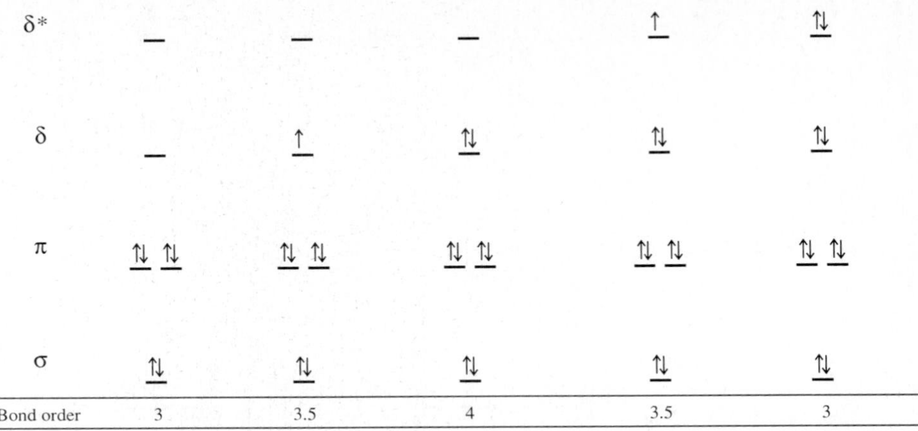

Examples: $[Mo_2(HPO_4)_4]^{2-}$ $[Mo_2(SO_4)_4]^{3-}$ $[Mo_2(SO_4)_4]^{4-}$

Mo—Mo = 223pm Mo—Mo = 217pm Mo—Mo = 211pm

$[Re_2Cl_4(PMe_2Ph)_4]^{2+}$ $[Re_2Cl_4(PMe_2Ph)_4]^{+}$ $Re_2Cl_4(PMe_2Ph)_4$

Re—Re = 221.5pm Re—Re = 221.8pm Re—Re = 224.1pm

FIGURE 14-19 Bond Order and Electron Count in Dimetal Clusters. (Sources: A. Bino and F. A. Cotton, *Inorg. Chem.*, **1979,** *18,* 3562; F. A. Cotton, *Chem. Soc. Rev.*, **1983,** *12,* 35.)

Similarly, fewer than 8 valence electrons would also give a bond order less than four. Examples of such complexes are shown in Figure 14-19.

Metal–metal multiple bonding can have dramatic effects on bond distances, as measured by X-ray crystallography. One way of describing the shortening of interatomic distances by multiple bonds is by comparing the bond distances in multiple bonds to the distances for single bonds. The ratio of these distances is sometimes called the *formal shortness ratio*. Values of this ratio are compared below for main group triple bonds and some of the shortest of the measured transition metal quadruple bonds:

Multiple bond distance/Single bond distance			
Bond	*Ratio*	*Bond*	*Ratio*
C≡C	0.783	Cr ≣ Cr	0.767
N≡N	0.786	Mo ≣ Mo	0.807
		Re ≣ Re	0.848

The ratios found for several quadruply bonded chromium complexes are the smallest ratios found to date for any compounds. Considerable variation in bond distances has been observed. Mo—Mo quadruple bonds, for example, have been found in the range of 203.7 to 230.2 pm.[20]

[20] F. A. Cotton and R. A. Walton, *Multiple Bonds between Metal Atoms,* New York, Wiley, 1982, pp. 161–65.

The effect of the population of δ and δ* orbitals on bond distances can sometimes be surprisingly small. For example, removal of δ* electrons on oxidation of $Re_2Cl_4(PMe_2Ph)_4$ gives only very slight shortening of the Re–Re distances:[21]

Complex	*Number of d electrons*	*Formal Re–Re bond order*	*Formal oxidation state of Re*	*Re–Re distance (pm)*
$Re_2Cl_4(PMe_2Ph)_4$	10	3	2	224.1
$[Re_2Cl_4(PMe_2Ph)_4]^+$	9	3.5	2.5	221.8
$[Re_2Cl_4(PMe_2Ph)_4]^{2+}$	8	4	3	221.5

A possible explanation for the small change in bond distance is that, with increasing oxidation state of the metal, the *d* orbitals contract. This contraction may cause overlap of *d* orbitals in pi bonding to become less effective. Thus as δ* electrons are removed, the pi interactions become weaker; the two factors (increase in bond order and increase in oxidation state of Re) very nearly offset each other.

14-3-8 ADDITIONAL COMMENTS ON CLUSTERS

As we have seen, transition metal clusters can adopt a wide variety of geometries and can involve metal–metal bonds of order as high as four. Clusters may also include much larger polyhedra than shown so far in this chapter; polyhedra linked through vertices, edges, or faces; and extended three-dimensional arrays. Examples of these types of clusters are given in Figure 14-20.

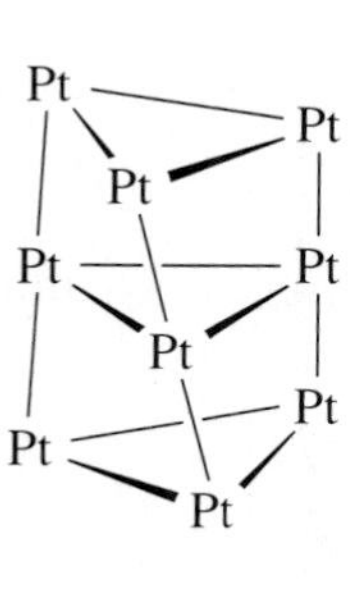

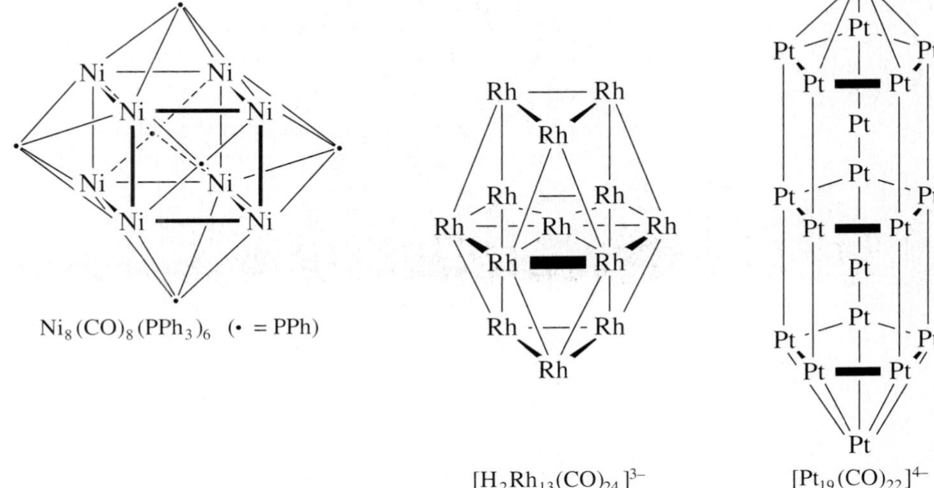

FIGURE 14-20 Examples of Large Clusters. CO and H ligands have been omitted for clarity.

[21] F. A. Cotton, *Chem. Soc. Rev.*, **1983,** *12,* 35.

GENERAL REFERENCES

The best reference on parallels between main group and organometallic chemistry is Roald Hoffmann's 1982 Nobel Lecture, "Building Bridges between Inorganic and Organic Chemistry," which describes in detail the isolobal analogy. This paper has been printed in *Angew. Chem. Int. Ed. Engl.*, **1982,** *21,* 711–724. Another very useful paper is John Ellis's "The Teaching of Organometallic Chemistry to Undergraduates," *J. Chem. Educ.*, **1976,** *53,* 2–6. K. Wade, *Electron Deficient Compounds,* Thomas Nelson, New York, 1971, provides detailed descriptions of bonding in boranes and related compounds. Topics related to multiple bonds between metal atoms are discussed in detail in F. A. Cotton and R. A. Walton, *Multiple Bonds between Metal Atoms,* Wiley, New York, 1982. Two articles in *Chemical and Engineering News* are recommended for further discussion of applications of cluster chemistry: E. L. Muetterties, "Metal Clusters," *Chem. Eng. News,* Aug. 20, **1982,** 28–41, and F. A. Cotton and M. H. Chisholm, "Bonds between Metal Atoms," *Chem. Eng. News,* June 28, **1982,** 40–46.

PROBLEMS

14-1 Predict the products:

a. $Mn_2(CO)_{10} + Br_2 \longrightarrow$

b. $HCCl_3$ + excess $Co(CO)_4^- \longrightarrow$

c. $Co_2(CO)_8 + (SCN)_2 \longrightarrow$

d. $Co_2(CO)_8 + C_6H_5{-}C{\equiv}C{-}C_6H_5$ (product has single Co—Co bond)

e. $Mn_2(CO)_{10} + [(\eta^5\text{-}C_5H_5)Fe(CO)_2]_2 \longrightarrow$

14-2 Find organic fragments isolobal with:

a. $Tc(CO)_5$ **b.** $[Re(CO)_4]^-$ **c.** $[Co(CN)_5]^{3-}$

d. $[CpFe(C_6H_6)]^+$ **e.** $[Mn(CO)_5]^+$

f. $Os_2(CO)_8$ (find organic molecule isolobal with this dimeric molecule)

14-3 Propose two organometallic fragments *not* mentioned in this chapter isolobal with:

a. CH_3 **b.** CH **c.** CH_3^+ **d.** CH_3^- **e.** $(\eta^5\text{-}C_5H_5)Fe(CO)_2$ **f.** $Sn(CH_3)_2$

14-4 Propose an organometallic molecule isolobal with each of the following:

a. Ethylene **b.** P_4 **c.** Cyclobutane **d.** S_8

14-5 Hydrides such as $NaBH_4$ and $LiAlH_4$ have been reacted with the complexes $[(C_5Me_5)Fe(C_6H_6)]^+$, $[(C_5H_5)Fe(CO)_3]^+$, and $[(C_5H_5)Fe(CO)_2(PPh_3)]^+$ (P. Michaud, C. Lapinte, and D. Astruc, *Ann. N.Y. Acad. Sci.*, **1983,** *415,* 97).

a. Show that these complexes are isolobal.

b. Predict the products of the reactions of these complexes with hydride reagents.

14-6 Hoffmann has described the following molecules to be composed of isolobal fragments. Subdivide the molecules into fragments, and show that the fragments are isolobal.

a. $H_2C{=}CH_2$, $(OC)_4Fe{=}CH_2$, and $(OC)_4Mn{=}PR_2$

b. $(OC)_4Fe{=}Fe(CO)_4$ and $(C_5H_5)(OC)Rh{=}Rh(CO)(C_5H_5)$

c. $[(OC)_4Re(\mu\text{-}Re(CO)_4)_2Re(CO)_4]^{2-}$ and $(RR'C)_2Re_2(CO)_8$

14-7 Verify that the following compounds are composed of isolobal fragments:

[Reference: G. A. Carriedo, J. A. K. Howard, and F. G. A. Stone, *J. Organomet. Chem.*, **1983,** *250,* C28.]

14-8 Calculations reported on the fragments $Mn(CO)_5$, $Mn(CO)_3$, $Cu(PH_3)$, and $Au(PH_3)$ have shown that the energies of their singly occupied hybrid orbitals are in the order $Au(PH)_3 > Cu(PH)_3 > Mn(CO)_3 > Mn(CO)_5$.

a. In the compound $(OC)_5Mn{-}Au(PH_3)$, would you expect the electrons in the Mn—Au bond to be polarized toward Mn or Au? Why? (HINT: Sketch an energy-level diagram for the molecular orbital formed between Mn and Au.)

b. The $Cu(PPh_3)$ fragment bonds to C_5H_5 in a manner similar to the isolobal $Mn(CO)_3$ fragment. However, the geometry of the corresponding $Au(PPh_3)$ complex is significantly different:

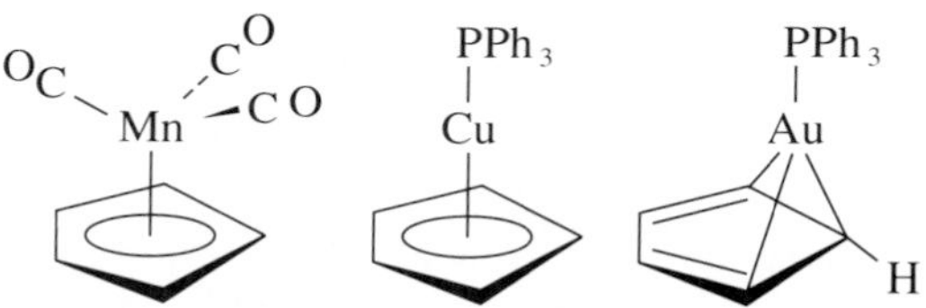

Suggest an explanation. (Reference: D. G. Evans and D. M. P. Mingos, *J. Organomet. Chem.*, **1982,** *232,* 171.)

14-9 **a.** A tin atom can bridge two $Fe_2(CO)_8$ groups in a structure similar to that of spiropentane. Show that these two molecules are composed of isolobal fragments.

b. Tin can also bridge two $Mn(CO)_2(\eta^5\text{-}C_5Me_5)$ fragments; the compound formed has Mn—Sn—Mn arranged linearly. Explain this linear arrangement. (HINT: Find a hydrocarbon isolobal with this compound. Reference: W. A. Herrmann, *Angew. Chem. Int. Ed. Engl.*, **1986,** *25,* 56.)

14-10 Using the coordinate system of Figure 14-4, for $B_6H_6^{2-}$:

a. Show that the p_z and s orbitals of the borons collectively have the same symmetry (obtain a representation for each).

b. Show that these representations reduce to $A_{1g} + E_g + T_{1u}$.

c. Show that the p_x and p_y orbitals of the borons form molecular orbitals of T_{2g} and T_{1u} symmetry.

d. Sketch a qualitative energy-level diagram for the molecular orbitals in $B_6H_6^{2-}$.

14-11 For the *closo* cluster $B_7H_7^{2-}$, which has D_{5h} symmetry, verify that there are eight framework bonding electron pairs.

14-12 Classify the following as *closo, nido,* or *arachno:*
a. $C_2B_3H_7$ **b.** B_6H_{12} **c.** $B_{11}H_{11}^{2-}$ **d.** $C_2B_7H_{13}$ **e.** $C_3B_5H_7$
f. $CB_{10}H_{13}^-$ **g.** $B_{10}H_{14}^{2-}$ **h.** $C_2B_8H_{10}$

14-13 Classify the following as *closo, nido,* or *arachno:*
a. $C_2B_4H_8$ **b.** $CB_9H_{10}^-$ **c.** $SB_{10}H_{10}^{2-}$ **d.** $NCB_{10}H_{11}$ **e.** $SiC_2B_4H_{10}$
f. $As_2C_2B_7H_9$ **g.** $PCB_9H_{11}^-$

14-14 Classify the following as *closo* or *nido:*
a. $B_3H_8Mn(CO)_3$ **b.** $B_4H_6(CoCp)_2$ **c.** $C_2B_7H_{11}CoCp$
d. $C_4B_8H_8Me_4(FeCp)_2$ **e.** $B_5H_{10}FeCp$ **f.** $C_2B_9H_{11}Ru(CO)_3$

14-15 The complex $Mo_2(NMe_2)_6$ (Me = methyl) contains a metal–metal triple bond. Would you predict this molecule to be more likely eclipsed or staggered? Explain.

14-16 In $Re_2Cl_8^{2-}$ the $d_{x^2-y^2}$ orbitals of rhenium interact strongly with the ligands. For one $ReCl_4$ unit in this ion, sketch the four group orbitals of the chloride ligands (assume one sigma donor orbital per Cl). Identify the group orbital of suitable symmetry to interact with the $d_{x^2-y^2}$ orbital of rhenium.

14-17 $[Tc_2Cl_8]^{2-}$ has a higher bond order than $[Tc_2Cl_8]^{3-}$; however, the Tc—Tc bond distance in $[Tc_2Cl_8]^{2-}$ is longer. Suggest an explanation. (HINT: see F. A. Cotton, *Chem. Soc. Rev.*, **1983,** *12,* 35).

15

Bioinorganic and Environmental Chemistry

Many biochemical reactions depend on the presence of metal ions. These ions may be present in specific coordination complexes or may act to facilitate or inhibit reactions in solution. In the first part of this chapter, we describe a few of these compounds and reactions.

Both organic and inorganic compounds released into the environment have serious consequences for plant, animal, and human life. The origin of some of these compounds, their reactions in the environment, and the technology being developed to remove them or prevent their formation are described later in this chapter.

Many metals are essential to plant and animal life, although in many cases their role is uncertain. The list includes all the first-row transition metals except scandium and titanium, but only molybdenum and perhaps tungsten from the heavier transition metals.[1] Table 15-1 lists several that are important in mammalian biochemistry. The importance of iron is obvious from the number of roles it plays, from oxygen carrier in hemoglobin and myoglobin to electron carrier in the cytochromes to detoxifying agent in catalase and peroxidase.

[1] E. Frieden, *J. Chem. Educ.*, **1985**, *62*, 917.

TABLE 15-1
Metal-containing enzymes and proteins

Fe (containing heme)	Hemoglobin, peroxidase, catalase, cytochrome P-450, tryptophan dioxygenase, cytochrome *c*
Fe (nonheme)	Pyrocatechase, ferredoxin, hemerythrin, transferrin, aconitase
Cu	Tyrosinase, amine oxidases, laccase, ascorbate oxidase, ceruloplasmin, superoxide dismutase, plastocyanin
Co (B_{12} coenzyme)	Glutamate mutase, dioldehydrase, methionine synthetase
Co(II) (noncorrin)	Dipeptidase
Zn(II)	Carbonic anhydrase, carboxypeptidase, alcohol dehydrogenase
Mg(II)	Activates phosphotransferases and phosphohydrases
K(I)	Activates pyruvate phosphokinase and K-specific ATPase
Na(I)	Activates Na-specific ATPase

15-1 PORPHYRINS AND RELATED COMPLEXES

One of the most important groups of compounds is the **porphyrins**, in which a metal ion is surrounded by the four nitrogens of a porphine ring in a square planar geometry and the axial sites are available for other ligands. Different side chains, metal ions, and surrounding species result in very different reactions and roles for these compounds. The parent porphine ring and some of the specific porphyrin compounds are shown in Figure 15-1.

15-1-1 IRON PORPHYRINS

Hemoglobin and myoglobin

The best-known iron porphyrin compounds are hemoglobin and myoglobin, oxygen transfer and storage agents in the blood and muscle tissue, respectively. Each of us has nearly 1 kg of hemoglobin in our body, picking up molecular oxygen in the lungs and delivering it to the rest of the body. Each molecule is made up of four globin protein subunits, two α and two β. In each of these, the protein molecule partially encloses the heme group, bonding to one of the axial positions through an imidazole nitrogen, as shown in Figure 15-2. The other axial position is vacant or has water bound to it. When dissolved oxygen is present, it can occupy this position, and subtle changes in the conformation of the proteins result. As one iron binds an oxygen molecule, the molecular shape changes to make binding of additional oxygen molecules easier. The four irons can each carry one O_2, with generally increasing equilibrium constants:

$$\mathrm{Hb} + \mathrm{O_2} \rightleftarrows \mathrm{HbO_2} \qquad K_1 = 5 - 60$$

$$\mathrm{HbO_2} + \mathrm{O_2} \rightleftarrows \mathrm{Hb(O_2)_2}$$

$$\mathrm{Hb(O_2)_2} + \mathrm{O_2} \rightleftarrows \mathrm{Hb(O_2)_3}$$

$$\mathrm{Hb(O_2)_3} + \mathrm{O_2} \rightleftarrows \mathrm{Hb(O_2)_4} \qquad K_4 = 3000 - 6000$$

Although there is still some disagreement over the details, the equilibrium constants increase, with the fourth many times larger (depending on the species

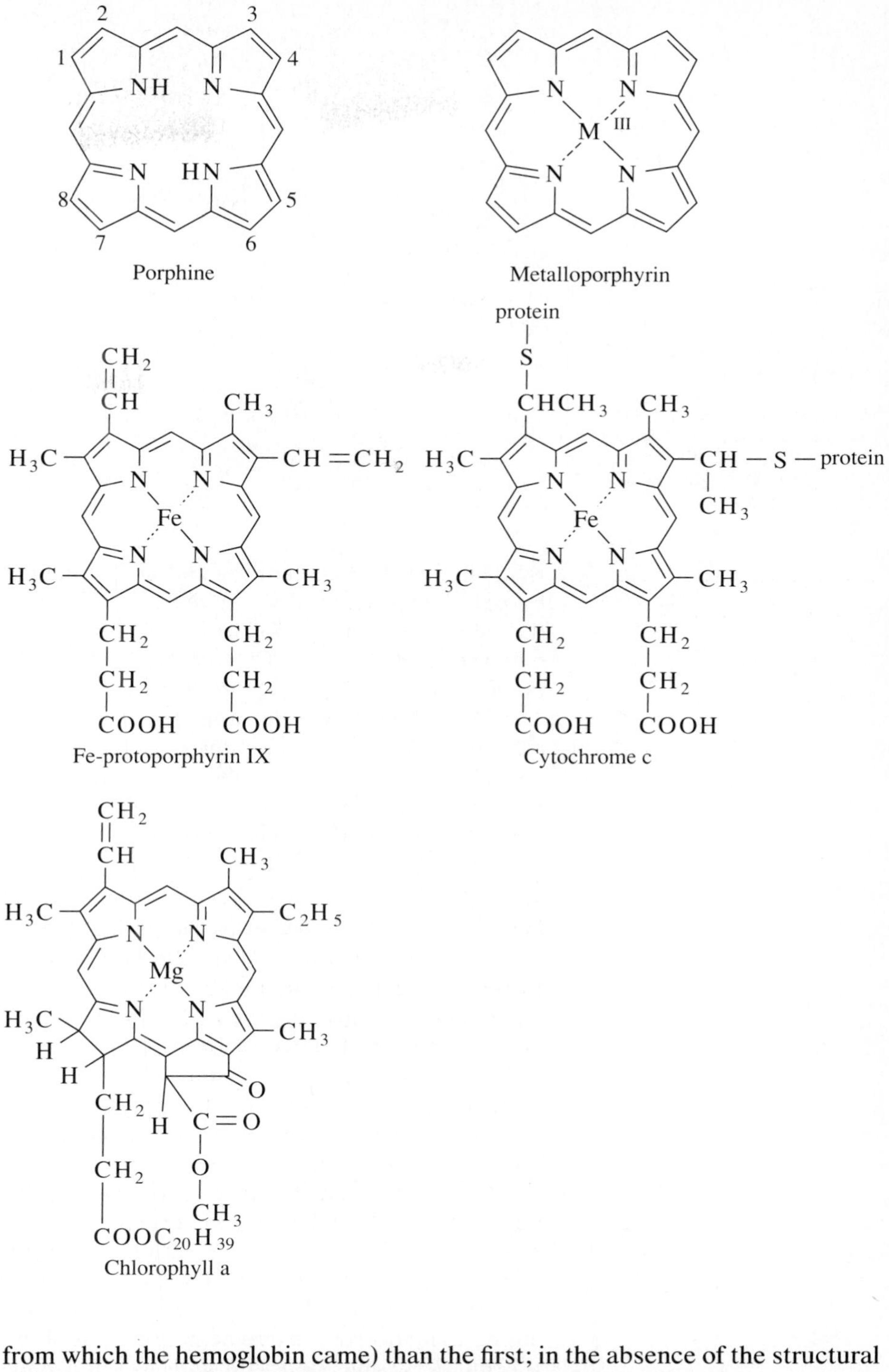

Porphine

Metalloporphyrin

Fe-protoporphyrin IX

Cytochrome c

Chlorophyll a

FIGURE 15-1 Porphine, Porphyrin, and Related Compounds.

from which the hemoglobin came) than the first; in the absence of the structural changes, it would be much smaller than the first. As a result, as soon as some oxygen has been bound to the molecule, all four irons are readily oxygenated. In a similar fashion, initial removal of oxygen triggers the release of the remainder, and the entire load of oxygen is delivered at the required site. The structural changes accompanying oxygenation are described thoroughly by

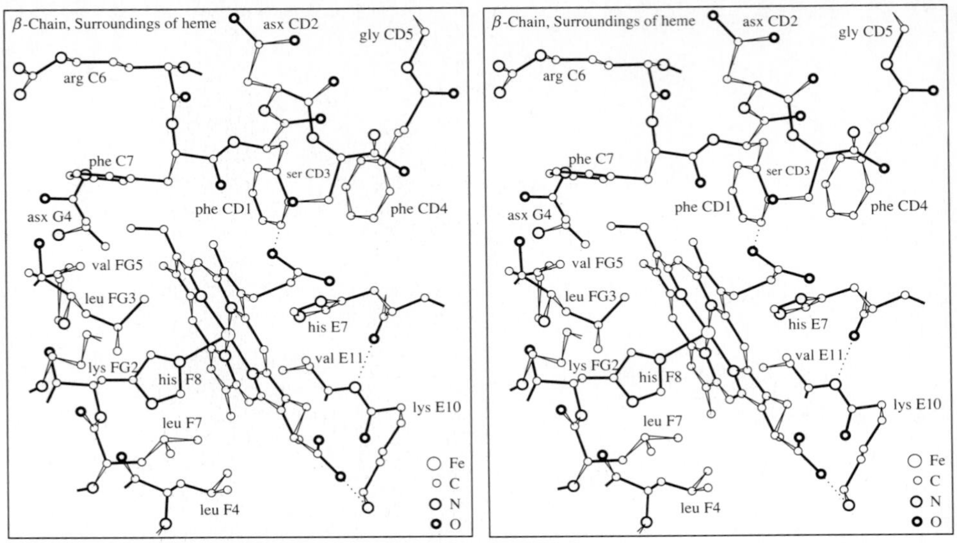

FIGURE 15-2 Heme Group Binding in Hemoglobin. A stereo drawing of the surroundings of the heme in the β chain of horse hemoglobin. Broken lines indicate hydrogen bonds. (Reprinted by permission from J. F. Perutz and H. Lehmann, *Nature*, **1968**, *219*, 902. Copyright © 1968 Macmillan Magazines Limited.)

Baldwin and Chothia[2] and by Dickerson and Geis.[3] This effect is also favored by pH changes caused by increased CO_2 concentration in the capillaries. As the concentration of CO_2 increases, the pH decreases due to formation of bicarbonate ($2\ H_2O + CO_2 \rightleftharpoons HCO_3^- + H_3O^+$), and the increased acidity favors release of O_2 from the oxyhemoglobin. This is called the Bohr effect.

Myoglobin has only one heme group per molecule, with a role as an oxygen storage molecule in the muscles. The myoglobin molecule is similar to a single subunit of hemoglobin, and bonding between the iron and the oxygen molecule is similar to that in hemoglobin, but the equilibrium is simpler because only one oxygen molecule is bound:

$$Mb + O_2 \rightleftharpoons MbO_2$$

When hemoglobin releases oxygen to the muscle tissue, myoglobin picks it up and stores it until it is needed. The Bohr effect and the cooperation of the four hemoglobin binding sites make the transfer more complete when the oxygen concentration is low and the carbon dioxide concentration is high; the opposite conditions in the lungs promote transfer of oxygen to hemoglobin and transfer of CO_2 to the gas phase in the lungs. As shown in Figure 15-3, myoglobin binds O_2 more strongly than the first O_2 of hemoglobin. However, the fourth equilibrium constant of hemoglobin is larger than that for myoglobin by about a factor of 50.

In hemoglobin, the iron is formally Fe(II), and bonding to oxygen does not oxidize it to Fe(III). However, when the heme group is removed from the protein, exposure to oxygen oxidizes the iron quickly to a μ-oxo dimer containing two Fe(III) ions. The presence of protein around the heme seems to prevent oxidation of Fe(II) in hemoglobin, while the presence of water alone allows oxidation of the free heme. In a test of this hypothesis, Wang[4] embedded a heme derivative saturated with CO in a polystyrene matrix and studied its equilibrium with CO and O_2. He was able to cycle the material between the

[2] J. Baldwin and C. Chothia, *J. Mol. Biol.*, **1979**, *129*, 175.
[3] R. E. Dickerson and I. Geis, *Hemoglobin*, Benjamin/Cummings, Menlo Park, Calif., 1983.
[4] J. H. Wang, *J. Am. Chem. Soc.*, **1958**, *80*, 3168.

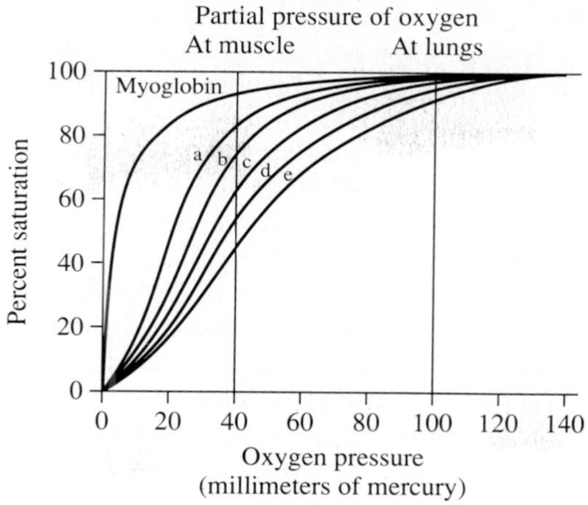

FIGURE 15-3 Myoglobin and Hemoglobin Binding Curves. Myoglobin, and hemoglobin at five different pH values: (a) 7.6; (b) 7.4; (c) 7.2; (d) 7.0; (e) 6.8. (Reproduced with permission from R. E. Dickerson and I. Geis, *Hemoglobin,* Benjamin/Cummings, Menlo Park, Calif., 1983, p. 24.)

Partial pressure of oxygen
At muscle At lungs
Myoglobin
a b c d e
Percent saturation
Oxygen pressure (millimeters of mercury)

oxygenated form, the CO-bound form, and the free heme with no oxidation to Fe(III). From this evidence, he concluded that a nonaqueous environment is required for reversible O_2 or CO binding. In hemoglobin, the protein surrounding the heme groups provides this nonaqueous environment and prevents oxidation. Others argue that oxidation results from one oxygen molecule simultaneously bonding to two hemes, which is effectively prevented by Wang's polystyrene matrix or the globin of native hemoglobin.

In hemoglobin, the Fe(II) is about 70 pm out of the plane of the porphyrin nitrogens in the direction of the imidazole nitrogen bonding to the axial position (Figure 15-2) and is a typical high-spin d^6 ion. When oxygen or carbon monoxide bonds to the sixth position, the iron becomes coplanar with the porphyrin and the resulting compound is diamagnetic. Carbon monoxide is a strong enough ligand to force spin pairing; the resulting π back-bonding stabilizes the complex. Oxygen bonds at an angle of approximately 130°, also with considerable π back-bonding. Ochiai has proposed a structure, shown in Figure 15-4, in which the triplet O_2 and the high-spin Fe(II) combine to form a spin-paired compound. The stronger σ interaction is between the d_{z^2} and the π_g^z orbitals. The weaker

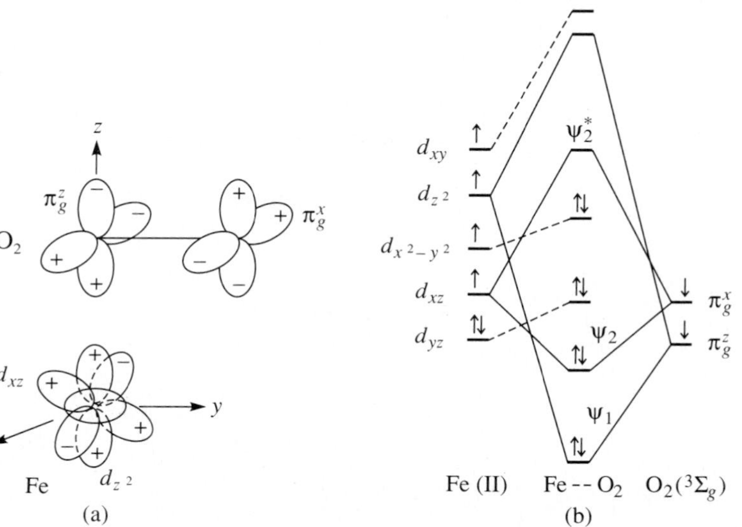

FIGURE 15-4 Electronic Structure of Oxyhemoglobin. (a) The most likely interaction between O_2 in the ground state ($^3\Sigma_g$) and Fe(II)-heme in the high spin state; *x* and *y* axes bisect the angle N—Fe—N. (b) The interaction between O_2 and Fe(II)-heme as expressed by an energy-level diagram. Fe in the high-spin state is located a little out of the porphyrin plane and as the reaction proceeds, it is thought to move to the center in the plane. This effect is shown by the broken lines. (Reprinted with permission from E. I. Ochiai, *J. Inorg. Nucl. Chem.,* **1974,** *36,* 2129. Copyright 1974, Pergamon Press PLC.)

π interaction is between d_{xz} and π_g^x. The increased ligand field results in pairing of the electrons and a weakened O—O bond. In hemoglobin, CO also forms bent bonds to Fe, probably because surrounding groups in the hemoglobin force it out of the linear form.

One method of studying hemoglobin (and many other complex biological systems) is through model compounds. Many heme derivatives have been synthesized and tested for oxygen binding with a more complete understanding of the process as a goal.[5] These compounds have been designed to protect the heme from the approach of another heme to prevent oxidation of the iron(II) and formation of an O_2 bridge between the two heme irons. In addition, some model compounds have an imidazole or pyridine nitrogen linked to the heme to hold it in a convenient location for binding to the iron. A few have reached the point of testing as synthetic hemoglobin substitutes.

Cytochromes, peroxidases, and catalases

Other heme compounds are also active biochemically. Cytochrome P-450 catalyzes oxidation reactions in the liver and adrenal cortex, helping to detoxify some substances by adding hydroxyl groups that make the compounds more water soluble and more susceptible to further reactions. Unfortunately, at times this process has the reverse effect; some relatively safe molecules are converted into potent carcinogens. Peroxidases and catalases are Fe(III)-heme compounds that decompose hydrogen peroxide and organic peroxides. The reactions seem to proceed through Fe(IV) compounds with another unpaired electron on the porphyrin, which is therefore a radical cation. Similar intermediates are known in simpler porphyrin molecules as well.[6]

TABLE 15-2
Rates of hydrogen peroxide decomposition

Catalyst	*Relative rate*
Catalase	10^8
$[Fe(trien)]^{3+}$	10^4
Methemoglobin [Fe(III) Hb]	1

A model compound that decomposes hydrogen peroxide rapidly has been made from Fe(III) and triethylenetetramine (trien).[7] Although the rate is not as high as that for catalase (Table 15-2), it is many times faster than that for hydrated iron oxide, which seems to have a large surface effect. The proposed mechanism for the $[Fe(trien)]^{3+}$ reaction is shown in Figure 15-5. Tracer studies using ^{18}O-labeled water have shown that the reaction produces oxygen gas in which all the oxygen atoms come from the peroxide; as a result, the steps forming O_2 must involve removal of hydrogen from H_2O_2. Formation of water as the other product requires breakage of the oxygen-oxygen bond.

A group of cytochromes (labeled a, b, and c, depending on their spectra) serve as oxidation–reduction agents, converting the energy of the oxidation process to synthesis of adenosine triphosphate (ATP), which makes the energy more available to other reactions. Copper is also involved in these reactions. The copper cycles between Cu(II) and Cu(I) and the iron cycles between Fe(III) and Fe(II) during the reactions. Details of the reactions are available in other sources.[8,9]

[5] K. S. Suslick and T. J. Reinert, *J. Chem. Educ.*, **1985**, *62*, 974.

[6] D. L. Hickman, A. Nanthakumar, and H. M. Goff, *J. Am. Chem. Soc.*, **1988**, *110*, 6384.

[7] J. H. Wang, *J. Am. Chem. Soc.*, **1955**, *77*, 822, 4715; *Acc. Chem. Res.*, **1970**, *3*, 90; R. C. Jarnagin and J. H. Wang, *J. Am. Chem. Soc.*, **1958**, *80*, 786.

[8] J. T. Groves, *J. Chem. Educ.*, **1985**, *62*, 928.

[9] E. I. Ochiai, *Bioinorganic Chemistry*, Allyn and Bacon, Boston, 1977, pp. 150–65; T. E. Meyer and M. D. Kamen, *Adv. Protein Chem.*, **1982**, *35*, 105–212; G. R. Moore et al., *Adv. Inorg. Bioinorg. Mech.*, **1984**, *3*, 1–96.

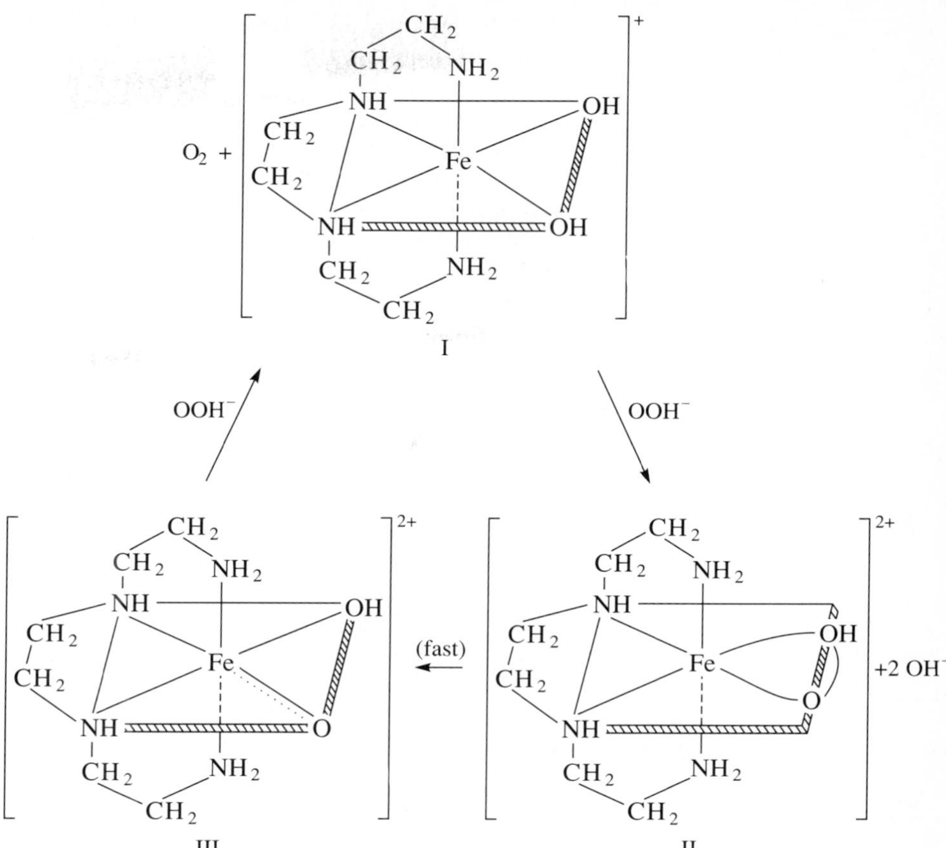

FIGURE 15-5 Mechanism of the $[Fe(trien)]^{3+}$—H_2O_2 Reaction. (Reproduced with permission from J. H. Wang, *J. Am. Chem. Soc.*, **1955,** *77,* 4715. Copyright 1955 American Chemical Society.)

15-1-2 SIMILAR RING COMPOUNDS

Chlorophylls

A porphine ring with one double bond reduced is called a chlorin. The chlorophylls (Figure 15-1) are examples of compounds containing this ring. They are green pigments found in plants, contain magnesium, and start the process of photosynthesis. They absorb light at the red end of the visible spectrum, transfer an electron to adjacent compounds, and, by a series of complex reactions, finally transfer the energy of the light to the metabolic processes of the plant. The overall process can be summarized in two reactions:

$$2\,H_2O \longrightarrow O_2 + 4\,H^+ + 4e^-$$

$$CO_2 + 4\,H^+ + 4e^- \longrightarrow [CH_2O] + H_2O$$

where $[CH_2O]$ represents sugars, carbohydrates, and cellulose synthesized in the plant. In effect, this process also reverses the oxidation process that produces the energy for animal life, in which the $[CH_2O]$ compounds are converted back to water and carbon dioxide. The entire process is very complicated and is far from being completely understood, but it includes a vital role for manganese in the first reaction.

Other compounds containing metal ions, such as ferredoxin,[10] are involved in the electron-transfer reactions of photosynthesis. Ferredoxin is an iron–sulfur compound whose active site is sometimes abbreviated as $Fe_2S_2(cys)_4$ (cys = cysteine). The structure is not completely determined, but seems to have Fe(II) and Fe(III) in tetrahedral sites bridged by sulfide ions and bound into the protein by Fe—S bonds to cysteine. There are also other more complex ferredoxins and related compounds in bacteria. In some of these, the Fe—S units are Fe_4S_4, again with tetrahedral iron and sulfide bridges. A similar structure has been suggested for a more uncertain Fe_6S_6 compound. Proposed structures for the Fe—S active sites of these compounds are shown in Figure 15-6.

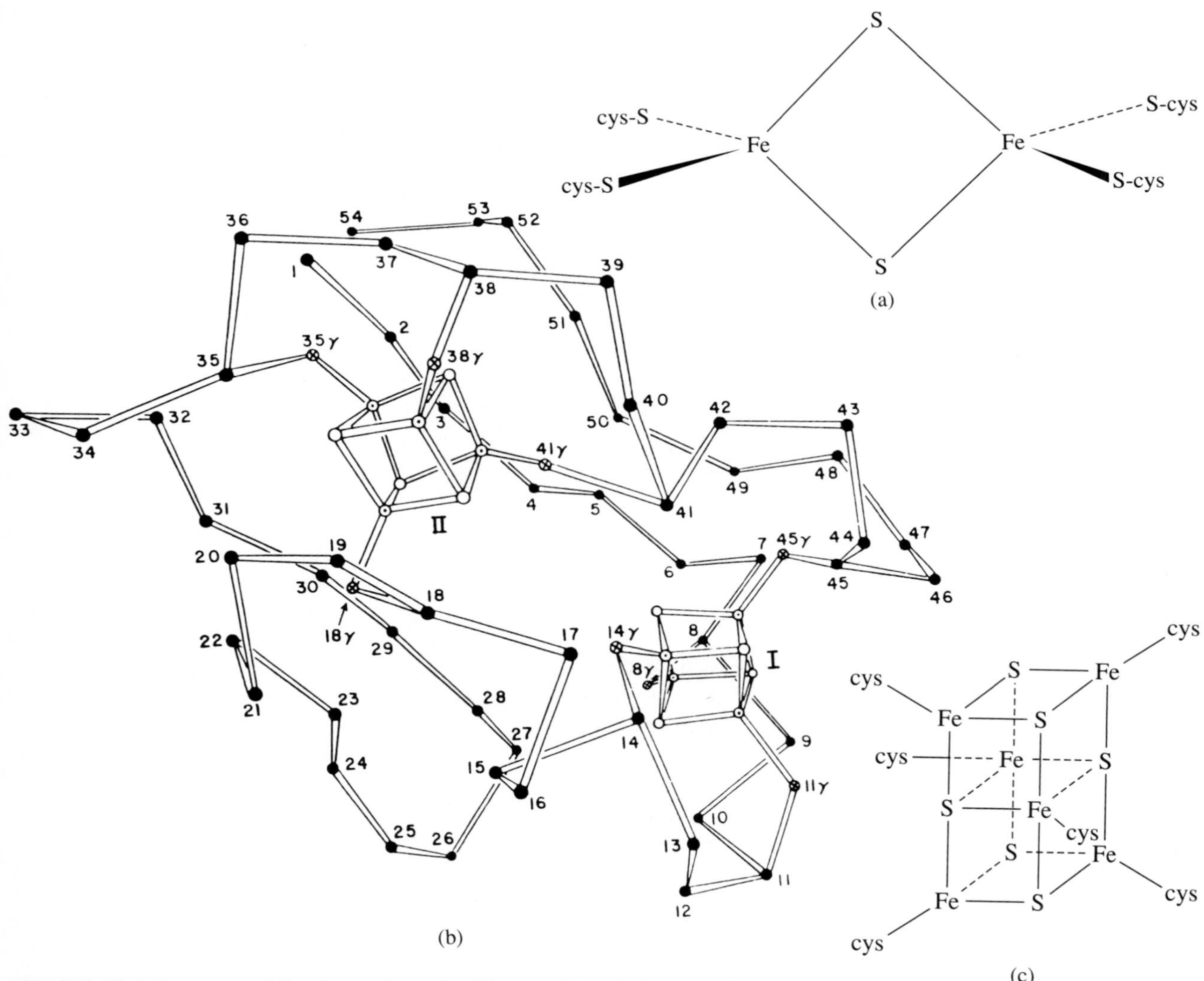

FIGURE 15-6 Structures of Fe—S Protein Active Sites. (a) Ferredoxin. (E.-I. Ochiai, *Bioinorganic Chemistry,* Allyn and Bacon, Boston, 1977, p. 184.) (b) Clostridial ferredoxin. Fe, ⊙; S_{inorg}, ○; S_{cys}, ⊗; C_α, ●. (Reproduced with permission from E. T. Adman, L. C. Sieker, and L. H. Jensen, *J. Biol. Chem.*, **1973,** *248,* 3987.) (c) A model for the structure of the Fe_6S_6 active unit. (Reproduced with permission from E.-I. Ochiai, *Bioinorganic Chemistry,* Allyn and Bacon, Boston, 1977, p. 192.)

[10] B. B. Buchanan, *Struct. Bonding,* **1966,** *1,* 109–148.

Coenzyme B_{12}

A vitamin known as coenzyme B_{12} is the only known organometallic compound in nature. It incorporates cobalt into a corrin ring structure, which has one less =CH— bridge between the pyrrole rings than the porphyrins (Figure 15-7). This compound is known to prevent anemia and has also been found to have many catalytic properties. During isolation of this compound from natural sources, the adenosine group is usually replaced by cyanide, and it is in this cyanocobalamin (Vitamin B-12) form that it is used medicinally. The cobalt can be counted as Co(III) in these compounds; the four corrin nitrogens contribute eight electrons and a charge of $2-$, the benzimidazole nitrogen contributes two electrons, and the cyanide or adenosine in the sixth position contributes two electrons and a charge of $1-$. Without the sixth ligand, it is called cobalamin.

Methylcobalamin (with methyl in the sixth position) can catalyze the methylation of metallic mercury, which makes the mercury much more dangerous. At one time it was thought that metallic mercury discarded into

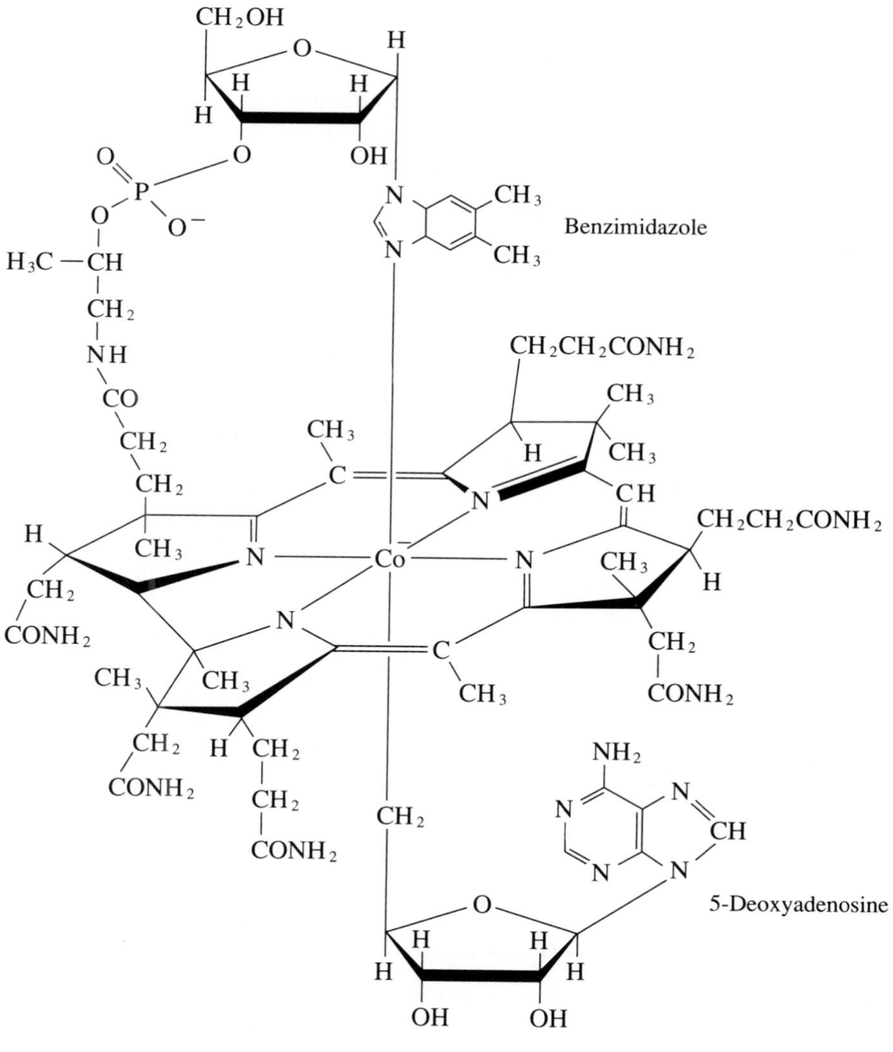

FIGURE 15-7 Coenzyme B_{12}.

rivers and lakes was inert and would simply remain on the bottom with no serious effects. More recently, it has been found that bacterial action, probably assisted by methylcobalamin, converts some of the mercury to methylmercury, a much more soluble form that is readily picked up by plant and animal life and concentrated in the food chain. As a result, discarded mercury poses a severe health threat. This problem is discussed in more detail later in this chapter.

The reactions of alkylcobalamins depend on cleavage of the alkyl—cobalt bond, which can result in Co(I) and an alkyl cation, Co(II) and an alkyl radical, or Co(III) and an alkyl anion. The alkyl products can then react in a number of ways. Some of the reactions include:[11]

Methylation or hydroxymethylation

$$\mathrm{HO_2C\underset{\displaystyle NH_2}{\underset{|}{C}}HCH_2CH_2SH \longrightarrow HO_2C\underset{\displaystyle NH_2}{\underset{|}{C}}HCH_2CH_2SCH_3}$$

$$\mathrm{H_2N{-}CH_2{-}CO_2H \longrightarrow H_2N{-}\underset{\displaystyle CH_2OH}{\underset{|}{C}H}{-}CO_2H}$$

Isomerization

$$\mathrm{HO_2C{-}\underset{\displaystyle H_2N}{\underset{|}{C}H}{-}CH_2{-}CH_2{-}CO_2H \longrightarrow HO_2C{-}\underset{\displaystyle H_2N}{\underset{|}{C}H}{-}\underset{\displaystyle CH_3}{\underset{|}{C}H}{-}CO_2H}$$

Isomerization and dehydration

$$\mathrm{HOCH_2{-}\underset{\displaystyle OH}{\underset{|}{C}H}{-}CH_3 \longrightarrow HO\underset{\displaystyle OH}{\underset{|}{C}H}{-}CH_2{-}CH_3 \longrightarrow H\underset{\displaystyle O}{\underset{\|}{C}}{-}CH_2{-}CH_3 + H_2O}$$

15-2 OTHER IRON COMPOUNDS

Ferritin and transferrin

Iron is stored in both plant and animal organisms in combination with a protein called apoferritin. The resulting ferritin contains a micelle of ferric hydroxide-oxide-phosphate surrounded by the protein and is present mainly in the spleen, liver, and bone marrow in mammals. Individual subunits of the apoferritin have a molecular weight of about 18,500, and 24 of these subunits combine to form the complex in Figure 15-8, with a protein molecular weight of about 445,000 and up to 4300 atoms of Fe in the iron core. The mechanisms for incorporation of iron into this complex and removal for use in the body are uncertain, but it appears that reduction to Fe(II) and chelation of the Fe(II) are required to remove iron from the core, and the reverse process moves it into the storage core of the complex. Another iron protein, called transferrin, serves to transport iron [as Fe(III)] in the blood. It has a molecular weight of 80,000, and titration evidence suggests that the iron is bound as Fe(III) by three tyrosine phenoxy groups, a carboxyl group, and either HCO_3^- or CO_3^{2-} as shown in Figure 15-8.[12]

[11] R. H. Abeles, "Current Status of the Mechanism of Action of B_{12} Coenzyme," in *Biological Aspects of Inorganic Chemistry*, A. W. Addison, W. R. Cullen, D. Dolphin, and B. R. James, eds., Wiley-Interscience, New York, 1977, p. 245–60.

[12] R. E. Feeney and S. K. Komatsu, *Struct. Bonding*, **1966**, *1*, 149–206; E. E. Hazan, cited in B. L. Vallee and W. E. C. Wacker, *Metalloproteins*, Academic Press, New York, 1969, p. 89.

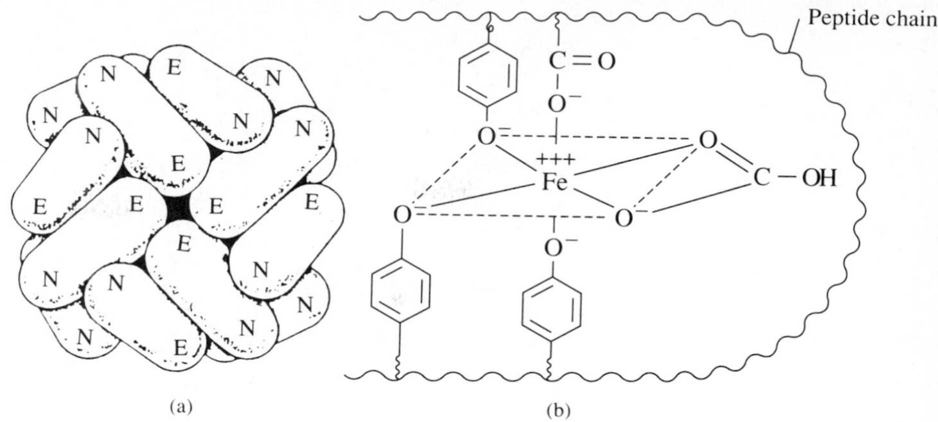

FIGURE 15-8 Ferritin and Transferrin. (a) A schematic diagram of the packing of the 24 subunits that form the protein sheath of horse spleen ferritin. The view is down one fourfold axis. (Reproduced with permission from F. A. Cotton and G. Wilkinson, *Advanced Inorganic Chemistry,* 5th ed., Wiley-Interscience, New York, 1988, p. 1338. Copyright © 1988 John Wiley & Sons, Inc. Reprinted by permission of John Wiley & Sons, Inc.) (b) Diagrammatic representation of the ferric ion, bicarbonate chelate of conalbumin, a transferrin. (Redrawn with permission from R. C. Warner, Transactions, New York Academy of Sciences, **1954,** *16,* 182.

Siderochromes

Bacteria and fungi also synthesize iron-transfer compounds, called siderochromes.[13] The common structures are complex hydroxamates (also called ferrichromes or ferrioxamines) or complex catechols, all shown in Figure 15-9. They have peptide backbones and are very strong chelating agents ($K \approx 10^{30}$), allowing the plant to extract iron from soil that contains little iron or is basic enough that the iron is present as the insoluble hydroxides or oxides. Some of these compounds act as growth factors for bacteria and others act as antibiotics. There are also examples in which the iron is bound by a mixture of phenolic hydroxyl, hydroxamate, amine, and alcoholic hydroxyl groups.

15-3 POTASSIUM AND SODIUM

Potassium ions are concentrated inside cells with sodium ions mainly in blood plasma and the interstitial fluids of the body, as shown in Table 15-3. The contrast in concentrations inside and outside the cells is also pronounced for magnesium and phosphate ions.

Since thermodynamic equilibrium requires equal concentration of ions on both sides of a membrane, energy must be expended to maintain the difference. Hydrolysis of adenosine triphosphate (ATP) to adenosine diphosphate (ADP) and phosphate ion is the source of energy, which is transferred to the transport reaction by coupling of the reactions, a common feature of metabolic cycles and other biochemical reactions. Maintaining the concentration difference for sodium or potassium ions requires nearly 6 kJ/mol (calculated by the equation $\Delta G = -RT \ln [Na^+]_{cell}/[Na^+]_{fluid}$); the ATP $\longrightarrow$ ADP + PO_4^{3-} reaction generates more than 33 kJ/mol. Overall, 3 moles of Na^+ and 2 to 3 moles of K^+ are transported per mole of ATP used in a very complex series of reactions.[14] At the same time, water must be pumped out of the cell to counter diffusion from the relatively dilute interstitial fluid into the cell, which has higher concentrations of proteins, sugars, and other compounds used in its metabolism.

Although the details of the transfer mechanisms are still uncertain, it appears that the ions are complexed by specific ligands that assist in the transfer through the hydrophobic cell wall. For example, valinomycin, a cyclic

[13] J. B. Neilands, in *Inorganic Biochemistry,* Vol. I, G. I. Eichhorn, ed., Elsevier, New York, 1973, pp. 167–202; *Struct. Bonding,* **1966**, *1*, 59–108; **1972**, *11*, 145–70.

[14] R. L. Post, C. D. Albright, and K. Dayani, *J. Gen. Physiol.*, **1967**, *50*, 1201.

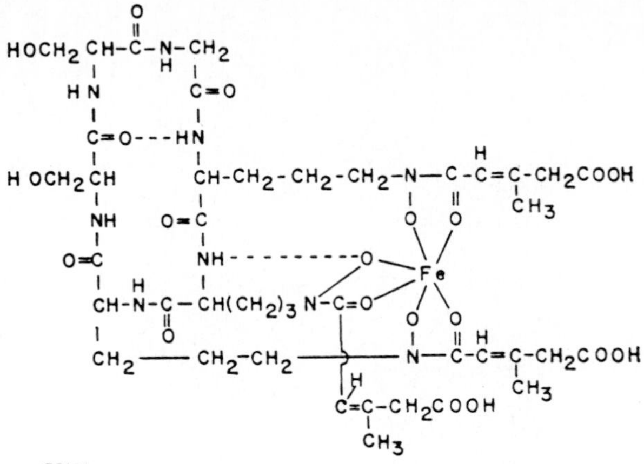

(a)

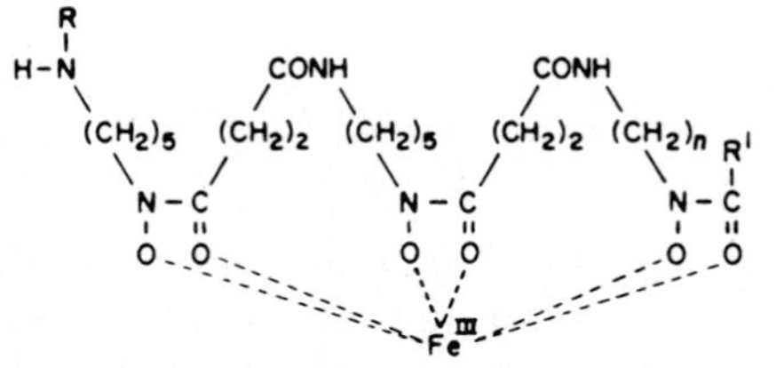

XV Ferrioxamines (linear)

	R	n	R′
XVa Ferrioxamine B	H	5	CH_3-
XVb Ferrioxamine D$_1$	CH_3CO-	5	CH_3-
XVc Ferrioxamine G	H	5	$HO_2C(CH_2)_2-$
XVd Ferrioxamine A$_1$	H	4	$HO_2C(CH_2)_2-$
XVe Ferrimycin A$_1$	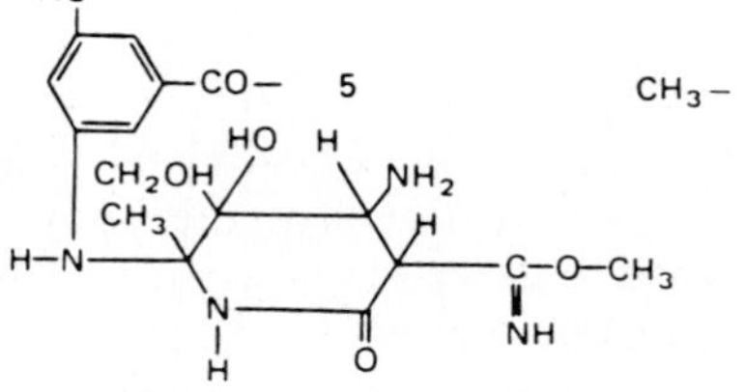	5	CH_3-

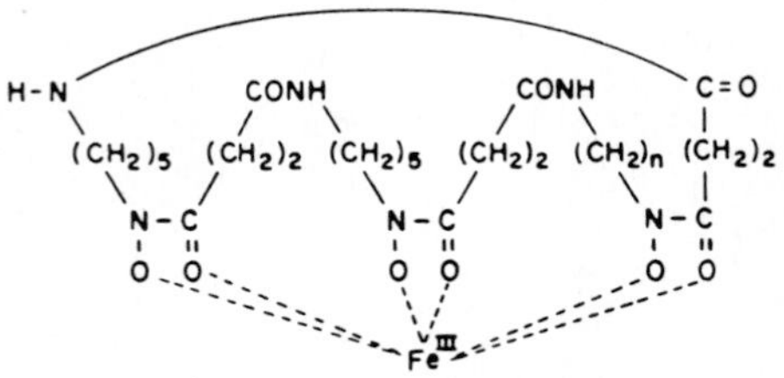

XVI Ferrioxamines (cyclic)

	n
XVIa Ferrioxamine E	5
XVIb Ferrioxamine D$_2$	4

(b)

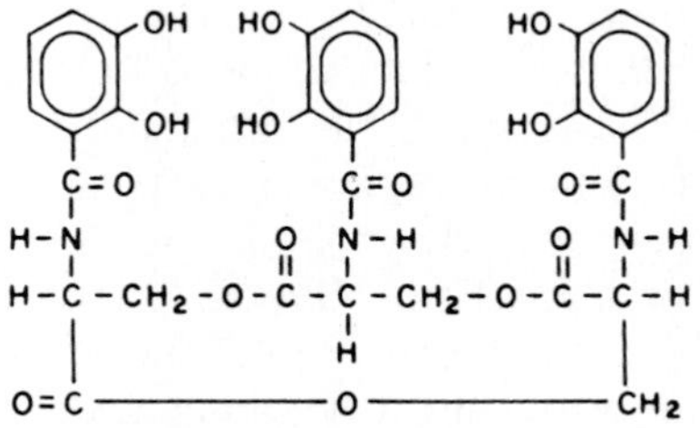

I Enterobactin

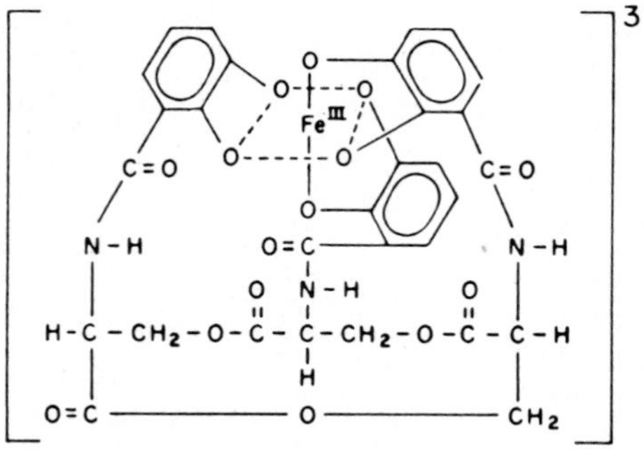

Ia Iron(III) enterobactin

(c)

FIGURE 15-9 Ferrichromes, Ferrioxamines, and Catechol Siderochromes. (a) Ferrichrome A. (Reproduced with permission from A. Zalkin, J. D. Forrester, and D. H. Templeton, *J. Am. Chem. Soc.*, **1966,** *88,* 1810.) (b) Ferrioxamines. (Reprinted by permission of the publisher from "Microbial Iron Transport Compounds (Siderochromes)" by J. B. Neilands in G. L. Eichhorn, ed., *Inorganic Biochemistry,* Vol. 1, Elsevier, New York, 1973, p. 196. Copyright 1973 by Elsevier Science Publishing Co., Inc.) (c) Fe(III) enterobactin, a catechol siderochrome. (Reprinted by permission of the publisher from J. B. Neilands, op. cit., p. 183. Copyright 1973 by Elsevier Science Publishing Co., Inc.)

oligopeptide with 12 carbonyl groups and a number of hydrophobic branching groups, can complex K^+ easily, but Na^+ is too small to form a good complex. The resulting potassium complex passes easily through cell walls that are usually impervious to the hydrated ions. However, the normal concentrations of these ions require a more active sequence of reactions. One suggested

TABLE 15-3
Distribution of ions in cells and body fluids

	Cations				*Anions*		
	Na^+	K^+	Ca^{2+}	Mg^{2+}	Cl^-	PO_4^{3-}	HCO_3^-
Blood plasma	152	5	2.5	1.5	113	1	27
Interstitial fluid	143	4	2.5	1.5	117	1	27
Cell	14	157	—	13		38	10

SOURCE: Data from A. Leaf and L. H. Newburgh, *Significance of Body Fluids in Clinical Medicine,* 2nd ed., Charles C Thomas, Springfield, Ill., 1955.
NOTE: All concentrations are in *mmoles/L*. Charge balance is maintained by charges on organic compounds, including the proteins.

sequence for the sodium pump has a phosphate ion bonding to a protein carrier at the same time Na^+ binds on the inside of the membrane, and removal of the phosphate and binding of K^+ on the outside. As a result, Na^+ and phosphate are picked up inside the cell, the protein rotates or moves to the outside of the cell where the Na^+ and phosphate are removed and K^+ bound, the protein rotates or moves to the inside of the cell where the K^+ is removed, and the cycle starts over. Figure 15-10 shows a schematic diagram of this process.

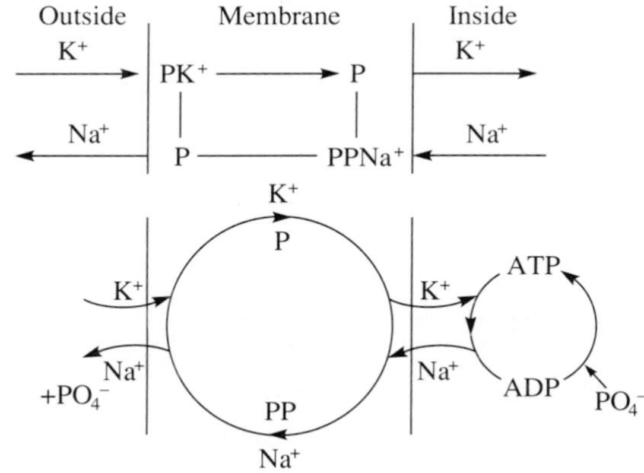

FIGURE 15-10 Schematic Representations of the Sodium Pump. (Reproduced with permission from M. N. Hughes, *The Inorganic Chemistry of Biological Processes,* Wiley, New York, 1972, p. 273. Copyright © 1972 John Wiley & Sons, Ltd. Reprinted by permission of John Wiley & Sons, Ltd.)

15-4 ZINC AND COPPER ENZYMES

Although zinc is found in more than 80 enzymes, only two, carboxypeptidase and carbonic anhydrase, will be discussed here.[15]

Carboxypeptidase

Carboxypeptidase is a pancreatic enzyme that catalyzes the hydrolysis of the peptide bond at the carboxyl end of proteins and peptides, with a strong preference for amino acids with an aromatic or branched aliphatic side chain. The zinc ion is bound in a nearly tetrahedral site by two histidine nitrogens, an oxygen from a glutamic acid carboxyl group, and a water molecule. A pocket in the protein structure accommodates the side chain of the substrate. Although details are still uncertain, current evidence indicates that the negative carboxyl

[15] I. Bertini, C. Luchinat, and R. Monnanni, *J. Chem. Educ.*, **1985**, *62*, 924.

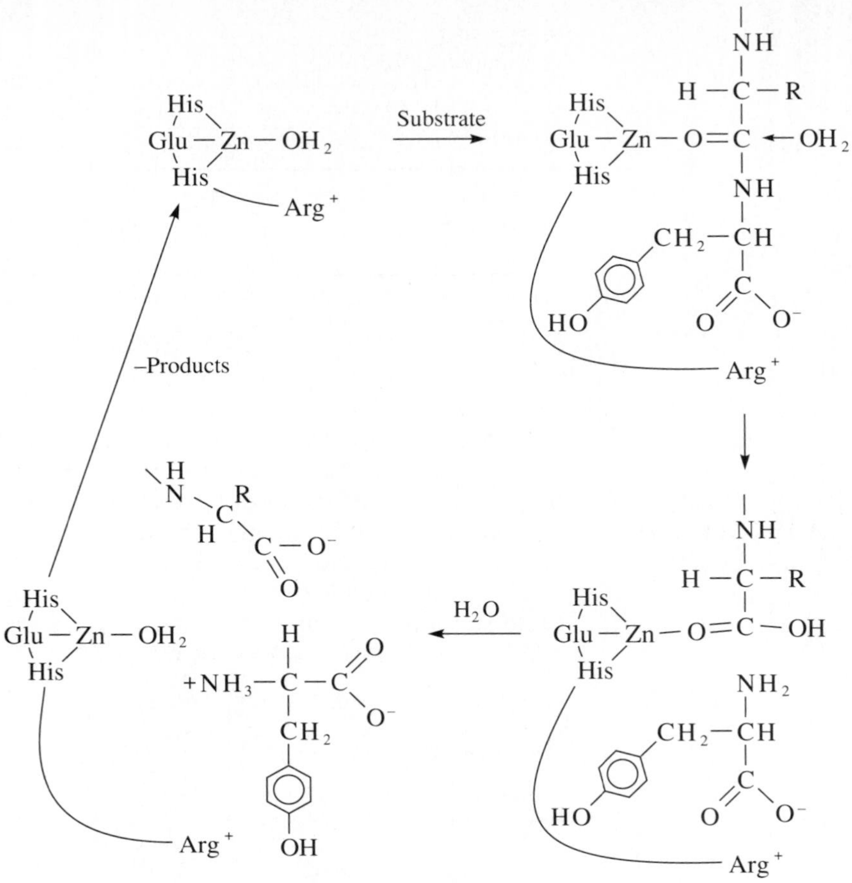

FIGURE 15-11 Proposed Mechanism of Carboxypeptidase Action.

group of the substrate hydrogen bonds to an arginine on the enzyme while the zinc bonds to the oxygen of the peptide carbonyl, as shown in Figure 15-11. A Zn—OH combination seems to be the group that reacts with the carbonyl carbon, with assistance of a glutamic acid carboxyl group from the enzyme.[16]

Carbonic anhydrase

Carbonic anhydrase is an essential enzyme for respiration. The hydration and dehydration equilibria of CO_2,

$$CO_2 + H_2O \rightleftharpoons H_2CO_3 \quad \text{and} \quad CO_2 + H_2O \rightleftharpoons HCO_3^- + H^+$$

are too slow to provide the rapid exchange needed for support of life. There are two similar forms of human carbonic anhydrase. In both, zinc is bound to three histidine imidazole nitrogens in a deep pocket of the protein. Enzyme activity is related to a group with $pK \sim 7$, thought to be a water molecule bonded to the zinc. The overall mechanism can be summarized in a four-step reaction:

$$E^- + CO_2 \rightleftharpoons E^- \cdot CO_2 \underset{H_2O}{\overset{H_2O}{\rightleftharpoons}} EH \cdot HCO_3^- \rightleftharpoons EH + HCO_3^- \rightleftharpoons E^- + H^+ + HCO_3^-$$

[16] M. W. Makinen et al., *Adv. Inorg. Biochem.*, **1984**, *6*, 1.

In this mechanism, E^- includes the Zn—OH required by the pH dependence. Possible structures involved are shown in Figure 15-12. Overall, the reaction is speeded by a factor of 10^6 or more, allowing the exchange of CO_2 between the blood and air in the lungs and between blood and other tissues of the body to be fast.[17,18]

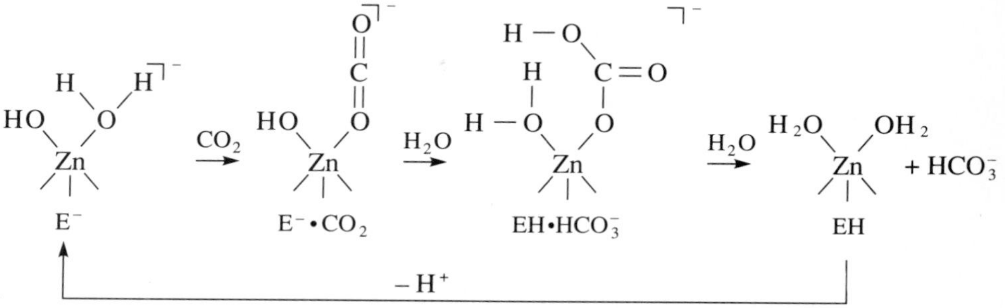

$$E^- \xrightarrow{CO_2} E^-\cdot CO_2 \xrightarrow{H_2O} EH\cdot HCO_3^- \xrightarrow{H_2O} EH + HCO_3^-;\quad EH \xrightarrow{-H^+} E^-$$

FIGURE 15-12 Proposed Carbonic Anhydrase Mechanism.

Superoxide dismutase and ceruloplasmin

Copper is present in mammals in superoxide dismutase and ceruloplasmin; it is also part of a number of enzymes in plants and other organisms, including laccase, ascorbate oxidase, and plastocyanin. In these compounds, it is present in four different forms, listed in Table 15-4.

TABLE 15-4
Forms of copper in proteins

	Absorption maximum (nm)	*Extinction coefficients*	
Type I	600	1000–4000	Responsible for the blue color of blue oxidases and electron-transfer proteins; L ⟶ M charge transfer spectrum of Cu—S bond
Type II	Near 600	300	Similar to ordinary tetragonal Cu(II) complexes, but more intensely colored
Type III	330	3000–5000	Paired Cu(II) ions, diamagnetic, associated with redox reactions of O_2, where it undergoes a 2-electron change, bypassing superoxide
Cu(I)			Colorless, diamagnetic, no *esr* spectrum

Ceruloplasmin[19] is an intensely blue glycoprotein of the α_2-globulin fraction of mammalian blood, whose physiological role is still somewhat uncertain. The structure is also uncertain, but it contains seven or eight copper atoms

[17] S. Lindskog, *Adv. Bioinorg. Chem.*, **1982**, *4*, 116.

[18] P. J. Stein, S. P. Merrill, and R. W. Henkens, *J. Am. Chem. Soc.*, **1977**, *99*, 3194; *Biochemistry*, **1985**, *24*, 2459.

[19] S. H. Lawrie and E. S. Mohammed, *Coord. Chem. Rev.*, **1980**, *33*, 279.

per mole (MW ~132,000), including all three types of Cu(II). It is believed to be part of the process of oxidizing Fe(II) to Fe(III) in the transfer of iron from ferritin to transferrin.

Bovine superoxide dismutase[20] contains one atom of Cu(II) and one of Zn(II) in each of two subunits of about 16,000 molecular weight. The copper is in a normal tetragonal site, bound to four histidine nitrogens and a water; the zinc is bound to three histidines (including a bridging imidazole ring bound to both metal ions) and an aspartate carboxyl oxygen, as shown in Figure 15-13. Cu is the more essential metal, which cannot be replaced while retaining activity. On the other hand, the Zn can be replaced by other divalent metals with retention of most of the catalytic activity. The enzyme catalyzes the conversion of superoxide to peroxide and oxygen:

$$2\,O_2^- + 2\,H^+ \longrightarrow H_2O_2 + O_2$$

In spite of the uncertain nature of the structure and action of the copper enzymes, their roles in electron transport, oxygen transport, and oxidation reactions have guaranteed continued interest in their study. In addition to studies of the natural compounds, there have been many attempts to design model structures of them, particularly of the binuclear species. Many of these include both nitrogen and oxygen donors built into macrocyclic ligands, although sulfur has been used as well.[21]

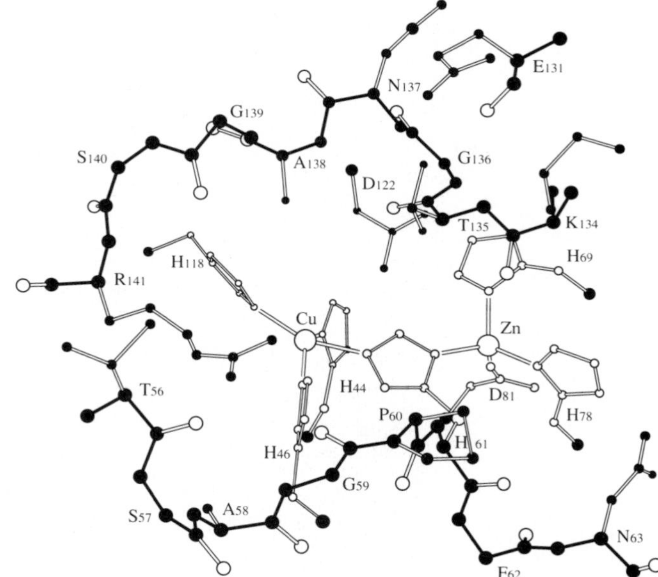

FIGURE 15-13 Active Site of Bovine Superoxide Dismutase. Drawing of the active site channel as viewed from the solvent. The main chain is shown in black, the ligand side chains as open circles and bonds, and the other side chains as solid atoms and open bonds. (Reprinted by permission from J. A. Tainer, E. D. Getzoff, J. S. Richardson, and D. C. Richardson, *Nature*, **1983,** *306*, 284. Copyright © 1983 Macmillan Magazines Limited.)

15-5 NITROGEN FIXATION

A very important sequence of reactions converts nitrogen from the atmosphere into ammonia, which can then be further converted into nitrate or nitrite or directly used in synthesis of amino acids and other essential compounds.

$$N_2 + 6\,H^+ + 6e^- \longrightarrow 2\,NH_3$$

[20] I. Fridovich, *Adv. Inorg. Biochem.*, **1979**, *1*, 67; J. S. Valentine and D. M. de Freitas, *J. Chem. Educ.*, **1985**, *62*, 990.

[21] K. D. Karlin and Y. Gultneh, *J. Chem. Educ.*, **1985**, *62*, 983.

This reaction takes place at 0.8 atm N_2 pressure and ambient temperatures in *Rhizobium* bacteria in nodules on the roots of legumes such as peas and beans, as well as in other independent bacteria. In contrast to these mild conditions, industrial synthesis of ammonia requires high temperatures and pressures with iron oxide catalysts and even then yields only 15% to 20% conversion of the nitrogen to ammonia. Intensive efforts to determine the bacterial mechanism and to improve the efficiency of the industrial process have so far been only moderately successful; the goal of approaching the enzymatic efficiency on an industrial scale is still only a goal.

The nitrogenase enzymes responsible for nitrogen fixation contain two proteins; one contains molybdenum, iron, and sulfur and the other contains iron and sulfur. Structures proposed for the iron–sulfur groups are similar to those of the ferredoxins, but the compounds are very difficult to work with and details of the structures are still uncertain. One method being used to study these compounds involves use of model compounds, in which essential features of the natural compounds are mimicked by simpler molecules that are easier to characterize and study. With nitrogenases, the great difficulty of reducing nitrogen ($\mathscr{E}^\circ = -3.40$ V) is a serious handicap; a moderately successful model may not be able to produce any ammonia. This problem can be somewhat relieved by using compounds such as acetylene, which is isoelectronic with nitrogen, for test purposes. Acetylene has a positive reduction potential ($\mathscr{E}^\circ = 0.73$ V), providing a much less difficult test reaction for model compounds. It is reduced easily by nitrogenases, as are HCN, azide, and other compounds with similar electronic structures:

$$C_2H_2 + 2\,H^+ + 2e^- \longrightarrow C_2H_4$$

The enzymes are difficult to study, and details of their structure are not yet available. The role of the iron and molybdenum centers is the subject of intense interest and activity, but information is still sketchy. These enzymes and other molybdenum enzymes, cofactors, and model systems are the subjects of a review.[22]

There is also considerable controversy over possible intermediates in the reduction of nitrogen; whether diimide (HN=NH, also called diazene) or hydrazine (H_2N—NH_2) is formed in the reactions is not known. If they are formed, the amounts are very small, since no direct evidence has been found for them. Suggestions have been made that two sites are involved in the reactions, one binding nitrogen and the other hydrogen, with the resulting combination providing a path for breaking the N—N bond. An alternative proposal has the nitrogen bound at both ends, weakening the N—N bond and facilitating transfer of hydrogen from other groups in the enzyme.[23]

Nitrification and denitrification

Oxidation of ammonia to nitrite, NO_2^-, and nitrate, NO_3^-, is called nitrification; the reverse reaction is ammonification. Reduction from nitrite to nitrogen is called denitrification. All these reactions, and more, occur in enzyme systems, many of which include transition metals. A molybdenum enzyme, nitrate reductase, reduces nitrate to nitrite. Further reduction to ammonia seems to

[22] S. J. N. Burgmayer and E. I. Stiefel, *J. Chem. Educ.*, **1985**, *62*, 943.

[23] M. E. Winfield, *Rev. Pure Appl. Chem.*, **1955**, *5*, 217.

go by 2-electron steps, through an uncertain intermediate with 1+ oxidation state (possibly hyponitrite, $N_2O_2^{2-}$) and hydroxylamine:

$$NO_2^- \longrightarrow N_2O_2^{2-} \longrightarrow NH_2OH \longrightarrow NH_3$$

Some nitrite reductases contain iron and copper; other enzymes active in these reactions contain manganese. Reactions giving NO, N_2O, and N_2 as products are also possible. Copper and iron enzymes are reported for these as well.

15-6 ENVIRONMENTAL CHEMISTRY

15-6-1 METALS

Mercury

Mercury and lead are two of the most prominent metallic environmental contaminants today. Although there have been continued efforts to prevent distribution of these metals and to clean up sources of contamination, they are still serious problems.

Because mercury has a significant vapor pressure, the pure metal can be as serious a problem as its compounds. Although the problem is usually less severe in present-day laboratories, mercury contamination and poisoning have been a problem in chemistry and physics laboratories for many years. When large amounts of the liquid are used in manometers, Toeppler pumps, and mercury diffusion pumps on vacuum lines, spills are inevitable. Since the liquid breaks into tiny drops, cleanup is extremely difficult and contamination remains even after strenuous efforts to remove it. As a result, a low level of mercury vapor is present in many laboratories and can result in toxic reactions. Mercury interferes with nerve action, causing both physical and psychological symptoms. Whether the Mad Hatter in Lewis Carroll's *Alice in Wonderland* had an origin in fact is uncertain, but mercury compounds were used in felt-making and some hatters were victims of mercury poisoning as a result.

Several industrial processes use mercury in large amounts, and the resulting potential for spills and loss to the environment is great. One of the largest is the chloralkali industry, in which mercury is used as an electrode for the electrolysis of brine to form chlorine gas and sodium hydroxide:

$$2\ H_2O + 2e^- \longrightarrow H_2 + 2\ OH^- \text{ and } 2\ Cl^- \longrightarrow Cl_2 + 2e^-$$

In one tragic incident, an entire community on Minamata Bay in Japan was poisoned, with extremely serious birth defects, very painful reactions, mental disorders, and many deaths. Only after lengthy research was the cause determined to be mercury compounds discarded into a river by a plastics factory. Whether it was inorganic salts or methylmercury seems uncertain, but the contamination was great and methylmercury compounds were found in the silt and in animals and humans. The methylmercury was readily taken up by the organisms living in the bay and, since the people of the community depended on fish and other seafood from the bay for much of their diet, the entire community was poisoned.

This incident showed the concentrating effect of the food chain and the need for extreme caution in predicting the outcome of dumping any material into the environment. The low concentration of methylmercury was readily taken up by the plants and microorganisms in the water. As these organisms were eaten by larger ones, each organism in the chain retained the mercury, and the concentration of mercury in these predators became larger, leading to damaging concentrations in the larger fish and other organisms eaten by the people.

During the research on mercury reactions, it was also discovered that insoluble metallic mercury can be converted to soluble methylmercury by bacterial action involving methylcobalamin. Earlier, it had been thought that elemental mercury was unreactive in lakes and rivers; now it is known to be dangerous. As a result, many more toxic metal sources are now known than had once been recognized. Historically, large amounts of metallic mercury have been discharged into the Great Lakes and other bodies of water in the belief that they were harmless; cleanup of these sources seems impossible, so the problem will remain with us forever.

While concern about mercury contamination is more visible and industries now use much safer practices, the increasing use of mercury in small batteries and other products results in a greater distribution of mercury into the environment as a whole. As a result, the problem is changing from one of a few large sources of contamination to many small ones, and the techniques for dealing with the problem must change as well. Concerns are being expressed about heavy metal contamination of the atmosphere by incinerators burning municipal garbage and trash, and it is likely that removal of these materials from the trash before burning or scrubbing of the flue gases to remove the volatile products will be needed.

Lead

Lead is another metal that is widespread in the environment, largely as a result of human activities. Two of the largest sources for environmental lead were paint pigments and leaded gasoline, both now much reduced in importance. White lead [basic lead carbonate, $2\ PbCO_3 \cdot Pb(OH)_2$] was used as a paint pigment for many years, and older buildings still have lead-containing paint, frequently under layers of more modern paint. If children living in these buildings eat paint chips, they are likely to pick up significant amounts of lead. In fact, in some cities lead poisoning of children is a very common problem.[24] As is the case with mercury, lead can affect nerve action and cause retardation and other mental problems, as well as cause acute illness. Unfortunately, the only solution is complete removal of the paint, a very time-consuming and expensive process.

Although commercial manufacturers of ceramics now avoid the use of heavy metal glazes, there are still reports of lead and other toxic heavy metals showing up in dishes imported from countries without similar controls or in ceramic items made by individuals who do not take the appropriate precautions. Since the glaze seems permanent and impervious to water and ordinary foods, it might seem that such materials would not be a hazard. However, acidic solutions can extract significant amounts of the heavy metals and result in chronic low-level poisoning.

Lead in gasoline is being phased out and should not be a continuing problem. Tetraethyl lead, $Pb(C_2H_5)_4$, has been used as an antiknock compound in gasoline for many years. When this compound is present, a low grade of gasoline burns as efficiently in automobile engines as does a higher grade

[24] M. W. Oberle, *Science,* **1969**, *165*, 991.

without the lead. Unfortunately, the lead in the gasoline has been distributed throughout the environment. Some studies have found increased lead levels in roadside plants and soil, and the population in general was exposed to higher levels of lead as a result of this use.

Laws requiring the use of nonleaded gasoline in newer cars have required other changes in the engines and in the refining of gasoline to compensate. The additional use of catalytic converters to reduce the amount of unburned hydrocarbons in exhaust emissions is an additional example of use of metals. (Reactions of these unburned hydrocarbons in the atmosphere are described later in the section on photochemical smog.) The catalyst used is platinum, in a very thin coating on ceramic spheres, which catalyzes the combustion of hydrocarbons in the exhaust gases to carbon dioxide and water. Platinum, palladium, and nickel are among the most reactive (and most used) catalytic materials. They are used in many different specific compounds and physical forms for reactions of surprising specificity in the petroleum and chemical industries. In another of the many interactions between problems and their solutions, catalysts in catalytic converters are poisoned by lead; for this reason, cars with catalytic converters are required to use only nonleaded gasoline.

Arsenic

Recent efforts to remove toxic materials from industrial sites, homes, and farms have unearthed other problems. For example, during the 1930s, farmers fought grasshopper infestations with bran poisoned with arsenic compounds. Fifty or more years later, burlap bags of arsenic-laced bran have been found in barns and storage sheds, where they are potentially serious hazards. Several states have begun programs to locate and remove these poisons for safe disposal, but since there is no way to detoxify this material, it will remain toxic forever. The only possible ways to remove the problem are to seal the material in a toxic waste dump and take every possible means to prevent leaching or other means of spreading the material or to find some other use for the heavy metal compounds that has a high enough value to make reprocessing profitable. So far, such uses have been very rare.

Other heavy metals are also toxic, but fortunately are less widely spread and are present in smaller amounts. Mine tailings (waste rock remaining after the valuable minerals have been removed) and waste material from processing plants are major sources of such metals. Many major rivers and lakes have sources of metal contamination from industries whose processes were developed and built before control of waste was recognized as a major problem.

Radioactive waste

Disposal of radioactive waste is a continuing controversial topic. Some argue that the technical problems have been solved and that only politics remain in the way of efficient permanent storage of such wastes. Others maintain that the technical problems are far from solution and, in addition, that the long half-lives of some of the isotopes will require protection of the disposal sites for hundreds or even thousands of years. At this time, it is impossible to predict the outcome, beyond noting that no location is perfect, either geologically or politically. Recent reports of contamination of water and land around processing sites have led to even more suspicion of any reported solution and have made the choices even more difficult.

As in the case of the heavy metals described earlier, the problem is the long lifetime of the atoms. Even though they are undergoing radioactive decay, the process is one that will leave some radioactive materials for thousands of years, and the radiation will be dangerous for that length of time. A related problem is the wide variety of elements in much radioactive waste. Spent fuel rods from nuclear reactors contain ^{238}U in large amounts, ^{235}U in small amounts (largely depleted by the chain reaction), fission and other decay products of a bewildering variety, and the metal cladding material, which has become radioactive because of the intense neutron flux of the reactor. Structural materials from decommissioned reactors and byproducts from ore-processing and isotope-enrichment plants are other examples of relatively high-level wastes. Low-level wastes from laboratories and hospitals provide different technical difficulties because of the relatively large volume and low radioactive level. For some purposes, concentration of such wastes would be desirable, but loss to the atmosphere during processing is an additional problem. As a relatively small, but very important, part of the overall problem of waste disposal, disposal of radioactive waste will be the subject of many fiercely fought battles for many years.

15-6-2 NONMETALS

Sulfur

Mine tailings are a source of both metal and nonmetal contamination. A common material in coal mines is iron pyrites, FeS_2. As a contaminant of coal, this compound and similar compounds contribute to the production of sulfur oxides in flue gases when coal is burned. As a material in mine tailings, it contributes both iron and sulfur to water pollution when the sulfide is oxidized in a series of reactions to sulfate and the Fe(II) to Fe(III):

$$4\ FeS_2 + 15\ O_2 + 6\ H_2O \longrightarrow 4\ Fe(OH)^{2+} + 8\ HSO_4^-$$

Since Fe(III) is a strongly acidic cation, the net result is a dilute solution of sulfuric acid containing Fe(II), Fe(III), and other heavy metal ions dissolved in the acidic solution (pH's of 2 to 3.5 have been measured). In areas with played-out mines, such solutions are common in the streams and rivers, effectively killing most plant and animal life in the water.

When coal containing sulfur compounds is burned, the resulting sulfur dioxide and sulfur trioxide can result in atmospheric contamination. There is much controversy worldwide over such contamination, since it travels across political and natural boundaries and the generators of the contamination are only rarely those who suffer the direct consequences from it. The sulfur oxides and nitrogen oxides from high-temperature combustion are readily dissolved in water droplets in the atmosphere and returned to the earth as acid rain. Although the evidence is still being debated, there seems little doubt that such acid rain has damaged forests and lakes worldwide, as well as attacked building materials and artistic works. Studies of the damage to limestone statues and building materials show an accelerating rate of destruction, with many carvings becoming completely unrecognizable over a relatively short time.

Although the amount of sulfur released by smelting is only about 10% of the total released into the atmosphere, even more dramatic effects of sulfur oxides can be seen around smelting industries, where nickel or copper are mined and purified. The major ores of these metals are sulfides, and the method of extracting the metal begins with roasting the ore in air to convert it to the oxide:

$$NiS + \frac{3}{2}\,O_2 \longrightarrow NiO + SO_2 \quad \text{and} \quad CuS + \frac{3}{2}\,O_2 \longrightarrow CuO + SO_2$$

The fraction of sulfur dioxide due to smelting operations is larger in Canada than in the United States, because more of Canada's power generation is hydroelectric and the total amount of power generated is smaller. Two sites that have been studied thoroughly are in Trail, British Columbia, and Sudbury, Ontario. When the area around Trail was studied in 1929–1936, after 30 to 40 years of smelter operation, no conifers were found within 12 miles, and damage to vegetation could be seen as far as 39 miles from the source.[25] Similar effects could also be seen around Sudbury, with evidence of acidified lakes up to 40 miles away. Current efforts to control the emission of SO_2 and SO_3 have reduced the contamination, but recovery of the environment is a very slow process.

One advantage of recovery of sulfur oxides from smelting is that the amounts are large enough to be economically useful; in most cases, the concentrations of sulfur dioxide and sulfur trioxide found in power plant flue gases are so small that it is simply an added expense to remove them. Two techniques are used, removal of the sulfur compounds from the coal before burning and scrubbing of the stack gases to remove the oxides. Since FeS_2 is much more dense than coal, much of it can be removed by reducing the coal to a powder and separating the two by gravitational techniques. Leaching with sodium hydroxide also removes much of the sulfide contaminant, but scrubbing of the stack gases with a substance such as an aqueous slurry of $CaCO_3$ is still required for complete removal. The resulting $CaSO_4$ must also be disposed of or used in some way. Other techniques require gasification of the coal (partial combustion in steam to CO and H_2) and scrubbing of the gas to remove the resulting H_2S, combustion of a fluidized bed of finely pulverized coal and limestone, or complete conversion of SO_2 to SO_3 on a V_2O_5 catalyst and removal of SO_3 and H_2SO_4.

Nitrogen oxides and photochemical smog

Nitrogen oxides are also major contaminants, primarily from automobiles. The combustion process in automotive engines takes place at a high enough temperature that NO and NO_2 are formed. In the air, NO is rapidly converted to NO_2, and both can react with hydrocarbons that are also released by cars. The resulting compounds are among the primary causes of smog seen in urban areas, particularly those where geography prevents easy mixing of the atmosphere and removal of contaminants. Cities such as Los Angeles and Denver are frequently in the news because of smog, and many others also have serious problems. The nitrogen oxides can also form nitric acid, which can contribute to acid rain:

$$3\,NO_2 + H_2O \longrightarrow 2\,HNO_3 + NO$$

Photochemical smog can form whenever air heavily laden with exhaust gases is trapped by atmospheric and topographic conditions and exposed to

[25] C. G. Down and J. Stocks, *Environmental Impact of Mining,* Wiley, New York, 1977, p. 63.

sunlight. Ozone and formaldehyde formed in the atmosphere from nitrogen oxides and hydrocarbons are also major contributors to the smog. The following are some of the major reactions in the sequence.[26]

Reactions during combustion of gasoline include

$$N_2 + O_2 \longrightarrow 2\,NO$$

$$C_nH_m + O_2 \longrightarrow CO_2 + CO + H_2O$$

Traces of ozone can be photolyzed, with hydroxyl radical the most important product:

$$O_3 + h\nu \longrightarrow O + O_2$$

$$O + H_2O \longrightarrow 2\cdot OH$$

Another important species, the hydroperoxyl radical, is formed by photolysis of formaldehyde:

$$HCHO + h\nu \longrightarrow H + HCO\cdot$$

$$H + O_2 + M \longrightarrow HO_2\cdot + M$$

(M is an unreactive molecule that removes kinetic energy from the products during this exothermic reaction.)

$$HCO\cdot + O_2 \longrightarrow HO_2\cdot + CO$$

Oxidation of NO at high concentration yields NO_2

$$2\,NO + O_2 \longrightarrow 2\,NO_2$$

and oxidation of NO by $HO_2\cdot$ at low NO concentrations, which is more common, also yields NO_2

$$NO + HO_2\cdot \longrightarrow NO_2 + \cdot OH$$

Photolysis of NO_2 gives

$$NO_2 + h\nu \longrightarrow NO + O$$

(This requires light with $\lambda < 395$ nm, at the ultraviolet edge of the visible region.)

Finally, production of ozone occurs:

$$O + O_2 + M \longrightarrow O_3 + M$$

Oxygen atoms and ozone react with NO and NO_2, forming NO_2, NO_3, and N_2O_5, which then react with water to form HNO_2 and HNO_3. These oxygen and nitrogen species also react with hydrocarbons to form aldehydes, oxygen-containing free radical species, and alkyl nitrites and nitrates, all of which are very reactive and contribute to eye and lung irritation and the

[26] B. J. Finlayson-Pitts and J. N. Pitts, Jr., *Atmospheric Chemistry: Fundamentals and Experimental Techniques,* Wiley, New York, 1986, pp. 29–37.

damaging effects on vegetation, rubber, and plastics. One of the most reactive is peroxyacetyl nitrate, formed by reaction of aldehydes with hydroxyl radical and NO_2:

$$CH_3\overset{\overset{\displaystyle O}{\|}}{C}H + \cdot OH \longrightarrow CH_3\overset{\overset{\displaystyle O}{\|}}{C}\cdot + H_2O$$

$$CH_3\overset{\overset{\displaystyle O}{\|}}{C}\cdot + O_2 \longrightarrow CH_3\overset{\overset{\displaystyle O}{\|}}{C}OO\cdot$$

$$CH_3\overset{\overset{\displaystyle O}{\|}}{C}OO\cdot + NO_2 \longrightarrow CH_3\overset{\overset{\displaystyle O}{\|}}{C}OONO_2$$

Photochemical reactions of the aldehydes and alkyl nitrites generate more radicals and continue the chain of reactions.

The ozone layer

Although it is an injurious pollutant in the lower atmosphere, ozone is an essential protective agent in the stratosphere. It is formed by photochemical dissociation of oxygen,

$$O_2 + h\nu \longrightarrow O + O^* \qquad \text{(requires light of } \lambda < 242 \text{ nm in the far UV)}$$

The activated oxygen atoms, O*, react with molecular oxygen to form ozone:

$$O^* + O_2 + M \longrightarrow O_3 + M$$

The ozone formed in this way absorbs ultraviolet radiation with $\lambda < 340$ nm, regenerating molecular oxygen:

$$O_3 + h\nu \longrightarrow O_2 + O \quad \text{followed by} \quad O + O_3 \longrightarrow 2\,O_2$$

This mechanism filters out much of the sun's ultraviolet radiation, protecting plant and animal life on the surface of the earth from other damaging photochemical reactions. Recent experiments have shown that this natural equilibrium is being affected by compounds added to the atmosphere by humans. The most publicized of these compounds are the chlorofluorocarbons, especially CF_2Cl_2 and CCl_3F (CFC 12 and 11, respectively, named for the number of carbons and fluorines in the molecule). These compounds are widely used as refrigerants, blowing agents for the manufacture of plastic foams, and, until recently, as propellants in aerosol cans. Since their damaging effects have been demonstrated conclusively, substitutes for chlorofluorocarbons are being sought, and nonessential uses (as in aerosols) are restricted.

The destruction of ozone by these compounds is caused, paradoxically, by their extreme stability and lack of reaction under ordinary conditions. Because they are so stable, they remain in the atmosphere indefinitely and finally diffuse to the stratosphere. The intense high-energy ultraviolet radiation in the stratosphere causes dissociation and forms chlorine atoms, which then undergo a series of reactions that destroy ozone:

$$CCl_2F_2 + h\nu \longrightarrow Cl\cdot + \cdot CClF_2 \qquad \text{(requires light with } \lambda \approx 200 \text{ nm)}$$

$$Cl\cdot + O_3 \longrightarrow ClO + O_2$$

$$ClO + O \longrightarrow Cl\cdot + O_2$$

(these two reactions remove O_3 and oxygen atoms without reducing the number of chlorine atoms, Cl·)

Other compounds, such as NO, also contribute to the chain of reactions:

$$NO + ClO \longrightarrow \cdot Cl + NO_2$$

The chains are terminated by reactions such as

$$\cdot Cl + CH_4 \longrightarrow HCl + \cdot CH_3 \quad \text{and} \quad \cdot Cl + H_2 \longrightarrow HCl + H\cdot$$

followed by combination of the new radicals to form stable molecules such as CH_4, H_2, and C_2H_6.

Whether reduction in use of these chlorofluorocarbons will be sufficient to prevent serious worldwide results caused by destruction of the ozone layer remains to be seen. Predictions based on the materials already in the atmosphere indicate that the damage will be significant even if production could be stopped immediately, but such predictions are based on computer models that are untested and subject to considerable error. Governments and industries are now acting to find substitutes, but the compounds proposed (primarily compounds containing C, H, Cl, and F with lower stability) are expensive and replacement will require years.[27] Methods for recycling CFCs from air conditioners and refrigeration units are also being developed.

The greenhouse effect

A related problem is the greenhouse effect. The major cause of the problem in this case is the carbon dioxide released by combustion and decomposition of organic matter. Other gases, including methane and CFCs, also contribute. In this effect, visible and ultraviolet radiation from the sun that is not absorbed in the stratosphere and upper atmosphere reaches the surface of the earth and is absorbed and converted to heat. This heat, in the form of infrared radiation, is transmitted out from the earth through the atmosphere. Molecules such as CO_2 and CH_4, which have low-energy vibrational energy levels, absorb this radiation and reradiate the energy, much of it toward the earth. As a result, the energy cannot escape from the earth, and the surface of the earth and the atmosphere are warmed. Whether this effect is already showing up or not is a source of controversy (largely due to inadequate computer models and lack of sufficient data for good projections), but there is general agreement that it will happen. Only the timing and the amount of warming are uncertain.

If there is a significant warming, even as much as 3° to 4°C increase in the average temperature over large portions of the earth, the consequences are expected to be extreme. Rainfall patterns will change drastically, the oceans will rise with significant melting of the polar ice caps, and every part of the earth will be affected. Efforts are being made to reduce the production of CO_2 and the release of hydrocarbons into the atmosphere, but the sources are so diffuse that it is difficult to have much effect.

Major sources of methane in the atmosphere are rice paddies, swamps, and animals. Methane is produced as a result of decay of vegetation under water in the paddies and swamps and as a result of the digestive processes in animals. Increasing population, coupled with increasing agriculture and more grazing animals, increases the amount of methane released.

[27] L. E. Manaer, *Science*, **1990**, *249*, 31.

GENERAL REFERENCES

Three short books with different emphases are R. W. Hay, *Bioinorganic Chemistry,* Halstead Press, New York, 1987, M. N. Hughes, *The Inorganic Chemistry of Biological Processes,* Wiley, New York, 1972, and D. R. Williams, *The Metals of Life,* Van Nostrand Reinhold, London, 1971. Hay is more recent and reviews many of the topics presented in this chapter, Williams concentrates on the solution chemistry of the metal ions and includes much more material on experimental methods, ligands, and other related topics. Hughes covers topics that parallel those of the first part of this chapter, with much more detail about the evidence for the reactions proposed.

J. E. Fergusson, *Inorganic Chemistry and the Earth,* Pergamon, Elmsford, N.Y., 1982, includes several chapters on environmental chemistry, and J. O'M. Bockris, ed., *Environmental Chemistry,* Plenum Press, New York, 1977, offers the viewpoints of many different authors. R. A. Bailey et al., *Chemistry of the Environment,* Academic Press, New York, 1978, covers a very broad range of environmental topics at an easily accessible level. B. J. Finlayson-Pitts and J. N. Pitts, Jr., *Atmospheric Chemistry: Fundamentals and Experimental Techniques,* Wiley, New York, 1986, offers very complete coverage of both the laboratory and field studies of all kinds of chemicals and their reactions. A more specific report on the greenhouse effect is *Changing Climate,* a National Academy of Sciences report by the Carbon Dioxide Assessment Committee, National Academy Press, Washington, D.C., 1983. Finally, two of the standard references used throughout this book must be mentioned again. F. A. Cotton and G. Wilkinson, *Advanced Inorganic Chemistry,* 5th ed., Wiley-Interscience, New York, 1988, includes a good review of bioinorganic chemistry, as does G. Wilkinson, R. D. Gillard, and J. A. McCleverty, eds., *Comprehensive Coordination Chemistry,* Pergamon, Elmsford, N.Y., 1987, in Volume 6, Applications.

PROBLEMS

15-1 Describe possible mechanisms for the B_{12}-catalyzed reactions given in the text.

15-2 The following curves show the degree of saturation of hemoglobin and myoglobin as a function of the pressure of oxygen. Explain how these two compounds are each best suited to their specific roles, with hemoglobin transferring oxygen from the lungs to the blood and myoglobin transferring oxygen from the blood to the other tissues. (Graph reprinted by permission of the publisher from "Hemoglobin and Myoglobin" by J. M. Rifkind in

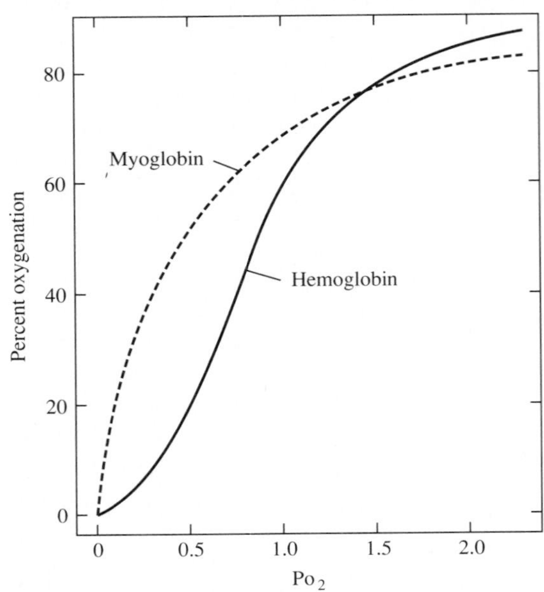

G. L. Eichhorn, ed., *Inorganic Biochemistry*, Vol. 2, Elsevier, New York, 1973, p. 853. Copyright 1973 by Elsevier Science Publishing Co., Inc.)

15-3 Acetohydroxamic acid (AcHA) is $CH_3\overset{\overset{\displaystyle O}{\|}}{C}NHOH$. Sketch the structure expected for a complex of the conjugate base of AcHA and Fe^{3+}, including the possible resonance structures.

15-4 Oxygen can bind to metal ions in several ways. Hemoglobin binds at one end of the oxygen molecule, with a Fe—O—O bond angle of 115° to 156° (solution and crystal). In other cases, oxygen bonds more like ethylene, with the O—O axis perpendicular to the bonding direction. Review the arguments for each of these, and the evidence for the structure of HbO_2 described in this chapter. [$Fe(III)-O_2^-$, J. J. Weiss, *Nature*, **1964**, *202*, 83; mix of $Fe(III)-O_2^-$ and $Fe(II)-O_2$, T. E. Tsai, J. L. Groves, and C. S. Wu, *J. Chem. Phys.*, **1981**, *74*, 4306; general review, L. Vaska, *Acc. Chem. Res.*, **1976**, *9*, 175.]

15-5 Dissociation of NO_2 to NO and O requires light with a wavelength less than 395 nm. Calculate the dissociation energy of NO_2, assuming all the energy is concentrated in this reaction. Dissociation of O_2 requires $\lambda < 242$ nm. Calculate the dissociation energy of O_2 with the same assumption. Do the results of these two calculations match the bonding of these molecules described in Chapters 3 and 5?

15-6 Methyl cobalamin is usually described as a Co(I) compound, which changes to Co(II) on dissociation of CH_3^-. Describe the probable electronic structure (splitting of *d* levels and number of unpaired electrons) of the cobalt in both cases.

15-7 Lead can accumulate in the bones and other body tissues unless removed soon after ingestion. In some cases, treatment with chelating agents such as EDTA has been used to remove lead, mercury, or other heavy metals from the body. Discuss the advantages and disadvantages of such treatment. Include both thermodynamic and kinetic arguments in your answer.

A

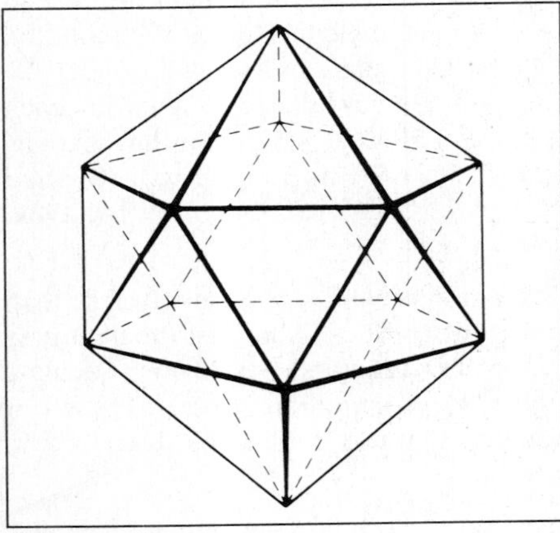

Answers to Exercises

CHAPTER 2

2-1 The nodal surface requires $2z^2 - x^2 - y^2 = 0$, so the angular nodal surface for a d_{z^2} orbital is the conical surface where $2z^2 = x^2 + y^2$.

2-2 The angular nodal surfaces for a d_{xz} orbital are the planes where $xz = 0$, which means that either x or z must be zero. The yz and xy planes satisfy this requirement.

2-3 If the $3p$ electrons all have the same spin, there are three possible exchanges (1 and 2, 1 and 3, and 2 and 3). There are no paired electrons, so the total energy is $3\Pi_e$. If there is one pair with opposite spin, there is only one possible exchange (one of the pair and the third electron), and there is Coulomb repulsion from the pair. Total energy is $\Pi_c + \Pi_e$.

2-4 Tin has $Z = 50$, and electron configuration $(1s^2)$ $(2s^2\ 2p^6)$ $(3s^2\ 3p^6)$ $(3d^{10})$ $(4s^2\ 4p^6)$ $(4d^{10})$ $(5s^2\ 5p^2)$.
For the $5p$ electron, $Z^* = Z - S = 50 - 2 - 8 - 8 - 10 - 8 \times 0.85 - 10 \times 0.85 - 3 \times 0.35 = 5.65$.
For the $5s$ electron, $Z^* = 5.65$ again, since $5s$ and $5p$ are in the same grouping.
For the $4d$ electron, $Z^* = Z - S = 50 - 2 - 8 - 8 - 10 - 8 - 9 \times 0.35 = 10.85$

2-5 Microstate table for d^2

M_L \ M_S	-1	0	$+1$
$+4$		$2^+\ 2^-$	
$+3$	$2^-\ 1^-$	$2^+\ 1^-$ $2^-\ 1^+$	$2^+\ 1^+$
$+2$	$2^-\ 0^-$	$2^+\ 0^-$ $2^-\ 0^+$ $1^+\ 1^-$	$2^+\ 0^+$
$+1$	$2^-\ -1^-$ $1^-\ 0^-$	$2^+\ -1^-$ $2^-\ -1^+$ $1^+\ 0^-$ $1^-\ 0^+$	$2^+\ -1^+$ $1^+\ 0^+$
0	$-2^-\ 2^-$ $-1^-\ 1^-$	$-2^+\ 2^-$ $-1^+\ 1^-$ $0^+\ 0^-$ $-1^-\ 1^+$ $-2^-\ 2^+$	$-2^+\ 2^+$ $-1^+\ 1^+$
-1	$-1^-\ 0^-$ $-2^-\ 1^-$	$-1^+\ 0^-$ $-1^-\ 0^+$ $-2^-\ 1^+$ $-2^+\ 1^-$	$-1^+\ 0^+$ $-2^+\ 1^+$
-2	$-2^-\ 0^-$	$-1^+\ -1^-$ $-2^+\ 0^-$ $-2^-\ 0^+$	$-2^+\ 0^+$
-3	$-2^-\ -1^-$	$-2^+\ -1^-$ $-2^-\ -1^+$	$-2^+\ -1^+$
-4		$-2^+\ -2^-$	

2-6

^2D $L = 2, S = \frac{1}{2}$
$M_L = -2, -1, 0, 1, 2$
$M_S = -\frac{1}{2}, \frac{1}{2}$

M_L \ M_S	$-\frac{1}{2}$	$+\frac{1}{2}$
$+2$	x	x
$+1$	x	x
0	x	x
-1	x	x
-2	x	x

^1P $L = 1, S = 0$
$M_L = -1, 0, 1$
$M_S = 0$

M_L \ M_S	0
$+1$	x
0	x
-1	x

^2S $L = 0, S = \frac{1}{2}$
$M_L = 0$
$M_S = -\frac{1}{2}, \frac{1}{2}$

M_L \ M_S	$-\frac{1}{2}$	$+\frac{1}{2}$
0	x	x

2-7

$L = 4,\ S = 0,\ J = 4$	^1G
$L = 3,\ S = 1,\ J = 4, 3, 2$	^3F
$L = 2,\ S = 0,\ J = 2$	^1D
$L = 1,\ S = 1,\ J = 2, 1, 0$	^3P
$L = 0,\ S = 0,\ J = 0$	^1S

Following Hund's rules:

1. The highest spin (S) is 1, so the ground state is ^3F or ^3P.
2. The highest L in step 1 is $L = 3$, so ^3F is the ground state.

2-8 The J values for each term are shown in the solution to Exercise 2-7. The full set of term symbols for a d^2 configuration is $^1G_4$, $^3F_4$, $^3F_3$, $^3F_2$, $^1D_2$, $^3P_2$, $^3P_1$, $^3P_0$, $^1S_0$. Hund's third rule (see following text in chapter) predicts which J value corresponds to the lowest energy state. This configuration has the d orbitals less than half-filled, so the minimum J value for ^3F, $J = 2$, is the ground state. Overall, it is $^3F_2$.

2-9 For $CaCl_2$, $\frac{r_+}{r_-} = \frac{114}{167} = 0.683$ For $CaBr_2$, $\frac{r_+}{r_-} = \frac{114}{182} = 0.626$

Both are predicted to have CN = 6, which fits the rutile structure found for both (see Figure 2-16).

CHAPTER 3

3-1 POF_3: The octet rule results in single P—F and P—O bonds; formal charge arguments result in a double bond for P═O. The actual distance is 143 pm, considerably shorter than the regular P—O bond (164 pm).

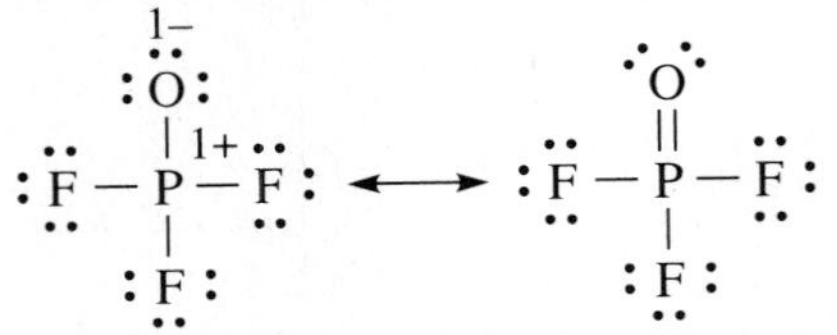

SOF_4: This is a distorted trigonal bipyramidal structure, with an S═O double bond and S—F single bonds required by formal charge arguments. The short 140 pm S═O bond length is in agreement.

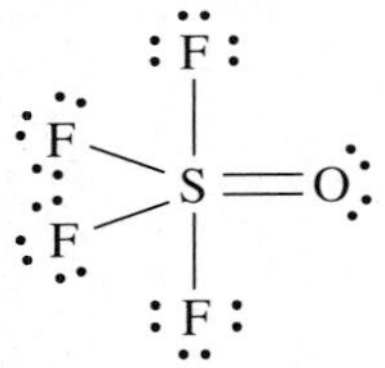

SO_3F^-: Basically tetrahedral, with two double bonds to oxygens and single bonds to fluorine and the third oxygen. The S—O bond order is then 1.67, and the bond length is 143 pm, shorter than the 149 pm of SO_4^{2-}, which has a bond order of 1.5.

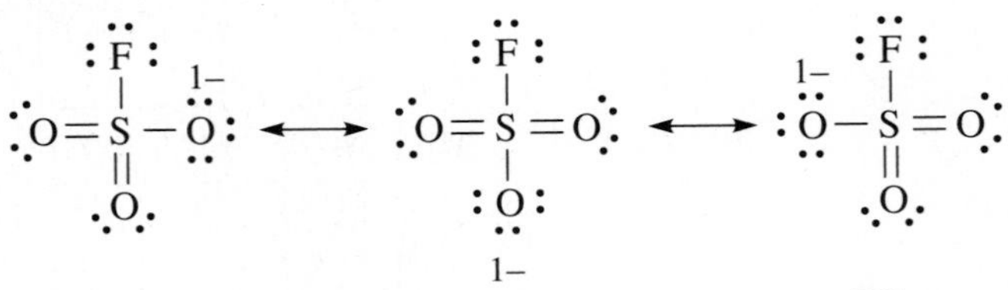

3-2

NH_2^-	NH_4^+	I_3^-	PCl_6^-
H–N–H < 109°	tetrahedral	linear	octahedral

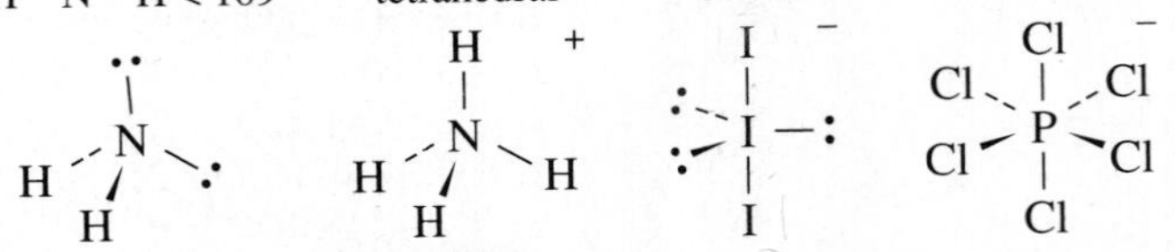

3-3 Total positions of bonds and lone pairs:

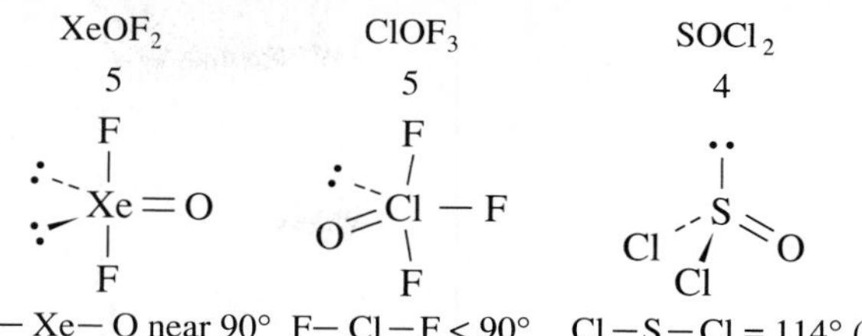

Angles: F—Xe—O near 90° F—Cl—F < 90° Cl—S—Cl = 114° (?)
F—Cl—O > 90° Cl—S—O = 106°

The Cl—S—Cl angle is probably about 94°. SOF_2 has an F—S—F angle of 92°, $SOBr_2$ has a Br—S—Br angle of 96°. See A. F. Wells, *Structural Inorganic Chemistry*, 5th ed., Oxford University Press, New York, 1984, p. 721.

3-4 Using electronegativity values from Appendix B-4, the H—O bond energy is

$$\frac{D(\text{H—H}) + D(\text{O—O})}{2} + \frac{\Delta\chi^2}{(0.102)^2} = \frac{436 + 213}{2} + \frac{(1.24)^2}{0.0104} = 472 \text{ kJ/mol}$$

CHAPTER 4

4-1 Show, using diagrams as necessary, that $S_2 = i$ and $S_1 = \sigma$.

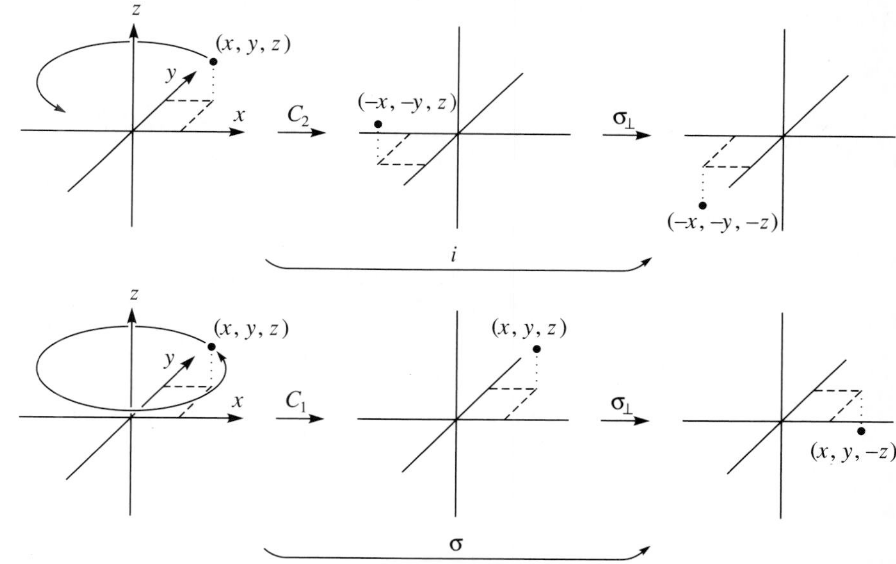

4-2 NH_3 has a three-fold axis through the N, perpendicular to the plane of the three hydrogens, and three mirror planes, each including the N and one H. C_3, 3σ

Cyclohexane (boat conformation) has a C_2 axis perpendicular to the plane of the lower four carbons and two mirror planes which include this axis and are perpendicular to each other. C_2, 2σ

Cyclohexane (chair conformation) has a C_3 axis perpendicular to the average plane of the ring, three perpendicular C_2 axes passing between carbons, and three mirror planes passing through opposite carbons and perpendicular to the average plane of the ring. It also contains a center of inversion and an S_6 axis collinear with the C_3 axis. (A model is very useful for this molecule.) C_3, $3C_2$, 3σ, i, S_6

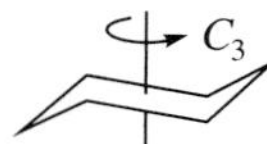

4-3 N_2F_2 has a mirror plane through all the atoms, which is the σ_h plane, perpendicular to the C_2 axis through the N=N bond. No other symmetry elements, so it is C_{2h}.

$B(OH)_3$ also has a σ_h mirror plane, the plane of the paper, perpendicular to the C_3 axis through the B. Again, no others, so it is C_{3h}.

H_2O has a C_2 axis in the plane of the drawing, through the O and between the two H's. It also has two mirror planes, one in the plane of the drawing and the other perpendicular to it. Overall, C_{2v}.

PCl_3 has a C_3 axis through the P and equidistant from the three Cl's. Like NH_3, it also has three σ_v planes, each through the P and one of the Cl's. Overall, C_{3v}.

BrF_5 has a C_4 axis through the Br and the F in the plane of the drawing, two σ_v planes, each through the Br, the F in the plane of the drawing, and two of the other F's, and two σ_v planes between the equatorial F's. Overall, C_{4v}.

HF, CO, and HCN all are linear, with the infinite rotation axis through the center of all the atoms. There are also an infinite number of σ_v planes, all of which contain the C_∞ axis. Overall, $C_{\infty v}$.

N_2H_4 has a C_2 axis perpendicular to the N—N bond and splitting the angle between the two lone pairs. No other elements, so it is C_2.

$P(C_6H_5)_3$ has only a C_3 axis, much like that in NH_3 or $B(OH)_3$. The twist of the phenyl rings prevents any other symmetry. C_3.

BF_3 has a C_3 axis perpendicular to the σ_h plane of the paper, and three C_2 axes, each through B and an F. Overall, D_{3h}. In addition, it has three mirror planes that include the C_3 axis, the B, and one H each, and an S_3 axis coincident with the C_3.

$PtCl_4^{2-}$ has a C_4 axis perpendicular to the σ_h plane of the paper. It also has four C_2 axes in the plane of the molecule, two through opposite Cl's and two splitting the Cl—Pt—Cl angles, making it D_{4h}. Additional symmetry elements are a C_2 coincident with the C_4, four mirror planes perpendicular to the plane of the molecule (two through opposite Cl—Pt—Cl atoms and two through Pt and splitting the angle between Cl's), an inversion center, and an S_4 axis coincident with the C_4.

$Os(C_5H_5)_2$ has a C_5 axis through the center of the two cyclopentadiene rings and the Os, five C_2 axes parallel to the rings and through the Os, and a σ_h plane parallel to the rings through the Os atom, for a D_{5h} assignment. It also has five σ_v planes, each including the Os atom, one of the C_2 axes, and the C_5 axis, in addition to an S_5 axis coincident with the C_5.

Benzene has a C_6 axis perpendicular to the σ_h plane of the ring, and six C_2 axes in the plane of the ring, three through two C atoms each and three between the atoms. These are sufficient to make it D_{6h}. In addition, it has C_2 and C_3 axes coincident with the C_6, three σ_d planes between the atoms, and three σ_v planes through opposite carbons, all perpendicular to the plane of the ring, along with an S_3 and an S_6 coincident with the C_6.

F_2, N_2, and H—C≡C—H are all linear, with a C_∞ axis through the atoms. There are also an infinite number of C_2 axes perpendicular to the C_∞ axis, and a σ_h plane perpendicular to the C_∞ axis, sufficient to make them $D_{\infty h}$. They also have an infinite number of σ_v planes, which include all the atoms, and an S_∞ axis coincident with the C_∞.

Allene, $H_2C{=}C{=}CH_2$, has a C_2 axis through the three C's, two C_2 axes perpendicular to the line of the C's, both at 45° angles to the planes of the H's. Two σ_d mirror planes through each H—C—H combination complete the assignment of D_{2d}. An additional S_4 axis is coincident with the C_2 through the three C's.

$Ni(C_4H_4)_2$ has a C_4 axis through the centers of the C_4H_4 rings and the Ni, four C_2 axes perpendicular to the C_4 through the Ni, and four σ_d planes, each including two opposite C's of the same ring and the Ni. Overall, D_{4d}. Additional S_8 and C_2 axes are coincident with the C_4.

$Fe(Cp)_2$ has a C_5 axis through the centers of the rings and the Fe, five C_2 axes perpendicular to the C_5 and through the Fe, and five σ_d planes including the C_5 axis. Overall, D_{5d}. An additional S_{10} coincident with the C_5 axis and an inversion center complete the symmetry.

$Ru(en)_3^{2+}$ has a C_3 axis perpendicular to the drawing through the Ru, and three C_2 axes in the plane of the paper, each intersecting an en ring at the midpoint and passing through the Ru. Overall, D_3.

4-4 **a.**
$$\begin{bmatrix}5 & 1 & 3\\4 & 2 & 2\\1 & 2 & 3\end{bmatrix} \times \begin{bmatrix}2 & 1 & 1\\1 & 2 & 3\\5 & 4 & 3\end{bmatrix}$$
$$= \begin{bmatrix}5\times2+1\times1+3\times5 & 5\times1+1\times2+3\times4 & 5\times1+1\times3+3\times3\\4\times2+2\times1+2\times5 & 4\times1+2\times2+2\times4 & 4\times1+2\times3+2\times3\\1\times2+2\times1+3\times5 & 1\times1+2\times2+3\times4 & 1\times1+2\times3+3\times3\end{bmatrix}$$
$$= \begin{bmatrix}26 & 19 & 17\\20 & 16 & 16\\19 & 17 & 16\end{bmatrix}$$

b.
$$\begin{bmatrix}1 & -1 & -2\\0 & 1 & -1\\1 & 0 & 0\end{bmatrix} \times \begin{bmatrix}2\\1\\3\end{bmatrix} = \begin{bmatrix}1\times2-1\times1-2\times3\\0\times2+1\times1-1\times3\\1\times2+0\times1+0\times3\end{bmatrix} = \begin{bmatrix}-5\\-2\\2\end{bmatrix}$$

c.
$$[1 \quad 2 \quad 3] \times \begin{bmatrix}1 & -1 & -2\\2 & 1 & -1\\3 & 2 & 1\end{bmatrix}$$
$$= [1\times1+2\times2+3\times3 \quad 1\times(-1)+2\times1+3\times2 \quad 1\times(-2)+2\times(-1)+3\times1]$$
$$= [14 \quad 7 \quad -1]$$

4-5 Chiral molecules may have only proper rotations. The C_1, C_n, and D_n groups, plus the rare T, O, and I groups, meet this condition.

4-6 $\Gamma_1 = A_1 + T_2$:

T_d	E	$8C_3$	$3C_2$	$6S_4$	$6\sigma_d$
Γ_1	4	1	0	0	2
A_1	1	1	1	1	1
A_2	1	1	1	−1	−1
E	2	−1	2	0	0
T_1	3	0	−1	1	−1
T_2	3	0	−1	−1	1

For A_1: $\frac{1}{24}[4 \times 1 + 8(1 \times 1) + 3(0 \times 1) + 6(0 \times 1) + 6(2 \times 1)] = 1$
For A_2: $\frac{1}{24}[4 \times 1 + 8(1 \times 1) + 3(0 \times 1) + 6(0 \times (-1)) + 6(2 \times (-1))] = 0$
For E: $\frac{1}{24}[4 \times 2 + 8(1 \times (-1)) + 3(0 \times 2) + 6(0 \times 0) + 6(2 \times 0)] = 0$
For T_1: $\frac{1}{24}[4 \times 3 + 8(1 \times 0) + 3(0 \times (-1)) + 6(0 \times 1) + 6(2 \times (-1))] = 0$
For T_2: $\frac{1}{24}[4 \times 3 + 8(1 \times 0) + 3(0 \times (-1)) + 6(0 \times (-1)) + 6(2 \times 1)] = 1$

$\Gamma_2 = A_1 + B_1 + E$:

D_{2d}	E	$2S_4$	C_2	$2C_2'$	$2\sigma_d$
Γ_2	4	0	0	2	0
A_1	1	1	1	1	1
A_2	1	1	1	−1	−1
B_1	1	−1	1	1	−1
B_2	1	−1	1	−1	1
E	2	0	−2	0	0

For A_1: $\frac{1}{8}[4 \times 1 + 2 \times 0 \times 1 + 0 \times 1 + 2 \times 2 \times 1 + 2 \times 0 \times 1] = 1$
For A_2: $\frac{1}{8}[4 \times 1 + 2 \times 0 \times 1 + 0 \times 1 + 2 \times 2 \times (-1) + 2 \times 0 \times (-1)] = 0$
For B_1: $\frac{1}{8}[4 \times 1 + 2 \times 0 \times (-1) + 0 \times 1 + 2 \times 2 \times 1 + 2 \times 0 \times (-1)] = 1$
For B_2: $\frac{1}{8}[4 \times 1 + 2 \times 0 \times (-1) + 0 \times 1 + 2 \times 2 \times (-1) + 2 \times 0 \times 1] = 0$
For E: $\frac{1}{8}[4 \times 2 + 2 \times 0 \times 0 + 0 \times (-2) + 2 \times 2 \times 0 + 2 \times 0 \times 0] = 1$

$\Gamma_3 = A_2 + B_1 + B_2 + 2E$:

C_{4v}	E	$2C_4$	C_2	$2\sigma_v$	$2\sigma_d$
Γ_3	7	−1	−1	−1	−1
A_1	1	1	1	1	1
A_2	1	1	1	−1	−1
B_1	1	−1	1	1	−1
B_2	1	−1	1	−1	1
E	2	0	−2	0	0

For A_1: $\frac{1}{8}[7 \times 1 + 2 \times (-1) \times 1 + (-1) \times 1 + (-1) \times 2 \times 1 + 2 \times (-1) \times 1] = 0$
For A_2: $\frac{1}{8}[7 \times 1 + 2 \times (-1) \times 1 + (-1) \times 1 + (-1) \times 2 \times (-1) + 2 \times (-1) \times (-1)] = 1$
For B_1: $\frac{1}{8}[7 \times 1 + 2 \times (-1) \times (-1) + (-1) \times 1 + (-1) \times 2 \times 1 + 2 \times (-1) \times (-1)] = 1$
For B_2: $\frac{1}{8}[7 \times 1 + 2 \times (-1) \times (-1) + (-1) \times 1 + (-1) \times 2 \times (-1) + 2 \times (-1) \times 1] = 1$
For E: $\frac{1}{8}[7 \times 2 + 2 \times (-1) \times 0 + (-1) \times (-2) + 2 \times (-1) \times 0 + 2 \times (-1) \times 0] = 2$

4-7 Vibrational analysis for NH_3:

C_{3v}	E	$2C_3$	$3\sigma_v$	
A_1	1	1	1	z
A_2	1	1	−1	R_z
E	2	−1	0	(x, y) (R_x, R_y)
Γ	12	0	2	

a. A_1: $\frac{1}{6}[(12 \times 1) + 2(0 \times 1) + 3(2 \times 1)] = 3$

A_2: $\frac{1}{6}[(12 \times 1) + 2(0 \times 1) + 3(2 \times (-1))] = 1$

E: $\frac{1}{6}[(12 \times 2) + 2(0 \times (-1)) + 3(2 \times 0)] = 4$

$\Gamma = 3A_1 + A_2 + 4E$

b. Translation: $A_1 + E$, based on the x, y, and z entries in the table.

Rotation: $A_2 + E$, based on the R_x, R_y, and R_z entries in the table.

Vibration: $2A_1 + 2E$ remaining from the total. (The A_1's are symmetric stretch and symmetric bend.)

c. All the vibrational modes are IR-active.

4-8 Taking only the CO stretching modes for the square pyramidal molecule:

C_{4v}	E	$2C_4$	C_2	$2\sigma_v$	$2\sigma_d$	
Γ	5	1	1	3	1	
A_1	1	1	1	1	1	z
A_2	1	1	1	−1	−1	R_z
B_1	1	−1	1	1	−1	
B_2	1	−1	1	−1	1	
E	2	0	−2	0	0	(x, y) (R_x, R_y)

$\Gamma = 2A_1 + B_1 + E$

$Mn(CO)_5Cl$ should have four IR-active stretching modes, the two A_1 and two from E. (The E modes are a degenerate pair; they give rise to a single infrared band.) The B_1 mode is IR-inactive.

CHAPTER 5

5-1

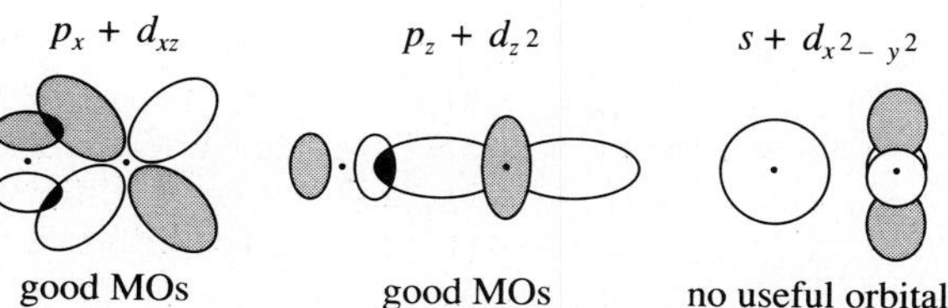

5-2 From top to bottom, the labels for the molecular orbitals in Figure 5-3 are σ_u, σ_g, σ_u, σ_g, π_g, π_u, δ_u, δ_g.

5-3 Bonding in the HF molecule.

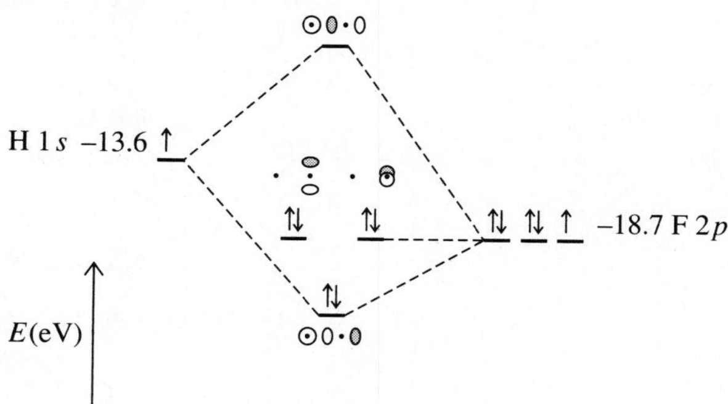

HF Molecular Orbitals

The energy match of the H 1*s* and the F 2*p* orbitals is fairly good, but that of the H 1*s* with the F 2*s* is poor. Therefore, molecular orbitals are formed between the H 1*s* and the F $2p_z$, as shown above (*z* is the axis through the nuclei). All the other F orbitals are nonbonding, either because of poor energy match or lack of useful overlap.

5-4 The oxygen *p* orbitals are at −15.9 eV; their group orbitals have essentially the same energy. The difference between this energy and that of the carbon 2*p* is only 5.2 eV; the difference between the oxygen 2*s* (and group orbitals 1 and 2) and the carbon 2*p* is 21.7 eV. The much closer match of C 2*p* and O 2*p* makes their combination much more likely. The diagram shows these relationships.

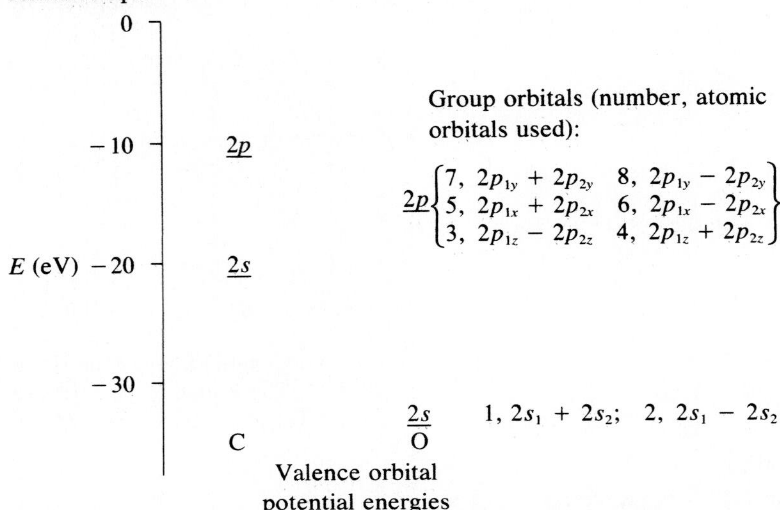

5-5 The molecular orbitals of N_3^- differ from those of CO_2 because all the atoms have the same initial orbital energies. Therefore, the best orbitals are formed by combinations of three 2*s* orbitals or the three 2*p* orbitals of the same type (*x*, *y*, or *z*). The resulting pattern of orbitals is shown in the diagram.

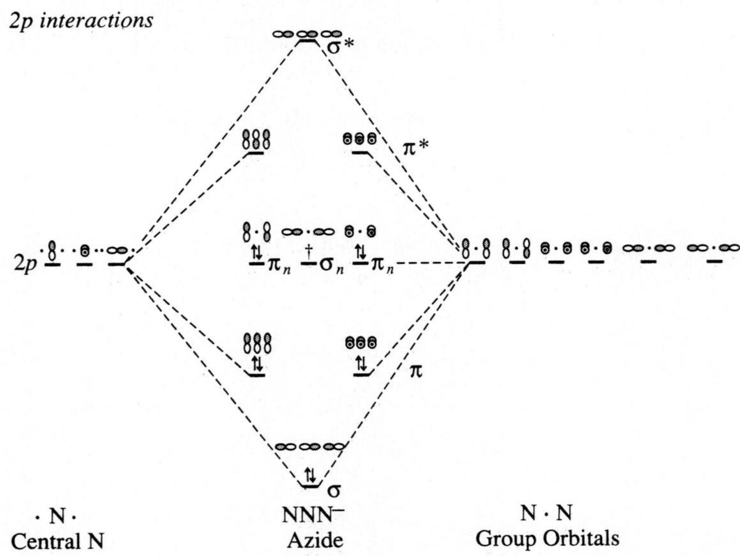

† This orbital is slightly higher in energy than the π_n orbitals as a consequence of an antibonding interaction with the 2*s* orbital of the central nitrogen.

2s interactions

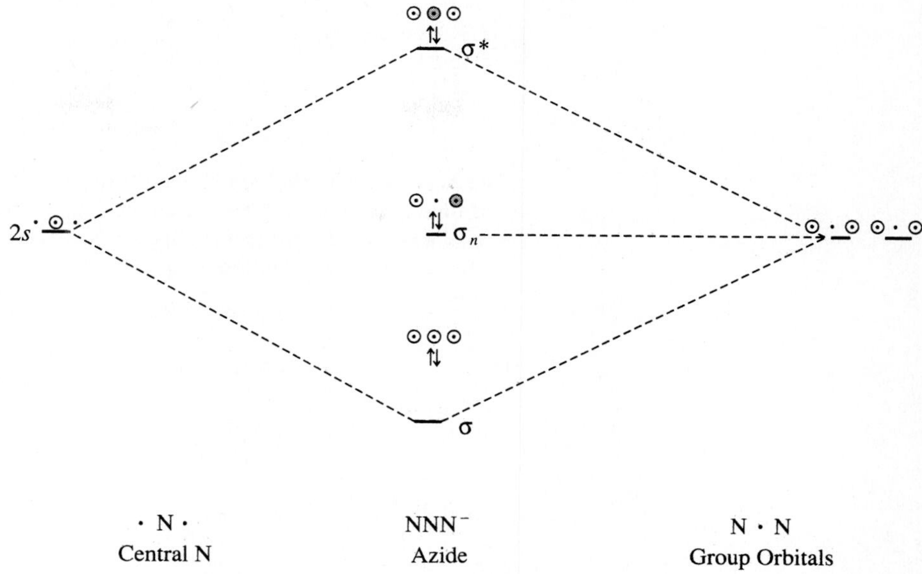

5-6 The group orbitals of a hexagonal ring of *s* orbitals:

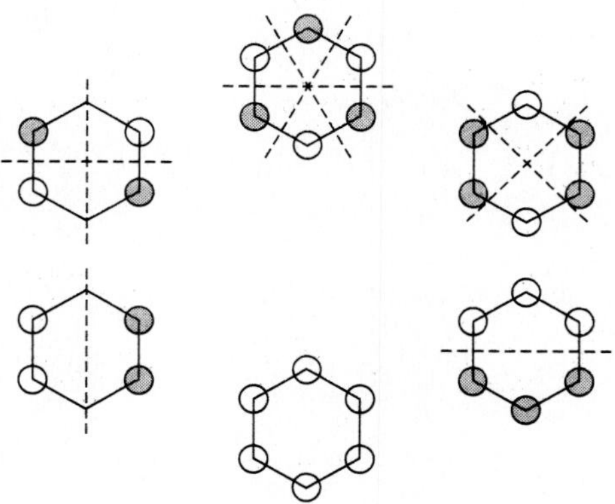

5-7 BF_3 is a planar triangle, with D_{3h} symmetry. The three F's each have *p* orbitals directed toward the B (along their respective *y* axes), *p* orbitals in the plane of the molecule (along their respective *x* axes), and *p* orbitals perpendicular to the plane of the molecule (along their respective *z* axes). (The F 2*s* orbitals are ignored; their energy is too low for effective orbital formation, and their overlap is small compared to the *p* orbitals. They are occupied by lone pairs.)

Each type of *p* orbital can be treated separately, as they cannot be interconverted by any of the symmetry operations of D_{3h}. The resulting group orbital symmetries are shown below:

D_{3h}	E	C_3	$3C_2$	σ_h	$2S_3$	$3\sigma_v$
Γ_y	3	0	1	3	0	1
Γ_x	3	0	−1	3	0	−1
Γ_z	3	0	−1	−3	0	1

Reduction leads to:

	Matches with B orbitals
$\Gamma_y = A_1' + E'$	A_1' matches $2s$, E' matches $(2p_x, 2p_y)$
$\Gamma_x = A_2' + E'$	A_2' does not match, E' matches $(2p_x, 2p_y)$
$\Gamma_z = A_2'' + E''$	A_2'' matches $2p_z$, E'' does not match

B $2s$ matches fairly well in energy with the F $2p$ (4.7 eV difference); the B $2p$ orbitals are 10.4 eV higher than the F $2p$'s, but still able to mix into molecular orbitals. The Γ_x group orbitals do not overlap well with the boron orbitals, and are treated as nonbonding orbitals. Overall, three σ bonds and a weak π bond.

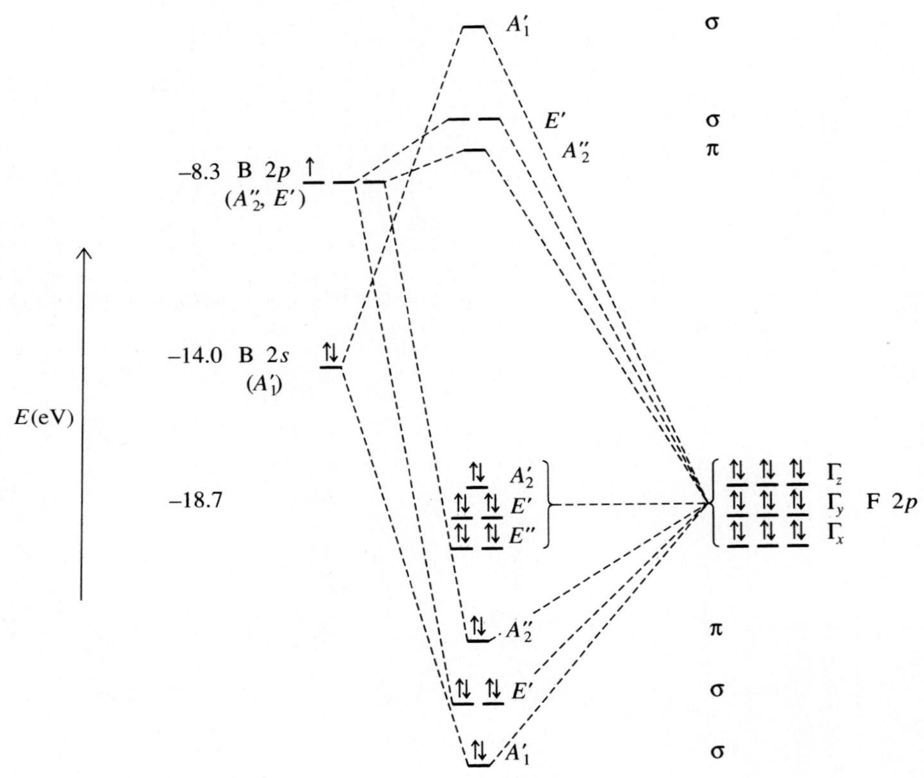

5-8 BeH_2 is linear, but simpler than CO_2 because there are no p orbitals on the hydrogens to be concerned. The result is sp hybridization on Be and single bonds each way.

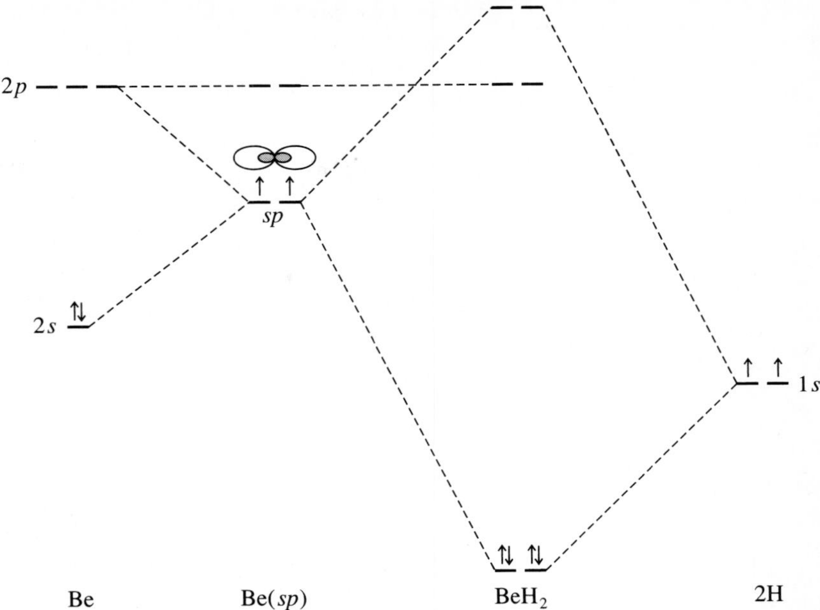

In group theory terms (Section 5-5), the molecule has $D_{\infty h}$ symmetry, which can be simplified to D_{2h}, as in the CO_2 example. The two H group orbitals have A_g and B_{1u} symmetry, matching the s and p_z orbitals of Be. The group orbitals of the hydrogens are

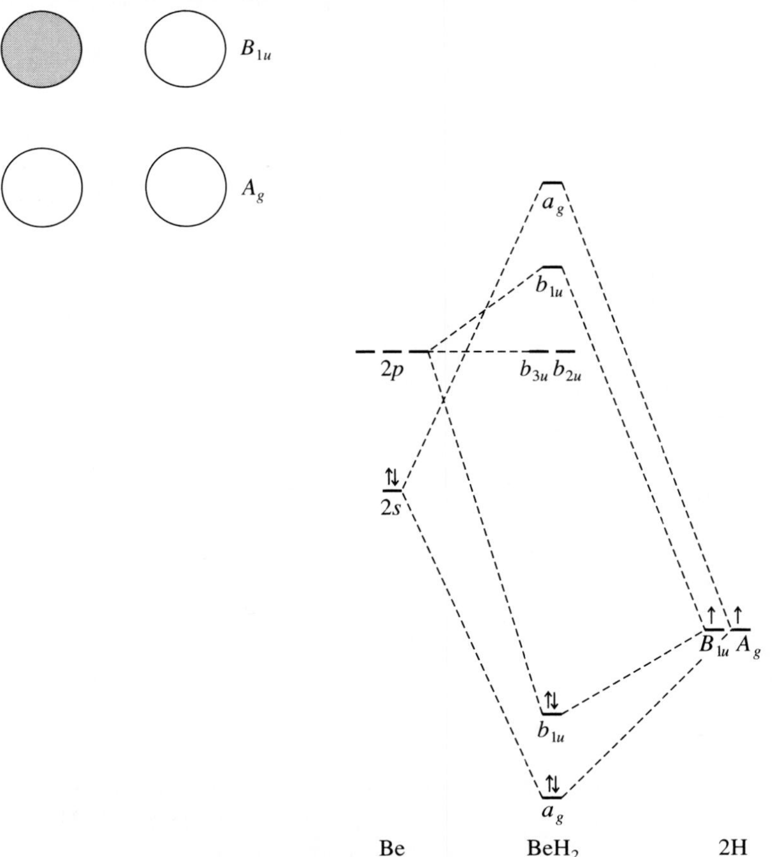

5-9 The four F atoms in XeF_4 are in a square planar structure, with D_{4h} symmetry. The four group orbitals by either method are

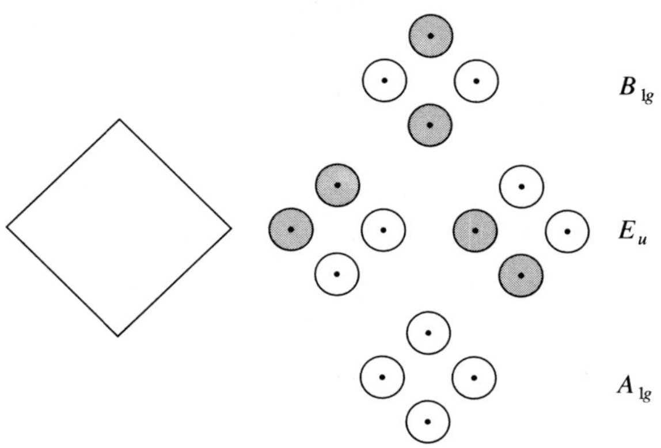

The reducible representation for all four and the component irreducible representations are

D_{4h}	E	$2C_4$	C_2	$2C_2'$	$2C_2''$	i	$2S_4$	σ_h	$2\sigma_v$	$2\sigma_d$		
Γ	4	0	0	2	0	0	0	4	2	0		
A_{1g}	1	1	1	1	1	1	1	1	1	1		$x^2 + y^2, z^2$
B_{1g}	1	−1	1	1	−1	1	−1	1	1	−1		$x^2 - y^2$
E_u	2	0	−2	0	0	−2	0	2	0	0	(x, y)	

Xenon orbitals used in bonding: s, $d_{z^2}(A_{1g})$; $d_{x^2-y^2}(B_{1g})$; p_x and $p_y(E_u)$.

CHAPTER 6

6-1 **a.** Dissociation of acetic acid: By Hess's law:

$$\Delta H^\circ = +55.9 - 56.3 = -0.4 \text{ kJ mol}^{-1}$$

$$\Delta S^\circ = -80.4 - 12.0 = -92.4 \text{ J K}^{-1} \text{ mol}^{-1}$$

b. By temperature dependence:

$$\Delta H^\circ = -2.8 \text{ kJ mol}^{-1}$$

$$\Delta S^\circ = -100 \text{ J K}^{-1} \text{ mol}^{-1}$$

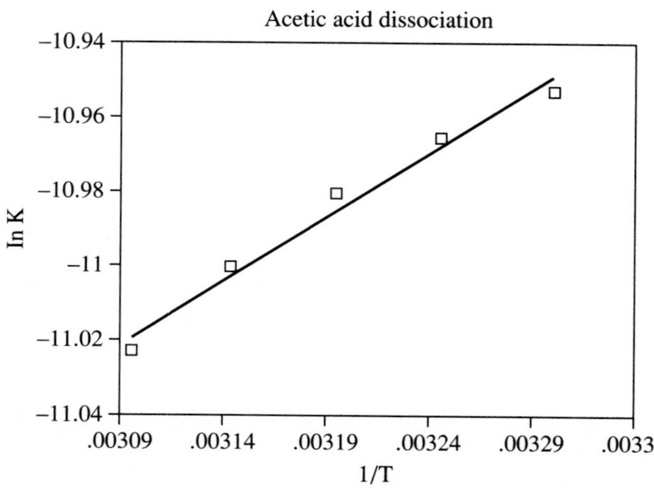

Even over this small temperature range, the data show the change in these functions with temperature and the different values obtained by different methods.

6-2

	H_2SO_3	HSO_3^-
$pK_a(9 - 7n)$	2	7
$pK_a(8 - 5n)$	3	8
pK_a(expt)	1.8	7.2

	H_3PO_3	$H_2PO_3^-$
$pK_a(9 - 7n)$	2	7
$pK_a(8 - 5n)$	3	8
pK_a(expt)	1.8	6.2

6-3 **a.** Acetic acid in water is only slightly dissociated according to the equation

$$HOAc \rightleftharpoons H^+ + OAc^-$$

Sodium hydroxide is completely dissociated into Na^+ and OH^-. During the titration, the primary reaction is

$$HOAc + OH^- \rightleftharpoons H_2O + OAc^-$$

At the midpoint, half the original acetic acid is present as HOAc and half as OAc^-. At the end point, all has been converted to OAc^-, and the solution contains almost entirely Na^+ and OAc^-. The next increment of OH^- added does not react, but remains as OH^-.

b. Acetic acid acts as a strong acid in pyridine, forming pyridinium ion and acetate:

$$HOAc + Py \rightleftharpoons HPy^+ + OAc^-$$

Tetramethylammonium hydroxide dissociates into $(CH_3)_4N^+ + OH^-$. During the titration, the hydroxide reacts with the pyridinium ion:

$$OH^- + HPy^+ \rightleftharpoons Py + H_2O$$

At the midpoint, half the pyridinium ion formed from reaction with acetic acid remains as HPy^+ and half has been converted to Py. At the end point, all the pyridine is converted to the free base Py and the remaining ions are $(CH_3)_4N^+$ and OAc^-. Any additional titrant added simply adds $(CH_3)_4N^+$ and OH^-.

6-4

	E_A	C_A		E_B	C_B	ΔH (E)	ΔH (C)	ΔH (total)
a. BF_3	9.88	1.62	NH_3	1.36	3.46	−13.44	−5.08	−18.52
			CH_3NH_2	1.30	5.88	−12.84	−9.53	−22.37
			$(CH_3)_2NH$	1.09	8.73	−10.77	−14.14	−24.91
			$(CH_3)_3N$	0.808	11.54	−7.98	−18.69	−26.67

	E_B	C_B		E_A	C_A	ΔH (E)	ΔH (C)	ΔH (total)
b. Py	1.17	6.40	Me_3B	6.14	1.70	−7.18	−10.88	−18.06
			Me_3Al	16.9	1.43	−19.77	−9.15	−28.92
			Me_3Ga	13.3	0.881	−15.56	−5.64	−21.20

c. The amine series shows a steady increase in C_B and decrease in E_B as methyl groups are added. The methyl groups push electrons onto the N, and the lone-pair electrons are then made more available to the acid BF_3. As a result, covalent bonds to BF_3 are more likely with more methyl groups.

The lone pairs of the molecules with fewer methyl groups are apparently more tightly held, requiring more nearly ionic type bonding and a larger E_B.

The B, Al, Ga series is less regular. Possible arguments for Al being the strongest in E_A:

1. As size increases, the electron density is reduced on the central atom.
2. The d electrons in Ga shield the outer electrons, so they are held less tightly. This makes them less likely to form an electrostatic-type bond. Again, the methyl groups push electrons onto the central atom, reducing the acidity. This is most effective for B, the smallest central atom. Ga is largest, but the filled d orbitals reduce the effect, so it is second. Al, with moderate size and no d electrons, is the best acid.

This series is counter to the expected simple HSAB argument for covalency. The larger Ga compound seems better suited to higher polarizability and softness. However, hard–hard interactions are most important.

CHAPTER 8

8-1 $S = n/2 \qquad \sqrt{4S(S+1)} = \sqrt{4(n/2)(n/2+1)} = \sqrt{n^2+2n} = \sqrt{n(n+2)}$

8-2 d^5, high spin. Start with electrons a and b in the upper e_g levels, c, d, and e in the lower t_{2g} levels. LFSE = 0.

Exchanges: Ten total exchanges, no pairing. Net energy: $+10\Pi_e$.

a–b
a–c b–c
a–d b–d c–d
a–e b–e c–e d–e

d^5 low spin: Start with a, b, and c with one spin (+), d and e with the other spin (−). LFSE = $-2\Delta_o$.

Exchanges: Four total, with two spin pairs. Net energy: $-2\Delta_o + 2\Pi_c + 4\Pi_e$.

a–b
a–c
b–c
d–e

Overall, the difference between low spin and high spin is $-2\Delta_o + 2\Pi_c - 6\Pi_e$, so when $\Delta_o > \Pi_c - 3\Pi_e$, the low spin configuration is more stable. From Table 8-4, Mn^{2+} has Δ_o = 7500, Π = 25,500 (high spin), and Fe^{3+} has Δ_o = 14,000, Π = 30,000 (high spin).

8-3 Square planar π bonding

	D_{4h}	E	$2C_4$	C_2	$2C_2'$	$2C_2''$	i	$2S_4$	σ_h	$2\sigma_v$	$2\sigma_d$
σ	Γ_{p_y}	4	0	0	2	0	0	0	4	2	0
$\pi_{\parallel}$	Γ_{p_x}	4	0	0	−2	0	0	0	4	−2	0
$\pi_{\perp}$	Γ_{p_z}	4	0	0	−2	0	0	0	−4	2	0

Γ_{p_y}: A_{1g} $\frac{1}{16}[4 \times 1 + 0 + 0 + 2 \times 2 \times 1 + 0 + 0 + 0 + 4 \times 1 + 2 \times 2 \times 1 + 0] = 1$
B_{1g} $\frac{1}{16}[4 \times 1 + 0 + 0 + 2 \times 2 \times 1 + 0 + 0 + 0 + 4 \times 1 + 2 \times 2 \times 1 + 0] = 1$
E_u $\frac{1}{16}[4 \times 2 + 0 + 0 + 0 + 0 + 0 + 0 + 4 \times 2 + 0 + 0] = 1$

Γ_{p_x}: A_{2g} $\frac{1}{16}[4 \times 1 + 0 + 0 + (-2) \times 2 \times (-1) + 0 + 0 + 0 + 4 \times 1 +$
$(-2) \times 2 \times (-1) + 0] = 1$

B_{2g} $\frac{1}{16}[4 \times 1 + 0 + 0 + (-2) \times 2 \times (-1) + 0 + 0 + 0 + 4 \times 1 + (-2) \times 2 \times (-1) + 0] = 1$

E_u $\frac{1}{16}[4 \times 2 + 0 + 0 + 0 + 0 + 0 + 0 + 4 \times 2 + 0 + 0] = 1$

Γ_{p_z}: A_{2u} $\frac{1}{16}[4 \times 1 + 0 + 0 + (-2) \times 2 \times (-1) + 0 + 0 + 0 + (-4) \times (-1) + 2 \times 2 \times 1 + 0] = 1$

B_{2u} $\frac{1}{16}[4 \times 1 + 0 + 0 + (-2) \times 2 \times (-1) + 0 + 0 + 0 + (-4) \times (-1) + 2 \times 2 \times 1 + 0] = 1$

E_g $\frac{1}{16}[4 \times 2 + 0 + 0 + 0 + 0 + 0 + 0 + (-4) \times (-2) + 0 + 0] = 1$

8-4 d_{xy} total for 7, 8, 9, 10 $= 1.33e_\sigma$
d_{xz} total for 7, 8, 9, 10 $= 1.33e_\sigma$
d_{yz} total for 7, 8, 9, 10 $= 1.33e_\sigma$
d_{z^2} total for 7, 8, 9, 10 $= 0$
$d_{x^2-y^2}$ total for 7, 8, 9, 10 $= 0$
Ligands go down in energy by e_σ each.

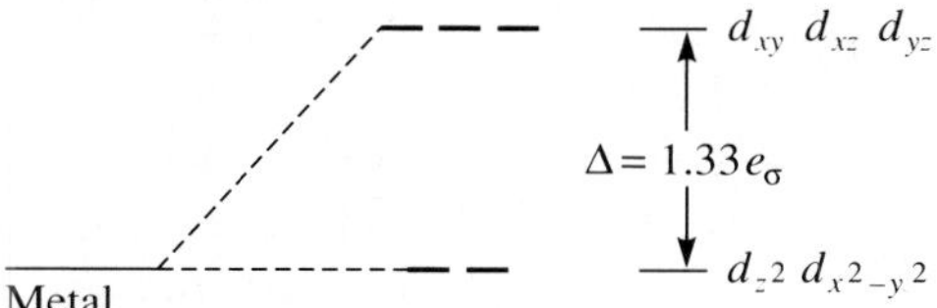

For octahedral, $\Delta = 3e_\sigma$
For tetrahedral, $\Delta = 4/9 \times 3e_\sigma = 1.33e_\sigma$

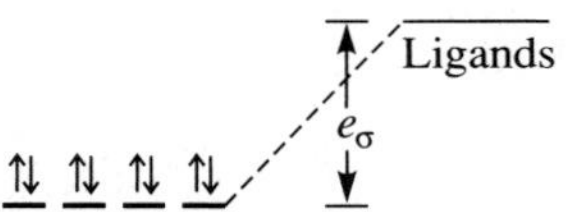

8-5 Adding π bonding to the results of Exercise 8-4: d_{xy}, d_{xz}, d_{yz} each total for 7, 8, 9, 10 $= 0.89e_\pi$. d_{z^2}, $d_{x^2-y^2}$ total for 7, 8, 9, 10 $= 2.67e_\pi$. Ligands go down in energy by $2e_\pi$ each.

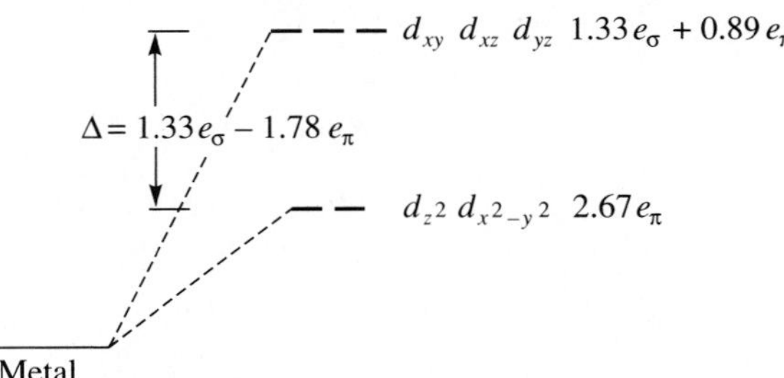

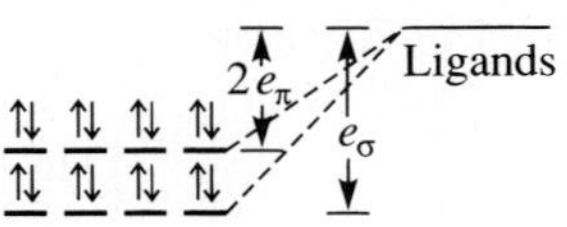

8-6 Square planar (2, 3, 4, 5)

σ only: d_{z^2} total for 2, 3, 4, 5, $= e_\sigma$
$d_{x^2-y^2}$ total for 2, 3, 4, 5 $= 3e_\sigma$

d_{xy}, d_{xz}, d_{yz} total for 2, 3, 4, 5 = 0

Ligands go down in energy by e_σ each.

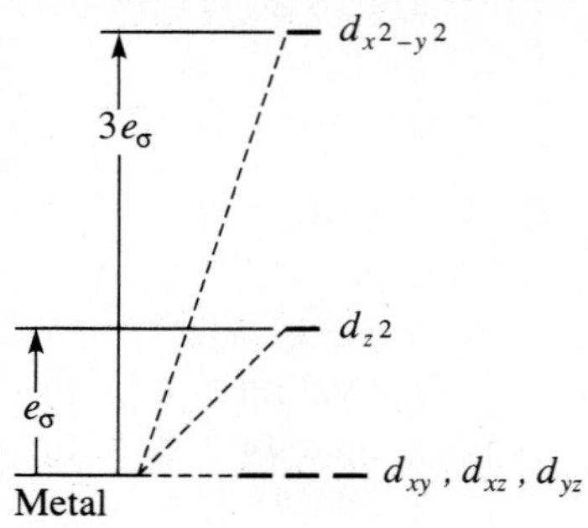

Adding π: d_{z^2}, $d_{x^2-y^2}$ total for 2, 3, 4, 5 = 0

d_{xz}, d_{yz} total for 2, 3, 4, 5 = $2e_\pi$

d_{xy} total for 2, 3, 4, 5 = $4e_\pi$

Ligands go up in energy by $2e_\pi$ each.

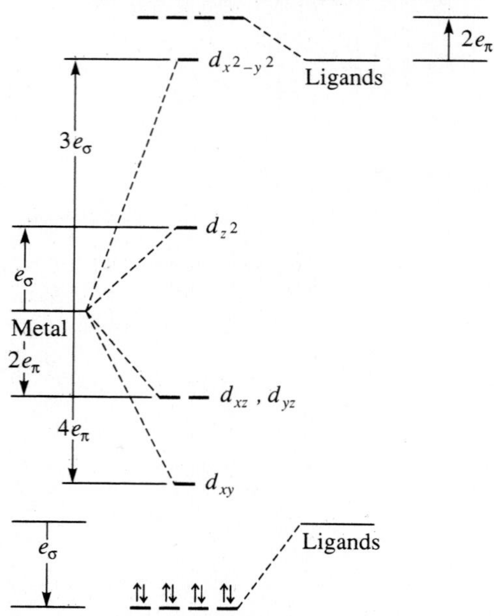

8-7 (See Table 8-3 for the complete high spin and low spin configurations.)

Number of electrons		1	2	3	4	5	6	7	8	9	10
High spin											
	e_g	0	0	0	1	2	2	2	2	3	4
	Jahn-Teller	w	w		s		w	w		s	
	t_{2g}	1	2	3	3	3	4	5	6	6	6
Low spin											
	e_g	0	0	0	0	0	0	1	2	3	4
	Jahn-Teller	w	w		w	w		s		s	
	t_{2g}	1	2	3	4	5	6	6	6	6	6

The weak J–T cases have unequal occupation of t_{2g} orbitals; the strong J–T cases have unequal occupation of e_g orbitals.

CHAPTER 9

9-1 High spin d^6: 1.

$$\begin{array}{c} \uparrow\ \uparrow \\ \uparrow\downarrow\ \uparrow\ \uparrow \end{array}$$

2. Spin multiplicity $= 4 + 1 = 5$
3. Maximum possible value of $M_L = 2 + 2 + 1 + 0 - 1 - 2 = 2$. Therefore, D term.
4. ^5D

Low spin d^6: 1.

$$\begin{array}{c} \_\ \_ \\ \uparrow\downarrow\ \uparrow\downarrow\ \uparrow\downarrow \end{array}$$

2. Spin multiplicity $= 0 + 1 = 1$
3. Maximum value of $M_L = 2 + 2 + 1 + 1 + 0 + 0 = 6$. Therefore, I term.
4. ^1I

9-2 $[Fe(H_2O)_6]^{2+}$ is a high-spin d^6 complex. The weak field (left) part of the Tanabe–Sugano diagram for d^6 shows that the only excited state with the same spin multiplicity (5) as the ground state is the ^5E. The transition is therefore $^5T_2 \longrightarrow {}^5E$. The excited state $t_{2g}^3e_g^3$ is subject to Jahn–Teller distortion; consequently, as in the d^1 complex, $[Ti(H_2O)_6]^{3+}$, the absorption band is split.

9-3

$$^4T_1 \rightarrow {}^4A_2 \quad 16{,}000\ \text{cm}^{-1} = \nu_1$$

$$^4T_1 \rightarrow {}^4T_1 \quad 19{,}800\ \text{cm}^{-1} = \nu_2$$

$$\nu_1/\nu_2 = 0.808$$

From the Tanabe–Sugano diagram, at $\Delta/B = 10$, $\nu_1 = 20$ and $\nu_2 = 25$, $\nu_1/\nu_2 = 0.80$. ν_1 rises at a slope of 2 and ν_2 at a slope of 1, so the value of Δ/B is 10.2 for a ratio of 0.81. B and Δ can then be calculated:

$$\nu_1 = 20.4 \quad B = E/\nu_1 = 16{,}000/20.4 = 784\ \text{cm}^{-1}$$

$$\nu_2 = 25.2 \quad B = E/\nu_2 = 19{,}800/25.2 = 786\ \text{cm}^{-1}$$

Average $B = 785\ \text{cm}^{-1}$

$$\Delta = \nu_1 \times B = 10.2 \times 785 = 8{,}000\ \text{cm}^{-1}$$

CHAPTER 10

10-1 **a.** d^9 σ donor, based on Figure 10-16

Tetrahedral:

$$5 \text{ electrons} \times 4/3e_\sigma + 4 \text{ electrons} \times 0 = 6.67e_\sigma$$

Square planar:

$$1 \text{ electron} \times 3e_\sigma + 2 \text{ electrons} \times e_\sigma + 6 \text{ electrons} \times 0 = 5e_\sigma$$

b. d^9 σ donor, π acceptor, based on Figure 10-17

Tetrahedral:

$$5 \text{ electrons} \times (4/3e_\sigma - 8/9e_\pi) + 4 \text{ electrons} \times (-8/3e_\pi) = 6.67e_\sigma - 15.11e_\pi$$

Square planar:

$$1 \text{ electron} \times 3e_\sigma + 2 \text{ electrons} \times e_\sigma + 4 \text{ electrons} \times (-2e_\pi) + 2 \text{ electrons} \times (-4e_\pi) = 5e_\sigma - 16e_\pi$$

Square planar is favored by $1.67e_\sigma$ and by $1.67e_\sigma + 0.89e_\pi$. d^8 favors square planar more in both cases.

10-2 Ma_2b_2cd has eight isomers, including two pairs of enantiomers, according to Table 10-3.

M⟨aa⟩⟨bb⟩⟨cd⟩ M⟨aa⟩⟨bc⟩⟨bd⟩ M⟨ac⟩⟨ad⟩⟨bb⟩ M⟨ab⟩⟨ab⟩⟨cd⟩

M⟨ab⟩⟨ad⟩⟨bc⟩ M⟨ab⟩⟨ac⟩⟨bd⟩ (both with enantiomers)

10-3 M(AA)bcde has 12 isomers, six pairs of enantiomers.

(all have enantiomers)

M⟨Ab⟩⟨Ac⟩⟨de⟩ M⟨Ab⟩⟨Ad⟩⟨ce⟩ M⟨Ab⟩⟨Ae⟩⟨cd⟩

M⟨Ac⟩⟨Ad⟩⟨be⟩ M⟨Ac⟩⟨Ae⟩⟨bd⟩ M⟨Ad⟩⟨Ae⟩⟨bc⟩

CHAPTER 11

11-1 The rate of product formation from the second reaction is

$$\frac{d[ML_5Y]}{dt} = k_2[ML_5XY]$$

The stationary state is based on the intermediate $[ML_5XY]$:

$$\frac{d[ML_5XY]}{dt} = k_1[ML_5X][Y] - k_{-1}[ML_5XY] - k_2[ML_5XY] = 0$$

from which the concentration of ML_5XY is $[ML_5XY] = \dfrac{k_1[ML_5X][Y]}{k_{-1} + k_2}$. Substituting into the first equation results in the overall rate equation:

$$\frac{d[ML_5Y]}{dt} = \frac{k_1k_2[ML_5X][Y]}{k_{-1} + k_2} = k[ML_5X][Y]$$

11-2 $PtCl_4^{2-} + NO_2^- \longrightarrow a$ $\quad a + NH_3 \longrightarrow b$

$a = PtCl_3(NO_2)^{2-}$ $\quad b = trans\text{-}PtCl_2(NO_2)(NH_3)^-$

NO_2^- is a better *trans* director than Cl^-

$PtCl_3NH_3^- + NO_2^- \longrightarrow c$ $\quad c + NO_2^- \longrightarrow d$

$c = cis\text{-}PtCl_2(NO_2)(NH_3)^-$ $\quad d = trans\text{-}PtCl(NO_2)_2(NH_3)^-$

Cl^- has a larger *trans* effect than NH_3, and NO_2^- has a larger *trans* effect than either Cl^- or NH_3

$PtCl(NH_3)_3^+ + NO_2^- \longrightarrow e$ $\quad e + NO_2^- \longrightarrow f$

$e = trans\text{-}PtCl(NH_3)_2(NO_2)$ $\quad f = trans\text{-}Pt(NH_3)_2(NO_2)_2$

Cl^- has a larger *trans* effect than NH_3, and NO_2^- has a still larger *trans* effect

$PtCl_4^{2-} + I^- \longrightarrow g$ $\quad g + I^- \longrightarrow h$

$g = PtCl_3I^{2-}$ $\quad h = trans\text{-}PtCl_2I_2^{2-}$

I^- has a larger *trans* effect than Cl^-

$$PtI_4^- + Cl^- \longrightarrow i \qquad i + Cl^- \longrightarrow j$$
$$i = PtI_3Cl^{2-} \qquad j = cis\text{-}PtI_2Cl_2^{2-}$$

I^- has a larger *trans* effect than Cl^-, and replacement of Cl^- in the second step would give no net change.

CHAPTER 12

12-1

	Method A		*Method B*	
a. $[Fe(CO)_4]^{2-}$	Fe^{2-}	10	Fe	8
	4 CO	8	4 CO	8
			2−	2
		18		18
b. $[(\eta^5\text{-}C_5H_5)_2Co]^+$	Co^{3+}	6	Co	9
	2 Cp^-	12	2 Cp	10
			1+	−1
		18		18
c. $(\eta^3\text{-}C_3H_5)(\eta^5\text{-}C_5H_5)Fe(CO)$	Fe	8	Fe	8
	$\eta^3\text{-}C_3H_3^+$	2	$\eta^3\text{-}C_3H_3$	3
	$\eta^5\text{-}Cp^-$	6	$\eta^5\text{-}Cp$	5
	CO	2	CO	2
		18		18

12-2

a. $[M(CO)_3PPh_3]^-$	3 CO	6	3 CO	6
	PPh_3	2	PPh_3	2
			1−	1
		8		9

Need 10 electrons for M^- or 9 for M, so Co is the metal.

b. $HM(CO)_5$	H^-	2	H	1
	5 CO	10	5 CO	10
		12		11

Need 6 electrons for M^+ or 7 for M, so Mn is the metal.

c. $(\eta^4\text{-}C_8H_8)M(CO)_3$	3 CO	6	3 CO	6
	$\eta^4\text{-}C_8H_8$	4	$\eta^4\text{-}C_8H_8$	4
		10		10

Need 8 electrons, so Fe is the metal.

d. $[(\eta^5\text{-}C_5H_5)M(CO)_3]_2$	3 CO	6	3 CO	6
	$\eta^5\text{-}C_5H_5^-$	6	$\eta^5\text{-}C_5H_5$	5
	M—M	1	M—M	1
		13		12

Need 5 for M^+ or 6 for M, so Cr is the metal.

12-3

$[Ni(CN)_4]^{2-}$	Ni(II)	8	Ni	10
	4 CN^-	8	4 CN	4
		16	2−	2
				16
$PtCl_2en$	Pt(II)	8	Pt	10
	2 Cl^-	4	2 Cl	2
	en	4	en	4
		16		16

	Method A		Method B	
$RhCl(PPh_3)_3$	Rh(I)	8	Rh	9
	Cl^-	2	Cl	1
	3 PPh_3	6	3 PPh_3	6
		16		16
$IrCl(CO)(PPh_3)_2$	Ir(I)	8	Ir	9
	Cl^-	2	Cl	1
	CO	2	CO	2
	2 PPh_3	4	2 PPh_3	4
		16		16

12-4 The N_2 σ and π levels are very close together in energy (see Chapter 5), and they are all symmetric. The CO levels are farther apart and skewed toward C. Therefore, the geometric overlap for CO is better, and CO has better σ donor and π acceptor qualities than N_2.

12-5 The two methods of electron counting are equivalent for these examples.

$M(CO)_4$ (M = Ni, Pd)	M	10
	4 CO	8
		18
$M(CO)_5$ (M = Fe, Ru, Os)	M	8
	5 CO	10
		18
$M(CO)_6$ (M = Cr, Mo, W)	M	6
	6 CO	12
		18
$M(CO)_6$ (M = V)	M	5
	6 CO	12
		17
$Co_2(CO)_8$ (soln)	Co	9
	4 CO	8
	Co—Co	1
		18
$Co_2(CO)_8$ (solid)	Co	9
	3 CO	6
	2μ_2-CO	2
	Co—Co	1
		18
$Fe_2(CO)_9$	Fe	8
	3 CO	6
	3μ_2-CO	3
	Fe—Fe	1
		18
$M_2(CO)_{10}$ (M = Mn, Tc, Re)	M	7
	5 CO	10
	M—M	1
		18

$Fe_3(CO)_{12}$,	Fe on left	Fe	8
		4 CO	8
		2 Fe—Fe	2
			18
	Other Fe's	Fe	8
		3 CO	6
		2-μ_2-CO	2
		2 Fe—Fe	2
			18
$M_3(CO)_{12}$	(M = Ru, Os)	M	8
		4 CO	8
		2 M—M	2
			18
$M_4(CO)_{12}$	(M = Co, Rh), M on top:	M	9
		3 CO	6
		3 M—M	3
			18
	Other M's	M	9
		2 CO	4
		2-μ_2-CO	2
		3 M—M	3
			18
$Ir_4(CO)_{12}$		M	9
		3 CO	6
		3 M—M	3
			18

12-6 Carbon and sulfur do not differ as much in electronegativity as carbon and oxygen. Consequently, the π^* orbitals (LUMO) are not skewed as much toward carbon in CS as in CO. This factor alone might make it seem that CS should be a weaker π acceptor. However, the greater bond distance in CS means that its π and π^* orbitals do not differ as much in energy as in the case of CO. As a result, the π^* orbitals of CS are slightly lower in energy and a better energy match for *d* orbitals of the metal; it is therefore better able to accept electron density from the metal. (This is a vastly oversimplified explanation; for additional information see P. V. Broadhurst, *Polyhedron,* **1985,** *4,* 1828 and references cited therein.)

12-7

2-Node group orbitals

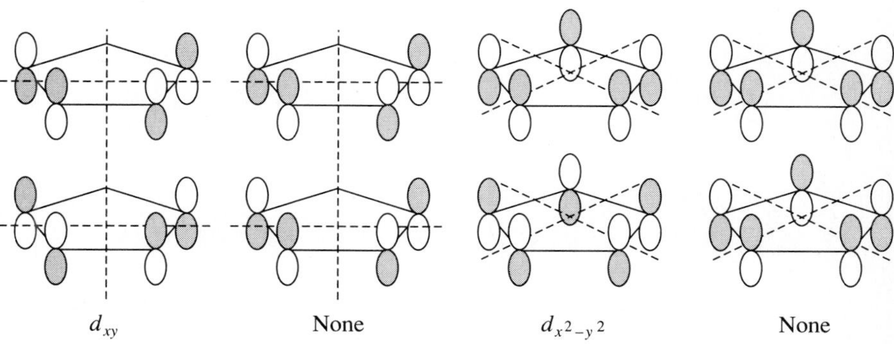

1-Node group orbitals

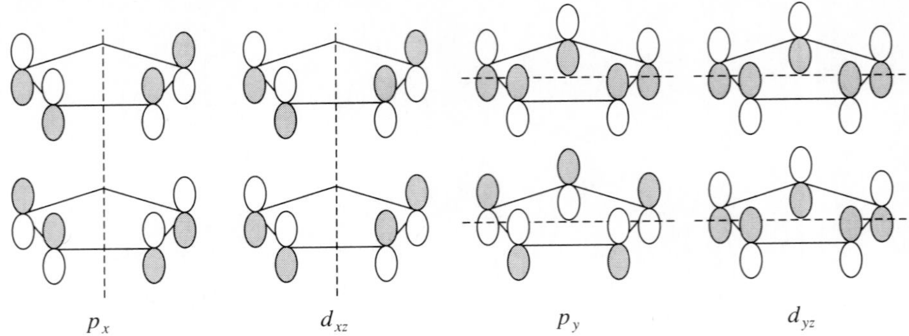

0-Node group orbitals

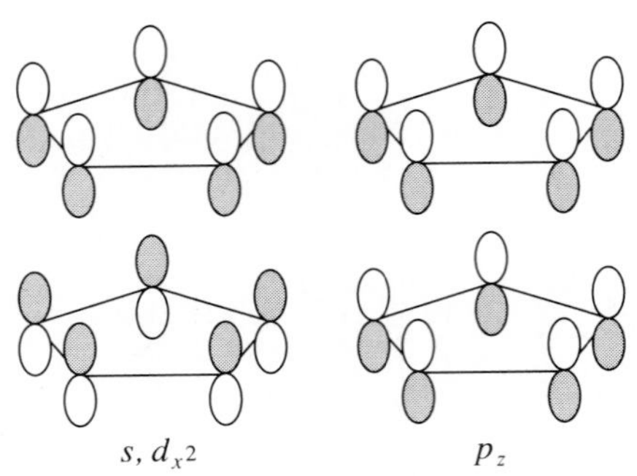

12-8 With just two bands, this is more likely the *fac* isomer.

12-9 II has three separate resonances in the CO range (at 194.98, 189.92, and 188.98 ppm) and is more likely the *fac* isomer. The *mer* isomer would be expected to have two peaks because two of the carbonyls have magnetically equivalent environments.

CHAPTER 13

13-1 The *cis* product is one with the labeled CO *cis* to CH_3. The reverse of mechanism 1 removes the acetyl ^{13}CO from the molecule completely, which means the product should have no ^{13}CO label at all.

13-2 The product distribution for the reaction of *cis*-$CH_3Mn(CO)_4(^{13}CO)$ with PR_3 ($R = C_2H_5$):

25% has ^{13}C in the CH_3CO.

$$(OC)_4Mn(^{\star}CO)(CH_3) \rightarrow (OC)_4Mn(^{\star}C(=O)CH_3) \leftarrow PR_3$$

($^{\star}CO = {}^{13}CO$)

25% has ^{13}CO *trans* to the CH_3CO.

$$(OC)_3Mn(^{\star}CO)(CH_3) \rightarrow (OC)_3Mn(^{\star}CO)(C(=O)CH_3) \leftarrow PR_3$$

50% has ^{13}CO *cis* to the CH_3CO.

All the products have PEt_3 *cis* to the CH_3CO.

13-3 The hydroformylation process for preparation of $(CH_3)_2CH—CH_2—CHO$ from $(CH_3)_2C═CH_2$ is exactly that of Figure 13-14, with R = CH_3.

CHAPTER 14

14-1 There are many possible answers. Examples:

a. $Re(CO)_4$ $(\eta^5\text{-}C_5H_5)Fe(CO)$

b. $Pt(CO)_3$ $(\eta^5\text{-}C_5H_5)Co^{2-}$

c. $Re(CO)_5$ $(\eta^5\text{-}C_5H_5)Mn(CO)_2^-$ $(\eta^6\text{-}C_6H_6)Mn(CO)_2$

14-2

a. This is a 15-electron species with three vacant positions, isolobal with CH.

b. This is a 16-electron species with one vacant position, isolobal with CH_3^+.

c. This is a 15-electron species with three vacant positions, isolobal with CH.

14-3

a. $B_{11}H_{13}^{2-}$ is derived from $B_{11}H_{11}^{4-}$, a *nido* species.

b. $B_5H_8^-$ is derived from $B_5H_5^{4-}$, a *nido* species.

c. $B_7H_7^{2-}$ is a *closo* species.

d. $B_{10}H_{18}$ is derived from $B_{10}H_{10}^{8-}$, a *hypho* species.

14-4

a. $C_3B_3H_7$ is equivalent to B_6H_{10}, derived from $B_6H_6^{4-}$, a *nido* species.

b. $C_2B_5H_7$ is equivalent to B_7H_9, derived from $B_7H_7^{2-}$, a *closo* species.

c. $C_2B_7H_{12}^-$ is equivalent to $B_9H_{14}^-$, derived from $B_9H_9^{6-}$, an *arachno* species.

14-5

a. SB_9H_9 is equivalent to $B_{10}H_{12}$, derived from $B_{10}H_{10}^{2-}$, a *closo* species.

b. $GeC_2B_9H_{11}$ is equivalent to $B_{12}H_{14}$, derived from $B_{12}H_{12}^{2-}$, a *closo* species.

c. $SB_9H_{12}^-$ is equivalent to $B_{10}H_{15}^-$, derived from $B_{10}H_{10}^{6-}$, an *arachno* species.

14-6

a. $C_2B_7H_9(CoCp)_3$ is equivalent to $B_9H_{11}(CoCp)_3$ or $B_{12}H_{14}$, derived from $B_{12}H_{12}^{2-}$, a *closo* species.

b. $C_2B_4H_6Ni(PPh_3)_2$ is equivalent to $B_6H_8Ni(PPh_3)_2$ or B_7H_9, derived from $B_7H_7^{2-}$, a *closo* species.

B

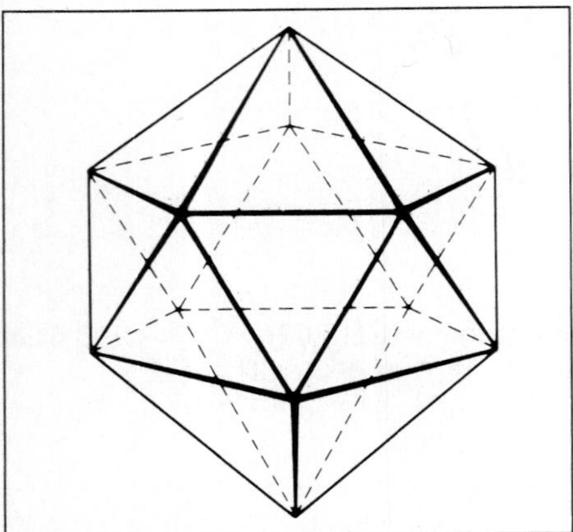

APPENDIX B-1 IONIC RADII

The values given are the crystal radii of Shannon, calculated using electron density maps and internuclear distances from X-ray data. Some of the trends that can be seen in these radii are:

1. Increase in size with increasing coordination number
2. Increase in size for a given coordination number with increasing Z within a periodic group
3. Decreasing size with increasing nuclear charge for isoelectronic ions
4. Decreasing size with increasing ionic charge for the same Z
5. Irregular, slowly decreasing size with increasing Z for transition metal, lanthanide, or actinide ions of the same charge
6. Larger size for high-spin ions than for low-spin ions of the same species and charge

Not shown in the table, but another apparent factor, is the decrease in anion size with increasing cation field strength, determined by the charge and size of the cation in the crystal. See O. Johnson, *Inorg. Chem.*, **1973,** *12,* 780 for the details.

Z	Coordination number 2	4	6	8	10	12	14
1 H	−4						
2 He							
3 Li^{+}		73	90	106			
4 Be^{2+}		41	59				
5 B^{3+}		25					
6 C^{4+}		29					
7 N^{3-}		132					
8 O^{2-}	121	124	126	128			
OH^{-}	118	121	123				
9 F^{-}	115	117	119				
10 Ne							
11 Na^{+}		113	116	132		153	
12 Mg^{2+}		71	86	103			
13 Al^{3+}		53	68				
14 Si^{4+}		40	54				
15 P^{3+}			58				
16 S^{2-}			170				
17 Cl^{-}			167				
18 Ar							
19 K^{+}		151	152	165	173	178	
20 Ca^{2+}			114	126	137	148	
21 Sc^{3+}			89	101			
22 Ti^{2+}			100				
Ti^{3+}			81				
Ti^{4+}		56	75	88			
23 V^{2+}			93				
V^{3+}			78				
24 Cr^{2+}			hs 94				
Cr^{2+}			ls 87				
Cr^{3+}			76				
25 Mn^{2+}		hs 80	hs 97				
Mn^{2+}			ls 81				
Mn^{3+}			hs 79				
Mn^{3+}			ls 72				
26 Fe^{2+}		hs 77	hs 92				
Fe^{2+}			ls 75				
Fe^{3+}		hs 63	hs 79				
Fe^{3+}			ls 69				
27 Co^{2+}		hs 72	hs 89				
Co^{2+}			ls 79				
Co^{3+}			hs 75				
Co^{3+}			ls 69				
28 Ni^{2+}		69	83				
Ni^{2+}		sq 63					
Ni^{3+}			hs 74				
Ni^{3+}			ls 70				
29 Cu^{+}	60	74	91				
Cu^{2+}		71	87				
30 Zn^{2+}		74	88	104			
31 Ga^{3+}		61	76				
32 Ge^{4+}		53	67				
33 As^{3+}			72				
As^{5+}		48	60				
34 Se^{2-}			184				
35 Br^{-}			182				
36 Kr							
37 Rb^{+}			166	175	180	186	197
38 Sr^{2+}			132	140	150	158	
39 Y^{3+}			104				
40 Zr^{4+}		73	86	98			
41 Nb^{3+}			86				
Nb^{4+}			82	93			
42 Mo^{3+}			83				
Mo^{4+}			79				
43 Tc^{4+}			79				

Z	Coordination number 2	4	6	8	10	12	14
44 Ru^{3+}			82				
Ru^{4+}			76				
45 Rh^{3+}			81				
Rh^{4+}			74				
46 Pd^{2+}		sq 78	100				
47 Ag^{+}	81	114	129	142			
Ag^{+}		sq 116					
48 Cd^{2+}		92	109	124		145	
49 In^{3+}		76	94	106			
50 Sn^{4+}		69	83	95			
51 Sb^{3+}			90				
52 Te^{2-}			207				
53 I^{-}			206				
54 Xe							
55 Cs^{+}			181	188	195	202	
56 Ba^{2+}			149	156	166	175	
57 La^{3+}			117	130	141	150	
58 Ce^{3+}			115	128	139	148	
59 Pr^{3+}			113	127			
60 Nd^{3+}			112	125		141	
61 Pm^{3+}			111	123			
62 Sm^{3+}			110	122		138	
63 Eu^{3+}			109	121			
64 Gd^{3+}			108	119			
65 Tb^{3+}			106	118			
66 Dy^{3+}			105	117			
67 Ho^{3+}			104	116	126		
68 Er^{3+}			103	114			
69 Tm^{3+}			102	113			
70 Yb^{3+}			101	113			
71 Lu^{3+}			100	112			
72 Hf^{4+}		72	85	97			
73 Ta^{3+}			86				
Ta^{4+}			82				
74 W^{4+}			80				
75 Re^{4+}			77				
76 Os^{4+}			77				
77 Ir^{3+}			82				
Ir^{4+}			77				
78 Pt^{2+}		sq 74	94				
Pt^{4+}			77				
79 Au^{+}			151				
Au^{3+}		sq 82	99				
80 Hg^{2+}	83	110	116	128			
81 Tl^{3+}		89	103	112			
82 Pb^{2+}		112	133	143	154	163	
Pb^{4+}		79	92	108			
83 Bi^{3+}			117	131			
84 Po^{4+}			108	122			
85 At^{7+}			76				
86 Rn							
87 Fr^{+}			194				
88 Ra^{2+}				162		184	
89 Ac^{3+}			126				
90 Th^{4+}			108	119	127	135	

SOURCE: Data from R. D. Shannon, *Acta Cryst.*, **1976,** *A32,* 751.
NOTE: hs = high spin, ls = low spin, sq = square planar
Values for CN = 4 are for tetrahedral geometry unless designated square planar. All values are in picometers.

APPENDIX B-2
IONIZATION ENERGY

Atomic no.	*Element*	*eV*	*kJ/mol*	*Atomic no.*	*Element*	*eV*	*kJ/mol*
1	H	13.598	1,312.0	49	In	5.786	558.3
2	He	24.587	2,372.8	50	Sn	7.344	708.6
3	Li	5.392	520.2	51	Sb	8.641	833.7
4	Be	9.322	899.4	52	Te	9.009	869.2
5	B	8.298	800.6	53	I	10.451	1,008.4
6	C	11.260	1,086.5	54	Xe	12.130	1,170.4
7	N	14.534	1,402.3	55	Cs	3.894	375.7
8	O	13.618	1,314.0	56	Ba	5.212	502.9
9	F	17.422	1,681.0	57	La	5.577	538.1
10	Ne	21.564	2,080.6	58	Ce	5.47	528
11	Na	5.139	495.8	59	Pr	5.42	523
12	Mg	7.646	737.8	60	Nd	5.49	530
13	Al	5.986	577.6	61	Pm	5.55	535
14	Si	8.151	786.5	62	Sm	5.63	543
15	P	10.486	1,011.7	63	Eu	5.67	547
16	S	10.360	999.6	64	Gd	6.14	592
17	Cl	12.967	1,251.1	65	Tb	5.85	564
18	Ar	15.759	1,520.5	66	Dy	5.93	572
19	K	4.341	418.8	67	Ho	6.02	581
20	Ca	6.113	589.8	68	Er	6.10	589
21	Sc	6.54	631	69	Tm	6.18	596
22	Ti	6.82	658	70	Yb	6.254	603.4
23	V	6.74	650	71	Lu	5.426	523.5
24	Cr	6.766	652.8	72	Hf	7.0	675
25	Mn	7.435	717.4	73	Ta	7.89	761
26	Fe	7.870	759.3	74	W	7.98	770
27	Co	7.86	758	75	Re	7.88	760
28	Ni	7.635	736.7	76	Os	8.7	839
29	Cu	7.726	745.5	77	Ir	9.1	878
30	Zn	9.394	906.4	78	Pt	9.0	868
31	Ga	5.999	578.8	79	Au	9.225	890.1
32	Ge	7.899	762.1	80	Hg	10.437	1,007.0
33	As	9.81	947	81	Tl	6.108	589.3
34	Se	9.752	940.9	82	Pb	7.416	715.5
35	Br	11.814	1,139.9	83	Bi	7.289	703.3
36	Kr	13.999	1,350.7	84	Po	8.42	812
37	Rb	4.177	403.0	85	At	7.289	703.3
38	Sr	5.695	549.5	86	Rn	10.748	1,037.1
39	Y	6.38	616	87	Fr		
40	Zr	6.84	660	88	Ra	5.279	509.3
41	Nb	6.88	664	89	Ac	6.9	666
42	Mo	7.099	684.9	90	Th		
43	Tc	7.28	702	91	Pa		
44	Ru	7.37	711	92	U		
45	Rh	7.46	720	93	Np		
46	Pd	8.34	805	94	Pu	5.8	560
47	Ag	7.576	731.0	95	Am	6.0	579
48	Cd	8.993	867.7				

SOURCE: Data from C. E. Moore, *Ionization Potentials and Limits Derived from the Analyses of Optical Spectra,* NSRDS-NBS 34, National Bureau of Standards, Washington, D.C., 1970.
NOTE: 1 eV = 96.4853 kJmol

APPENDIX B-3
ELECTRON AFFINITY

Atomic no.	*Element*	*eV*	*kJ/mol*	*Atomic no.*	*Element*	*eV*	*kJ/mol*
1	H	0.754	72.8	45	Rh	1.137	109.7
2	He	−0.5*	−50	46	Pd	0.557	53.7
3	Li	0.618	59.6	47	Ag	1.302	125.6
4	Be	−0.5*	−50	48	Cd	−0.7*	−68
5	B	0.277	26.7	49	In	0.3	29
6	C	1.263	121.9	50	Sn	1.2	116
7	N	−0.07	−7	51	Sb	1.07	103
8	O	1.461	141.0	52	Te	1.971	190.2
9	F	3.399	328.0	53	I	3.059	295.2
10	Ne	−1.2*	−116	54	Xe	−0.8*	−77
11	Na	0.548	52.9	55	Cs	0.472	45.5
12	Mg	−0.4*	−39	56	Ba	−0.3*	−29
13	Al	0.441	42.6	57	La	0.5	48
14	Si	1.385	133.6	58	Ce	<0.5[a]	<48
15	P	0.747	72.0	59	Pr	<0.5[a]	<48
16	S	2.077	200.4	60	Nd	<0.5[a]	<48
17	Cl	3.617	349.0	61	Pm	<0.5[a]	<48
18	Ar	−1.0*	−97	62	Sm	<0.5[a]	<48
19	K	0.501	48.4	63	Eu	<0.5[a]	<48
20	Ca	−0.3*	−29	64	Gd	<0.5[a]	<48
21	Sc	0.188	18.1	65	Tb	<0.5[a]	<48
22	Ti	0.079	7.6	66	Dy	<0.5[a]	<48
23	V	0.525	50.7	67	Ho	<0.5[a]	<48
24	Cr	0.666	64.3	68	Er	<0.5[a]	<48
25	Mn	<0	0.0	69	Tm	<0.5[a]	<48
26	Fe	0.163	15.7	70	Yb	<0.5[a]	<48
27	Co	0.661	63.8	71	Lu	<0.5[a]	<48
28	Ni	1.156	111.5	72	Hf	~0	~0
29	Cu	1.228	118.5	73	Ta	0.322	31.1
30	Zn	−0.6*	−58	74	W	0.815	78.6
31	Ga	0.3	29	75	Re	0.15	14.5
32	Ge	1.2	115.8	76	Os	1.1	106.1
33	As	0.81	78	77	Ir	1.565	151.0
34	Se	2.021	195.0	78	Pt	2.128	205.3
35	Br	3.365	324.7	79	Au	2.309	222.8
36	Kr	−1.0*	−97	80	Hg	−0.5*	−48
37	Rb	0.486	46.9	81	Tl	0.2	19
38	Sr	−0.3*	−29	82	Pb	0.364	35.1
39	Y	0.307	29.6	83	Bi	0.946	91.3
40	Zr	0.426	41.1	84	Po	1.9	183
41	Nb	0.893	86.2	85	At	2.8	270
42	Mo	0.746	72.0	86	Rn	−0.7*	−68
43	Tc	0.55	53.1	87	Fr	0.6*	58
44	Ru	1.05	101.3	88	Ra	−0.3*	−29

SOURCE: All data from W. Hotop and W. C. Lineberger, *J. Phys. Chem. Ref. Data*, **1985,** *14,* 731, except those marked * from S. G. Bratsch and J. J. Lagowski, *Polyhedron*, **1986,** *5,* 1763.
NOTE: Many of these data are known to greater accuracy than shown in the table, some to 10 significant figures.
[a] Estimated values.

APPENDIX B-4
PAULING ELECTRONEGATIVITIES

Element	Electronegativity	Element	Electronegativity
H	2.20	Y	1.22
He	5.2[b]	Zr	1.33
Li	0.98	Nb	
Be	1.57	Mo	2.16
B	2.04	Tc	1.9[a]
C	2.55	Ru	2.2[a]
N	3.04	Rh	2.28
O	3.44	Pd	2.20
F	3.98	Ag	1.93
Ne	4.5[b]	Cd	1.69
Na	0.93	In	1.78
Mg	1.31	Sn(IV)	1.96
Al	1.61	Sn(II)	1.80
Si	1.90	Sb	2.05
P	2.19	Te	2.1[a]
S	2.58	I	2.66
Cl	3.16	Xe	2.4[b]
Ar	3.2[b]	Cs	0.79
K	0.82	Ba	0.89
Ca	1.00	La	1.10
Sc	1.36	Hf	1.3[a]
Ti	1.54	Ta	1.5[a]
V	1.63	W	2.36
Cr	1.66	Re	1.9[a]
Mn	1.55	Os	2.2[a]
Fe	1.83	Ir	2.20
Co	1.88	Pt	2.28
Ni	1.91	Au	2.54
Cu	1.90	Hg	2.00
Zn	1.65	Tl(III)	2.04
Ga	1.81	Tl(I)	1.62
Ge	2.01	Pb(IV)	2.33
As	2.18	Pb(II)	1.87
Se	2.55	Bi	2.02
Br	2.96	Po	2.0[a]
Kr	2.9[b]	At	2.2[a]
Rb	0.82	Rn	2.1[b]
Sr	0.95		

SOURCE: Data from A. L. Allred, *J. Inorg. Nucl. Chem.*, **1961,** *17,* 215, except

[a] L. Pauling, *The Nature of the Chemical Bond,* 3rd ed., Cornell University Press, Ithaca, N.Y., 1960, p. 93

[b] L. C. Allen and J. E. Huheey, *J. Inorg. Nucl. Chem.*, **1980,** *42,* 1523.

C

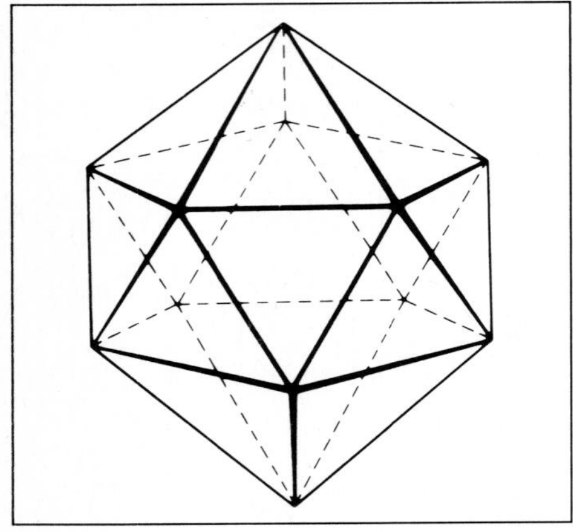

Character Tables†

1. GROUPS OF LOW SYMMETRY

C_1	E
A	1

C_s	E	σ_h		
A'	1	1	x, y, R_z	x^2, y^2, z^2, xy
A''	1	-1	z, R_x, R_y	yz, xz

C_i	E	i		
A_g	1	1	R_x, R_y, R_z	$x^2, y^2, z^2\ xy, xz, yz$
A_u	1	-1	x, y, z	

2. C_n, C_{nv}, AND C_{nh} GROUPS

The C_n groups

C_2	E	C_2		
A	1	1	z, R_z	x^2, y^2, z^2, xy
B	1	-1	x, y, R_x, R_y	yz, xz

† $i = \sqrt{-1}$; $\epsilon^* = \epsilon$ with $-i$ substituted for i.

C_3	E	C_3	C_3^2		
A	1	1	1	z, R_z	$x^2 + y^2, z^2$
E	1	ϵ	ϵ^*	$(x, y), (R_x, R_y)$	$(x^2 - y^2, xy), (yz, xz)$
	1	ϵ^*	ϵ		

$\epsilon = e^{(2\pi i)/3}$

C_4	E	C_4	C_2	C_4^3		
A	1	1	1	1	z, R_z	$x^2 + y^2, z^2$
B	1	-1	1	-1		$x^2 - y^2, xy$
E	1	i	-1	$-i$	$(x, y), (R_x, R_y)$	(yz, xz)
	1	$-i$	-1	i		

C_5	E	C_5	C_5^2	C_5^3	C_5^4		
A	1	1	1	1	1	z, R_z	$x^2 + y^2, z^2$
E_1	1	ϵ	ϵ^2	ϵ^{2*}	ϵ^*	$(x, y), (R_x, R_y)$	(yz, xz)
	1	ϵ^*	ϵ^{2*}	ϵ^2	ϵ		
E_2	1	ϵ^2	ϵ^*	ϵ	ϵ^{2*}		$(x^2 - y^2, xy)$
	1	ϵ^{2*}	ϵ	ϵ^*	ϵ^2		

$\epsilon = e^{(2\pi i)/5}$

C_6	E	C_6	C_3	C_2	C_3^2	C_6^5		
A	1	1	1	1	1	1	z, R_z	$x^2 + y^2, z^2$
B	1	-1	1	-1	1	-1		
E_1	1	ϵ	$-\epsilon^*$	-1	$-\epsilon$	ϵ^*	(x, y), (R_x, R_y)	(xz, yz)
	1	ϵ^*	$-\epsilon$	-1	$-\epsilon^*$	ϵ		
E_2	1	$-\epsilon^*$	$-\epsilon$	1	$-\epsilon^*$	$-\epsilon$		$(x^2 - y^2, xy)$
	1	$-\epsilon$	$-\epsilon^*$	1	$-\epsilon$	$-\epsilon^*$		

$\epsilon = e^{(\pi i)/3}$

C_7	E	C_7	C_7^2	C_7^3	C_7^4	C_7^5	C_7^6		
A	1	1	1	1	1	1	1	z, R_z	$x^2 + y^2, z^2$
E_1	1	ϵ	ϵ^2	ϵ^3	ϵ^{3*}	ϵ^{2*}	ϵ^*	(x, y), (R_x, R_y)	(xz, yz)
	1	ϵ^*	ϵ^{2*}	ϵ^{3*}	ϵ^3	ϵ^2	ϵ		
E_2	1	ϵ^2	ϵ^{3*}	ϵ^*	ϵ	ϵ^3	ϵ^{2*}		$(x^2 - y^2, xy)$
	1	ϵ^{2*}	ϵ^3	ϵ	ϵ^*	ϵ^{3*}	ϵ^2		
E_3	1	ϵ^3	ϵ^*	ϵ^2	ϵ^{2*}	ϵ	ϵ^{3*}		
	1	ϵ^{3*}	ϵ	ϵ^{2*}	ϵ^2	ϵ^*	ϵ^3		

$\epsilon = e^{(2\pi i)/7}$

C_8	E	C_8	C_4	C_2	C_4^3	C_8^3	C_8^5	C_8^7		
A	1	1	1	1	1	1	1	1	z, R_z	$x^2 + y^2, z^2$
B	1	-1	1	1	1	-1	-1	-1		
E_1	1	ϵ	i	-1	$-i$	$-\epsilon^*$	$-\epsilon$	ϵ^*	(x, y), (R_x, R_y)	(xz, yz)
	1	ϵ^*	$-i$	-1	i	$-\epsilon$	$-\epsilon^*$	ϵ		
E_2	1	i	-1	1	-1	$-i$	i	$-i$		$(x^2 - y^2, xy)$
	1	$-i$	-1	1	-1	i	$-i$	i		
E_3	1	$-\epsilon$	i	-1	$-i$	ϵ^*	ϵ	$-\epsilon^*$		
	1	$-\epsilon^*$	$-i$	-1	i	ϵ	ϵ^*	$-\epsilon$		

$\epsilon = e^{(\pi i)/4}$

The C_{nv} groups

C_{2v}	E	C_2	$\sigma_v(xz)$	$\sigma'_v(yz)$		
A_1	1	1	1	1	z	x^2, y^2, z^2
A_2	1	1	−1	−1	R_z	xy
B_1	1	−1	1	−1	x, R_y	xz
B_2	1	−1	−1	1	y, R_x	yz

C_{3v}	E	$2C_3$	$3\sigma_v$		
A_1	1	1	1	z	$x^2 + y^2, z^2$
A_2	1	1	−1	R_z	
E	2	−1	0	$(x, y), (R_x, R_y)$	$(x^2 - y^2, xy), (xz, yz)$

C_{4v}	E	$2C_4$	C_2	$2\sigma_v$	$2\sigma_d$		
A_1	1	1	1	1	1	z	$x^2 + y^2, z^2$
A_2	1	1	1	−1	−1	R_z	
B_1	1	−1	1	1	−1		$x^2 - y^2$
B_2	1	−1	1	−1	1		xy
E	2	0	−2	0	0	$(x, y), (R_x, R_y)$	(xz, yz)

C_{5v}	E	$2C_5$	$2C_5^2$	$5\sigma_v$		
A_1	1	1	1	1	z	$x^2 + y^2, z^2$
A_2	1	1	1	−1	R_z	
E_1	2	2 cos 72°	2 cos 144°	0	$(x, y), (R_x, R_y)$	(xz, yz)
E_2	2	2 cos 144°	2 cos 72°	0		$(x^2 - y^2, xy)$

C_{6v}	E	$2C_6$	$2C_3$	C_2	$3\sigma_v$	$3\sigma_d$		
A_1	1	1	1	1	1	1	z	$x^2 + y^2, z^2$
A_2	1	1	1	1	−1	−1	R_z	
B_1	1	−1	1	−1	1	−1		
B_2	1	−1	1	−1	−1	1		
E_1	2	1	−1	−2	0	0	$(x, y), (R_x, R_y)$	(xz, yz)
E_2	2	−1	−1	2	0	0		$(x^2 - y^2, xy)$

The C_{nh} groups

C_{2h}	E	C_2	i	σ_h		
A_g	1	1	1	1	R_z	x^2, y^2, z^2, xy
B_g	1	−1	1	−1	R_x, R_y	xz, yz
A_u	1	1	−1	−1	z	
B_u	1	−1	−1	1	x, y	

C_{3h}	E	C_3	C_3^2	σ_h	S_3	S_3^5		
A'	1	1	1	1	1	1	R_z	$x^2 + y^2, z^2$
E'	$\begin{Bmatrix} 1 \\ 1 \end{Bmatrix}$	$\begin{matrix} \epsilon \\ \epsilon^* \end{matrix}$	$\begin{matrix} \epsilon^* \\ \epsilon \end{matrix}$	$\begin{matrix} 1 \\ 1 \end{matrix}$	$\begin{matrix} \epsilon \\ \epsilon^* \end{matrix}$	$\begin{matrix} \epsilon^* \\ \epsilon \end{matrix}$	(x, y)	$(x^2 - y^2, xy)$
A''	1	1	1	−1	−1	−1	z	
E''	$\begin{Bmatrix} 1 \\ 1 \end{Bmatrix}$	$\begin{matrix} \epsilon \\ \epsilon^* \end{matrix}$	$\begin{matrix} \epsilon^* \\ \epsilon \end{matrix}$	$\begin{matrix} -1 \\ -1 \end{matrix}$	$\begin{matrix} -\epsilon \\ -\epsilon^* \end{matrix}$	$\begin{matrix} -\epsilon^* \\ -\epsilon \end{matrix}$	(R_x, R_y)	(xz, yz)

$\epsilon = e^{(2\pi i)/3}$

C_{4h}	E	C_4	C_2	C_4^3	i	S_4^3	σ_h	S_4		
A_g	1	1	1	1	1	1	1	1	R_z	$x^2 + y^2, z^2$
B_g	1	−1	1	−1	1	−1	1	−1		$x^2 - y^2, xy$
E_g	1	i	−1	$-i$	1	i	−1	$-i$	(R_x, R_y)	(xz, yz)
	1	$-i$	−1	i	1	$-i$	−1	i		
A_u	1	1	1	1	−1	−1	−1	−1	z	
B_u	1	−1	1	−1	−1	1	−1	1		
E_u	1	i	−1	$-i$	−1	$-i$	1	i	(x, y)	
	1	$-i$	−1	i	−1	i	1	$-i$		

C_{5h}	E	C_5	C_5^2	C_5^3	C_5^4	σ_h	S_5	S_5^7	S_5^3	S_5^9		
A'	1	1	1	1	1	1	1	1	1	1	R_z	$x^2 + y^2, z^2$
E_1'	1	ϵ	ϵ^2	ϵ^{2*}	ϵ^*	1	ϵ	ϵ^2	ϵ^{2*}	ϵ^*	(x, y)	
	1	ϵ^*	ϵ^{2*}	ϵ^2	ϵ	1	ϵ^*	ϵ^{2*}	ϵ^2	ϵ		
E_2'	1	ϵ^2	ϵ^*	ϵ	ϵ^{2*}	1	ϵ^2	ϵ^*	ϵ	ϵ^{2*}		$(x^2 - y^2, xy)$
	1	ϵ^{2*}	ϵ	ϵ^*	ϵ^2	1	ϵ^{2*}	ϵ	ϵ^*	ϵ^2		
A''	1	1	1	1	1	−1	−1	−1	−1	−1	z	
E_1''	1	ϵ	ϵ^2	ϵ^{2*}	ϵ^*	−1	$-\epsilon$	$-\epsilon^2$	$-\epsilon^{2*}$	$-\epsilon^*$	(R_x, R_y)	(xz, yz)
	1	ϵ^*	ϵ^{2*}	ϵ^2	ϵ	−1	$-\epsilon^*$	$-\epsilon^{2*}$	$-\epsilon^2$	$-\epsilon$		
E_2''	1	ϵ^2	ϵ^*	ϵ	ϵ^{2*}	−1	$-\epsilon^2$	$-\epsilon^*$	$-\epsilon$	$-\epsilon^{2*}$		
	1	ϵ^{2*}	ϵ	ϵ^*	ϵ^2	−1	$-\epsilon^{2*}$	$-\epsilon$	$-\epsilon^*$	$-\epsilon^2$		

$\epsilon = e^{(2\pi i)/5}$

C_{6h}	E	C_6	C_3	C_2	C_3^2	C_6^5	i	S_3^5	S_6^5	σ_h	S_6	S_3		
A_g	1	1	1	1	1	1	1	1	1	1	1	1	R_z	$x^2 + y^2, z^2$
B_g	1	−1	1	−1	1	−1	1	−1	1	−1	1	−1		
E_{1g}	1	ϵ	$-\epsilon^*$	−1	$-\epsilon$	ϵ^*	1	ϵ	$-\epsilon^*$	−1	$-\epsilon$	ϵ^*	(R_x, R_y)	(xz, yz)
	1	ϵ^*	$-\epsilon$	−1	$-\epsilon^*$	ϵ	1	ϵ^*	$-\epsilon$	−1	$-\epsilon^*$	ϵ		
E_{2g}	1	$-\epsilon^*$	$-\epsilon$	1	$-\epsilon^*$	$-\epsilon$	1	$-\epsilon^*$	$-\epsilon$	1	$-\epsilon^*$	$-\epsilon$		$(x^2 - y^2, xy)$
	1	$-\epsilon$	$-\epsilon^*$	1	$-\epsilon$	$-\epsilon^*$	1	$-\epsilon$	$-\epsilon^*$	1	$-\epsilon$	$-\epsilon^*$		
A_u	1	1	1	1	1	1	−1	−1	−1	−1	−1	−1	z	
B_u	1	−1	1	−1	1	−1	−1	1	−1	1	−1	1		
E_{1u}	1	ϵ	$-\epsilon^*$	−1	$-\epsilon$	ϵ^*	−1	$-\epsilon$	ϵ^*	1	ϵ	$-\epsilon^*$	(x, y)	
	1	ϵ^*	$-\epsilon$	−1	$-\epsilon^*$	ϵ	−1	$-\epsilon^*$	ϵ	1	ϵ^*	$-\epsilon$		
E_{2u}	1	$-\epsilon^*$	$-\epsilon$	1	$-\epsilon^*$	$-\epsilon$	−1	ϵ^*	ϵ	−1	ϵ^*	ϵ		
	1	$-\epsilon$	$-\epsilon^*$	1	$-\epsilon$	$-\epsilon^*$	−1	ϵ	ϵ^*	−1	ϵ	ϵ^*		

$\epsilon = e^{(\pi i)/3}$

3. D_n, D_{nd}, AND D_{nh} GROUPS

The D_n groups

D_2	E	$C_2(z)$	$C_2(y)$	$C_2(x)$		
A	1	1	1	1		x^2, y^2, z^2
B_1	1	1	−1	−1	z, R_z	xy
B_2	1	−1	1	−1	y, R_y	xz
B_3	1	−1	−1	1	x, R_x	yz

D_3	E	$2C_3$	$3C_2$		
A_1	1	1	1		$x^2 + y^2, z^2$
A_2	1	1	−1	z, R_z	
E	2	−1	0	$(x, y), (R_x, R_y)$	$(x^2 - y^2, xy), (xz, yz)$

D_4	E	$2C_4$	$C_2(= C_4^2)$	$2C_2'$	$2C_2''$		
A_1	1	1	1	1	1		$x^2 + y^2, z^2$
A_2	1	1	1	−1	−1	z, R_z	
B_1	1	−1	1	1	−1		$x^2 - y^2$
B_2	1	−1	1	−1	1		xy
E	2	0	−2	0	0	$(x, y), (R_x, R_y)$	(xz, yz)

D_5	E	$2C_5$	$2C_5^2$	$5C_2$		
A_1	1	1	1	1		$x^2 + y^2, z^2$
A_2	1	1	1	−1	z, R_z	
E_1	2	2 cos 72°	2 cos 144°	0	$(x, y), (R_x, R_y)$	(xz, yz)
E_2	2	2 cos 144°	2 cos 72°	0		$(x^2 - y^2, xy)$

D_6	E	$2C_6$	$2C_3$	C_2	$3C_2'$	$3C_2''$		
A_1	1	1	1	1	1	1		$x^2 + y^2, z^2$
A_2	1	1	1	1	−1	−1	z, R_z	
B_1	1	−1	1	−1	1	−1		
B_2	1	−1	1	−1	−1	1		
E_1	2	1	−1	−2	0	0	$(x, y), (R_x, R_y)$	(xz, yz)
E_2	2	−1	−1	2	0	0		$(x^2 - y^2, xy)$

The D_{nd} groups

D_{2d}	E	$2S_4$	C_2	$2C_2'$	$2\sigma_d$		
A_1	1	1	1	1	1		$x^2 + y^2, z^2$
A_2	1	1	1	−1	−1	R_z	
B_1	1	−1	1	1	−1		$x^2 - y^2$
B_2	1	−1	1	−1	1	z	xy
E	2	0	−2	0	0	$(x, y), (R_x, R_y)$	(xz, yz)

D_{3d}	E	$2C_3$	$3C_2$	i	$2S_6$	$3\sigma_d$		
A_{1g}	1	1	1	1	1	1		$x^2 + y^2, z^2$
A_{2g}	1	1	−1	1	1	−1	R_z	
E_g	2	−1	0	2	−1	0	(R_x, R_y)	$(x^2 - y^2, xy), (xz, yz)$
A_{1u}	1	1	1	−1	−1	−1		
A_{2u}	1	1	−1	−1	−1	1	z	
E_u	2	−1	0	−2	1	0	(x, y)	

D_{4d}	E	$2S_8$	$2C_4$	$2S_8^3$	C_2	$4C_2'$	$4\sigma_d$		
A_1	1	1	1	1	1	1	1		$x^2 + y^2, z^2$
A_2	1	1	1	1	1	−1	−1	R_z	
B_1	1	−1	1	−1	1	1	−1		
B_2	1	−1	1	−1	1	−1	1	z	
E_1	2	$\sqrt{2}$	0	$-\sqrt{2}$	−2	0	0	(x, y)	
E_2	2	0	−2	0	2	0	0		$(x^2 - y^2, xy)$
E_3	2	$-\sqrt{2}$	0	$\sqrt{2}$	−2	0	0	(R_x, R_y)	(xz, yz)

D_{5d}	E	$2C_5$	$2C_5^2$	$5C_2$	i	$2S_{10}^3$	$2S_{10}$	$5\sigma_d$		
A_{1g}	1	1	1	1	1	1	1	1		$x^2 + y^2, z^2$
A_{2g}	1	1	1	−1	1	1	1	−1	R_z	
E_{1g}	2	2 cos 72°	2 cos 144°	0	2	2 cos 72°	2 cos 144°	0	(R_x, R_y)	(xz, yz)
E_{2g}	2	2 cos 144°	2 cos 72°	0	2	2 cos 144°	2 cos 72°	0		$(x^2 - y^2, xy)$
A_{1u}	1	1	1	1	−1	−1	−1	−1		
A_{2u}	1	1	1	−1	−1	−1	−1	1	z	
E_{1u}	2	2 cos 72°	2 cos 144°	0	−2	−2 cos 72°	−2 cos 144°	0	(x, y)	
E_{2u}	2	2 cos 144°	2 cos 72°	0	−2	−2 cos 144°	−2 cos 72°	0		

D_{6d}	E	$2S_{12}$	$2C_6$	$2S_4$	$2C_3$	$2S_{12}^5$	C_2	$6C_2'$	$6\sigma_d$		
A_1	1	1	1	1	1	1	1	1	1		$x^2 + y^2, z^2$
A_2	1	1	1	1	1	1	1	−1	−1	R_z	
B_1	1	−1	1	−1	1	−1	1	1	−1		
B_2	1	−1	1	−1	1	−1	1	−1	1	z	
E_1	2	$\sqrt{3}$	1	0	−1	$-\sqrt{3}$	−2	0	0	(x, y)	
E_2	2	1	−1	−2	−1	1	2	0	0		$(x^2 - y^2, xy)$
E_3	2	0	−2	0	2	0	−2	0	0		
E_4	2	−1	−1	2	−1	−1	2	0	0		
E_5	2	$-\sqrt{3}$	1	0	−1	$\sqrt{3}$	−2	0	0	(R_x, R_y)	(xz, yz)

The D_{nh} groups

D_{2h}	E	$C_2(z)$	$C_2(y)$	$C_2(x)$	i	$\sigma(xy)$	$\sigma(xz)$	$\sigma(yz)$		
A_g	1	1	1	1	1	1	1	1		x^2, y^2, z^2
B_{1g}	1	1	−1	−1	1	1	−1	−1	R_z	xy
B_{2g}	1	−1	1	−1	1	−1	1	−1	R_y	xz
B_{3g}	1	−1	−1	1	1	−1	−1	1	R_x	yz
A_u	1	1	1	1	−1	−1	−1	−1		
B_{1u}	1	1	−1	−1	−1	−1	1	1	z	
B_{2u}	1	−1	1	−1	−1	1	−1	1	y	
B_{3u}	1	−1	−1	1	−1	1	1	−1	x	

D_{3h}	E	$2C_3$	$3C_2$	σ_h	$2S_3$	$3\sigma_v$		
A_1'	1	1	1	1	1	1		$x^2 + y^2, z^2$
A_2'	1	1	−1	1	1	−1	R_z	
E'	2	−1	0	2	−1	0	(x, y)	$(x^2 - y^2, xy)$
A_1''	1	1	1	−1	−1	−1		
A_2''	1	1	−1	−1	−1	1	z	
E''	2	−1	0	−2	1	0	(R_x, R_y)	(xz, yz)

D_{4h}	E	$2C_4$	C_2	$2C'_2$	$2C''_2$	i	$2S_4$	σ_h	$2\sigma_v$	$2\sigma_d$		
A_{1g}	1	1	1	1	1	1	1	1	1	1		$x^2 + y^2, z^2$
A_{2g}	1	1	1	-1	-1	1	1	1	-1	-1	R_z	
B_{1g}	1	-1	1	1	-1	1	-1	1	1	-1		$x^2 - y^2$
B_{2g}	1	-1	1	-1	1	1	-1	1	-1	1		xy
E_g	2	0	-2	0	0	2	0	-2	0	0	(R_x, R_y)	(xz, yz)
A_{1u}	1	1	1	1	1	-1	-1	-1	-1	-1		
A_{2u}	1	1	1	-1	-1	-1	-1	-1	1	1	z	
B_{1u}	1	-1	1	1	-1	-1	1	-1	-1	1		
B_{2u}	1	-1	1	-1	1	-1	1	-1	1	-1		
E_u	2	0	-2	0	0	-2	0	2	0	0	(x, y)	

D_{5h}	E	$2C_5$	$2C_5^2$	$5C_2$	σ_h	$2S_5$	$2S_5^3$	$5\sigma_v$		
A'_1	1	1	1	1	1	1	1	1		$x^2 + y^2, z^2$
A'_2	1	1	1	-1	1	1	1	-1	R_z	
E'_1	2	2 cos 72°	2 cos 144°	0	2	2 cos 72°	2 cos 144°	0	(x, y)	
E'_2	2	2 cos 144°	2 cos 72°	0	2	2 cos 144°	2 cos 72°	0		$(x^2 - y^2, xy)$
A''_1	1	1	1	1	-1	-1	-1	-1		
A''_2	1	1	1	-1	-1	-1	-1	1	z	
E''_1	2	2 cos 72°	2 cos 144°	0	-2	-2 cos 72°	-2 cos 144°	0	(R_x, R_y)	(xz, yz)
E''_2	2	2 cos 144°	2 cos 72°	0	-2	-2 cos 144°	-2 cos 72°	0		

D_{6h}	E	$2C_6$	$2C_3$	C_2	$3C'_2$	$3C''_2$	i	$2S_3$	$2S_6$	σ_h	$3\sigma_d$	$3\sigma_v$		
A_{1g}	1	1	1	1	1	1	1	1	1	1	1	1		$x^2 + y^2, z^2$
A_{2g}	1	1	1	1	-1	-1	1	1	1	1	-1	-1	R_z	
B_{1g}	1	-1	1	-1	1	-1	1	-1	1	-1	1	-1		
B_{2g}	1	-1	1	-1	-1	1	1	-1	1	-1	-1	1		
E_{1g}	2	1	-1	-2	0	0	2	1	-1	-2	0	0	(R_x, R_y)	(xz, yz)
E_{2g}	2	-1	-1	2	0	0	2	-1	-1	2	0	0		$(x^2 - y^2, xy)$
A_{1u}	1	1	1	1	1	1	-1	-1	-1	-1	-1	-1		
A_{2u}	1	1	1	1	-1	-1	-1	-1	-1	-1	1	1	z	
B_{1u}	1	-1	1	-1	1	-1	-1	1	-1	1	-1	1		
B_{2u}	1	-1	1	-1	-1	1	-1	1	-1	1	1	-1		
E_{1u}	2	1	-1	-2	0	0	-2	-1	1	2	0	0	(x, y)	
E_{2u}	2	-1	-1	2	0	0	-2	1	1	-2	0	0		

D_{8h}	E	$2C_8$	$2C_8^3$	$2C_4$	C_2	$4C'_2$	$4C''_2$	i	$2S_8$	$2S_8^3$	$2S_4$	σ_h	$4\sigma_d$	$4\sigma_v$		
A_{1g}	1	1	1	1	1	1	1	1	1	1	1	1	1	1		$x^2 + y^2, z^2$
A_{2g}	1	1	1	1	1	-1	-1	1	1	1	1	1	-1	-1	R_z	
B_{1g}	1	-1	-1	1	1	1	-1	1	-1	-1	1	1	1	-1		
B_{2g}	1	-1	-1	1	1	-1	1	1	-1	-1	1	1	-1	1		
E_{1g}	2	$\sqrt{2}$	$-\sqrt{2}$	0	-2	0	0	2	$\sqrt{2}$	$-\sqrt{2}$	0	-2	0	0	(R_x, R_y)	(xz, yz)
E_{2g}	2	0	0	-2	2	0	0	2	0	0	-2	2	0	0		$(x^2 - y^2, xy)$
E_{3g}	2	$-\sqrt{2}$	$\sqrt{2}$	0	-2	0	0	2	$-\sqrt{2}$	$\sqrt{2}$	0	-2	0	0		
A_{1u}	1	1	1	1	1	1	1	-1	-1	-1	-1	-1	-1	-1		
A_{2u}	1	1	1	1	1	-1	-1	-1	-1	-1	-1	-1	1	1	z	
B_{1u}	1	-1	-1	1	1	1	-1	-1	1	1	-1	-1	-1	1		
B_{2u}	1	-1	-1	1	1	-1	1	-1	1	1	-1	-1	1	-1		
E_{1u}	2	$\sqrt{2}$	$-\sqrt{2}$	0	-2	0	0	-2	$-\sqrt{2}$	$\sqrt{2}$	0	2	0	0	(x, y)	
E_{2u}	2	0	0	-2	2	0	0	-2	0	0	2	-2	0	0		
E_{3u}	2	$-\sqrt{2}$	$\sqrt{2}$	0	-2	0	0	-2	$\sqrt{2}$	$-\sqrt{2}$	0	2	0	0		

4. LINEAR GROUPS

$C_{\infty v}$	E	$2C_\infty^\phi$	$\dots$	$\infty\sigma_v$		
$A_1 \equiv \Sigma^+$	1	1	$\dots$	1	z	$x^2 + y^2, z^2$
$A_2 \equiv \Sigma^-$	1	1	$\dots$	-1	R_z	
$E_1 \equiv \Pi$	2	$2\cos\phi$	$\dots$	0	$(x, y), (R_x, R_y)$	(xz, yz)
$E_2 \equiv \Delta$	2	$2\cos 2\phi$	$\dots$	0		$(x^2 - y^2, xy)$
$E_3 \equiv \Phi$	2	$2\cos 3\phi$	$\dots$	0		
$\dots$	$\dots$	$\dots$	$\dots$	$\dots$		

$D_{\infty h}$	E	$2C_\infty^\phi$	$\dots$	$\infty\sigma_v$	i	$2S_\infty^\phi$	$\dots$	∞C_2		
Σ_g^+	1	1	$\dots$	1	1	1	$\dots$	1		$x^2 + y^2, z^2$
Σ_g^-	1	1	$\dots$	-1	1	1	$\dots$	-1	R_z	
Π_g	2	$2\cos\phi$	$\dots$	0	2	$-2\cos\phi$	$\dots$	0	(R_x, R_y)	(xz, yz)
Δ_g	2	$2\cos 2\phi$	$\dots$	0	2	$2\cos 2\phi$	$\dots$	0		$(x^2 - y^2, xy)$
$\dots$	$\dots$	$\dots$	$\dots$	$\dots$	$\dots$	$\dots$	$\dots$	$\dots$		
Σ_u^+	1	1	$\dots$	1	-1	-1	$\dots$	-1	z	
Σ_u^-	1	1	$\dots$	-1	-1	-1	$\dots$	1		
Π_u	2	$2\cos\phi$	$\dots$	0	-2	$2\cos\phi$	$\dots$	0	(x, y)	
Δ_u	2	$2\cos 2\phi$	$\dots$	0	-2	$-2\cos 2\phi$	$\dots$	0		
$\dots$	$\dots$	$\dots$	$\dots$	$\dots$	$\dots$	$\dots$	$\dots$	$\dots$		

5. S_{2n} GROUPS

S_4	E	S_4	C_2	S_4^3		
A	1	1	1	1	R_z	$x^2 + y^2, z^2$
B	1	-1	1	-1	z	$x^2 - y^2, xy$
E	1	i	-1	$-i$	$(x, y), (R_x, R_y)$	(xz, yz)
	1	$-i$	-1	i		

S_6	E	C_3	C_3^2	i	S_6^5	S_6		
A_g	1	1	1	1	1	1	R_z	$x^2 + y^2, z^2$
E_g	1	ϵ	ϵ^*	1	ϵ	ϵ^*	(R_x, R_y)	$(x^2 - y^2, xy)$, (xz, yz)
	1	ϵ^*	ϵ	1	ϵ^*	ϵ		
A_u	1	1	1	-1	-1	-1	z	
E_u	1	ϵ	ϵ^*	-1	$-\epsilon$	$-\epsilon^*$	(x, y)	
	1	ϵ^*	ϵ	-1	$-\epsilon^*$	$-\epsilon$		

$\epsilon = e^{(2\pi i)/3}$

S_8	E	S_8	C_4	S_8^3	C_2	S_8^5	C_4^3	S_8^7		
A	1	1	1	1	1	1	1	1	R_z	$x^2 + y^2, z^2$
B	1	-1	1	-1	1	-1	1	-1	z	
E_1	1	ϵ	i	$-\epsilon^*$	-1	$-\epsilon$	$-i$	ϵ^*	(x, y), (R_x, R_y)	
	1	ϵ^*	$-i$	$-\epsilon$	-1	$-\epsilon^*$	i	ϵ		
E_2	1	i	-1	$-i$	1	i	-1	$-i$		$(x^2 - y^2, xy)$
	1	$-i$	-1	i	1	$-i$	-1	i		
E_3	1	$-\epsilon^*$	$-i$	ϵ	-1	ϵ^*	i	$-\epsilon$		(xz, yz)
	1	$-\epsilon$	i	ϵ^*	-1	ϵ	$-i$	$-\epsilon^*$		

$\epsilon = e^{(\pi i)/4}$

6. TETRAHEDRAL, OCTAHEDRAL, AND ICOSAHEDRAL GROUPS

T	E	$4C_3$	$4C_3^2$	$3C_2$		
A	1	1	1	1		$x^2 + y^2 + z^2$
E {	1	ϵ	ϵ^*	1		$(2z^2 - x^2 - y^2,$
	1	ϵ^*	ϵ	1		$x^2 - y^2)$
T	3	0	0	-1	$(R_x, R_y, R_z), (x, y, z)$	(xy, xz, yz)

$\epsilon = e^{(2\pi i)/3}$

T_d	E	$8C_3$	$3C_2$	$6S_4$	$6\sigma_d$		
A_1	1	1	1	1	1		$x^2 + y^2 + z^2$
A_2	1	1	1	-1	-1		
E	2	-1	2	0	0		$(2z^2 - x^2 - y^2,$ $x^2 - y^2)$
T_1	3	0	-1	1	-1	(R_x, R_y, R_z)	
T_2	3	0	-1	-1	1	(x, y, z)	(xy, xz, yz)

T_h	E	$4C_3$	$4C_3^2$	$3C_2$	i	$4S_6$	$4S_6^5$	$3\sigma_h$		
A_g	1	1	1	1	1	1	1	1		$x^2 + y^2 + z^2$
A_u	1	1	1	1	-1	-1	-1	-1		
E_g {	1	ϵ	ϵ^*	1	1	ϵ	ϵ^*	1 }		$(2z^2 - x^2 - y^2,$
	1	ϵ^*	ϵ	1	1	ϵ^*	ϵ	1 }		$x^2 - y^2)$
E_u {	1	ϵ	ϵ^*	1	-1	$-\epsilon$	$-\epsilon^*$	-1 }		
	1	ϵ^*	ϵ	1	-1	$-\epsilon^*$	$-\epsilon$	-1 }		
T_g	3	0	0	-1	3	0	0	-1	(R_x, R_y, R_z)	(xy, xz, yz)
T_u	3	0	0	-1	-3	0	0	1	(x, y, z)	

$\epsilon = e^{(2\pi i)/3}$

O	E	$6C_4$	$3C_2(= C_4^2)$	$8C_3$	$6C_2$		
A_1	1	1	1	1	1		$x^2 + y^2 + z^2$
A_2	1	-1	1	1	-1		
E	2	0	2	-1	0		$(2z^2 - x^2 - y^2,$ $x^2 - y^2)$
T_1	3	1	-1	0	-1	$(R_x, R_y, R_z), (x, y, z)$	
T_2	3	-1	-1	0	1		(xy, xz, yz)

O_h	E	$8C_3$	$6C_2$	$6C_4$	$3C_2(=C_4^2)$	i	$6S_4$	$8S_6$	$3\sigma_h$	$6\sigma_d$		
A_{1g}	1	1	1	1	1	1	1	1	1	1		$x^2 + y^2 + z^2$
A_{2g}	1	1	−1	−1	1	1	−1	1	1	−1		
E_g	2	−1	0	0	2	2	0	−1	2	0		$(2z^2 - x^2 - y^2, x^2 - y^2)$
T_{1g}	3	0	−1	1	−1	3	1	0	−1	−1	(R_x, R_y, R_z)	
T_{2g}	3	0	1	−1	−1	3	−1	0	−1	1		(xy, xz, yz)
A_{1u}	1	1	1	1	1	−1	−1	−1	−1	−1		
A_{2u}	1	1	−1	−1	1	−1	1	−1	−1	1		
E_u	2	−1	0	0	2	−2	0	1	−2	0		
T_{1u}	3	0	−1	1	−1	−3	−1	0	1	1	(x, y, z)	
T_{2u}	3	0	1	−1	−1	−3	1	0	1	−1		

I	E	$12C_5$	$12C_5^2$	$20C_3$	$15C_2$		
A	1	1	1	1	1		$x^2 + y^2 + z^2$
T_1	3	$\frac{1}{2}(1 + \sqrt{5})$	$\frac{1}{2}(1 - \sqrt{5})$	0	−1	$(x, y, z), (R_x, R_y, R_z)$	
T_2	3	$\frac{1}{2}(1 - \sqrt{5})$	$\frac{1}{2}(1 + \sqrt{5})$	0	−1		
G	4	−1	−1	1	0		
H	5	0	0	−1	1		$(xy, xz, yz, x^2 - y^2, 2z^2 - x^2 - y^2)$

I_h	E	$12C_5$	$12C_5^2$	$20C_3$	$15C_2$	i	$12S_{10}$	$12S_{10}^3$	$20S_6$	15σ		
A_g	1	1	1	1	1	1	1	1	1	1		$x^2 + y^2 + z^2$
T_{1g}	3	$\frac{1}{2}(1 + \sqrt{5})$	$\frac{1}{2}(1 - \sqrt{5})$	0	−1	3	$\frac{1}{2}(1 - \sqrt{5})$	$\frac{1}{2}(1 + \sqrt{5})$	0	−1	(R_x, R_y, R_z)	
T_{2g}	3	$\frac{1}{2}(1 - \sqrt{5})$	$\frac{1}{2}(1 + \sqrt{5})$	0	−1	3	$\frac{1}{2}(1 + \sqrt{5})$	$\frac{1}{2}(1 - \sqrt{5})$	0	−1		
G_g	4	−1	−1	1	0	4	−1	−1	1	0		
H_g	5	0	0	−1	1	5	0	0	−1	1		$(2z^2 - x^2 - y^2, x^2 - y^2, xy, xz, yz)$
A_u	1	1	1	1	1	−1	−1	−1	−1	−1		
T_{1u}	3	$\frac{1}{2}(1 + \sqrt{5})$	$\frac{1}{2}(1 - \sqrt{5})$	0	−1	−3	$-\frac{1}{2}(1 - \sqrt{5})$	$-\frac{1}{2}(1 + \sqrt{5})$	0	1	(x, y, z)	
T_{2u}	3	$\frac{1}{2}(1 - \sqrt{5})$	$\frac{1}{2}(1 + \sqrt{5})$	0	−1	−3	$-\frac{1}{2}(1 + \sqrt{5})$	$-\frac{1}{2}(1 - \sqrt{5})$	0	1		
G_u	4	−1	−1	1	0	−4	1	1	−1	0		
H_u	5	0	0	−1	1	−5	0	0	1	−1		

D

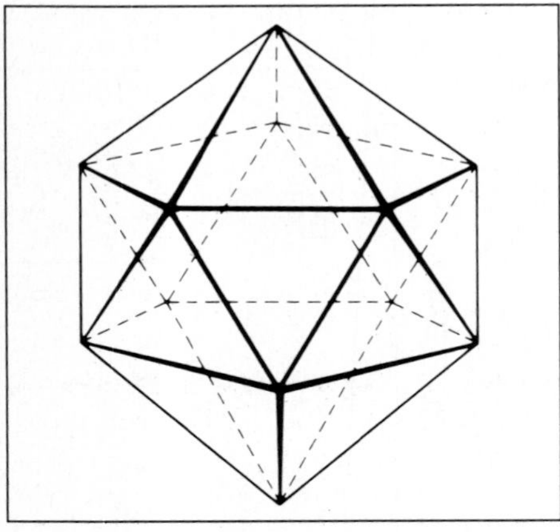

Electron-dot Diagrams and Formal Charge

DRAWING ELECTRON-DOT DIAGRAMS

Lewis electron-dot diagrams show the number of bonds between specific atoms and the resonance possibilities. The method described here, developed by Miller[1] and summarized by Malerich,[2] is slightly different from those presented in many general chemistry texts. The general approach is to calculate the number of bonds in the molecule, draw them in, and then add the lone-pair electrons. The procedure is as follows:

1. Calculate the number of electrons needed to satisfy the normal valence structure of the atoms if each were totally independent of the others. Hydrogen needs 2 electrons; all other atoms need 8 electrons. Using NH_3 as an example,

3 hydrogens need 2 electrons each	$3 \times 2 = 6$	
1 nitrogen needs 8 electrons	$1 \times 8 = 8$	
Total		14 electrons needed

2. Calculate the number of valence electrons available in the atoms, counting only those outside any noble gas core. If the molecule has a charge, add

[1] G. T. Miller, Jr., *Chemistry: Principles and Applications,* Wadsworth, Belmont, Calif., 1976, Chapter 4, Supplement 1.

[2] C. J. Malerich, *J. Chem. Educ.*, **1987,** *64,* 403.

an electron for each negative charge, and subtract an electron for each positive charge. For NH_3,

3 hydrogens have 1 valence electron each	$3 \times 1 = 3$	
1 nitrogen has 5 valence electrons	$1 \times 5 = 5$	
Total		8 electrons available

3. Find the difference between the number of electrons needed and the number available. This is the number of bonding electrons, which must be shared by two atoms and counted twice, once for each atom. Since each bond uses two electrons, the number of bonds is half the number of bonding electrons. For NH_3,

$$\begin{array}{rl} 14 & \text{electrons needed} \\ -\ 8 & \text{electrons available} \\ \hline 6 & \text{electrons are shared, for 3 bonds} \end{array}$$

4. Sketch the molecule with the number of bonds calculated. Several additional rules useful in determining how to draw the molecule are given below. In this case, there are three bonds, just enough for one connecting each hydrogen to the nitrogen.
5. Fill in electron pairs around the atoms up to the total number of electrons available and the maximum of 8 around each atom (2 on hydrogen) to complete the structure. In the ammonia example, one lone pair is added to the nitrogen. In summary, for NH_3,

	N 3 H	
Electrons needed	$8 + 3 \times 2 =$	14
− Electrons available	$-(5 + 3 \times 1) =$	− 8
	Shared electrons	6 = 3 bonding pairs

Net, 3 bonding pairs, 1 lone pair, for a total of 8 electrons. For ease in drawing, a line frequently designates a pair of electrons, as in Figure D-1.

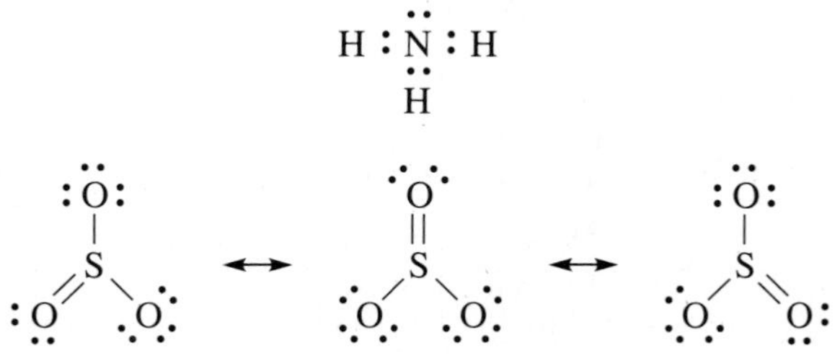

H : N̈ : H
H

FIGURE D-1 Lewis Diagrams for NH_3 and SO_3.

6. If the number of bonding pairs exceeds the minimum needed to form single bonds between the atoms, double or triple bonds are used. For example, for SO_3,

	S 3 O	
Electrons needed	$8 + 3 \times 8 =$	32
− Electrons available	$-(6 + 3 \times 6) =$	− 24
	Shared electrons	8 or 4 bonds

Net, 8 bonding electrons or 4 bonds, and 16 other electrons or 8 lone pairs.

The structure must have a double bond and two single bonds from the central S to the O's. The remaining 16 electrons are lone pairs around the oxygen atoms, three pairs on each singly bonded oxygen and two pairs on the doubly bonded oxygen. Figure D-1 shows the completed structure. The section on resonance in Chapter 3 completes the discussion of structures like this.

7. If step 3 gives fewer electron pairs than the number of atoms surrounding the central atom, *expand the octet* on the central atom to give enough bonds to connect all the atoms together. Examples are given later in this appendix.

DRAWING THE MOLECULE

In drawing the molecule, several general principles can be helpful. There are exceptions to all of them, so they should be taken only as guides. These principles are as follows:

1. If one atom is different from the others, it is positioned in the center of the molecule with the others arranged around it (for example, NH_3, SO_3, CH_4, SO_4^{2-}). Hydrogen and oxygen atoms are usually found on the outside of the molecule.
2. If there are single atoms of two elements, the one with larger atomic number is in the center of the molecule with the others arranged around it (for example $POCl_3$, $SOCl_2$).
3. The carbon family usually has four bonds, the nitrogen family three bonds, and the oxygen family two bonds, and the halogens usually have one bond in neutral molecules.
4. When oxygen and hydrogen are in the same molecule, they usually form the combination H—O—X, where X is whatever other atom is in the molecule.
5. Three-membered rings are unlikely for most molecules. Larger rings are possible, but still not as common as other structures.

FORMAL CHARGE

Formal charge examples (shown in Figure D-2)

In CH_4, NH_3, and H_2O, the formal charges of all atoms are zero. In CH_4, carbon initially had 4 valence electrons and in the compound it shares 8 electrons with the hydrogens. Half the shared electrons are assigned to carbon, so it has 4 electrons in the molecule and a net formal charge of zero. Each hydrogen initially has one electron, and in methane each shares 2 electrons with carbon. One of the shared electrons is assigned to each hydrogen, so the net formal charge for each hydrogen is also zero. The ammonia and water cases are left as examples for the reader.

In each SO_3 resonance structure, the doubly bonded oxygen has a formal charge of zero, the singly bonded oxygens have formal charges of $1-$, and the sulfur has a formal charge of $2+$. Since each resonance structure contributes equally to the Lewis description of SO_3, we average the three structures to

Molecule	Atom	*In the free atom* *Valence electrons*	*In the molecule* *Electrons shared*		*Lone pair electrons*		*Formal charge*
CH_4	C	4	$-\left(\frac{8}{2}\right.$	+	$\left.0\right)$	=	0
	H	1	$-\left(\frac{2}{2}\right.$	+	$\left.0\right)$	=	0
NH_3	N	5	$-\left(\frac{6}{2}\right.$	+	$\left.2\right)$	=	0
	H	1	$-\left(\frac{2}{2}\right.$	+	$\left.0\right)$	=	0
H_2O	O	6	$-\left(\frac{4}{2}\right.$	+	$\left.4\right)$	=	0
	H	1	$-\left(\frac{2}{2}\right.$	+	$\left.0\right)$	=	0
SO_3	S	6	$-\left(\frac{8}{2}\right.$	+	$\left.0\right)$	=	2+
	—O	6	$-\left(\frac{2}{2}\right.$	+	$\left.6\right)$	=	1−
	=O	6	$-\left(\frac{4}{2}\right.$	+	$\left.4\right)$	=	0
SO_2	S	6	$-\left(\frac{6}{2}\right.$	+	$\left.2\right)$	=	1+
	−O	6	$-\left(\frac{2}{2}\right.$	+	$\left.6\right)$	=	1−
	=O	6	$-\left(\frac{4}{2}\right.$	+	$\left.4\right)$	=	0

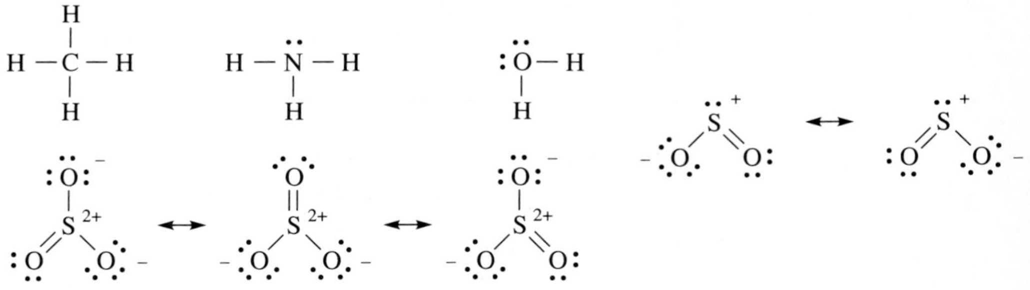

FIGURE D-2 Formal Charge Diagrams for CH_4, NH_3, H_2O, SO_3, and SO_2.

give sulfur a formal charge of 2+ and each oxygen a formal charge of $\frac{2}{3}-$. In SO_2, the doubly bonded oxygen has a formal charge of zero, the singly bonded oxygen has a formal charge of 1−, and the sulfur has a formal charge of 1+. Averaging again, sulfur has a formal charge of 1+ and each oxygen a formal charge of $\frac{1}{2}-$.

Some further rules about formal charge can make it more useful in deciding between different possible structures.

1. Structures with small formal charges (2+, 2−, or less) are more likely than those with larger formal charges.
2. Nonzero formal charges on adjacent atoms are usually of opposite sign.

3. More electronegative atoms (those in the upper-right corner of the periodic table) should have negative rather than positive formal charges.
4. Formal charges of opposite signs separated by large distances are unlikely.
5. The most stable structures have the largest sum of the electronegativity differences for adjacent atoms. For example, HOCl is more stable than HClO:

	H	O	Cl	H	Cl	O
Electronegativities	2.20	3.44	3.16	2.20	3.16	3.44
Differences	1.24	0.28		0.96	0.28	
Sum of differences		1.52			1.24	

In other words, a bond with large polarity is more stable than a bond with smaller polarity, and atoms with large electronegativity difference are likely to be bonded to each other.

EXAMPLES

NO_3^-

	N O −1	
Electrons needed	4 × 8	= 32
− Electrons available	−(5 + 3 × 6 + 1) =	−24
	Shared electrons	8

Net, 4 bonding pairs, 8 lone pairs, 3 structures, with resonance.

CO_3^{2-}

	C O −2	
Electrons needed	4 × 8	= 32
− Electrons available	−(4 + 3 × 6 + 2) =	−24
	Shared electrons	8

Net, 4 bonding pairs, 8 lone pairs, 3 structures, with resonance.

ClF_3

Normal	3F Cl		Expanded shell	3F Cl	
Needed	3 × 8 + 8 =	32	Needed	3 × 8 + 10 =	34
− Available	−(3 × 7 + 7) =	−28	− Available	−(3 × 7 + 7) =	−28
	Bonding electrons	4		Bonding electrons	6

Not enough for the 3 bonds needed. 3 bonds can be formed.

Since at least three bonds, or 6 electrons, are needed, the number of electrons that can be around the chlorine must be increased from the usual octet to 10. This is justified by the closeness of the 3*d*-orbital energy to that of the 3*s* and 3*p* energies; one of the 3*d* orbitals of chlorine is used in the bonding. After the 6 bonding electrons are drawn in, 6 electrons (three pairs) are added to each fluorine and 4 (two pairs) are added to the chlorine as lone pairs. The chlorine may thus be viewed as having an expanded octet of 10 electrons.

EXAMPLE

SF_6

	Normal			Expanded shell	
	S 6 F			S 6 F	
Needed	$8 + 6 \times 8 =$	56	Needed	$12 + 6 \times 8 =$	60
− Available	$-(6 \times 7 + 6) =$	−48	− Available	$-(6 + 6 \times 7) =$	−48
Bonding electrons		8	Bonding electrons		12

Only 4 bonds possible, too few. 6 bonds are possible.

Since at least six bonds are needed, the number of electrons around the sulfur must be increased from 8 to 12. Two 3*d* orbitals are used in the bonding. (The molecular orbital discussion in Chapter 5 explains this further.) Three lone pairs on each fluorine complete the picture. Octets may generally be exceeded for elements of atomic number 14 (Si) or higher, where the next higher *d* orbitals are near the energy of the *s* and *p* orbitals.

In expanding the octet, the number around the central atom is expanded two electrons at a time until the number of bonds is sufficient for all the atoms. There are exceptions, but this procedure handles most molecules.

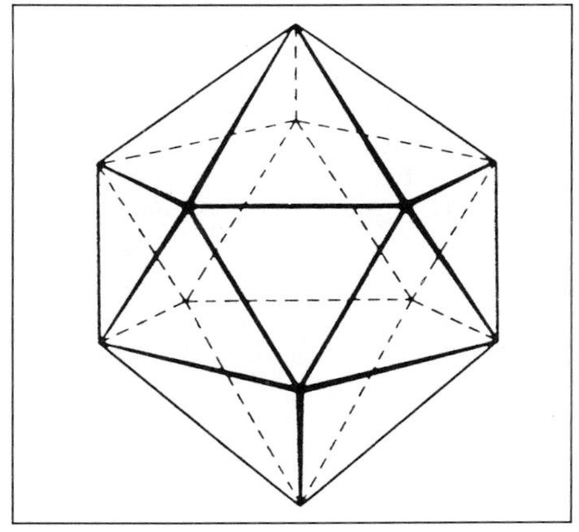

Index